Hydrodynamic Instabilities and Turbulence

Turbulent mixing induced by hydrodynamic instabilities is found in many high- and low-energy-density regimes, ranging from supernovae to inertial confinement fusion to scramjet engines. While these applications have long been recognized, unprecedented advances in both computational and experimental tools have provided novel, critical insights to the field. Incorporating the most recent theoretical, computational, and experimental results, this title provides a comprehensive yet accessible description of turbulent mixing driven by Rayleigh–Taylor, Richtmyer–Meshkov, and Kelvin–Helmholtz instabilities. An overview of core concepts and equations is provided, followed by detailed descriptions of complex and turbulent flows. The influences of stabilizing mechanisms, rotations, magnetic fields, and time-dependent accelerations on the evolution of hydrodynamic instabilities are explained. This book is ideal for advanced undergraduates as well as graduates beginning research in this exciting field, while also functioning as an authoritative reference volume for researchers in the wide range of disciplines for which it has applications.

Dr. Ye Zhou is a Physicist at Lawrence Livermore National Laboratory in California, having previously worked as a Senior Staff Scientist at the Institute for Computer Applications in Science and Engineering, NASA Langley Research Center. He has published over 130 peer-reviewed publications and is a regular invited speaker at major international conferences. He is a Fellow of the American Physical Society and currently serves as an associate editor of *Computers & Fluids* and an editor of *Physica D: Nonlinear Phenomena.*

Hydrodynamic Instabilities and Turbulence

Rayleigh–Taylor, Richtmyer–Meshkov, and Kelvin–Helmholtz Mixing

YE ZHOU
Lawrence Livermore National Laboratory

CAMBRIDGE
UNIVERSITY PRESS

CAMBRIDGE
UNIVERSITY PRESS

Shaftesbury Road, Cambridge CB2 8EA, United Kingdom

One Liberty Plaza, 20th Floor, New York, NY 10006, USA

477 Williamstown Road, Port Melbourne, VIC 3207, Australia

314–321, 3rd Floor, Plot 3, Splendor Forum, Jasola District Centre, New Delhi – 110025, India

103 Penang Road, #05–06/07, Visioncrest Commercial, Singapore 238467

Cambridge University Press is part of Cambridge University Press & Assessment, a department of the University of Cambridge.

We share the University's mission to contribute to society through the pursuit of education, learning and research at the highest international levels of excellence.

www.cambridge.org
Information on this title: www.cambridge.org/9781108489645

DOI: 10.1017/9781108779135

First published 2024

A catalogue record for this publication is available from the British Library

Library of Congress Cataloging-in-Publication Data
Names: Zhou, Ye (Physicist), author.
Title: Hydrodynamic instabilities and turbulence : Rayleigh-Taylor, Richtmyer-Meshkov, and Kelvin-Helmholtz Mixing / Ye Zhou, Lawrence Livermore National Laboratory, California.
Description: Cambridge, United Kingdom ; New York, NY : Cambridge University Press, 2024. | Includes bibliographical references and index.
Identifiers: LCCN 2023035316 | ISBN 9781108489645 (hardback) | ISBN 9781108779135 (ebook)
Subjects: LCSH: Turbulence. | Hydrodynamics. | Stability.
Classification: LCC QA913 .Z46 2024 | DDC 532/.0527–dc23/eng/20231122
LC record available at https://lccn.loc.gov/2023035316

ISBN 978-1-108-48964-5 Hardback

To my wife Lily
and to our sons
David and Kevin

Contents

Part 3 From the microscopic to cosmic scales

Preface

The turbulent mixing of fluids induced by Rayleigh–Taylor (RT), Richtmyer–Meshkov (RM), and Kelvin–Helmholtz (KH) instabilities is ubiquitous in nature, with widespread applications including astrophysics, geophysics, and engineering. In the astrophysical fluid dynamics and space plasma context, such mixing plays a crucial role in both core-collapse supernovae and the majestic extrusions of solar prominences in the Sun's corona. Meanwhile, in our efforts to bring the Sun's power down to the terrestrial sphere, this phenomenon inhibits our ability to obtain clean energy through inertial confinement fusion (ICF). On the other hand, if these instabilities can be harnessed and controlled, they will be extremely helpful in building more efficient scramjet engines. Thus, the advancement of our understanding and control of instability-induced turbulent mixing has the potential to both deepen our knowledge of the universe and improve engineering applications.

Throughout the twentieth and twenty-first centuries, researchers have faced a myriad of obstacles in their quest to describe instability-driven turbulent flows. These flows are complex because they evolve from linear to nonlinear regimes as they grow from a small initial perturbation into full-scale turbulence. They also vary dramatically in size, ranging from perturbations measuring just a few microns at the onset of instability to geophysical and astrophysical flows spanning thousands of kilometers, or more, after the flow has transitioned to turbulence. Nobel laureate Richard Feynman famously stated that the study of turbulence was the last unresolved problem of classical physics. In addition, external factors such as magnetic fields, rotation, and high-energy flux can have considerable influence on the evolution of hydrodynamic instabilities. Hence, in describing these phenomena, the scientist must synthesize several challenging physics topics – fluid instabilities, nonlinearity, and fluid- and magneto-hydrodynamic (MHD) turbulence – all while encompassing vast spatial and temporal scales.

In the last several decades, the development of supercomputers and powerful diagnostic tools for experimental measurements in both intensive laser platforms and traditional laboratories has opened the door to major advancements in the study of hydrodynamic instability-driven turbulent mixing. National laboratories and universities in the United States, United Kingdom, France, Japan, China, Russia, and other countries, have devoted substantial time and resources to studying this subject. The National Ignition Facility (NIF), Sandia's Z machine, and Omega laser are actively working on ICF in the United States, and several large experimental facilities are currently under construction abroad, including the Laser Megajoule (LMJ) and ShenGuang facilities in France and China, respectively. In addition, a number of institutions are carrying out intensive numerical and experimental studies on hydrodynamic instabilities. The exciting developments in the study of instability-driven turbulent mixing are reflected in the thousands of papers being published in the field every year.

The RT, RM, and KH instabilities, which are the focus of this book, are not the only fluid instabilities that exist in nature. In fact, scientists have observed more than 20 distinct kinds of fluid instabilities, all of which fall under the broad umbrella of "hydrodynamic instabilities." Many of the instabilities not covered in this book arise in the fields of aerospace or mechanical engineering, where they affect the design of aircraft and other equipment. However, these instabilities are completely different than the RT, RM, and KH instabilities, both in their characteristics and their effects on the flow of surrounding fluids. This book represents the first complete reference guide to RT, RM, and KH instabilities, and the turbulent mixing they induce. Furthermore, several astrophysical and fusion-related applications connected to these instabilities require the use of an MHD treatment, which is irrelevant to other kinds of instabilities, but the discussions of which in this book also set it apart from other published works.

It is the author's hope that this book will serve as an authoritative reference on the RT, RM, and KH instabilities for scientists and engineers who work at universities and national laboratories, as well as those in industries such as aerospace, defense, and chemical engineering. Indeed, this book arose out of the observation that no publication previously available comprehensively addresses all aspects of this extremely active, interdisciplinary field of study. As a result, a researcher wanting to learn about the field was previously required to painstakingly piece together information from several different textbooks, including many that are out of date and omit key developments. The primary resources for hydrodynamic instabilities, for example, remain the classic texts published decades ago by S. Chandrasekhar, *Hydrodynamic and Hydromagnetic Stability* (Oxford University Press, 1961) and P.G. Drazin and W.H. Reid, *Hydrodynamic Stability* (Cambridge University Press, 1981). Those books, however, do not incorporate important recent discoveries regarding the nonlinear phase of instabilities and the transition to turbulent flow. Meanwhile, the established texts on turbulence are *A First Course in Turbulence* (MIT Press, 1972) by H. Tennekes and J. Lumley, *Turbulent Flows* (Cambridge University Press, 2000) by S. Pope, *Turbulence: An Introduction for Scientists and Engineers* (Oxford University Press, 2015) by P. Davidson, and *Buoyancy Effects in Fluids* (Cambridge University Press, 1979) by J.S. Turner. None of these texts, however, consider the time-dependent nature of instability-driven turbulent flows. Finally, specialized books on ICF (J.D. Lindl, *Inertial Confinement Fusion: The Quest for Ignition and Energy Gain using Indirect Drive*, AIP Press, 1998; S. Atzeni and J. Meyer-ter-Vehn, *The Physics of Inertial Fusion: Beam Plasma Interaction, Hydrodynamics, Hot Dense Matter*, Clarendon, Oxford, 2004), high-energy-density physics (HEDP) (R.P. Drake, *High Energy Density Physics*, Springer, 2nd ed., 2018), and supernova explosion (D. Branch and J.C. Wheeler, *Supernova Explosion*, Springer, 2017) have given very limited coverage to the turbulent mixing induced by hydrodynamic instabilities.

This book has been deliberately structured to be maximally useful to both active researchers – physicists, astronomers, engineers, and mathematicians alike – and as a textbook for graduate or advanced undergraduate courses. In particular, the 24 chapters are organized into 3 broad parts. Part 1 provides a logical and systematic introduction to the foundations of hydrodynamic instabilities and turbulence, and naturally lends itself to a classroom presentation. Part 2 details the many factors that make the flows induced by the RT, RM, and KH instabilities so rich and complex. Part 3 presents applications from the laboratory to the cosmos, as well as the engineering modeling and numerical simulation methods required to describe them. The broad set of topics discussed in Parts 2 and 3 render the text an excellent reference, while also giving classroom instructors a wide range of options in designing a course syllabus around selected topics relevant to the program mission and specialization of the department offering the course.

Acknowledgments

I am deeply indebted to Professor D.L. Youngs for his many insightful suggestions and constructive feedback. I am also extremely grateful to Drs. P.D. Powell and J.D. Sadler for their thorough reviews and detailed comments.

I very much appreciate the thoughtful discussions with Drs. S. Balachandar, D.S. Clark, M. Groom, A. Hillier, J.W. Jacobs, K.O. Mikaelian, B.E. Morgan, W.D. Mostert, B. Mueller, O. Soulard, K. Peterson, P. Ramaprabhu, B.A. Remington, B.B. Rollin, O. Schilling, B. Thornber, R.J.R. Williams, E.P. Yu, and Q. Zhang.

I am much obliged to many previous and recent discussions with Drs. H. Aluie, A. Banerjee, R. Bonazza, A.C. Buckingham, G.C. Burton, W.H. Cabot, B. Cheng, T.T. Clark, A.W. Cook, S.B. Dalziel, G. Dimonte, P.E. Dimotakis, F.W. Doss, R.P. Drake, D. Drikakis, K.A. Flippo, S. Gauthier, S.G. Glendinning, J. Glimm, F.F Grinstein, S.W. Haan, C.M. Huntington, O.A. Hurricane, A.L. Kuhl, C.C. Kuranz, S. Kurien, D. Livescu, S.A. MacLaren, E.E. Meshkov, S.R. Nagel, K.P. Prestridge, D. Ranjan, H.F. Robey, D.D. Ryutov, C.-W. Shu, L. Welser-Sherrill, and G.B. Zimmerman. I also acknowledge useful conversations that I held with Drs. P. Amendt, C. Batha, J.D. Bender, G. Boffetta, W.T. Buttler, P. Carlès, A. Casner, A.M. Dimits, A. Do, D. Eliason, R.A. Gore, B.J. Gréa, J. Greenough, G. Gregori, J. Griffond, B.M. Haines, R. Hill, M. Koenig, E. Leinov, M. Lombardini, L.J. Perkins, A. Mahalov, W.H. Matthaeus, P.L. Miller, B.J. Olson, K.S. Raman, M.M. Scase, J.D. Schwarzkopf, M.A. Skinner, K.L. Stalsberg-Zarling, P. Wang, C.R. Weber, and G.P. Zank.

Additionally, I would like to express my sincere gratitude to the late Drs. M.J. Andrews, J.R. Herring, R.H. Kraichnan, C.E. Leith, C.G. Speziale, and W.C. Reynolds.

I would like to offer special thanks to Drs. M.R. Baer, A. Banerjee, D. Frost, D. Jewitt, and A. Lawrie for providing their original photographs and Drs. H. Aluie, X. Bian, T.T. Clark, M. Groom, A. Krygier, D. Livescu, D. Martinez, B.E. Morgan, S. Nagel, and G. Rigon for their assistance regarding the figures. I would also like to acknowledge Drs. A. Hillier, T. Janka, T. Kaneko, T. Sano, and T. Shigeyama for granting me permission to use their published figures.

Finally, I would like to express my deep and sincere appreciation to the editorial and production teams at Cambridge University Press for their invaluable assistance and support throughout the preparation of this book.

PART 1

FUNDAMENTALS

A journey of a thousand miles begins with a single step.

Lao Tzu, ancient Chinese philosopher

1 A first glimpse of Rayleigh–Taylor, Richtmyer–Meshkov, and Kelvin–Helmholtz instabilities

A common feature during the evolutions of Rayleigh–Taylor (RT) and Richtmyer–Meshkov (RM) instabilities is the formation of a characteristic mushroom shape. Figure 1.1 shows cirrus clouds, where such beautiful structures can be observed along the lower edge of the cloud. In fact, the concept of the Rayleigh–Taylor instability (RTI) was born when Lord Rayleigh's interest was sparked by observations of cirrus cloud formation (Jevons, 1857). He derived the properties of an interface instability that occurs when a dense fluid is supported by a lighter one (Rayleigh, 1883).

Lord Rayleigh, John William Strutt, 3rd Baron Rayleigh, made seminal contributions to many areas of physics and introduced a number of fluid dynamics concepts bearing his name. More than a century ago, it became clear to scientists in general and applied mathematicians in particular, that it would be prudent, before ever publishing any of their research, to first check whether it had not already been done by Lord Rayleigh (Gillis, 1962). He was awarded the 1904 Nobel Prize in Physics "for his investigations of the densities of the most important gases and for his discovery of argon in connection with these studies."

According to an account by Professor J.S. Turner, the modern contribution of Sir Geoffrey I. Taylor arose from his observation and interpretation of the surface effect of an underwater explosion: the sudden appearance, then equally sudden disappearance, of an expanding dark circle (Turner, 1997). After a discussion with Professor William George Penney (Lord Penney) who had previously investigated instabilities in submarine explosions (Penney and Price, 1942), Taylor concluded that it must be due to an instability, with drops being thrown off the water surface as it is disturbed and then returns to its equilibrium position. Only a few days later, Taylor informed Lord Penney that he had solved the problem of the behavior of a modulated plane interface between two fluids in a field of acceleration, which became the original calculation of the Taylor instability (Taylor, 1950). The analytical study of the Taylor instability at a fluid interface accelerating in the direction of the denser fluid was published along with supporting experiments carried out by his research student (Lewis, 1950). This should not be a surprise, as Taylor was a superb experimentalist himself and a strong advocate of the interaction between experiment and theory in fluid mechanics (Taylor, 1974). There is, of course, an equivalence between gravity and acceleration which Taylor recognized – therefore this phenomenon has rightly become known as the "Rayleigh–Taylor instability."

Sir Geoffrey was described as one of the most notable scientists of the twentieth century (Batchelor, 1994), with many fluid dynamics concepts also bearing his name. "To fluid dynamicists in the world at large, Taylor undoubtedly remains a giant. The centrality of fluid mechanics in Taylor's research cannot be debated," according to Sreenivasan (2011).

Fig. 1.1 Rayleigh–Taylor cirrus clouds. Photograph courtesy of Prof. David Jewitt, University of California at Los Angeles.

Among other things, his contribution is so vast that it provided sufficient material for an entire course at Harvard University, covering topics ranging from hydrodynamic stability and turbulence to electrohydrodynamics and the locomotion of small organisms (Brenner and Stone, 2000).

Early work in the Soviet Union, circa 1950, used a very interesting approach. Mixing at a gravitationally unstable interface between fluids of different density was modeled as turbulent diffusion due to buoyancy-generated turbulence (Belen'kii and Fradkin, 1965). This research was carried out at the Lebedev Institute in Moscow and published posthumously after the death of Belen'kii in 1956. Andronov et al. (1976) considered deceleration by multiple-shocks, rather than by RT, but the Schlieren image was the first image of a "turbulent mixing zone" in this area (Andronov et al., 1976). A shock wave is a discontinuous surface propagating in any elastic medium (such as air, water, or a solid substance) at which density and velocity experience abrupt changes (Courant and Friedrichs, 1948; Zel'dovich and Raizer, 1966; Ben-Dor et al., 2000; Anderson, 2007; Krehl, 2008).

Very little scientific research can be directly linked to the creative world of art, so we should not miss this opportunity to discuss some of these topics as well. Several exquisite paintings of well-known Mexican muralist David Alfaro Siqueiros are created by the artist's "accidental painting" method (Siqueiros, 1936a,b, 1969; Folgarait, 1987; White, 2009). Unbeknown to him, Siqueiros' technique of absorption of two or more superimposed colors which infiltrate one another is firmly rooted in fluid physics of the RTI – such "infiltration" or mixing of colors is a direct consequence of an imbalance between the paints' densities, subject to gravity (Fig. 1.2). Indeed, Zetina et al. (2015) and de la Calleja et al. (2014) have demonstrated that the formation of spotted, patchy areas of contrasting color can be reproduced in a controlled manner (Fig. 1.3). Thus, Siqueiros' accidental painting technique was not accidental at all (Zenit, 2019).

Fig. 1.2 Image sequence of the preparation of an "accidental painting" experiment. Fig. 3 of Zetina et al. (2015). Cited in accordance with the Creative Commons Attribution (CC BY) license.

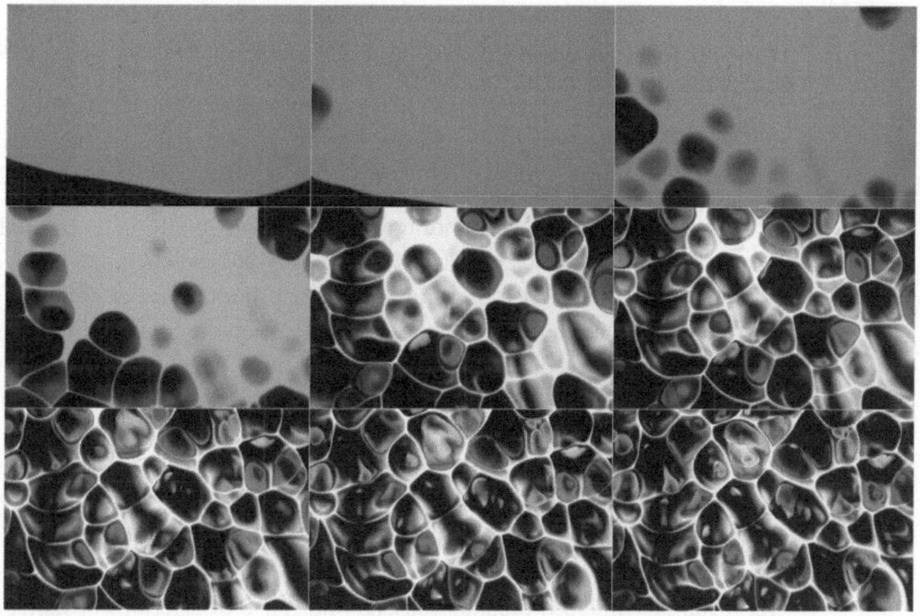

Fig. 1.3 Image sequence, white over black paint. The time difference in between frames is 63 s. Time progresses from left to right and from top to bottom. Fig. 5 of Zetina et al. (2015). Cited in accordance with the Creative Commons Attribution (CC BY) license.

At very high-density ratios, for example, water/air, as used in the early RT experiments, the dense fluid forms spikes as it penetrates the lighter fluid (sometimes these structures are described as "fingers"). The lighter fluid "rises" up through the heavier fluid, which in turn "drops" through the lighter fluid. The characteristic structures of the lighter fluid penetrating the heavier fluid are usually referred to as "bubbles," and the heavier fluid forms "spikes" as it penetrates the lighter fluid. Most notably, "mushroom" shapes form on the spike tips at lower density ratios and for small density differences there is little asymmetry. However, using the terms bubble and spike at all density ratios has persisted in the RT/RM literature.

Fig. 1.4 Crater produced by the vertical impact of a liquid drop onto a less dense liquid pool. A spherical RT instability develops around the crater when it decelerates, which results in mushroom-shaped plumes growing radially outward. Lherm et al. (2022).

The RTI is one of the most observed physical processes in nature and in people's daily life. Pouring cream into your morning coffee results in RTI-induced mixing (see the cover image of the book). The spherical RT instability also enters the dynamics of drop impacts when a denser-than-water drop impacts a deep liquid pool of water. Figure 1.4 shows a snapshot from one such experiment, at a time when the impact of the drop has produced a sizeable crater. We see on this picture that the liquid from the drop, which covers the crater floor, has developed mushroom-shaped structures penetrating radially into the water pool (Lherm et al., 2022).

The RT instability of a self-gravitating fluid sphere is important as a mechanism for corona formation on Venus (Hoogenboom and Houseman, 2006) (Fig. 1.5) and in the context of Earth's core formation (e.g., Ida et al., 1989; Mondal and Korenaga, 2017; Sasaki and Abe, 2007; Roy et al., 2024). In a much more violent scenario, asteroid (and/or meteor) RT-instability-induced breakup may be one cause of asteroid (or meteor) swarm impacts in the Venusian atmosphere (Korycansky et al., 2000) (Fig. 1.6) and Earth's atmosphere (Svetsov et al., 1995). The RT-induced perturbation growth and subsequent breakup are dependent on the material strength.

Granular dynamics not only occur in geological applications but also enable technologies ranging from carbon capture to pharmaceutical production (McLaren et al., 2019; Kobayashi and Kurita, 2022). For example, the combination of vibration and upward gas flow through a granular material can produce an analogous process to the RTI. Figure 1.7 shows a time series of optical images of the RTI forming when denser particles are placed initially above lighter particles. Lighter grains start to rise through the heavier grains in fingers (a long thin shape) and "granular bubbles," leaving a trail of particles in their wake (McLaren et al., 2019).

The oceans are dangerous for many creatures. A shrimp (the species *Alpheus heterochaelis*; ~5.5 cm in size, Fig. 1.8) must adopt some unique capabilities to stun and even kill prey animals or to defend a shelter or territory. More specifically, the shrimp issues

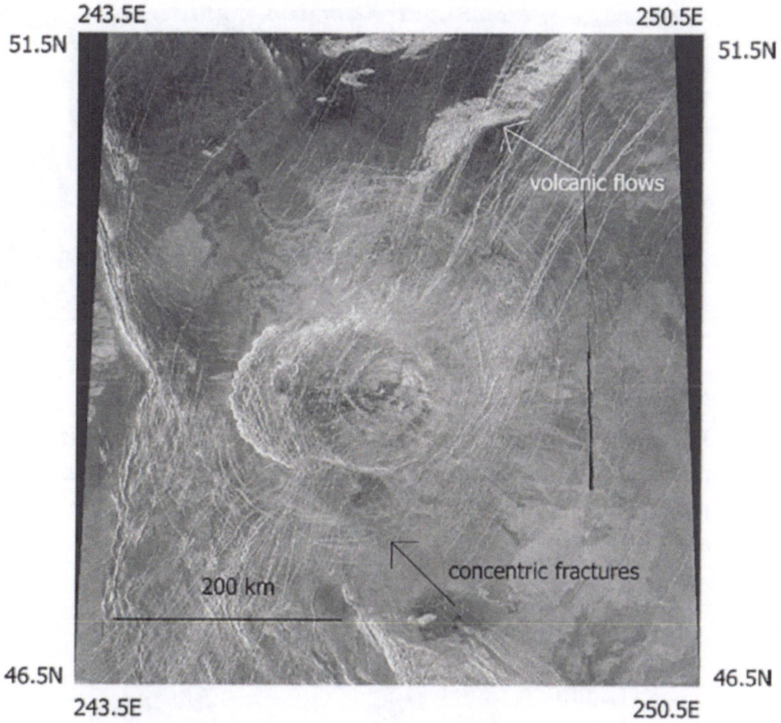

Fig. 1.5 Magellan synthetic aperture radar image of Nalwomga corona on Venus (48.7°N, 247°E). Nalwomga is a concentric, depression-shaped corona (380 km in diameter) located in the plains. Black regions indicate data gaps. Image from http://pdsmaps.wr.usgs.gov.

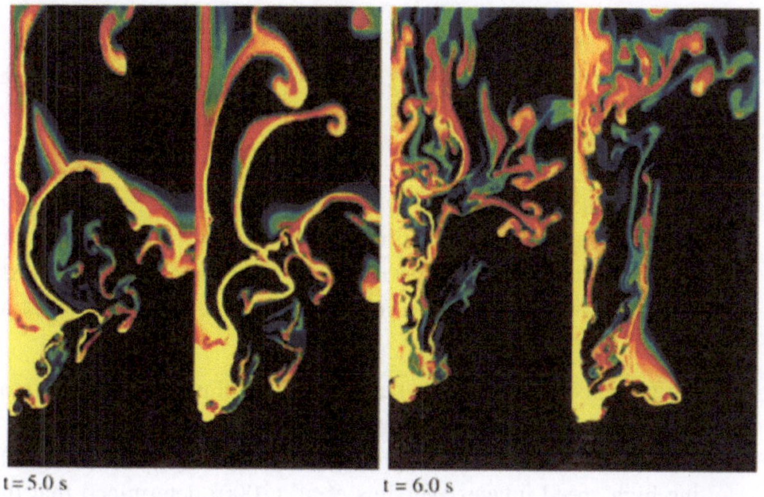

Fig. 1.6 Asteroid (and/or meteor) Rayleigh–Taylor-instability-induced breakup may be one cause of asteroid (or meteor) swarm impacts. From Korycansky et al. (2000). Reprinted with permission from Elsevier.

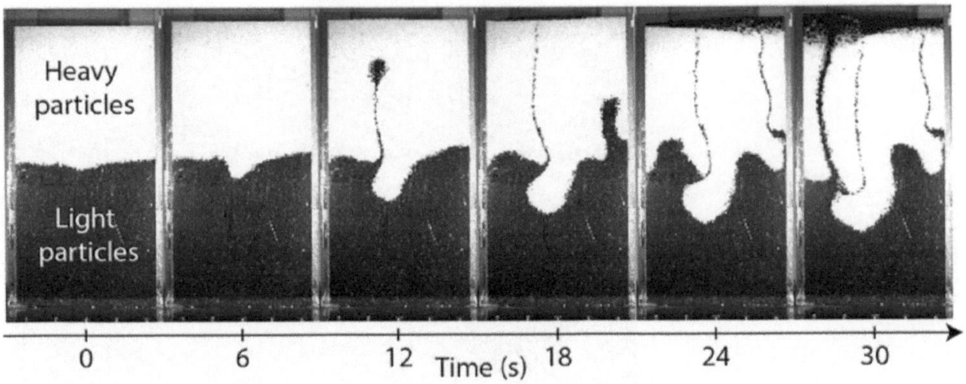

Fig. 1.7 Time series of optical images of the RT instability forming when denser particles are placed initially above lighter particles. Diameter of light particles: 1.70 $\pm$ 0.07 mm; density of light particles: 2,500 kg/m^3; ratio of heavy-to-light particle diameters and densities: 0.69 $\pm$ 0.03 and 2.40, respectively. Vibration frequency: 188 rad/s. Fig. 1 of McLaren et al. (2019), with permission.

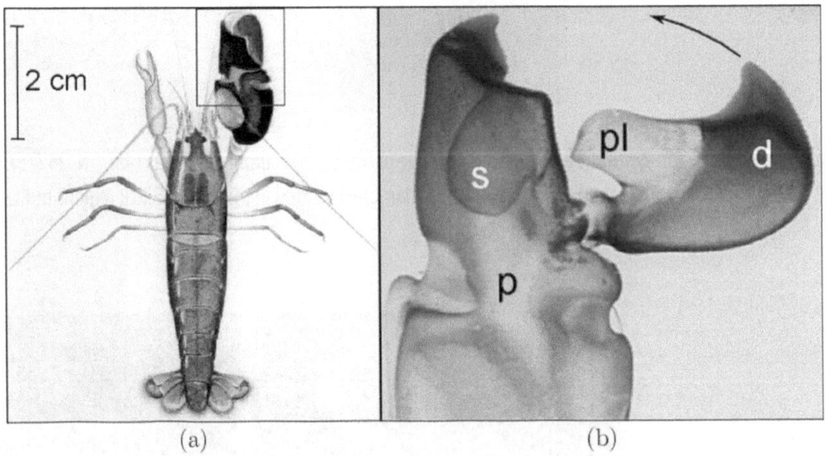

Fig. 1.8 (a) *Alpheus heterochaelis*, one of the largest snapping shrimp. The large snapper claw may be either on the right or the left in both sexes. (b) Close-up of the snapper claw in its cocked position. The claw is made transparent by the use of methyl salicylate. The claw has a protruding plunger (pl) on the dactyl (d) and a matching socket (s) in the immobile propus (p) (photograph by B. Seibel). Versluis et al. (2000). Reprinted with permission from AAAS.

a high-velocity water jet when the dactyl plunger is driven into the propus socket, displacing water. A cavitation bubble is also created by a water jet formed by an extremely rapid closure of its large snapper claw, which may reach about half of its body size. Studying high-speed images, Versluis et al. (2000) determined that the loud snap is produced through collapse of a cavitation bubble, destroyed through the RTI, instead of that previously attributed to the mechanical contact made when the dactyl and the propus edges hit each other as the claw closes.

The Richtmyer–Meshkov instability (RMI) shares a great many features with the RTI. The RMI arises when a shock wave passes through an interface between two fluids and can be viewed as the impulsive counterpart of the RT instability (i.e., RT with high acceleration for a short time). Recognizing this important insight, Professor Robert D. Richtmyer published the original theoretical analysis of the RMI while he was in residence at Los Alamos National Laboratory (Richtmyer, 1960), during which period he also served as the head of its theoretical division for more than two years. During his next residence at The Courant Institute of Mathematical Sciences, New York University, he moved on to prove the Lax–Richtmyer theorem with Peter Lax, a fundamental theorem in numerical analysis (Lax and Richtmyer, 1956). He eventually settled down at The University of Colorado at Boulder.

Professor Evgeny Evgrafovich Meshkov carried out the first successful isolated experiments of the RMI (Meshkov, 1969) at the All-Russian Scientific Research Institute of Experimental Physics, in Sarov, Russia (also known as Arzamas-16, a closed city long omitted from official maps during the Soviet Union era). Fortunately, Meshkov's important work became well known worldwide, and the appropriately coined term "Richtmyer–Meshkov instability" has been found in many scientific and engineering applications.

Filling a balloon with water and then throwing it at something (or someone) is a game enjoyed by children both young and old throughout the world (Lund and Dalziel, 2014). The balloon rupture following a collision with a rigid surface, shown in Fig. 1.9, provides an example of the RMI in action. At later times following the balloon rupture, Lund and Dalziel observed a large-scale growth of the interfacial amplitude, where the generation mechanism is momentum in the water due to the preburst waves, a manifestation of the same mechanism that drives the RMI.

(a) (b)

Fig. 1.9 The rupturing of a water-filled balloon. (a) The capillary-like waves following a collision with a rigid surface. (b) The Richtmyer–Meshkov-like instability following the removal of the restoring force (i.e., gravity). Lund and Dalziel (2014).

The shock-bubble interaction (SBI) has been considered as a finite-mass, high-interface-curvature analog to the classical RMI (Abd-El-Fattah et al., 1976; Abd-El-Fattah and Henderson, 1978a,b; Bryson and Gross, 1961; Catherasoo and Sturtevant, 1983; Haas and Sturtevant, 1987; Layes and Le Métayer, 2007; Layes et al., 2003, 2005, 2009; Markstein, 1957a,b; Ranjan et al., 2011; Rudinger and Somers, 1960; Singh et al., 2021; Sturtevant and Kulkarny, 1976). A single-shocked (Ranjan et al., 2005, 2007, 2008) or a twice-shocked (Haehn et al., 2010, 2011) spherical gas inhomogeneity has been studied extensively with both shock tube experiments and numerical simulations.

Of particular interest is the fact that depending on the shock strength, the SBI may lead to ignition of chemically reactive gases inside the bubble (Haehn, 2012; Haehn et al., 2012). The reacting shock–bubble interaction (RSBI) allows one to study the interaction between the RMI and the reaction waves inside the bubble that are initiated by the shock (Diegelmann et al., 2016a). Haehn (2012) and Haehn et al. (2012) performed first-of-its-kind RSBI experiments for a range of Mach numbers (Fig. 1.10). Shock focusing leads to an ignition of the bubble gas. Different behaviors from the introduction of reactive gases and variation of the shock strength illustrate why hydrodynamic instability–driven flows are so complex. The numerical work of Diegelmann et al. (2016a,b, 2017) nicely complements that of Haehn et al. These investigations, quite innovative in that they include

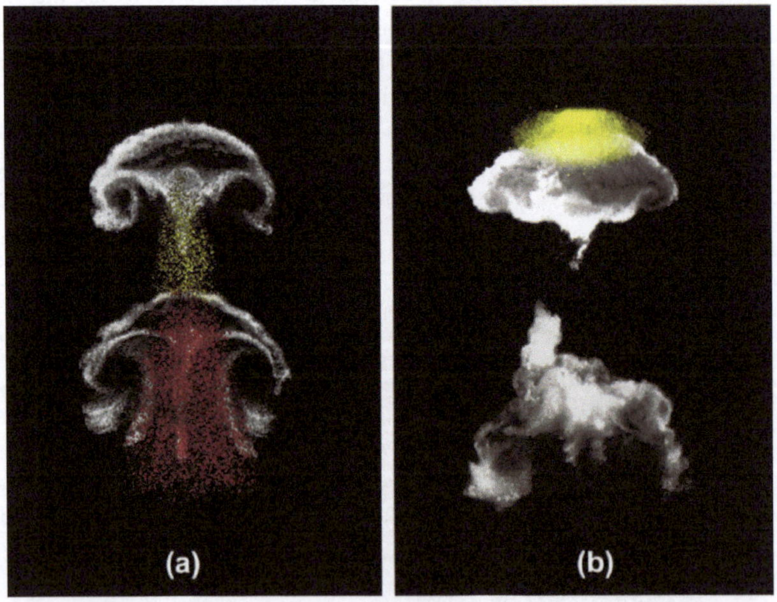

Fig. 1.10 Composite images showing the development of a shock-accelerated bubble filled with a reactive mixture initially free-falling in N_2. Each image consists of two chemiluminescence exposures in yellow and red overlaid upon two planar Mie scattering images in grayscale. For clarity, the signal levels have been adjusted and do not reflect the intensity of the chemiluminescence signal. In each image, the incident shock wave travels downward. Haehn et al. (2012). Reprinted with permission from Elsevier.

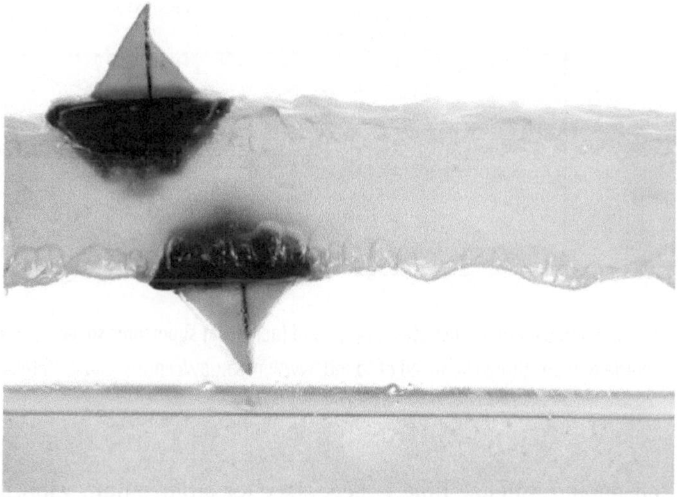

Fig. 1.11 Apffel et al. (2020) levitated a liquid above a layer of air in a container by vertically shaking the container. They observed that objects can float upside down from the lower side of the air–liquid interface. Reprinted with permission from Springer Nature.

chemical reactions, have established an experimental database and a numerical framework for further studies.

Most hydrodynamic instabilities are unwelcome, a prominent example of which are those ones encountered in inertial confinement fusion (ICF). Thus, significant efforts have been devoted to finding mechanisms for suppressing and controlling these instabilities (e.g., Sterbentz et al., 2022; Schill et al., 2024). In a fascinating study that flipped reality on its head, Apffel et al. (2020) succeeded in inducing small boats to float "upside down" based on what has been called "weird physics" (Fig. 1.11). As the authors noted, "[w]hen placed over a less dense medium, a liquid layer will typically collapse downwards if it exceeds a certain size, as gravity acting on the lower liquid interface triggers a destabilising effect called a Rayleigh–Taylor instability." Impressively, the RTI can be mitigated. The authors demonstrated that vertical shaking can produce a steady rhythm of compressions that maintains the levitating liquid intact and causes bubbles injected into the liquid to be pushed downward, forming an air cushion below the levitating liquid. At the lower interface of the liquid, the vertical shaking also creates stable buoyancy positions, which behave as though the gravitational force were inverted.

While hydrodynamic instabilities are typically detrimental to experimental or engineering designs, enhanced RMI is highly desirable for improving the performance of scramjets (Fig. 1.12), where supersonic air flow into the combustion chamber limits the residence time to a few milliseconds (Marble et al., 1989, 1990; Jacobs, 1992; Yang et al., 1993; Yu et al., 2020; Cao and Michaels, 2022). For flame holdings in supersonic combustion,

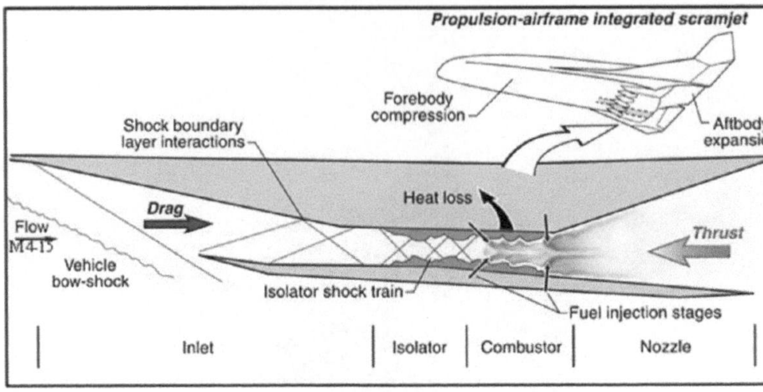

Fig. 1.12 This simplified graphic illustrates how air and fuel mix at supersonic speeds inside a scramjet engine to propel the vehicle to many times the speed of sound. www.nasa.gov/centers-and-facilities/armstrong/x-43a/ Credit: NASA.

adequate residence time is necessary for atomization, vaporization, mixing and chemical reaction (e.g., Nakaya et al., 2015). The numerical results of Yang et al. (2014) indicate that the RMI phenomenon markedly enhances the mixing efficiency (up to 43%) and is necessary to initiate chemical reactions.

Perhaps the most photogenic examples of hydrodynamic instabilities are the mesmerizing images obtained by the Hubble Space Telescope and other astronomical observational facilities. In fact, RT instabilities are the proposed origin for the filaments of the Crab Nebula; the heavy fluid forms fingers (spikes) that stream downwards while the light fluid forms bubbles that rise between the fingers (Fig. 1.13).

The evolution of the Crab Nebula's bubbles and fingers provides an excellent example of how the RTI or RMI can lead to the development of the Kelvin–Helmholtz instability (KHI), the name given to the instability of two adjacent fluid layers in relative motion. Originally, Lord Kelvin (Sir William Thomson) sought to understand the KHI in the context of the generation of ocean surface waves by wind (Thomson, 1855), while von Helmholtz used this instability to explain the banded clouds (von Helmholtz, 1868).

Figure 1.14 is a photo showing what are known as Kelvin–Helmholtz (KH) clouds over mountains. The KHI has also been observed in the solar atmosphere (Li et al., 2018a; Yuan et al., 2019), and KH billows can be visible at the turbulent boundary between two latitudinal bands in Saturn's atmosphere, which curl repeatedly along the edge (Fig. 1.15).

For a challenging measurement, carried out at a depth of half a kilometer in the Atlantic some 700 miles south of the Azores, Van Haren and Gostiaux (2010) used a network of temperature sensors to observe KH billows ripping down the sides of an underwater mountain (Fig. 1.16(a)). While KH instabilities are well observed in the laboratory and nature, the authors still could not resist the temptation to use a definitely nonscientific word "beautiful" to describe this amazing image. Finally, in the deep ocean, van Haren (2015) reported that the RTI penetrates stable density structures under high-frequency internal waves. It arises at the same time as KHI billows emerge at some distance above the slope – see Fig. 1.16(b).

Fig. 1.13 This Hubble image gives the most detailed view of the entire Crab Nebula. It was assembled from 24 individual exposures taken with the Hubble Space Telescope. www.spacetelescope.org/images/heic0515a/. Credit: NASA.

Fig. 1.14 Kelvin–Helmholtz clouds over the Rocky Mountains. Image: University Corporation for Atmospheric Research, with permission. Photo by Benjamin Foster.

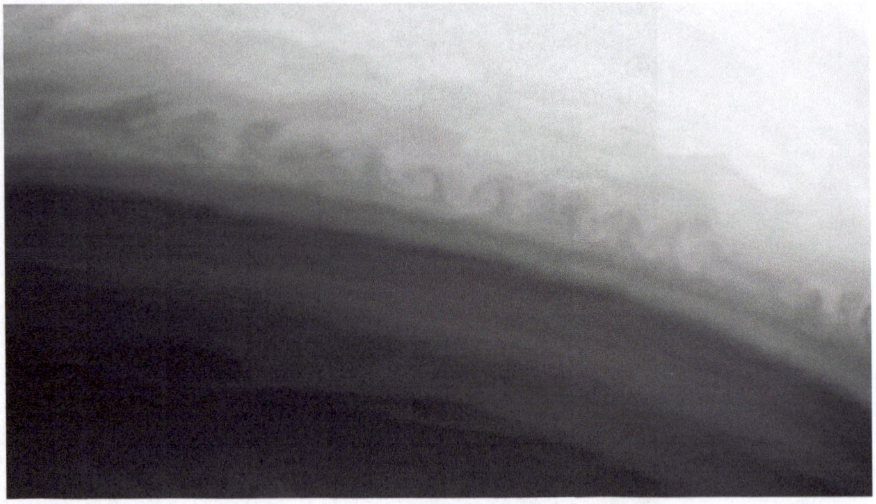

Fig. 1.15 This image of Saturn was taken with the Cassini spacecraft narrow angle camera on Oct. 9, 2004, at a distance of 5.9 million kilometers (3.7 million miles) from Saturn. It used a filter sensitive to wavelengths of infrared light centered at 889 nm. www.jpl.nasa.gov/spaceimages/details.php?id=PIA06502.

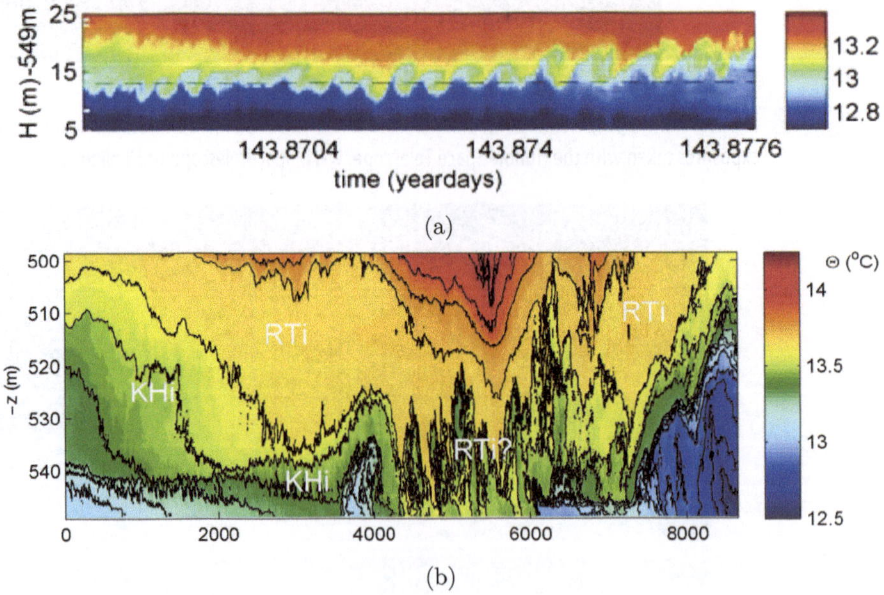

Fig. 1.16 (a) A deep-ocean Kelvin–Helmholtz billow train. Fig. 3 of Van Haren and Gostiaux (2010). Reprinted with permission from AGU & John Wiley and Sons. (b) High-resolution moored temperature observations above Great Meteor Seamount, Van Haren (2015). Reprinted with permission from Elsevier.

As we will illustrate in Chapters 10–17, the evolution of the RTI and RMI depends on flow properties such as the viscosity, surface tension, and compressibility, as well as on several external factors such as rotation and magnetic field. Moreover, the development of these hydrodynamic instabilities may be strongly affected by the initial conditions. To make the situation more complicated, multiple instabilities appear at the same time in many important scientific and engineering applications.[1] ICF is one such example, the details of which will be investigated in Chapter 19.

From the microscopic to cosmic scales, this manuscript will show that the RT, RM, and KH instability-induced flows and mixing obey the same physical laws and exhibit similar structures and behaviors. The main feature of this work is to strive to treat the RT and RM instabilities in a uniform framework that clearly delineates their similarities and differences.

[1] Over a thousand articles are published each year that are focused on either the physics or applications of RT, RM, and KH instabilities.

2 The linear stage for a single mode

In this chapter, we focus on some simple cases in which analytical treatments of the RT and RM instabilities can be carried out. This treatment requires the introduction of the key fluid equations.

Much of the early work in the United States and United Kingdom, after G.I. Taylor's 1950 paper, focused on the growth of a single sinusoidal mode, following its development from the linear (small amplitude) to nonlinear (large-amplitude) regimes (Youngs, 2013).

The single-mode RT or RM instability is a rather special case. While such a perturbation profile is introduced strictly for analytical convenience, surfaces with regular grooves are a common result of machining processes in many practical applications. Despite its simplicity, the asymptotic behavior of instabilities growing from such regular surfaces may still be a useful guide to behaviors found in experimental studies at intermediate times. Small irregularities will inevitably grow and break the repetitive symmetry of the initial conditions in the long term.

Mathematically, we assume that the initial interface perturbation can be described by a single sine function of known amplitude (a) and wavelength (λ). For these so-called "single-mode" flows, any perturbation at a plane boundary can be represented as a sum of Fourier modes, $\exp[i(k_x x + k_y y)]$, where k_x and k_y are the Cartesian components of the total wavenumber $k = 2\pi/\lambda = \sqrt{k_x^2 + k_y^2}$. The linear theory is fully general due to the principle of superposition.

In the linear stage, the bubble (lighter material) and spike (denser material) heights h_b and h_s are both equal to the sinusoidal amplitude $a(t)$, where $a \ll \lambda$. The total (peak-to-valley) width of the interface is $h(t) = h_b(t) + h_s(t)$. Thus, in the linear stage, $h = 2a$. In the nonlinear stage, the interface is no longer sinusoidal and so $h_b \neq h_s$. However, the total width of the interface is still $h = h_b + h_s$.

Here, it is important to introduce a nondimensional parameter to characterize the mass density difference between two fluids separated by an initial interface. One option is the density ratio ρ_2/ρ_1, where ρ_2 and ρ_1 are the "heavy" and "light" fluid densities, respectively. An alternative parameter, intimately linked to the RTI and RMI, is known as the *Atwood number*:

$$\mathcal{A} = \frac{\rho_2 - \rho_1}{\rho_1 + \rho_2}. \tag{2.1}$$

The Atwood number (also denoted A or A_t in the literature) is zero for density matched fluids and reaches unity for a fluid to vacuum interface.

We also draw a distinction between whether fluids of the same phase are miscible or immiscible, as each case typically requires a different numerical approach. The existence of a sharp interface is characteristic of immiscible fluids. In this chapter, a sharp interface between two fluids is considered. The effects of viscosity, diffusivity and surface tension are assumed to be negligible here, but do need to be taken into account when modeling experiments.[1] The distinction between the mixing of miscible fluids and immiscible fluids, with surface tension, is particularly important for RT as many experiments use immiscible fluids to study turbulent mixing. In the case of miscible fluids, there is no distinct boundary, at least on a microscopic scale (Corey, 1994).

2.1 Navier–Stokes equations and Reynolds number

Claude-Louis Navier was one of the best known French scientists of several generations. Given his towering reputation, Navier was put in charge of constructing the first monumental suspension bridge in Paris over the Seine River (Fig. 2.1). Unfortunately, he was later blamed when major defects were discovered that led to the dismantling of the entire bridge. Accusations were made that Navier was "too much of a theoretical mathematician" (Cannone and Friedlander, 2003).

It is interesting that Sir George Gabriel Stokes was so little known at the time of his appointment to the Lucasian Chair of Mathematics of Cambridge University in 1849 that *The Times* newspaper and two of three Cambridge newspapers incorrectly stated that his brother, W.H. Stokes of Caius College, had been appointed (Larmor, 1907; Wilson, 1990; Craik, 2005). The list of Lucasian Professors of Mathematics include Sir Isaac Newton, Nobel laureate Paul Dirac, and, more recently, "the No 1 celebrity scientist" Stephen Hawking.[2] Of course, Stokes later became famous for deriving the most important equations of fluid dynamics in 1845, the Navier–Stokes equations. Originally proved

Fig. 2.1 Pont des Invalides, taken from the original drawing by Navier. Rapport à Monsieur Becquey et Mémoire sur les Ponts Suspendus, Imprimerie Royale, Paris (1823).

[1] These issues will be addressed in Chapter 10.

[2] "Mind over matter: Stephen Hawking" – obituary by Nobel laureate Sir Roger Penrose, www.theguardian.com/science/2018/mar/14/stephen-hawking-obituary.

by Claude-Louis Navier, in 1822, the equations were rediscovered or re-derived at least three other times, by Augustin-Louis Cauchy in 1823, by Siméon Denis Poisson in 1829, and by Jean Claude Barré de Saint-Venant in 1837 (Darrigol, 2002).

Applying momentum conservation to an infinitesimal fluid element gives rise to the well-known relation for a single fluid:

$$\frac{\partial \rho u_i}{\partial t} + \frac{\partial \rho u_i u_j}{\partial x_j} = -\frac{\partial p}{\partial x_i} + \frac{\partial \tau_{ij}}{\partial x_j} + F_i, \tag{2.2}$$

where ρ is the fluid density, $\mathbf{u}$ is the fluid velocity, p is the pressure, and $F(t)$ is the external source term.

The deviatoric stress tensor τ_{ij} is given by

$$\tau_{ij} = \mu \left[\frac{\partial u_i}{\partial x_j} + \frac{\partial u_j}{\partial x_i} - \frac{2}{3}\delta_{ij}\frac{\partial u_n}{\partial x_n} \right], \tag{2.3}$$

where μ is the dynamic viscosity. The kinematic viscosity of the fluid is defined as $\nu = \mu/\rho$. The Euler equations are a special case of the Navier–Stokes equations in the limit of $\mu = 0$.

The equation governing the density transport and mass conservation is

$$\frac{\partial \rho}{\partial t} + \nabla \cdot (\rho \mathbf{u}) = 0. \tag{2.4}$$

Equation 2.4 can also be written in the form

$$\frac{\mathrm{D}\rho}{\mathrm{D}t} + \rho \nabla \cdot \mathbf{u} = 0, \tag{2.5}$$

where the total derivative

$$\frac{\mathrm{D}}{\mathrm{D}t} = \frac{\partial}{\partial t} + \mathbf{u} \cdot \nabla, \tag{2.6}$$

describes the time evolution for a fluid parcel following the flow. If a fluid parcel is incompressible, its density must be constant $D\rho/Dt = 0$, leading to the result $\nabla \cdot \mathbf{u} = 0$ for incompressible flows.

When discussing a Navier–Stokes fluid, there is little if any dispute regarding either the dynamical equation (see Eq. 2.2) or the physical parameters appearing in it. Thus, if one is describing water close to standard conditions, the mass density ρ is nearly 1 g/cm^3 and the kinematic viscosity ν is roughly 1 centistoke (Bird et al., 2006).[3] The parameters of other fluids, when not calculable from first principles, can be readily obtained from engineering handbooks. For dilute gases, the calculations can, in fact, be carried out from first principles (Chapman-Enskog theory) and have better than qualitative agreement with measurements for monatomic gases (Chapman and Cowling, 1990; Balescu, 1975).

It should be stressed that, unlike many other branches of physics, turbulence researchers do not have to deal with uncertainty in the underlying governing equations (Nelkin, 1994). The Navier–Stokes equations are partial differential equations that are widely believed to contain all the physics necessary to describe fluid motion at the macroscale, including turbulent flows (Frisch, 1995).

[3] The physical unit for kinematic viscosity is the stokes (St), named after Sir George Stokes.

The properties of compressible fluid flow may be shown to depend on the Mach number M. In general, the Mach number is defined as the ratio of the bulk flow speed to the local speed of sound (e.g., Gombosi, 1994; Rotty, 1962). The speed of sound is the speed at which small disturbances, especially acoustic disturbance, propagate relative to the movement of the substance (Laney, 1998). Note that the sonic velocity is not constant but varies with the thermodynamic state of the fluid. Flows with $M \gg 1$ exhibit fundamentally different behaviors from the incompressible case $M \ll 1$ and must be described using modified governing equations and numerical methods (e.g., Gauthier, 2017; Livescu, 2020).

By necessity, experimental investigation has played a key role in turbulence research for many years. Observations from the data collected at the University of Manchester's tank resulted in Osborne Reynolds' discovery of a parameter, now bearing his name, for distinguishing a turbulent flow from a laminar one (Reynolds, 1883).

The Navier–Stokes equations, when placed in an appropriate nondimensional form, contain a key dimensionless parameter, the Reynolds number (*Re*). For a viscous, incompressible fluid with constant properties, Eqs. 2.2 and 2.4 reduce to

$$\frac{\partial \mathbf{u}}{\partial t} + \mathbf{u} \cdot \nabla \mathbf{u} = \nu \nabla^2 \mathbf{u} - \frac{\nabla p}{\rho_0}, \tag{2.7}$$

where ρ_0 is the uniform density, and the incompressibility condition is $\nabla \cdot \mathbf{u} = 0$.

Let us consider a flow with characteristic length and velocity scales, $\mathcal{L}$ and $\mathcal{U}$, respectively. By defining the dimensionless velocity field $\mathbf{u}^* = \mathbf{u}/\mathcal{U}$ and gradient $\nabla^* = \mathcal{L}\nabla$, Eq. 2.7 can be recast in the dimensionless form

$$\frac{\partial \mathbf{u}^*}{\partial t^*} + \mathbf{u}^* \cdot \nabla^* \mathbf{u}^* = \frac{\nu}{\mathcal{L}\mathcal{U}} \nabla^{*2} \mathbf{u}^* - \nabla^* \left(\frac{p}{\rho_0 \mathcal{U}^2} \right), \tag{2.8}$$

where $t^* = \mathcal{U}t/\mathcal{L}$ is a dimensionless time. In this form, it is clear that the pressure scale is the familiar $\rho_0 \mathcal{U}^2$, while a single dimensionless parameter, the Reynolds number (Re), is defined as the ratio between inertial and viscous forces,

$$\mathrm{Re} = \frac{\mathcal{L}\mathcal{U}}{\nu}. \tag{2.9}$$

As pointed out by Jackson and Launder (2007), it is no exaggeration to assert that Reynolds' two principal papers on turbulent flow (Reynolds, 1883, 1895), which were both published in the *Philosophical Transactions of the Royal Society*, essentially provided a marker for the direction of research in fluid mechanics for the next century.

2.2 The classical linear RT problem

Chandrasekhar (1961), as well as Drazin and Reid (1981), present detailed derivations of the RTI linear stability for the case of a sinusoidal perturbation in the absence of diffusion. Chandrasekhar also includes analyses for the cases with and without surface tension, and for inviscid fluids as well as viscous fluids with differing viscosities. In addition, he extends the analysis to include hydromagnetic effects.

Naturally, the Navier–Stokes equations are nonlinear differential equations for describing flow fields. It is important to show that the governing equations can recover the linear instability phase. We will carry out this analysis explicitly for the RT instability and then show that the corresponding RM case is its impulsive limit (e.g., Zhou et al., 2021).

The classical RT problem consists of two semi-infinite incompressible ($\nabla \cdot \mathbf{u} = 0$) and inviscid ($\mu = 0$) fluids in a gravitational field, with the denser fluid at a higher gravitational potential (Rayleigh, 1883). The question posed is then: How does an arbitrarily small perturbation to the fluid interface evolve, based on the Navier–Stokes equations, from a static initial condition?

The key to solving the classical RT problem is to first expand the governing equations about the static initial condition and then to linearize the system with small deviations in ρ, p, and $\mathbf{u}$. The smallness of the perturbation allows one to neglect high-order terms in the governing equations and thereby determine the decay or growth in time; that is, whether the system will return to or depart from equilibrium. The latter outcome indicates an instability. While this picture seems neat and qualitatively well-defined, in many cases, it is very difficult not only to determine whether a system is stable but also to define what we really mean by stability! For example, the vanishing derivatives of ρ in the bulk of either fluid (i.e., anywhere *except* at the two-fluid interface) reduces the continuity equation to

$$\nabla \cdot \mathbf{u} = 0. \tag{2.10}$$

Further, continuity of the velocity field requires that, while this condition was derived away from the fluid interface, it must nonetheless be maintained there. However, the density discontinuity means that its derivatives certainly *cannot* be ignored there. Thus, expanding about the unperturbed state ($\rho = \rho_0$, $\mathbf{u} = \mathbf{0}$) by writing $\rho = \rho_0 + \delta\rho$, we find the lowest order in the perturbation variables

$$\frac{\partial(\delta\rho)}{\partial t} + u_z \frac{\partial \rho_0}{\partial z} = 0. \tag{2.11}$$

For the following equations, we have adopted a Cartesian coordinate system and selected z as n, the direction normal to the interface.

Turning to the Navier–Stokes equations, they reduce in the present case to the Euler equations with the gravitational force $F = -\rho g\hat{\mathbf{z}}$, where g is the gravitational acceleration. The momentum equation becomes

$$\rho \frac{\mathrm{D}\mathbf{u}}{\mathrm{D}t} = -\nabla p - \rho g\hat{\mathbf{z}}, \tag{2.12}$$

where the constant gravitational field is taken to point along $-\hat{\mathbf{z}}$. Expanding about the unperturbed state (now also writing $p = p_0 + \delta p$), while requiring the resulting equations to be satisfied for the unperturbed equilibrium, yields

$$\nabla p_0 + \rho_0 g\hat{\mathbf{z}} = 0, \tag{2.13}$$

$$\rho_0 \frac{\partial \mathbf{u}}{\partial t} + \nabla(\delta p) + (\delta\rho) g\hat{\mathbf{z}} = 0. \tag{2.14}$$

The $\mathcal{O}(1)$ equation produces the usual result for hydrostatic pressure in a fluid:

$$p_0(z) = p_0(0) - \rho_0 g z, \tag{2.15}$$

while the next order equation of perturbation expansion constitutes the linearized Navier–Stokes equation for the Rayleigh–Taylor problem.

Fourier transforming Eqs. 2.10, 2.11, and 2.14 in x, y, and t, then algebraically eliminating u_x, u_y, δp, and $\delta\rho$ yields:

$$\frac{\partial}{\partial z}\left(\rho_0 \frac{\partial u_z}{\partial z}\right) - k^2 \rho_0 u_z = \frac{k^2 g}{\omega^2}\frac{\partial \rho_0}{\partial z} u_z, \tag{2.16}$$

where k_x, k_y, and ω are the Fourier transform variables corresponding to x, y, and t, respectively, and $k^2 = k_x^2 + k_y^2$. In the bulk of either fluid, we have $\partial\rho_0/\partial z = 0$, so that Eq. 2.16 reduces to

$$\frac{\partial^2 u_z}{\partial z^2} - k^2 u_z = 0, \tag{2.17}$$

which admits the solution

$$u_z(z) = u_{z0} e^{-k|z|}, \tag{2.18}$$

where u_{z0} is a constant, and we have imposed both continuity of u_z at the interface $z = 0$, as well as the boundary condition that u_z remains finite as $z \to \pm\infty$. In order to determine the dispersion relation $\omega = \omega(k)$, we next integrate Eq. 2.16 across the interface to obtain

$$\left[\!\left[\rho_0 \frac{\partial u_z}{\partial z}\right]\!\right] = \frac{k^2 g}{\omega^2} [\![\rho_0]\!]\, u_z(0). \tag{2.19}$$

The square brackets indicate the difference across the interface. Finally, substituting Eq. 2.18 into this expression yields the result

$$\omega = \sqrt{\left(\frac{\rho_1 - \rho_2}{\rho_1 + \rho_2}\right) kg}, \tag{2.20}$$

where ρ_1 and ρ_2 are the "bottom" and "top" fluid densities, respectively. Thus, we find that if $\rho_1 > \rho_2$ (the denser fluid is on the bottom) then ω is real and the system is stable, while if $\rho_1 < \rho_2$ (the denser fluid is on top) the frequency is imaginary, and the velocity grows exponentially in time, $u_z(t) = u_z(0)e^{\gamma t}$, with a growth rate $\gamma = i\omega$.

Integrating $u_z(t)$ with respect to time, we obtain the perturbation amplitude in the (early time) linear regime:

$$a(t) = a_0 e^{\gamma t}, \tag{2.21}$$

where a_0 is the initial amplitude. This demonstrates that the RTI significantly enhances interfacial perturbations at early times. The RT growth rate,

$$\gamma = \sqrt{kg\mathcal{A}}, \tag{2.22}$$

increases with the perturbation wavenumber, density ratio, and acceleration. The Atwood number of the system, $\mathcal{A}$, has been defined in Eq. 2.1.

More generally, the RT interface perturbation amplitude $a(t)$ evolves as (Lamb, 1932)

$$\frac{d^2 a}{dt^2} = kg\mathcal{A}a, \tag{2.23}$$

with any time-dependent acceleration of the interface $g(t)$. Constant gravitational acceleration, leading to Eq. 2.22, is just a convenient special case. For the RT instability $\mathcal{A}g$

is positive; if $\mathcal{A}g$ is negative, Eq. 2.23 describes oscillations at a stable interface, as considered by Lamb (1932).

Other RTI analyses have been based on an initial velocity distribution at the interface, see references Mikaelian (1983); Jacobs and Sheeley (1996); Mikaelian (2019). The linear theory gives modes $\exp(\pm\gamma t)$. This means there is a choice of an $a = a_0 \cosh(\gamma t)$ (initial amplitude) mode or an $a = a_0 \sinh(\gamma t)$ mode (which assumes an initial velocity distribution; see, e.g., Eq. (11) of Mikaelian (1983)). Historically, before RM was considered, most researchers adopted the natural choice $\dot{a}(0) = 0$. For example, Taylor explicitly stated that if the initial amplitude is finite and the initial velocity is zero then the solution is $a_0 \cosh(\gamma t)$. Indeed, one had no incentive to consider a finite initial $\dot{a}(0)$, when RTI is the only consideration (Mikaelian, 2019). For an initial vorticity and an initial velocity (in the sense of $\dot{a}$), the two can be related. This was done by Jacobs and Sheeley (1996) (see their Eq. (17)). If both the initial amplitude (a) and velocity ($\dot{a}$) are provided, however, then the two may become degenerate and cancel each other at late times, as Mikaelian (2019) discussed recently.

2.3 The classical linear RM problem

Despite the fact that the RMI canonically involves a shock-interface interaction, and hence compressible bulk fluids, the linear growth phase in many practical applications closely approximates the incompressible, impulsive RTI solution at early times (Richtmyer, 1960). The RM instability may be viewed as a version of the RTI in which the gravitational force is impulsive in the weak shock limit. Richtmyer (1960) recognized this connection with RTI and applied the linear theory of Taylor (1950), but modeling the shock-interface interaction as the impulsive acceleration of two incompressible fluids. Specifically, the acceleration is given by $g(t) = \Delta V\,\delta(t)$ (i.e., applied instantaneously at $t = 0$) on an initial amplitude of a_0, such that $\int g(t)dt = \Delta V$ is the interfacial velocity imparted by the impulse, and $\delta(t)$ is the Dirac delta function. In light of these expressions, Eq. 2.23 becomes

$$\frac{d^2a}{dt^2} = k\mathcal{A}\Delta V\delta(t)a, \tag{2.24}$$

which can be immediately integrated to yield

$$\frac{da}{dt} = k\mathcal{A}\Delta V a_0. \tag{2.25}$$

One complication in the RMI case is that the compression due to the passing shock changes both the Atwood number, $\mathcal{A}$, and the initial perturbation amplitude a_0. Richtmyer (1960) chose to use the post-shock values to arrive at

$$\frac{da}{dt} = k\mathcal{A}^+\Delta V a_0^+. \tag{2.26}$$

The superscript $+$ is introduced to distinguish the post-shock quantities from the pre-shock quantities (denoted with superscript $-$).

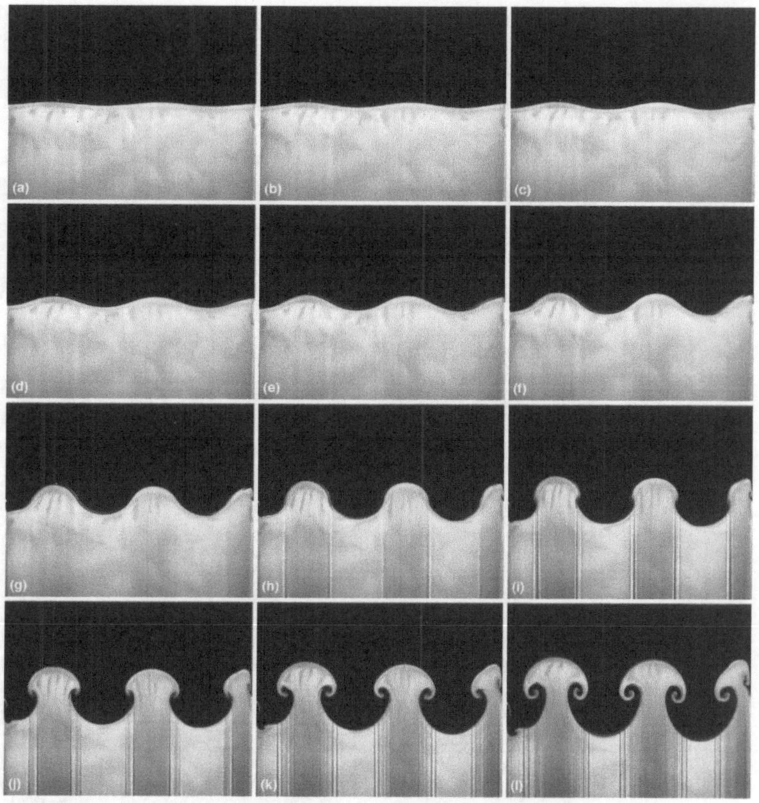

Fig. 2.2 A sequence of planar laser-induced fluorescence (PLIF) images showing the development of a miscible RTI system with A = 0.155 accelerated at 0.74 g. The initial perturbation wavelength was 54 mm. The first frame (a) was taken immediately after the test sled was released and there is a 0.033 s increment between each subsequent image. Fig. 4 of Waddell et al. (2001).

While we have emphasized the similarity of RT and RM instabilities, Eq. 2.26 reveals an RM amplitude growth that is linear in time, rather than the exponential growth of RTI. Another key distinction is that the shock wave alters the Atwood number and several other important parameters.

Various modifications to the linear RMI growth velocity have been proposed, and the interested reader is referred to the reviews by Brouillette (2002) and Zhou (2017a). In the present context, it is worthwhile mentioning two examples.

Meyer and Blewett (1972) carried out a 2D Lagrangian simulation of the single-mode RM flow in Meshkov's experiments and found that improved agreement can be achieved with

$$\frac{da}{dt} = k\Delta V \mathcal{A}^{+}\frac{1}{2}\left[a_0^{+} + a_0^{-}\right]. \tag{2.27}$$

Heuristically, Vandenboomgaerde et al. (1998) proposed another modified Richtmyer model,

$$\frac{da}{dt} = k\Delta V\left\{\frac{1}{2}\left[a_0^{+}\mathcal{A}^{+} + a_0^{-}\mathcal{A}^{-}\right] - \frac{1}{6}\left[(\mathcal{A}^{+} - \mathcal{A}^{-})(a_0^{+} - a_0^{-})\right]\right\}. \tag{2.28}$$

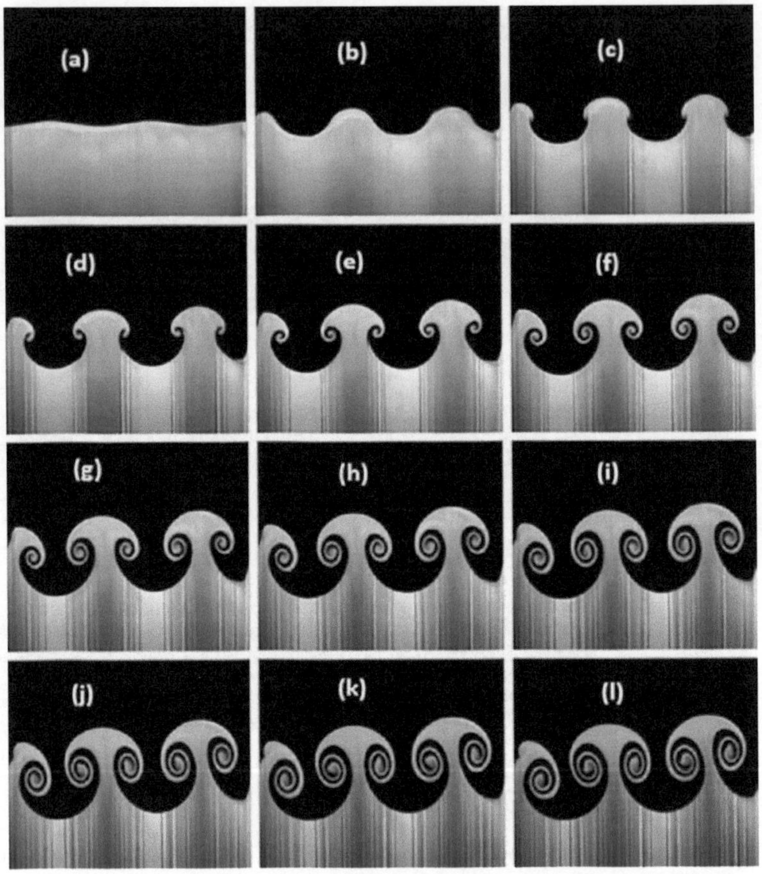

Fig. 2.3 A sequence of images from a RMI experiment with $2\frac{1}{2}$ waves and a dimensionless initial amplitude of $ka_0 = 0.16$. Times relative to the midpoint of spring impact are (a) 25 ms, (b) 92ms, (c) 175 ms, (d) 259 ms, (e) 342 ms, (f) 426 ms, (g) 509 ms, (h) 592 ms, (i) 676 ms, (j) 759 ms, (k) 843 ms, and (l) 909 ms. Fig. 6 of Niederhaus and Jacobs (2003).

If the compressibility of the two fluids is similar, $\mathcal{A}^+$ and $\mathcal{A}^-$ are likely to be similar. In this case, Eq. 2.28 reduces to Eq. 2.27. Several additional works have discussed the linear phase of the RM instability (e.g., Fraley, 1986; Yang et al., 1994b; Nishihara et al., 2010; Wouchuk and Nishihara, 1996, 1997; Cobos-Campos and Wouchuk, 2014, 2016, 2017).

The linear RTI and RMI theory is fully general due to the principle of superposition. However, eventually the perturbation grows large enough to invalidate the linearizing assumption, and nonlinear effects begin to dominate (namely, when $a \simeq \lambda$ so that $ka(t) > 1$). This will be the subject of the next chapter.

The visualizations from single-mode RT and RM experiments, such as those in Collins and Jacobs (2002), are often used for verifying and validating numerical simulation codes (e.g., Henry de Frahan et al., 2015b,c; Kaman et al., 2021; Kaman and Holley, 2022). Figures 2.2 and 2.3 show the time-dependent evolutions of RTI and RMI from laboratory experiments, from the linear to nonlinear phases. The characteristic bubble-spike configurations and the peculiar mushroom-shapes are due to the effects of higher harmonics, which come into play beyond the range of validity of the linear analysis. In particular, the linear growth phase may be very short-lived (and the utility of linear theory limited) in experiments and applications which evolve from finite amplitude (rather than infinitesimal) perturbations. In any case, small irregularities will inevitably grow and break the repetitive symmetry of the initial single-mode conditions in the long term.

2.4 The classical Kelvin–Helmholtz instability

The Kelvin–Helmholtz Instability (KHI) occurs at the shearing interface between two counter-flowing streams of fluid. If the interface between the two fluids is absolutely smooth, the fluids will, in principle (at least mathematically!) maintain laminar flow. If the interface is perturbed, however, the perturbation will grow, under the influence of the KHI (Chandrasekhar, 1961; Drazin and Reid, 1981). The development of KHI often constitutes the critical step toward the onset of turbulent mixing observed both in the atmosphere and in the oceans (Woods, 1968; Peltier and Caulfield, 2003). Also, the KHI is the reason for the creation of mushroom structures during the RT and RM nonlinear regimes (Yabe et al., 1991; Collins and Jacobs, 2002; Zabusky, 1999).

Lord Kelvin (Sir William Thomson) made fundamental contributions to fluid mechanics (for a summary, see Craik, 2012), with several concepts named after him (e.g., Kelvin circulation theorem, Kelvin waves and others). When Thomson was appointed Professor of Natural Philosophy at the University of Glasgow at the tender age of 22, he already had twenty-six published papers (Lekner, 2017). He was most famous for his work in the laws of thermodynamics and the SI unit of thermodynamic temperature was posthumously named after him. He also contributed significantly as an advisor to the transatlantic telegraph project, making a key improvement to the receiver (Schils, 2012), for which he was knighted by H.M Queen Victoria. Hermann Ludwig Ferdinand von Helmholtz, a great scientist himself, maintained that "the gift to translate real facts into mathematical equations, and vice versa, is by far rarer than that to find the solution of a given mathematical problem," adding, "And in this direction Sir William Thomson is most eminent and original" (Thompson, 1907).

Helmholtz first met Lord Kelvin at Bad Kreuznach in 1855; Thomson's lively intellect made a deep impression on him and they cemented a close personal friendship (Wade, 1994). Helmholtz had a broad range of interests, from astronomy to physics, physiology, and the relationship between the physical world and human sensory perception. As pointed out by Gruber and Gruber (1956), few scholars in any branch of knowledge have left so varied a record of research and discovery as Helmholtz. His scientific papers (Wissenschaftliche Abhandlungen) were later published in three volumes.

Lord Kelvin was deeply inspired by Helmholtz's famous paper "Ueber Integrale der hydrodynamischen Gleichungen, welche den Wirbelbewegungen entsprechen," translated by P.G. Tait and published in English under the title "On integrals of the hydrodynamical equations, which express vortex-motion." Apart from the KHI, the legacy of Helmholtz and

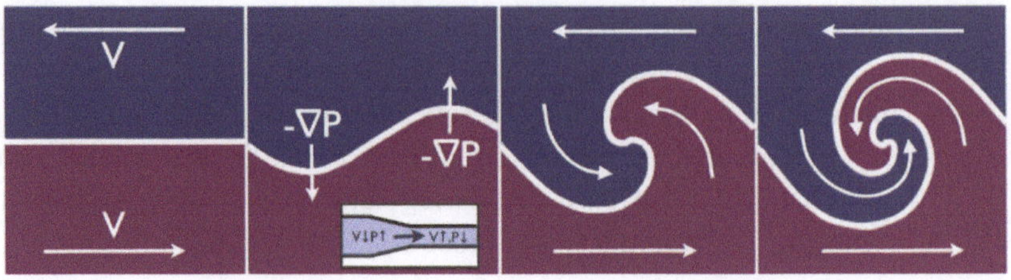

Fig. 2.4 The KH instability occurs at the interface between two fluids with velocity shear. The deformation into the lower fluid leads to a reduction of pressure below the deformation and an increased pressure above the deformation so the pressure gradient is upward, and the pressure force pushes the deformation downward. Similarly the upward deformation is pushed upward by the pressure force. The deformation rolls up as it is caught up in the flow, leading to the formation of a vortex. Fig. 1 of Johnson et al. (2014). Reprinted with permission from Springer.

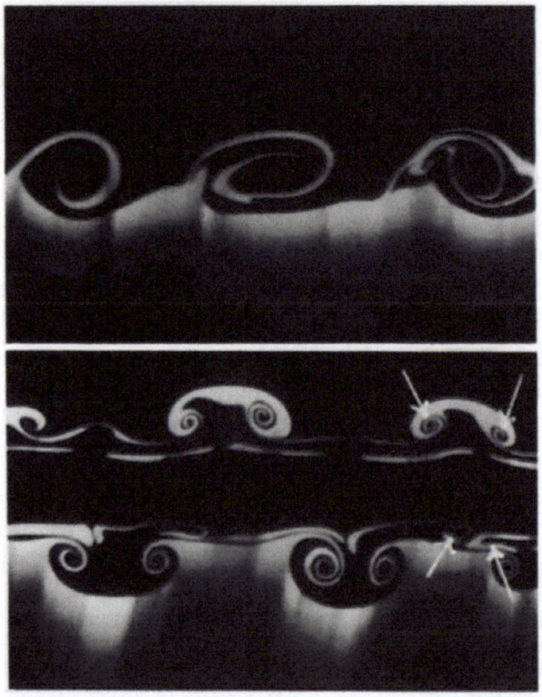

Fig. 2.5 Images from the experiments showing a side view (top) and end view (bottom) of a KH shear layer. The end view is from an experiment without buoyancy effects, yet it still exhibits the mushroom shaped structures. Reproduced with permission from Schowalter et al. (1994).

Kelvin is their pioneering work on vortex dynamics (Moffatt, 2008), a subject fundamental to the topics discussed in this manuscript.

As in the case of the RTI and RMI, the classical KHI analysis assumes a sinusoidal perturbation. The analyses presented by Chandrasekhar (1961) and by Drazin and Reid (1981) also include the effects of buoyancy, surface tension, rotation and a diffuse (rather

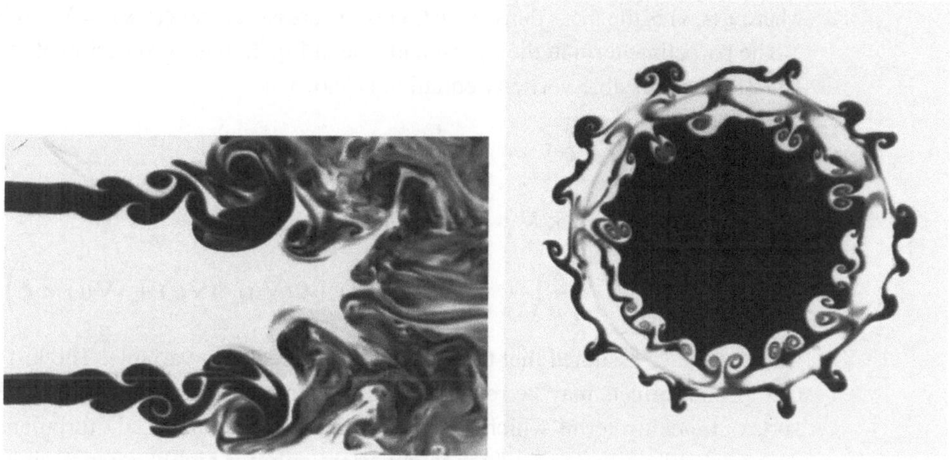

Fig. 2.6 Images from the experiments of Villermaux et al. showing a side view (left) and the an end view (right) of circular jet. Both the inner shear layer and the outer shear layer exhibit the mushroom shapes in the end view. Reproduced from Villermaux et al. (1998). Reprinted with permission from AIP Publishing.

than discontinuous) interface. For the case of buoyancy, and with a discontinuous interface, the KHI exponential growth rate is given by

$$\gamma = \pm \left\{ gk\mathcal{A} + \rho_1 \rho_2 k^2 \left[\frac{V_1 - V_2}{\rho_1 + \rho_2} \right]^2 \right\}^{1/2} - ik \left\{ \frac{\rho_1 V_1 + \rho_2 V_2}{\rho_1 + \rho_2} \right\}, \tag{2.29}$$

where g is the acceleration normal to the interface, V_1 and V_2 are the fluid velocities along the interface, k is the wavenumber of the disturbance, and ρ_1 and ρ_2 are the fluid densities on each side of the interface (Fig. 2.4).

The KHI plays a significant role in the nonlinear RTI and RMI. The characteristic mushroom shapes are induced by a shear instability. The mushroom shape is, in fact, physically associated with the shear, rather than the RTI or RMI – see Fig. 2.5 adapted from Schowalter et al. (1994). The mushroom features also appear in other sheared flows, for example, around the jet shown in Fig. 2.6, where both fluids are the same ($\mathcal{A} = 0$) (Villermaux et al., 1998). The fluids are distinguished in the experiment via a dye.

2.5 Vorticity paradigm

The *vorticity paradigm* provides a physically intuitive perspective (Hawley and Zabusky, 1989; Zabusky, 1999). The RTI and RMI can be viewed as the result of baroclinic torque created by a mismatch between the density gradient and the pressure gradient (Taylor, 1950), as given by the inviscid 2D vorticity equation (Kundu and Cohen, 2004),

$$\frac{D\zeta}{Dt} = \frac{1}{\rho^2} \nabla\rho \times \nabla p, \tag{2.30}$$

where $\rho(t,\mathbf{x})$ is the mass density, $p(t,\mathbf{x})$ is the pressure, and $\boldsymbol{\zeta}(t,\mathbf{x}) = \nabla\times\mathbf{u}$ is the vorticity.[4]

The baroclinic term in the right-hand side of Eq. 2.30 is most pertinent to our discussion. The 3D compressible vorticity equation is shown as

$$\begin{aligned}\frac{D\boldsymbol{\zeta}}{Dt} &= \boldsymbol{\zeta}\cdot\nabla\mathbf{u} - \boldsymbol{\zeta}(\nabla\cdot\mathbf{u}) + \frac{1}{\rho^2}\nabla\rho\times\nabla p \\ &+ \frac{\mu}{\rho^2}(\nabla\rho\times\nabla\times\boldsymbol{\zeta}) - \frac{4\mu}{3\rho^2}[\nabla\rho\times\nabla(\nabla\cdot\mathbf{u})] + \frac{\mu}{\rho}\nabla^2\boldsymbol{\zeta} \\ &+ \left\{\nabla\times\left[\frac{1}{\rho}\left(-\frac{2}{3}(\nabla\cdot\mathbf{u})(\nabla\mu) + 2(\nabla\mathbf{u})\cdot(\nabla\mu) + (\nabla\mu)\times\boldsymbol{\zeta}\right)\right]\right\},\end{aligned} \tag{2.31}$$

where we have assumed that the viscosity, μ, is spatially variable. The last two lines show that viscous effects may be relevant. The remaining terms on the right of Eq. 2.31 are the vortex stretching term, which is essential for discussions of 3D turbulence and mixing; the vortex dilatation term, which is important only for highly compressible fluids; and the baroclinic term (Hewett and Madnia, 1998; Kida and Orszag, 1990b; Porter et al., 2015; Zabusky, 1999).

As mentioned previously, Rayleigh's original interest was the formation of cirrus clouds in the presence of gravity, but in Taylor's problem the light fluid was accelerated into the heavy fluid without gravity. The equivalence of the two problems studied by Rayleigh and Taylor becomes clear when one considers that in both cases the pressure gradient (∇p) points from the heavy fluid to the light fluid; the unstable stratification is defined by the condition $\nabla p\cdot\nabla\rho < 0$ (Dalziel et al., 2008; Tryggvason and Unverdi, 1990; Banerjee, 2020).

In the RTI, the pressure gradient is supplied by the acceleration field (e.g., gravity) which imposes hydrostatic pressure on the fluid. At the interface, if the pressure gradient opposes the density gradient, but is not perfectly aligned to it, vorticity will be gradually generated, leading to the instability (Fig. 2.7).

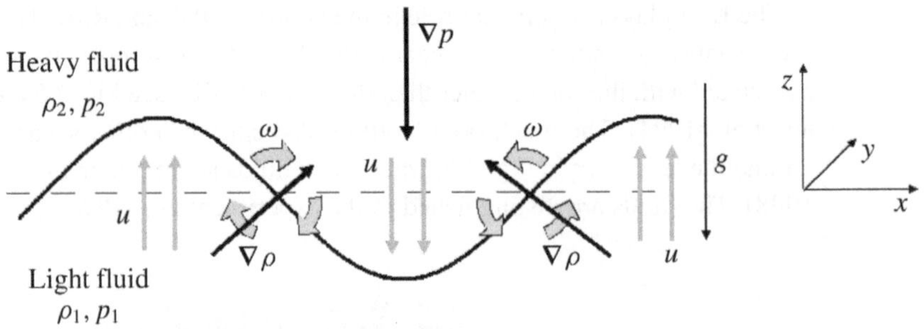

Fig. 2.7 Visualization of an unstable RT configuration. Baroclinic torque at the interface creates vorticity and induces a velocity field that increases the baroclinic torque. Here, p is pressure, ρ is density, u is velocity, and g is gravity. The vorticity is denoted by ω, instead of ζ used in this book and the atmospheric sciences community. Fig. 1 of Roberts and Jacobs (2016).

[4] Here, we shall follow the notation convention of the vorticity from the atmospheric sciences community, ζ (e.g., Orszag, 1970b), instead of ω used by many authors. In this book, ω is the Fourier transform variables corresponding to t, see Fig. 2.16.

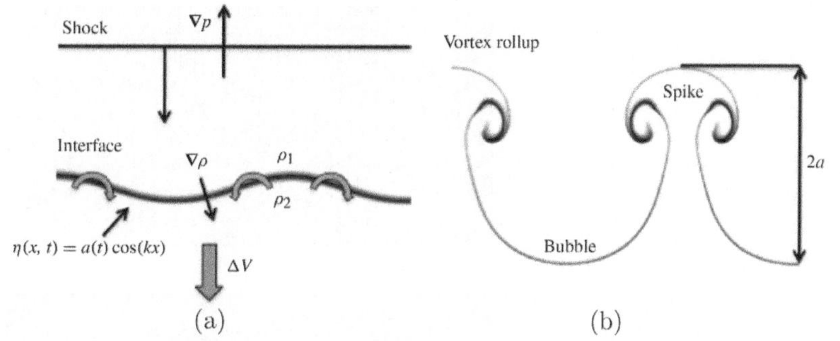

Fig. 2.8 Explanation of the RMI deposition of baroclinic vorticity by a shock. (a) The shock wave travels downward toward the interface and imposes a pressure gradient across the density gradient, generating baroclinic vorticity on the interface. (b) As the interface travels along with the mean post-shock flow, this baroclinic vorticity rolls up into spikes of heavy fluid between bubbles of light fluid. Note that the interface velocity jump, in Richtmyer's formalism, is denoted as ΔV here. Fig. 1 of Morgan et al. (2012).

A similar mechanism operates in the RMI. The shock wave travels through the interface and applies a pressure gradient across the perturbed density gradient, generating baroclinic vorticity on the interface. Vorticity deposited on an interface will cause it to curl up under the influence of the induced rotational velocity, **u**. This is illustrated in Fig. 2.8, adapted from Morgan et al. (2012).

Matsuoka et al. (2003) summarized the existence of the vorticity on the interface at $t = 0$ for the three instabilities:

- RTI: no vorticity
- RMI: an inhomogeneous (2D) weak vorticity
- KHI: a homogeneous (1D) strong vorticity

As the instability evolves, the vorticity begins to concentrate at points midway between the crests and troughs. These correspond to the points of maximum initial interface slope (Zabusky, 1999). The resulting vortices produce the symmetric mushroom pattern typical of the RT and RM instability (see Fig. 2.9 for an RMI example).

Unlike the RTI, RMI can be induced regardless of the direction of the fluid accelerations. While Richtmyer only considered the light-to-heavy configuration, his derived equation applies equally to both. Meshkov (1969) pointed out the applicability of Richtmyer's impulsive model to both configurations and carried out shock tube experiments with the acceleration direction from both sides (Collins and Jacobs, 2002). Figure 2.10 shows how the RM shock causes a phase inversion for the heavy/light configuration. It also illustrates that the vorticity direction depends upon the shock direction (Puranik et al., 2004). The first configuration has a similar orientation to that of RTI. When the shock is traveling from light to heavy fluid, ($\nabla p \cdot \nabla \rho < 0$) results in a direct amplification of the initial interface perturbations. Most studies in this field focus on this case because of its direct applicability to ICF (Jourdan and Houas, 2005; Collins and Jacobs, 2002; Mikaelian, 1985, 1998; Sadot et al., 1998; Chen et al., 2023). In the opposite case of a shock traveling from heavy to light

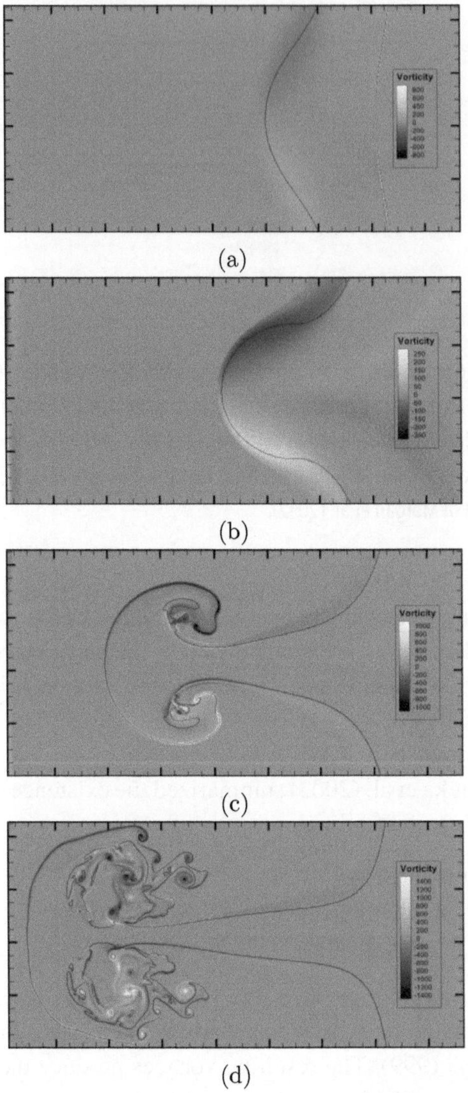

Fig. 2.9 Out-of-plane vorticity arising from the 2D RM instability between air and SF_6, produced using implicit large eddy simulations. Also shown is the contour line denoting the volume fraction $f_1 = 0.5$. Upon shock interaction, baroclinic vorticity is deposited at the interface, causing perturbations to grow and the interface to roll up into the classic mushroom shape. The shock is visible near the right boundary in the first frame. Zhou et al. (2021).

fluid ($\nabla p \cdot \nabla \rho > 0$), there is a phase reversal. The perturbations diminish and pass through zero, after which they continue to grow, maintaining the π phase change (Brouillette, 2002; Jourdan and Houas, 2005). The RMI evolution may be classified into two distinct phases: the deposition of vorticity during shock contact, and the vorticity evolution coupled with secondary baroclinic effects after the shock has passed (Samtaney and Zabusky, 1994).

The strength of a vortex, in a domain ϱ enclosed by contour ℓ, can be represented by the circulation Γ, which is the summation of vorticity ζ in the domain,

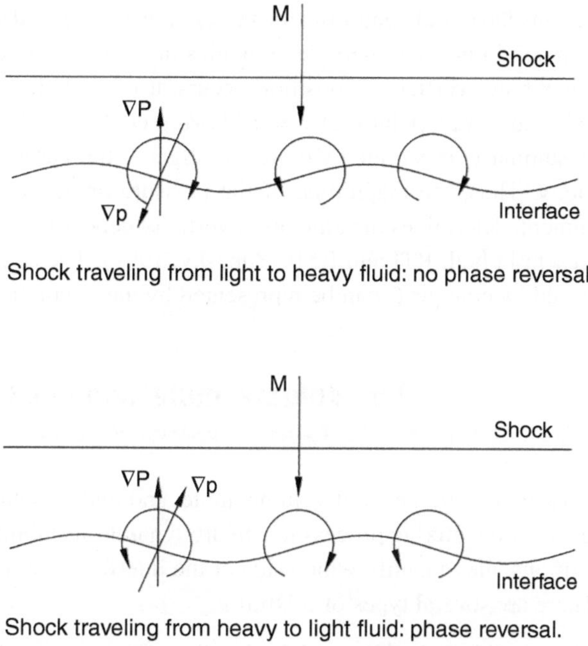

Fig. 2.10 Schematic showing the shock-interface interaction for the case of a shock travelling from a light to a heavy gas and vice versa, illustrating whether or not the vorticity deposition results in perturbation phase reversal. Fig. 10 of Puranik et al. (2004). Reprinted with permission from Springer.

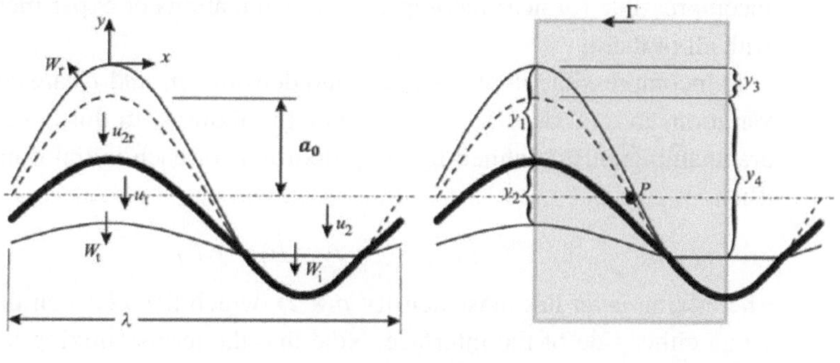

Fig. 2.11 Schematic of a shock wave interaction with a sinusoidal perturbation. There are two vortices (of equal strength but of opposite sign) per wavelength. The circulation of the vortex at point P can be found by integrating the velocity around the gray rectangle perimeter. Fig. 1 of Motl et al. (2009). Reprinted with permission from AIP Publishing.

$$\Gamma = \int_{\varrho} \zeta d\varrho = \int_{\ell} \mathbf{u} \cdot d\ell. \tag{2.32}$$

The integral is taken along the boundary (ℓ) that exactly encloses the perturbed region (Sun and Takayama, 2003). Indeed, circulation deposition on interfaces is a global quantification of the vorticity generation in the early-time RMI process and is the subject of significant modeling efforts (Zabusky, 1999 and references therein).

Circulation is deposited during the shock wave interaction with the perturbation. For the single wavelength shown in Fig. 2.11, the thick line is the interface, while the dashed line represents the initial condition; the initial, transmitted, and reflected shock waves are shown as thin solid lines; and particle velocities in a region are labeled u. Considering Eq. 2.30, the maximum vorticity deposition occurs at the location of the maximum misalignment (maximum slope), point P at $x = \pi/2$ (Motl et al., 2009).

In summary, if ∇p and $\nabla \rho$ are misaligned, the baroclinic term generates or destroys vorticity. Thus, the magnitude of the pressure or density gradients, and their degree of alignment, determines the amount of vorticity deposition on the interface. This mechanism is present in both RTI and RMI (Zabusky, 1999). The strength of a vortex, in a domain ϱ enclosed by contour ℓ, can be represented by the circulation.

2.6 Volume, mole, and mass fractions

The basic measurements of volume, mole, and mass fractions are the starting points for an analysis of the mixing process in a hydrodynamic-instability-induced flow. Using different definitions, they identify which side of the interface that the fluid came from.

There are several types of mixtures:

(a) Immiscible fluids (liquid-liquid or liquid-gas),
(b) Miscible fluids,
(c) Two different gases at the same temperature and pressure,
(d) Two regions of the same fluid but at different temperatures.

Incompressible (or near-incompressible) simulations or experiments have been performed with all of these.

In incompressible flow, the pure-fluid densities ρ_1 and ρ_2 are constant and uniform, and variations in total density ρ arise solely in mixtures. In this case, volume fractions, f_1,f_2, are unambiguously defined by the proportions of each initial fluid present in the mixture. Then

$$\rho = f_1\rho_1 + f_2\rho_2. \tag{2.33}$$

The mixing layer has mass density $\rho(\mathbf{x}, t)$, which lies between the uniform values ρ_1 and ρ_2 on either side of the interface. Note that the terms "mixing layer" and "mixing zone" are used interchangeably in this book.

For random mixtures of immiscible fluids, volume fractions can more generally be defined as the probabilities of a point being in each fluid and then

$$\rho = f_1\overline{\rho}_1 + f_2\overline{\rho}_2, \tag{2.34}$$

where $\overline{\rho}_1$, $\overline{\rho}_1$ need not be constants.

The mixing layer growth can also be robustly defined in terms of entrainment of pure-fluid mole fractions. Assuming $\rho_2 > \rho_1$, the mole fractions of each fluid are defined as (e.g., Cabot and Cook, 2006)

$$X_1(\mathbf{x}, t) = \frac{\rho_2 - \rho}{\rho_2 - \rho_1}, \tag{2.35}$$

$$X_2(\mathbf{x}, t) = \frac{\rho - \rho_1}{\rho_2 - \rho_1} = 1 - X_1. \tag{2.36}$$

By this definition, $0 \le X_1 \le 1$, $0 \le X_2 \le 1$ and $\rho = X_1\rho_1 + X_2\rho_2$.

Now, Avogadro's hypothesis is true for gases – equal volumes of different gases at the same temperature and pressure contain the same number of molecules. Hence for case (c) f_i and X_i are the same – but not for the other cases. For alcohol/water mixtures equal volumes alcohol contains many fewer molecules because of the lower density and higher molecular weight. For case (d) there may well be only one molecular species, that is, the same fluid at two different temperatures.

The mass fractions Y_1 and Y_2 are defined as the fraction of the total density ρ that is contributed by each fluid (Zhou and Cabot, 2019)

$$Y_1(\mathbf{x}, t) = \frac{X_1\rho_1}{\rho} = \frac{X_1\rho_1}{X_1\rho_1 + (1 - X_1)\rho_2}, \tag{2.37}$$

$$Y_2(\mathbf{x}, t) = \frac{X_2\rho_2}{\rho} = \frac{X_2\rho_2}{(1 - X_2)\rho_1 + X_2\rho_2}. \tag{2.38}$$

After some algebra, this leads to

$$Y_1(\mathbf{x}, t) = \frac{\rho^{-1} - \rho_2^{-1}}{\rho_1^{-1} - \rho_2^{-1}}, \tag{2.39}$$

$$Y_2(\mathbf{x}, t) = \frac{\rho_1^{-1} - \rho^{-1}}{\rho_1^{-1} - \rho_2^{-1}} = 1 - Y_1. \tag{2.40}$$

and (Cook and Dimotakis, 2001)

$$\frac{1}{\rho(\mathbf{x}, t)} = \frac{Y_1(\mathbf{x}, t)}{\rho_1} + \frac{Y_2(\mathbf{x}, t)}{\rho_2}. \tag{2.41}$$

The volume fraction of fluid i (f_i) is calculated from the mass fractions (e.g., Youngs, 1994):

$$f_i = \frac{Y_i/\rho_i}{Y_1/\rho_1 + Y_2/\rho_2}. \tag{2.42}$$

The mass fraction profile will be different than the volume fraction profile. Figure 2.12 shows contour plots of the volume and mass fractions in direct numerical simulations of

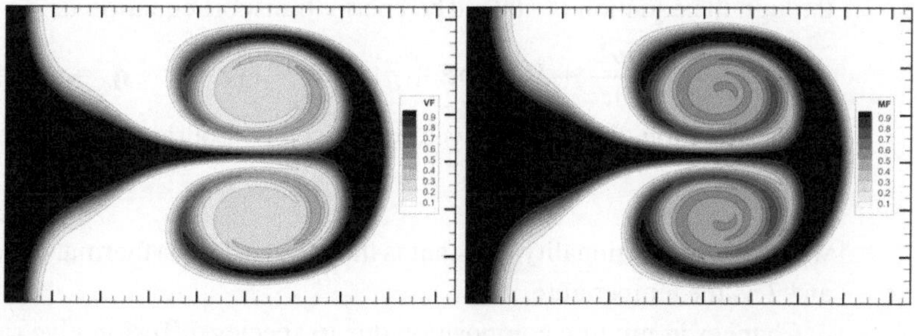

Fig. 2.12 Contours of heavy-fluid volume fractions (VF) and mass fractions (MF) in the 2D single-mode RM instability at an Atwood number of $\mathcal{A} = 0.5$, taken from direct numerical simulations. Note the subtle differences in the locations of contour lines for equal values of volume and mass fractions. Zhou et al. (2021).

the 2D single-mode RM instability. If the constituent species of the mixture are intimately mixed at the molecular level and the species temperatures are locally equal, the volume fraction and mole fraction of each species are identical. Figure 2.12 also shows the subtle differences in the locations of contour lines for equal values of volume and mass fractions for a RM mixing layer.

2.7 Interspecies diffusion and energy transport

If acoustic effects are important, as quantified by the Mach number, then a fully compressible system description must be employed, which allows for the possibility of discontinuous changes in the flow properties, at least when only considering the macroscale and not looking at internal shock structure (which is not well described by the Navier–Stokes equations for shock Mach number > 2). Equations 2.2 and 2.4 must be supplemented by the transport equation for the energy, which is derived by applying the first law of thermodynamics, and species mass fraction. The transport equation for the mass fraction of species m, Y_m, is given by

$$\frac{\partial Y_m \rho}{\partial t} + \nabla \cdot (Y_m \rho \mathbf{u}) = \nabla \cdot \mathbf{J}_m. \tag{2.43}$$

For a two-fluid mixture, the mass flux $\mathbf{J}_m$ allows each species to have a separate velocity. If $\mathbf{J}_m = 0$, the mass fraction Y_m is a constant advected with the flow velocity.

The mass flux for species 1 is found via the application of Fick's law of diffusion, giving

$$\mathbf{J}_1 = \rho D_{12} \nabla Y_1, \tag{2.44}$$

where D_{12} is the binary diffusion coefficient. While often in fundamental studies D_{12} (as well as the viscosity μ) is assumed to be constant, this need not be the case in general. For mixtures of more than two species, Fick's law is often still used but with an additional corrections to ensure mass conservation (Poinsot and Veynante, 2005), whilst a more accurate representation is the Hirschfelder–Curtiss approximation (Hirschfelder et al., 1964).

The total energy per unit mass is given by the sum of internal and kinetic energy, $E = e + \frac{1}{2}\mathbf{u} \cdot \mathbf{u}$. The internal energy per unit mass e is related to the state variables ρ and p through the equation of state. The energy transport equation is

$$\frac{\partial \rho E}{\partial t} + \nabla \cdot \left(\left[\rho E + p\right]\mathbf{u}\right) = \nabla \cdot \left(\boldsymbol{\tau} \cdot \mathbf{u} - \mathbf{q}_c - \mathbf{q}_d\right) + \rho \mathbf{g} \cdot \mathbf{u}. \tag{2.45}$$

In this equation, the heat flux vector is given by Fourier's law of conductivity,

$$\mathbf{q}_c = -\kappa_{\mathrm{T}} \nabla T, \tag{2.46}$$

where the proportionality constant is the (dimensional) thermal conductivity of the material and T is the temperature.

Changes in mixture composition due to species diffusion give rise to changes in energy, for which we must also account. This is done via the enthalpy diffusion flux, defined as:

$$\mathbf{q}_d = -\sum_m h_m \mathbf{J}_m, \tag{2.47}$$

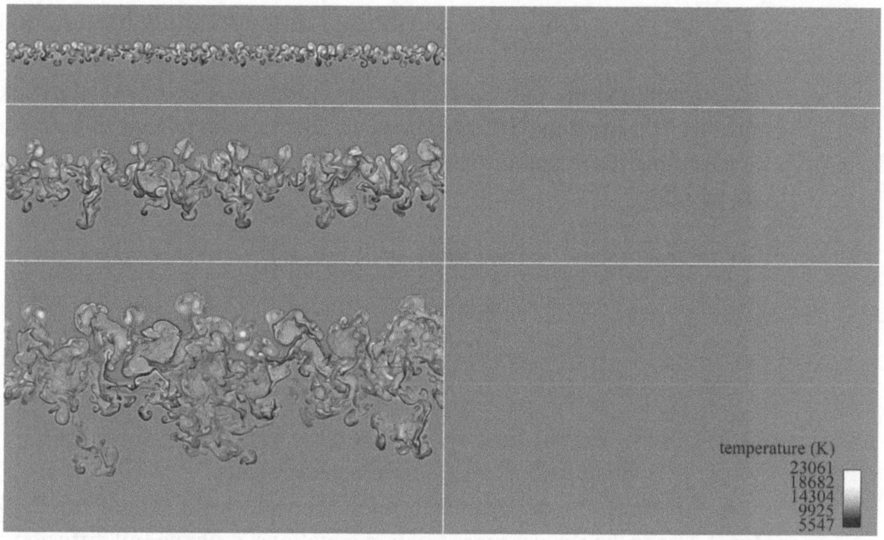

Fig. 2.13 Temperature at 20 (top), 40 (middle), and 60 (bottom) ms from Rayleigh–Taylor simulations neglecting (left) and including (right) enthalpy diffusion. Fig. 8 of Cook (2009). Reprinted with permission from AIP Publishing.

where $h_m = e_m + p_m/\rho_m$ is the enthalpy of species m. The enthalpy diffusion flux in the multicomponent energy equation accounts for energy changes associated with compositional changes resulting from species diffusion. The enthalpy diffusion also enhances the acoustic field of mixing layers. As an example, Cook (2009) showed that neglecting the enthalpy diffusion term in an RTI simulation leads to large gradients (Fig. 7 of Cook, 2009, not shown). In simulations with enthalpy diffusion, the temperature variation was over 100 times less than in simulations without enthalpy diffusion (see Fig. 2.13). The enthalpy diffusion term prevents local violations of the entropy condition in flows with significant mixing between species of dissimilar molecular weight. The term is furthermore responsible for smoothing density gradients in diffusion regions by indirectly causing local dilatations of the velocity field.

2.8 Reduced models of the Navier–Stokes equations

The Navier–Stokes equations have two interesting special cases (in order of decreasing generality): incompressible variable density, and incompressible Boussinesq Navier–Stokes. Two nondimensional parameters already introduced, the Mach and Atwood numbers, are the keys to their validity.

As we move through the section, I will focus particular attention on the following questions: In what regimes are any of these descriptions valid? What are the assumptions made, and how does the choice of equations constrain the system evolution?

2.8.1 Variable density incompressible formulations

The incompressible limit of the Navier–Stokes may be obtained by taking $p \to \infty$ (Sandoval, 1995; Livescu, 2013; Saenz et al., 2021), which leads to uniform density in regions of pure fluid (as opposed to a non-constant background density). The governing equations are therefore:

$$\frac{\partial \rho}{\partial t} + \nabla \cdot (\rho \mathbf{u}) = 0, \tag{2.48a}$$

$$\frac{\partial \rho \mathbf{u}}{\partial t} + \nabla \cdot (\rho \mathbf{u}\mathbf{u}) = -\nabla p + \nabla \cdot \boldsymbol{\tau} + \rho \mathbf{g}, \tag{2.48b}$$

$$\nabla \cdot \mathbf{u} = -\nabla \cdot \left(\frac{D}{\rho}\nabla \rho\right), \tag{2.48c}$$

where D is the mass diffusivity (assumed constant).

In this limit, the divergence term on the right-hand side of Eq. 2.48c is obtained from Eq. 2.41 and the species mass fraction equations, with the full diffusion operator in the limit of an infinite speed of sound. This non-zero velocity divergence is a direct consequence of the change in specific volume that occurs when the two fluids mix.

While this formulation allows accurate results for mixtures of miscible fluids at arbitrary $\mathcal{A}$, its applicability is restricted to low-speed flows. In general, this formulation is appropriate for the RTI but it precludes the study of the RMI – although the RMI development at late times is virtually incompressible (for small to moderate shock Mach numbers), the shock wave itself is inherently compressible. In order to circumvent the need to initialize a shock wave, an alternative approach is to use an equivalent velocity perturbation (Thornber et al., 2010) in conjunction with compressible solvers or incompressible solvers if only the solenoidal velocity field is of interest (Thornber et al., 2010, 2011).

The allowance for variable density introduces a non-zero divergence of the velocity field. Issues also remain in the cubic nonlinearities in the momentum equations. These two issues result in additional challenges not present in the incompressible constant-density (Eq. 2.7) or Boussinesq formulations (next subsection). First, the variable-density formulation, after combining Eq. 2.48c and the divergence of Eq. 2.48b, leads to a Poisson equation with a factor of $1/\rho$ in the coefficients, making its solution more complicated and computationally demanding. Second, the triple velocity products in the momentum equations correspond to fourth-order nonlinearities in the energy transfer equation. This makes established analysis in the spectral space almost impractical without introducing a variable transformation (Zhou, 2021).

2.8.2 Boussinesq incompressible limit

For incompressible flows, the pressure and velocity fields may be considered to remain smooth and continuous; needless to say, this reduces the validity range of the corresponding predictions. In the limit $\mathcal{A} \to 0$, density variations can be assumed to be small and the governing equations may be simplified (Boussinesq, 1903). The "Boussinesq approximation," as it is known in the community, was first termed by Lord Rayleigh (1916). This may be considered the simplest formulation that still permits the study of buoyancy-driven turbulence (Zeytounian, 2003).

The Boussinesq approximation – the basic formulation for buoyancy-driven incompressible flow – is given in Batchelor et al. (1992), who first introduced the approach, and by Livescu and Ristorcelli (2007) and Chung and Pullin (2010). These works, however, focus on statistically stationary or decaying homogeneous variable-density turbulence and they do not investigate the seeding instability (Burlot et al., 2015a,b).

From the fully compressible Navier–Stokes equations, Schneider and Gauthier (2015) derived the Boussinesq approximation as one of three zero- and low-Mach number models. Under the Boussinesq approximation (e.g., Tritton, 1988; Batchelor et al., 1992; Landau and Lifshitz, 1959), fluctuations in density (ρ') are assumed to be small relative to the mean density (ρ_0). The resulting buoyancy force is destabilizing when gravity and density gradients are of opposite sign.

The Boussinesq fluid equations evolve the fluctuating velocity field $\mathbf{u}$ (which is incompressible, $\nabla \cdot \mathbf{u} = 0$) and a scalar C (such as the concentration), via (Gréa and Soulard, 2019)

$$\frac{\partial \mathbf{u}}{\partial t} + \mathbf{u} \cdot \nabla \mathbf{u} = \nu \nabla^2 \mathbf{u} - \nabla p + 2\mathcal{A} C g(t) \hat{\mathbf{z}}, \tag{2.49a}$$

$$\frac{\partial C}{\partial t} + \mathbf{u} \cdot \nabla C = D \nabla^2 C - \frac{1}{L} u_3, \tag{2.49b}$$

where $g(t)\hat{\mathbf{z}}$ is the time-varying acceleration vector aligned with the vertical direction, z, and D is the diffusion coefficient. Fluctuating flow quantities have zero mean.

Griffond et al. (2014) stressed that unstably stratified homogeneous (USH) flows are not intended to mimic RT flows, as is seen from their differing kinetic energy growth: exponential for USH, quadratic for RT. By removing all phenomena related to inhomogeneity and transport from RT flows, USH trained the spotlight on the interplay between buoyancy production, nonlinearities and dissipation.

As one might expect, the outer scales of USH flows differ due to inhomogeneous effects (Carnevale et al., 1995). Yet, it appears that the energetic and smaller scales of turbulence are qualitatively well reproduced in USH (Soulard et al., 2015; Burlot et al., 2015a,b).

Equations 2.49a and 2.49b show that the dynamics of the velocity and buoyancy fields are coupled through two linear terms involving the destabilizing buoyancy force. The dynamics of the advecting velocity field now depend on the transported field (Eq. 2.49b), which is then termed active (Celani et al., 2002).

Another line of inquiry is to study Boussinesq flows with some additional simplification to the passive scalar equation. In this case,

$$\frac{\partial \mathbf{u}}{\partial t} + \mathbf{u} \cdot \nabla \mathbf{u} = \nu \nabla^2 \mathbf{u} - \nabla p - \beta_B T \mathbf{g}, \tag{2.50a}$$

$$\frac{\partial T}{\partial t} + \mathbf{u} \cdot \nabla T = \kappa \nabla^2 T. \tag{2.50b}$$

Here, $\mathbf{g} = (0, 0, -g)$ is the gravitational field, κ is the thermal diffusivity, and β_B is the thermal expansion coefficient (e.g., Boffetta and Mazzino, 2017). Situations where the feedback of the scalar field is negligible and the velocity determines the properties of the scalar, but not vice versa, are dubbed passive. This ideal case is well approximated

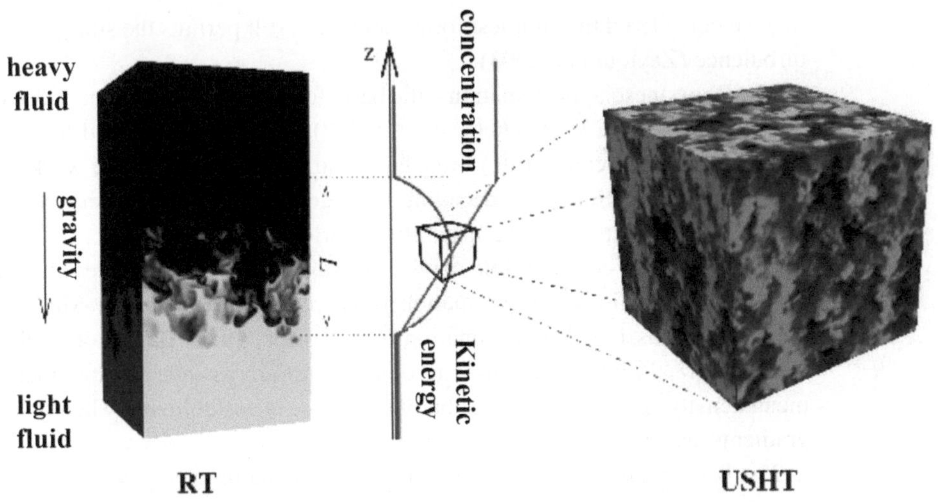

Fig. 2.14 Sketch of the Boussinesq fluids configuration with the underlying gradient length and reference profiles, as schematized from low Atwood number RT configurations. Fig. 1 of Griffond et al. (2015). Reprinted with permission from Taylor and Francis.

by the use of fluorescent dye (the scalar field) in laboratory experiments to mark fluid parcels (Celani et al., 2004).

The density and temperature are linearly related as

$$\rho(T) = \rho(T_0)[1 - \beta_B(T - T_0)], \tag{2.51}$$

where T_0 is a reference temperature. From this perspective, the role of gravity in a classical hydrodynamical system is simply to act as a time-dependent pumping scale (Chertkov, 2003).

The utilization of the Boussinesq approximation not only removes a major roadblock for performing any thorough theoretical analysis, but also allows researchers to take advantage of accurate and efficient codes developed for studying homogeneous turbulence – they may be applied with little modification to study buoyancy-driven effects. In Batchelor et al. (1992), Eqs. 2.50 were solved numerically using the Fourier pseudo-spectral code of Rogallo (1981), developed for homogeneous turbulence.

Figure 2.14 shows the idealized profiles of RT mixing zones used to derive the Boussinesq fluid equations (Griffond et al., 2015). As noted by Burlot et al. (2015a,b), the Boussinesq approximation is valid when the integral length scale of the flow is much smaller than the characteristic scale of the mixing region. This requirement is shown by Vladimirova and Chertkov (2009) and more recently, by Zhou and Cabot (2019) to be valid in well-developed RT mixing zones.

3 The nonlinear stage for a single mode

This chapter first studies the nonlinear growth of the interface amplitude, $a(t)$, and its associated terminal velocity with potential flow models. Next, one describes several models intended to predict the evolution of the bubble and spike heights $h_{b/s}(t)$, and the corresponding velocities $v_{b/s}(t)$, for the nonlinear stage with $h_{b/s}(t) > \lambda/10$. These models depend on the acceleration g, the wavelength λ (or $k = 2\pi/\lambda$), the Atwood number $\mathcal{A}$, and the initial perturbation amplitude a_0. The RMI models also depend on the initial growth velocity v_0.

3.1 Threshold or "visual" width

In order to quantify the nonlinear stage of hydrodynamic instability growth, we must mathematically define the spike and bubble heights. The most common spike height (h_s) definition for miscible fluids (for experimental measurements or numerical simulations) is the distance along z to the position with $X_2 = 1\%$. This distance is measured from the interface position z_0, which is where the interface would be in an equivalent experiment with a flat interface. Similarly, the bubble height (h_b) is the distance from the interface position to the position with $X_1 = 1\%$.

These definitions will be denoted as $h_{b/s}(n\%)$, where n is the threshold mole fraction level. This definition tends to give the greatest overall widths, the greatest spike-to-bubble ratios, and a good deal of statistical noise at late times, when only a few large structures determine the threshold values. Note that some researchers use volume fractions instead of the mole fractions used in the original definitions, while others use similar definitions based on mass fractions.

Note that the words height and width are often used interchangeably for these quantities. The spike and bubbles together are referred to as the mixing layer and the combined length, $h(n\%) = h_s(n\%) + h_b(n\%)$, is the full width of the mixing layer. Note that only h_s and h_b depend on the interface position, while h does not. The bubble velocity is $v_b = dh_b/dt$ and the spike velocity is $v_s = dh_s/dt$.

In images of laboratory experiments, light transmission thresholds are commonly used to determine the mixing layer widths, so in a loose sense one can think of these as visual experimental widths. However, light transmission can be difficult to relate precisely to material concentrations (Roberts and Jacobs, 2016).

3.2 Terminal velocities with a potential flow

3.2.1 Governing equations for a potential flow

In the simplest case, both fluids are inviscid and incompressible, and they extend to infinity on either side of the interface (Layzer, 1955). Potential flow modeling efforts, pioneered by Layzer for the RTI, and subsequently developed by Goncharov (2002) and other researchers, all assume that the perturbed surface height is singled–valued, an assumption that breaks down when the KH roll-up begins.

For an incompressible, irrotational, and inviscid system, a potential flow is described by a scalar velocity potential of fluid i, ϕ_i, where $\mathbf{u}_i = \nabla\phi_i$. The governing equations for RTI and RMI are given by the following kinematic condition (Laplace equation) and jump conditions (Bernoulli equation) at the fluid interface:

$$\nabla^2\phi_i = 0, \tag{3.1}$$

$$\frac{\partial a}{\partial t} - \nabla\phi_i \cdot \nabla a + \frac{\partial \phi_i}{\partial z} = 0, \quad \text{at } z = a, \tag{3.2}$$

$$\sum_{i=1}^{2}(-1)^i\rho_i\Big(- ga + \frac{\partial\phi_i}{\partial t} - \frac{1}{2} \mid \nabla\phi_i \mid^2 \Big) = \tilde{F}(t), \quad \text{at } z = a. \tag{3.3}$$

Here, $z = a(t)$ is the vertical position of the material interface at time t, and depends only on the time, and the horizontal variables (i.e., $\partial a/\partial z = 0$, see Guo and Zhang, 2020). $\tilde{F}(t)$ is an arbitrary function which depends on the temporal variable t.

The above equations are obtained from the mass-conservation equation, the continuity condition for the velocity component normal to the fluid interface, and the so-called Bernoulli's equation.

3.2.2 The Layzer model for $\mathcal{A} = 1$

Layzer (1955) attempts a solution to the governing equations in the form of a local expansion near the bubble tip. The nonlinear boundary value problem for the Euler equations was simplified to a set of ordinary differential equations. The solution demands knowledge of the velocity field, which Layzer assumed to be a single Fourier mode, and is consistent with the results of Rayleigh and Taylor for short and long times, respectively. This model is appropriate only to the fluid-vacuum interfaces (Atwood number $\mathcal{A} = 1$).

Layzer's equations for the bubble velocity, $v_b = dh_b/dt$, can be written in the following forms. For 2D bubbles in the form of archways (Fourier mode)

$$(2 + E)\frac{dv_b}{dt} = (1 - E)g - 6\pi\frac{v_b^2}{\lambda}, \tag{3.4}$$

with

$$E = \exp\Big(- \frac{6\pi h_b}{\lambda}\Big), \tag{3.5}$$

and for 3D bubbles rising in a cylindrical tube of diameter $\mathfrak{D}$ (Bessel function mode)

$$(1+E)\frac{dv_b}{dt} = (1-E)g - 2\beta_1\frac{v_b^2}{\mathfrak{D}}, \tag{3.6}$$

with

$$E = \exp\left(-\frac{2\beta_1 h_b}{\mathfrak{D}}\right), \tag{3.7}$$

where $\beta_1 = 3.83171$ is the first zero of the Bessel function $J_1(r)$.

These equations can be interpreted as having the form:

$$\text{mass} \times \text{acceleration} = \text{buoyancy force} - \text{drag} \tag{3.8}$$

and have provided inspiration for the commonly used buoyancy-drag model (see Chapter 22 on mix models).

If h_b is small, the equations reduce to the required linear behavior

$$\text{2D:}\ \frac{d^2h_b}{dt^2} = \frac{2\pi g}{\lambda}h_b \qquad \text{3D:}\ \frac{d^2h_b}{dt^2} = \frac{2\beta_1 g}{\mathfrak{D}}h_b. \tag{3.9}$$

The equations are, if fact, valid for time varying $g(t)$. If g is constant, the limiting bubbles velocities are

$$\text{2D:}\ v_b = \sqrt{\frac{g\lambda}{6\pi}} \qquad \text{3D:}\ v_b = \sqrt{\frac{g\mathfrak{D}}{2\beta_1}}. \tag{3.10}$$

Therefore, for a single-mode with fixed wavelength λ, the nonlinear RT growth is independent of the initial amplitude (Ramaprabhu et al., 2005). Also, the perturbation wavenumber, k, is given by

$$\text{2D:}\ k = \frac{2\pi}{\lambda} \qquad \text{3D:}\ k = \frac{2\beta_1}{\mathfrak{D}}. \tag{3.11}$$

The theory of Davies and Taylor (1950) was also based on the idea of approximately satisfying the equations of motion in the neighborhood of the bubble vertex (Zufiria, 1988). There is a reference earlier than Davies and Taylor (1950) giving the limiting velocity for a bubble rising in a cylindrical tube (Dumitrescu, 1943). This paper gives $v = 0.351\sqrt{g\mathfrak{D}}$, compared to $v = 0.361\sqrt{g\mathfrak{D}}$ for the Layzer model in Eq. 3.10.

Layzer's method suggests a fundamental difference between 2D and 3D motions (Inogamov and Oparin, 2003). While the vortices are rectilinear in planar geometry and ring-shaped in 2D axisymmetric geometry, the vortices are found to be very complicated in the 3D case.

To preserve continuity in amplitude and growth rate, the nonlinear transition may occur when the linear velocity equals the terminal velocity (attributed to Nobel laureate physicist Enrico Fermi, see Layzer, 1955; Dimonte, 2004).

Quantitatively, the linear stage growth is $a(t) = a_0 e^{\gamma t}$, with bubble velocity $v(t) = a_0\gamma e^{\gamma t} = \gamma a$ and $\gamma = \sqrt{\mathcal{A}gk}$. Under the assumption that the terminal velocity is reached when the perturbation becomes nonlinear, with $ak = 1$, the velocity is then $v_b = \gamma a = \sqrt{k\mathcal{A}g}/k$. For the case $\mathcal{A} = 1$, this becomes $v_b = \sqrt{g\lambda/(2\pi)}$, quite close to the Layzer value in Eq. 3.10.

The "terminal bubble velocity," which varies as $\sqrt{g\lambda}$, was first measured in Lewis' 1950 paper (Lewis, 1950). In the early research, it was assumed that bubbles reached a terminal velocity and spikes grew as $gt^2/2$. This picture of single mode behavior at $\mathcal{A} = 1$ is the origin of the use of the terms "bubble" and "spike" that we still use today.

Hecht et al. (1994) extended the work of Layzer and predicted that the 2D asymptotic velocity of a single-mode RM bubble decays as λt^{-1}, in contrast with the constant velocity attained in the RT case. The model was also applied to the evolution of 3D modes with an initial rectangular geometry, revealing that the bubbles lost memory of their initial geometry and morphed into the fastest growing shape, exhibiting a circular tip. Zhang (1998) extended the Layzer approach to spikes and derived the first analytical prediction for the asymptotic RMI spike growth. All these studies were restricted to systems with infinite density ratios (this just means that one "fluid" is zero density).

The bubble curvature is an important parameter, as it sets a length scale and plays a key role in the bubble merger process. The Layzer model incorrectly predicts a constant asymptotic bubble curvature (Sohn, 2004b; Sohn and Baek, 2017). Other limitations of the Layzer model were discussed by Mikaelian (2008).

3.2.3 The Goncharov model for $\mathcal{A} < 1$

Goncharov (2002) extended Layzer's model to arbitrary Atwood numbers for both the RTI and RMI. He obtained an accurate approximation of the conservation equations (valid at the bubble tip) in the form of a convergent Fourier series. The governing equation for the velocity potential, ϕ, obeys the Laplace equation and satisfies the jump conditions at the interface.

It is reasonable to start with two irrotational, incompressible, inviscid fluids in a 2D geometry (Goncharov, 2002). The fluids are subject to an external acceleration pointing from the heavier to the lighter fluid. The Laplace equation, velocity potential, and jump conditions at the fluid interface are provided by the 2D version of Eqs. 3.1–3.3. Here, the velocity potential function must satisfy the jump conditions at the fluid interface $y = a(x,t)$.

The velocity potential near the bubble tip takes the form

$$\phi^h = \tilde{a}_1(t)\cos(kx)\exp[-k(y-a_0)], \tag{3.12a}$$

$$\phi^L = \tilde{b}_1(t)\cos(kx)\exp[k(y-a_0)] + \tilde{b}_2(t)y, \tag{3.12b}$$

where a_0 is the initial sinusoidal perturbation amplitude.

Finally, the interface amplitude, a, is expanded to second order near the bubble tip at $(x,y) = (0, a(0,t))$,

$$a = a_0(t) + a_2(t)x^2, \tag{3.13}$$

where $a_0(t)$ is the time-dependent vertex amplitude with initial value $a_0(0)$. The system is therefore described by five ordinary differential equations, which can be solved for the functions $a_0(t)$, $a_2(t)$, $\tilde{a}_1(t)$, $\tilde{b}_1(t)$, and $\tilde{b}_2(t)$.

Of particular interest, we have (Goncharov, 2002)

$$\frac{da_2}{dt} = -\left(\frac{da_0}{dt}\right)\frac{k}{2}(k + 6a_2), \tag{3.14}$$

and

$$\frac{d^2a_0}{dt^2}\left\{\frac{k^2 - 4\mathcal{A}ka_2 - 12\mathcal{A}a_2^2}{2(k - 6a_2)}\right\} + \left(\frac{da_0}{dt}\right)^2 k^2\left\{\frac{(4\mathcal{A} - 3)k^2 + 6(3\mathcal{A} - 5)ka_2 + 36\mathcal{A}a_2^2}{2(k - 6a_2)^2}\right\} + \mathcal{A}ga_2 = 0. \tag{3.15}$$

Equation 3.14 can be integrated directly to yield,

$$a_2 = -\frac{k}{6} + \left[\frac{k}{6} - a_0(0)\frac{k^2}{2}\right]\exp\left[-3k\big(a_0 - a_0(0)\big)\right]. \tag{3.16}$$

Following Layzer, the procedure described above for the 2D geometry can be extended to treat the bubble evolution of a 3D flow in a cylindrical symmetry. Letting the z-axis point in the direction of the density gradient, the velocity potential in the heavy and light fluids is given by (Goncharov, 2002)

$$\phi^h = \tilde{a}(t)J_0(kr)\exp[-kz], \tag{3.17a}$$

$$\phi^L = \tilde{b}_1(t)J_0(kr)\exp[kz] + \tilde{b}_2(t)z. \tag{3.17b}$$

Here, $J_0(x)$ is the Bessel function of zero order. Next, the velocity potential and the jump conditions across the fluid interface are expanded, also up to the second order in r, resulting in

$$a_2 = -\frac{k}{8} + \left[\frac{k}{8} + a_2(0)\right]\exp\left[-2k\big(a_0 - a_0(0)\big)\right], \tag{3.18}$$

and

$$\frac{d^2a_0}{dt^2}\left\{\frac{k^2 - 4\mathcal{A}ka_2 - 32\mathcal{A}a_2^2}{4(k - 8a_2)}\right\} + \left(\frac{da_0}{dt}\right)^2 k^2\left\{\frac{(5\mathcal{A} - 4)k^2 + 16(2\mathcal{A} - 3)ka_2 + 64\mathcal{A}a_2^2}{8(k - 8a_2)^2}\right\} + \mathcal{A}ga_2 = 0. \tag{3.19}$$

This model provides a continuous bubble evolution from the earlier exponential growth to the nonlinear regime, when the asymptotic RTI bubble velocity saturates at

$$v_b = \begin{cases} 1.025\sqrt{\frac{2\mathcal{A}g}{3k(1+\mathcal{A})}} & \text{for 2D} \\ 1.02\sqrt{\frac{2\mathcal{A}g}{k(1+\mathcal{A})}} & \text{for 3D} \end{cases}. \tag{3.20}$$

Note that these reduce to Eq. 3.10 for the case with $\mathcal{A} = 1$. The asymptotic RMI bubble velocity becomes

$$v_b = \begin{cases} \frac{3+\mathcal{A}}{3kt(1+\mathcal{A})} & \text{for 2D} \\ \frac{2}{kt(1+\mathcal{A})} & \text{for 3D} \end{cases}. \tag{3.21}$$

Note again that Fourier mode (2D case) and Bessel function (3D case) expansions were used.

Using different forms of the velocity potentials, Sohn (2003) also generalized the 2D and 3D Layzer-type model to the case of an arbitrary density ratio.

Mikaelian (2008) offers a critical assessment of Goncharov's theory. In particular, Goncharov's model underestimated the RTI and RMI spike growth with large a_0, but overestimated RMI spike growth with small a_0 (see also, Mikaelian, 2009b). The reasons for this are as follows: in an infinite density ratio system, one of the phases is vacuum so that the spike and the bubble are uncoupled. However, for finite density ratio systems, the spike and the bubble are coupled, and their description should be symmetric under interchange of ρ_1 and ρ_2. This symmetry should be preserved not only in the final solution, but also in the entire derivation and approximation. Goncharov introduced a linear term $b_2(t)y$ in one of the phases, but not in the other, thus breaking the symmetry (Zhang and Guo, 2016; Zhou, 2017a; Guo and Zhang, 2020). Fortunately, the bubble curvature is not sensitive to the Atwood number (Sohn, 2004a), so Goncharov's theory gives a good approximation for bubbles. However, the situation is very different for spikes, whose asymptotic curvatures vary from around 0.177 to infinity.

For single-mode cases, Mikaelian (2014a) used a Layzer-type model to investigate the suitability of the Boussinesq approximation. The approximation is good in the linear and incompressible regime because in setting $\mathcal{A} \approx 0$ (except in $g\mathcal{A}$ or $\Delta V\mathcal{A}$), one takes $\rho_1 \approx \rho_2$ and $a_b \approx a_s$. Bubbles are much less sensitive to $\mathcal{A}$ and the approximation may be acceptable even in the extreme limit $\mathcal{A} = 1$. The terminal velocity in the Boussinesq approximation is $v_b^{\text{Boussinesq}} \sim \sqrt{2\mathcal{A}g/k}$, and thus overpredicts the true bubble velocity by $v_b^{\text{Boussinesq}}/v_b \sim \sqrt{1+\mathcal{A}}$. This approximation is a much poorer one, however, at high $\mathcal{A}$ in the nonlinear regime, when the spikes become substantially longer than bubbles.

3.2.4 Layzer model generalized

To account for uniform time varying densities and accelerations, $\rho_1(t), \rho_2(t), g(t)$, Mikaelian (2009b), with the detailed derivation given in the Appendix of his paper, extended the Layzer model, assuming potential flow within each fluid. This resulted in two coupled ordinary differential equations:

$$\frac{d}{dt}(\rho_2 a_2) = -\frac{c_m k}{4}\left[k + \frac{4(1+c_m)}{c_m}a_2\right]\frac{d}{dt}(\rho_2 a), \tag{3.22}$$

$$\rho_2\left\{\frac{c_m k}{4}\frac{d}{dt}\left[\frac{1}{\rho_2}\frac{d}{dt}(\rho_2 a)\right] + \frac{(c_m k)^2}{8}\left[\frac{1}{\rho_2}\frac{d}{dt}(\rho_2 a)\right]^2\right\}$$
$$+\rho_1\left\{\frac{c_m k}{4}\dot{b}_1 - \frac{(c_m k)^2}{8}b_1^2\right\} + (\rho_2 - \rho_1)(g + \ddot{a})a_2 = 0, \tag{3.23}$$

where $a(t)$ is the bubble amplitude, $a_2(t)$ is related to the bubble curvature, and

$$b_1 = \left[\frac{1}{k\rho_1}\frac{d}{dt}(\rho_1 a_2)\right]\left[(1+c_m)a_2 - \frac{c_m k}{4}\right]^{-1}. \tag{3.24}$$

Here, c_m is a model constant which is equal to 2 for 2D Cartesian perturbations, while $c_m = 1$ for axisymmetric perturbations (Morgan et al., 2018c).

Mikaelian (2009b) noted that such a method can be traced back to Bell (1951) who pointed out that irrotational flow model, Eqs. 3.1–3.3, can be extended to uniformly compressing fluids if the velocity potentials satisfy $\nabla^2\phi = -\dot{\rho}/\rho = \tilde{F}(t)$, where $\tilde{F}(t)$ is a function of time.

3.2.5 Higher order nonlinear perturbation solutions for the RM instability

A significant effort has been expended to derive the RMI nonlinear solutions beyond the second order for incompressible fluids (e.g., Jacobs and Catton, 1988a). As usual, the expansion is in terms of the amplitude of the disturbance at the initial interface. A general formula has been obtained for n-th order solutions (Zhang and Sohn, 1997).

As an example, the bubble or spike growth velocity, explicitly through the fourth order, is

$$v = v_0 - kv_0^2(\pm\mathcal{A} + ka_0)t + k^2v_0^3\left(\mathcal{A}^2 - \frac{1}{2}\pm 2\mathcal{A}ka_0\right)t^2 \pm \frac{4}{3}k^3v_0^4(\mathcal{A} - \mathcal{A}^3)t^3 + O(t^4), \tag{3.25}$$

where we have rewritten the small expansion parameter in terms of the time, t, and the initial RMI growth velocity, v_0. In Eq. 3.25, the $+/-$ sign corresponds to the expression for the growth rate of bubbles/spikes.

Of course, the number of terms can increase rapidly with increasing order. Even at 4th order, Zhang and Sohn (1999) remarked, "at this order, the expansion is rather complicated and inclusion of higher order terms is avoided." An approximate perturbation expansion approach (Vandenboomgaerde et al., 2002) retains only the terms with the highest power in time, which are those with the largest unbounded part (Nayfeh, 1973; Bender and Orszag, 1978). Expressions up to $(a_0k)^{11}$ are then easily obtainable thanks to the major simplification provided. However, due to the simplification, the derived expressions are not exact.

The perturbation expansion method gives solutions in terms of polynomials of time, which are divergent series. In general, the Taylor series does not allow one to approximate a function in a given interval (Pozzi, 1994). The higher the order of the polynomial, the faster it diverges. Zhang and Sohn (1997) only carried out the full perturbation expansion up to $(a_0k)^4$, which gives them the small-time information needed to construct the single-point Padé approximation that we now discuss.

3.3 Single-point Padé approximation

The Padé approximation is a powerful tool designed to render a divergent asymptotic expansion (e.g., a Taylor expansion) convergent. As detailed in Bender and Orszag (1978); Pozzi (1994); Lagerstrom (2013) the Padé approximant of order [i/j] is the ratio of a polynomial of degree i to a polynomial of degree j, P_j^i. Pozzi (1994) stressed that the principal advantage of Padé approximants with respect to the generating Taylor series is that they provide an extension of the interval of convergence.

Zhang and Sohn (1996, 1997, 1999) derived the perturbation solution Eq. 3.25 for the purpose of using it as a generating series for Padé approximations. They constructed a Padé approximant for the bubble and spike growth rates in compressible fluids. The unknown Padé coefficients are determined by expanding the approximant for small t and matching the results with the small-time perturbation solution up to $O(a_0^4)$ (i.e., Eq. 3.25). The overall growth rate of RM fingers, for example, is given by

$$v = \frac{v_0}{1 + v_0 a_0 k^2 t + \max\left\{0, a_0^2 k^2 - \mathcal{A}^2 + \frac{1}{2}\right\}[v_0 k t]^2}. \tag{3.26}$$

The single-point Padé approximation has been carried out to high orders by Velikovich and Dimonte (1996) for the case of $\mathcal{A} = 1$ and by Velikovich et al. (2014) for arbitrary Atwood number. These works showed that the sequence of Padé approximants P_6^6, P_8^8, and P_{10}^{10} gives a much better prediction than the Taylor series obtained via from perturbation theory.

Building on a previous work by Wouchuk and Nishihara (1997), Matsuoka et al. (2003) developed a model describing the nonlinear evolution of a non-uniform vortex sheet with a density discontinuity across it. Using 2D hydrodynamic simulations, the authors validated the model by comparing its nonlinear dynamics with that of a shocked interface. Following the usual procedure, the model was constructed in both Taylor and Padé form.

The range of validity of equations such as Eq. 3.26 is not limited to the range of validity of the generating series. The expression 3.26 is valid over a much larger time domain than is the Taylor expansion given by Eq. 3.25. Nevertheless, it is not valid at asymptotic large times: as pointed out by Mikaelian (1998), this models decays too quickly at asymptotically large times – as $1/t^2$ instead of $1/t$ (except for the spike at $\mathcal{A} = 1$ (Zhang, 1998)). This is due to the fact that the single-point Padé approximation incorporates the small-time behavior, but no large-time behavior is included. To overcome this problem, one needs a two-point Padé approximation with information on large-time behavior.

3.4 Two-point Padé approximation

To develop a model valid for asymptotically large times, one should determine the behavior of bubbles and spikes analytically, for any Atwood number. Such information cannot be extrapolated from the small-time information and has only become available recently (Zhang and Guo, 2016, 2022; Guo and Zhang, 2020; Liu et al., 2023a). Indeed, the knowledge at asymptotic large times presented in Zhang (1998) and Zhang and Guo (2016) plays a key role in constructing the two-point Padé approximant valid for all times. These works showed that when one applies the Layzer approach properly, the methodology works for both spikes and bubbles in systems with arbitrary density ratios.

3.4.1 Signed Atwood number, asymptotic behavior and universality

Zhang and Guo studied scaling laws and universality for the asymptotic large-time behavior of RTI and RMI by considering the governing equations for incompressible, inviscid and irrotational fluids with arbitrary density ratio (Zhang and Guo, 2016; Guo and Zhang, 2020). To overcome the problem of incorrect predictions for the behavior of spikes in Goncharov's model, Zhang and Guo (2016) pointed out an important symmetry between spikes and bubbles, namely when one switches ρ_2 and ρ_1, a theory for the bubble should become a theory for the spike. Such a switching on the conventional Atwood number $\mathcal{A} = \frac{\rho_2-\rho_1}{\rho_2+\rho_1} > 0$ leads to $\frac{\rho_1-\rho_2}{\rho_1+\rho_2} = -\mathcal{A} < 0$. Therefore, they introduced a signed Atwood number $\mathcal{A} = \frac{\rho_{\text{top}}-\rho_{\text{bottom}}}{\rho_{\text{top}}+\rho_{\text{bottom}}}$. A positive $\mathcal{A}$ is for bubbles and a negative $\mathcal{A}$ is for spikes. According to Zhang and co-workers, the advantage of the signed Atwood number is that a single function describes both bubbles and spikes, with the only differences being the signs of $\mathcal{A}$ and g (Zhang and Sohn, 1997; Guo and Zhang, 2020). Such a property is important since it incorporates the mapping between bubbles and spikes. In this notation, the lower fluid is always penetrating the upper fluid (see Fig. 3.1). In their approach, the heavy fluid and the light fluid always have the same functional forms to preserve the symmetry. This is very different from Goncharov's approach, in which ϕ^h in Eq. 3.12a and ϕ^L in Eq. 3.12b have different functional forms, as they do in Eqs. 3.17a and 3.17b.

As both the velocity and the curvature of the finger tip are insensitive to time in the quasi-steady state (Glimm et al., 2002; Sohn, 2004a), an equation for the velocity of the finger for RT instability in the quasi-steady stage can be obtained (Guo and Zhang, 2020):

$$\frac{dv}{dt} = -\tilde{\alpha}k(v^2 - v_\infty^2), \tag{3.27}$$

where

$$\tilde{\alpha}(d,\mathcal{A}) = \frac{1}{4}\frac{d+1}{d-1}\frac{(1+\mathcal{A})(3+\mathcal{A})}{3+\mathcal{A}+\sqrt{2}(1+\mathcal{A})^{1/2}} \left(\frac{4(3+\mathcal{A})+\sqrt{2}(9+\mathcal{A})(1+\mathcal{A})^{1/2}}{(3+\mathcal{A})(d+1-(d-3)\mathcal{A})+2\sqrt{2}(1+\mathcal{A})^{1/2}(d+1-(d-1)\mathcal{A})}\right), \tag{3.28}$$

and the asymptotic behavior of the velocity field is given by

$$v_\infty(d,\mathcal{A}) = \left(\frac{8(d-1)^2}{d+1}\frac{[3+\mathcal{A}+\sqrt{2}(1+\mathcal{A})^{1/2}]^2}{(1+\mathcal{A})(3+\mathcal{A})[4(3+\mathcal{A})+\sqrt{2}(9+\mathcal{A})(1+\mathcal{A})^{1/2}]}\frac{\mathcal{A}g}{k}\right)^{1/2}. \tag{3.29}$$

Here, $\tilde{\alpha}(d,\mathcal{A})$ and $v_\infty(d,\mathcal{A})$ are functions of two parameters: the signed Atwood number $\mathcal{A}$ and the dimension d. The expressions are applicable to systems not only with different density ratios and different finger types (spikes and bubbles) but also with different dimensions.

Considering the finger transitions into the asymptotic stage at time t^*, a scaling law can be derived for the RTI similarly. In particular, an integration from t^* to t yields

$$v = v_\infty \frac{(v_\infty + v^*) - (v_\infty - v^*)\exp[-2\tilde{\alpha}kv_\infty(t-t^*)]}{(v_\infty + v^*) + (v_\infty - v^*)\exp[-2\tilde{\alpha}kv_\infty(t-t^*)]}. \tag{3.30}$$

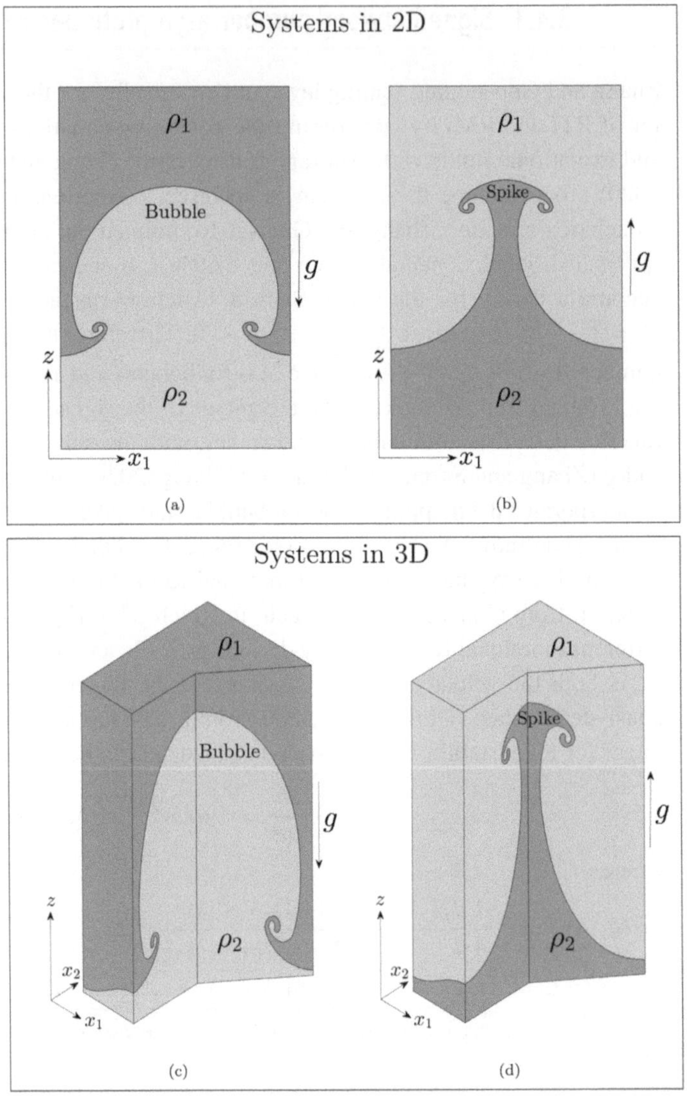

Fig. 3.1 Schematic figures of the nonlinear bubble-spike configuration in two dimensions, (a) and (b), and those in three dimensions, (c) and (d). The horizontal boundaries are periodic. The vertical direction is unbounded with no flux at $z = \pm\infty$. Gravity points from the heavy fluid (in dark color) to the light fluid (in light color). Fig. 1 of Guo and Zhang (2020). Reprinted with permission from Elsevier.

Here v^* is the finger growth rate at t^*. For RMI, we take the limit $g \to 0$ in Eq. 3.27 and obtain the asymptotic dynamic equation

$$\frac{dv}{dt} = -\tilde{\alpha} k v^2. \tag{3.31}$$

After an integration from t^* to t, we obtain a solution for the growth of RM fingers in the long-time limit

$$v = \frac{v^*}{1 + \tilde{\alpha}(d, \mathcal{A})kv^*(t - t^*)}. \tag{3.32}$$

By introducing the scaled dimensionless variables

$$u_{RM} = \frac{v}{v^*}, \quad \tau_{RM} = \tilde{\alpha}(d, \mathcal{A})kv^*(t - t^*), \tag{3.33}$$

for RM instability and

$$u_{RT} = \left(\frac{v}{v_\infty}\right)^{\mathrm{sgn}(\tilde{r})}, \quad \tau_{RT} = \tilde{\alpha}(d, \mathcal{A})kv_\infty(t - t^*) - \frac{1}{2}\ln|\tilde{r}|, \tag{3.34}$$

with $\tilde{r} = (v_\infty - v^*)/(v_\infty + v^*)$ for RT instability.[1] Equations 3.30 and 3.32 can be expressed as

$$u_{RT} = \frac{1 - e^{-2\tau_{RT}}}{1 + e^{-2\tau_{RT}}}, \tag{3.35}$$

for RT fingers, and

$$u_{RM} = \frac{1}{1 + \tau_{RM}}, \tag{3.36}$$

for RM fingers. In terms of the scaled dimensionless variables τ and u, the expressions Eqs. 3.35 and 3.36 are independent of dimensionality (the parameter d), the density ratio of the two fluids (the magnitude of Atwood number $\mathcal{A}$) and the finger type (the sign of $\mathcal{A}$). This means that the relation given by Eqs. 3.35 and 3.36 is universal, namely, it holds for all systems. These scaling laws and universalities have been verified by the data from various numerical simulations. Figure 3.2 is for RM fingers and shows that the data from 29 different finger curves agree well with the universality curve given by Eq. 3.36. Figure 3.3 shows a comparison between the data from 23 different finger curves and the prediction of the universality curve (Eq. 3.35) for RT fingers. The curves in Figs. 3.2 and 3.3 correspond to different dimensions (2D and 3D), different density ratios, and different finger types (spikes and bubbles). This universality property suggests that through the derived scaling laws, the data for 2D fingers can be used to predict the dominant behavior of 3D fingers (see also, Guo and Zhang, 2022).

3.4.2 Two-point Padé approximant and compressible RMI theory

The perturbation expansion series given by Eq. 3.25 provides information about the small-time behavior of fingers. The asymptotic solution given by Eq. 3.32 provides information about fingers at large times. Based on such information, Zhang et al. (2018b) constructed a two-point Padé approximant for the growth rates of RM fingers from early to late times. Furthermore, they applied an asymptotic matching technique to develop a theory for RM fingers in compressible fluids. The result is

$$v = v_{lin}(t)P_3^2(t), \tag{3.37}$$

[1] The sign of a real number, also called signum or sgn, is defined as follows:

$$\mathrm{sgn}(x) = \begin{cases} 1, & \text{for } x > 0, \\ 0, & \text{for } x = 0, \\ -1, & \text{for } x < 0. \end{cases}$$

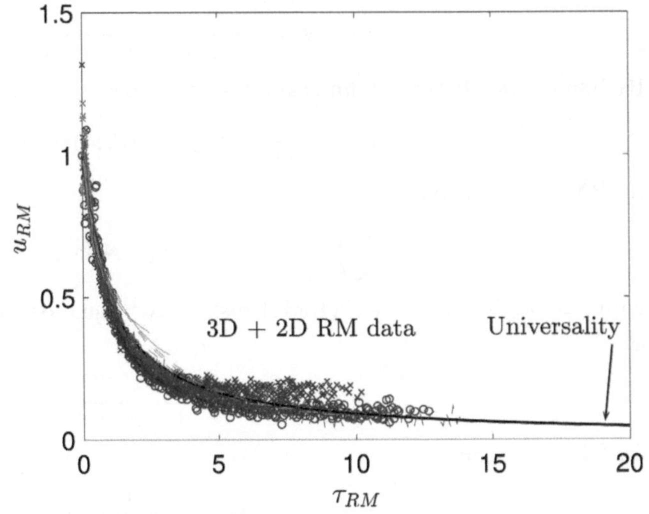

Fig. 3.2 Comparison of the prediction of universality curve (black solid curve) and data from experiments and numerical simulations in both two and three dimensions in terms of scaled growth rate u_{RM} and scaled time τ_{RM}. There are 29 different finger curves. Fig. 12 of Guo and Zhang (2020). Reprinted with permission from Elsevier.

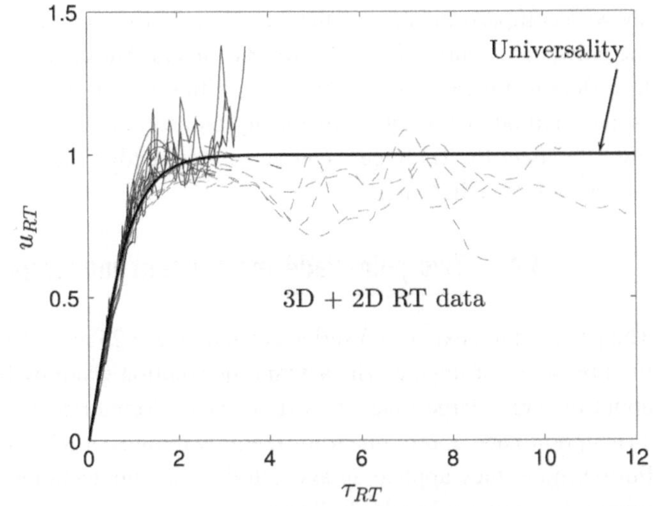

Fig. 3.3 Scaled growth rates u_{RT} versus scaled time τ_{RT} for RT fingers from numerical simulations at different signed Atwood numbers $\mathcal{A}$, in two and three dimensions. There are 23 different finger curves. The black curve is the universality curve. Fig. 16 of Guo and Zhang (2020). Reprinted with permission from Elsevier.

where

$$P_3^2(t) = \frac{1 + a_1\tilde{t} + a_2\tilde{t}^2}{1 + b_1\tilde{t} + b_2\tilde{t}^2 + b_3\tilde{t}^3}. \tag{3.38}$$

Here $\tilde{t} = kv_\infty^{cmp}t$, and v_∞^{cmp} is the asymptotic growth rate of fingers in the compressible linear theory. The coefficients a_1, a_2, b_1, b_2, and b_3 are determined by expanding Eq. 3.38 in terms of small t to match Eq. 3.25 and in terms of large t to match Eq. 3.32. The explicit expressions for these coefficients can be found in Eqs. (2) and (3) of Zhang et al. (2018b). Consequently, the two-point Padé approximant given by Eq. 3.37 has the following properties (Zhang et al., 2018b):

- For $t \ll 1$, the growth rate agrees with the compressible linear theory (Richtmyer, 1960; Fraley, 1986; Yang et al., 1994a);
- At the early nonlinear stage, the growth rate for incompressible fluids agrees with the solution from the perturbation expansion for small t (Zhang and Sohn, 1997)

$$v = v_0(t)\Big[1 - (\mathcal{A} + ka_0)(kv_0t) + \Big(\mathcal{A}^2 - \frac{1}{2} + 2\mathcal{A}ka_0\Big)(kv_0t)^2 + \frac{4}{3}(\mathcal{A} - \mathcal{A}^3)(kv_0t)^3\Big] + O(t^4); \tag{3.39}$$

- The bubble growth rate tends to zero when $\mathcal{A} = 1$ (Hecht et al., 1994; Zhang, 1998; Sohn, 2003);
- For a system with $\mathcal{A} = -1$, the model agrees with the known result that the spike growth rate asymptotically approaches a constant $v_0[(3 + 3ka_0)/(1 + 3ka_0)]^{1/2}$ (Zhang, 1998);
- At asymptotically large times, that is, $t \gg 1$, the asymptotic growth rate for arbitrary $\mathcal{A}$ is (Zhang and Guo, 2016)

$$v_{asym} \sim \frac{1}{\tilde{\alpha}kt}, \tag{3.40}$$

where $\tilde{\alpha}$ is given by Eq. 3.28. For above properties, $a_0 = a_0^+$ when the reflected wave is a shock and $a_0 = \frac{1}{2}(a_0^- + a_0^+)$ when the reflected wave is a rarefaction wave. Here a_0^- and a_0^+ are pre- and post-shock initial amplitudes, respectively. The signed Atwood number, $\mathcal{A}$, is positive (negative) for bubbles (spikes).

The model of Zhang and co-workers is valid for compressible RMI. Although the dynamics of the system at late times is nearly incompressible, the solution in that regime still has strong dependence on compressibility. This is because the functional form of the solutions of nonlinear and incompressible dynamics contains effective initial conditions. These effective initial conditions are determined by asymptotic matching with the linear compressible theory in the transition regime. Such matching leads to a strong compressibility dependence of the solution, even though the dynamics of the system is nearly incompressible. For this reason, the theoretical predictions of Zhang et al. model (Zhang and Guo, 2016; Zhang et al., 2018b) are in excellent agreement with numerical data of both weak and strong shocks.

An example that shows a surprisingly accurate prediction of Eq. 3.37 for the overall growth rate of an RM-unstable interface in compressible fluids is displayed in Fig. 3.4, for the case of an incident shock with Mach number 15.3. Even for such a strong shock, which contains strong compressibility effects, the theoretical prediction matches the data from

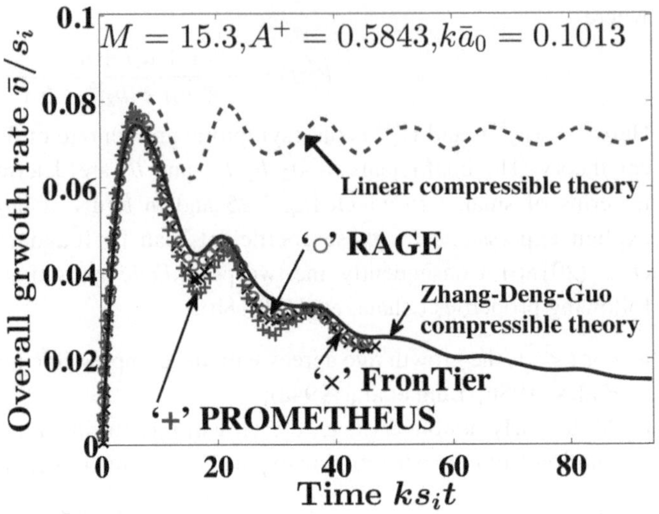

Fig. 3.4 Comparison of the dimensionless overall growth rates between the compressible nonlinear theory by Eq. 3.37 (solid), the compressible linear theory (dashed) (Yang et al., 1994b), and three numerical simulations ('×' for FronTier, '+' for PROMETHEUS and 'o' for RAGE) (Holmes et al., 1999). Here, s_i is the incident shock speed. Fig.1(d) of Zhang et al. (2018b). Reprinted with permission from APS.

numerical simulations quite well. There are two features shown in this figure: oscillation and decay. The oscillation is due to the compressibility effect and the decay is due to the nonlinear effect. The oscillation phenomenon is pronounced at early times, which indicates that the compressibility effect dominates at early times. Figure 3.4 also shows that the magnitude of the oscillation progressively diminishes, with the decay phenomenon becoming dominant at late times. This confirms an important physical picture about the dynamics of an RM unstable interface, pointed out by Zhang and his collaborators: "the system gradually changes from a linear and compressible stage at early times to a nonlinear and incompressible stage at late times" (Zhang et al., 2018b).

Expression 3.37 provides a solution for the topic originated by Richtmyer (1960) in his path-breaking work, namely to derive an analytical formula which gives accurate predictions of the single-mode RM fingers for compressible fluids, for arbitrary density ratios, for both spikes and bubbles, and for all time. This is achieved through several essential steps in the theoretical approaches (Zhang and Sohn, 1997; Zhang, 1998; Zhang and Guo, 2016; Zhang et al. 2018b; Guo and Zhang, 2020): (1) an accurate Taylor expansion to obtain small-time behavior; (2) a correct application of the Layzer approximation to obtain large-time behavior of fingers in systems with arbitrary density ratio; (3) an introduction of a signed Atwood number, which preserves the symmetry between spikes and bubbles; (4) an application of the two-point Padé approximation to make the theory valid for all times; (5) an application of an asymptotic matching technique to make the theory applicable for compressible fluids. It is noted that this procedure keeps the Taylor expansion up to the fourth order only, for the purposes of avoiding the expressions being too complicated to display explicitly and avoiding potential singularities in high-order Padé approximants. Since the Taylor expansion is only used for extracting small-time behavior, the four leading orders should be sufficient, as demonstrated in Fig. 3.4.

3.5 Heuristic models

Heuristic models such as those developed by Mikaelian, Sadot et al., and Dimonte and Ramaprabhu essentially join the linear to the nonlinear regime by making an abrupt change in the growth rate. The model coefficients are then chosen to be consistent with relevant simulations and experiments, based on the developers' expert judgment.

3.5.1 Mikaelian model – Joining the linear and nonlinear solutions

Mikaelian's method consists of joining the linear and nonlinear solutions of the RTI and RMI equations at a specific value of the amplitude (Mikaelian, 2003), for which explicit analytic expressions have been given previously for $\mathcal{A} = 1$ (Mikaelian, 1998),

$$v = \frac{v_0}{1 + v_0/v_\infty}, \tag{3.41}$$

where v_∞ is evaluated by Mikaelian using Goncharov's late time expression.

It should be noted that Mikaelian considers his model to be valid only for bubble growth. The early and late time models are then matched at $ka = 0.5$ (for 3D), assuring continuity of a and v yielding the following result for RMI (Long et al., 2009):

$$ka = \begin{cases} ka_0 + kv_0 t & \text{for } ka_0 + kv_0 t \le \frac{1}{2} \\ \frac{1}{2} + \frac{2}{1+\mathcal{A}} \ln\left[1 + \frac{1+\mathcal{A}}{2}\left(ka_0 + kv_0 t - \frac{1}{2}\right)\right] & \text{for } ka_0 + kv_0 t > \frac{1}{2}. \end{cases} \tag{3.42}$$

Mikaelian also presented another explicit model attributed to Fermi and compared with numerical simulations.

3.5.2 Sadot model – Fit to the early and late evolution

Sadot et al. (1998) proposed an RM model that captures the initial weakly nonlinear behavior, yet also provides the correct late-time asymptotic form. It is given by

$$\frac{v_{b/s}}{v_0} = \frac{1 + v_0 kt}{1 + D_{b/s}(v_0 kt) + E_{b/s}(v_0 kt)^2}, \tag{3.43}$$

where $D_{b/s}$ and $E_{b/s}$ are model coefficients (Niederhaus and Jacobs, 2003).

Sadot et al. (1998) chose a value for $D_{b/s}$ to match the 2D weakly nonlinear series solution of Zhang and Sohn (1996, 1997) at early times, and an $E_{b/s}$ value that approximates the 2D late-time behavior observed in the computation of Alon et al. (1995):

$$D_{b/s} = 1 \pm \mathcal{A}, \tag{3.44}$$

$$E_{b/s} = \frac{1 \pm \mathcal{A}}{1 + \mathcal{A}} \frac{1}{2\pi \tilde{C}}. \tag{3.45}$$

In these expressions, the plus sign denotes the bubble velocity while the minus sign denotes the spike velocity. Meanwhile, $\tilde{C}$ is a function of the asymptotic velocity of the bubbles and spikes, $\tilde{C} = 1/(3\pi)$ for $\mathcal{A} \geq 0.5$ and $1/(2\pi)$ when $\mathcal{A} \to 0$ (Alon et al., 1995; Niederhaus and Jacobs, 2003).

In the small time limit, $t \to 0$, $v_{b/s}/v_0 \to 1 + (1 - D_{b/s})(kv_0 t) + \ldots$. The bubble and spike velocities also possess the desirable $1/t$ dependence at large time.

Finally, Dimonte and Ramaprabhu (2010) advanced an empirical nonlinear RMI model that produces simulation results consistent with potential flow for a broader range of a_0 and $\mathcal{A}$.

3.6 Models compared with experiments

3.6.1 Amplitudes

The empirical models of Mikaelian (2003), Dimonte and Ramaprabhu (2010), and Sadot et al. (1998) (Eq. 3.43) offer reasonable agreement with data, as they were constructed specifically to match such data at both early and late-time (Fig. 3.5).

The Padé approximation of Zhang and Sohn (1997), Vandenboomgaerde et al. (2002), and Matsuoka et al. (2003) produces an improved agreement with data at early times. However, a single-point Padé approximation will not be able to predict the asymptotic behavior correctly, since it contains no information about late times. These approximations give a late-time dependence of $da/dt \propto 1/t^2$, instead of the correct asymptotic behavior $da/dt \propto 1/t$ (Fig. 3.6(a,b)).

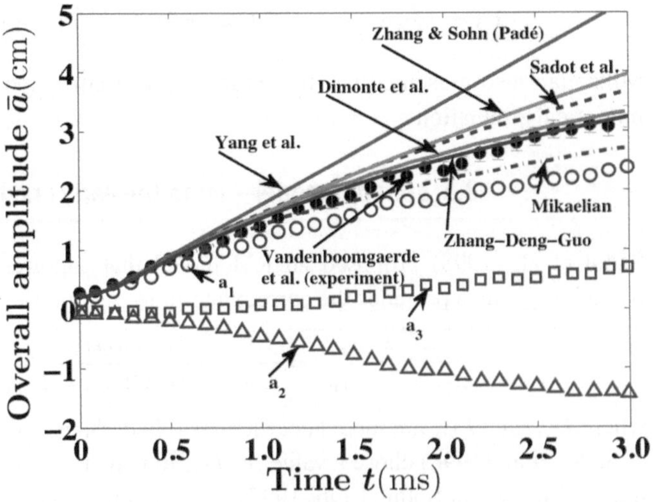

Fig. 3.5 Comparison of the RMI amplitude with several nonlinear solutions and models. The experimental data from Fig. 6 of Vandenboomgaerde et al. (2014).

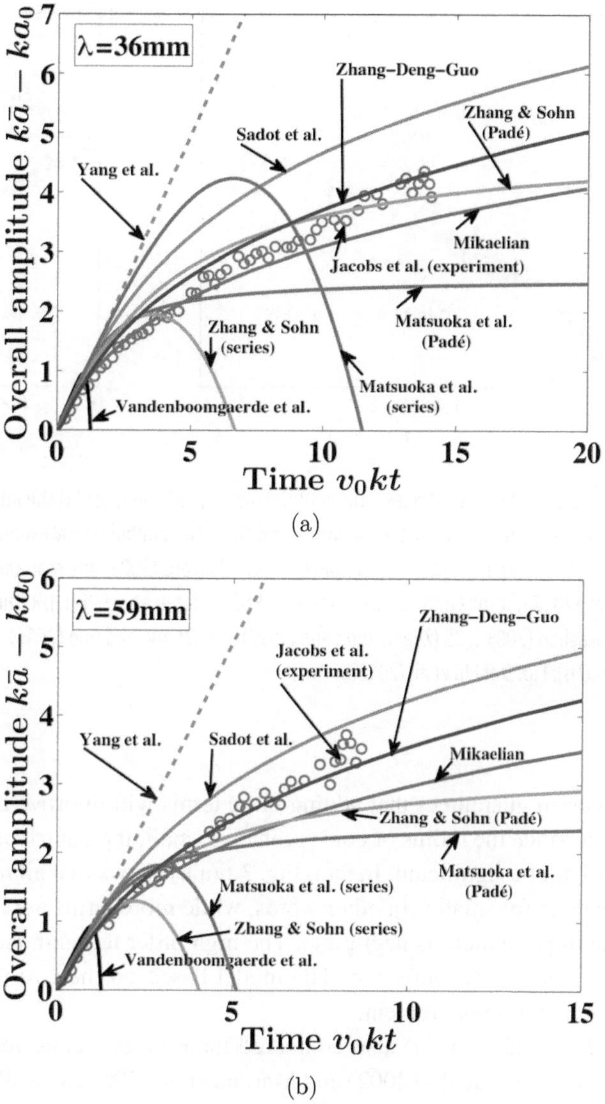

Fig. 3.6 Comparison of RMI dimensionless amplitude measurements with several nonlinear solutions and models. The experimental data from Jacobs and Krivets (2005) are represented by o.The curve labeled as Yang et al. is the prediction from the linear compressible theory (Yang et al., 1994b). The curve labeled as Zhang–Deng–Guo is the prediction from the Zhang–Deng–Guo model (2018b). (a) The pre-shock initial conditions are $\lambda = 36$ mm, $a_0^- = 1.47$ mm, $\mathcal{A}^- = 0.656$, Mach number $= 1.274$, and $\Delta V = 90.1$ m/s. (b) The pre-shock initial conditions are $\lambda = 59$ mm, $a_0^- = 2.90$ mm, $\mathcal{A}^- = 0.606$, Mach number $= 1.292$, and $\Delta V = 92.6$ m/s. Fig. 7, Jacobs and Krivets (2005), with the permission of AIP Publishing.

For a single-point Padé approximant, the series should converge when t is smaller than its radius of convergence. Therefore, within the radius of convergence, the mean value

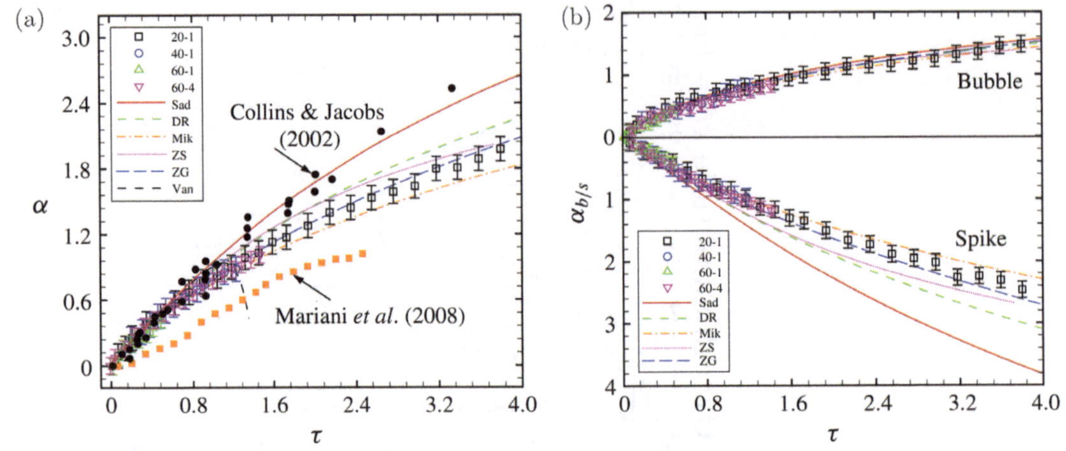

Fig. 3.7 Comparisons between RMI experiments and model predictions for (a) the interface amplitude and (b) the bubble and spike amplitudes, as a function of normalized time. The symbols show experimental results and the lines show model predictions. Experimental results from Collins and Jacobs (2002) and Mariani et al. (2008) are compared with the Liu et al. (2018a) results and nonlinear models [Sad (Sadot et al. (1998)), DR (Dimonte and Ramaprabhu (2010)), Mik (Mikaelian (2003)), ZS (Zhang and Sohn (1997)), ZG (Zhang and Guo (2016)) and Van (Vandenboomgaerde et al. (2002))]. Fig. 3 of Liu et al. (2018a).

theorem guarantees that adding more terms will improve the accuracy of the model. However, since the radius of convergence is small, the contribution from these high order terms will not be significant. In fact, Fig. 3.6(a,b) shows that all models give almost the same predictions for small t. In other words, while more terms are better for small time, in practice, the improvement is negligible. The high order terms induce large errors when t is outside its radius of convergence. The model based on the two-point Padé approximant showed significant improvement.

Liu et al. (2018a) also compared their experimental results (Fig. 3.7) with those from Collins and Jacobs (2002) and Mariani et al. (2008) as well as nonlinear model predictions. The models shown are Sad (Sadot et al., 1998), DR (Dimonte and Ramaprabhu, 2010), Mik (Mikaelian, 2003), ZS (Zhang and Sohn, 1997), ZG (Zhang and Guo, 2016), and Van (Vandenboomgaerde et al., 2002).

3.6.2 Froude number

It is useful to have a dimensionless measure of RTI spike and bubble velocities in the nonlinear stage. The dimensionless Froude number, Fr, is defined as the ratio of the flow inertia to the external force (in the current RT case, gravity) (e.g., Ramaprabhu et al., 2006; Wilkinson and Jacobs, 2007),

$$Fr_{b/s} = v_{b/s}/\sqrt{\mathcal{A}g\lambda/(1 \pm \mathcal{A})}. \tag{3.46}$$

The single-mode RTI growth has four stages: (1) the linear growth stage for $ka \ll 1$, where the amplitude and velocity of the perturbation grow exponentially with time; (2) the terminal velocity growth stage, where the perturbation undergoes a nonlinear saturation toward a terminal bubble/spike velocity – and therefore a constant Froude number for $ka_0 > 1$; (3) the reacceleration stage, where the spike rolls up along the sides to form a mushroom structure due to KHI and interactions among different bubbles (e.g., Glimm et al., 2002; Ramaprabhu et al., 2006; Wilkinson and Jacobs, 2007) and (4) the chaotic development stage, where the bubble and spike features show a proliferation of small-scale structure, resulting in turbulent and chaotic mixing of the fluids (Lewis, 1950; Sharp, 1984; Ramaprabhu et al., 2012; Wei and Livescu, 2012). For instance, the Goncharov model Eq. 3.20 predicts constant late-time velocity with $Fr_b \approx 0.56$.

Using the apparatus of Waddell et al. (2001), Wilkinson and Jacobs (2007) studied the 3D RTI in a miscible fluid system. Their results are shown in Fig. 3.8. The experimental fluids are the same miscible liquid combination as used by Waddell et al. at a low Atwood number of $\mathcal{A} = 0.15$. The 3D initial perturbation had a surface deflection of the form $a_0 \sin(kx) \sin(ky)$, where the x and y axes are oriented along the tank diagonals. Figure 3.8 (reproduced from Fig. 5 of Wilkinson and Jacobs) shows an example PLIF image of this type of perturbation (with 2.5 wavelengths) across the diagonal. Figures 2.2 and 3.8 illustrate the difference between 2D and 3D initialized RTI flows.

The authors showed that, after an initial growth stage, Fr_b plateaus at intermediate times, and then increases at late times (Fig. 3.9(a)). This reacceleration provides experimental

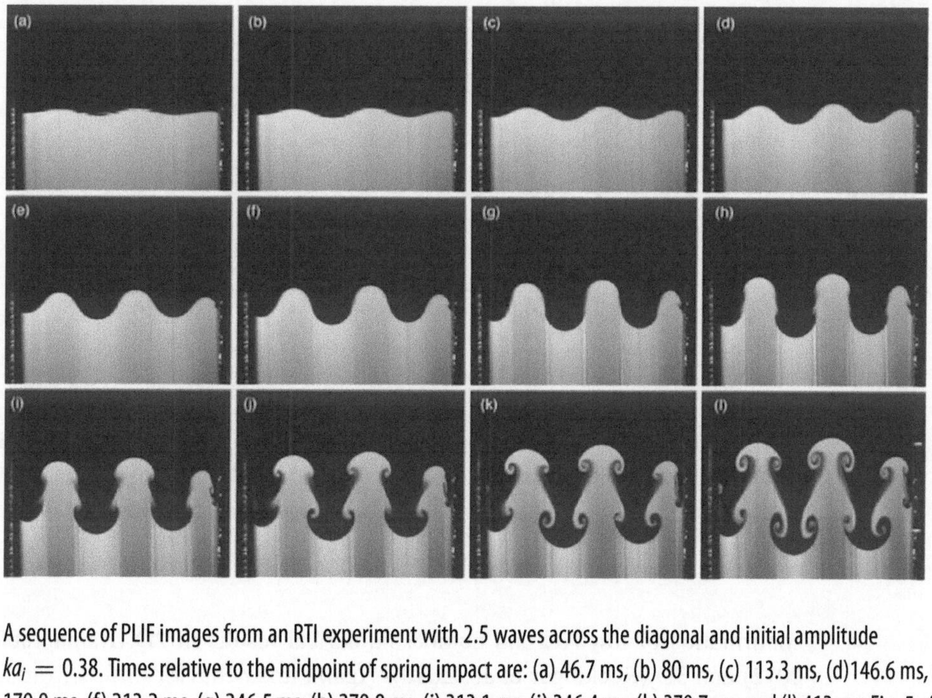

Fig. 3.8 A sequence of PLIF images from an RTI experiment with 2.5 waves across the diagonal and initial amplitude $ka_i = 0.38$. Times relative to the midpoint of spring impact are: (a) 46.7 ms, (b) 80 ms, (c) 113.3 ms, (d)146.6 ms, (e) 179.9 ms, (f) 213.2 ms, (g) 246.5 ms, (h) 279.8 ms, (i) 313.1 ms, (j) 346.4 ms, (k) 379.7 ms, and (l) 413 ms. Fig. 5 of Wilkinson and Jacobs (2007), with the permission of AIP Publishing.

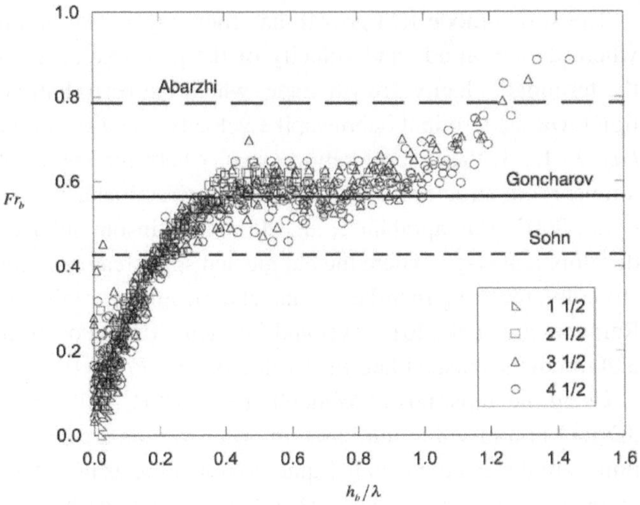

(a)

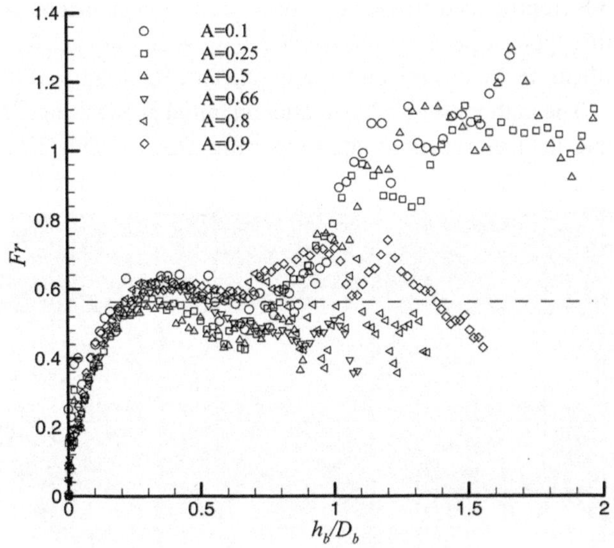

(b)

Fig. 3.9 RTI bubble Froude number vs the normalized bubble amplitude. (a) Experimental data from Fig. 15 of Wilkinson and Jacobs (2007), with the permission of AIP Publishing. (b) 3D numerical simulations with various Atwood numbers ranging from 0.1 to 0.9. Fig. 6 of Ramaprabhu et al. (2006). Reprinted with permission from APS.

confirmation of low Atwood number numerical results in 2D (Ramaprabhu et al., 2006, 2012) and 3D (Reckinger et al., 2010). Again, this behavior differs from the result of the Goncharov model, which assumes constant late-time velocity of $Fr_b \approx 0.56$. The models of Sohn (2003) and Abarzhi et al. (2003) appear to have worse agreement with the data.

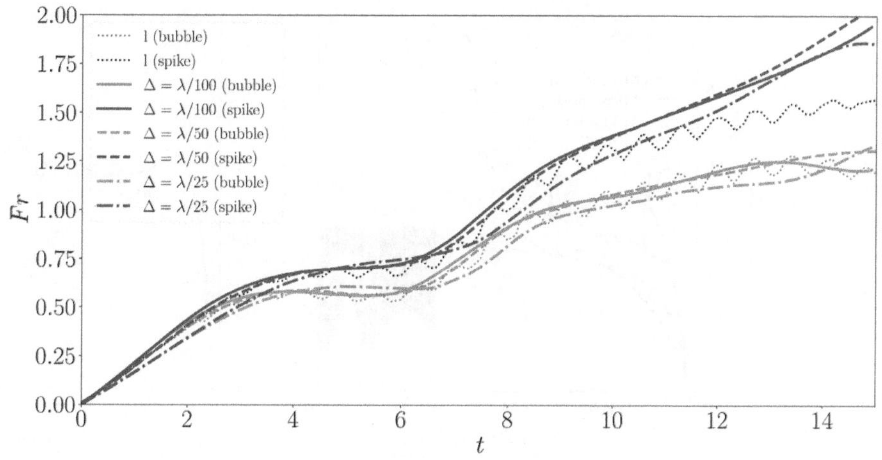

Fig. 3.10 Numerically simulated RTI bubble (gray) and spike (black) Froude numbers as a function of normalized time. The constant Froude number and reacceleration stages are visible. The dotted lines are the results from Liang et al. and are denoted as "l" in the legend. Fig. 23 of Jain et al. (2020). Reprinted with permission from Elsevier.

Figure 3.9(b) shows results of 3D implicit large-eddy simulations (ILES) obtained for a single mode, $\cos(kx) + \cos(ky)$ perturbation. The numerical simulations indeed level off at a terminal velocity for all $\mathcal{A}$ values. Although the Fr value remains near 0.56 for $\mathcal{A} \sim 1$, bubbles accelerate away from terminal velocity for $\mathcal{A} < 1$.

Figure 3.10 illustrates the bubble and spike Froude numbers as a function of time. Results from three grid resolutions show the achieved grid convergence. These results are in reasonable agreement with the results of Liang et al. (2016) (dotted gray) for Fr_b; however, there is a small disagreement for the Fr_s (dotted black) at the late-time chaotic mixing stage.[2] The four distinct growth stages exhibited by the RTI induced flow can be clearly seen as (a) the linear growth stage for $t \leq 2$, (b) the terminal velocity growth stage for $2 < t \leq 6$, (c) the reacceleration stage for $6 < t \leq 10$, and (d) the chaotic mixing stage for $t > 10$ (Jain et al., 2020).

In order to extend experiments into higher Atwood number regime, Morgan et al. (2016, 2018c) studied the single-mode, nonlinear, gas phase RT instability using a rarefaction tube. With this apparatus, RT unstable interfaces can be generated by using the acceleration provided by a rarefaction wave. It should be stressed that the rarefaction wave generates a large but non-constant acceleration, and results in time varying densities (i.e., time varying Atwood numbers).

Figure 3.11(a) shows their results for experiments with air/SF_6 and a perturbation of 3.5 wavelengths.[3] The experimental behavior is consistent with the previous discussion.

[2] The results from Liang et al. exhibit an oscillatory behavior throughout all four growth stages in the simulation, which could be a numerical artifact (Jain et al., 2020).

[3] The scatter in these plots is caused by the numerical derivatives used. Additionally, since the data are produced from images, pixel aliasing can exacerbate this scatter.

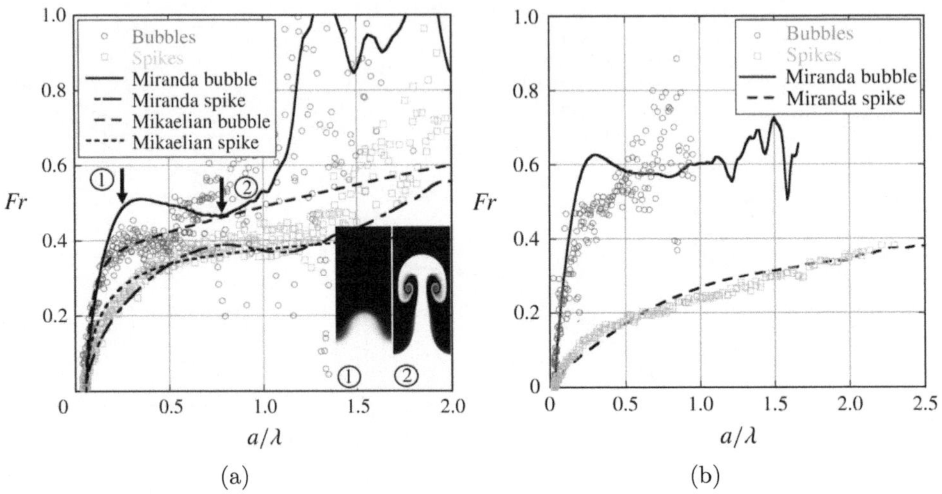

Fig. 3.11 Experimental bubble (open circle) and spike (open square) Froude numbers of 2D single-mode perturbations, as a function of their normalized amplitude. Experiments were repeated for (a) Low Atwood number and (b) High Atwood number. Lines show the prediction of two nonlinear models. Figs. 20 and 26 of Morgan et al. (2018c).

The Froude number increases until it reaches an approximate plateau, before increasing again at late times. The Froude number plateau is observed at a different value than that measured in Wilkinson's experiments and predicted by the Goncharov model. Morgan et al. (2018c) argued that this disagreement is likely due to the variable acceleration (Morgan and Jacobs, 2020) and compressibility, which the model neglects and which were not present in the liquid experiments. The model of Mikaelian (2009b) seems to predict spike Froude numbers well.

Figure 3.11(b) indicates that for He/SF_6 2D experiments with large $\mathcal{A}$, bubble and spike Froude numbers do not appear to approach constant values.

There have been relatively few experiments exploring the 3D, single-mode instability, with the exception of Wilkinson and Jacobs (2007) who studied moderate Atwood number incompressible RTI systems (Figs. 3.8 and 3.9 (a)). Figure 3.12, reproduced from Fig. 4 of Chapman and Jacobs (2006), illustrates the evolution of a single-mode 3D RM instability. Once again, experimental images revealed a developing array of vortex rings, alternatively traveling upward and downward. Their measured amplitude had good agreement with linear stability theory up until $k\dot{a}_0 t \approx 1$. This is significantly longer than has been observed in 2D experiments, which showed good agreement only until $k\dot{a}_0 t \approx 0.7$.

Figure 3.13 shows the late-time $Fr_{b/s}$ behavior of the 3D RTI experiments with air/SF_6 (Morgan et al., 2018c). The bubbles are seen to grow more slowly in experiments than in the simulations, with the experiments saturating very quickly and appearing to reach the asymptotic Froude number, $Fr = 1/\sqrt{\pi} \approx 0.56$, predicted by the incompressible theory

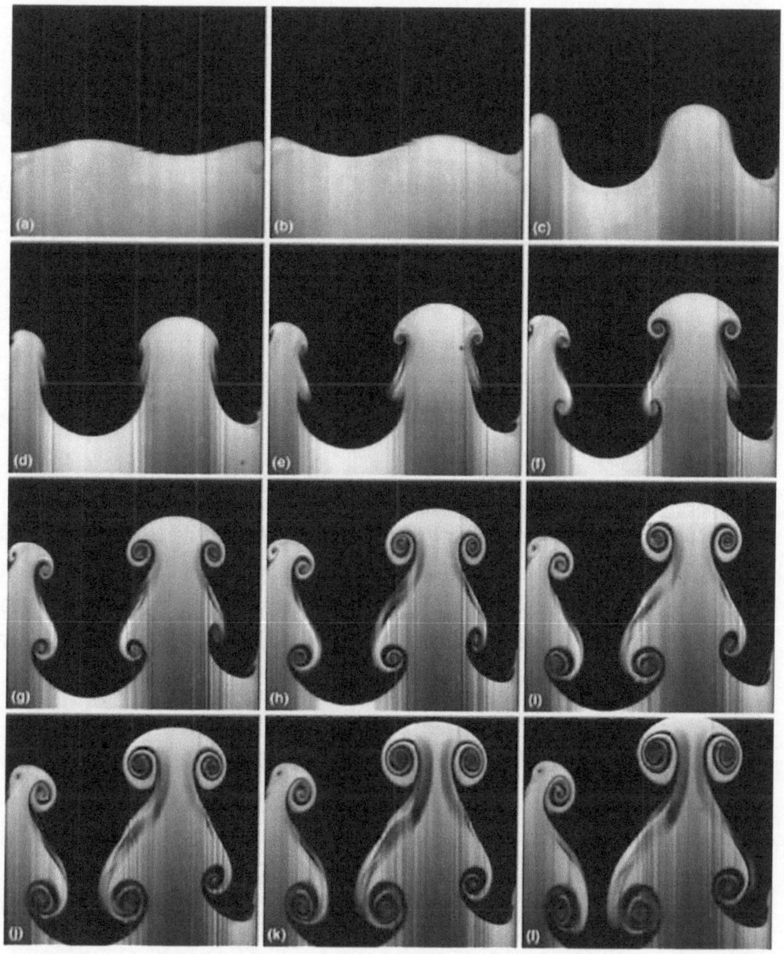

Fig. 3.12 A sequence of images from an RMI experiment with 1.5 waves across the diagonal and an initial amplitude $ka_i = 0.38$. Times relative to the midpoint of spring impact are: (a) −33 ms, (b) 33 ms, (c) 100 ms, (d) 167 ms, (e) 233 ms, (f) 300 ms, (g) 367 ms, (h) 433 ms, (i) 500 ms, (j) 567 ms, (k) 633 ms, and (l) 700 ms. Fig. 4 of Chapman and Jacobs (2006), with the permission of AIP Publishing.

of Goncharov (Eq. 3.20). Indeed, the simulated bubble Froude number agrees only briefly with the experiments. The simulated spike Froude number agrees well with the experiments (for both Atwood numbers), but then diverges at late times.[4] Good agreement between the simulations and the theory of Mikaelian (2009b) is observed.

[4] Simulated flows were computed using the compressible research hydrodynamics code Miranda of LLNL.

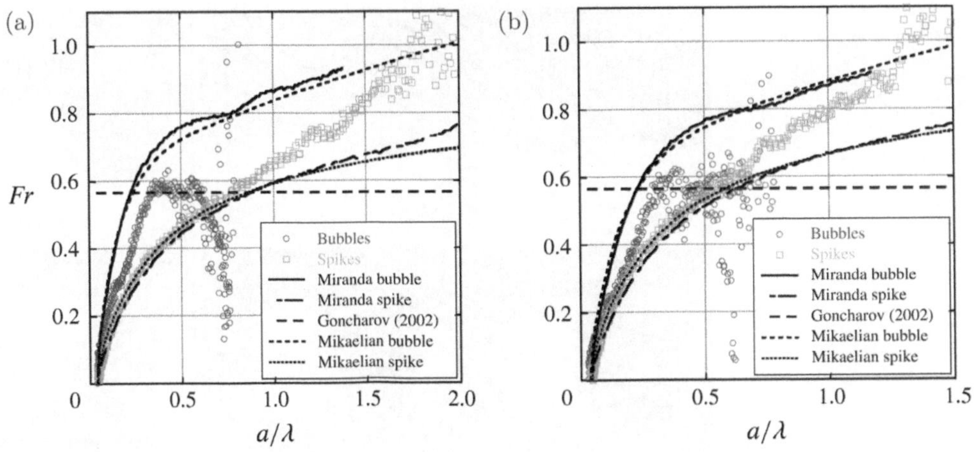

Fig. 3.13 Measured experimental RTI bubble and spike Froude numbers for 3D single-mode perturbations, as a function of their normalized amplitude: air/SF_6 (a, $\mathcal{A} = 0.63$) and CO_2/SF_6 (b, $\mathcal{A} = 0.49$). Good agreement with simulations (Miranda code) and theory is seen for the spikes at early times. The bubbles reach late-time constant velocity much more quickly than in simulations, and appear to approach the asymptotic Froude number found by Goncharov (2002). Good agreement between the simulations and the theory of Mikaelian (2009b) is observed. Fig. 25 of Morgan et al. (2018c).

3.7 Sensitivity of RT flows to the Atwood, Reynolds, and Mach numbers

3.7.1 The Atwood and Reynolds numbers

For a 3D single mode, $\cos(kx) + \cos(ky)$ initial perturbation, Ramaprabhu et al. (2006) used 3D ILES to show that the bubble's late-time acceleration may be suppressed when the values of the Atwood number are high.

Wei and Livescu (2012) have confirmed many of these results, but emphasized the important role of the perturbation Reynolds number (Re_p),

$$Re_p = \frac{\lambda}{\nu}\sqrt{\frac{\mathcal{A}}{1+\mathcal{A}}g\lambda}, \tag{3.47}$$

on the single-mode RTI late-time growth. Here, Re_p is the ratio of inertial forces to viscous actions. Above a threshold Re_p value, the flow undergoes re-acceleration followed by what they termed a "chaotic development" stage where the bubble front's mean acceleration is constant, corresponding to quadratic growth in mixing layer width.

Hu et al. (2019) considered a 2D single-mode Rayleigh–Taylor instability with low Atwood number ($\mathcal{A} = 0.15$) for a wide range of Reynolds numbers, with a special focus on intermediate Re_p numbers. After the reacceleration stage, the authors found that the vorticity near the bubble head is first decreased due to the inhibition of viscosity and

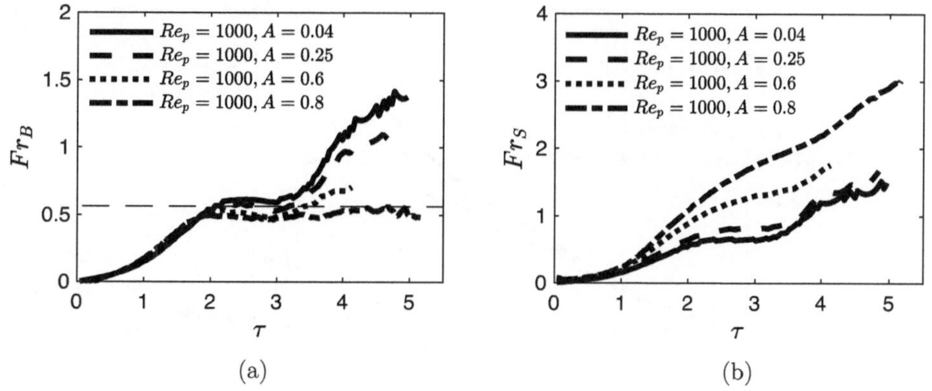

Fig. 3.14 Simulated time evolutions of the 3D RTI bubble (a) and spike (b) Froude numbers, at different values of $\mathcal{A}$ and Re_p. The dashed line shows the Goncharov value $Fr_B = \sqrt{1/\pi}$. The normalized time is smaller than the definition used in Ramaprabhu et al. by a factor of $\sqrt{2\pi}$. Fig. 9 of Bian et al. (2020). Reprinted with permission from Elsevier.

then increased again when a pair of new KHI-induced vortices approach the bubble head. Consequently, the bubble velocity decelerates and accelerates repeatedly.

Bian et al. (2020) carried out a systematic analysis of the effects of the Atwood number ($\mathcal{A}$) and the perturbation Reynolds number (Re_p), which is helpful to quantify the role of viscosity. Contrary to the common belief that single-mode RTI reaches a terminal bubble velocity, Bian et al. (2020) once again showed (Fig. 3.14) that the bubble re-accelerates when Re_p is sufficiently large and $\mathcal{A}$ is sufficiently small. This is consistent with Ramaprabhu et al. (2006) and Wei and Livescu (2012).

A related study was carried out for the immiscible single-mode RTI with high Reynolds numbers in a long tube (Hu et al., 2020). The authors found that increasing Re_p reduces the terminal bubble velocity below the potential model prediction.

3.7.2 The Atwood and Mach numbers for compressible RTI

For single-mode compressible RTI, the Mach number is given by[5]

$$M = \sqrt{\frac{g\lambda}{p_r/\rho_r}}, \tag{3.48}$$

where p_r and ρ_r are the reference (interfacial) pressure and density, respectively. Physically, the Mach number is the ratio of the gravity wave speed, $\sqrt{g\lambda}$, to the isothermal speed of sound $\sqrt{p_r/\rho_r}$ (e.g., Wieland et al., 2019).

Figure 3.15 displays the simulated time evolutions of bubble height at different Atwood numbers $\mathcal{A}$ and at two Mach numbers: (a) $M = 0.1$ and (b) $M = 0.5$. Luo et al. (2020a) also considered the Mach number effect on the time-evolution of bubble velocity. Figure 3.16 presents the evolutions of spike velocity and bubble velocity at $\mathcal{A} = 0.1$, and $M = 0.1$,

[5] The Mach number, M, should not be confused with the mixed mass, $\mathcal{M}$ (defined in Eq. 5.19).

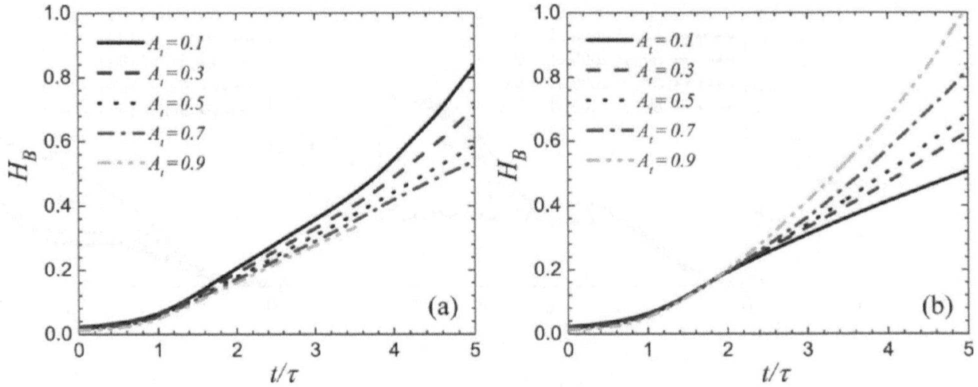

Fig. 3.15 The simulated time evolutions of bubble height H_B at different Atwood numbers A_t and at two Mach numbers: (a) $M = 0.1$ and (b) $M = 0.5$. Fig. 2 of Luo et al. (2020a), with the permission of AIP Publishing.

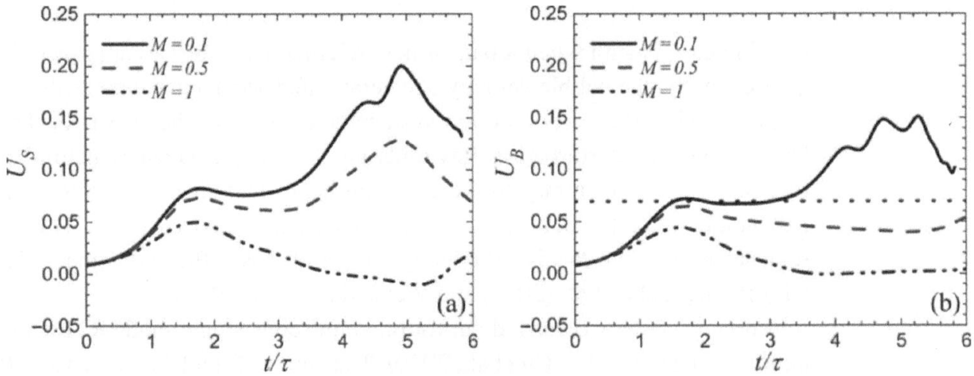

Fig. 3.16 The simulated time evolutions of (a) spike velocity and (b) bubble velocity for Atwood number 0.1 and various Mach numbers, M. The dotted line represents the asymptotic bubble velocity of the potential flow model. Fig. 4 of Luo et al. (2020a), with the permission of AIP Publishing.

0.5, and 1.0. The dotted line represents the asymptotic bubble velocity predicted by incompressible potential flow theory. The suppression of the mixing layer growth is apparent with increased compressibility (Mach number).

3.8 Reacceleration and vorticity

Vortex dynamics offer a perspective on the physical process of the reacceleration, as described by Ramaprabhu et al. (2012). From the simulation images shown in Fig. 3.17, a sinusoidal interface ($\tau = 3.8$) went through the linear stage, followed by the formation of bubble and spike fingers corresponding to the onset of nonlinearity ($\tau = 5$). It is evident

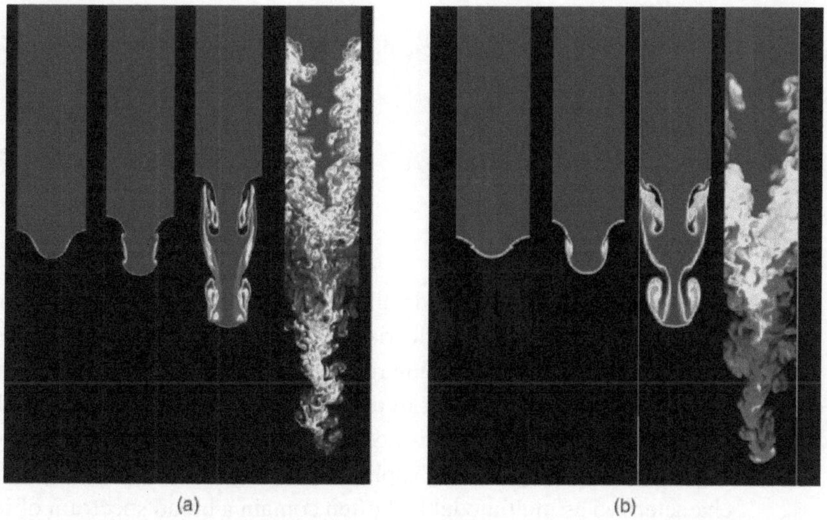

Fig. 3.17 Simulated evolution of single-mode RTI. The gray scale shows species volume fraction in (a) PPM and (b) RTI-3D simulations at $\mathcal{A} = 0.15$, with frames at dimensionless times $\tau = 3.8, 5.3, 9$, and 15. Fig. 5 of Ramaprabhu et al. (2012), with the permission of AIP Publishing.

that the reacceleration of $Fr_{b/s}$ corresponds to the formation of "secondary" KH roll-ups on the primary RT plume. The secondary KH vortex doublet (vortex ring in 3D) induces a net velocity that augments v_b and v_s, increasing the Froude number beyond the potential flow value. Finally, at $\tau = 15$, the bubbles and spikes show a proliferation of small scale structures. This further instability likely contributes to increased atomic mixing, thereby reducing the Froude number associated with the fundamental mode.

The strong correlation between vorticity and bubble velocity suggests that re-acceleration and deceleration of the bubble front is determined by vorticity accumulation inside the bubble (Betti and Sanz, 2006; Ramaprabhu et al., 2012; Zhou et al., 2022). Bian et al. (2020) quantitatively showed that the vortices which propel the bubble front are not generated inside the bubble but are generated near the spike interface as it penetrates into the lighter fluid. The vortices then propagate towards the bubble tip.

Finally, Fu et al. (2022, 2023) studied bubble re-acceleration behaviors in compressible RTI with isothermal stratification. Unlike the bubble re-acceleration behavior in quasi-incompressible RTI with uniform background density, the authors found that characteristics in the stratified compressible RTI are driven not only by vorticity accumulation inside the bubble, but also by flow compressibility resulting from the stratification (Fu et al., 2023).

4 Multimode instabilities: Linear and nonlinear regimes

In this chapter, we provide a detailed treatment of the various stages of development of the RT and RM instabilities which arise from *multimodal* initial perturbations. Mathematically, the pathway to turbulent mixing requires that the initial interface be multimodal, to permit cross-mode coupling leading to turbulence. Moreover, it is practically difficult to experimentally produce a non-multi-mode initial interface. Indeed, hydrodynamic instabilities in experiments and engineering applications are typically seeded by perturbations that can be characterized as multimodal and often contain a broad spectrum of modes.

For multimodal random initial perturbations, the initial conditions need to be described by a random surface, which is characterized by a 2D power spectrum, $P_2(k_x, k_y)$, where k_x and k_y are the wavenumbers in the x and y directions and $P_2(k_x, k_y)dk_x dk_y$ is the variance of the Fourier modes in the interval $dk_x dk_y$. The standard deviation, $\hat{\sigma}$, of the height distribution is then given by

$$\hat{\sigma}^2 = \int_{-\infty}^{\infty}\int_{-\infty}^{\infty} P_2(k_x, k_y)dk_x dk_y. \tag{4.1}$$

If the surface is assumed to be isotropic, the power spectrum will only depend on the total wavenumber $k = \sqrt{k_x^2 + k_y^2}$. The standard deviation is then given by

$$\hat{\sigma}^2 = \int_0^{\infty} P(k)dk \quad \text{where } P(k) = 2\pi k P_2(k_x, k_y). \tag{4.2}$$

In the following sections, each distinct stage of the development in multimodal RTI and RMI will be discussed in greater detail alongside corresponding flow visualizations. To illustrate the interactions between the modes, we will also present the simulated and experimental results that clearly exhibit the mode competition and the bubble merger processes in the dual-mode RMI.

A key element of this chapter is the discussion of the late-time self-similar scaling for the RT and RM growth factors, α and θ. These growth factors are not only important as a theoretical consideration, but also due to their essential use in engineering and complex applications.

If viscous effects are small, short wavelength perturbations grow quickly and then saturate. Long wavelength perturbations grow at a slower rate and saturate at a larger amplitude. This implies that the dominant length scale should be monotonically increasing (in time) for the RT, RM, and KH instabilities. If the dominant length scale increases in proportion to the width of the mixing zone, the process tends to become self-similar.

4.1 Linear regime

For simplicity, we consider perturbations consisting of a range of initial wavelengths $\{\lambda_1, \ldots, \lambda_N\}$, all independently growing in the linear regime. In other words, the amplitude a_k of each mode with wavenumber $k = 2\pi/\lambda_k$ is growing at a rate that is well described by linear stability analysis (Drazin and Reid, 1981).

Figure 4.1 provides a visualization of early time evolution of the volume fraction distributions of a multimodal RMI. Figure 4.1(a) shows the initial condition used in a recent fundamental study of the Richtmyer–Meshkov mixing layer (RMML) (Thornber et al., 2017). While the image sequence is taken from that study, the same general trends also apply to Rayleigh–Taylor mixing layers (RTML) (see Youngs (2017) for visualizations of an analogous RTI case). The initial surface perturbation shown in Fig. 4.1(a) contains a superposition of modes in the interval $k_{\min} = 4$ to $k_{\max} = 8$. To ensure that all modes are initially linear, the total standard deviation of the perturbation is taken as one tenth of the minimal wavelength. The interface also has an initial diffusive thickness, which modifies the initial growth rate but has a negligible impact on the subsequent evolution, see Duff et al. (1962).

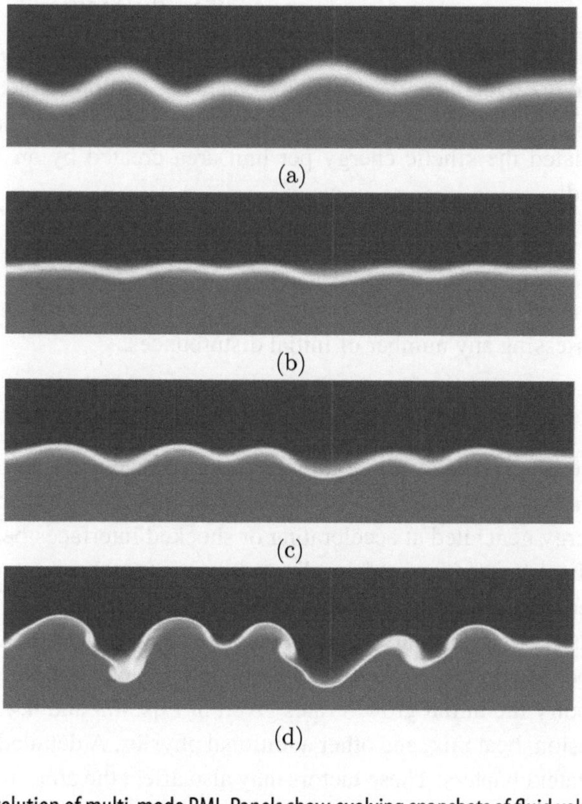

Fig. 4.1 Simulated early time evolution of multi-mode RMI. Panels show evolving snapshots of fluid volume fraction, where black indicates unmixed heavy fluid and gray indicates unmixed light fluid. Zhou et al. (2021).

Shortly after initiation of the instability (in this case, by a Mach 1.84 shock wave), the layer growth is in the linear regime, as shown in Fig. 4.1(b). For this particular case, the incident shock travels from the heavy to the light fluid, resulting in a phase inversion of the initial perturbation. The initial diffusive thickness of the interface has also been compressed by the shock. While the evolution of RTI from the same initial perturbation would be different, at an equivalent early time the layer would look qualitatively similar.

If individual modes in the initial perturbation wavepacket are of sufficiently small amplitudes, their early evolution can be described by the linear theories discussed earlier. It is implicitly assumed that the interface separating the two fluids is initially smooth and continuous, such that it can be described as an expansion in some orthonormal basis (e.g., as a Fourier series). This allows for analysis in terms of amplitudes and wavelengths of individual modes (i.e., harmonics) in the perturbation.

For the RTI, the initial growth in a_k is given by rewriting Eq. 2.23 for mode k:

$$\ddot{a}_k = \mathcal{A} g k a_k, \tag{4.3}$$

the solution to which is $a_k = a_{k,0} \exp[t\sqrt{\mathcal{A}gk}]$, where $a_{k,0}$ is the initial amplitude of mode k.

Similarly, the RM growth rate due to the initial impulse is given by rewriting Eq. 2.26 as:

$$\dot{a}_k = \mathcal{A}^+ \Delta V k a_{k,0}^+. \tag{4.4}$$

As noted already, Richtmyer proposed using post-shock Atwood numbers and amplitudes to agree with his numerical solutions to the compressible, linearized Euler equations.

Following a previous suggestion by Leith (1986), Saffman and Meiron (1989) formulated the kinetic energy per unit area created by an RM shock of an incompressible fluid:

$$E_i = \frac{(\rho_1 + \rho_2)}{4} \frac{2\pi}{\lambda_0} a_0^2 (\mathcal{A} \Delta V)^2. \tag{4.5}$$

Besnard et al. (1989) extended this single-mode relationship to the case of an interface possessing any number of initial disturbances,

$$E_i = c_B \frac{(\rho_1 + \rho_2)}{4} \sqrt{\langle a_0^2 \rangle} (\mathcal{A} \Delta V)^2, \tag{4.6}$$

where $\langle a_0^2 \rangle$ denotes the total interface variance and $c_B \approx 0.2$ was estimated by numerical simulations. Mikaelian (1990d) also developed a methodology for estimating the kinetic energy generated at accelerating or shocked interfaces between two fluids.

Each mode is considered linear when its amplitude a_k satisfies $ka_k < 1$ (Wilkinson and Jacobs, 2007; Chapman and Jacobs, 2006), provided $ka_k \ll 1$ initially. Typically a more conservative estimate is taken, such as $a_k \leq 0.1\lambda_k$ (Brouillette, 2002), beyond which the mode is considered nonlinear or saturated. Note that there are a wide range of factors that modify the initial growth rates given in Eqs. 4.3 and 4.4. These include viscosity, surface tension, heat flux, and other additional physics. A detailed discussion of these can be found in later chapters. These factors may also affect the amplitude at which each mode saturates. However, for simplicity, it will be assumed that a multi-mode perturbation is in the linear regime if all modes satisfy $a_k \leq 0.1\lambda_k$.

4.2 Nonlinear regime

When a hydrodynamic instability has evolved to the point $a_{k_{\max}} > 0.1\lambda_{\min}$, the layer can be considered to have entered the weakly nonlinear regime. In this regime, the growth of the highest amplitude modes is no longer well described by linear theory; however, the interface remains smooth and simply connected. Figure 4.1(c) depicts a visualization of the previously shown RMI mixing layer in this regime.

A distinction is generally made between weakly-nonlinear and strongly-nonlinear regimes. An RTI/RMI mixing layer is considered strongly nonlinear when the interface has become multi-valued, as shown in Fig. 4.1(d). This occurs due to the intense vorticity generation at the interface, causing it to roll up into saturated vortex structures dominated by localized KH instabilities. These roll-ups are also the sites of intense molecular mixing between the two fluids, due to the stirring motions and increased surface area. The interface may also no longer be simply connected, as regions of strong circulation cause material to pinch off and advect away from the main layer. This is more pertinent to RMI than RTI, due to the former's initial impulsive acceleration.

4.2.1 Multimode saturation models

Haan (1989) constructed a model built on assuming that the spectral amplitudes of 3D, broadband modulations grow quickly until reaching a saturation amplitude, $S_k = 2/(Lk^2)$ (L is the size of the analysis box), after which their amplitude is given by a logarithmic construction depending on how far into the nonlinear regime they have grown. For classical RT, this amounts to assuming that each mode grows exponentially with its own RT growth rate, until reaching the saturation amplitude, after which it grows linearly in time with the velocity $V_s(k)$ with which the mode was growing when it became nonlinear (Birkhoff, 1954; Smalyuk, 2012).

More specifically, short-wavelength modes grow initially most rapidly and quickly saturate at levels S_k, while very long-wavelength modes have a very slow exponential growth rate. As a result, the mid-wavelength modes have the largest growth factors, producing a mid-wavelength peak in the spectrum. As the evolution continues, this peak moves to longer wavelengths (Smalyuk et al., 2006).

As described by the Haan saturation model, for a broad spectrum of Fourier modes, nonlinear saturation can occur earlier due to the phase mixing of adjacent modes in Fourier space. Some further explanation of the model is now appropriate – the amplitude, a_k, at given wavenumber is taken to be

$$a_k = \sqrt{\int_{k-\delta k}^{k+\delta k} P(k')dk'} \quad \text{with } \delta k \sim k. \tag{4.7}$$

One can now argue that saturation occurs when $a_k \sim 0.1\lambda$. Self-similarity also occurs if amplitude $\propto$ wavelength, that is $a_k \propto \lambda$, which implies $P(k) \propto k^{-3}$. This should be contrasted with the case of single-mode modulations, in which the RTI grows exponentially

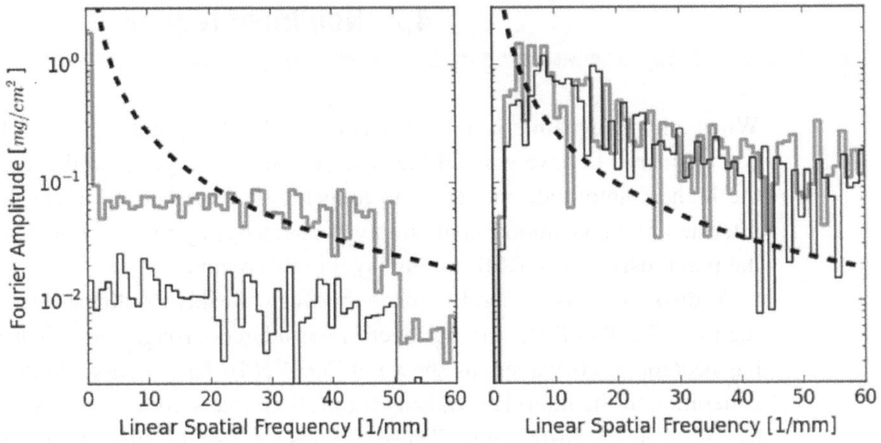

Fig. 4.2 Comparing the absolute amplitude per mode of the RT broadband modulations at 15.5 ns for initial rms amplitudes of 2.8 and 0.5 μm. The initial spectra (solid) are plotted on the left with the Haan saturation limit (dashed) and then the data is plotted on the right with the saturation limit once again (dashed line). The gray and black lines correspond to the appropriate profile. Fig. 3 of Martinez et al. (2005). Reprinted with permission from APS.

until the amplitude is $\sim 0.1\lambda$, after which nonlinear effects become important (Smalyuk et al., 1998).

Even in the extreme high energy density conditions created by the Omega laser system, the measured RTI nonlinear saturation and post-saturation growth were found to be in excellent agreement with Haan's model. Furthermore, the spectra and growth velocities were insensitive to the initial conditions (Smalyuk et al., 1998, 2005). Since the initial conditions are mostly "forgotten" in the nonlinear phase, small uncertainties in the initial seeds will not have a significant impact on the resulting perturbation growth (Welser-Sherrill et al., 2007). Ofer et al. (1996) also proposed a model that includes a description of nonlinear low-order mode coupling, mode growth saturation, and post-saturation mode coupling. This model was shown to significantly extend the range of applicability of Haan's model.

Martinez et al. (2015) went on to obtain several interesting results from their indirectly driven planar RT experiments on the National Ignition Facility. Figure 4.2 illustrates the insensitivity of RT growth to the initial conditions. Foils with different initial amplitudes show similar saturated amplitudes for $k > 10$ mm^{-1}. The RT growth is slower once the Haan saturation limit is exceeded. In Fig. 4.3, the data is seen to be in reasonable agreement with the Haan model in both cases. The spectral peak is visible at $k \simeq 5-10$ mm^{-1}.

4.2.2 Second order mode-coupling analysis

Performing an asymptotic study provides a means of scrutinizing the early nonlinear evolution of hydrodynamic instabilities (e.g., Ingraham, 1954; Jacobs and Catton, 1988a,b). A second-order analysis by Haan (1991) was constructed for the case of an interface $z = Z(\mathbf{x}, t)$ between two materials in an imploding ICF capsule. For a slightly perturbed

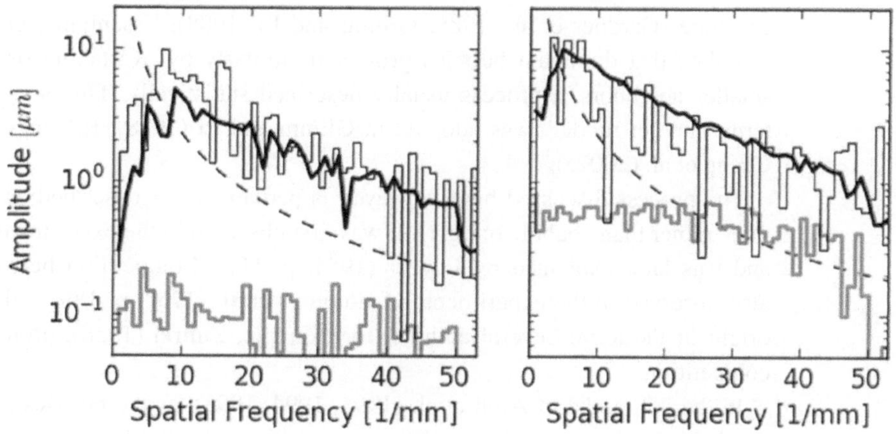

Fig. 4.3 Experimental multi-mode RT amplitudes on the NIF laser, obtained from a spatial Fourier transform of the measured areal density. The amplitude per mode of the broadband region at 15.5 ns (thin black lines), the Haan saturation amplitude (dashed), and the initial modulation spectrum (gray) are shown for a (left) 0.5 μm and (right) 2.8 μm rms initial amplitude. Overlapped with the data is the expected spectrum from the Haan model (thick black lines). Fig. 4 of Martinez et al. (2005). Reprinted with permission from APS.

interface, an expansion was carried out in spherical harmonics to derive a second-order mode-coupling equation:

$$\ddot{Z}_k = \gamma^2(k)Z_k + \mathcal{A}k \sum_{k_2} \left[\ddot{Z}_{k_2} Z_{k_2'} (1 - \hat{\mathbf{k}}_2 \cdot \hat{\mathbf{k}}) + \dot{Z}_{k_2} \dot{Z}_{k_2'} \left(\frac{1}{2} - \hat{\mathbf{k}}_2 \cdot \hat{\mathbf{k}} - \frac{1}{2} \hat{\mathbf{k}}_2 \cdot \hat{\mathbf{k}}_2' \right) \right], \tag{4.8}$$

where the surface modes are $Z_k(t) = L^{-2} \int d\mathbf{x} \exp[-i\mathbf{k} \cdot \mathbf{x}] Z(\mathbf{x}, t)$, $\mathbf{k}_2' = \mathbf{k} - \mathbf{k}_2$, and $\hat{\mathbf{k}}$ is the unit vector.

Mode coupling generates large scale structures that dominate the late-stage evolution. The importance of long-wavelength components in the initial perturbation spectrum has also been illustrated, for example, in Shvarts et al. (1995).

The second-order analysis has proven to be valuable for estimating the character and magnitude of weakly nonlinear effects (Haan, 1991). The individual modes may become nonlinear even when their amplitudes are below the single-mode threshold, owing to their interaction with nearby modes (Ramaprabhu et al., 2005). This picture also found applications in relating the nonlinear saturation of the single-mode and multi-mode beam-plasma instability (DeNeef, 1975; Dimonte, 1982).

4.2.3 Competition and merger models

Bubble competition, which is believed to determine the multi-mode front evolution, was investigated for both RTI and RMI by Hecht et al. (1994). Previously, Sharp and Wheeler (1961) proposed a 3D model for bubble rise and merger (see also, Sharp, 1984). This model is defined by a simplified set of equations that are solved to find the bubble envelope (Gardner et al., 1988; Glimm and Li, 1988). Essentially, competition models postulate that dominant bubbles grow self-similarly by overtaking or assimilating their smaller neighbors, a process usually described statistically (Dimonte et al., 2005). The term "merger model" was adopted in Glimm and Li (1988), Rikanati et al. (2000), and Cheng et al. (2002a).

The process discussed here, however, is perhaps better described as "bubble competition" rather than "bubble merger." It was first observed in the experiments of Lewis (1950) and was later explained by Layzer (1955, p. 11). "Competition between bubbles" was also observed in the experiments of Emmons et al. (1960, p. 189) – this is probably the origin of the term. Several authors, for example, Zufiria (1988), prefer the term bubble competition.

In the 2D model of Alon et al. (1993, 1994, 1995), as the bubbles rise and merge, the surviving bubbles expand to fill the space vacated by the bubbles swept away from the front. Two adjacent bubbles of wavelengths λ_i and λ_{i+1} merge (at a rate provided by either a potential flow model or a vortex model) to generate a new bubble of wavelength $\lambda_i + \lambda_{i+1}$. Neglecting near-neighbor correlations, the bubble size distribution is obtained by solving a differential-integral equation. Using a mean-field approach, Alon et al. (1995) showed that the bubble size distribution reaches an asymptotically scale-invariant regime. Shvarts et al. (1995) compared and validated the performance of the merger models with the model of Haan (1989).

Sadot et al. (1998) performed Mach 1.3 air-to-SF_6 shock tube experiments to study the two-bubble interaction in both early- and late-nonlinear stages. The results agreed with both the model prediction and numerical simulations.

Shvarts et al. (2000), Oron et al. (2001), Cheng et al. (2002b), and Kartoon et al. (2003) argued that the discrepancies between the 2D bubble competition model and 3D linear electric motor (LEM) experimental results (Dimonte, 2000; Dimonte and Schneider, 1996, 1997, 2000) are due to dimensionality effects. Oron et al. (2001) used the mean-field approximation in calculating the average bubble velocity. Their 3D model included three main changes with respect to 2D: (1) the terminal bubble velocity increases in 3D by a factor 1.5–2 due to the reduced drag per unit volume; (2) in 3D the conserved quantity when two bubbles merge is an area rather than a length; (3) the average number of neighboring bubbles is ≈ 6. It is interesting to note (Chapman and Jacobs, 2006; Mikaelian, 2003) that their asymptotic bubble velocities agree with Goncharov (2002), shown in Eq. 3.20.

The bubble-merger model suggests that the mixing region eventually behaves as a single array of bubbles with a wavelength equal to the width originally occupied by both bubbles. To further verify the key elements of the approach, Wadas and Johnsen (2020) conducted numerical simulations initialized with a dual-mode perturbation from the above mentioned shock tube (Sadot et al., 1998). Figure 4.4 shows a general agreement with the experiments. As described in Wadas and Johnsen (2020), both large and small bubbles grow, and the larger bubble begins to overtake the smaller bubble. Beyond a point, approximately 1.4 ms in this case, the larger bubble expands to fill the space originally occupied by both (Fig. 4.5(a)).

The influence of the initial condition is a major issue in RT and RM studies. In the case of RMI with bubbles of equal initial amplitude, Fig. 4.5(b) illustrates that both bubbles

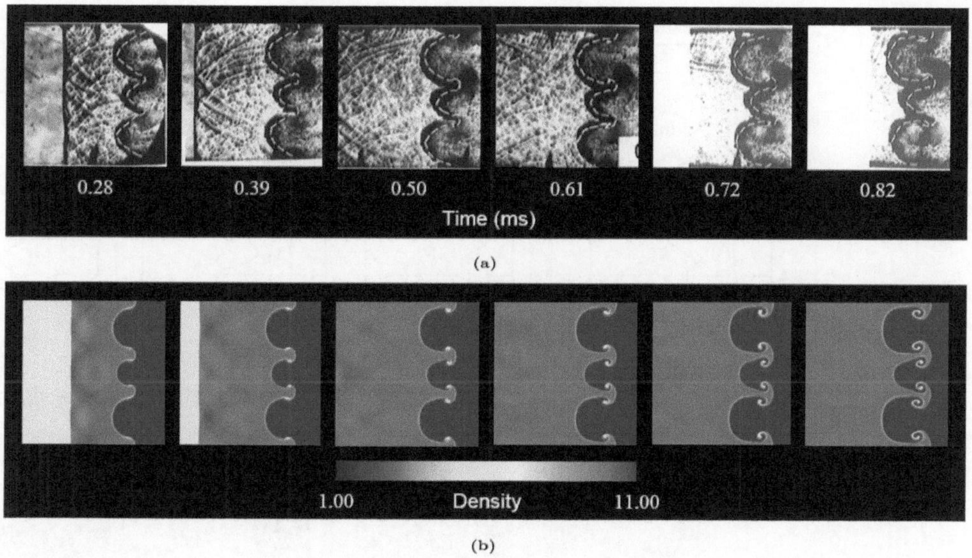

Fig. 4.4 (a) Schlieren images showing the bubble competition process observed in the experiments of Sadot et al. (1998) (white dashed lines are simulations performed by Sadot et al. (1998)), reproduced with permission, and (b) density plots of a simulation performed for this work matching the experimental conditions of Sadot et al. (1998), where Mach = 1.3, $\mathcal{A} = -0.67$. $\lambda_1/\lambda_2 = 1.59$, and $a_{0,1}/a_{0,2} = \lambda_1/\lambda_2$. Fig. 2 of Wadas and Johnsen (2020). Reprinted with permission from Elsevier.

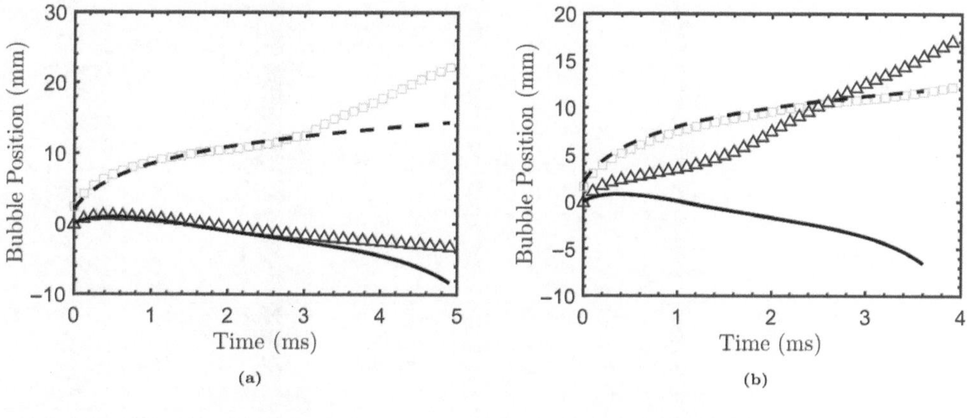

Fig. 4.5 The larger (squares/dashed line) and smaller (triangles/solid line) bubble tip positions from simulation/bubble-merger model for Mach = 1.3, $\mathcal{A} = -0.67$, $\lambda_1/\lambda_2 = 1.59$, and (a) $a_{0,1}/a_{0,2} = \lambda_1/\lambda_2$ and (b) $a_{0,1}/a_{0,2} = 1$. Fig. 3 of Wadas and Johnsen (2020). Reprinted with permission from Elsevier.

begin to grow after the shock passes. While the large bubble evolution is well predicted by the bubble merger model, the smaller bubble immediately diverges from the model.

Using an extended soap-film technique and high-speed Schlieren photography, Luo et al. (2020b) performed dual-mode RMI shock-tube experiments (Fig. 4.6) emphasizing the mode coupling effects. Figure 4.7 depicts results from four well-controlled dual-mode

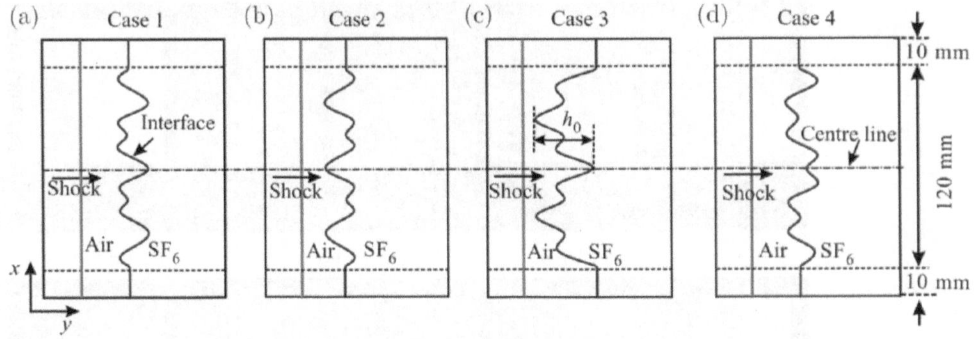

Fig. 4.6 Schematics of the four initial experimental interface configurations studied in the present work (not drawn to scale). Here h_0 is the initial total width of the interface. Fig. 2 of Luo et al. (2020b).

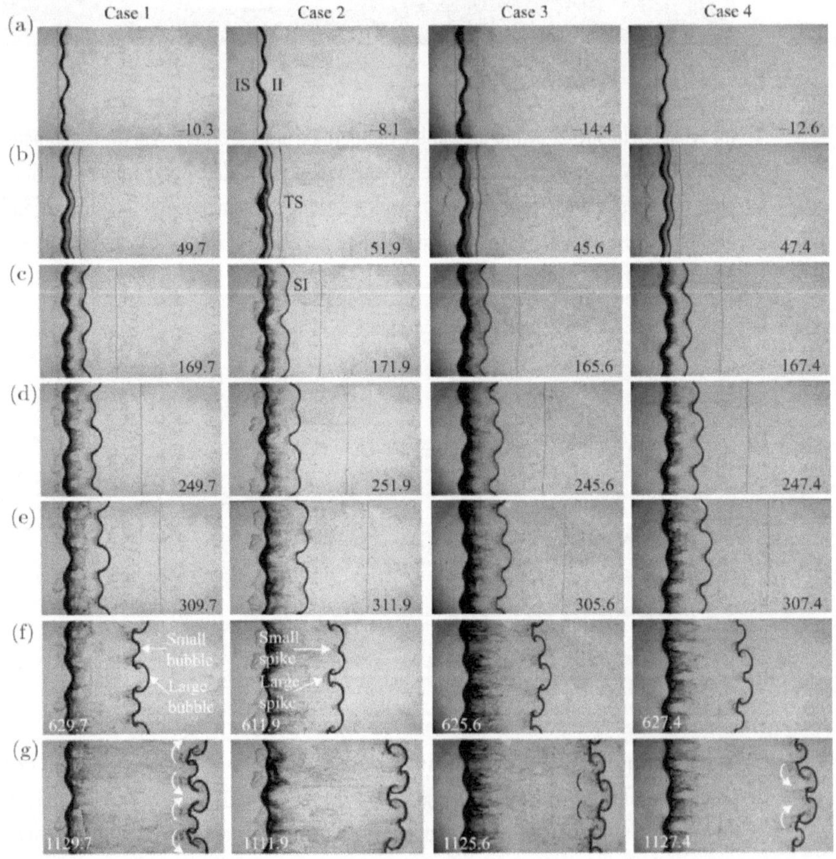

Fig. 4.7 Experimental Schlieren pictures illustrating the dual-mode interface evolution for the four well-controlled initial conditions detailed in Fig. 4.6. Numbers denote the time in μs, and $t = 0$ is defined as the moment when the incident shock arrives at the mean position of the initial interface. IS, incident shock; II, initial interface; SI, shocked interface; TS, transmitted shock. Arrows in the last row indicate the tilt of the spike during the evolution. Fig. 3 of Luo et al. (2020b).

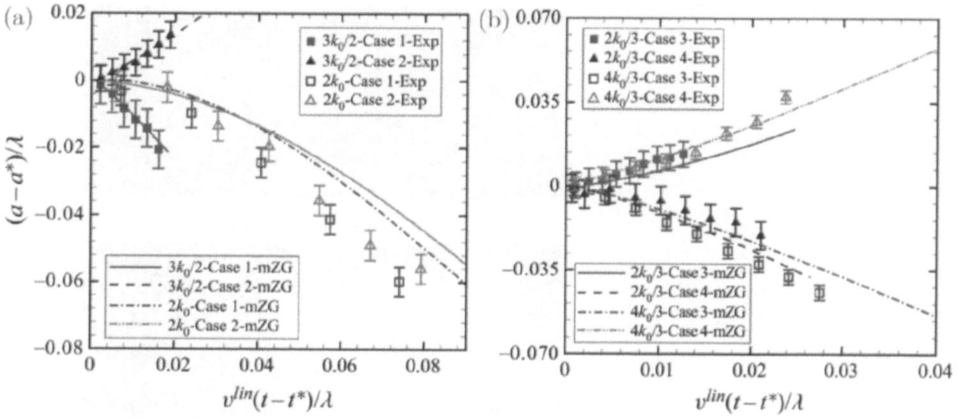

Fig. 4.8 Comparison of experimental results (markers) with the predictions of the mZG model (lines) for the growth of the second harmonics in (a) cases 1 and 2 and (b) cases 3 and 4 shown in Fig. 4.7. Fig. 7 of Luo et al. (2020b).

interfaces with different initial mode compositions and phases. The initial bubble sizes and distribution significantly affect the evolution.

The amplitudes of the basic (first harmonic) and second harmonic modes are obtained by a serial Fourier analysis of the interface in high-quality Schlieren images (Luo et al., 2020b). A noticeable difference between the basic mode growth and the corresponding single-mode RMI is observed, which suggests evident mode coupling effects in dual-mode RMI. Figure 4.8 shows that a model ("mZG") combining some elements from Haan (1991), Ofer et al. (1996), and Zhang and Guo (2016) agrees well with the data, illustrating the importance of incorporating the nonlinear growth rate reduction.

4.3 Strongly nonlinear and transition regimes

In addition to the primary roll-ups that occur due to baroclinic vorticity, the mixing layer begins to develop secondary instabilities within the roll-ups and elsewhere. This is due to shearing motion, meaning that there is a jump in tangential velocity. In regions of strongly sheared flow, the interface becomes KH unstable, resulting in the production of increasingly fine-scale vortical motions. It is these motions that drive the transition from a laminar to turbulent state, often generating modes outside the range present in the initial perturbation. Since the interfacial surface area also rapidly increases, mixing occurs much more effectively. Figure 4.9(a) shows a visualization of the RMML in this transitional regime. Figure 4.9(b) shows the RMML after the transition to the turbulent regime.

4.4 Late-time behavior and self-similarity

With notable exception from the Russian literature (e.g., Anuchina et al., 1978), the vast majority of early experimental studies of hydrodynamic instabilities concentrated

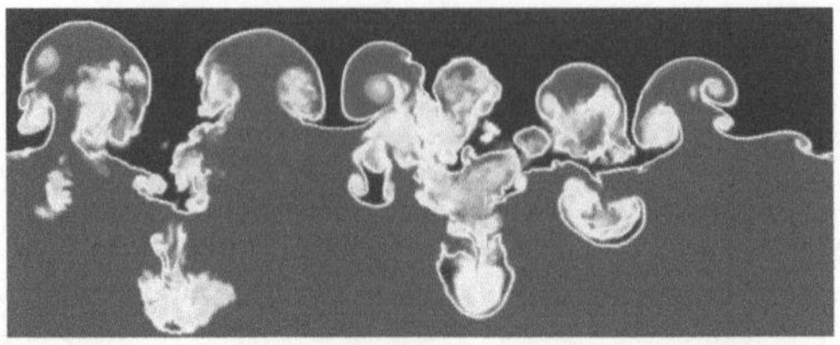

(a) Transitional regime.

(b) Turbulent regime.

Fig. 4.9 Transition to turbulence in a multi-mode RMI mixing layer. Zhou et al. (2021).

on studying the growth of a single wavelength perturbation, rather than turbulent mixing (Lewis, 1950; Allred et al., 1953; Emmons et al., 1960; Duff et al., 1962; Cole and Tankin, 1973; Popil and Curzon, 1979).

Enrico Fermi, together with John von Neumann, was probably the first to perform a single-mode analysis for the nonlinear stage of the Rayleigh-Taylor instability (Fermi and Von Neumann, 1955). Actually written in 1951 and 1953, this work first suggested free-fall for the spikes. Several more recent studies have also argued that spikes will free fall when $\mathcal{A} \to 1$, that is, $a_s(t) \propto z$, where $z = gt^2/2$ is the interface displacement in the direction of gravity or acceleration (e.g., Baker et al., 1980; Menikoff and Zemach, 1983; Zhang, 1998).

For the circumstance of a multimodal initial interface perturbation, the RT or RM mixing layer grows, after the initial linear growth regime, into a chaotic flow of interacting "fingers" or "mushrooms" of penetrating fluids (also often referred to as "bubbles" of lighter fluid and "spikes" of heavier fluid). At late times, as the kinetic energy of the flow grows, the flow in the core of the mixing layer becomes increasingly chaotic. However, the edges may still be characterized by isolated fingers penetrating a relatively quiescent fluid, and thus may not appear to be classically turbulent. For RM flows there can be highly energetic spikes that get ejected far from the main layer (e.g., Thornber and Zhou, 2012, 2015; Zhou et al., 2020a; Wadas et al., 2023).

As the velocities grow and the mixing widths increase, the nonlinearities begin to grow and eventually play a dominant role in the mixing process. As an example of the nonlinear

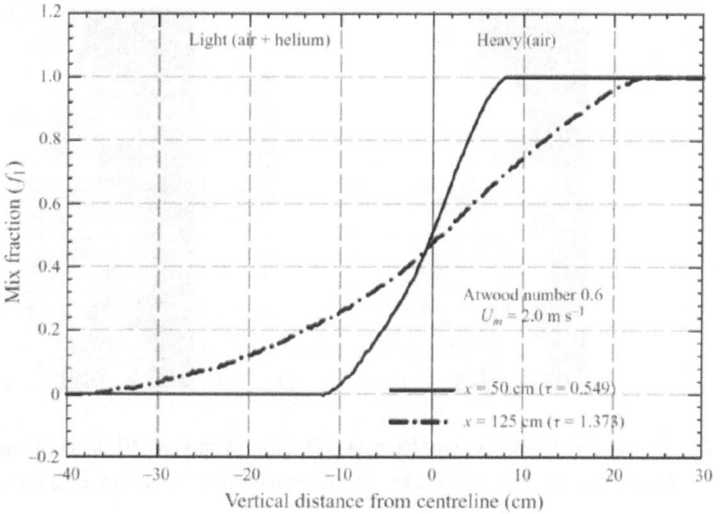

Fig. 4.10 Volume fraction profiles from a gas channel experiment at Atwood number $\mathcal{A} = 0.5$. Fig. 11 of Banerjee et al. (2010a).

stage, Fig. 4.10 shows two volume fraction profiles from a statistically steady RT mixing layer gas channel experiment.[1] The data reveal the mixing layer asymmetry with $h_b \sim$ 22 cm and $h_s \sim$ 33 cm (at a downstream location 1.25 m from the splitter plate). This asymmetry (meaning $h_b \neq h_s$) is not predicted by the linear theory in the previous chapter, since a sinusoid has $h_b = h_s$. The mixture fraction profiles illustrate that the RT mixing layer centerline, that is, the position that corresponds to a 50% mix fraction, no longer corresponds to the initial position of the density interface, for $\mathcal{A} \geq 0.47$.

As noted previously, the mass fraction profile will be different from the volume fraction profile. Figure 4.11 shows contour plots of the volume and mass fractions in direct numerical simulations of the 3D multi-mode RM instability.

Experiments with relatively low viscosity, surface tension and initial perturbations have shown that the dominant length scale increases as mixing evolves (Youngs, 2004, 2013). The higher saturation amplitude of long wavelength modes allows them to continue growing after the short wavelengths have saturated, thereby dominating the overall layer growth at late times. In this late-time limit, such flows tend to exhibit approximately self-similar behavior, the scaling of which can generally be deduced from fairly simple dimensional arguments.

4.4.1 Self-similarity scaling for RTI

It is widely believed that many "simple" turbulent flows, such as RTI, RMI, and KHI mixing layers, evolve toward self-similarity. A criterion for transition to self-similarity

[1] In this experimental setup, heavy gas (on top) and light gas (on bottom), initially flowed parallel to each other at equal velocity, separated by a thin splitter plate (Banerjee et al., 2010a). The streams met at the end of the splitter plate, with the downstream formation of an unstable interface, and thereafter RTI induced mixing.

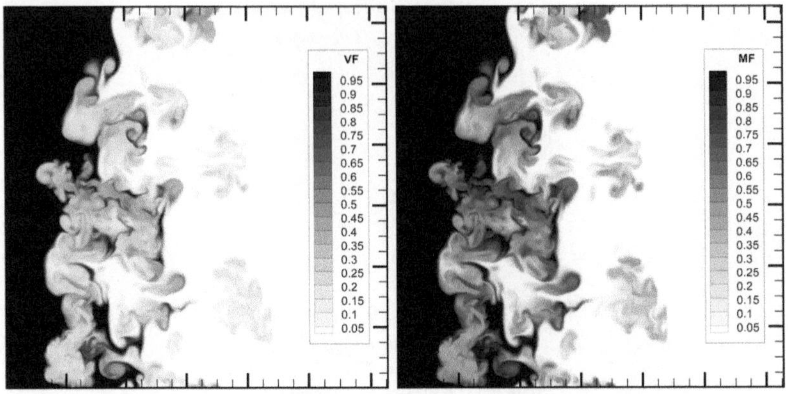

Fig. 4.11 Contours of heavy-fluid volume fractions (VF) and mass fractions (MF) in the 3D multi-mode RM instability at an Atwood number of $\mathcal{A} = 0.5$, taken from direct numerical simulations. Zhou et al. (2021).

(which is related to the transition to turbulence) should also be mentioned. A factor ~ 10 increase in dominant length scale is often given for this transition (Youngs, 1984).

In an early work, Sharp and Wheeler (1961) (see also, Sharp, 1984) constructed a bubble amalgamation model for the RTI that predicted an average velocity of bubble rise,

$$v_{av} \propto gt, \tag{4.9}$$

where g is the acceleration. Youngs (1984) exploited a self-similar modal growth of the RTI interface to arrive at the formula for mixing layer growth

$$h_i(t) = \alpha_i \mathcal{A} g t^2. \tag{4.10}$$

The total mixed width is $h = h_b + h_s$.

In fact, the $\propto t^2$ scaling for the RTML width was previously reported in Birkhoff (1954). The significance of the Birkhoff (1955) reference is generally underestimated. His analysis applied the single mode theory (exponential growth to terminal velocity) for bubbles to multi-mode random perturbations with $a_k \sim \lambda$. He derived $\alpha_b = 0.06$ at $\mathcal{A} = 1$. In the earliest Russian publication on the subject, Belen'kii and Fradkin (1965) determined that the dimensional argument suggests the length scale should be proportional to gt^2 if the mixing is self-similar. Also, Neuvazhaev and Yakovlev (1976a,b) treated RTI with g as a function of time.

Using the Euler–Lagrange equations for the variation of the potential and kinetic energy, a simplified Fermi model resulted in the following RTML width (also recalled in Boffetta et al. (2011)):

$$h_i(t) = \alpha_i \mathcal{A} g (t + t_{i,0})^2, \tag{4.11}$$

where $t_{i,0}$ is the time origin of spikes and bubbles (Clark and Zhou, 2003), sometimes dubbed the "virtual origin." As remarked by Cabot and Cook (2006), the RTI virtual origin depends on how long it takes for the flow to become self-similar, which, in turn, depends on the spectrum of initial perturbations.

Of course, the value of α_i can be obtained directly from Eq. 4.11,

$$\alpha_i = \frac{h_i(t)}{\mathcal{A}g(t + t_{i,0})^2}. \tag{4.12}$$

However, recent results show that the two procedures introduced below provide comparable, or in many cases, better solutions.

It is straightforward to show that the asymptotic growth formula of Eq. 4.11 can be written in the form

$$[h_i(t)]^{1/2} = [\alpha_i \mathcal{A}g]^{1/2}(t + t_{i,0}) = [\alpha_i \mathcal{A}g]^{1/2} t + [h_{i,0}]^{1/2}, \tag{4.13}$$

for some measure of mixing width h_i, where $h_{i,0}$ are constants depending on earlier transient stages.

The value of α_i can thus be easily obtained by simply plotting $\sqrt{h_i(t)}$ versus $\sqrt{\mathcal{A}g}t$ and fitting a line to the late time portion of the curve, where the slope is $\sqrt{\alpha_i}$ (see Olson and Jacobs, 2009; Zhou and Cabot, 2019).

Upon taking the derivative of Eq. 4.13, we obtain

$$\alpha_i(h) = [d(h_i^{1/2})/dt]^2/g\mathcal{A} = [dh_i/dt]^2/(4g\mathcal{A}h_i) \tag{4.14}$$

(Glimm et al., 2001; Ristorcelli and Clark, 2004; Cook et al., 2004; Jacobs and Dalziel, 2005).

An equivalent formula,

$$\dot{h}_i(t)^2 = 4\alpha_i \mathcal{A}gh_i, \tag{4.15}$$

has been rederived more recently though different means. For instance, Ristorcelli and Clark (2004) and Jacobs and Dalziel (2005) used a similarity assumption, whereas Cook et al. (2004) used a mass flux and energy balance argument. The value of α can be estimated from experiments, as exhibited by Dimonte and Schneider (2000), Dimonte et al. (2004), Banerjee et al. (2010a), and Akula and Ranjan (2016) (Fig. 4.12), or from numerical simulation (e.g., Cabot and Cook, 2006; Youngs, 2013).

The available experimental and numerical estimates yield a reasonably constant late-time value of α from either Eq. 4.13 or Eq. 4.14. Both methods are also independent of the virtual origin (Zhou and Cabot, 2019).

Cabot and Cook (2006) attempted to shed some light on the somewhat ironic state of affairs with α, in that the agreement between simulations and experiments is worse today than it was several decades ago. This is due to the availability of ever more powerful computers. Experimental values of α are usually much higher than those obtained by simulations, most likely due to the presence of miniscule long wavelength initial perturbations.

Although many papers have focused on the ideal case described above, some researchers have endeavored to obtain better agreement with experimental data by using equivalent initial conditions (see Dalziel et al., 1999; Latini et al., 2007b; Mueschke and Schilling, 2009a; Mueschke and Schilling, 2009b; Glimm et al., 2013, see more detailed discussion in Chapter 9). Inogamov (1978) proposed that random perturbations with amplitude $a_k \propto \lambda_k$ give self-similar growth at an enhanced rate. This corresponds to a k^{-3} power spectrum for the initial perturbations, used in several papers including Youngs (2013) and Ramaprabhu et al. (2005).

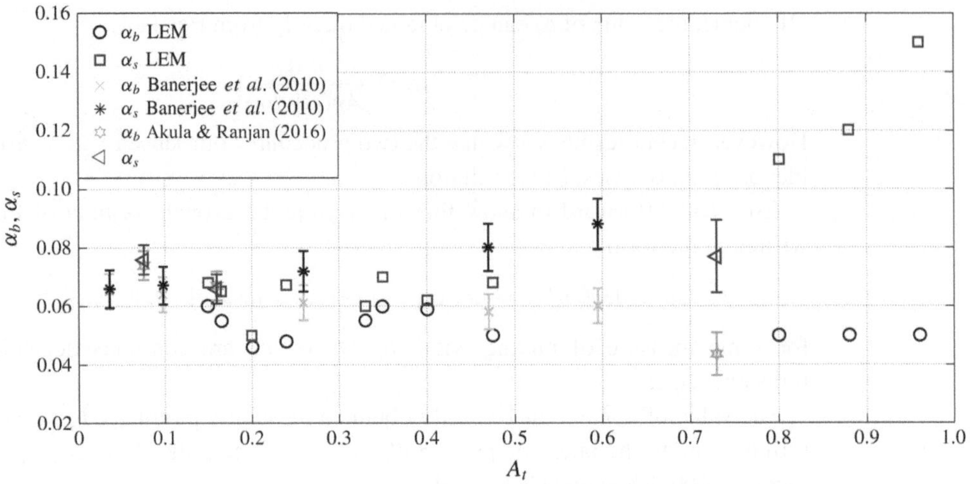

Fig. 4.12 Self-similar power-law exponents of RT bubbles and spikes, α_b and α_s, vs. Atwood number. Markers show results from Akula and Ranjan (2016), Banerjee et al. (2010b), and the Linear Electric Motor (LEM) experimental data of Dimonte and Schneider (2000). Fig. 5 of Akula and Ranjan, (2016).

Much attention has been paid to growth factors α, for RTI, and θ, for RMI (see Section 4.4.2). This is because the mixed width is sometimes the only measurement that can be obtained from experiments. Moreover, the self-similar RTI and RMI scaling laws have been used to constrain a few adjustable mix model parameters (see Chapter 22). Table 4.1 lists some recent values of α obtained from experiments or simulations, and the earlier results can be found in Dimonte et al. (2004) and Zhou (2017a).

4.4.2 Self-similarity scaling for RMI

It is reasonably well established that, like the RT case, a multi-mode RM perturbation will transition to a turbulent mixing layer. The precise behavior of the layer, however, is complicated by a strong dependence on initial conditions and chaotic turbulence physics. While gt^2 is the only length scale for RTI induced mixing (Youngs, 1984), its RMI counterpart is more complicated. The mixing layer is a decaying flow field (Zhou, 2001), so it is inappropriate to invoke the velocity jump, ΔV, in its dimensional analysis.

One might tempted to write down the RMML scaling as

$$h_i(t) \propto (t + t_{i,0})^{\theta_i}, \tag{4.16a}$$

where a proportionality constant is introduced on the right hand side. But this would not be correct dimensionally – since such a "constant" is actually not a constant at all – as it will be a function of both the value of the exponent, θ, and the physical unit of the time, t. Therefore, a proper scaling law should read

$$h_i(t) \propto (\tau + \tau_0)^{\theta_i}, \tag{4.16b}$$

Table 4.1 Growth ratios of RT "spike" and "bubble."

Authors	α_b	α_s	Atwood number	Methodology
Akula and Ranjan (2016)	0.040	0.078	0.73	Exp.
Burton (2011)	0.02		0.50, 0.75–0.96	Sim.
Liang et al. (2019)	0.025	0.027	0.1	Sim.
Roberts and Jacobs (2016)	0.047 ± 0.007	0.059 ± 0.002	0.48	Exp. unforced immiscible
Roberts and Jacobs (2016)	0.044 ± 0.009	0.057 ± 0.014	0.48	Exp. forced immiscible
Roberts and Jacobs (2016)	0.030 ± 0.008	0.047 ± 0.023	0.48	Exp. unforced miscible
Roberts and Jacobs (2016)	0.023 ± 0.007	0.017 ± 0.008	0.48	Exp. forced miscible
Yilmaz (2020)	≈ 0.027	≈ 0.04	0.5	Sim.
Yilmaz (2020)	≈ 0.027	≈ 0.06	0.9	Sim.
Zhou and Cabot (2019)	0.031	0.030	0.8	Sim.
Zhou and Cabot (2019)	0.021	0.025	0.5	Sim.
Zhou and Cabot (2019)	0.021	0.025	0.2	Sim.

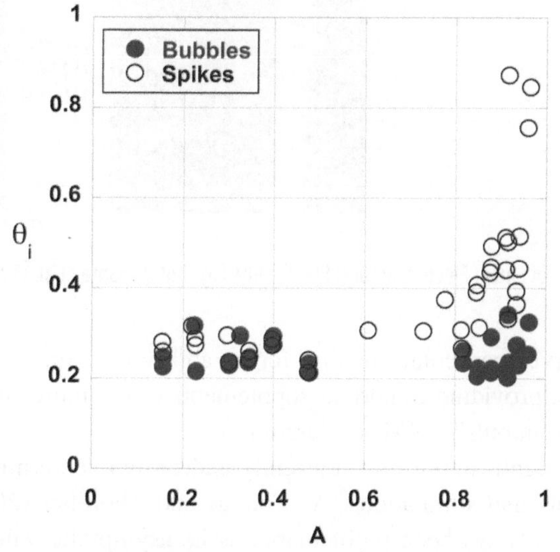

Fig. 4.13 Self-similar power-law exponents of RM bubbles and spikes, θ_b and θ_s, versus Atwood number. Figure 26 of Dimonte and Schneider (2000), with the permission of AIP Publishing.

where τ and τ_0 are dimensionless time and virtual origin, respectively. The scaling exponent, from numerical simulations and experiments, has a value $\theta \in (0, 1)$.

The variation of θ_i with Atwood number is shown in Fig. 4.13. Linear Electric Motor (LEM) experiments indicate that the bubble response is insensitive to $\mathcal{A}$ with a sample average of $\theta_b = 0.25 \pm 0.05$. For the spikes, $\theta_s \sim \theta_b$ out to $\mathcal{A} \sim 0.8$, then it increases dramatically to $0.35 < \theta_s < 0.85$ for $\mathcal{A} \sim 0.92 - 0.94$ (Dimonte and Schneider, 2000). The LEM is powered by 16 independent capacitor banks and can provide arbitrary acceleration profiles up to 1,000 times Earth's gravity (Fig. 4.14) (Dimonte et al., 1996a). This facility

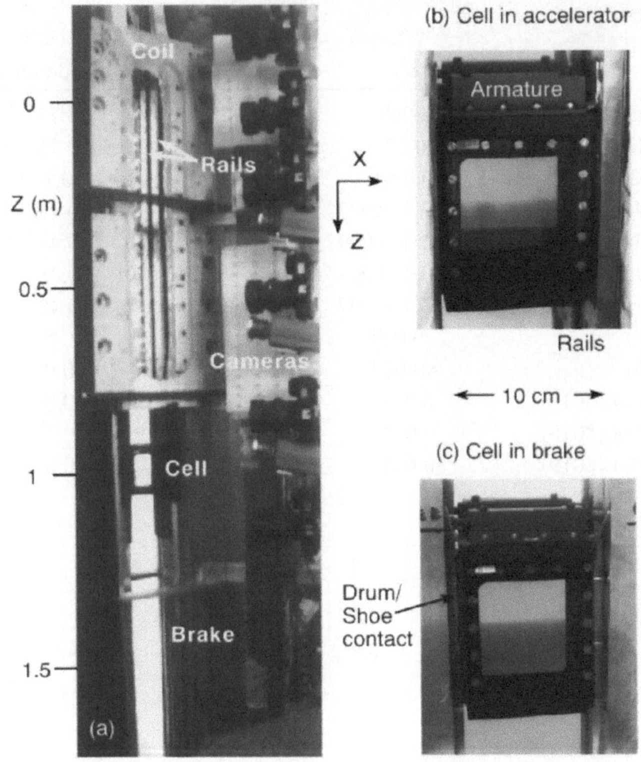

Fig. 4.14 Sketch of Linear Electric Motor (LEM) facility. Fig. 1 of Dimonte et al. (1996a), with the permission of AIP.

deserves particular mention for its ability to realize experiments with variable accelerations, providing important supplements to the more standard investigations of a constant RT or impulsive RM acceleration.

A significant effort is currently underway to determine the value of θ using both experiments and simulations. As Youngs and Thornber (2020b) stressed recently, estimating θ has always been problematic, as its asymptotic value is strongly influenced by initial conditions. Four methods have been utilized in an attempt to obtain θ. The first was to exclude early times and then fit the remaining data with Eq. 4.16b (Thornber et al., 2017). For narrowband layers, we typically start the nonlinear curve fit at the point of peak mix velocity. As one might expect, the value of θ is sensitive to the choice of the point where self-similarity begins (e.g., Thornber, 2016). It also depends on the quality of the statistics, for example, having sufficient separation between the integral length scale and the domain length. Another approach is to utilize the derivative of the integral width to directly determine

$$\theta = \frac{t\dot{h}}{h}, \tag{4.17}$$

if τ_0 is assumed to be small. In the case when a virtual time offset cannot be neglected, we need both the first and second derivatives (e.g., Oggian et al., 2015)

$$\theta = \frac{1}{1 - \ddot{h}h/\dot{h}^2}. \tag{4.18}$$

This expression can also be derived from a simple buoyancy-drag model (Youngs and Thornber, 2020a). Finally, it would be worthwhile to mention the idea put forward by Youngs and Thornber (noted in Thornber et al., 2010) about how θ can be related to the decay rate of turbulent kinetic energy (TKE, see Chapter 6). This is the method that Jacobs and co-workers have been using in their recent work on turbulent RMI. Experiments that use particle image velocimetry (PIV) measurements (e.g., Sewell et al., 2021) define a width based on threshold values of the TKE profile, for example, cutoff at 5% of peak TKE.

A summary of previous estimates of θ up to 2017, for both narrowband and broadband initial conditions, is given in the review paper of Zhou (2017a, Table 6.4). Using the Eddy Damped Quasi-Normal Markovian theory (EDQNM) (Orszag, 1970a; Lesieur, 1987; Sagaut and Cambon, 2019; Zhou, 2021), Soulard et al. (2018) and Soulard and Griffond (2022) found that $\theta = 1/4$ when long wave initial perturbations are absent. If significant long wavelength perturbations are present with amplitude spectrum $P(k) \sim k^m$, then Inogamov (1999), Youngs (2004), and Thornber et al. (2010) have found

$$\theta = 2/(m+5). \tag{4.19}$$

Recently, Elbaz and Shvarts (2018) evaluated a new formulation based on the well known mode-coupling models of Haan (1989, 1991), Shvarts et al. (1995), and Ofer et al. (1996). In addition to providing a self-consistent extension of previous RTI theoretical studies, Elbaz and Shvarts obtained results for the RMI that $\theta = 2/5$ in 2D and $\theta = 1/3$ in 3D for mixing layers in which the growth rate is due solely to mode coupling, for example, narrowband layers. In Elbaz and Shvarts (2018), the value $\theta = 1/3$ is the only one mentioned in the abstract and conclusion, yet $\theta = 1/(\min(m, 0) + 3)$ is the actual formula derived in the main body of the article (see Fig. 6b of this reference which shows the variation of θ as a function of m). The discrepancy between the minimum value of $1/3$ of this expression and the minimum of $1/4$ otherwise found in the literature is discussed in Soulard and Griffond (2022).

4.4.3 Group-theoretic analysis

The fact that the mixing zone width is the largest observable feature of the flow suggests the importance of a dominant large length scale. This, in turn, is associated with the low-wave number power-law of the energy spectrum. Such a viewpoint is similar to the analytical and experimental findings in isotropic turbulence, beginning with Batchelor (1953), Batchelor and Proudman (1956), Saffman (1967), and productively exploited by others, for example, Comte-Bellot and Corrsin (1966).

The self-similar scalings of RTI, RMI, and KHI are directly related to the assumption of time-translation-invariance (Ibragimov, 1999; Olver, 2000). Such approaches are based on a mathematical definition of self-similarity – solutions invariant under an appropriate symmetry group. For the case of RTI/RMI induced flows, the appropriate group consists of temporal scaling and translation,

$$t' = \tau(t + t_0) - t_0, \tag{4.20}$$

and a simple scaling of length;

$$\mathcal{L}' = \varrho \mathcal{L}. \tag{4.21}$$

The existence of a virtual time origin, t_0, is a statement of time-translation invariance–the behavior of the system should be independent of the absolute time. Clark and Zemach (1998) found that the same group was also appropriate for decaying weakly anisotropic turbulence.

This two-parameter group was restricted to a simple power-law subgroup,

$$\varrho = \tau^{\vartheta}. \tag{4.22}$$

This subgroup (τ, ϑ) implies that changes in lengths are related to changes in time by a power law. Using this subgroup, in addition to the above scaling, the generally accepted power-law behavior for decaying isotropic turbulence was derived. This simple argument also showed that initially anisotropic flow never becomes fully isotropic.

Self-similarity requires that the scaled function h is equal to the unscaled function of the scaled argument, t:

$$\varrho h(t) = h\left[\tau(t + t_0) - t_0)\right]. \tag{4.23}$$

The power-law subgroup, Eq. 4.22, yields

$$\tau^{\vartheta} h(t) = h\left[\tau(t + t_0) - t_0)\right]. \tag{4.24}$$

It is generally true that Eq. 4.24 is implied for all group elements if it holds for the element $1 + \delta\tau$, where $\delta\tau$ is infinitesimal. Thus, we may differentiate Eq. 4.24 with respect to τ and set τ to unity to yield the so-called determining equation,

$$\vartheta h(t) = (t + t_0)\frac{dh(t)}{dt}. \tag{4.25}$$

Solving for $h(t)$ yields

$$h(t) = h_0 \left[\frac{t + t_0}{t_0}\right]^{\vartheta}. \tag{4.26}$$

Self-similarity of a physical law implies that the scaled functions of the unscaled arguments are equal to the unscaled functions of the scaled arguments. It also implies that physical parameters (e.g., viscosity, acceleration et cetera) scale onto themselves, that is, that the relevant dimensional parameters remain invariant under the group symmetry transformations. Thus, the parameter ϑ may be determined by requiring invariance of the physical parameters.

For the RTML, Clark and Zhou (2003) required the acceleration, g (dimensions [Length/Time^2]) be made invariant,

$$g' = g\frac{\varrho}{\tau^2} = g\tau^{\vartheta - 2}. \tag{4.27}$$

From this, we see that $\vartheta - 2 = 0$ (or $\vartheta = 2$) makes g invariant and gives a length-scale (e.g., the mixing layer width) that grows as t^2 at late times. The growth will look almost linear when $0 < t \approx t_0$, and only when $t \gg t_0$ will the growth be quadratic. This alone might account for some variations in reported measurements of the scaling exponent.

For the KH mixing layer, the operative physical parameter is the velocity difference, ΔU. Following the same procedure above,

$$\Delta U' = \Delta U \frac{\varrho}{\tau} = \Delta U \tau^{\vartheta - 1}, \tag{4.28}$$

and ΔU is invariant for $\vartheta = 1$, giving a mixing layer width that grows as t^1 (Clark and Zhou, 2003).

The selected values of the RMI growth exponent are summarized in Table 4.2. As noted previously, the asymptotic value of θ is strongly influenced by the initial conditions (Youngs and Thornber, 2020b). Zhou (2001) has presented an analysis that draws a connection between the RM exponents and the growth of the length-scale in a decaying, weakly anisotropic turbulence. The physical picture is an initially quiescent flow field subject to an impulsive energy injection, perhaps at all scales, that is subsequently allowed to freely decay. Despite significant corrections due to lack of isotropy and equilibration at early times, the analogy with the RM induced flow improves at later times (Clark and Zhou, 2006).

Zhou's description of the RMML is remarkably reminiscent of isotropic turbulence with permanent large-scale structures. For isotropic turbulence, Clark and Zemach considered two physical effects – viscous dissipation and the observed persistent large-scale eddies. They assumed the low wave number ("infrared") energy spectrum

$$\lim_{k \to 0} E(k,t) = e_0 k^{\mathfrak{s}}. \tag{4.29}$$

For this "permanent large eddies" assumption, the value of e_0 is a time-independent constant for decay with $2 \leq \mathfrak{s} \leq 4$ (Batchelor and Proudman, 1956; Saffman, 1967), and has dimensions of [length$^{3+\mathfrak{s}}$/time^2]. Specifically, the invariance of e_0 requires that (Clark and Zhou, 2006)

$$\tau^{(3+\mathfrak{s})\vartheta - 2} e_0 = e_0, \tag{4.30}$$

which has a solution for ϑ (=θ) of

$$\vartheta = \frac{2}{\mathfrak{s} + 3}. \tag{4.31}$$

Thus, permanence of large-eddies – or equivalently, invariance of the infrared energy spectrum – establishes a connection between the RMI ϑ and the spectral power-law at low wavenumber (Soulard et al., 2020; Zhou, 2021). The initial energy spectrum at the center of the RM mixing layer is related to the perturbation amplitude spectrum by $E(k,0) \sim k^2 P(k)$. Hence, this result is equivalent to Eq. 4.19 with $\mathfrak{s} = m + 2$. This is also what Soulard et al. (2018) showed for $\mathcal{A} = 0$ and more recently for non-zero Atwood number in Soulard and Griffond (2022).

Table 4.3 highlights the documented relationship between θ and initial conditions predicted in the literature.[2] The particulars are complicated and have been omitted here for brevity, but we refer the interested reader to the original articles for further details.

[2] The parameters of Table 4.3 are defined as follows:

- m is the infrared exponent of the initial interfacial perturbation height spectrum.
- s_{3D} is the infrared exponent of the 3D velocity spectrum.

Table 4.2 RMI growth exponent values for various models, simulations, as well as experiments.

Authors	θ	Atwood number	Methodology
Barenblatt (1983)	2/3	Boussinesq (i.e., small Atwood number)	Self-similar flow with constant kinetic energy
Alon et al. (1994)	0.40	$\mathcal{A} \sim 1$	Bubble merger, 2D
Alon et al. (1995)	0.40	small Atwood number	Bubble merger, 2D
Oron et al. (2001)	0.2–0.25	weak Atwood dependence	Bubble merger, 3D
Weber et al. (2013)	0.24	$-0.73 \leq \mathcal{A} \leq -0.35$ $0.22 \leq \mathcal{A} \leq 0.73$	Shock remains planar as it passes through the interface
Youngs and Thornber (2020b)	0.31–0.36	0.5	Model and data analysis
Oggian et al. (2015)	0.244	0.5	Compressible sim.
	0.225		Hybrid sim.
Thornber et al. (2017) (θ-group)	0.291	0.5	Sim.
Mansoor et al. (2020)	0.36	0.67	Exp.
Reese et al. (2018)	0.34 ± 0.01	0.7	Exp.
Sewell et al. (2021)	0.51 ± 0.04 (high-amplitude) 0.45 ± 0.08 (low-amplitude)	0.67	Exp. (TKE width)
	0.51 ± 0.06 (high-amplitude) 0.68 ± 0.11 (low-amplitude)	0.67	Exp. (TKE decay)

- s_{2D} is the infrared exponent of the 2D velocity spectrum at the center of the mixing zone.
- s_{fol} is the infrared exponent of the foliated velocity spectrum.
- n_{cor} is the decay exponent of two-point longitudinal correlation of velocity at large separation.
- s^*_{3D} is a bounded expression of the infrared exponent of the 3D velocity spectrum:

$$s^*_{\mathrm{3D}} = \begin{cases} -1 & \text{if } s_{\mathrm{3D}} < -1 \\ s_{\mathrm{3D}} & \text{if } -1 < s_{\mathrm{3D}} < 4 - \eta^* \\ 4 - \eta^* & \text{if } 4 - \eta^* < s_{\mathrm{3D}} \end{cases} .$$

The parameter η^* is approximately 0.5. Here, σ^*_{so} is a bounded expression of the infrared exponent σ_{so} of the 3D spectrum of the solenoidal component of the linear momentum:

$$\sigma^*_{\mathrm{so}} = \begin{cases} -1 & \text{if } \sigma_{\mathrm{so}} < -1 \\ \sigma_{\mathrm{so}} & \text{if } -1 < \sigma_{\mathrm{so}} < 4 - \eta^* \\ 4 - \eta^* & \text{if } 4 - \eta^* < \sigma_{\mathrm{so}} \end{cases}$$

Indeed, one has:

$$\sigma_{\mathrm{so}} = m + 1, \text{ and only for } \mathcal{A} \ll 1, \; s_{3D} = n_{\mathrm{cor}} - 1 = \sigma_{\mathrm{so}}.$$

Dr. Olivier Soulard of Commissariat à l'Énergie Atomique provided key assistance in compiling this table.

Table 4.3 Relations between the RMI growth factor, θ, and initial conditions predicted in the literature. Several symbols are given in footnote 2.

Authors	θ	Atwood number	Methodology
Inogamov (1999)	$2/(m+5)$	$0 < \mathcal{A} < 1$	Just-saturated modes
Youngs (2004)	$2/(\min(m,3)+5)$	$0 < \mathcal{A} < 1$	Just-saturated modes
Thornber et al. (2010)	$2/(m+5)$	$0 < \mathcal{A} < 1$	Just-saturated modes
Elbaz and Shvarts (2018)	$2/(\min(m,1)+5)$	$0 < \mathcal{A} < 1$	Haan's model + saturation
Chasnov (1997)	$2/(s_{3D}+4)$	$\mathcal{A} \ll 1$	Perm. large eddies
Llor (2006)	$2/(n_{\text{cor}}+3)$	$\mathcal{A} \ll 1$	Perm. large eddies
Clark and Zhou (2006)	$2/(s_{2D}+3)$	$\mathcal{A} \ll 1$	Perm. large eddies
Poujade and Peybernes (2010)	$2/(s_{\text{fol}}+4)$	$\mathcal{A} \ll 1$	Perm. large eddies
Lombardini et al. (2012)	$2/(s_{3D}+3)$	$\mathcal{A} \ll 1$	Perm. large eddies
Soulard et al. (2018)	$2/(s^*_{3D}+4) \in [\frac{1}{4}, \frac{2}{3}]$	$\mathcal{A} \ll 1$	Perm. large eddies (EDQNM + Sim.)
Soulard and Griffond (2022)	$2/(\sigma^*_{\text{so}}+4) \in [\frac{1}{4}, \frac{2}{3}]$	$0 < \mathcal{A} < 1$	Perm. large eddies (EDQNM + Sim.)

4.5 Reynolds number for self-similar RT, RM, and KH flows

It is important to note that diffusive effects (viscosity, diffusivity, and heat conductivity) will act to moderate fluid instabilities, most strongly at small scales. This has potentially important ramifications for the mixing layer evolution. The Reynolds number, Eq. 2.9, characterizes the viscous damping. To estimate it, appropriate choices for the length scale $\mathcal{L}$ and velocity scale $\mathcal{U}$ must be made, so that they can capture the dominant flow dynamics. One such choice is to set $\mathcal{L} = h$ and $\mathcal{U} = \dot{h}$, which results in the outer-scale Reynolds number that represents the large scales dominating the flow (Zhou et al., 2019). One further assumes the self-similar regime discussed in this chapter and $t_{s,0} = t_{b,0} = t_0$ for simplicity's sake.

In the RT self-similar regime, it is trivial to show

$$Re = \frac{h\dot{h}}{\nu} = \frac{2(\alpha \mathcal{A} g)^2 (t+t_0)^3}{\nu}. \tag{4.32}$$

Recall from Eq. 4.10 that $h_i(t) = \alpha_i \mathcal{A} g t^2$, $i = b, s$, and $\alpha = \alpha_s + \alpha_b$. Thus, the Re growth is asymptotically proportional to t^3, a result that has been demonstrated in Fig. 4.15.

Figure 4.16 illustrates the difference between the RTML and RMML (Tritschler et al., 2014a). As will be discussed in the next chapter, several researchers (Clark and Zhou, 2006; Thornber et al., 2010, 2017) have argued in favor of a single value of θ for the layer as a whole, implying $\theta_s = \theta_b$ (except for the singular case $\mathcal{A} = 1$, Williams (2016)). In this case, the Re of the RMML is given by

$$Re = \frac{\tilde{h}^2}{\nu} \frac{\theta}{|t_0|} \left[\frac{t+t_0}{|t_0|} \right]^{2\theta - 1}, \tag{4.33}$$

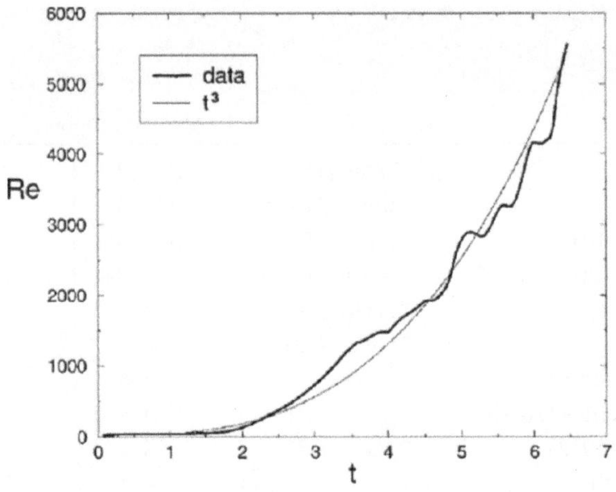

Fig. 4.15 Outer-scale Reynolds number, based on the extent and rate of growth of the RT mixing region Cook and Zhou (2002). Reproduced with permission from Cook and Zhou (2002).

where $h = \tilde{h}\tau^{\theta}$ and $\tilde{h} = \tilde{h}_{s,0}+\tilde{h}_{b,0}$. Thus in this "self-similar" regime, the RMML Reynolds number grows as $t^{2\theta-1}$, and since $\theta < 1$, the growth in the Reynolds number is sublinear in time.

Given that the value of θ can be related to the initial conditions, so too can the outer-scale Reynolds number. We point out the results of Barenblatt (1983) who showed that constant kinetic energy implies $\theta = 2/3$, which is a stricter constraint than $\theta < 1$. Note also that *Re* decreases with time if $\theta < 0.5$. For sufficiently low values of the outer-scale Reynolds number, the growth of secondary instabilities will be suppressed and the mixing layer will not reach a state of fully developed turbulence.

For completeness, we note that in the self-similar growth regime, the width of the KH mixing layer at late-time grows as

$$h = \beta_u \Delta_u (t + t_0), \tag{4.34}$$

where Δ_u is the velocity difference across the layer (e.g., Akula et al., 2017). Meanwhile, the Reynolds number of the KHI mixing layer can be represented as

$$Re = \frac{(\beta_u \Delta_u)^2 (t + t_0)}{\nu}. \tag{4.35}$$

Thus, in this case, the *Re* grows as a linear function of *t*.

4.6 The role of experiment and numerical simulation

By necessity, experimental investigation has played a key role in hydrodynamic instabilities and turbulence research for many years. It worth noting that experiments on the time evolving processes RT and RM are more difficult than for many of the statistically steady

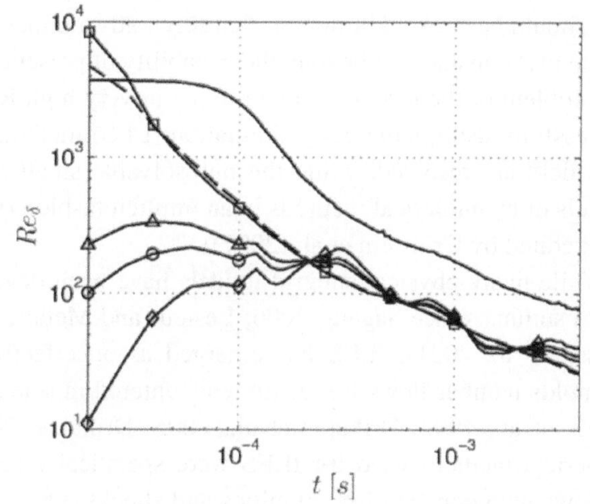

Fig. 4.16 Outer-scale Reynolds number, based on extent and rate of growth of the RM mixing region (solid line) for a shock Mach number of $M = 1.5$. Temporal evolutions of different Reynolds number approximations, as defined in Pope (2000), are given by symbols. For details, see Table II of Tritschler et al. (2014a). Fig. 13 of Tritschler et al. (2014a). Reprinted with permission from APS.

shear flow processes and that a fundamental understanding of the nonlinear and turbulent aspects of RM and RT has, to a very large extent, depended on advances in numerical simulation.

Over the past decades, carefully designed numerical simulations have emerged as an accepted surrogate for experiments when detailed information is not readily measurable in the laboratory. The pioneering paper of Orszag and Patterson Jr (1972) may be considered the first high-quality "numerical experiment." The authors noted that numerical simulation is feasible and even economical at Reynolds numbers of the order achieved in wind-tunnel turbulence experiments. They also stressed that the value of the simulations lies not only in the completeness of the data they provide but also in the opportunity they give for the assessment of the accuracy of turbulence theories under controlled and known conditions. With the exponential growth of supercomputing capabilities, numerical simulations have proven capable of following the three phases of turbulent mixing identified by Eckart (1948): entrainment, stirring, and molecular mixing.

As noted by Moin and Mahesh (1998), the wide range of scales present in turbulent flows requires that care must be taken in their developing numerical solution. The benefit of direct numerical simulations (DNS) is that the calculation can faithfully resolve all the scales that can be handled by the computational resources available to the researcher. DNS also provides a complete, 3D, time-accurate description of the flow field. As an example, Youngs (2017) noted that the early stages of the mixing process are influenced by viscosity and diffusivity and DNS is essential for understanding the influence of Reynolds number on the mixing process. For quantities such as enstrophy, which are determined by the small scales present in the flow, DNS is required to get meaningful results (typically with higher grid resolution or with higher-order numerics being required to obtain accurate results).

It should be stressed, however, that Reynolds numbers for practical flows of interest are extremely high and still beyond the capability of present-day supercomputers. Fortunately, the problem of the turbulent flow mixing at very high Reynolds number has been tackled successfully using large-eddy simulation (LES) methods. In LES, the large scales of the flow field are resolved, while the unresolvable small scales are represented by subgrid models or by numerical methods in an implicit fashion (implicit LES or ILES, see e.g., the book edited by Grinstein et al. (2007)).

While many physical subgrid models have been developed for large-eddy simulations (for a summary, see Sagaut, 2006; Lesieur and Metais, 1996; Meneveau and Katz, 2000; Moser et al., 2021), ILES has emerged as an effective approach for calculating high Reynolds number flows frequently encountered in scientific and engineering applications (Boris et al., 1992; Drikakis et al., 2005; Drikakis, 2003; Grinstein et al., 2007). The numerical methods used for ILES were specifically devised for the accurate simulation of flows with contact discontinuities and shocks – key aspects of the RT and RM mixing processes. Arbitrary numerics will not work for LES: good or bad subgrid scale physics can be built into the simulation model depending on the choice of numerics and its particular implementation.

Our guiding principle in the chapters that follow is to pick the highest quality data available to illustrate various points, regardless of whether the figures are from simulations or experiments. As appropriate, some comparisons between the methodologies will be presented in Chapter 23.

5 Global features from the lens of integrated mixing measurements

In the sections to follow, we focus on three integrated measures of the mixing: the mixed-width, mixedness, and mixed mass.

5.1 Mixed-width

5.1.1 Integrated mixing widths

In addition to the threshold or "visual" width, $h(n\%)$, there are several other methods for quantifying the width of the mixing layer. *The integral mixing width* (Andrews and Spalding, 1990; Youngs, 1994) is given by

$$W = \int_{-\infty}^{\infty} 6\overline{f_1 f_2}\, dz, \tag{5.1}$$

where f_1 and f_2 are the volume fractions of the two fluids. The overbar represents the spatial average taken over the x-and y-directions tangent to the original interface. The z-direction is normal to the interface.

Note that many authors have used compressible codes to calculate weakly compressible flows [case (c) of Section 2.6] and it is then appropriate to use

$$W = \int_{-\infty}^{\infty} 6\bar{X}_1\bar{X}_2\, dz. \tag{5.2}$$

In Eqs. (5.1) and (5.2), the factor 6 ensures the correct mixing width for a linear transition profile (Andrews and Spalding, 1990). For example, a linear profile with $\rho(z) = \rho_1 + (\rho_2 - \rho_1)z/L$ for $0 < z < L$, yields $X_1X_2 = z/L - z^2/L^2$ for $0 < z < L$, and $X_1X_2 = 0$ elsewhere. Integration of Eq. 5.2 then yields $W = L$.

The definition of W above can be further decomposed into the heights for the bubble and spike. With respect to the interface position z_0, the integral bubble and spike heights are defined as (e.g., Krivets et al., 2017)

$$W_b(z_0) = \int_{z_0}^{\infty} 6\overline{X}_1\overline{X}_2\, dz\,, \quad W_s(z_0) = \int_{-\infty}^{z_0} 6\overline{X}_1\overline{X}_2\, dz\,. \tag{5.3}$$

This definition is consistent with $W = W_b + W_s$.

The product width is analogous to the product in fast reactions (Cook and Dimotakis, 2001), and it is defined here with respect to a "stoichiometric" mole fraction of 1/2. It is similar to the integral mixing width except that, rather than X_1X_2, it uses a different measure of mixedness X_p.

The mole fraction of heavy fluid is (e.g., Cabot and Cook, 2006)

$$X = \frac{\rho - \rho_1}{\rho_2 - \rho_1}. \tag{5.4}$$

Assuming a passive, equilibrium chemical reaction between fluids, the chemical product is

$$X_p(X) = \begin{cases} 2X & \text{if } X \leq X_s \\ 2(1-X) & \text{if } X > X_s \end{cases}, \tag{5.5}$$

where X_s is the heavy-fluid mole fraction for a stoichiometric mixture, here taken as $X = 1/2$.

The product width is therefore defined as

$$H = \int_{-\infty}^{\infty} X_p(\langle X \rangle)\, dz. \tag{5.6}$$

In a similar way to the integral width W, this can be decomposed into the bubble and spike heights via

$$H_b(z_0) = \int_{z_0}^{\infty} X_p(\langle X \rangle)\, dz\,, \quad H_s(z_0) = \int_{-\infty}^{z_0} X_p(\langle X \rangle)\, dz. \tag{5.7}$$

The height of the mixing region is defined as the product thickness that would result if the entrained fluids were perfectly mixed in x and y.

Rather than the interface position z_0, another definition measures the spike and bubble from the physical midplane position z_s where $X_1 = X_2 = 1/2$. According to this definition we have

$$H_b(z_s) = \int_{z_s}^{\infty} X_p(\langle X \rangle)\, dz\,, \quad H_s(z_s) = \int_{-\infty}^{z_s} X_p(\langle X \rangle) dz. \tag{5.8}$$

5.1.2 RT growth rate, α_s and α_b

We now focus on understanding the time-dependent anisotropy, mixing, and scaling of flows induced by RTI. In particular, we utilize three large datasets from well-resolved direct numerical simulations (DNS) with moderate Reynolds number (Zhou and Cabot, 2019). The three cases used density ratios $\rho_2/\rho_1 = 1.5, 3$, and 9, corresponding to Atwood numbers $0.2, 0.5$, and 0.8, respectively. The grid sizes were 2048^3, 3072^3, and $2048^2 \times 2304$, respectively (Cabot and Cook, 2006; Cabot and Zhou, 2013).

We emphasize that the RT self-similar α (see Chapter 4) is by no means a universal quantity because it scales as h. Any quoted value of α should always be accompanied by the precise method used to estimate it (Section 4.4.1), as well as the precise definition of h (Sections 3.1 and 5.1.1), if there is to be any hope of making useful comparisons.

Figure 5.1 shows the time histories of α, using Eq. 4.14 for the total product width H, the 1% threshold bubble height, and the 1% threshold spike height. The other frequently used threshold values in the literature are 2%, 5%, and 10%. At late times the values for α level out somewhat and oscillate roughly between 0.020 and 0.025 for all $\mathcal{A}$. However, the 1% threshold heights are very noisy, making it difficult to determine an asymptotic mean

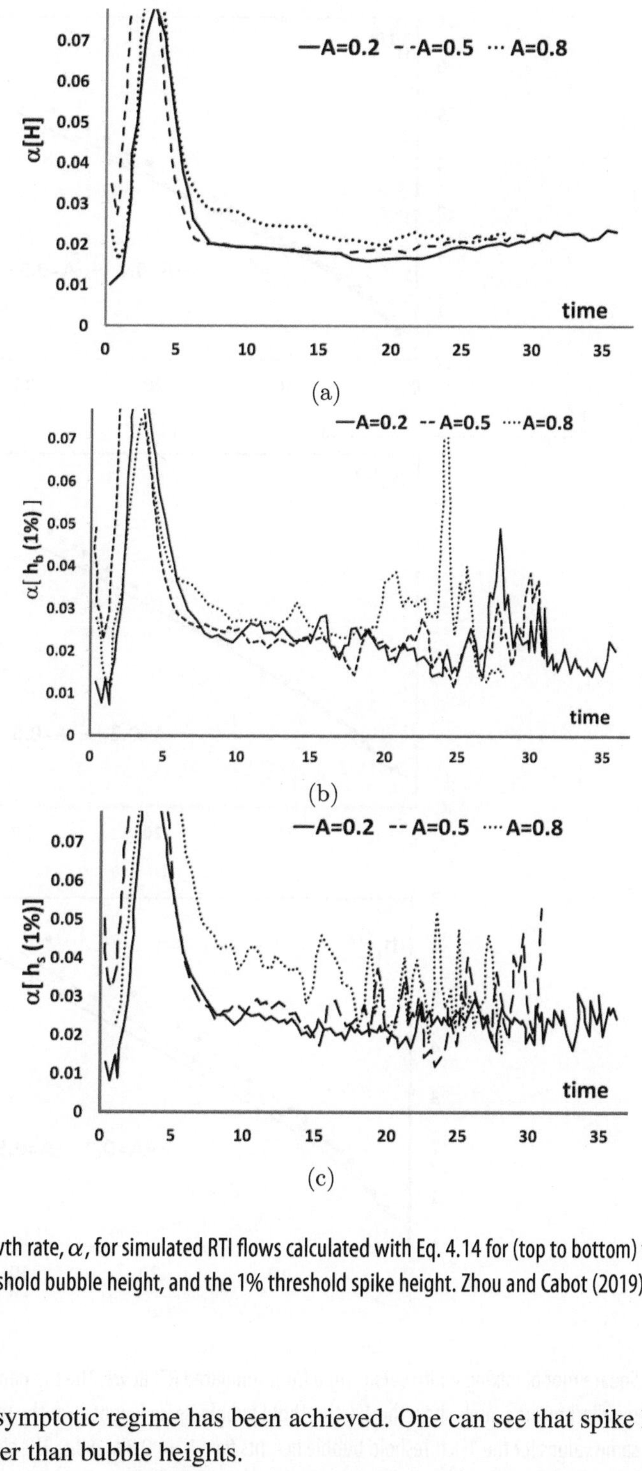

Fig. 5.1 Instantaneous growth rate, α, for simulated RTI flows calculated with Eq. 4.14 for (top to bottom) the total product width, the 1% threshold bubble height, and the 1% threshold spike height. Zhou and Cabot (2019).

or even if an asymptotic regime has been achieved. One can see that spike heights tend to be slightly larger than bubble heights.

For the same measures of height, Fig. 5.2 shows the square root of mixing width versus time, along with some late-time linear fits whose slopes equal $\alpha^{1/2}$, using Eq. 4.13. For

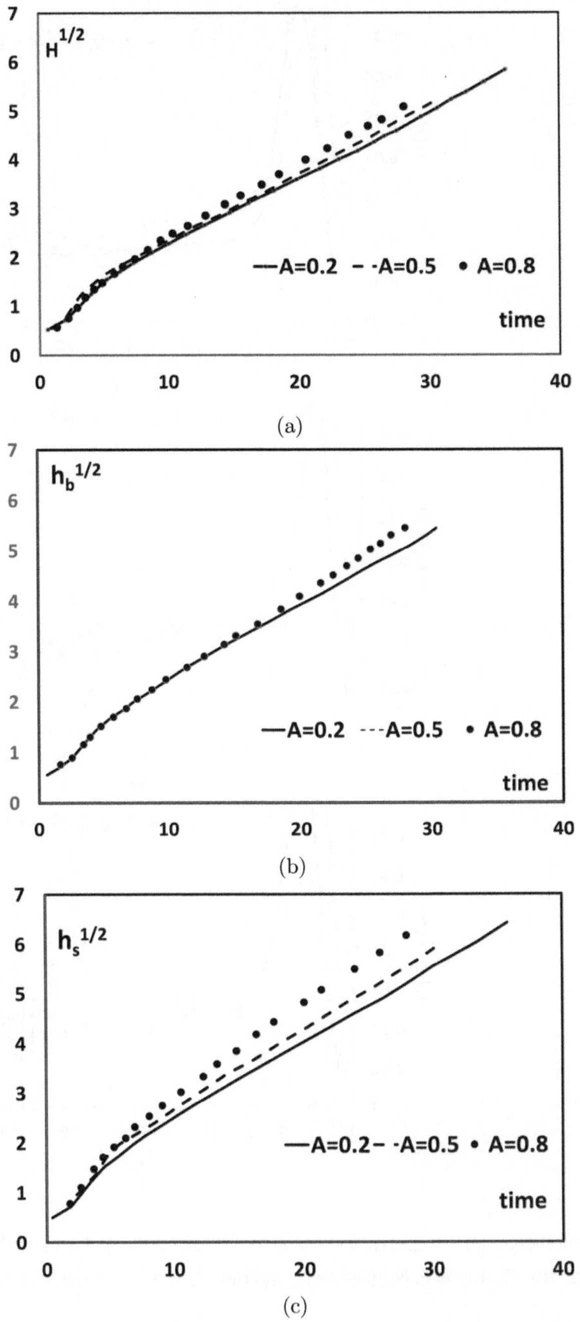

Fig. 5.2 Square root of mixing width versus time from simulated RTI flows. The top panel shows the total product width; the middle/bottom panels show the 1% threshold bubble/spike heights. For the middle panel, we find essentially the same values for the 1% threshold bubble heights for $\mathcal{A} = 0.2$ and 0.5. The slope of the late-time linear fits give approximations for $\alpha^{1/2}$. (a) $\alpha = 0.021$ for all $\mathcal{A}$, (b) $\alpha_b = 0.021, 0.021, 0.031$ ($\mathcal{A} = 0.2, 0.5, 0.8$) and (c) $\alpha_s = 0.025, 0.025, 0.030$ ($\mathcal{A} = 0.2, 0.5, 0.8$). Zhou and Cabot (2019).

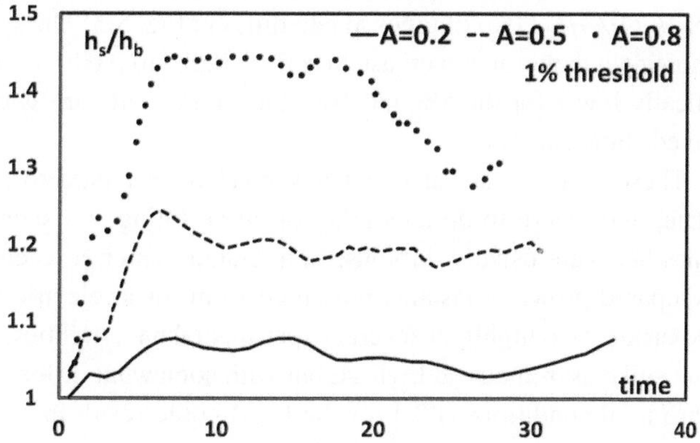

(a)

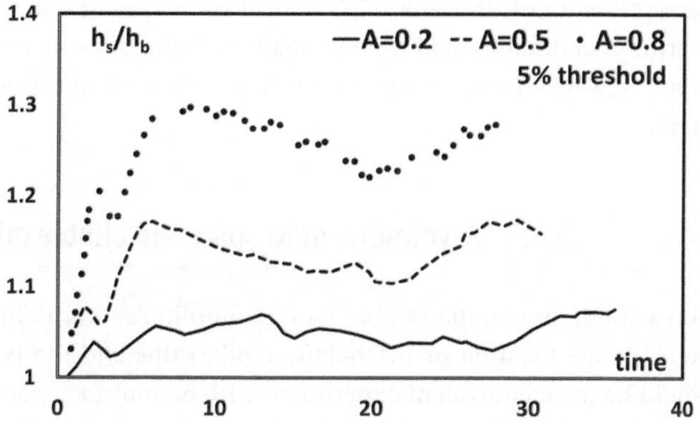

(b)

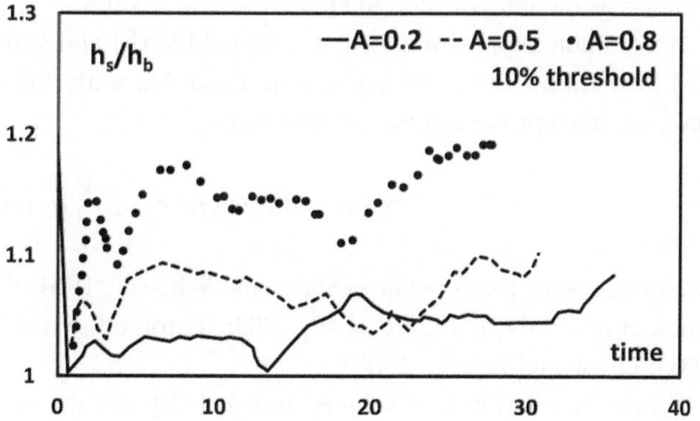

(c)

Fig. 5.3 Ratio of spike height to bubble height using (from top to bottom) 1%, 5%, and 10% thresholds for simulated RTI flow with several Atwood numbers. Zhou and Cabot (2019).

total product width H, we find $\alpha = 0.021$ for all $\mathcal{A}$ by this method. We find similar values for h_b at $\mathcal{A} = 0.2$ and 0.5; however, the height for $\mathcal{A} = 0.8$ is seen at late times to take a jog to a higher growth rate that we suspect is anomalous. This is the main reason that $h_s(1\%)/h_b(1\%)$ plunges at late times in Fig. 5.3. The spike heights are better behaved, showing a monotonic increase in α from 0.025 to 0.030 in Fig. 5.2. Values of α are systematically lower for the 5% and 10% thresholds, but vary widely as result of the previously noted statistical noise.

These methods are also rather subjective in regard to applicability of the asymptotic state, and where to do averaging or curve fitting. As such, quoted values of α (like the ones here) are usually dubious. In a related matter, a recent publication (Banerjee, 2020) compared growth-constants measured from linear electric motor (LEM) and gas channel experiments to highlight several experimental uncertainties. These datasets reported bubble and spike asymmetry at high $\mathcal{A}$, but with somewhat different magnitudes. First, the different initial conditions (ICs) for the LEM could result in a $\sim$ 20% variation of α; second, the use of surfactants in the LEM immiscible experiments might yield different growth parameters; third, the lack of statistical convergence due to insufficient flow sample size; fourth, uncertainties with image analysis techniques or with the experimental configurations; finally, the use of the Taylor hypothesis, of questionable validity at large Atwood numbers.

5.1.3 Asymmetry of RT spike and bubble mixing layer widths

For a single mode, the bubble (spike) amplitudes are defined by computing the distance between the location of the bubble (spike) tips and the position where the flat interface would be in an equivalent experiment with no interface perturbation. The Rayleigh–Taylor mixing layer (RTML) for a multi-mode center at $X_1 = X_2 = 1/2$ only gives the same location as the unperturbed interface for low $\mathcal{A}$ values.

A common supposition is that both bubble (b) and spike (s) growth for a multi-mode RTI flow follow the same asymptotic form, but with different values of α. Were this the case, the asymptotic solution should satisfy

$$h_s/h_b = (\alpha_s/\alpha_b)(t + t_{s,0})^2/(t + t_{b,0})^2, \tag{5.9}$$

where α_b and α_s are constants (Sharp and Wheeler, 1961; Sharp, 1984; Glimm et al., 2001; Cook et al., 2004; Dimonte et al., 2004; Ristorcelli and Clark, 2004; Jacobs and Dalziel, 2005; Olson and Jacobs, 2009).

While the relation $h = h_b + h_s$ always holds (by definition), the relations $\alpha = \alpha_b + \alpha_s$ and $h_s/h_b = \alpha_s/\alpha_b$ generally do not. The exception is when $h_{b/s} \approx \alpha_{b/s} g \mathcal{A} t^2$ at very late times ($t \gg \max(|t_{s,0}|, |t_{b,0}|)$) that are not easily achievable with numerical simulations and experiments (Zhou, 2017b). From this perspective, measurements of h_s/h_b in the early asymptotic state, contaminated by the virtual origin, are not particularly useful, since they will vary in time.

The consensus of the late-time RTML was an asymptotic, approximately self-similar state, wherein the width of the mixing layer for $t \gg t_{b/s,0}$ grows as

$$h_{b/s}(t) = \alpha_{b/s} \mathcal{A} g t^2. \tag{5.10}$$

In this limit one obtains

$$h_s(t)/h_b(t) \to \alpha_s/\alpha_b. \tag{5.11}$$

While the static values of α_s/α_b could in principle be determined at any point in the asymptotic stage, unfortunately, α_s/α_b is notoriously difficult to measure without a very long range of time in the asymptotic regime. Nevertheless, several publications have reported the bubble and spike ratios from either experiments or modeling (e.g., Dimonte and Schneider, 2000; Cheng et al., 2000; Ofer et al., 1996). Bubble growth is observed to be nearly independent of Atwood number, while spike growth is not.

Of particular note is the fact that studies have obtained higher values for α_s/α_b than for h_s/h_b. Dimonte and Schneider (2000) found greater asymmetry based on measuring α separately for bubbles and spikes, $\alpha_s/\alpha_b = (\rho_1/\rho_2)^{0.33\pm0.05}$. Youngs (2013) noted that the results of Dimonte and Schneider (2000), at high Atwood number, do appear to disagree with his ILES simulations. Including the uncertainties due to choosing the correct normalization factor for the volume fraction calculations, Akula and Ranjan (2016) performed an experiment at a moderately high Atwood number ($\mathcal{A} = 0.73 \pm 0.02$) and determined that α_s/α_b =1.95 or 1.93, depending on whether one adopts the virtual origin method (Snider and Andrews, 1994; Bell and Mehta, 1990) or the derivative method (Glimm et al., 2001; Ristorcelli and Clark, 2004; Cook et al., 2004; Jacobs and Dalziel, 2005). Akula and Ranjan reported their measured asymmetry, $h_s/h_b \approx 1.6 \pm 0.07 < \alpha_s/\alpha_b$.

For the reasons outlined above, care must be taken when quoting RTI experimental results. For the rocket-rig experiment, results are given for h_s/h_b, not α_s/α_b as suggested

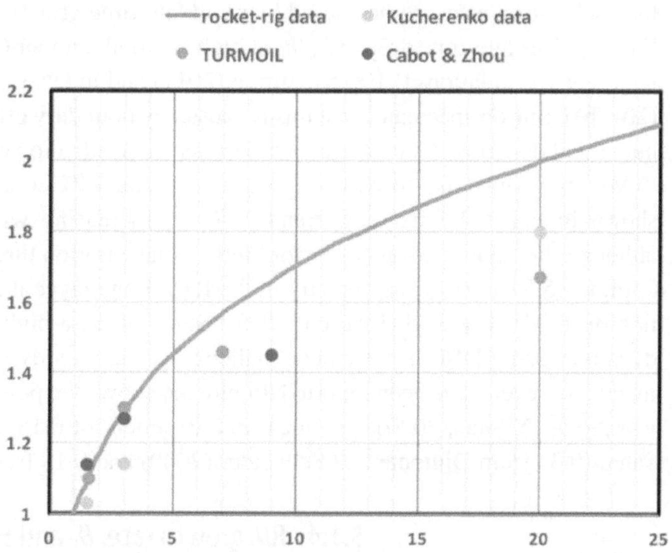

Fig. 5.4 Plot of simulated RTI spike/bubble mix width ratio, $h_s(1\%)/h_b(2\%)$, versus ρ_1/ρ_2. Fig. 5 of Youngs (2013). Additional data from DNS discussed in Cabot and Zhou (2013) added (Zhou (2017b).) UK Ministry of Defence © Crown Owned Copyright 2022/AWE.

Table 5.1 Mixing width asymmetry (ratio of the spike height to the bubble height) from numerical simulations with various Atwood numbers $\mathcal{A}$ (rows). Columns show data for various definitions of the spike and bubble heights – three threshold methods, the integral width and the product width. The $\mathcal{A} = 0.8$ value for $h(1\%)$ is poorly determined.

$\mathcal{A}$	$h_s/h_b(1\%)$	$h_s/h_b(5\%)$	$h_s/h_b(10\%)$	H_s/H_b	W_s/W_b
0.2	1.06	1.07	1.06	1.001	1.03
0.5	1.19	1.17	1.09	1.004	1.06
0.8	*1.30*	1.26	1.18	1.013	1.12

in the caption of Fig. 2 of Zhang et al. (2016b). Also, the LEM results are quoted in α_s/α_b, not h_s/h_b, as indicated in Eq. (31) of Burton (2011).

The spike-bubble asymmetry ratio $h_s(1\%)/h_b(2\%)$ has been plotted in Fig. 5.4 for DNS with Atwood numbers 0.2, 0.5, and 0.8 (Cabot and Zhou, 2013; Zhou, 2017b). Conditions were chosen to emulate the experiment of Kucherenko et al. (1991) and the rocket-rig (Youngs, 1989). Comparisons were also made to the ILES of Youngs (2013), which, at high $\mathcal{A}$, found lower spike-bubble ratios than the corresponding experiments. Higher threshold levels reduce the degree of asymmetry, but as was shown in Zhou and Cabot (2019), do not appreciably reduce the statistical noise.

The spike-bubble asymmetry for different threshold values ($h_b(n\%)$ or $h_s(n\%)$) is given in Table 5.1. Spike-bubble ratios are shown in Fig. 5.3 for different Atwood numbers and thresholds. The ratios for $\mathcal{A} = 0.2$ and 0.5 level out at late times, but the ratio for $\mathcal{A} = 0.8$, after leveling at about 1.45 at intermediate times, plunges to 1.3 at late times. This anomalous behavior appears to be caused by rapid late-time growth of the bubble height, seen in Fig. 5.2. This late-time drop in h_s/h_b at high Atwood number (> 0.75) was also observed in large-eddy simulations (LES) by Burton (2011) and in DNS by Livescu (Fig. 9.14 in Zhou (2017b)). Burton speculates that this is caused by boundary effects from insufficiently large numerical domains. Much larger simulations are needed to verify this hypothesis.

Mixing width spike-to-bubble ratios for 5% and 10% concentration thresholds are also shown in Fig. 5.3. The same trend of h_s/h_b increasing with $\mathcal{A}$ is generally observed, although the asymmetry is lower for higher concentration thresholds.

Figure 5.5 shows the asymmetry of the RT mixing layer at $\mathcal{A} = 0.75$ from 3D visualization of the density field (Livescu, 2013). Study of ultra-high Atwood number RT mixing dynamics using DNS is extremely challenging. Large-eddy simulations and LEM experiments, however, have been reported. Figure 5.6 shows the power law scaling between h_s/h_b and ρ_2/ρ_1 (Yilmaz, 2020). The measured exponent of 0.18 should be compared with the values 0.33 from Dimonte and Schneider (2000) and 0.18 from Burton (2011).

5.1.4 RM growth rate, θ_s and θ_b

The usual scaling assumptions regarding late-time RM growth rates are based on empirical observation: h_b and h_s grow as power-laws in time, with the bubble-side exponent θ_b having a somewhat smaller value than the spike-side exponent θ_s,

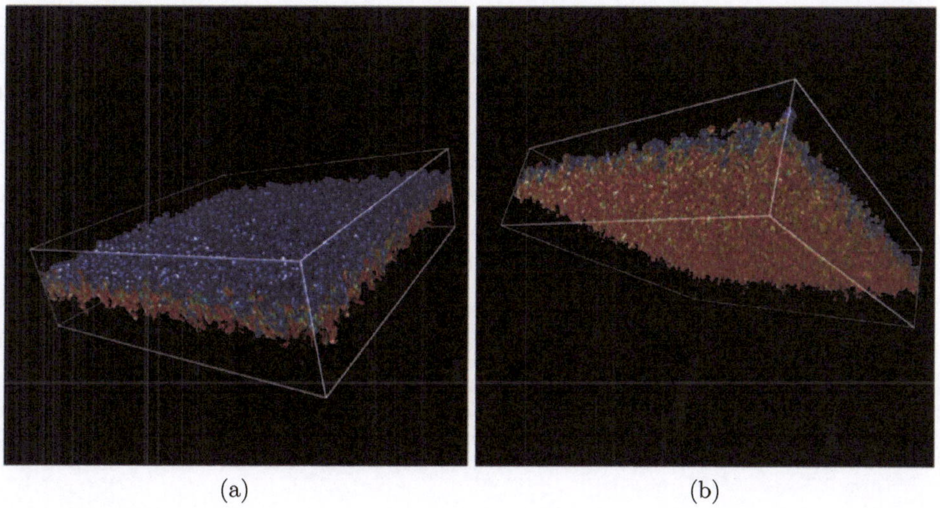

Fig. 5.5 (a,b) 3D visualization of the density field showing the asymmetry of the RT mixing layer at $\mathcal{A} = 0.75$, from a $4096^2 \times 4032$ simulation. Fig. 2 of Livescu (2013), with permission.

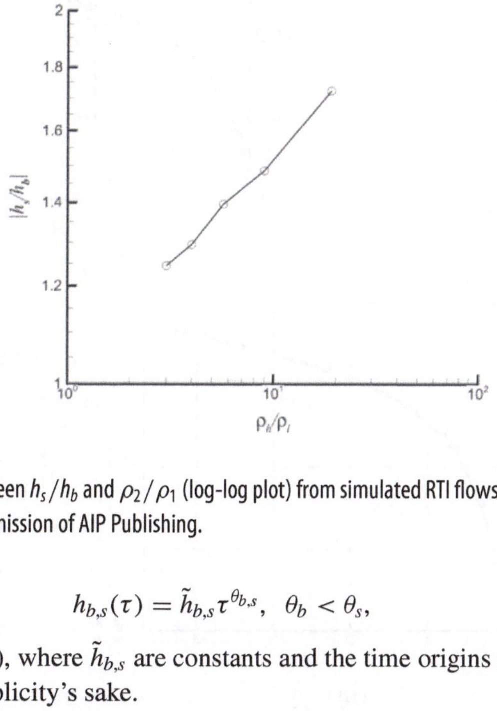

Fig. 5.6 The power law scaling between h_s/h_b and ρ_2/ρ_1 (log-log plot) from simulated RTI flows of various density ratios. Yilmaz (2020), with the permission of AIP Publishing.

$$h_{b,s}(\tau) = \tilde{h}_{b,s}\tau^{\theta_{b,s}}, \quad \theta_b < \theta_s, \tag{5.12}$$

for $\mathcal{A} \to 1$ (Fig. 4.13), where $\tilde{h}_{b,s}$ are constants and the time origins of bubbles and spikes are neglected for simplicity's sake.

In the opinion of Dimonte and Schneider (2000), $\theta_s > \theta_b$ implies that h_s/h_b grows in time without bound, implying a non-self-similar internal structure. However, using the analogy with isotropic turbulence, Clark and Zhou (2003, 2006) studied the late-time evolution of the RM mixing layer and argued analogy indicates that both h_b and h_s should evolve as power-laws in time, with the same exponent for both sides. This is a direct conclusion from turbulence theory, which implies that the bubble-spike correlation length at large separation should be the same. This suggests that the bubble and spike asymmetry should be instead attributed to the virtual time origins and coefficients for each side.

Youngs and co-workers (Thornber et al., 2010, 2017; Youngs and Thornber, 2020a,b) have also argued in favor of $\theta_s = \theta_b$, which gives constant h_s/h_b at late times. This is analogous to the self-similar behavior of RT mixing layers.[1] Furthermore, recent ILES results for $0.5 < \mathcal{A} < 0.9$ show that, while h_s/h_b varies initially, it eventually approaches a constant value, implying that $\theta_b = \theta_s$ (Youngs and Thornber, 2020a).

5.1.5 Asymmetry of RM bubble and spike mixing layer widths

Using a DNS with $\mathcal{A}^+ = 0.487$, Groom and Thornber (2021) show the time evolution of W_s, W_b, and the ratio W_s/W_b. Figure 5.7 shows that the asymmetry is largest at early time,

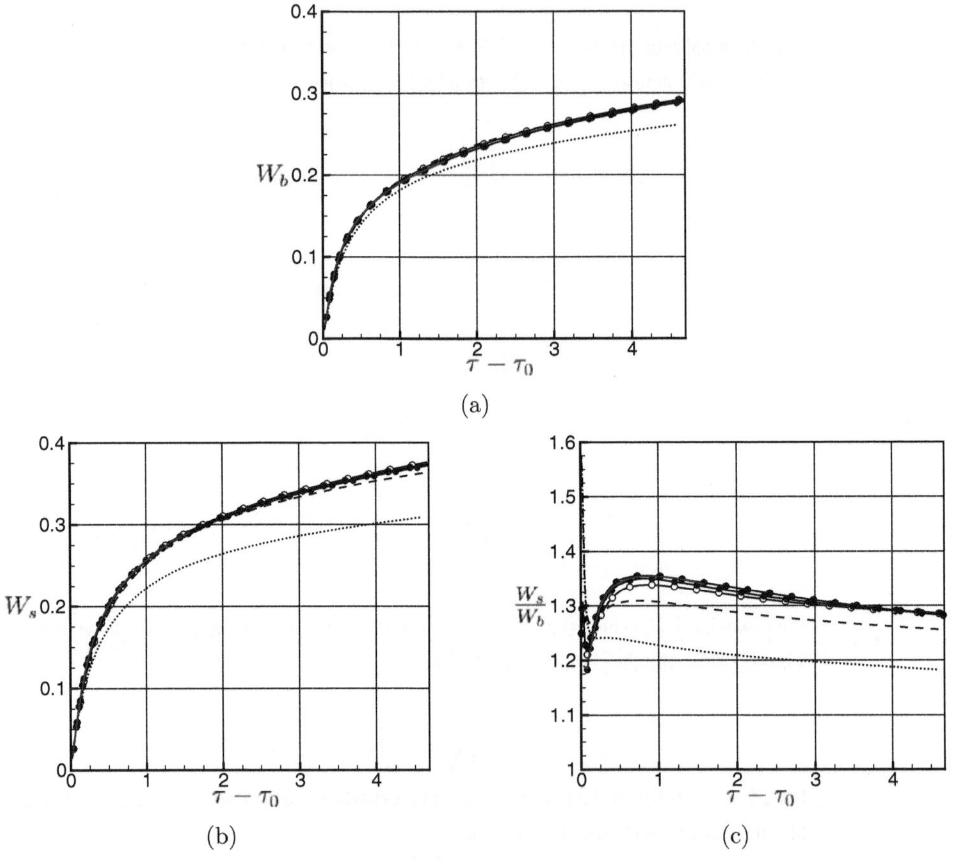

Fig. 5.7 Temporal evolution of integral bubble and spike widths as well as the ratio of spike and bubble widths from DNS of RM mixing layers. The legend of the original figure used W instead of h. Fig. 4.17 of Groom (2020), with permission.

[1] The authors used a Buoyancy–Drag model, which was described in detail in Section 3.2.2.

after which it decreases gradually. In all of the simulations the ratio appears to approach a constant value, although they do not run long enough to show this definitively.

Figure 5.8 shows the spike-to-bubble ratio in ILES simulations, as a function of W/λ. Spike-to-bubble ratios derived from the 1% and 99% volume fraction limits, h_s/h_b (1%), are compared with values of h_s^n/h_b^n derived from the integral measures (Youngs and Thornber, 2020a),

$$\bar{h}_b^{(n)} = \left[\frac{(n+1)(n+2)}{2} \frac{\int_{-\infty} |x|^n (1-f_1)dx}{\int_{-\infty} (1-f_1)dx}\right]^{\frac{1}{n}}, \tag{5.13}$$

$$\bar{h}_s^{(n)} = \left[\frac{(n+1)(n+2)}{2} \frac{\int_{-\infty} |x|^n f_1 dx}{\int_{-\infty} f_1 dx}\right]^{\frac{1}{n}}. \tag{5.14}$$

Integral bubble and spike distances for $n = 1, 2, 3$ are considered here. The integral measures behave in a very similar way to the distances derived from volume fraction cut-offs (which are affected by fluctuations). At early time the asymmetry is high. This is attributed

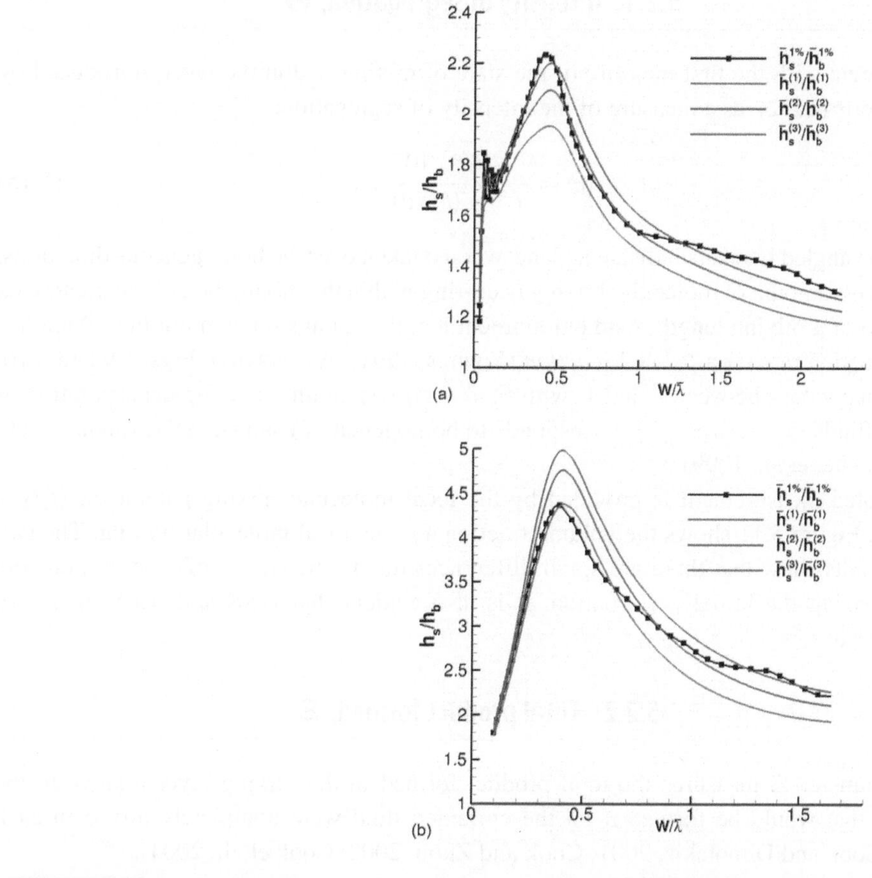

Fig. 5.8 The spike to bubble ratio, h_s/h_b, versus normalized mixed width, W/λ, from ILES for RMI mixing at (a) density ratio of 3 and (b) density ratio of 20. Fig. 3 of Youngs and Thornber (2020a).

to the formation of vortex rings on the spike side, which propagate with near constant velocity and persist to late times before breaking up (Youngs and Thornber, 2020a). It should be noted that 2D single-mode calculations show a similar degree of asymmetry (vortex pair rather than vortex ring).

5.2 Mixedness

The mixing-layer widths, either from integral or threshold measurements, do not differentiate between atomically mixed fluid and fluid that is penetrating the interface while remaining unmixed. For example, the mixing-layer width in 2D grows faster than in 3D while being less atomically mixed (Cabot, 2006; Zhao et al., 2022). It is often useful to quantify the degree to which the two fluids are atomically mixed by considering a dimensionless measure of the mixing state.

5.2.1 Intensity of segregation, Θ

Let us begin with the first measure of the state of mixing within the layer, introduced by Danckwerts (1952), as a measure of the intensity of segregation:

$$\Theta = \frac{\int \langle f_1 f_2 \rangle \, \mathrm{d}V}{\int \langle f_1 \rangle \langle f_2 \rangle \, \mathrm{d}V}, \tag{5.15}$$

where the angled brackets indicate a plane average taken over the homogeneous directions. The relative amount of molecular mixing occurring within the mixing layer is characterized as a ratio of a mixing length to an entrainment length. Youngs pointed out that Θ applies to the reaction rate for a *"slow" reaction* (Youngs, 1991). As shown in Figs. 5.9 and 5.10, Θ can take values between 0 and 1, with $\Theta = 0$ corresponding to complete separation of the two fluids, whereas $\Theta = 1$ corresponds to homogeneity (Youngs, 1994; Linden et al., 1994; Dalziel et al., 1999).

A related measurement is provided by the local molecular mixing parameter, $\langle f_1 f_2 \rangle / \langle f_1 \rangle \langle f_2 \rangle$. Figure 5.11 shows the volume fraction and the local molecular mixing. The two ILES results show that there are small differences due to the choice of random numbers characterizing the initial perturbation. It is also evident that DNS and ILES give very similar results.

5.2.2 Total product formed, Ξ

The parameter Ξ measures the total product formed in the mixing layer relative to the product that would be formed if all the entrained fluid were completely mixed in each plane (Cook and Dimotakis, 2001; Cook and Zhou, 2002; Cook et al., 2004):

$$\Xi = \frac{\int_{-\infty}^{\infty} \langle X_p \rangle \, \mathrm{d}z}{H}. \tag{5.16}$$

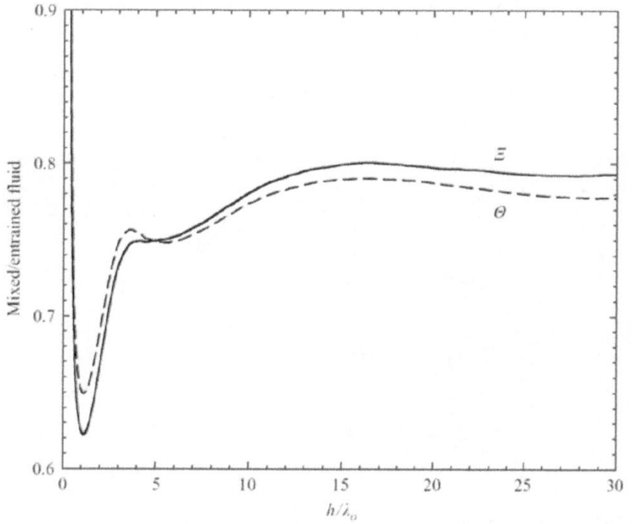

Fig. 5.9 Ratio of simulated mixed to entrained fluid for the RTI mixing region: Θ, $\Xi = 1$, corresponds to completely mixed fluid (no horizontal variation), while Θ, $\Xi = 0$ corresponds to complete segregation (immiscible case). Cook et al. (2004).

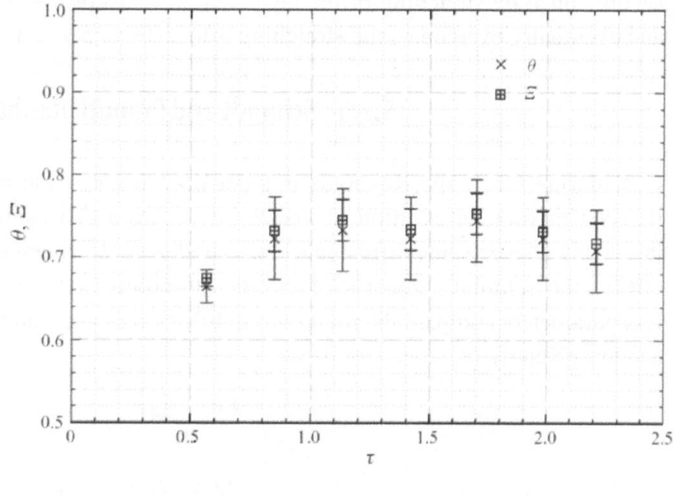

Fig. 5.10 The evolution of mix parameters measured at the centreline, Θ, Ξ, from a RT experiment Banerjee et al. (2010a).

Note that with homogeneity in x and y, ensemble averages (denoted by angle brackets) are equivalent to horizontal integrals. The value of Ξ is limited by the amount of lean reactant (light fluid) in the mixture. Thus, Ξ gives the equivalent result for a *"fast" reaction* for which either the reactant in fluid 1 or fluid 2 is fully depleted (Cook and Dimotakis, 2001).

For a typical simulation of a conventional RTI/RMI configuration {Θ, Ξ} will have an initial value close to 1, as the only heterogeneity present in the flow is due to the surface perturbation. As the instability develops, {Θ, Ξ} will decrease and eventually reach

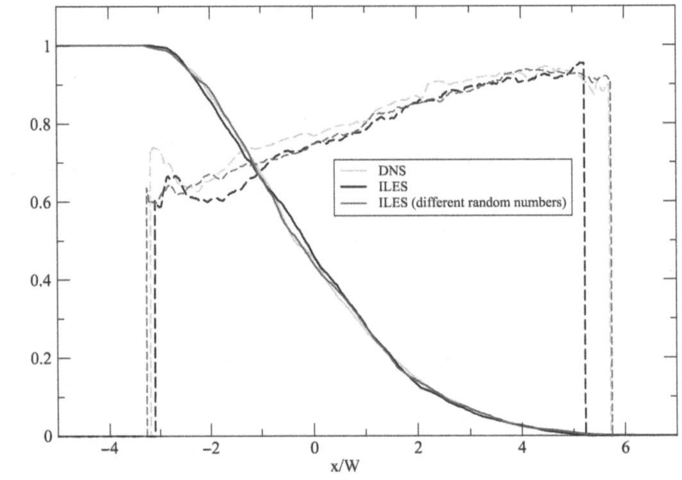

Fig. 5.11 Profiles of dense fluid volume fraction (solid lines) and local molecular mixing parameter (dotted lines). Fig. 3 of Youngs (2017). Reproduced with permission. UK Ministry of Defence © Crown Owned Copyright 2022/AWE.

a minimum, beyond which it increases again and tends towards some asymptotic value. The decrease in $\{\Theta, \Xi\}$ is due to the inter-penetration of bubbles and spikes, a process accompanied by stretching of the layer. As the roll-ups develop, fine-scale mixing occurs and eventually overcomes the stretching effect (Figs. 5.9 and 5.10).

5.2.3 Schmidt and Prandtl numbers

Considering solely the Reynolds number does not provide a full description of RT/RM induced mixing. In addition to viscous dissipation (diffusion of momentum) there will invariably also be some diffusion of mass and heat. To characterize the magnitude of these effects, the Schmidt number (Sc) and Prandtl number (Pr) are required. These dimensionless parameters compare the momentum diffusion to mass and heat diffusion, respectively, and are given by

$$\mathrm{Sc} = \frac{\nu}{\mathcal{D}}, \tag{5.17}$$

$$\mathrm{Pr} = \frac{\nu}{\mathcal{D}_\mathrm{T}}. \tag{5.18}$$

Here $\mathcal{D}$ is the mass diffusivity between two species, and $\mathcal{D}_\mathrm{T} = \kappa_\mathrm{T}/(\rho c_p)$ is the thermal diffusivity, with κ_T being the thermal conductivity, and c_p the specific heat capacity at constant pressure. For many gases, $\mathrm{Sc} \approx \mathrm{Pr} \approx 0.7$, and it is common in fundamental turbulent mixing studies to set $\mathrm{Sc} = \mathrm{Pr} = 1$ (Youngs, 2017). Thus at very low Reynolds numbers, for Pr and $\mathrm{Sc} \sim \mathcal{O}(1)$, there will be significant amounts of heat and mass transfer by diffusion, which in such cases will constitute the primary mixing mechanism (rather than turbulent stirring) between the two fluids.

As recalled in Banerjee et al. (2010a), the brine/water experiments of Dalziel et al. (1999) ($\mathrm{Sc} \sim 600$) resulted in a late-time value of $\Theta \sim 0.6-0.7$. Banerjee and Mutnuri

(2012) carried out an experimental study of mixing induced by RTI at $\mathcal{A} \sim 7.5 \times 10^{-4}$ and Sc $\sim 10^3$ with a transient experimental facility developed on the working principles of the draw-tank facility at Cambridge (Dalziel et al., 1999).

At Sc $\sim 10^3$, Mueschke et al. (2007) and Banerjee and Mutnuri (2012) found that the molecular mixing parameters (Θ, Ξ) were increasing at the end of the experiment. They also observed a higher ($2\times$) Reynolds number regime than that of Dalziel et al. (1999). The desired asymptotic Reynolds number could not be reached due to limitations in the test-section's dimensions. Well diagnosed experiments at higher Reynolds number are needed to confirm that the same asymptotic value is reached.

5.3 Mixed mass and normalized mixed mass

The actual amount of mixed mass could be viewed as a more direct measure of the evolution of the mixing layers than the width of the mixing zone. The mixed mass is directly relevant to recent high-energy-density-physics (HEDP) measurements.

5.3.1 Mixed mass, $\mathcal{M}$

To frame the characterization based on how much the heavy fluid is mixed into the light fluid, the mixed mass $\mathcal{M}$ can be defined by Zhou et al. (2016),

$$\mathcal{M} = \int_{-\infty}^{\infty} 4\rho Y_1 Y_2 \, \mathrm{d}V. \tag{5.19}$$

The mixed mass, $\mathcal{M}$, should not be confused with the Mach number, M.

An especially attractive feature of the mixed mass is the relation

$$\frac{\partial Y_1 Y_2 \rho}{\partial t} + \nabla \cdot (Y_1 Y_2 \rho \mathbf{u}) = Y_2 \nabla \cdot \mathbf{J}_1 + Y_1 \nabla \cdot \mathbf{J}_2, \tag{5.20}$$

$$\frac{\partial \mathcal{M}}{\partial t} = \int_{-\infty}^{\infty} 4(Y_2 \nabla \cdot \mathbf{J}_1 + Y_1 \nabla \cdot \mathbf{J}_2) \mathrm{d}V, \tag{5.21}$$

which has been obtained with use of Eq. 2.43. This shows that the mixed mass can only be changed by interspecies diffusion $\mathbf{J}_m$, either physical or numerical. In addition, with $Y_2 = 1 - Y_1$ and Fick's law Eq. 2.44, it can be shown that $d\mathcal{M}/dt \geq 0$.

The conversion from the mixed width to the mixed mass is nontrivial, as the nonlinearity in Eq. 5.19 (in terms of the density and mass fraction) has risen from quadratic to cubic. With the density incorporated, Eq. 5.19 now describes the time evolution of the mixing of materials within the mixing layer.

Figure 5.12 shows the temporal evolution of the RTI mixed mass for three Atwood numbers (solid lines). Also, we have presented the corresponding results for the mixed width (dashed lines) multiplied by arbitrary constants. Clearly, the $\mathcal{M}$ measurements closely track the scaling of the mixed width for all $\mathcal{A}$ (Zhou et al., 2016).

For the single-shock RMI case, it is interesting to compare two specific Atwood numbers, $\mathcal{A} = 0.5$ and $\mathcal{A} = 0.9$. The corresponding incident shock wave has a Mach number of

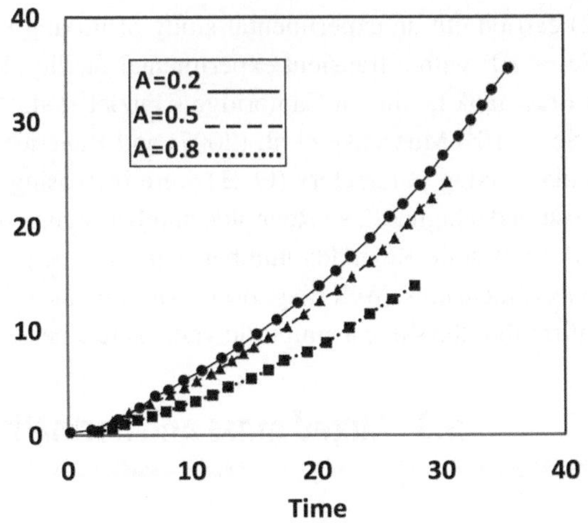

Fig. 5.12 Temporal evolution of simulated RTI mixed mass for three Atwood numbers (Solid, dashed & dotted Lines). Symbols (circle, triangle & square) show the corresponding results for the mixed width, multiplied by arbitrary constants. Zhou et al. (2016).

1.843 and 1.91, respectively, where the variation in shock Mach ensures that both cases have the same initial impulse (growth rate). Figure 5.13(a) indicates that the late-time behavior of mixed mass for $\mathcal{A} = 0.5$ is consistent with that of RTI. The scaling of the mixed mass is consistent with that of the mixing layer width. However, for $\mathcal{A} = 0.9$, the growth rate of the mixed mass is much lower than that of the scaled width (Fig. 5.13(b)), which experiences sustained growth due to the spike inertia. For the same scaled time, the high Atwood number case has substantially greater anisotropy, another indicator of a slower transition toward self-similarity.

The main thesis is that the amount of mixed mass can be viewed as a more direct measure of mixing than the mixed width. The mixed mass is also more useful for distinguishing between results at different Reynolds numbers (Fig. 5.14). The importance of the mixed mass is evident from recent measurements using the HEDP platform (Ma et al., 2013, 2017; Regan et al., 2013; Smalyuk et al., 2014). Figure 5.15 shows the measured total fusion neutron yield across the ensemble of cryogenic low-foot deuterium-tritium experiments against their inferred CH(Si) mix mass. While there is large variability in the amount of measured mix, there is clearly a trend of neutron yield dropping with increasing mix (Ma et al., 2013, 2017).

5.3.2 Normalized mixed mass, Ψ

The normalized mixed mass is given by (Zhou et al., 2016)

$$\Psi = \frac{\int_{-\infty}^{\infty} \rho Y_1 Y_2 \, dV}{\int_{-\infty}^{\infty} \langle\rho\rangle\langle Y_1\rangle\langle Y_2\rangle \, dV}. \tag{5.22}$$

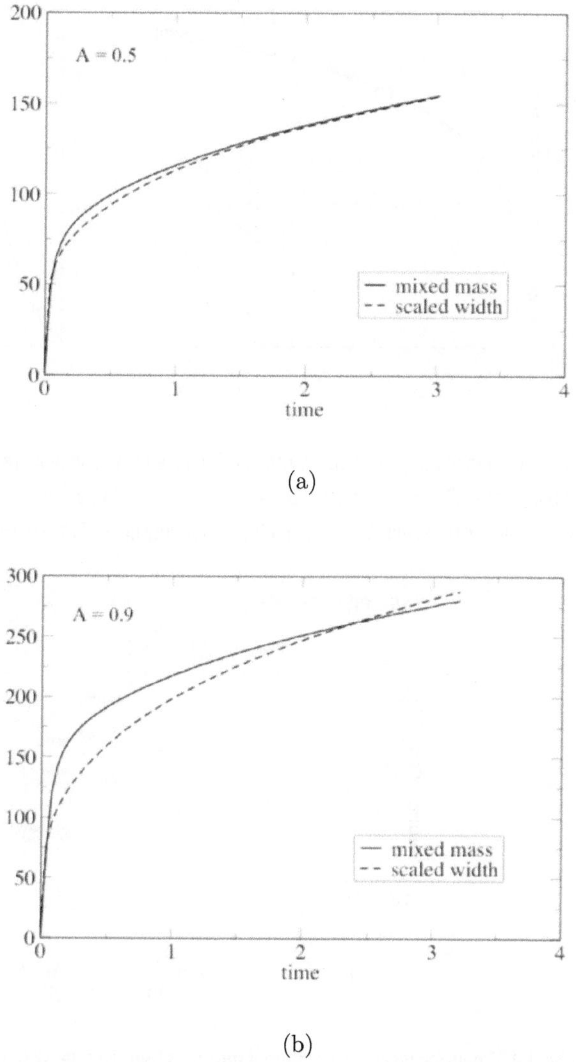

Fig. 5.13 Temporal evolution of simulated mixed mass for two RMI single shock cases: Atwood numbers (a) 0.5 and (b) 0.9. Dashed lines show the corresponding results for the mixed width, multiplied by arbitrary constants. Zhou et al. (2016).

Figures 5.16 and 5.17 illustrate that Θ, Ξ, and Ψ follow similar evolutions for all Atwood numbers under consideration. At early times, the entrainment rates exceed the mixing rates and the curves decrease for all three Atwood numbers. Later, the curves are enhanced significantly due to the Kelvin–Helmholtz instabilities in the shearing regions along the mushroom necks; the interfacial surface area then rapidly increases, the mixing rates overtake the entrainment rates, and the curves reverse direction toward their asymptotic values.

Despite a fairly significant span of Atwood numbers, Θ and Ξ are essentially the same. Their late-time values are around 0.80 and 0.78, respectively. Yet, the normalized mixed

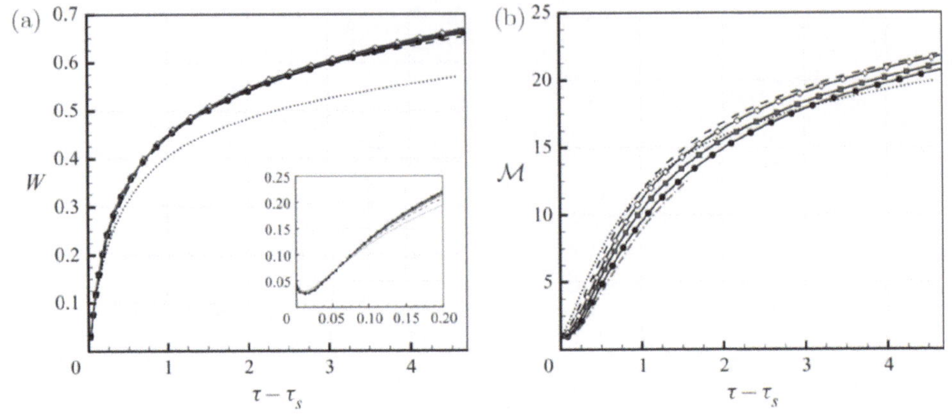

Fig. 5.14 Temporal evolution of the (a) integral width and (b) mixed mass. Shown are simulated RMI data for $Re_0 = 43$ (dotted black lines), $Re_0 = 86$ (dashed black lines), $Re_0 = 174$ (white diamonds), $Re_0 = 348$ (grey squares), $Re_0 = 697$ (black circles) and $Re_0 = 1395$ (dash-dot grey lines). Fig. 3 of Groom and Thornber (2021).

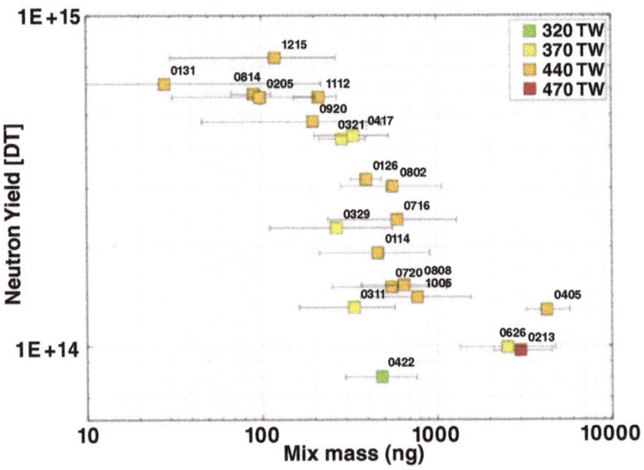

Fig. 5.15 Experimental DT neutron yield versus inferred mix mass (which scales with the X-ray enhancement factor) for layered low-foot implosions. Points are coded by peak laser power. Fig. 3 of Ma et al. (2013). Reprinted with permission from APS.

mass, Ψ, informs us how efficiently the mixed mass is generated for a given mixed width. While the values of Θ, Ξ, and Ψ are essentially indistinguishable for $\mathcal{A} = 0.2$, the Ψ curves are progressively lower for $\mathcal{A} = 0.5$ and $\mathcal{A} = 0.8$. It is evident, therefore, that the relative mixing (as measured by the mixed mass) becomes less efficient as the Atwood number increases (Fig. 5.16).

The normalized mixed mass Ψ for single shock RMI is shown in Fig. 5.17(a,b) for $\mathcal{A} = 0.5$ and $\mathcal{A} = 0.9$. While for $\mathcal{A} = 0.5$, Ψ reaches its asymptotic value, for $\mathcal{A} = 0.9$, Ψ still has not settled down completely, even after significant late-time computations. Apparently, $\mathcal{A} = 0.9$ requires much longer time to decay to a constant self-similar value. This phenomenon, in turn, suggests that the flow is still undergoing an active transient period,

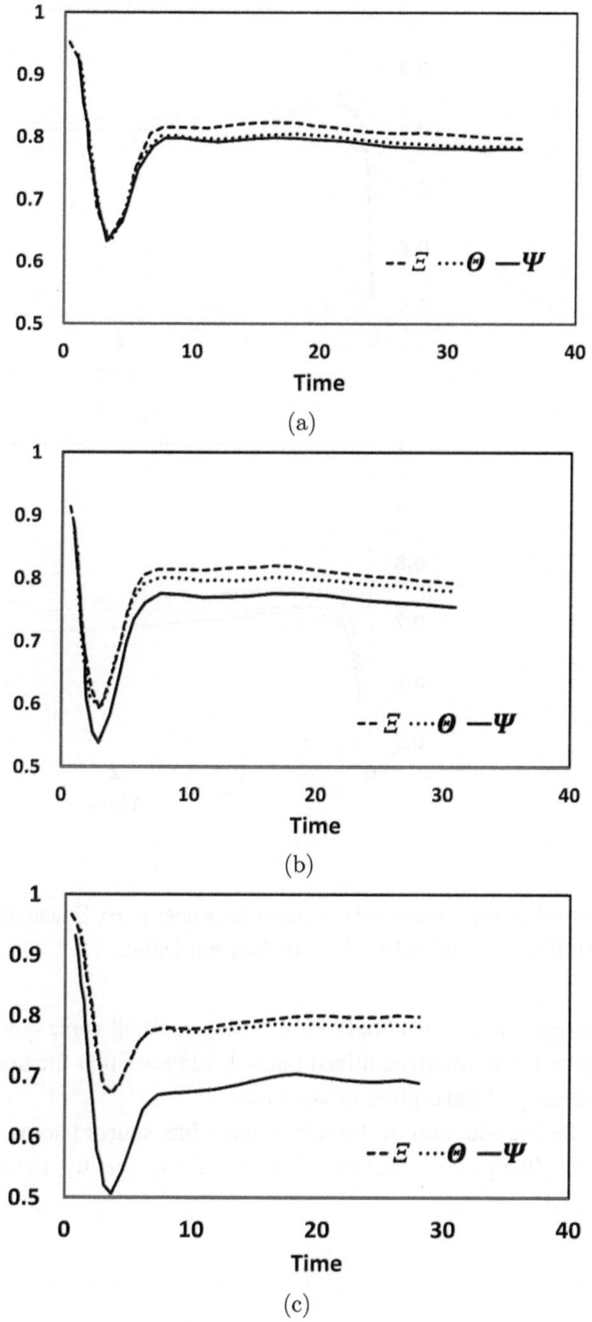

Fig. 5.16 Simulated temporal evolution of the mixedness parameters (Θ, Ξ) and normalized RTI mixed mass (Ψ) for several Atwood numbers: (a) 0.2, (b) 0.5, and (c) 0.8. Zhou et al. (2016).

possibly caused by physical phenomena, which do not collapse under a single scaling (elongated spikes for example). Nevertheless, the late-time values of Ψ are lower for high

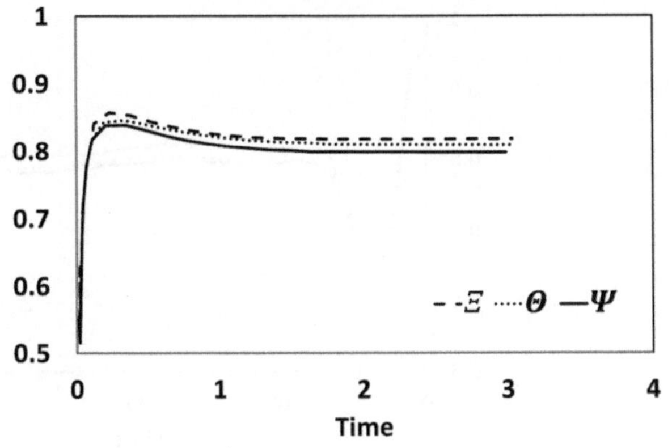

(a)

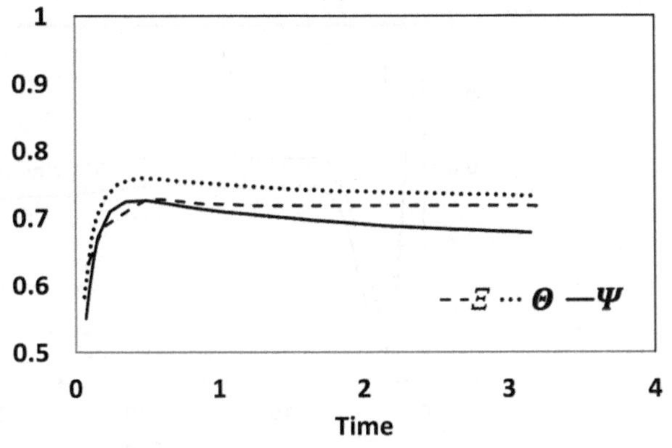

(b)

Fig. 5.17 Simulated temporal evolution of the mixedness parameters (Θ, Ξ) and normalized mixed mass (Ψ) for single-shock RMI with (a) $\mathcal{A} = 0.5$ and (b) $\mathcal{A} = 0.9$. Zhou et al. (2016).

Atwood number than that of lower $\mathcal{A}$ case (0.68 verse 0.80), consistent with the RTI flows. Again, the normalized mixed mass Ψ suggests that the higher Atwood number reduces the efficiency of generating mixed mass.

The introduction of the mixed mass has spurred some additional research (Ruan et al., 2020; Zhang et al., 2020c,d). For the sake of brevity, these results are not repeated here.

6 Internal dynamics from the lens of statistical mixing measurements

In this chapter, we will focus on the statistical spectral dynamics which are paramount to understanding the development of the integrated quantities described in the previous chapter.

The insight for invoking a statistical methodology to the study of turbulence may even trace its roots back to Leonardo da Vinci, who was a good scientist and engineer, in addition to being one of the greatest painters of all time (Keele, 2014; Marusic and Broomhall, 2021; Taylor, 1974). Figure 6.1, circa 1500, depicts turbulent flows generated by a jet from a square pipe into a pool.[1] Perhaps the world's first use of visualization as a scientific tool to study a turbulent flow (Gad-el Hak, 1998), da Vinci observed "the motion of the surface of the water, which resembles that of hair, which has two motions, of which one is caused by the weight of the hair, the other by the direction of the curls; thus, the water has eddying motions, one part of which is due to the principal current, the other to the random and reverse motion".[2]

Beginning with the pioneering work of Osborne Reynolds, mathematical descriptions of turbulent flows have been statistical, representing, for example, the velocity, u_i as the sum of an averaged value, $\langle u_i \rangle$, and a fluctuating value. Only "grudgingly" accepted for publication by two eminent referees (Launder, 2015), Sir George Stokes and Sir Horace Lamb,[3] Reynolds (1895) published his classical approach to turbulence, in which he began with the Navier–Stokes equation and then decomposed a flow variable Υ (e.g., velocity, pressure) into a mean component and a fluctuation about that mean.

One might wonder and speculate, as Lumley (1992) did, that Leonardo da Vinci might have developed his physical understanding of the so-called Reynolds decomposition four centuries beforehand. It is fascinating that da Vinci also carried out another fluid mechanics study (circa 1510) that focused on cardiac mechanics and the vascular system, unveiled in his more than fifty surviving pages of drawings (Keele, 1952; Gharib et al., 2002). Leonardo's heart studies integrate structural form and dynamic function in a way that called upon his extensive experience and knowledge of hydrology, engineering, mathematics, and architectural design (Wells, 2014).

[1] Recently, a French-Italian study has been performed aiming at numerically reproducing the physics behind da Vinci's drawings (Colagrossi et al., 2021).

[2] Translation credit: Ugo Piomelli (Lumley, 1992).

[3] Among his other achievements, Sir Horace is known to generations of fluid dynamic researchers for his book (Lamb, 1932), simply entitled "Hydrodynamics." The book ran to six editions; the sixth, which weighed in at nearly 740 pages, appeared in 1932 (Launder, 2012). G. I. Taylor (1935) paid homage to Lamb and his book: "The long-continued supremacy of this book in a field where much development has taken place is very remarkable and is evidence of the complete mastery which its author retained over this subject throughout his life." The book review of the sixth edition by H. Jeffreys (1933) declared that "a book on hydrodynamics capable of competing with Lamb's is unlikely to be produced for some considerable time."

Fig. 6.1 Turbulent flows. Leonardo da Vinci, circa 1500.

6.1 Statistics of the flow fields

Even before the transition of instability induced flows to fully developed turbulence, it is necessary to utilize statistical treatments. This is an important issue, and it will be addressed in Chapter 7.

For general turbulent flows, the mean of any flow variable is taken to be the ensemble average

$$\overline{\Upsilon}(\mathbf{x}, t) = \langle \Upsilon(\mathbf{x}, t) \rangle = \lim_{N \to \infty} \frac{1}{N} \sum_{i=0}^{N} \Upsilon^{(i)}(\mathbf{x}, t), \tag{6.1}$$

where an average is taken over N repeated experiments. A flow is statistically steady if the ensemble average does not change with time. The average over a long time period, then, gives an estimate of the ensemble average. A flow is homogeneous if the ensemble average does not vary with position. The average over space is then equal to the ensemble average

$$\overline{\Upsilon}(t) = \lim_{V \to \infty} \frac{1}{V} \int_V \Upsilon(\mathbf{x}, t) dV, \tag{6.2}$$

as well as the time average

$$\overline{\Upsilon}(\mathbf{x}) = \lim_{T \to \infty} \frac{1}{2T} \int_{-T}^{T} \Upsilon(\mathbf{x}, t) dt. \tag{6.3}$$

In the present case, where we are concerned with the temporal evolution of a strongly inhomogeneous field, neither temporal nor spatial averages seem appropriate, so we assume an ensemble average over a number of realizations. This means the supposedly larger-scale details of the field, such as the geometry, may be identical across each realization. However, the Navier–Stokes equations are deterministic, so repeated experiments will give identical results. The finer-scale details, for example the perturbations at an interface, must therefore vary across individual realizations.

More specifically, time-averaging of turbulent flows is inappropriate due to the flows' rapid time evolution. Spatial averaging may be appropriate in circumstances wherein (a) the scale of fluctuations is much smaller than the bulk motions of the fluid or (b) when there is at least one homogeneous direction over which to average. It must be noted that the assumption of homogeneity is not equivalent to an assumption of periodicity, and that assuming homogeneity may mask effects of large scale flow features in the presumed homogeneous directions. The ensemble average provides the most generally applicable approach but is difficult to realize, both experimentally and via DNS. However, for the task of deriving fundamental evolution equations, and for theoretical study, the ensemble average provides the most succinct description. Hence, from a fundamental point of view, the ensemble average is the appropriate definition of the averaging process. For example, for mixing at a plane boundary, if initial random periodic perturbations are suitably chosen, the ensemble average will be the same for all points in planes parallel to the initial interface. The planar average may then be used to estimate the ensemble average. Therefore, for most time-evolving RT/RM applications, the ensemble average is approximated by averaging over any available statistically homogeneous directions, for example, parallel to the interface. Due to the above considerations, many publications use the angle brackets or overlines to represent not only the ensemble average, but also any other appropriate (and approximately equivalent) averages.

Given the above considerations, we may write the average of the velocity field in the form

$$\langle u_i(\mathbf{x}) \rangle = \frac{1}{N_e} \sum_{n=1}^{N_e} u_i^{(n)}(\mathbf{x}), \tag{6.4}$$

where N_e is the number of elements in the ensemble and the superscript (n) represents the nth realization in the ensemble. The fluctuational part, denoted with a prime, is simply

$$u_i'(\mathbf{x}) = u_i(\mathbf{x}) - \langle u_i(\mathbf{x}) \rangle, \tag{6.5}$$

and obviously, $\langle u_i'(\mathbf{x}) \rangle = 0$ by construction. Under these restrictions, the evolution equation will have a quadratic nonlinearity in the fluctuating velocity,

$$R_{ij}(\mathbf{x}) = \langle u_i'(\mathbf{x})\, u_j'(\mathbf{x}) \rangle, \tag{6.6}$$

where $R_{ij}(\mathbf{x})$ is the so-called Reynolds stress tensor, which is related to the turbulent kinetic energy, K, by

$$K(\mathbf{x}) = \frac{1}{2} R_{nn}(\mathbf{x}), \tag{6.7}$$

where the repeated index implies summation. The notion of the Reynolds stress can be generalized by considering the correlation of velocities at two different points in space,

$$R_{ij}(\mathbf{x}_1, \mathbf{x}_2) = \langle u_i'(\mathbf{x}_1)\, u_j'(\mathbf{x}_2) \rangle. \tag{6.8}$$

Defining the average and relative coordinates

$$\mathbf{x}_c = \frac{1}{2}\left[\mathbf{x}_1 + \mathbf{x}_2\right], \tag{6.9a}$$

$$\mathbf{r} = \left[\mathbf{x}_1 - \mathbf{x}_2\right], \tag{6.9b}$$

allows Eq. 6.8 to be rewritten as

$$R_{ij}\left(\mathbf{x}_c, \mathbf{r}\right) = \left\langle u'_i\left(\mathbf{x}_c + \frac{1}{2}\mathbf{r}\right) u'_j\left(\mathbf{x}_c - \frac{1}{2}\mathbf{r}\right)\right\rangle. \tag{6.10}$$

If we assume that the field is homogeneous (Batchelor, 1953), then the two-point Reynolds stress tensor is independent of $\mathbf{x}_c$ and can be written as;

$$R_{ij}\left(\mathbf{x}_c, \mathbf{r}\right) \rightarrow R_{ij}\left(\mathbf{r}\right). \tag{6.11}$$

Fourier transforming the Reynolds stress tensor yields the 3D spectrum

$$R_{ij}\left(\mathbf{k}\right) = \iiint R_{ij}\left(\mathbf{r}\right) e^{i\mathbf{k}\cdot\mathbf{r}} \mathbf{dr}. \tag{6.12}$$

Next, one can average over all angles in Fourier space to arrive at

$$E_{ij}\left(k\right) = \frac{1}{2}\int R_{ij}\left(\mathbf{k}\right) \frac{\mathrm{d}\Omega_k}{\left(2\pi\right)^3}, \tag{6.13}$$

where $k = |\mathbf{k}|$. The turbulent kinetic energy spectrum is thus $E(k) = E_{nn}(k)$ and the turbulent kinetic energy per unit of mass is

$$K = \langle u'_n u'_n \rangle = \int_0^\infty E\left(k\right) \mathrm{d}k, \tag{6.14}$$

where the Einstein summation convention is applied over n, and

$$K_\alpha = \langle u'_\alpha u'_\alpha \rangle = \int_0^\infty E_\alpha\left(k\right) \mathrm{d}k, \tag{6.15}$$

where $E_\alpha(k)$ is the component energy spectrum (without summation over α).

Finally, the so-called velocity autocorrelation (Hinze, 1975; Champagne et al., 1970), $\mathcal{R}_{ii}(r)$, is defined as

$$\mathcal{R}_{ii}(r) = \frac{R_{ii}(r)}{R_{ii}(0)}. \tag{6.16}$$

By applying this decomposition (or the related mass-averaged statistical decomposition) to the Navier–Stokes equations, it is possible to derive an exact equation for the mean fields. This leads to the so-called Reynolds Averaged Navier–Stokes (RANS) turbulence models (e.g., Speziale, 1991; Wilcox, 1998), which will be covered in Chapter 22. Unfortunately, the mean flow equation will have second-order moments which must be modeled using an additional evolution equation.

Finally, studies of the Rayleigh–Taylor and Richtmyer–Meshkov instabilities typically require either compressible or variable density equations. The Favre decomposition of an arbitrary scalar, $\mathfrak{f}$, is denoted by (Favre, 1965; Bilger, 1975)

$$\mathfrak{f} = \bar{\mathfrak{f}} + \mathfrak{f}' = \tilde{\mathfrak{f}} + \mathfrak{f}'', \tag{6.17}$$

where

$$\tilde{f} = \overline{\rho f}/\bar{\rho}. \tag{6.18}$$

Thus, the velocity fluctuation is taken as

$$u_i'' = u_i - \tilde{u}_i, \tag{6.19}$$

where $\tilde{u}_i = \overline{\rho u_i}/\bar{\rho}$.

6.2 Energy budget

Fundamentally, hydrodynamic instabilities rely on a free energy source which drives the perturbation beyond the restraining ability of any restoring forces. The sources of the above instabilities differ: (i) for the RTI, the source is gravity (or, equivalently, surface acceleration), which tends to drive denser material through lighter material that might otherwise support it; (ii) for the RMI, a sudden intense acceleration, such as a shock, sets the material in motion; (iii) for the KH instability, by contrast, the source of free energy is the sideways relative motion of two components across a shearing surface.

In the case of classical RTI, the Navier–Stokes equations are driven by gravitational forces. Figure 6.2 shows snapshots of density field evolution from DNS (Cabot and Cook, 2006). The flows were initialized with a stationary interface between the high and low-density fluids, perturbed with random variations of characteristic wavelength λ_0. The simulations proceeded until the mixing layer filled 60%–70% of the domain in the vertical direction. The simulation used $\mathcal{A} = 0.5$ and 3072^3 grid points.

Taking the dot product of velocity with the momentum equation, Eq. 2.2, and dropping the negligible volume-expansion work term, one obtains

$$\frac{\partial}{\partial t}\left(\frac{1}{2}\rho u_i u_i\right) + \frac{\partial}{\partial x_j}\left(\frac{1}{2}\rho u_i u_i u_j + p u_j - u_i \tau_{ij}\right) = -\rho g u_3 - \tau_{ij}\frac{\partial u_i}{\partial x_j}. \tag{6.20}$$

Because there is no net flux through the boundaries, integrating Eq. 6.20 over the domain volume removes all the pure divergence terms on the left-hand side, yielding

$$\begin{aligned} \frac{dK}{dt} &= -\frac{d}{dt}\int_V \rho g z dV - \int_V \tau_{ij}\frac{\partial u_i}{\partial x_j} dV \\ &= -\frac{d\mathcal{P}}{dt} - \varepsilon, \end{aligned} \tag{6.21}$$

where the kinetic energy, potential energy, and energy dissipation rate are given by

$$K(t) = \frac{1}{2}\int_V \rho u_i u_i dV, \tag{6.22}$$

$$\mathcal{P}(t) = \int_V \rho g z dV, \tag{6.23}$$

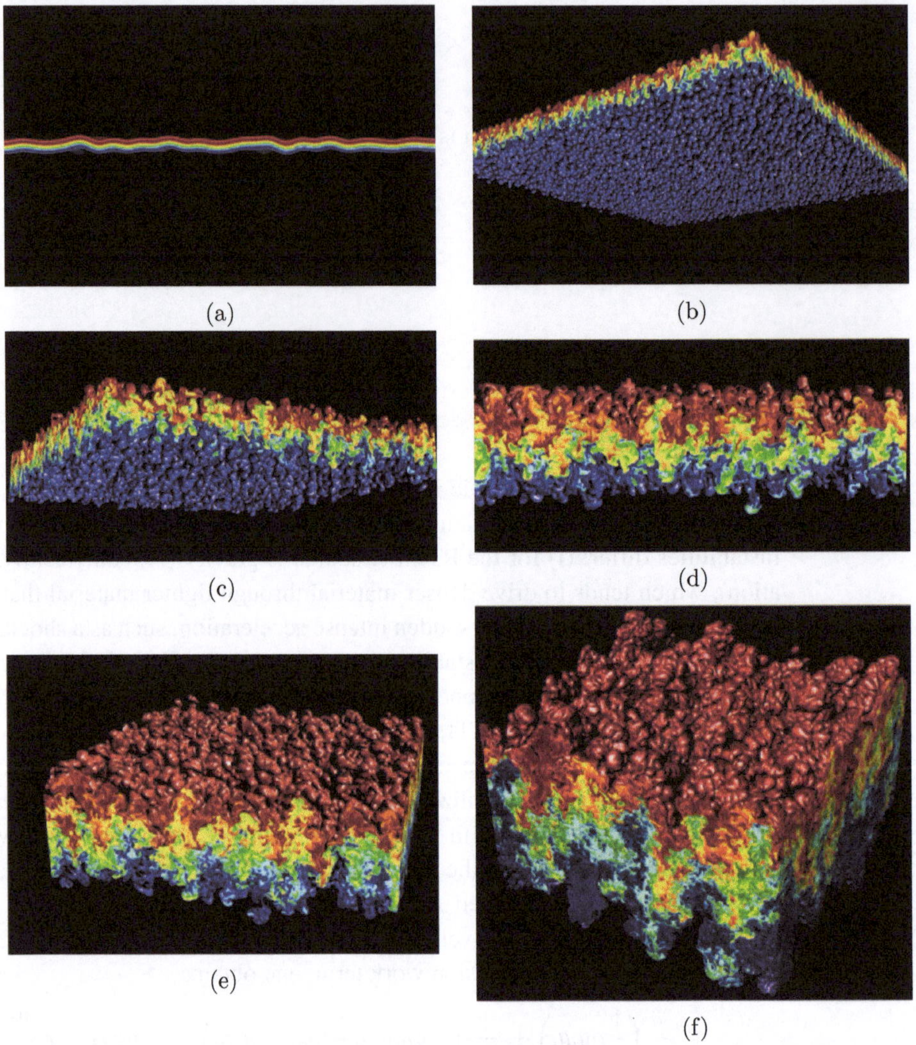

Fig. 6.2 (a–f) Snapshots of density field from DNS of RTI. Images, proceeding from upper to lower, were taken from a movie at dimensionless time 1, 6, 10, 12, 16, 31 (courtesy Dr. A.W. Cook, Lawrence Livermore National Laboratory). The heavy fluid is red ($\rho = 3$), the light fluid is blue ($\rho = 1$), and the mixed fluid is green ($\rho = 2$). The other colors represent mixed fluid of various compositions. The later panels show RTI in the fully turbulent regime. The last figure also appeared in Cabot and Cook (2006).

$$\varepsilon = \int_V \tau_{ij} \frac{\partial u_i}{\partial x_j} dV. \tag{6.24}$$

Further integrating Eq. 6.21 over the time from 0 to t, with stationary domain boundaries, yields (Cabot and Cook, 2006),

$$K(t) = \delta \mathcal{P}(t) - \mathcal{E}(t), \tag{6.25}$$

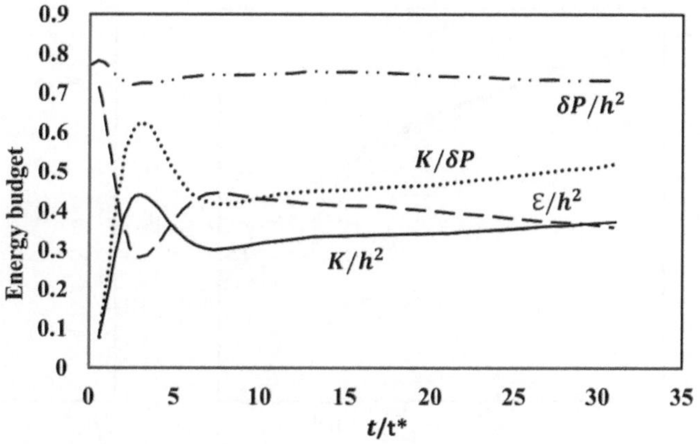

Fig. 6.3 RTI energy budget. Released potential energy scales with the square of the mixed width (h^2); however, kinetic energy and dissipated energy do not exhibit the same scaling. Consequently, the ratio of kinetic to released potential energy exhibits a gradual rise at late time. All dimensional terms in this figure are scaled in terms of gravity (g), the density of the light fluid, and the horizontal area of the computational domain. Fig. 5 of Cabot and Cook (2006).

where the potential energy released into the flow is

$$\delta \mathcal{P}(t) = \int_V [\rho(\mathbf{x}, 0) - \rho(\mathbf{x}, t)] gz dV, \tag{6.26}$$

and the energy dissipated into heat is

$$\mathcal{E}(t) = \int_0^t \int_V \left[\tau_{ij} \frac{\partial u_i}{\partial x_j}\right] dV dt' \tag{6.27}$$

Figure 6.3 depicts each of the three terms in Eq. 6.25, normalized by the square of the mixing width h^2 for the RTI. This normalization is motivated by a scaling relationship suggested by Dimonte et al. (2004), $\delta P = (\rho_2 - \rho_1) g h^2/6$. While the energy production does indeed follow h^2 scaling, the turbulent kinetic energy (TKE) and dissipation do not. Since the dissipation lags the production, the kinetic energy grows faster than h^2 and serves as the free energy driving perturbation growth. Cabot and Cook (2006) suggest this state of affairs is a result of a non-equilibrium process caused by the time lag between energy injection at large scales and dissipation at small scales. If so, it casts doubt on attaining self-similarity in all RTI statistics. Zhao et al. (2022) presented evidence in support of an increasing time lag, where they found that the energy flux across scales exhibits a self-similar evolution and peaks at a scale $\ell_{\text{peak}} \sim h(t)$ that grows in direct proportion to the mixing-layer width.

By way of contrast, for the RMI, vorticity is deposited on the perturbed interface. The amount scales directly with the shock wave pressure gradient and the density gradient at the interface. Instead of a separate forcing term, such as is present in the case of RTI, the amount of TKE created by the RMI impulsive acceleration is given by Eq. 6.22. Taking advantage of the freely decaying and incompressible flow after shock passage (Zhou, 2001), the energy budget (Eq. 6.25) now reads $K(t) = -\mathcal{E}(t)$, as the gravitational acceleration is $g = 0$.

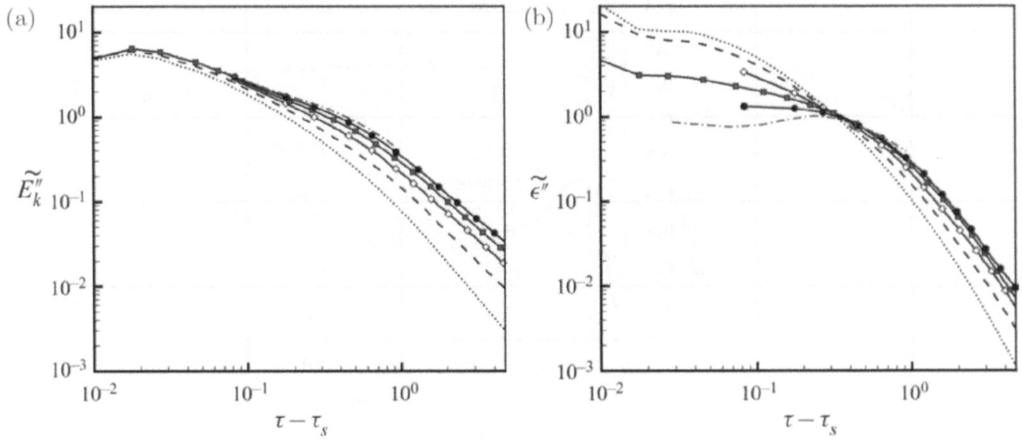

Fig. 6.4 Temporal evolution of (a) turbulent kinetic energy and (b) dissipation rate at the RMI mixing layer center plane. τ_s denotes the (dimensionless) arrival time of the incidental shock at the perturbed interface. Plots show the initial Reynolds numbers: 43 (dotted black lines), 86 (dashed black lines), 174 (white diamonds), 348 (gray squares), 697 (black circles) and 1395 (dash-dot gray lines). Fig. 5, Groom and Thornber (2021).

Figure 6.4 shows the evolution of the TKE and dissipation rate at the mixing layer center plane, which is defined as the position of equal mixed volumes (Groom and Thornber, 2021; Walchli and Thornber, 2017). Both quantities exhibit a decay in time as the initial kinetic energy is converted into internal energy by irreversible processes.

6.3 Energy transfer

We now turn our attention to how the free energy delivered to RT and RM flows, via acceleration or a shock wave, respectively, is transferred in inhomogeneous and homogeneous directions.

For the RTI, the difference in forcing between vertical and horizontal velocity components is demonstrated in Fig. 6.5, which shows the (horizontal) scale distribution of terms in the vertical and horizontal kinetic energy equations,[4]

$$\left(\frac{1}{2}\rho w^2\right)_{,t} + \left(\frac{1}{2}\rho w^2 u_j - w\tau_{zj}\right)_{,j} = \rho w g - w p_{,z} - w_{,j}\tau_{zj}\,, \tag{6.28}$$

$$\left(\frac{1}{2}\rho u_i^2\right)_{,t} + \left(\frac{1}{2}\rho u_i^2 u_j - u_i\tau_{ij}\right)_{,j} = -u_i p_{,i} - u_{i,j}\tau_{ij}\,, \tag{6.29}$$

whose global averages, denoted by angle brackets, are given by

[4] In this section, we have relabeled the vertical velocity u_3 to w to emphasize its distinctive role. In the meantime, we have also denoted the horizontal velocity components $(u_i, i = 1, 2)$ as u and v, respectively (Fig. 6.5).

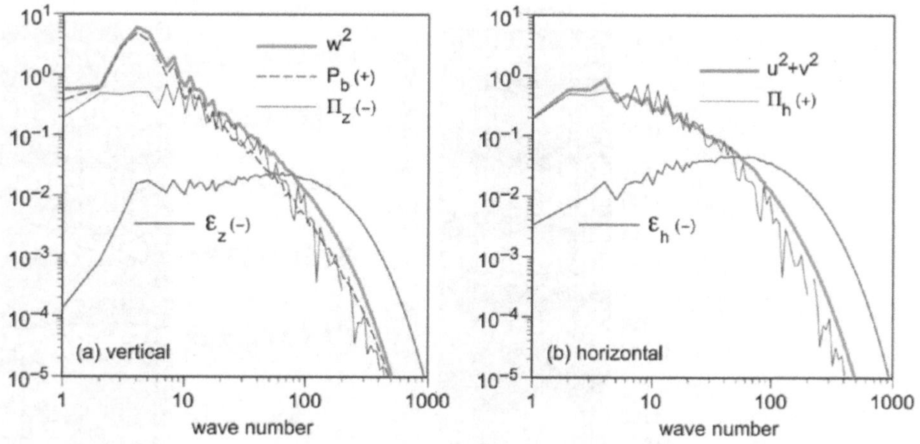

Fig. 6.5 Horizontal midplane spectra for (a) vertical and (b) horizontal kinetic energy rate terms: buoyancy production (dashed line), pressure-strain (thin solid line), and dissipation (medium solid line). The pressure-strain term is a sink (−) for the vertical kinetic energy and a source (+) for the horizontal kinetic energy. The thick curves show the corresponding vertical and horizontal velocity spectra. Cabot and Zhou (2013).

$$\left\langle \frac{1}{2}\rho w^2 \right\rangle_{,t} = +\langle \rho w \rangle g - \langle -pw_{,z} \rangle - \langle w_{,j}\tau_{zj} \rangle \equiv +P_b - \Pi_z - \mathcal{E}_z \,, \tag{6.30}$$

$$\left\langle \frac{1}{2}\rho u_i^2 \right\rangle_{,t} = +\langle p u_{i,i} \rangle - \langle u_{i,j}\tau_{ij} \rangle \equiv +\Pi_h - \mathcal{E}_h, \tag{6.31}$$

where the index i is summed over only x and y in Eqs. 6.29 and 6.31. [Note also that the standard notations for temporal and spatial partial derivatives are $(\ldots)_{,t}$, $\langle\ldots\rangle_{,t}$ and $(\ldots)_{,j}$, $\langle\ldots\rangle_{,j}$, respectively.]

Buoyancy production, pressure-strain, and dissipation terms appear on the right-hand sides of Eqs. 6.30 and 6.31 with + and − signs to indicate if they are sources or sinks of the energy balance equations. Two-dimensional horizontal cospectra are calculated near the midplane for their constituent terms (g being constant in the buoyancy production term). Because the flow is nearly incompressible at late times, the pressure-strain terms nearly balance, that is, $\Pi_h \approx \Pi_z$. The vertical component is forced at the largest scales by buoyancy production (P_b). The horizontal scales, on the other hand, have energy injected at intermediate scales via the pressure-strain (Π_h), at the expense of the vertical kinetic energy ($-\Pi_z$). The dynamic range in the simulation is probably inadequate to allow the horizontal scales to achieve a Kolmogorov-like spectrum. Figure 6.5 overlays spectra for vertical and horizontal velocity components, which have a strong resemblance. As a result of the turbulence cascade, the velocity spectra have somewhat greater small-scale energy content than the production spectra.

For the RM-instability-induced flow, Fig. 6.6 illustrates the simultaneous density and velocity field data for Mach 1.36 experiments, performed at the Los Alamos horizontal gas shock tube (Orlicz et al., 2013, 2015b). The Atwood number for this mixture composition in the shock-driven heavy gas curtain is 0.49. As time progresses, there are noticeable

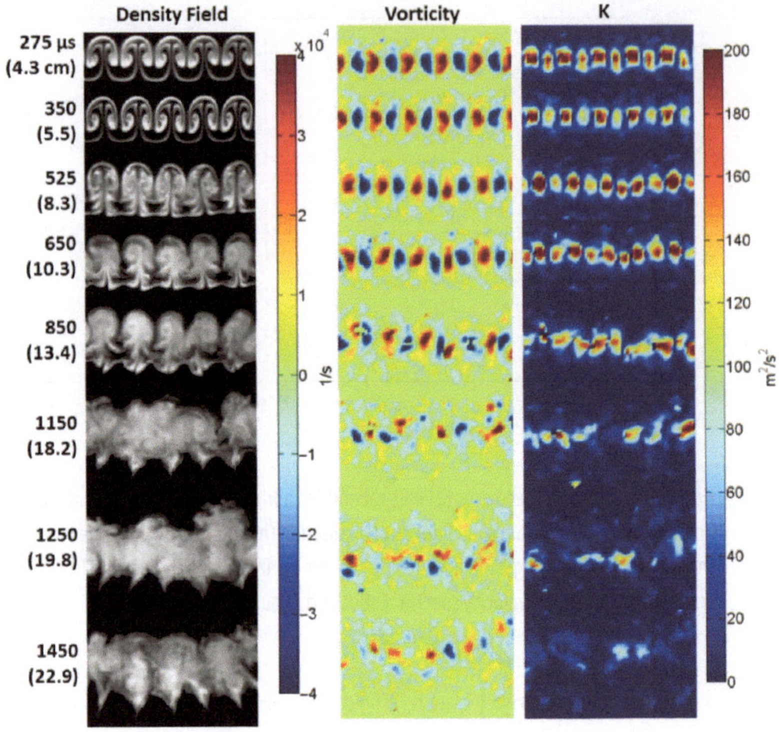

Fig. 6.6 Simultaneous PIV/PLIF data for Mach 1.36 experiments of RMI-induced flows. From left to right, the columns are PLIF SF_6 volume fraction maps, vorticity fields, and estimations of turbulent kinetic energy (K). Fig. 2 of Orlicz et al. (2015), with the permission of AIP Publishing.

changes in the density and vorticity profiles of the flow. The array of alternating blobs of positive and negative vorticity begin to lose their symmetry (left column, the density field) and the vorticity field evolves to a more disordered state, with a break up of the main vortices into smaller ones (middle column). In the meantime, the estimated kinetic energy reduces its magnitude and organization (right column).

6.4 Transfer spectra and scale interactions

Let us now move one step deeper to investigate the role of each term in the Navier–Stokes equation in the energy transfer process. Unlike the idealized homogeneous, isotropic turbulence, we must work in both the spectral and physical space to treat the homogeneous and inhomogeneous directions, respectively.

The variable density Navier–Stokes equations are the appropriate starting point for our present analysis. In order to extend the methodology of constant-density energetics to the variable-density case, a new variable is introduced:

$$v_i \equiv \rho^{1/2} u_i. \tag{6.32}$$

This variable has been used for similar purposes by various authors (Yih, 1960; Rodriguez Azara and Emanuel, 1989; Kida and Orszag, 1990a; Cabot et al., 2004). Another approach to incorporate density variations into the scale energetics was introduced in Aluie (2011, 2013) by spatial filtering (or coarse-graining), which we shall discuss briefly below.

The Navier–Stokes equation can be rewritten in terms of v_i as

$$\frac{\partial v_i}{\partial t} = \rho^{1/2} g_i - \frac{\partial v_i u_j}{\partial x_j} - \rho^{-1/2}\frac{\partial p}{\partial x_i} + \frac{1}{2} v_i \frac{\partial u_k}{\partial x_k} + 2\rho^{-1/2}\frac{\partial}{\partial x_j}\left[\mu\left(S_{ij} - \frac{1}{3}\delta_{ij}\frac{\partial u_k}{\partial x_k}\right)\right] \tag{6.33}$$

The variable-density energy equation has been derived in Fourier space by Cook and Zhou (2002). This permits the analysis of the kinetic energy budget as a function of wave number and position, where the wave number has been reduced through integration over Fourier annuli of radius $k = \sqrt{k_x^2 + k_y^2}$.

For simplicity, Eq. 6.33 may be rewritten as

$$\frac{\partial v_i}{\partial t} = \mathscr{F}_i + N_i + \mathscr{D}_i, \tag{6.34}$$

where the forcing is

$$\mathscr{F}_i = \begin{cases} \rho^{1/2} g_i & \text{RT}, \\ 0 & \text{RM}, \end{cases}$$

the nonlinear (quadratic, pressure, and dilatation) contribution is

$$N_i = \sum_m N_{m,i} \quad (m = q, p, d), \tag{6.35}$$

where

$$N_{q,i} = -\frac{\partial v_i u_j}{\partial x_j}, \tag{6.36}$$

$$N_{p,i} = -\rho^{-1/2}\frac{\partial p}{\partial x_i}, \tag{6.37}$$

$$N_{d,i} = \frac{1}{2} v_i \frac{\partial u_k}{\partial x_k}, \tag{6.38}$$

and the viscous diffusion is

$$\mathscr{D}_i = 2\rho^{-1/2}\frac{\partial}{\partial x_j}\left[\mu\left(S_{ij} - \frac{1}{3}\delta_{ij}\frac{\partial u_k}{\partial x_k}\right)\right]. \tag{6.39}$$

Next, we let $\hat{v}_i$, $\hat{\mathscr{F}}_i$, $\hat{N}_i$, and $\hat{\mathscr{D}}_i$, be horizontal Fourier transforms (denoted $FT_{x,y}$) of v_i, $\mathscr{F}_i$, N_i, and $\mathscr{D}_i$, respectively; for example, $\hat{v}_i(\mathbf{k}, z, t) \equiv FT_{x,y}\{v_i(\mathbf{x}, t)\}$, where $\mathbf{k} = (k_x, k_y)$ is the horizontal wave vector. If $\hat{v}_i^*(\mathbf{k}, z, t)$ denotes the complex conjugate of $\hat{v}_i(\mathbf{k}, z, t)$, then multiplying the transform of Eq. 6.34, and its conjugate, by $\hat{v}_i^*$ and $\hat{v}_i$, respectively, and adding the equations together yields

$$\frac{\partial \hat{v}_i^* v_i}{\partial t} = \hat{v}_i^* \hat{\mathscr{F}}_i + \hat{v}_i^* \hat{N}_i + \hat{v}_i^* \hat{\mathscr{D}}_i + v_i \hat{\mathscr{F}}_i^* + v_i \hat{N}_i^* + v_i \hat{\mathscr{D}}_i^*. \tag{6.40}$$

Integrating Eq. 6.40 over Fourier annuli of radius $k = \sqrt{k_x^2 + k_y^2}$ leads to the energy transfer equation

$$\frac{\partial E(k,z,t)}{\partial t} = \mathcal{P}(k,z,t) + \mathcal{T}(k,z,t) + \mathcal{E}(k,z,t), \tag{6.41}$$

where the kinetic energy is

$$E(k,z,t) = \frac{1}{2}\oint v_i^* v_i \, d\sigma, \tag{6.42}$$

the production is

$$\mathcal{P}(k,z,t) = \frac{1}{2}\oint (v_i^* \mathscr{F}_i + v_i \mathscr{F}_i^*) d\sigma, \tag{6.43}$$

and the viscous dissipation is

$$\mathcal{E}(k,z,t) = \frac{1}{2}\oint (v_i^* \mathscr{D}_i + v_i \mathscr{D}_i^*) d\sigma, \tag{6.44}$$

where $d\sigma$ is a differential element of a wave space annulus.

There are three contributions to the nonlinear energy transfer. The first is from the quadratic term in Eq. (6.35), which is responsible for passive-vector advection, while the second and third contributions are from pressure and dilatation effects. In order to ascertain the relative importance of each process, the total nonlinear transfer is subdivided into individual components:

$$\mathcal{T}(k,z,t) = \mathcal{T}_q(k,z,t) + \mathcal{T}_p(k,z,t) + \mathcal{T}_d(k,z,t), \tag{6.45}$$

where

$$\mathcal{T}_m = \oint (v_i^* N_{m,i} + v_i N_{m,i}^*) d\sigma \quad (m = q, p, d). \tag{6.46}$$

We now offer a few remarks on the RM energy transfer analysis (Thornber and Zhou, 2012), since there are clear parallels with a previous investigation for the RTI (Cook and Zhou, 2002).

As shown in Fig. 6.7(a–c), the quadratic terms are largely negative in the energy-containing scales in the layer core. The transfer spectra are asymmetric, with greater activity on the spike side. The quadratic and pressure components are of opposite sign and almost cancel each other out in the spikes. The dilatational terms are negligible in comparison to the difference between the quadratic and pressure transfer terms.

To elucidate the source of strongly varying transfer at high z, the total nonlinear transfer is shown in Fig. 6.7(d), alongside a close-up of the escaping vortex ring that highlights the intense vorticity transport (Fig. 6.8). The velocity vectors within this vortex ring are larger than in the mixing layer itself, and the peaks in nonlinear transfer are associated with these spikes.

In related studies, Noble et al. (2020b, 2023) focused on the evolution of RMI spectral quantities, the scalar power spectrum, and the scalar structure functions. Wang et al. (2023c) investigated scale-to-scale energy transfer in rarefaction-driven RTI-induced transitional mixing. Luo and Wang (2022) considered the effect of flow compressibility on the

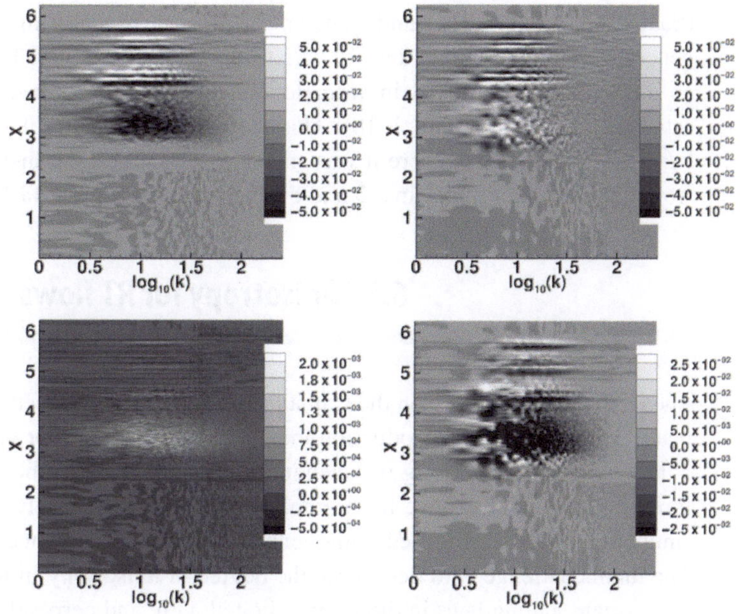

Fig. 6.7 Simulated transfer due to quadratic (upper left), pressure (upper right), and dilatational (lower left) components, and the total of the nonlinear RMI transfer (lower right), at dimensionless time 37. Thornber and Zhou (2012).

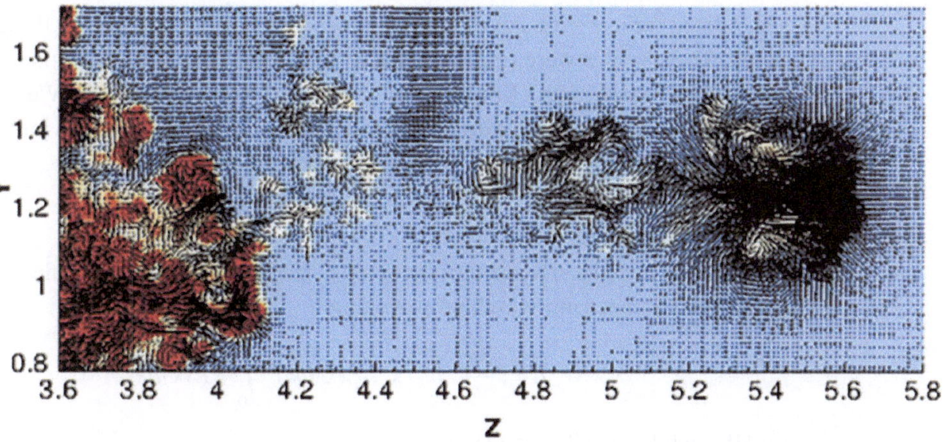

Fig. 6.8 A close-up of the simulated mass fraction contours and velocity vectors around a spike, at dimensionless time 37. Thornber and Zhou (2012).

mixing and energy transfer in 3D compressible RT turbulence for isothermal background stratification.

An alternate *coarse-graining* approach to analyzing the energetics across scales relies on filtering in physical space (Eyink, 2005; Aluie and Eyink, 2009) rather than in Fourier space. This allows its application to inhomogeneous flows to diagnose the energy spectra

(Sadek and Aluie, 2018) and transfer across scales locally in space. The approach was generalized to incorporate density variations (Aluie, 2011, 2013), where it was shown that an inertial range can exist in variable density flows if the so-called *inviscid criterion* is satisfied (Zhao and Aluie, 2018). This methodology was implemented to analyze in RTI by Zhao et al. (2022), where it was shown that the energy transfer across scales proceeds from large to small scales in 3D and evolves in a self-similar fashion.

6.5 Anisotropy for RT flows

The impact of anisotropy on the statistical theory of RTI induced flows can be studied from many perspectives. The production induced by acceleration or gravity acts predominantly at low wave numbers (long wavelengths). As a result, it can be viewed as an analogue of forcing functions that have been employed extensively in fully resolved direct numerical simulations (DNS) of forced homogeneous isotropic turbulence (HIT) (Batchelor, 1953). The main challenge is to determine the degree of anisotropy in RTI-induced flow, and the appropriate scaling laws in directions aligned with, and perpendicular to, gravity.

6.5.1 Length scales

In this section, we discuss a variety of turbulence length scales that can be, to some extent, generalized for use with anisotropic RTI flow (see Cabot and Zhou, 2013; Zhou and Cabot, 2019). These scales were originally based on velocity fluctuation statistics in HIT (Batchelor, 1953; Tennekes and Lumley, 1972; Frisch, 1995; Pope, 2000). We will proceed from large outer scales (expected to be comparable to the mixing layer width) to intermediate Taylor microscales, and finally to small viscous scales.

The integral scales

The integral scale is a measure of the scale of the most energetic turbulent motions, defined as the energy-spectrum-weighted wavelength:

$$L = \frac{3\pi}{4} \int_0^\infty k^{-1} E(k) dk \Big/ \int_0^\infty E(k) dk, \tag{6.47}$$

for HIT, while for RTI induced flows,

$$L_i = \int_0^\infty k^{-1} E_i(k) dk \Big/ \int_0^\infty E_i(k) dk, \tag{6.48}$$

where $E_i(k)$ is the 2D spectral energy density for ith spatial component and k is the 2D wavenumber in horizontal planes. Zhou and Cabot (2019) showed that the 2D integral scale for horizontal velocity components at the midplane agree (within 10%) with the correlation lengths, which are defined by integrating to the first zero crossing (Λ_0). Another definition involves taking the half-widths at $1/e$ maximum (Λ_H). Given the approximate equivalence of these terms, we will treat Λ and L interchangeably.

The inhomogeneous vertical direction causes some challenges in the spectral analysis of RTI-induced turbulent flows. In order to obtain the information unavailable from the spectral analysis, we turn our attention to the correlation function. Just as with the velocity derivatives, when discussing correlations and spectra in one dimension, "longitudinal" is taken to be aligned with the velocity component, while "transverse" is normal to the velocity component. In the nonperiodic vertical direction in RTI flow, where power spectra are not available, L_z is represented by the vertical correlation length, whilst Λ_z is calculated from integrals of vertical autocorrelation functions of the vertical velocity (Cabot and Zhou, 2013). Because of inhomogeneity in the vertical direction, the correlation functions may also be height-dependent, depending on the spatial averaging domain. The vertical correlation lengths are shown in Fig. 6.9 to scale reasonably well with H, going roughly as $\Lambda_z/H \approx 0.3$ at late times. The horizontal integral scale is still gradually decreasing at late times, but may level off at a steady value of $L_x/H < 0.08$.

Taylor microscales

The Taylor microscale is one of the fundamental turbulence scales, namely, the scale for which viscous dissipation begins to affect the eddies (Townsend, 1980; Tennekes and Lumley, 1972). It is also an effective measure of the anisotropy of the flow.

A standard way to compute the Taylor microscale, λ_T, is from the curvature of the velocity autocorrelation (Champagne et al., 1970), $\mathcal{R}(r)$. Note that $\mathcal{R}(r)$ is an even function: $\mathcal{R}(r) = \mathcal{R}(-r)$. Performing a Taylor expansion, the leading terms may be written as

$$\mathcal{R}(r) = 1 - \frac{r^2}{\lambda_T^2}, \tag{6.49}$$

where the Taylor microscale is defined as

$$\lambda_T = \left[-\frac{1}{2}\frac{d^2\mathcal{R}(0)}{dr^2} \right]^{-1/2}. \tag{6.50}$$

Another way of estimating the Taylor microscale (O'Neill et al., 2004; Reese et al., 2018) from experimental data utilizes the variance and the first derivative of the velocity to yield

$$\lambda_T = \left[\frac{2\langle u^2\rangle}{\langle(\partial u/\partial x)^2\rangle} \right]^{1/2}. \tag{6.51}$$

A third method for estimating the Taylor microscale (see pages 66–68 of Tennekes and Lumley (1972) for more details) is given by

$$\lambda_T = \sqrt{10\nu\frac{K}{\mathcal{E}}}. \tag{6.52}$$

The component form of the Taylor microscale is given by

$$\lambda_{T,i} = \lambda_i = u'_i/u'_{i,i}, \tag{6.53}$$

where u'_i is the rms of the ith component of velocity and $u'_{i,i}$ is the rms of the spatial derivative in the ith direction of the ith component of velocity, that is, the longitudinal velocity derivative. To good approximation $\mathcal{E}_i = 15\nu u'^2_{i,i}$ (Cabot and Zhou, 2013), hence

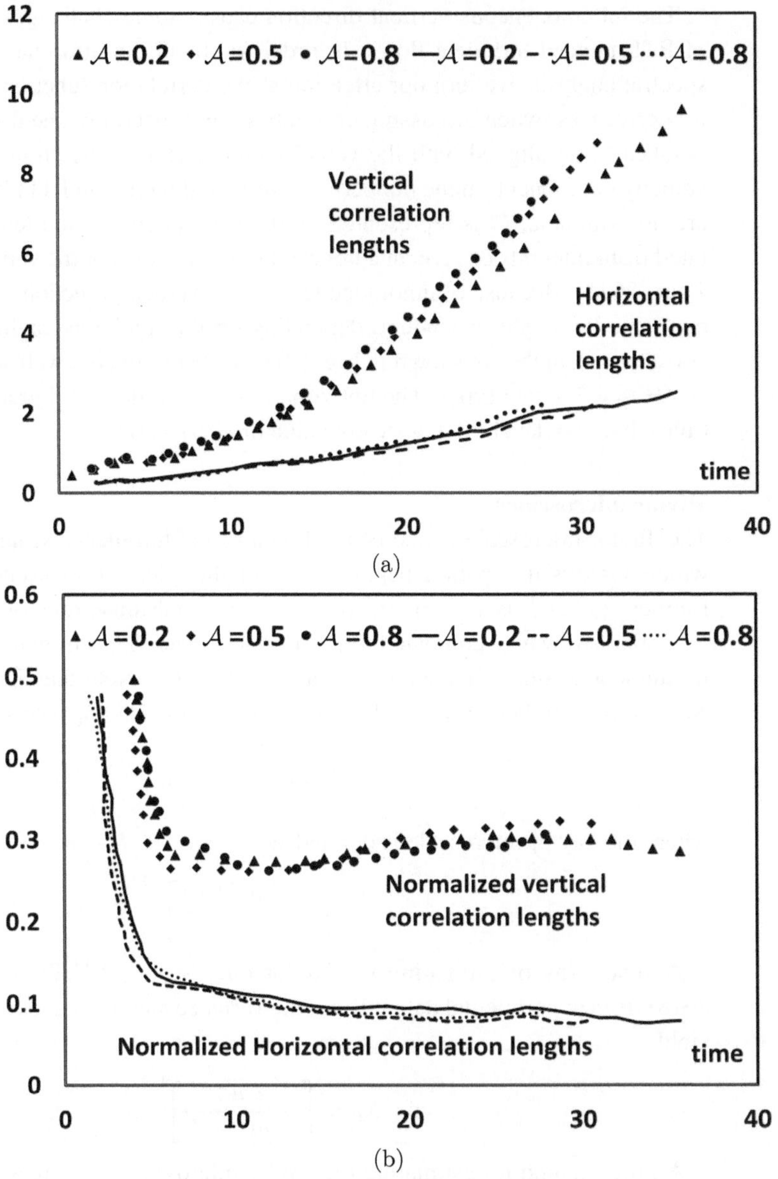

Fig. 6.9 Vertical correlation lengths, Λ_z, and horizontal integral length, L_{ix}, from Eq. 6.48, for RTI flow for several Atwood numbers. Correlation lengths are scaled by (top) initial perturbation lengths and (bottom) the mixing layer width H. Zhou and Cabot (2019).

$\lambda_{T,i} \approx (15\nu/\varepsilon_i)^{1/2}u'_i$. In HIT all components are equal, namely $\lambda_{T,i} = \lambda_T$. In RTI flow, the rms velocity component u'_i is a macroscopic factor sensitive to the anisotropy of the flow, while the velocity derivative is not. The temporal development of the anisotropy in $\lambda_{T,i}$ is shown in Fig. 6.10(a) for three Atwood numbers. It is seen in all cases that the

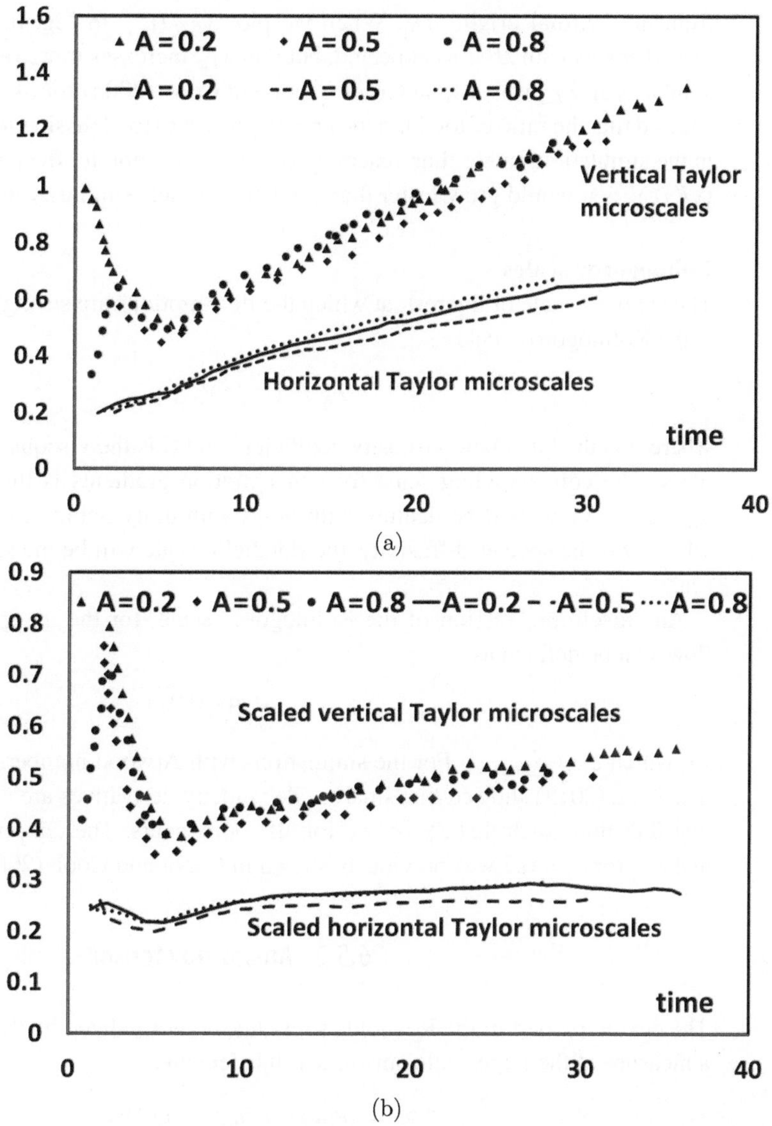

Fig. 6.10 Midplane Taylor microscales, λ_i, for longitudinal velocity variations in RTI flow for several Atwood numbers in the horizontal (x and y) directions (dashed lines) and the vertical (z) direction (solid lines). The time development of λ_i is shown in the top panel and, as a test of self-similarity, $\lambda_i/H^{1/4}$ is shown in the bottom panel. Zhou and Cabot (2019).

vertical component $\lambda_{T,z}$ grows as a linear function of time, remaining twice the size of the horizontal component $\lambda_{T,x}$. The latter appears to have a slightly sublinear growth rate.

The self-similarity analysis of Ristorcelli and Clark (2004) suggested that λ_T scales as $H^{1/4}$. Yet it would not be appropriate to utilize the $t^{1/2}$ asymptotic growth to normalize the Taylor microscale. Indeed, Fig. 5.2(a) shows that the RTI flow in these simulations is still early enough in development that there is a significant offset in $H \sim (H_0^{1/2} + \alpha^{1/2} t)^2$

from the "virtual origin" H_0. When we plot $\lambda_{T,i}/H^{1/4}$ in Fig. 6.10(b), we see that $\lambda_{T,x}$ indeed scales with $H^{1/4}$ as expected, but that $\lambda_{T,z}$ increases more rapidly. This lack of self-similarity in $\lambda_{T,z}$ was also noted in Cabot and Cook (2006) for $\mathcal{A} = 0.5$. In addition, they showed that the ratio of total kinetic energy growth to total dissipation continually increases in the simulation, rather than reaching a constant asymptote. Because λ_T scales roughly as $(\nu K/\varepsilon)^{1/2}$, it would grow faster than predicted by self-similarity in this situation.

Kolmogorov scales

The smallest scale of interest, at which the flow motions are strongly damped by viscosity, is the Kolmogorov scale

$$\lambda_K = (\nu^3/\mathcal{E})^{1/4}, \tag{6.54}$$

where ν is the kinematic viscosity coefficient and $\mathcal{E}$ is the viscous dissipation rate per unit mass. The corresponding scale for concentration gradients is the Batchelor scale $\lambda_B = \lambda_K/\mathrm{Sc}^{1/2}$. As we will be dealing with flows with unity Schmidt number $\mathrm{Sc} = \nu/D = 1$, where D is the species diffusivity, the Batchelor scale will be the same as the Kolmogorov scale.

An anisotropic version of the Kolmogorov scale, for the ith spatial component in 3D flow, can be defined as

$$\lambda_{Ki} = (\nu^3/3\mathcal{E}_i)^{1/4}, \tag{6.55}$$

for which $\mathcal{E} = \sum_{i=1}^{3} \mathcal{E}_i$. For the simulations with Atwood numbers from 0.2 to 0.8, Cabot and Zhou (2013) showed that small-scale velocity derivatives are nearly isotropic in turbulent RTI flow, such that $\lambda_{Ki} \approx \lambda_K$ for all components. The temporal development of λ_{Kx} and λ_{Kz} for $A = 0.5$ was previously shown in Cabot and Cook (2006).

6.5.2 Anisotropy tensor

The deviatoric part of the Reynolds stress tensor, normalized by the kinetic energy term, is a measure of the large-scale anisotropy. Its elements

$$b_{ij} = \langle u_i u_j \rangle / \langle u_n u_n \rangle - (1/3)\delta_{ij}, \tag{6.56}$$

are very important, especially in terms of the second and third principal invariants (Lumley and Newman, 1977). Here, the Einstein summation convention is used. The tensor b_{ij} is bounded, that is, $-1/3 \le b_{ij} \le 2/3$. The value $b_{ij} = 0$ (for all i) corresponds to the isotropic turbulence limit. In contrast (Lumley, 1978; Pope, 2000), the value $b_{ii} = 2/3$ means the flow field contains all of the turbulent kinetic energy (1D flow), and the value $b_{ii} = -1/3$ means no energy at all in this component (2D distributions of the turbulent kinetic energy) (no sum over i).

Banerjee et al. (2010b) measured the late time anisotropy tensor across the RT mixing layer at a location downstream from the splitter plate. As shown in Fig. 6.11, most of the turbulent transport appears in the vertical direction. The velocity fluctuation statistics are fairly constant across the RT mixing layer, consistent with the finding of Ramaprabhu

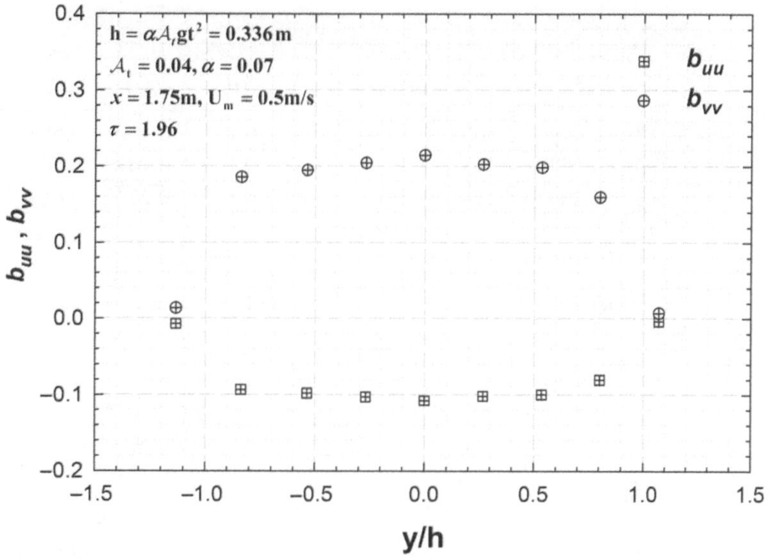

Fig. 6.11 Profiles of the anisotropic tensor (b_{uu} & b_{vv}) across the RT mixing layer for an experiment with $\mathcal{A} = 0.04$. Fig. 27 of Banerjee et al. (2010a).

and Andrews (2004a). Near the edges of the RT mixing layer, it appears that $b_{ii} \rightarrow 0$, suggesting isotropy. The profiles and peak values of these components are similar to those of a recent gas channel experiment (Akula and Ranjan, 2016) and small Atwood number numerical simulations (Livescu et al., 2009; Mueschke and Schilling, 2009). Using two large DNS datasets, Livescu et al. (2009, 2010) found that b_{33} appears to become almost constant across the layer, with an asymptotic value around 0.3. This corresponds to almost 65% of the energy being contained in the vertical velocity component, along the direction of the acceleration.

Figure 6.12 shows that the probability distribution function (pdf) of the vertical velocity fluctuation v' is more or less flat, with two peaks observed on the left and right sides (Akula and Ranjan, 2016). The kurtosis values are 1.9 and 3.2 for v' and w', respectively, similar to those in the small-Atwood-number experiments of Banerjee et al. (2010). The disparity between these kurtosis values suggests significant anisotropy inside the mixing layer. The low kurtosis value (< 3.0) also implies that the structure of rising bubbles and falling spikes generates a broad spectrum of scales (Banerjee et al., 2010).

Using direct numerical simulations, Cabot and Zhou (2013) also showed that the fluctuation intensity (square of the rms) of the velocity components displays a high degree of anisotropy (Table 6.1). Other papers have shown similar results, for example, ILES results at various $\mathcal{A}$ in Youngs (2013). Vladimirova and Chertkov (2009), Boffetta et al. (2009), and Schneider and Gauthier (2016a) considered the anisotropy in numerical Rayleigh–Taylor Boussinesq turbulence. For the large-scales, the anisotropy (defined as the ratio between the two rms velocity components) is approximately 1.8. Zhao et al. (2022) showed that this high anisotropy at large scales in 3D is absent in 2D simulations, where the large

Table 6.1 Late-time turbulence statistics for several Atwood numbers with "mid-slab" average.

$\mathcal{A}$	$\langle u_1^2 \rangle$	$\langle u_3^2 \rangle$	$\langle u_{1,1}^2 \rangle$	$\langle u_{3,3}^2 \rangle$	$\langle \varepsilon_1 \rangle$	$\langle \varepsilon_3 \rangle$	λ_1	λ_3	$Re_{\lambda 1}$	$Re_{\lambda 3}$
0.2	1.477	5.565	3.168	3.091	0.132	0.150	0.683	1.342	95	362
0.5	1.257	4.403	3.402	3.360	0.126	0.147	0.608	1.145	87	307
0.8	1.434	4.473	3.298	3.342	0.177	0.211	0.659	1.157	70	216

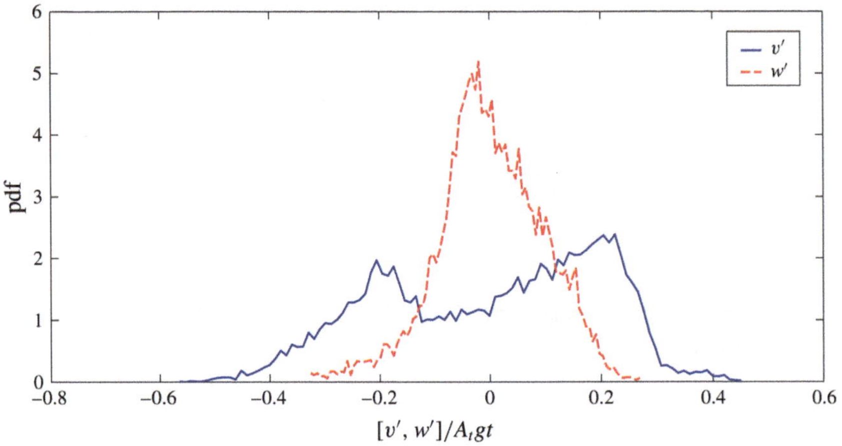

Fig. 6.12 Vertical velocity fluctuation v and spanwise velocity fluctuation w pdfs are normalized with velocity RT anisotropy. Fig. 18 of Akula and Ranjan (2016).

scales are influenced by an inverse cascade from small scales, which generate a large-scale overturning circulation in 2D RTI that is approximately isotropic.

Table 6.1 presents several additional late-time statistics of interest. The ratio between the vertical and horizontal Taylor microscales is approximately 2. The Taylor microscale Reynolds number scales as $\langle u_\alpha^2 \rangle$; hence, the vertical value will be 3–4 times the horizontal one. The anisotropy is therefore quite pronounced at intermediate length scales.

Toward smaller scales, the velocity derivatives of $u_{1,1}^2$ and $u_{3,3}^2$ indicate a return to isotropy. Furthermore, the component Kolmogorov length scales are essentially identical (within 5%, see Cabot and Zhou, 2013). Finally, the vertical dissipation rate is about 20% larger than the horizontal rate.

6.6 Anisotropy for RM flows

A strong case can be made for an analysis that draws a connection between the exponents of the infrared spectra and the length-scale growth in decaying, weakly anisotropic turbulence. The physical picture is one of an initially quiescent flow field subject to an impulsive injection of energy, perhaps at all scales, which is subsequently allowed to freely decay (Zhou, 2001). Although there are undoubtedly significant corrections due to strong anisotropy and lack of equilibration at early time, the analogy between RM induced flow and decaying isotropic turbulence becomes better and better as the shock-induced anisotropy decays. As a result, one can anticipate that this analogy will be acceptable at a late time, long after the shock passage. At this late time, one expects the RM mixing zone width to grow as a power-law in time.[5]

The mixing layer width is the largest observable feature of the flow. This suggests the importance of a dominant large length scale, which, in turn, is associated with the low-wave number power-law of the energy spectrum (Clark and Zhou, 2006; Soulard et al., 2018; Soulard and Griffond, 2022). Such a viewpoint is very similar to the analytical and experimental findings in isotropic turbulence, beginning with Batchelor and Proudman (1956), Saffman (1967), and productively exploited by others, for example, Comte-Bellot and Corrsin (1971).

Such a physical picture appears to have been realized in the LES of multi-mode RMI by Lombardini et al. (2012), who investigated five Mach numbers, 1.05, 1.25, 1.56, 3.0, and 5.0. For all Mach numbers considered, the late-time flow resembles homogeneous decaying turbulence of Batchelor type.[6] In particular, the authors considered an anisotropy measure of the velocity power spectra for Mach 1.05 and 3.0 (the authors reported that similar results are also observed at the three other Mach numbers). Here, the axial velocity component is computed across a transverse plane slicing through the center of the mixing-layer at three different times. Figure 6.13 indicates that, at early times, the axial velocity power spectrum contains more than a third of the total energy at almost every scale. As the mixing layer grows, isotropization of the flow is evident. However, perfect TKE isotropy is never completely reached, even at the highest Mach number considered.

Another anisotropy measure is provided by the ratio of the turbulent kinetic energy components[7] $TKR = (2 \times TKX)/(TKY + TKZ)$. Figure 6.14(a) shows the temporal evolution of this ratio for the DNS and ILES of a multi-mode narrowband RMI. At the latest time, the ratio of turbulent kinetic energy components is 1.692 for the DNS (Groom and Thornber, 2019), compared to 1.423 for the ILES (Thornber et al., 2010; Oggian et al., 2015). Sewell et al. (2021) also plot this quantity based on their shock tube experiments ($TKR \approx 1.52$ after the incident shock wave). Therefore, the mixing zone remains anisotropic. This also suggests that viscosity has suppressed secondary instabilities that transfer kinetic energy from the x direction to the y- and z-directions. Nevertheless, Groom and Thornber (2021) noted that the anisotropy tensor suggests that the isotropization occurs due to increasingly turbulent flow, especially in the core of the mixing layer. Figure 6.14(b) shows the decay of turbulent kinetic energy (TKE) at different grid resolutions.

Lombardini et al. (2012) remarked that this spectral measure of anisotropy, which relies solely on the TKE components, must be complemented with anisotropic Taylor microscale

[5] The effect of anisotropy in RMI was discussed in Mikaelian (1990d), where he argued that anisotropy is necessary to overcome Barenblatt's $t^{2/3}$ growth (Barenblatt, 1983).

[6] Batchelor-type decaying turbulence is associated with a turbulent kinetic energy decay exponent of 1.4 and large-scale energy spectrum $\sim k^4$.

[7] The direction of shock propagation is x for this shock tube configuration.

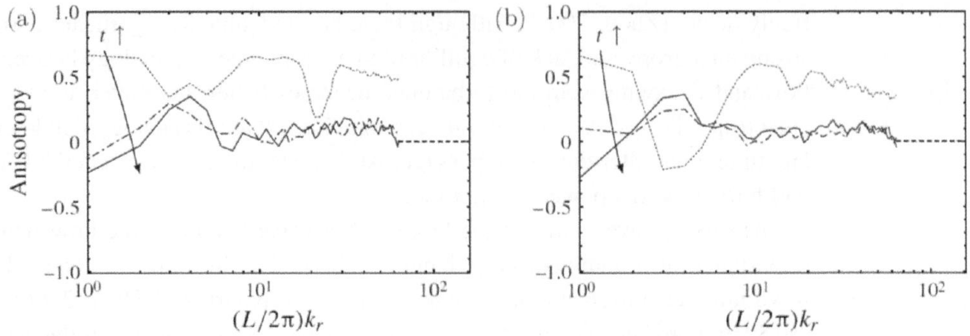

Fig. 6.13 Anisotropy of simulated RMI velocity power spectra at three different normalized times, 10 (dotted line), 800 (dash-dotted line) and 1600 (solid line). (a) Mach number 1.05, (b) Mach number 3. Fig. 13 of Lombardini et al. (2012).

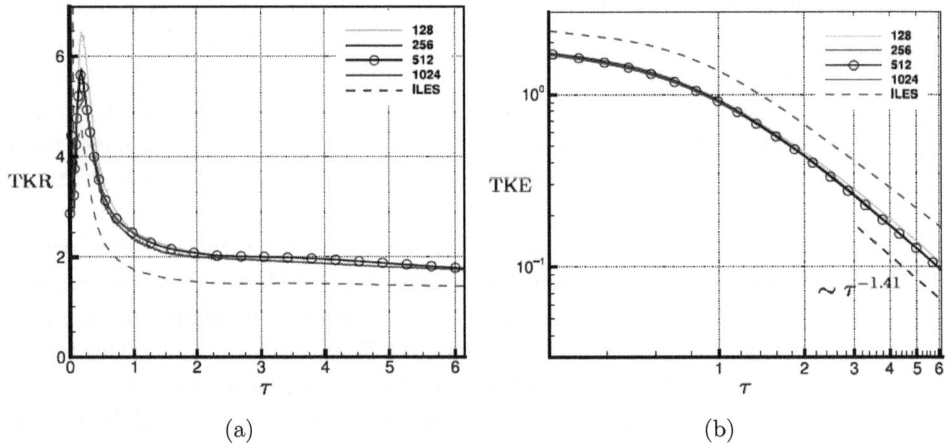

Fig. 6.14 (a) Time histories of the ratios of TKE components for a simulated RMI mixing layer. (b) Time histories of decaying turbulent kinetic energy from the DNS and ILES results. Figures 10a and 9a of Groom and Thornber (2019).

measures. For DNS with an initial outer Reynolds number[8] of 697, Fig. 6.15 indicates that the Taylor microscale and Reynolds number anisotropies persist until the end of the simulation (Groom and Thornber, 2021). Experimental measurements of the Taylor microscale components have been obtained by Weber et al. (2014a), but there exists a sensitivity to which definition is used. In the variance-based method, according to Weber et al., the scale appears to be slightly larger than the horizontal scale by the latest time, while the autocorrelation-based method produces a vertical scale that stays below the horizontal scale. This sensitivity certainly requires additional consideration to be more fully understood.

[8] The initial Reynolds number, Re_0, is defined by the estimated initial growth rate, a weighted average wavenumber $\bar{k} = \sqrt{\int_0^\infty k^2 P(k) dk} / \sqrt{\int_0^\infty P(k) dk}$, and the viscosity (Groom and Thornber, 2021).

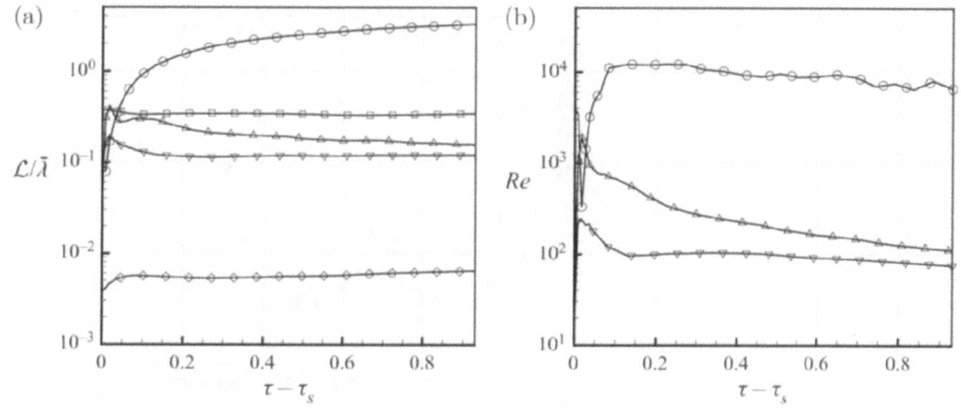

Fig. 6.15 Evolution of (a) length scales (b) and Reynolds numbers for a simulated RMI mixing layer with initial Reynolds number $Re_0 = 1385$. Plots show the outer scale h, the associated Reynolds number (circles), the integral length (squares), the axial (upward triangles) and transverse (downward triangles) Taylor microscales, the associated Reynolds numbers and the Kolmogorov microscale (diamonds). Fig. 18 of Groom and Thornber (2021).

The generalized mass-weighted Reynolds stresses in a flow are defined by Favre averaged velocity fluctuations,

$$R_{ij} = \overline{\rho u_i'' u_j''}. \tag{6.57}$$

The evolution of each Reynolds stress component in shock tube experiments (Reese et al., 2018) is shown in Fig. 6.16. The experiments detailed in the studies allow for the all components of the Reynolds stress to be obtained at the initial condition (IC) and four post-shock times (PS1-PS4). Following passage of the shock wave, the streamwise component R_{22} dominates the transverse stresses, R_{11}. At early times, the interface tends toward zero away from the mixing zone, indicating the greatest forces from turbulent fluctuations are occurring near the center of the mixing zone. As time progresses, stresses near the top of the layer begin to grow, while a decay in Reynolds stress is seen at the center of the mixing layer.

As RM slows down, the width doubling time gets longer and it is very difficult to run simulations long enough to obtain self-similarity. It is even more difficult to run experiments long enough. The differences between the values of the RMI growth exponent can be attributed to the influence of initial conditions and to the duration of the simulation/experiment being insufficient to establish self-similarity (Youngs, private communication, 2017). An associated theoretical argument has been presented by Elbaz and Shvarts (2018), based on the number of mode-coupling generations required to reach self-similarity. For RMI, each generation takes longer and longer to simulate, whereas the opposite is true for RTI. Furthermore, discrepancies are also caused by initial conditions, the presence of reshock, and shock Mach number (Prestridge, private communication, 2017). The initial conditions strongly affect the time for transition to anisotropy, while the incident Mach number affects the mixing as well, in ways that are difficult to scale. Finally, the dimensionality (2D versus 3D) nature of the initial conditions is very important, as well.

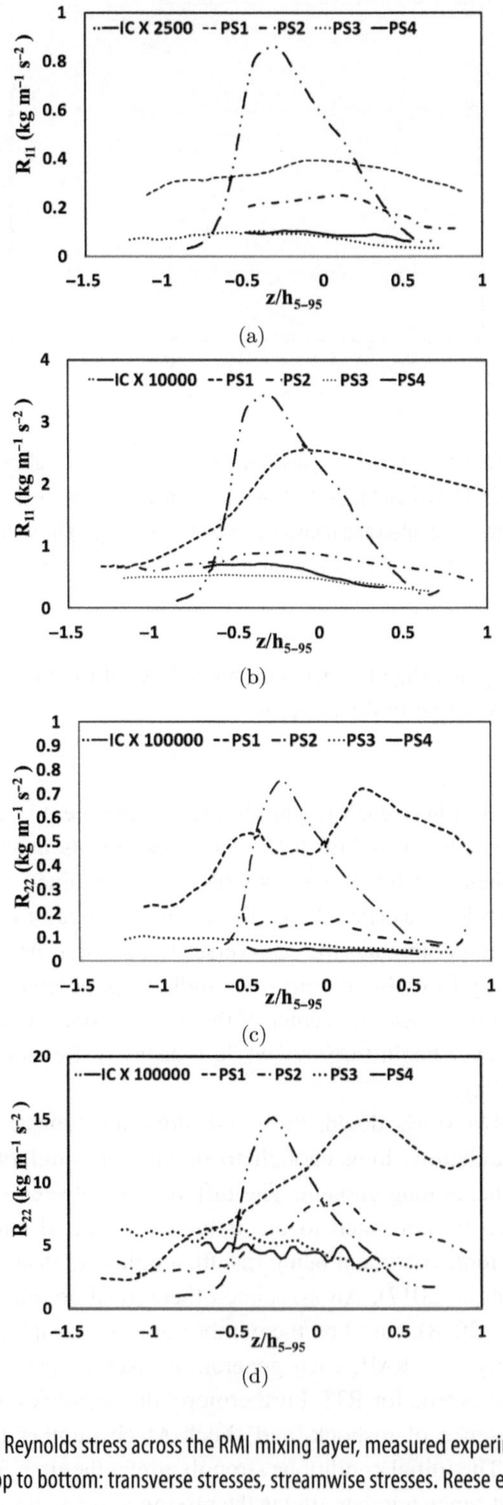

Fig. 6.16 The two components of the Reynolds stress across the RMI mixing layer, measured experimentally for Mach 1.6 (a,c) and Mach 2.2 (b,d). From top to bottom: transverse stresses, streamwise stresses. Reese et al. (2018).

7 Elementary aspects of turbulent flows

Fluid flows as varied as steady laminar flows, nonlinear instabilities, and turbulence are all described by the Navier–Stokes equations (Hinze, 1975; Monin and Yaglom, 1971, 1975). Prior to 1887 (Schmitt, 2017), the word "turbulence" was used for describing the agitation and disorder of crowds or children. It was not used by fluid dynamics pioneers such as Joseph Boussinesq, Osborne Reynolds, Lord Rayleigh, and Horace Lamb. Popularized later by Lord Kelvin (William Thomson, Thomson (1887a,b)), the term "turbulence" introduced a new concept, a new understanding of a scientific phenomenon identified as distinct from laminar flow. Turbulent evolution is characterized by the randomness of the flow field in both time and space (Fig. 7.1(a) and (b)). A laminar flow, on the other hand, is one that moves as if comprised of layers with different velocities, smooth and steady (Munson et al., 1998; White, 1994) (Fig. 7.1(c)).

Turbulence as a distinct branch of physics is relatively new, founded when Lev Landau placed hydrodynamics firmly in the physics corpus (Falkovich, 2011) with the publication of "Fluid Mechanics" as Volume 6 of his authoritative "Course of Theoretical Physics" series (Landau and Lifshitz, 1959). Turbulence as a physical phenomenon is everywhere in nature and has been observed and documented for centuries by scientists, writers, and artists. In his manifestations of water paintings, the famous Chinese artist Ma Yuan (1160–1225) of the Song Dynasty contrasted some very turbulent flows with that of a much calmer laminar flow. For an interesting book on this famous artist and his extensive body of work, the reader is referred to *The Heart of Ma Yuan: The Search for a Southern Song Aesthetic* (Edwards, 2011).

Several hundred years later, Vincent van Gogh painted the essence of turbulence with a high degree of realism in "The Starry Night" and other paintings. "The Starry Night" was painted by van Gogh in 1889 at the asylum in Saint-Remy, France during one of his periods of prolonged psychotic agitation (Bailey, 2018). Aragón et al. (2008) used digital images of these paintings of van Gogh to show that the statistics of luminance contains the characteristic fingerprint of turbulent flow. Indeed, these authors showed that Starry Night, and other impassioned van Gogh paintings, transmitted the essence of turbulence with a high degree of realism.

Why do we care about turbulence? Usually, due to its effect on mixing. Mixing of the mass, momentum, and energy is significantly more rapid in turbulent flows than in laminar flows. Based on the role of mixing, turbulent flows can be grouped into three categories (Dimotakis, 2005). The first category is composed of those cases where turbulent flow drives mixing, but the mixing itself plays essentially no role in the dynamics of the turbulent velocity field; a good example is the mixing of dye in a liquid, where the scalar dye concentration (Rudyak and Minakov, 2014) does not influence flow dynamics. The second

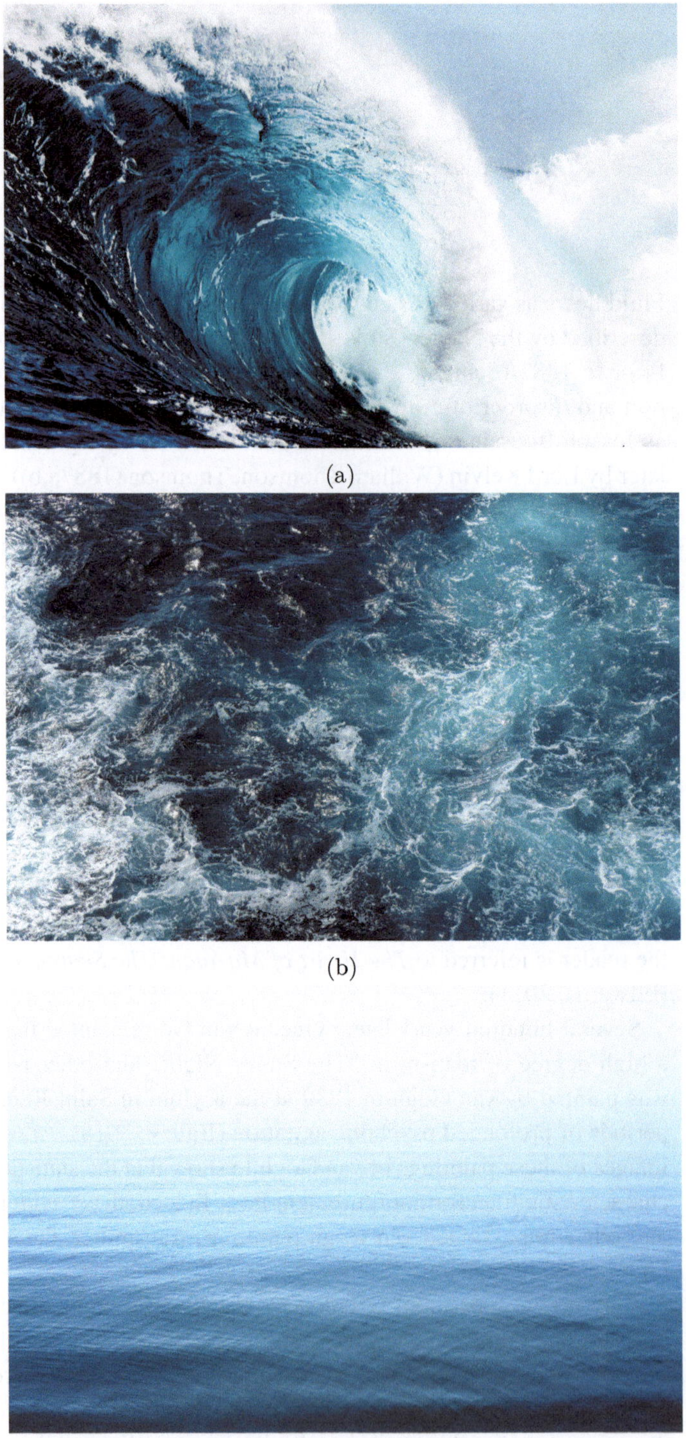

Fig. 7.1 Turbulent flows (a) and (b) versus laminar flow (c). Getty Images.

category is where the scalar being mixed plays an intrinsic role in driving the velocity dynamics; an example is RTI, wherein the scalar is the density field, and the momentum forcing is due to the coupling of this scalar with the acceleration. For the third and last category, the scalar being mixed has the dominant active role in the behavior of the overall flow field; examples for this category include flows driven by combustion and/or by detonation and deflagration phenomena.

7.1 The last unresolved classical physics problem

When I meet God, I'm going to ask him two questions: why relativity? And why turbulence? This probably apocryphal quote is attributed to Nobel Laureate Werner Heisenberg (Castelvecchi, 2017).

Richard Feynman, another Nobel Laureate, once described turbulence as "the most important unsolved problem of classical physics" (Feynman, 1963). The field of turbulence research has attracted some of the most prominent scientific minds of the nineteenth and twentieth centuries – Lord Rayleigh, Werner Heisenberg, Lev Landau, Subrahmanyan Chandrasekhar, Tsung-Dao Lee, Lars Onsager, and Richard Feynman – all Nobel Laureates. Despite their efforts, even these great minds faced challenges and experienced setbacks.

Turbulence presents a significant challenge to researchers because it is both multiscale and highly nonlinear. Analytical solutions for even the simplest turbulent flows do not exist (Moin and Mahesh, 1998), even though the fundamental equations of motion have been known to scientists for over 170 years (Hinze, 1975). The Navier–Stokes equations (Navier, 1827; Stokes, 1845) are partial differential equations that likely contain all the physics (Frisch, 1995) necessary to describe fluid motion. In many problems in fundamental physics, we must simultaneously deal with uncertainty in the underlying equations of motion and uncertainty in our ability to solve them. In turbulence, we have only the latter (Nelkin, 1994).

Yet these equations are so challenging that the Clay Mathematics Institute selected them in 2000 as one of seven "Millennium Prize Problems," and offered a $1 million reward to the first individual who could solve them.[1] To date, no one has won this award. For the record, only one of the seven Millennium Prize problems has been solved in 20 years. On March 18, 2010, The Clay Mathematics Institute announced that Dr. Grigoriy Perelman of the St. Petersburg branch of The Steklov Mathematical Institute, Russia, was the recipient of the Millennium Prize for resolution of the Poincaré conjecture.[2]

Professors Lord Julian Hunt, Peter Bradshaw, and Steven A. Orszag, all leading turbulence experts, have expressed their thoughts about the challenges posed by turbulence. Orszag (1970a) said that "[i]t must be admitted that the principal result of fifty years of

[1] www.claymath.org/millennium-problems/navier-stokes-equation.
[2] www.claymath.org/sites/default/files/millenniumprizefull.pdf.

turbulence research is the recognition of the profound difficulties of the subject." Bradshaw once humorously opined that turbulence might be the "invention of the Devil on the seventh day of creation" (Bradshaw, 1994). Lord Hunt added his observation that turbulence is "like certain diseases" (Hunt and Kevlahan, 1993).

The key property that makes understanding turbulence possible is the remarkable fact that a wide range of processes can be described by simple self-similar patterns. The most well known is of course "Kolmogorov's Law." There are also a range of turbulent shear flows, the plane mixing layer, the round jet, the turbulent wake, the turbulent boundary layer and also buoyant flows such as turbulent plumes. Much of the book deals with self-similar RT, RM, and KH. Also, with the major advances in computing there is much scope for understanding turbulence from first principle using 3D simulation (DNS, LES, ILES).

7.2 Cascade

When a fluid flow is laminar, the viscous action plays an important role. Any perturbations from smooth motion are damped away. When the flow is fully turbulent, however, it contains eddy motions of all sizes admissible within the flow geometry. In such a case, the large eddies reflect the inhomogeneity characteristics of the forces and boundary conditions which produce them. The energy injection takes place at the largest eddy length-scales.

Since the characteristic eddy *timescales* decrease geometrically with eddy size, a large number of eddies are generated during the lifetime of the large anisotropic eddies. The flow will evolve into a quasi-stationary self-similar hierarchy of eddies, frequently termed a cascade process, whose wavenumbers form a geometric progression (Rose and Sulem, 1978; Frisch et al., 1978; Frisch, 1995). The smallest eddies will be limited by the viscous action at the smallest length scales, the Kolmogorov dissipation scales, $\lambda_K = 2\pi/k_d$, a precise definition of which was previously given in Eq. 6.54.

The sketch in Fig. 7.2 is a good starting point for study of high *Re* 3D turbulent flows. The well-known poem by Richardson (1922) vividly illustrates some key features of turbulent flows:

Big whirls have little whirls
Which feed on their velocity,
And little whirls have lesser whirls
And so on to viscosity.

Formation of eddies requires a large part of the mechanical energy in a turbulent flow, and the effects of viscosity eventually dissipate their energy as heat (Warhaft, 1997). Thus, the energy transfer in the 3D turbulent cascade proceeds from large to small scales. The time required for an eddy to be distorted and, in this distortion process, generate smaller eddies, is called an *eddy-turnover-time*. Finally, the energy is dissipated from the smallest eddies due to viscous effects.

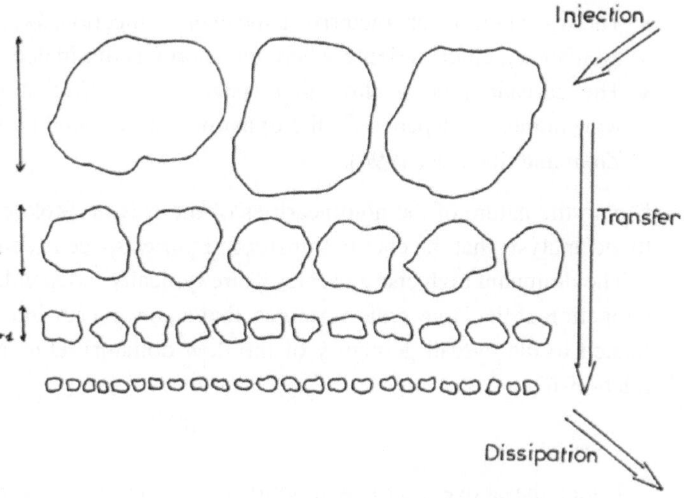

Fig. 7.2 The energy cascade according to the 1941 Kolmogorov theory. Notice that at each step the eddies are space filling, Frisch et al. (1978).

7.3 Kolmogorov phenomenology

While Sir G.I. Taylor pioneered the statistical treatment of turbulence, it was Kolmogorov (1941) who made the turbulent energy spectrum the centerpiece of modern turbulence theory.

Andrey Nikolaevich Kolmogorov has been described as the founder of modern probability theory and one of the 20th century's most eminent mathematicians. It is fascinating how Kolmogorov's breakthrough became known internationally during "the darkest hour" of the Second World War. George Keith Batchelor, who later became a revered turbulence researcher himself, made an amazing discovery during his first year in Cambridge – he came upon the landmark papers by Kolmogorov from the English-language editions of the 1941 issues of Doklady, the Comptes Rendus of the USSR Academy of Sciences. The bound volumes of Doklady and other Soviet journals were used as ballast for supply ships on their return voyage from northern Russian ports having carried armaments for besieged Russian cities on the outward voyage (Barenblatt, 2001; Moffatt, 2002, 2010; Zhou, 2021).

Kolmogorov's 1941 "phenomenology," which is basically a dimensional argument, remains the single most important theory of turbulence. The phenomenology makes the following assumptions:

- The turbulence is locally isotropic, that is, at small scales the turbulence statistics are invariant under translation and rotation.
- The turbulence is near "equilibrium," that is, the energy injection (the forcing on the turbulence), $\mathcal{F}$ that drives the turbulence is approximately matched by the rate, $\mathcal{E}$ at which the viscous effects dissipate the energy, that is, $\mathcal{F} \approx \mathcal{E}$.

- The wave number characterizing the energy injection, k_F, is much smaller than the wave number, k_d, characterizing where the energy is dissipated, $k_F \ll k_d$
- The cascade process through k-space is "local," that is, that the transfer through wavenumber k depends on the dynamics in the neighborhood near k (Zhou, 1993a,b; Zhou and Speziale, 1998).

Due to the nature of the nonlinearities of the Navier–Stokes equations, it is also essential to the analysis that the energy transfer in Fourier space is conservative.

The dominant high energy scales $\mathcal{L}$ are typically independent of viscous effects, and are a function of the large-scale variables, that is, the energy injection rate $\mathcal{F}$ (which is usually related to the overall geometry of the flow domain). Dimensional analysis produces the relationship

$$\mathcal{L} = \frac{K^{3/2}}{\mathcal{F}}. \tag{7.1}$$

Under the above set of assumptions, the energy in the cascade is to be independent of the details of both the forcing and the dissipation processes and is simply a function of the energy moving from small to large wave numbers. However, since the transfer is conservative, this energy transfer rate is equal to the injection rate, $\mathcal{F}$, which is equal to the dissipation rate $\mathcal{E}$. Under these conditions, one can choose to express $E(k)$ either in terms of the injection rate or in terms of the dissipation rate. Thus, for the near-equilibrium case, we more commonly see Eq. 7.1 expressed as

$$\mathcal{L} = \frac{K^{3/2}}{\mathcal{E}}, \tag{7.2}$$

which is generally preferred since it does not depend on a variable that is extrinsic to the description of the turbulent field itself.

7.3.1 Structure function

The structure function of order $\mathfrak{m}$, constructed from the velocity field, and defined over separation $\mathbf{r}$, is given by

$$S_{\mathfrak{m}}(\mathbf{r}) = \langle |u_i(\mathbf{x}+\mathbf{r}) - u_i(\mathbf{x})|^{\mathfrak{m}} \rangle. \tag{7.3}$$

The main results of Kolmogorov (1941) can be stated in plain language as follows (Davidson, 2015):

- Kolmogorov's *First Similarity Hypothesis* states that:

When Re is large enough, and $r \ll L$, the statistical properties of the structure function have a universal form which depends on only $\mathcal{E}$, r, and ν. Recall that $\mathcal{E}$ is the fundamental energy transfer rate and has units of energy/time.

- Kolmogorov's *Second Similarity Hypothesis* states that:

When Re is large, and in the inertial range $\lambda_K \ll r \ll L$, the statistical properties of the structure function have a universal form which is uniquely determined by $\mathcal{E}$ and r alone. Therefore,

$$S_{\mathfrak{m}}(r) \propto (\mathcal{E}r)^{\mathfrak{m}/3} \quad (\lambda_K \ll r \ll L). \tag{7.4}$$

In particular, one obtains for $\mathfrak{m} = 2$,

$$S_2(r) \propto \mathcal{E}^{2/3} r^{2/3} \quad (\lambda_K \ll r \ll L). \tag{7.5}$$

The second-order structure function is the energy contained in all Fourier modes with wavenumbers larger than $1/r$.

7.3.2 Kolmogorov energy spectrum

The structure function and energy spectrum are related in a standard form (Batchelor, 1953; Hou et al., 1998), and Eq. 7.5 can be reformulated into the Kolmogorov energy spectrum. Figure 7.3 shows the relevant features of the Kolmogorov 1941 theory in the spectral space. Essentially, a flow field always has an energy-containing scale, which corresponds to the large-eddies discussed in the previous subsection. The "energy spectrum," $E(k)$, defined above, provides a useful mathematical object for understanding Kolmogorov's theory. The contributions to the energy spectrum from the energy-containing scale are located in the vicinity of low wavenumbers. At the other end of the energy spectrum is the so-called dissipative scale, where $E(k)$ decreases exponentially as a result of the viscous action. Moreover, Kolmogorov predicted that if the Reynolds number is sufficiently large, the energy-containing scale will be separated from the dissipation scale by an inertial range in which $E(k)$ scales as $k^{-5/3}$. The larger the Reynolds number, the wider the inertial range will be.

The longest experimentally measured inertial ranges are especially noteworthy: Grant et al. collected their data from the Discovery Passage, a tidal channel of the west coast of

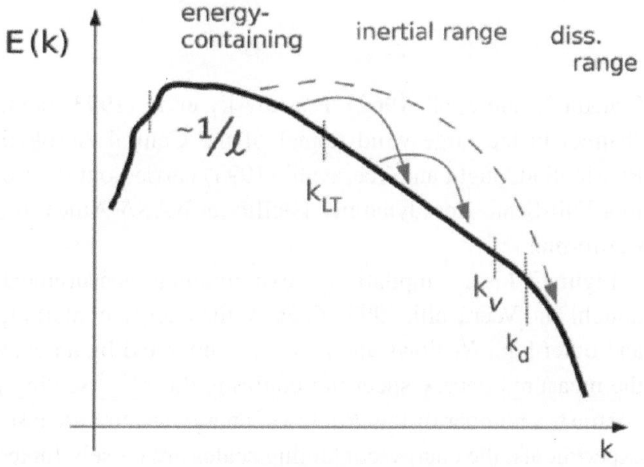

Fig. 7.3 Sketch of a kinetic energy spectrum indicating the energy-containing, inertial, and dissipation ranges and their wavenumber boundaries. The idea behind the minimum state is that the inertial range should be long enough so that direct interactions between modes in the energy-containing and dissipation ranges are energetically weak, indicated by the dashed arrow. Some "strong" interactions are indicated via the solid arrows. Zhou and Oughton (2011).

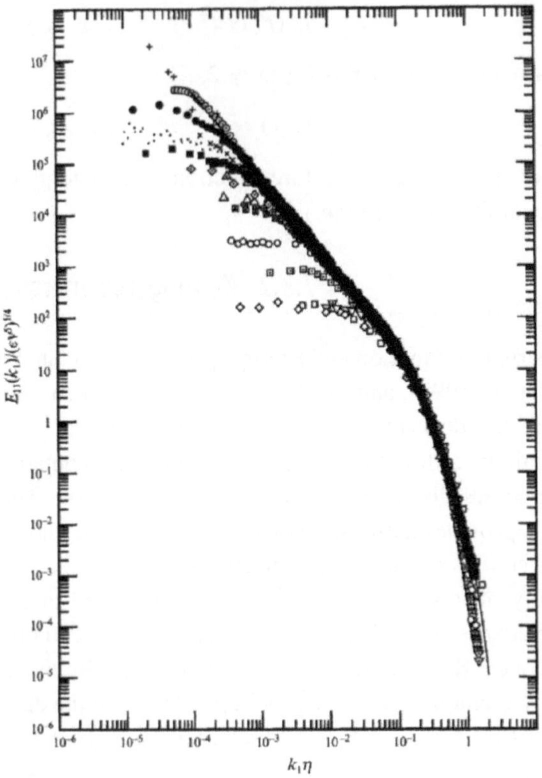

Fig. 7.4 Kolmogorov's universal scaling for 1D longitudinal power spectra Saddoughi and Veeravalli (1994). Reproduced with permission from Saddoughi and Veeravalli (1994).

Canada (Grant et al., 1962). Praskovsky et al. (1993) obtained their results from a return channel in the large wind tunnel of the Central Aerohydrodynamic Institute, Moscow, Russia. Saddoughi and Veeravalli (1994) carried out their experiments with the 80 × 120-foot Full-Scale Aerodynamics Facility at NASA Ames Research Center, Mountain View, California.

Figure 7.4 is a compilation of experimental measurements of the energy spectrum (Saddoughi and Veeravalli, 1994). Clearly, the energy-containing and dissipation scales in these and other high *Re* flows are very well separated by an extended inertial range. Moreover, the measured energy spectrum confirms the $k^{-5/3}$ scaling predicted by Kolmogorov. For contrast, a number of low *Re* flows' energy spectra are also included in Fig. 7.4. For these experiments, the energy-containing scales are closely linked to the dissipation scales. As a result, no inertial range can be discerned.

In terms of the dissipation rate, the energy spectrum can be written generally in the form

$$E(k) = C_K \mathcal{E}^p k^q, \tag{7.6}$$

where C_K is the so-called Kolmogorov Constant, and the exponents p and q are determined by dimensional arguments. Since $E(k)$ has units of [Length^3/Time^2], $\mathcal{E}$ has units of [Length^2/Time^3] and k has units of [1/Length], the time scaling of E can only be reconciled if the exponent of $\mathcal{E}$ is $p = 2/3$, where now $\mathcal{E}^{2/3}$ has units of [Length$^{4/3}$/Time^2]. Thus, the exponent of k must be $q = -5/3$ to produce the correct length scaling for E. The result is the Kolmogorov scaling law:

$$E(k) = C_K \mathcal{E}^{2/3} k^{-5/3}. \tag{7.7}$$

There is no consensus regarding whether C_K is a universal constant, but it appears to be approximately $C_K \approx 1.5$ (Sreenivasan, 1995; Yeung and Zhou, 1997; Donzis and Sreenivasan, 2010; Ishihara et al., 2016).

The finding of the $k^{-5/3}$ spectrum has been called a "simultaneous discovery" (Battimelli, 1986), because it was also suggested independently by Heisenberg (1948a,b), Weizsäcker (1948), Prandtl (Bodenschatz and Eckert, 2011), and Onsager (1945, 1949).

Kolmogorov's hypothesis essentially states the statistical independence of high and low wavenumbers in large Reynolds number turbulence. Fundamentally, he suggested that the inertial range scales are not influenced by the large scales directly and the large scales evolve very slowly by comparison with the eddies in the inertial range. The $-5/3$ scaling has received excellent confirmations (Grant et al., 1962; Chapman, 1979; Saddoughi and Veeravalli, 1994) from high Reynolds number measurements in geophysical flows and in the world's largest wind tunnels.[3] An asymptotic expansion analysis offers support that Kolmogorov's hypothesis is equivalent to the locality assumption (Branover et al., 1999).

7.3.3 Scalar variance spectrum

The Kolmogorov theory addresses the behavior of the statistics of the covariance of the velocity field but says nothing about the density statistics. For those flows where density variations are intrinsic to the behavior (RTI, RMI, and KHI with buoyancy), the density behaves as an active scalar. Obukhov (1949) and Corrsin (1951) applied/extended the Kolmogorov 1941 theory to the case of a passive scalar, for example, weak temperature fluctuations, and deduced that the scalar spectrum, $T(k)$ behaves as

$$E_\vartheta(k) = C_\vartheta \mathcal{E}_\vartheta \mathcal{E}^{-1/3} k^{-5/3}, \tag{7.8}$$

where $\mathcal{E}_\vartheta$ is the dissipation rate of the scalar due to diffusion, and is assumed to be equal to the injection rate, and C_ϑ is a constant. There are also potentially different "subranges" in the scalar spectrum depending on the Schmidt number, $Sc = \mathcal{D}/\nu$, where $\mathcal{D}$ is the diffusivity of the scalar. This topic is beyond our current scope and the reader is referred to Tennekes and Lumley (1972) for further details. For RTI flows, $Sc \sim 600$ for the brine/water experiments of Dalziel et al. (1999) and the follow-up experimental work by Banerjee and Mutnuri (2012) has $Sc \sim 10^3$. Molecular mixing measurements were

[3] Just to show how robust the Kolmogorov spectrum is, Düring et al. (2019) found the $k^{-5/3}$ scaling in a totally different physical system: the vibrations of elastic plates in the full nonlinear limit.

reported (Mueschke et al., 2007) for a high-Schmidt-number ($Sc \sim 10^3$) buoyancy-driven turbulent RT mixing layer in a water channel facility with a weakly buoyant flow.

7.4 RTI and RMI modified energy spectra

Having discussed the energy transfer in Eq. 6.41, we are now prepared to discuss the spectral evolution. Zhou (2001) argued that the extended Kolmogorov–Kraichnan phenomenology, developed for turbulent flows with an external agent (Kraichnan, 1965; Matthaeus and Zhou, 1989; Zhou and Matthaeus, 1990), should also be applicable to instability-driven turbulent flows. In this framework, excitation in the energy-containing range does not affect energy transfer within the inertial range.

Due to the above considerations, the average rate of energy dissipation is identified with the rate of spectral energy transfer and the rate of energy production (denoted here as $\mathcal{E}$). In order to infer the inertial-range spectrum, it is necessary to estimate the magnitude of the transfer function correlations,

$$\mathcal{E} \propto \tau_T(k) k^4 E^2(k). \tag{7.9}$$

The timescale τ_T for decay of transfer function correlations, which are responsible for inducing turbulent spectral transfer, may depend on any relevant turbulence parameters.

The well-known Kolmogorov spectrum can be demonstrated within this framework for homogeneous, isotropic, statistically steady turbulence. In this case, there is no excitation of the energy-containing range due to external agents. Therefore, the local dynamical timescale, $\tau_{nl} = [k^3 E(k)]^{-1/2}$, is the only available timescale of homogeneous, isotropic, statistically steady turbulence. A direct substitution of $\tau_T(k) = \tau_{nl}$ into Eq. 7.9 reproduces the Kolmogorov spectrum,

$$E(k) = C_K \mathcal{E}^{2/3} k^{-5/3}, \tag{7.10}$$

where C_K is the Kolmogorov constant (Kolmogorov, 1941; Saddoughi and Veeravalli, 1994; Yeung and Zhou, 1997). However, for flows induced by RT and RM instabilities, the external timescales are introduced (see Eqs. 2.23 and 2.25) and given by $\tau_{RT}(k) = (kg\mathcal{A})^{-1/2}$ and $\tau_{RM}(k) = (k\Delta V\mathcal{A})^{-1}$.

We are now in a position to determine the energy spectrum for both RTI and RMI driven turbulent flows. For turbulent flow induced by RTI, we find that a direct application of $\tau_T = \tau_{RT}$ or τ_{RM} results in the scaling of the kinetic energy spectra for the RTI (Zhou, 2001)

$$E(k, z) = C_{RT} (\mathcal{A}g)^{1/4} \mathcal{E}^{1/2} k^{-7/4}, \tag{7.11}$$

and for the RMI

$$E(k, z) = C_{RM} (\mathcal{A}\Delta V \mathcal{E})^{1/2} k^{-3/2}. \tag{7.12}$$

The extended phenomenology suggests that the lifetime of transfer function correlations τ_T is more accurately treated by considering the external agent and turbulent nonlinear interactions. Here, the external timescale, τ_{EX}, represents any timescale other than the nonlinear cascade timescale (in our applications here, τ_{RT} or τ_{RM}). Matthaeus and Zhou (1989), Zhou (1995), and Zhou and Matthaeus (1990) have developed a framework in which both timescales, τ_{nl} and τ_{EX}, coexist in a fashion analogous to the composition of triple decay times in the EDQNM closures (Pouquet et al., 1976)

$$\frac{1}{\tau_T} = \frac{1}{\tau_{nl}} + \frac{1}{\tau_{EX}}. \tag{7.13}$$

Combining Eqs. 7.9 and 7.13, the extended energy spectrum $E(k) \propto k^{\mathfrak{s}}$ is obtained, where

$$\mathfrak{s} = \begin{cases} -7/4 \text{ (RTI) or } -3/2 \text{ (RMI)}, & \text{for } k < k_Z, \\ -5/3 \text{ (Kolmogorov)} & \text{for } k \geq k_Z. \end{cases} \tag{7.14}$$

A controlling parameter, k_Z, indicates the extended range of influences exerted by the external agents in wave number space (Zhou, 2001). It changes, in a reasonably quasi-steady state, from instability induced turbulence to the fully developed turbulent state. When τ_{EX} is much shorter than τ_{nl}, we should reproduce the external agent modified spectra "$\mathfrak{s} = -7/4$" (RTI driven turbulent flow) or "$\mathfrak{s} = -3/2$" spectrum (RMI driven turbulent flow). The inertial range of the RTI or RMI spectrum will exhibit a split spectrum that varies from the classical Kolmogorov spectrum ($\mathfrak{s} = -5/3$) at higher wave numbers to one with a modified slope at lower wave numbers.

For RT flow, Cook et al. (2004) plotted the evolution of the vertical velocity spectrum computed in the $z = 0$ plane. Past an early time period, Fig. 7.5(a) shows that the peak of the perturbation spectrum begins migrating to lower wavenumbers, eventually forming an inertial range. The movement of the peak of the spectrum is important because it is inversely proportional to the width of the mixing layer. Figure 7.5(b) reproduces the density and velocity spectra of Cabot and Cook (2006) at the end of their simulation. A fiducial line corresponding to famous Kolmogorov scaling is drawn for reference. However, the simulated spectra of Zhao et al. (2020c) are consistent with the $-7/4$ inertial range power law predicted by Zhou (2001). At any rate, the inertial range must be sufficiently long to exhibit this split. Indeed, Cabot and Zhou (2013) found that the narrowness of the inertial range and the noise in numerical simulations make it difficult to clearly distinguish the split.

The evolution of the energy spectrum of a simulated RMI-induced flow is also given in Fig. 7.6 (Oggian et al., 2015). Yet, unlike the RTI, the RMML is not a driven turbulent field (Zhou, 2001). Once the shock has passed, there is no continued source of energy for the flow field and the initially deposited energy begins to decay similar to those of decaying turbulence. Therefore, the energy spectrum decreases in magnitude with time.

Yet as in the case of RTML, the peak of the spectrum also migrates to lower wavenumbers as time evolves. Hence, the mixing layer width also increases with time. For the given initial condition, the late-time energy scales as $k^{-3/2}$, an RM-modified energy spectrum first suggested in Zhou (2001).

The fluctuating RM kinetic energy spectra are shown in Fig. 7.7 at several snapshots in time. At early times, the spectrum is peaked around the highest perturbation frequency $k_{\max}$. At later times, a clear $k^{-3/2}$ range can be seen from nondimensionalized time 100 to 500, consistent with the theory of Zhou (2001). The spectral peak moves from $k/k_{\max} = 1$ at the start of the simulation to $k/k_{\max} = 0.2$ ($k \approx 7$) by the end, in accordance with the increased layer width (Thornber et al., 2010).

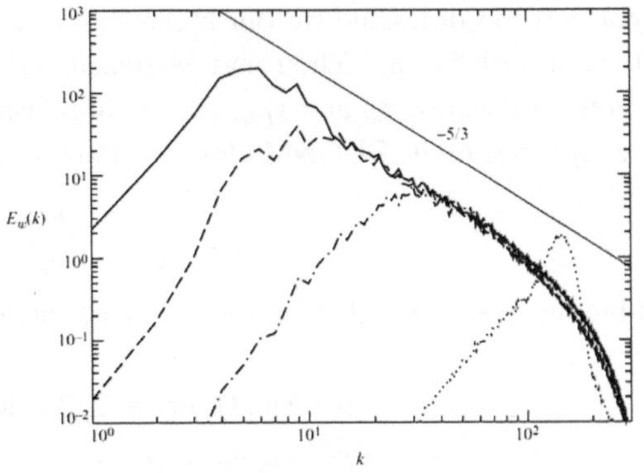

(a)

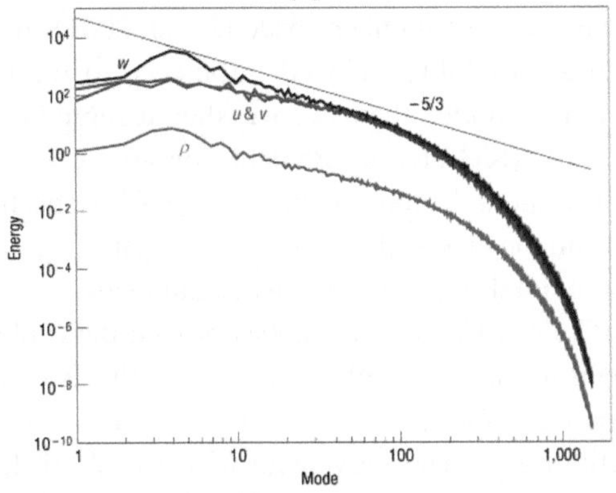

(b)

Fig. 7.5 RT spectrum from numerical simulations with the Miranda code. (a) Evolution of the vertical velocity spectrum computed in the $z = 0$ plane from a simulation with a 1152^3 grid resolution. A fiducial corresponding to the Kolmogorov $k^{-5/3}$ spectrum is drawn for reference. Fig. 14 of Cook et al. (2004). (b) Rayleigh–Taylor energy spectra at the end of the 3072^3 simulation. Data are from the $z = 0$ plane at a dimensionless time 31 (end of simulation). The spectra are computed by taking 2D Fourier transforms of density and velocity components, multiplying by the complex conjugate, and summing over each wavenumber annulus. A fiducial corresponding to Kolmogorov scaling is drawn for reference. The energy spectra are scaled in terms of gravity, computational grid spacing, and the density of the light fluid. Fig. 2 of Cabot and Cook (2006), with permission from Nature Physics, © SpringerNature.

Peng et al. (2021a) studied this issue from the vortex dynamics perspective. The effect of the secondary baroclinic vorticity (SBV) (Peng et al., 2021b) causes the slope of the total energy spectrum at intermediate wavenumbers to evolve towards the $-3/2$ scaling at the critical time. Subsequently, the SBV effect diminishes and the energy spectrum decays to the $-5/3$ law, as shown in Fig. 7.8.

More recently, researchers from the University of Science and Technology of China performed high-fidelity simulations of RM turbulence with two different initial perturbation spectra (Zhou et al., 2023). A visible $-3/2$ spectrum of the turbulent kinetic energy is observed, which confirms the prediction of Zhou (2001).

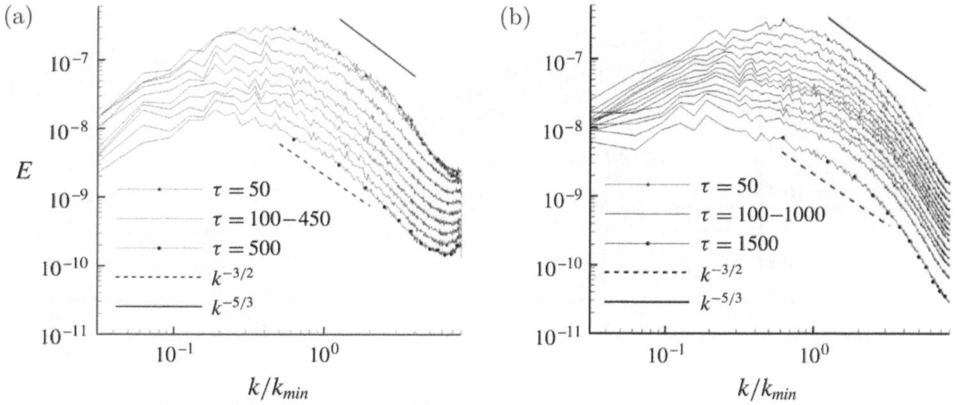

Fig. 7.6 The hybrid spectra of the radial turbulent kinetic energy averaged on the y- and z-planes in the bulk of the mixing layer for the 512 cross-section grid and the $k^{-3/2}$ guideline analytically predicted by Zhou (2001). The spectra have been obtained by an averaging process over 10 slices of the domain along the x direction, which is the direction of the shock propagation. (a) Compressible solution and (b) Hybrid solution. Fig. 9 of Oggian et al. (2015).

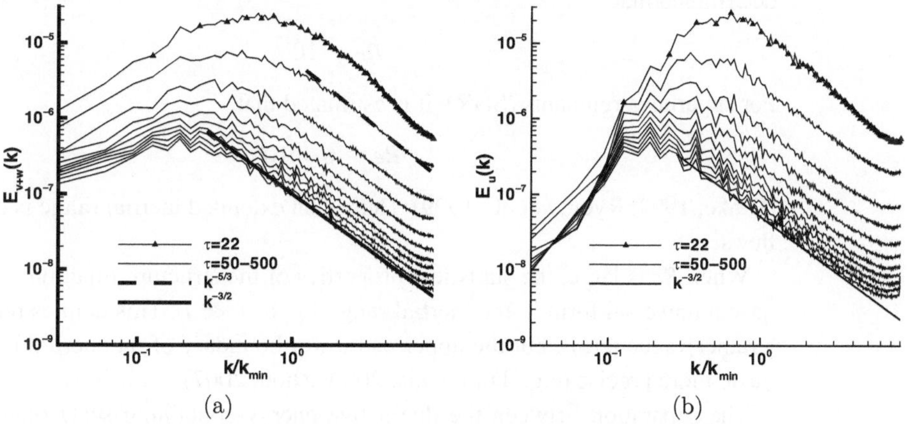

Fig. 7.7 RM energy spectrum with a narrow band initial condition. (a) The homogeneous plane. (b) The spectrum of the energy component along the direction of the shock passage. At later times, a clear −3/2 range of Zhou (2001) can be seen. Thornber et al. (2010).

7.5 Inertial ranges

The Reynolds numbers for astrophysical flows are very high. For the degenerate carbon–oxygen white dwarf explosion (Type Ia),[4] Kuhlen et al. (2006) and Woosley et al. (2004) found

[4] The basic supernova classification scheme is based on early- and late-time spectra and other features. See Section 21.1.

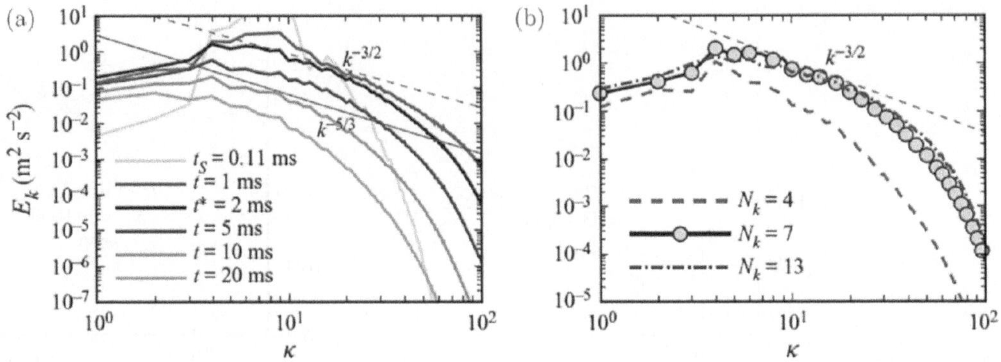

Fig. 7.8 Comparisons of kinetic energy spectra. (a) At different times with the intermediate-band perturbation and (b) With different initial perturbations at $t = 2$ ms. Fig. 4 of Peng et al. (2021a).

$$Re \sim 10^{14}, \tag{7.15}$$

while for the core-collapse of a massive star (Type II), Abdikamalov et al. (2015) determined that

$$Re \sim 10^{17}. \tag{7.16}$$

For supernova remnants (SNR), it is estimated that

$$Re \sim 6 \times 10^{8}. \tag{7.17}$$

(Drake, 1999; Ryutov et al., 1999). Hence, an extended inertial range is expected for these flows.

When Re is large, the statistical properties of the structure function or energy spectrum have a universal form in the inertial range $\lambda_K \ll r \ll L$. This defines the inertial range in a rather vague terms, but the upper and lower boundary of the inertial range can (now) be made more precise (e.g., Dimotakis, 2000; Zhou, 2007).

The separation between the distinctive energy-containing scale range and universally independent inertial range is the Liepmann–Taylor scale which can be defined in terms of the outer-scale $\mathcal{L}$ and Reynolds number,

$$\lambda_{LT} = 5\mathcal{L}Re^{-1/2}. \tag{7.18}$$

The other end of the inertial range is the so-called inner viscous scale, which is also determined by the Re and $\mathcal{L}$,

$$\lambda_{\nu} = 50\mathcal{L}Re^{-3/4}. \tag{7.19}$$

An inertial range exists, therefore, when the following condition is met,

$$\frac{\lambda_{LT}}{\lambda_{\nu}} \gg 1. \tag{7.20}$$

The still unanswered question is: How large must the Reynolds number be for a flow field to qualify as fully developed turbulence? Chapter 8 will address this question.

8 Transition to turbulence

Having laid the necessary groundwork in the previous chapter, we are now ready to determine the necessary and sufficient conditions for a time-dependent flow to transition to turbulent flow. The reader should not be surprised that we will start our discussion by investigating the role of the Reynolds number, *Re*, as it is the key nondimensional number capturing the relative importance of the inertial and viscous forces in the flow.

In the low *Re* limit, inertial forces are so low as to be negligible. In this case, the velocities of objects are essentially governed by the applied force and viscous drag. In an intriguing essay, Purcell (1977) probed small things moving and living in low *Re* fluids (see also, Mogilner and Manhart, 2018).

In many other cases, one has to confront the more challenging problem of turbulent flows where the Reynolds number is significantly higher. For avian flight, as an example, air flow in birds' respiratory systems must achieve its high *Re* limit; more specifically, birds rely on their unique lung structures for efficient oxygen uptake, particularly when they fly at high altitudes or long distances (Fig. 8.1(a)). The respiratory system of birds is designed to optimize efficient exchange of oxygen and carbon dioxide (Nguyen et al., 2021). This is achieved with air sacs providing a continuous unidirectional airflow through the lungs. It does not require any valve-like structures, typical of circulatory systems (Fig. 8.1(b)).

Nguyen et al. (2021) offered some insights from an experimental study. In particular, when the fluid in the setup encountered a T junction where it could either continue straight or turn, the fluid preferentially continued straight. This is because flows at high Reynolds number are dominated by inertial forces. Moreover, vortices (another ramification of flows at high *Re* number) created at the junction corner served as blockades to the orthogonal branch of the junction (Fig. 8.1(c)).

8.1 Reynolds number criterion for transition

8.1.1 Flux rate of energy transfer

The most basic measure of the energy transfer process is the energy flux across a scale k. According to the Kolmogorov theory of the universal equilibrium range, it is the only link between the energetic and dissipative scales of motion (Batchelor, 1953).

In the classical Kolmogorov inertial range, where energy injection is absent and dissipation is negligible, energy conservation implies that the 3D energy flux across the spectrum is uniform. For homogeneous isotropic turbulence (HIT), the spectral energy

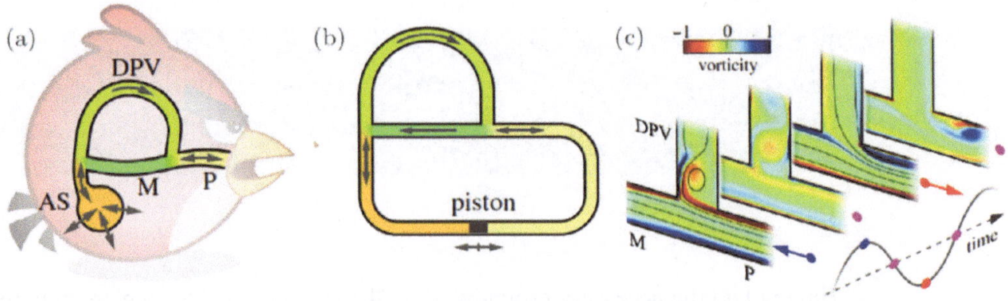

Fig. 8.1 (a) Network model of bird lungs. Reciprocal expansion and contraction of air sacs (AS) drives oscillatory inhalation and exhalation flows in the primary bronchus ("P"). Directed flow is observed in dorso-, para-, ventrobronchi ("DPV") that span the mesobronchus ("M"). (b) A "spherical bird" simplification with a single AS and DPV, with a closed network formed by connecting P to AS, the latter replaced by a reciprocating piston. (c) Vorticity and streamlines near the junction, at four instances in the oscillation cycle. At peak velocity during inhalation (first image), fluid is injected from P and predominantly goes straight past the junction down M, with little turning up the DPV (inertial effect). A stagnation streamline impinges on the far (left) corner of the junction and divides the straight and turning flows, and a large vortex shed from the near (right) corner "plugs" the side branch. Nguyen et al. (2021).

balance equation reads

$$\left(\frac{\partial}{\partial t} + 2\nu k^2\right)E(k) = T(k), \tag{8.1}$$

where $T(k)$ is the energy transfer function (Lin, 1956). Meanwhile, contributions to the total flux from the various scale interactions can be written as

$$\Pi(k) = \int_k^\infty T(k')dk' = -\int_0^k T(k')dk'. \tag{8.2}$$

To provide a direct visualization of the Kolmogorov energy transfer process in a finite section of the "ideal" inertial range, we can resort to a suitable analogy of an infinitely long "pipe" without leaks. To illustrate the interaction of scales, we "cut" a finite section from this "pipe" and view its inflow and outflow. The "ideal" $T(k)$, constructed from several simulations at different resolutions, is shown in Fig. 8.2. The negative and positive peaks correspond to inflow and outflow. Since the flow is statistically steady and $\int dkT(k) = 0$, we have shifted the peaks, so that the three mesh sizes can overlap. Because the "ideal" pipe does not leak, its length is not important. This is a direct visualization of the Kolmogorov energy transfer process in a finite section of the "ideal" inertial range (Zhou, 1993a,b).

For a developing time-dependent flow, Fig. 8.3(a,b) provides sketches of the energy transfer and flux functions at an early and late time, respectively. Note that the energy containing scale and the dissipation scale are indicated by the "ideal" inflow and outflow profiles, as indicated by negative and positive peaks (shaded regions). More specifically, at an early time, the Re of the flow is low and the energy-containing and dissipation scales are not separated. In this case, the energy flux across the spectrum is not uniform. However, an inertial range will be formed at a late-time as Re increases. At a sufficiently late time

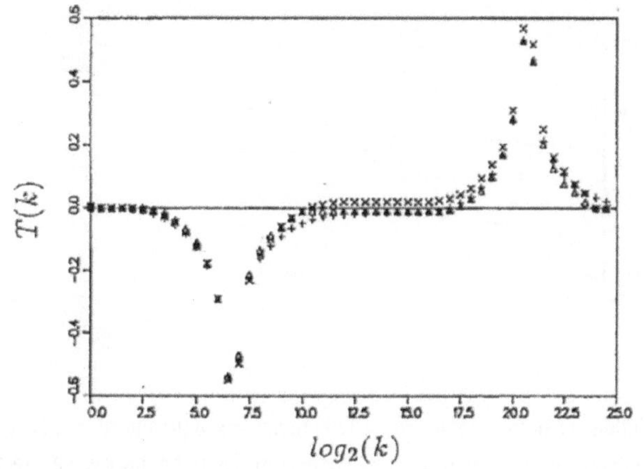

Fig. 8.2 The "ideal" self-similar transfer $T(k)$, computed from three numerical simulations at grid resolution 64^3 to 256^3. Zhou (1993b).

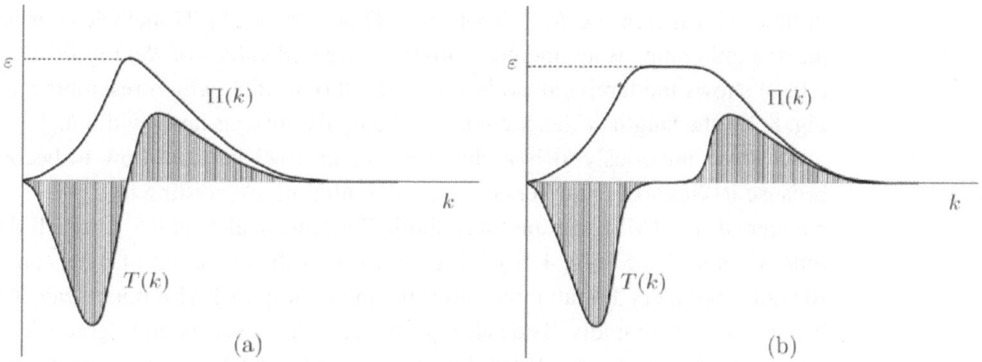

Fig. 8.3 Sketch of the early- and late-time energy transfer processes. Schematic views of the energy flux and energy transport spectrum as functions of wave number. Reproduced with permission from McComb (2008).

(Section 8.2), the energy-containing and dissipation scales will be separated by an extended inertial range that will satisfy the fundamental criterion of constant flux, $\Pi(k) = \mathcal{E}$.

8.1.2 Evolution of the length scales for time-dependent flows

The development of a separation between large- and small-length scales provides another means of characterizing the evolution of the time-dependent flow fields. Figure 8.4 shows the temporal evolution of the outer scale, integral length, Taylor microscale and Kolmogorov scale from DNS at $Re_O = 174$ (a), $Re_O = 348$ (b) and $Re_O = 697$ (c). The initial Re_O is defined by a weighted average wavenumber and the initial growth rate. Because of the choice of nondimensionalization, the outer scales are essentially the same, except for a small difference mostly at late time. Comparing results across the three simulations, the

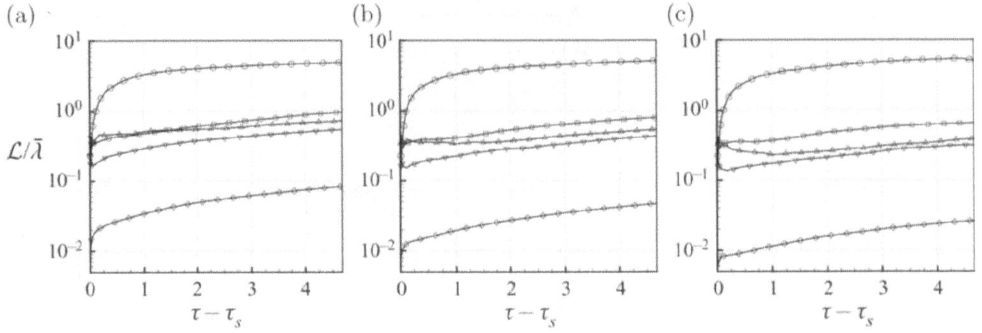

Fig. 8.4 Evolution of length scales for $Re_0 = 174$ (a), $Re_0 = 348$ (b) and $Re_0 = 697$ (c). Markers show the outer scale h (circles), the integral length (squares), the Taylor microscales in the shock wave direction (upward triangles), the Taylor microscales in the homogeneous direction (downward triangles) and the Kolmogorov microscale (diamonds). Fig. 16 of Groom and Thornber (2021).

integral lengths, Taylor microscales, and Kolmogorov microscales all decrease uniformly in time with increasing *Re* (Groom and Thornber, 2021). Nonetheless, it is evident that the dynamic range is augmented with the increased values of *Re*. Figure 11 in Zhao et al. (2022) shows the temporal evolution of the flux in RTI, which resembles the sketches in Fig. 8.3 if the length-scales are normalized by the mixing-layer width h.

As noted previously, it is a challenge for an RMI-induced flow to become turbulent, because its Reynolds number could decay with time. According to Eq. 4.33, the Reynolds number of the RMML in the "self-similar" regime scales as $t^{2\theta-1}$, which decreases with time when $\theta < 0.5$ (Fig. 4.16). Figure 8.5 shows the evolution of outer- and Taylor-scale Reynolds numbers. For all three cases, the initial drop in Taylor microscale Reynolds numbers occurs very rapidly. The subsequent decay, however, is quite gradual. Moreover, the ratio of Taylor microscale Reynolds numbers indicates significant anisotropy is present in the velocity field and persists even to the latest times.

Finally, Taylor-microscale Reynolds numbers for vertical velocity components in the DNS results are about 200–400 at the latest times, which is marginally near the level needed to form an inertial range, but those for the horizontal components are 3–4 times lower at the sub-100 level. As a result, strong anisotropy is present at the end of calculations.

8.1.3 Mixing transition

Dimotakis (2000) suggested the notion of a mixing transition. This concept arose from his observation that a variety of experiments showed a distinct transition from a low-mix state to a high mix-state. Such a mixing transition is evident in the evolution of the probability density function of the mole-fraction, which indicates an increase in mixed fluid (Cook et al., 2004).

An inertial range will start to form when the following condition is met:

$$\frac{\lambda_{LT}}{\lambda_\nu} \approx 1, \tag{8.3}$$

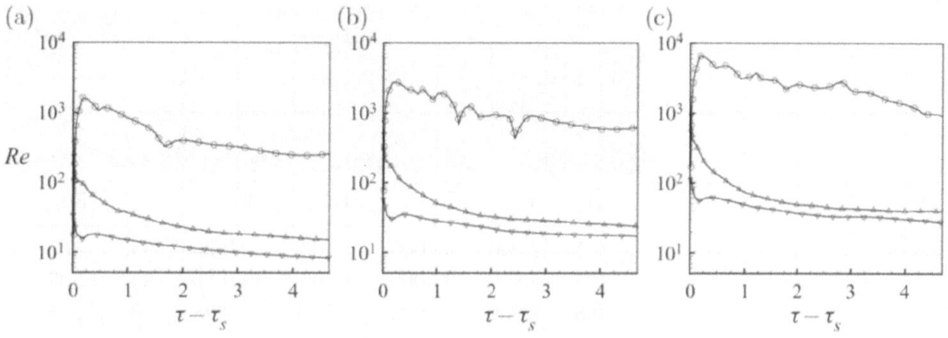

Fig. 8.5 Evolution of Reynolds numbers for $Re_0 = 174$ (a), $Re_0 = 348$ (b) and $Re_0 = 697$ (c). Plotted are the outer-scale Reynolds number (circles), the Taylor microscales in the shock wave direction (upward triangles), and the Taylor microscales in the homogeneous direction (downward triangles). Fig. 17 of Groom and Thornber (2021).

where λ_{LT} and λ_ν are the Liepmann–Taylor and inner viscous scales, respectively (see Section 7.5). The *Re* associated with the mixing transition, $Re \sim 10^4$, corresponds to the starting point of the scale separation. Thus, for time-dependent RT and RM flows, there is no scale separation between the energy-containing and dissipation scales at early time. However, an inertial range does develop at late times. When the flow field's Reynolds number is around the mixing transition $Re \sim 10^4$, the separation between the energy-containing and dissipation scales is just beginning. In other words, the inertial range does not exist at this *Re*.

It is important to note that one should not confuse a flow which has just achieved the mixing transition with one exhibiting fully-developed turbulence. Indeed, a turbulent flow, according to any standard textbook, mandates an established inertial range (see Chapter 7 and Fig. 7.4) and the *Re* should be much greater than that associated with the mixing transition.

For RTI flows, the values of $\mathrm{Re}_{h1\%} = h(1\%)\dot{h}(1\%)/\nu$, using a 1% concentration threshold, are sometimes quoted (e.g., Cabot and Cook, 2006), giving somewhat inflated late-time simulation values that greatly exceed the required 10^4 value (Fig. 4.15) for a mixing transition state (Dimotakis, 2000). However, the values of their $\mathrm{Re}_H = H\dot{H}/\nu$ counterparts, defined in terms of the product width H, are several times lower, and fall well short of the *Re* for mixing transition (see Table 8.1).

8.1.4 Minimum state turbulence

A developing flow reaches a fully developed turbulent state when its inertial range is sufficiently broad (Zhou, 2007; Zhou et al., 2019). However, a more narrowly defined criterion is required to guide experimental designs in both the laboratory and HEDP platforms. The so-called *"minimum state"* is significant because it implies that one doesn't need to perform experimental measurements or numerical simulations of extremely high *Re* flows. Specifically, the significant work required to perform such difficult studies would simply extend the length of the inertial range, which is already well understood via self-similar

Table 8.1 Approximate late-time Reynolds numbers for simulations with different Atwood numbers. The minima for development of an inertial range and "minimum state" in HIT are $\mathrm{Re}_\ell \approx 10^4$ and 1.6×10^5, and $\mathrm{Re}_\lambda \approx 250$ and 1,000, respectively.

A	Re_H	$\mathrm{Re}_{h1\%}$	$\mathrm{Re}_{\ell x}$	$\mathrm{Re}_{\ell z}$	$\mathrm{Re}_{\lambda x}$	$\mathrm{Re}_{\lambda z}$
0.2	7,000	38,000	1,400	20,000	95	362
0.5	5,400	32,000	1,100	14,000	87	307
0.8	3,500	24,000	700	7,000	70	216

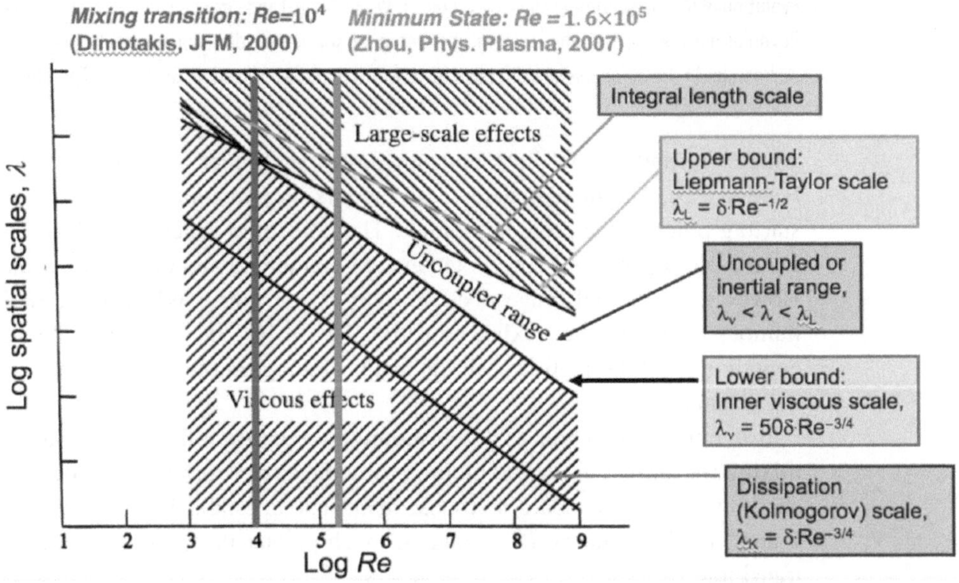

Fig. 8.6 *Spatial Transition.* This figure indicates the Reynolds numbers required for achieving the mixing transition of Dimotakis (2000) and minimum state turbulence of Zhou (2007). Reproduced from Dimotakis (2000). See also, Zhou (2007).

analysis and has little impact on the actual physical processes of interest. In other words, once the *"minimum state"* is achieved, the most important physics of the turbulence has already been captured.

For flows that have achieved this minimum state criterion, there is a complete decoupling between the energy-containing and dissipation scales, so that the statistics of the energy-containing scales becomes independent of Reynolds number.

The resolution of large-scale dynamics refers to the direct computation or measurement of the 3D, time-dependent evolution at large scales. Measurements include the energy-containing velocity fluctuations and scalar variances, which are known to govern the dynamics of the flow. To ensure the integrity of the physical picture, the large-scale modes should not be contaminated by interaction with the non-universal dissipation range. This is

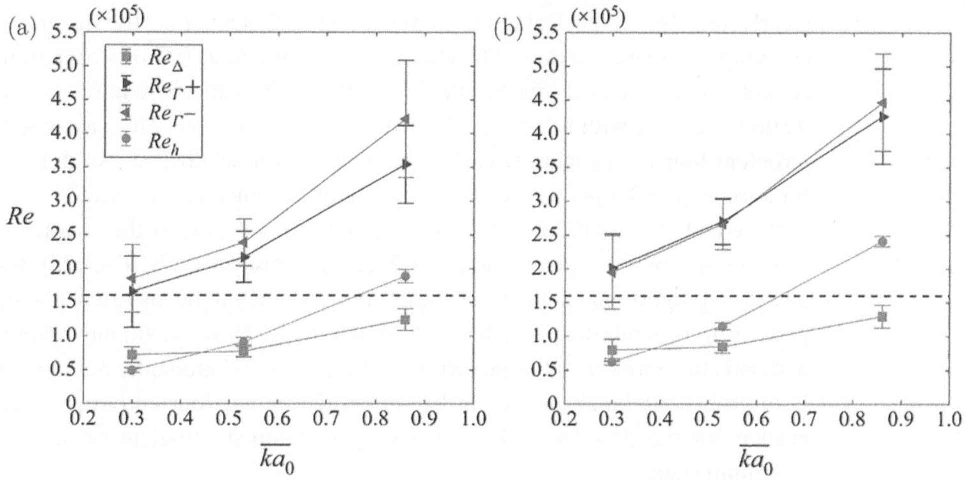

Fig. 8.7 Physical significance of the minimum state criterion in the context of RMI studies. Reynolds numbers based on local and global information, plotted against the initial $\overline{ka_0}$, at times (a) $t = 2.65$ ms and (b) $t = 5.65$ ms. The dashed lines indicate the minimum state criterion of $Re = 1.6 \times 10^5$ in Ref. Zhou (2007). The authors found that, compared to other methods, the Reynolds number defined by Eq. 2.9 better characterizes the flow in terms of the mixing transition. (see Mansoor et al. (2020) for detailed definitions). Fig. 14 of Mansoor et al. (2020).

true for both experiments and numerical simulations. Studying the energy transfer process and interacting scales, Zhou (2007) determined that the lowest Re satisfying this condition is around 1.6×10^5. Figure 8.6 is a sketch summarizing the spatial transition criterion described above.

To demonstrate the physical significance of the minimum state criterion in the context of RMI studies, Mansoor et al. (2020) performed single-mode RM experiments at the Los Alamos National Laboratory (LANL) Vertical Shock Tube (VST) facility experiments (see also, Mejia-Alvarez et al., 2015). Figure 8.7 shows the variously defined Reynolds numbers as functions of the initial conditions at two times. The Reynolds numbers increase with a_0 and t. Given the physical formation of mixing regions, $Re = h\dot{h}/\nu$ characterizes the flow better than methods that are not based on the mixing transition.

8.2 Temporal criterion of transition

While the spatial transition criterion can be used to analyze turbulence experiments, the time-dependence of RTI/RMI flows suggests that, in addition to high Re, a state of strong mixing requires a time-dependent criterion. In particular, a sufficient amount of time (or distance if the turbulence is spatially evolving) is required for the turbulence to establish a full cascade that enhances the mixing at fine scales.

Zhou et al. (2003a,b), Robey et al. (2003), and Zhou (2007) focused on developing a temporal criterion of transition for time-dependent RTI and RMI. These studies showed clearly that both high Reynolds number and sufficient time (or distance, in the spatial evolving cases) were required to achieve a transition to a state of strong molecular mixing. As noted in Drake et al. (2008), the Zhou–Robey hypothesis requires that uncoupled fluctuations develop within laminar-flow regions in the viscous diffusion boundary layers. A turbulent transition occurs when the boundary layer widths,[1] $\lambda_D = 5(\nu t)^{1/2}$, have grown for a long enough time to become greater than the inner viscous scale.

Rose and Sulem (1978) defined the eddy-turn-over-time as the characteristic size of an eddy divided by the typical velocity difference across the eddy. The eddy-turn-over-time is therefore proportional to the time required for the eddy to be distorted, and in this distortion process, to generate smaller eddies. As remarked by Drake et al., the mixing transition corresponds to approximately one turn-over-time for the dominant, energy-containing eddies. Using the same language, it can be shown that about four eddy-turn-over-times will be needed for the flow to qualify as a fully developed turbulent flow, as defined by the minimum state.

An intuitive understanding can be gained by considering the TKE evolution of simple isotropic turbulence, as predicted by closure theory (Zhou, 2021). The time required to establish a turbulent cascade from an initial non-cascading state was found by Lesieur and

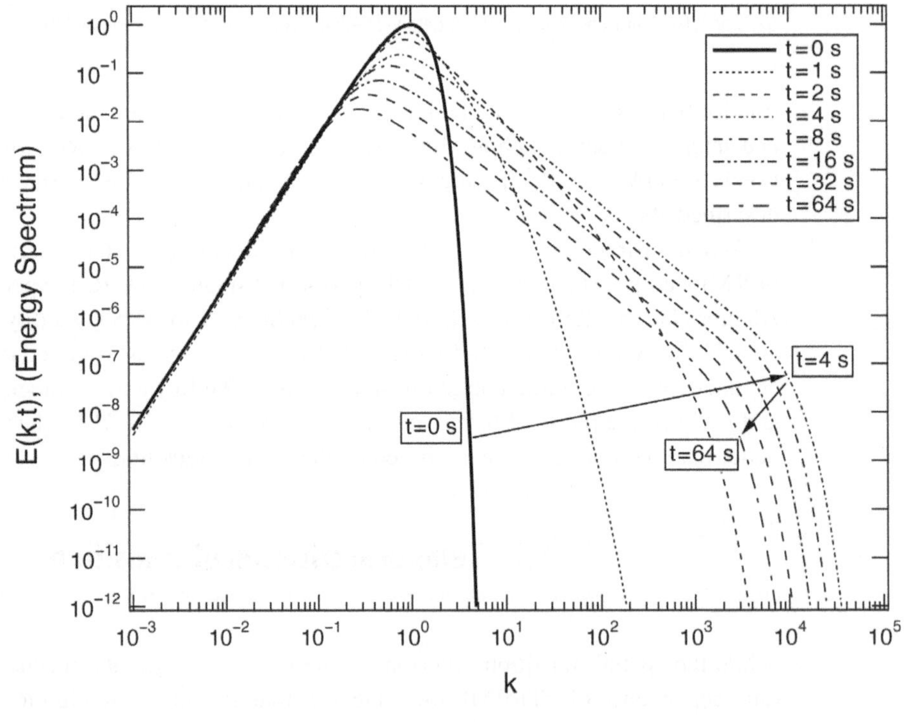

Fig. 8.8 Evolution of the energy spectrum for an isotropic decaying turbulence as predicted by the EDQNM model. Note that the spectrum establishes an inertial range by $t = 4$. Zhou et al. (2019).

[1] Is it worth mentioning that this is the same coefficient as in the Liepmann–Taylor scale.

Schertzer (1978) using the Eddy Damped Quasi-Normal Markovian (EDQNM) closure. The EDQNM model is a "spectral model"; that is, it models the turbulent energy spectrum dynamics for isotropic turbulence, and represents the non-local Fourier-space transfer terms with reasonable fidelity (Orszag, 1970a, 1977; Lesieur, 1987; Sagaut and Cambon, 2019; Zhou, 2021). Similar behavior was also observed by Clark and Zemach (1998) (who found the same phenomenon in EDQNM) and in a simple spectral model due to Leith (1967), which represents the cascade as a purely local transfer via nonlinear diffusion. Both models exhibited the finite time onset of the cascade. The time of onset was referred to as the "catastrophe time" by Lesieur and Schertzer, and denoted by t^*. Both models also found a weak dependence of this time on the viscosity (or equivalently, the initial turbulent Reynolds number). Clark and Zemach observed that EDQNM established a cascade in approximately $4\tau_e$, whereas the Leith model established a cascade in approximately $2.3\tau_e$. The eddy-turn-over-time τ_e was defined as

$$\tau_e = \left(Kk_{0,\max}^2\right)^{-1/2}, \tag{8.4}$$

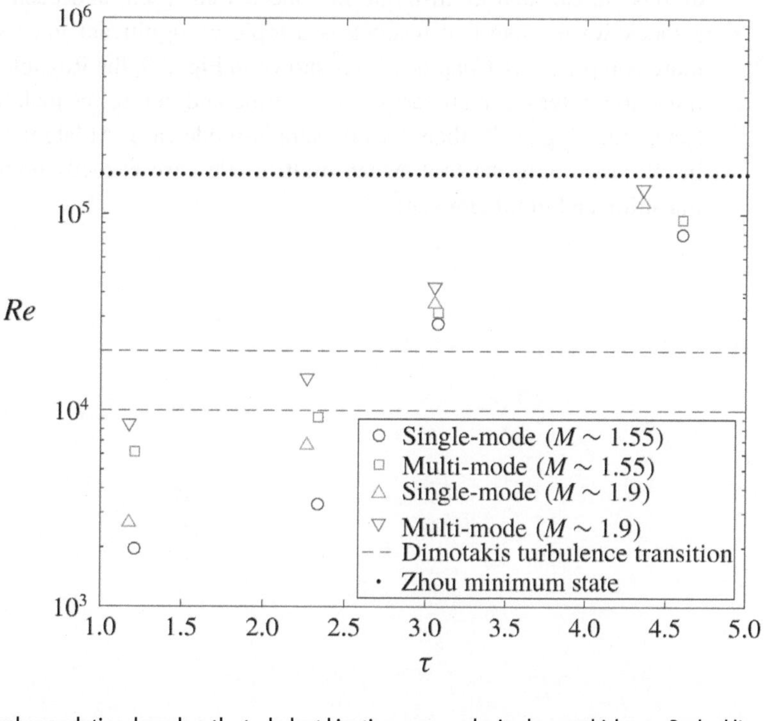

Fig. 8.9 Reynolds number evolution, based on the turbulent kinetic energy and mixed-mass thickness. Dashed lines indicate a threshold for turbulent mixing (Dimotakis, 2000) and the dotted line indicates the minimum state criterion ($Re \approx 1.6 \times 10^5$, Zhou (2007)). After reshock, there is a sharp increase in the Reynolds number, and the mixing transition occurs in the flow. However, this flow is still not a fully developed turbulent flow, and does not meet the minimum state criterion to reproduce the spectral range corresponding to an astrophysical event. Mohaghar et al. (2019).

where K is the turbulent kinetic energy and $k_{0,\max}$ is the wave number at which the initial energy spectrum has its maximum value. Again, note that the wave number of the spectral maximum is inversely proportional to the integral scales. Figure 8.8 shows the EDQNM-predicted spectral evolution from a state exhibiting no inertial range to a state exhibiting a strong cascade.

Neither Lesieur and Schertzer (1978) nor Clark and Zemach (1998) commented on the significance of this finite time onset of the cascade in relation to a transition process. However, Zhou et al. (2003a,b), Robey et al. (2003), Zhou (2007), and Zhou et al. (2019), in their efforts to develop the temporal criterion of transition, connected this onset with a new notion. Strong mixing should occur only when the difference between the dissipative scale wave number, $k_d \approx 1/\lambda_K$, and the integral scale wave number, $k_L \approx 1/L$, is sufficiently larger than the nonlocality associated with the Fourier-space energy transfer.

The temporal transition criterion and its corresponding Reynolds number transition criterion (discussed in the last subsection) allow us to nicely correlate a number of important physics concepts. Using an incident shock at Mack number 1.9 (Mohaghar et al., 2019), along with previous work at Mach 1.55, Mohaghar et al. (2017) examined the effects of incident shock strength on the RMI mixing transition. To quantify the effect of initial conditions, single- and multi-mode interfaces were used, and each case was followed by a reshock wave. Note that reshock is a topic of significant interest and will be discussed more completely in Chapter 13. As shown in Fig. 8.9, the Reynolds numbers are below the transition criterion in all cases at early time and before reshock, but the multi-mode case has a much higher *Re* than does the single-mode case. At late time, although *Re* increases in all cases, only the high-Mach multi-mode case exceeds both criteria for the mixing transition and minimum state.

PART 2

HYDRODYNAMICS OF COMPLEX FLOWS

One learns to hope that Nature possesses an order that one may aspire to comprehend.

Chen-Ning Yang, Nobel laureate

9 Influence of initial conditions

The temporal evolution of RTI, RMI, and KHI induced flows described in the preceding chapters can be stalled at any stage through the action of dissipative or restoring forces. For the KH instability, for instance, the action of a stabilizing gravitational force can lead to the wave amplitude growing very slowly, if at all. Viscosity, surface tension, material strength and magnetic fields can also act to stifle growth, or in some cases arrest it completely.

The parameter space of possible initial conditions (ICs) is extremely large. The most prominent parameter relevant to instability growth is the ratio between the initial amplitude and wavelength. Other parameters that could potentially change the course of instability growth include Atwood and Mach numbers, viscosity, heat flux, radiation transport, surface tension, rotation, stratification, and magnetic field. Unfortunately, only limited cases have been investigated in sufficient detail to assess the impact of ICs on flow development (e.g., Morgan and Black, 2020; Motl et al., 2009; Rasteiro dos Santos et al., 2022, 2023; Suchandra and Ranjan, 2023).

There are significant indications that many of the mix measurements in previous chapters are influenced by ICs. Much attention has been paid to whether the late-time flow depends on the specific ICs and other properties. These issues will be the subject of this and the next several chapters. Whether the flows remember their IC remains a major unknown, since almost all relevant studies have had insufficient Reynolds number and temporal scales to produce a fully developed turbulent state.

9.1 Initial conditions of RT

Early papers on experiments were performed with immiscible liquids, such as the well-known Lewis's 1950 RTI experiment that was published along with Taylor's landmark paper. He obtained an approximately sinusoidal initial interface perturbation by oscillating the blade of a small paddle. There were two important conclusions from Lewis' paper: (1) the experimental data agreed with linear stability theory until the interface amplitude reached $a/\lambda \sim 0.4$, and (2) a constant late time bubble velocity $v_b \propto \sqrt{g}$ was attained. In their follow up works, Emmons et al. (1960), Cole and Tankin (1973), and Jacobs and Catton (1988b) further improved the experiments of Lewis. Popil and Curzon (1980) highlighted one of the problems with experiments using immiscible fluids – the edge effects due to the meniscus.

Roberts and Jacobs (2016) performed experiments using miscible and immiscible liquid combinations (Fig. 9.1). The experiments were initiated either with parametrically

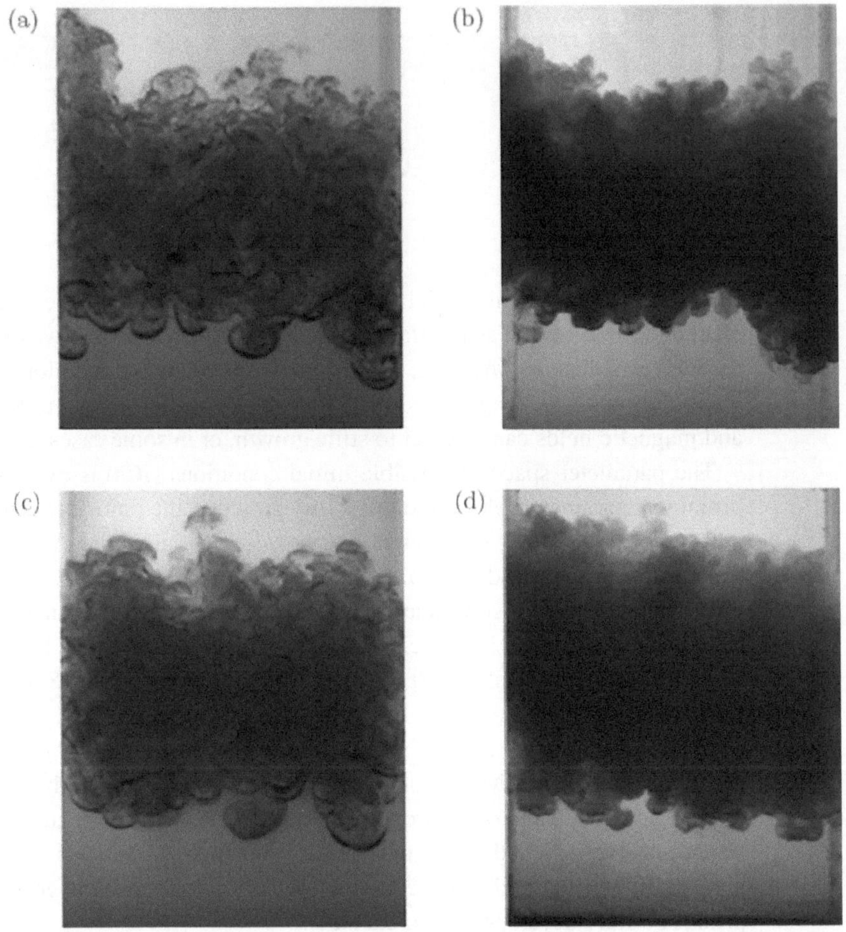

Fig. 9.1 Comparison of forced, unforced, miscible, and immiscible experiments performed on a weight-and-pulley apparatus for Atwood number 0.48. Images are shown at (a) $t = 320$ ms for immiscible unforced, (b) 380 ms for miscible unforced, (c) 275 ms for immiscible forced, and (d) 300 ms for miscible forced. Images were chosen at times in which the mixing-layer width is approximately matched between experiments. Fig. 13 of Roberts and Jacobs (2016).

forced initial perturbations or left unforced. The unforced experiments were seeded by random interfacial fluctuations, resulting from background noise. The forced experiments had an initial perturbation imposed by vertically oscillating the liquid-filled tank to produce Faraday waves at the interface.

The experiments of Roberts and Jacobs (2016) achieved the self-similar regime, as evidenced by the increase of the dominant scales in proportion to the mixing-layer width. The scaling parameter α was significantly smaller for the miscible experiments than for the immiscible experiments, indicating that miscibility plays a role in the late-time turbulent instability growth – see Figs. 9.2 and 9.3.

By using a lattice Boltzmann (LB) method (e.g., Chen and Doolen, 1998), Tavares et al. (2021a) compared simulated 2D RTI immiscible and miscible flows. Figure 9.4(a) shows

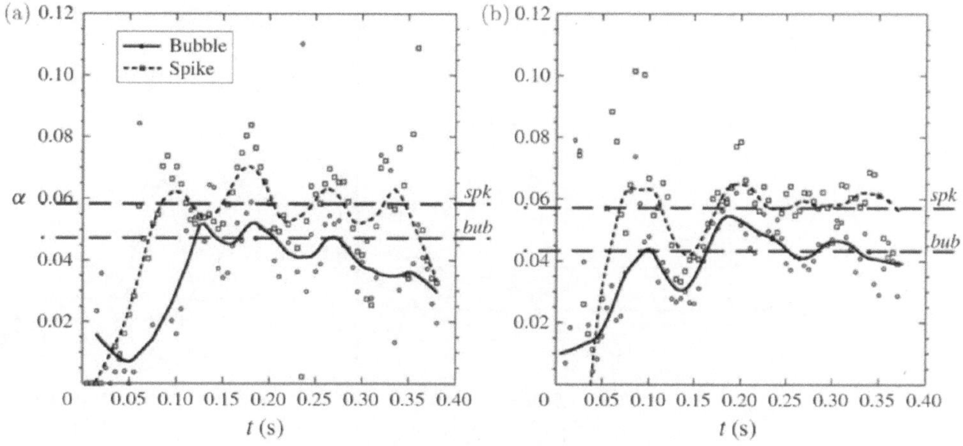

Fig. 9.2 Measurements of α, determined by the derivative method, for ensemble-averaged immiscible experiments. The mixing-layer width cutoff was taken as 80% of the pure liquid intensity. Here, $\dot{h}_{b/s}$ was estimated using a central difference approximation. Data are for the two cases: (a) immiscible unforced and (b) immiscible forced. The horizontal line represents the value of α obtained from the square-root method and is drawn to show the consistency of the methods. For both plots, the top line represents the spike and the bottom represents the bubble. Solid and dashed lines show a weighted running average of 20% of the points in each experiment. For the immiscible unforced experiments, it was found that $\alpha_s = 0.059 \pm 0.002$ and $\alpha_b = 0.047 \pm 0.007$, while for the immiscible forced experiments $\alpha_s = 0.057 \pm 0.014$ and $\alpha_b = 0.044 \pm 0.009$. Roberts and Jacobs (2016).

the width and speed of the RTML, averaged with respect to ensembles of realizations. The extracted α values are shown in Fig. 9.4(b), tending to constant values in the well-developed mixing regions. The estimated values of α are 0.027 ± 0.005 for immiscible and 0.033 ± 0.004 for miscible flows, respectively. Finally, based on statistical similarities between miscible and immiscible RT flows, Tavares et al. (2021a) suggested (as yet to be fully confirmed) that the tendency towards isotropy of small-scale fluctuations occurs for the immiscible case in a similar manner to the previously simulated miscible case (Biferale et al., 2010; Boffetta et al., 2009, 2010). Surface tension, meniscus effects and 3D effects could cause differences between miscible and immiscible flows. These will be examined in later chapters.

Two distinctive types of ICs were previously recognized by Birkhoff (1955) and Sharp and Wheeler (1961). First, the initial interface may be populated by "small random perturbations," in which high-wavenumber modes dominate the energy content. In this case, long wavelengths in the flow are only generated by nonlinear mode coupling (Haan, 1989; Dimonte et al., 2004). It seems reasonable to assume that flows will then lose memory of the ICs, implying self-similar mixing (Youngs, 2013).

A second type of IC is one in which long-wavelength perturbations appear in the initial spectral content, the amplitudes of which govern the late-time self-similar growth (Youngs, 2013). As an example, Inogamov (1978) argued that random perturbations with wavelengths up to the domain size, and amplitude proportional to wavelength, would give

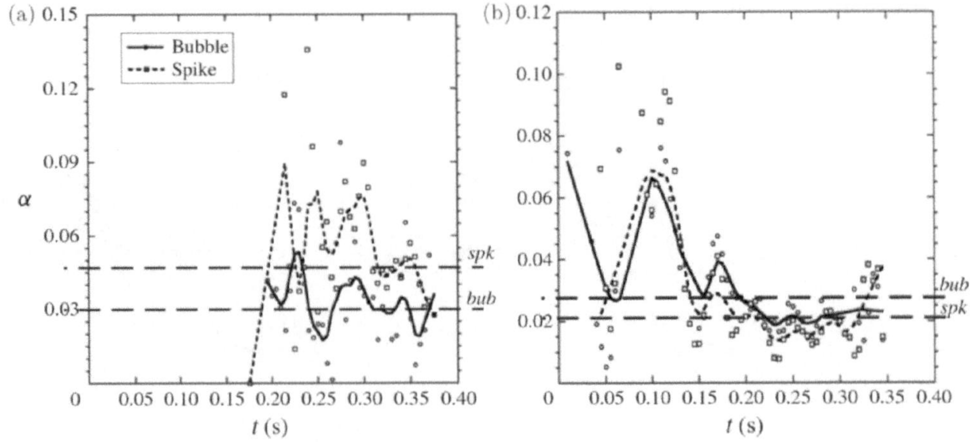

Fig. 9.3 Measurements of α, determined by the derivative method, for ensemble-averaged miscible experiments. Here $\dot{h}_{b/s}$ was estimated using a central difference approximation. Data are for the cases: (a) miscible unforced and (b) miscible forced. The horizontal line here represents the value of α obtained from the square-root method and it is drawn to show the consistency between the methods. The top line is the spike and the bottom is the bubble, while it is opposite for the right panel. Solid and dashed lines show a weighted running average of 20% of the points in each experiment. For the miscible unforced experiments, it was found that $\alpha_s = 0.047 \pm 0.023$ and $\alpha_b = 0.030 \pm 0.008$, while for the miscible forced experiments $\alpha_s = 0.017 \pm 0.008$ and $\alpha_b = 0.023 \pm 0.007$. Roberts and Jacobs (2016).

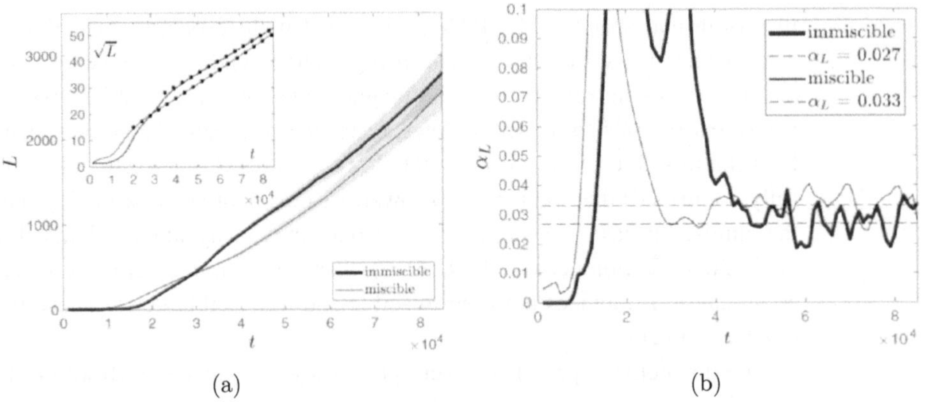

Fig. 9.4 (a) Width of the mixing layer, $L(t)$, as a function of time for immiscible (bold line) and miscible (thin line) flows. Shaded areas indicate standard deviations. The inset compares the $\sqrt{L}$ graphs in the region of turbulent mixing. (b) Measurement of the dimensionless prefactors, α, for the immiscible (bold line) and miscible (thin line) flows for the mixing layer width. Constant values (dashed lines) are estimated in the regions of turbulent mixing. Note also that the authors used α_L instead of α. Tavares et al. (2021a).

self-similar mixing at an enhanced rate. In terms of the initial perturbations power spectrum, $P(k)$, this is equivalent to $P(k) \sim k^{-3}$, where the standard deviation of the initial perturbation is given by

$$\hat{\sigma}^2 = \int P(k)dk, \tag{9.1}$$

where k is the wavenumber and

$$P(k) = \begin{cases} k^m & \text{if } k_{\min} < k < k_{\max}, \\ 0 & \text{otherwise}, \end{cases} \tag{9.2}$$

with $m \leq 0$ (Youngs, 2013). Youngs (2013), Dimonte et al. (2004), and Ramaprahhu et al. (2005) have all used such initial perturbations.

Desirable initial perturbations can be implemented in numerical simulations and a number of calculations have been designed to show the impact of "small random perturbations" and random long-wavelength perturbations (k^{-3} spectrum). Ramaprabhu et al. (2005) reported that with long wavelengths present in the ICs, α_b increased logarithmically with the initial amplitudes (Fig. 9.5(a)). On the other hand, in the mode-coupling dominated limit, no such dependence is observed, and α_b takes a universal lower-bound value of $\sim 0.03 \pm 0.003$ (Fig. 9.5(b)). This may explain the low values of α_b reported by most numerical simulations initialized with annular spectra of short-wavelength modes, which consequently evolve purely through mode-coupling. Moreover, Ramaprabhu et al. found α_b to be independent of h_0k in RTI computations of the mode-coupling regime.

Performing a further numerical study, Banerjee and Andrews (2009) concluded that the overall growth of RTI mixing is strongly dependent on ICs. In Fig. 9.6, Youngs (2013) showed that higher values of α can be obtained if long-wavelength perturbations with a k^{-3} spectrum are added. By varying the magnitude of the long-wavelength perturbations, the value of α can increase from 0.025 (the value frequently found from idealized DNS) to about 0.06, typical of experimental results. Other publications studying this issue include those by Dimonte (2004), Kuchibhatla and Ranjan (2013), Miles et al. (2004), and Rozanov et al. (2015).

The data reported in Burrows et al. (1984) and Smeeton and Youngs (1987) are part of an extensive series of "rocket rig" experiments investigating the mixing of incompressible fluids by RT instability (Read and Youngs, 1983; Read, 1984; Youngs, 1989). As shown in Fig. 9.7, two fluids were enclosed in a rectangular tank, the less dense fluid initially resting on top of the denser fluid. A novel technique was used to generate the desired acceleration between 25 and 75 times that of gravity. One or more rocket motors, attached to the top of the tank, were used to drive the tank vertically downwards. Read (1984) and Youngs (1989) found that progressively larger structures appear as time proceeds (Fig. 9.8) and concluded that the mixed-width is described well by a t^2 scaling law.

Glimm et al. (2013) study the RT mixing layer, presenting simulations in agreement with experimental data. For example, the authors reconstructed initial data from Smeeton–Youngs (1987) rocket rig experiments, noting that these experiments had significantly weaker low-mode perturbations than the water channel experiments at Texas

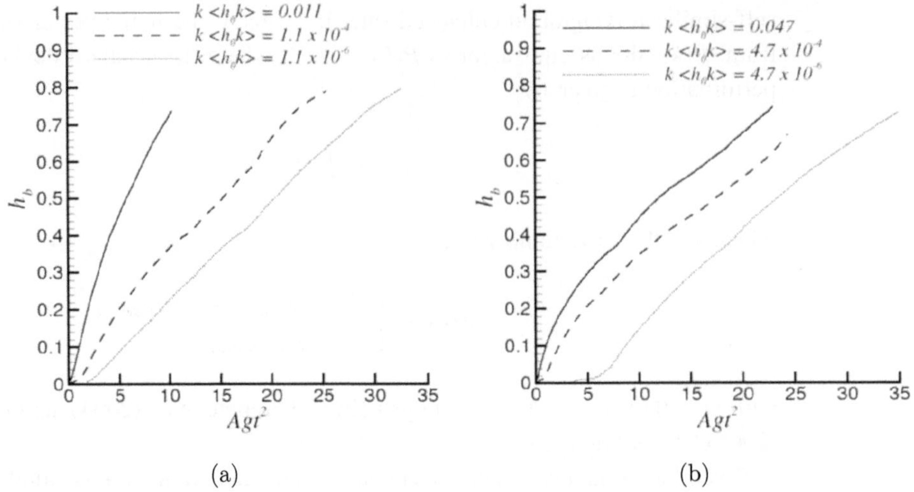

Fig. 9.5 Effect of kh_0 on the evolution of bubble amplitude, h_b: (a) three cases with long wavelengths present in the initial conditions; (b) evolution of bubble amplitude for mode-coupling cases. Fig. 13 of Ramaprabhu et al. (2005).

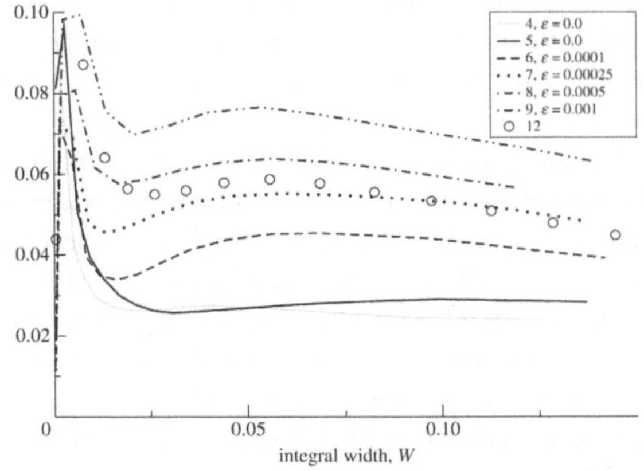

Fig. 9.6 Variation of α for RTI with integral mix width at $\rho_2/\rho_1 = 3$ and various values of the long wavelength parameter (denoted ϵ). Fig. 3 of Youngs (2013). A. UK Ministry of Defence © Crown Owned Copyright 2022/AWE.

A&M (Mueschke, 2010). Moreover, Zhang et al. (2018a) analyzed the role of the meniscus in these immiscible experiments, finding that it delays the achievement of self-similar scaling, but leaves the growth rate, α_b, largely unchanged.

The data of Burrows et al. (1984) and Smeeton and Youngs (1987) is notable due to the long-wavelength noise in the initial perturbation profile. Dimonte (2004) estimated the initial perturbations in the LEM experiments and used a model to show that they should enhance α. He argued that growth by mode-coupling is a very slow process and likely to be enhanced by long wavelength perturbations. He also noted that there is a weak logarithmic dependence of α on the magnitude of the long wavelength perturbations. Glimm et al. (2020) disagrees, instead arguing that its strength is insufficient to appreciably modify α_b.

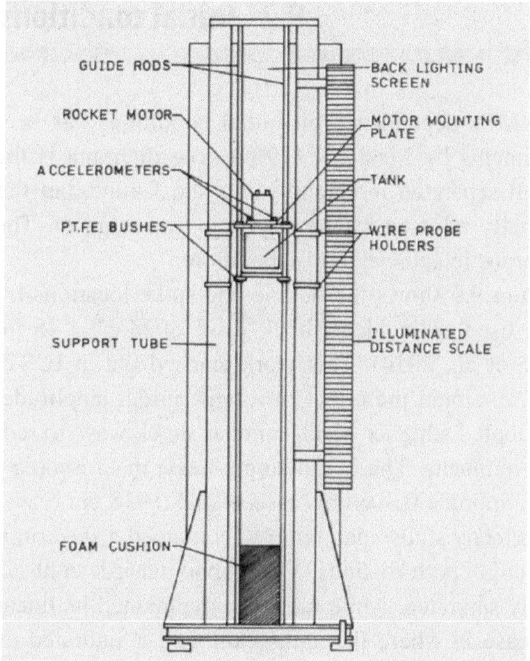

Fig. 9.7 Sketch of the rocket-rig facility. Rocket-rig experiments are also described in Youngs (1989). Fig. 1 from Read (1984).

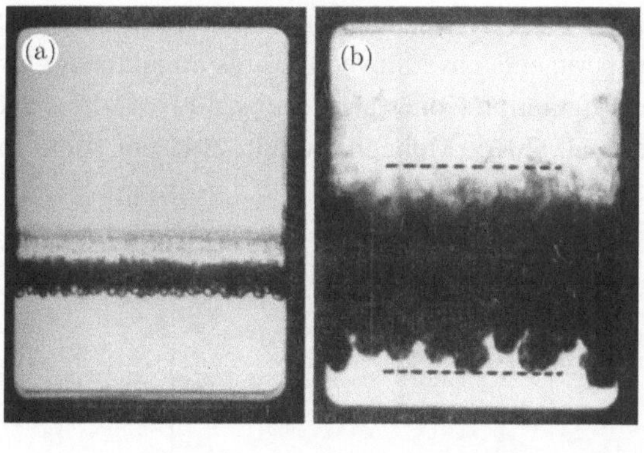

Fig. 9.8 Experimental figures from (a) third experimental plate and (b) sixth experimental plate for experiment no. 105. Smeeton and Youngs (1987), with permission. UK Ministry of Defence © Crown Owned Copyright 2022/AWE.

The authors further question the analysis in Dimonte (2004) and Dimonte and Schneider (1996), wherein long-wavelength noise was asserted to be significant.

Additionally, Liu et al. (2006) presented a set of 3D RT multi-mode mixing simulations to model the experimental data of Banerjee and Andrews (2006). They used improved physics and an improved front tracking code FronTier. In a follow-up work, Glimm et al. (2013) used estimates of experimental ICs to match observed values of the RTI scaling prefactor α in the range 0.05–0.85.

9.2 Initial conditions of RM

The RMI's dependence on initial conditions has been recognized since the pioneering experiments by Meshkov (1969). The dilemma is that the heavy and light fluids must be well separated incipiently in a shock tube, but the resulting mixing layer should be minimally affected by the separation mechanism. The earliest RMI experiments used a membrane to achieve fluid separation.

Figure 9.9 shows the bubble and spike locations from single-mode RMI experiments, which used an incident shock wave of Mach 1.45 in an air-SF_6 system (Vandenboomgaerde et al., 2014). This work, carried out at IUSTI, Marseille, France (Houas et al., 2003), explored the effect of a large initial amplitude on the growth of the perturbation and adopted a higher-Mach-number shock wave to reduce the deleterious effects of membrane remnants. The initial single-mode macroscopic perturbation used wavelength 8 cm with amplitude 0.306 cm (case 1) and 0.918 cm (case 2). Nevertheless, the interface was not perfectly sinusoidal, but also contained a spectrum of residual small-amplitude, short-wavelength perturbations (Vandenboomgaerde et al., 2014). In case 1, the bubble velocity quickly saturated, while the spike maintained its linear growth. This should be compared with case 2, where the spike exhibited a saturated evolution, departing from the linear regime around $t = 1$ ms.

The presence of membrane fragments impedes the use of advanced visualization techniques (Collins and Jacobs, 2002). Moreover, the remaining fragments have a damping effect on the ensuing RMI, due to their high inertia. The fragments also inhibit the free mixing of the experimental gases (Sewell et al., 2021). As a result, most modern shock tube experiments have moved towards a membrane-less interface, using either a counterflow configuration (Jones and Jacobs, 1997; Collins and Jacobs, 2002; Weber et al., 2012; Reese et al., 2014; Mohaghar et al., 2019) or some form of a gas curtain (Jacobs et al., 1995; Balakumar et al., 2008).

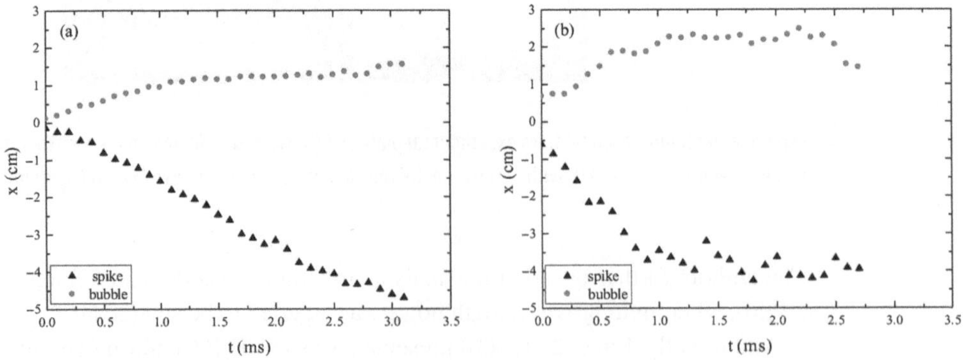

Fig. 9.9 Experimental temporal evolution of the locations of the RMI bubble and the spike. (a) Case 1. (b) Case 2., Fig. 7 of Vandenboomgaerde et al. (2014), with the permission of AIP Publishing.

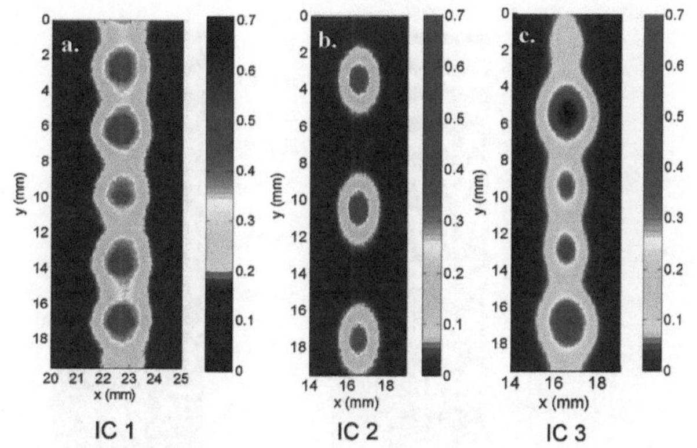

Fig. 9.10 Images of the initial concentration: (a) single-mode short wavelength (3.6 mm, IC 1), (b) single-mode long wavelength (7.2 mm, IC 2), and (c) multi-mode initial condition (IC 3). The intensity bar shows the volume fraction of SF_6. Fig. 3 of Balasubramanian et al. (2013), with permission.

Continuing their previous experimental work (e.g., Balakumar et al., 2008; Prestridge et al., 2013), Balasubramanian et al. (2013) studied the effects of initial amplitude and wavenumber. Images of the RMI evolution after transmission of a planar shock (Mach $= 1.21$) were captured using simultaneous planar laser-induced fluorescence (PLIF) and particle image velocimetry (PIV) diagnostics. The single- and multi-mode membrane-free ICs are shown for the light–heavy–light gas curtain experiments in Fig. 9.10.

For an air-SF_6-air configuration ($\mathcal{A} = 0.67$), there is a correlation between the ICs and small-scale mixing at late times (Fig. 9.11). Single-mode RMI growth rate depends on both initial amplitude and wavelength. In contrast, multi-mode RMI depends only weakly on the ICs.[1]

Mansoor et al. (2020) studied the late-time RMI growth of sinusoidal perturbations on an air/SF_6 interface subjected to a Mach 1.2 planar shock wave. Interface perturbations are established using a novel membrane-less technique where cross-flowing air and SF_6 are separated by an oscillating splitter plate. It is evident that larger initial perturbations give rise to greater RM growth (Fig. 9.12), with the increase being significant for the highest $\overline{ka_0}$. Figure 9.12 shows a comparison of the experimental data with some nonlinear and Padé models. Several models agree with the data for $\overline{ka_0} = 0.30$ and 0.53, but then deviate for $\overline{ka_0} = 0.86$, the highest initial amplitude. A variant of Sadot and co-workers' empirical rational function has been suggested to remedy this shortfall.

In a related work, Rasteiro dos Santos et al. (2022) identified different regimes corresponding to single- or multi-mode initial interfaces. The flows were generated by rotating several thin plates that initially separated two gases of different densities.

The dimensionless RM growth of the 3D single-mode perturbation (Chapman and Jacobs, 2006) is higher than that of the 2D single-mode (Niederhaus and Jacobs, 2003), as shown in Fig. 9.13. Luo et al. (2013) also compared these two datasets with the results

[1] Several interesting numerical works should also be noted. Recently, Pereira et al. (2021) carry out ILES of an air-SF_6-air gas curtain subjected to two shock waves at Mach 1.2, while Romero et al. (2021) perform simulations of the shock-driven Kelvin–Helmholtz instability in inclined gas curtains.

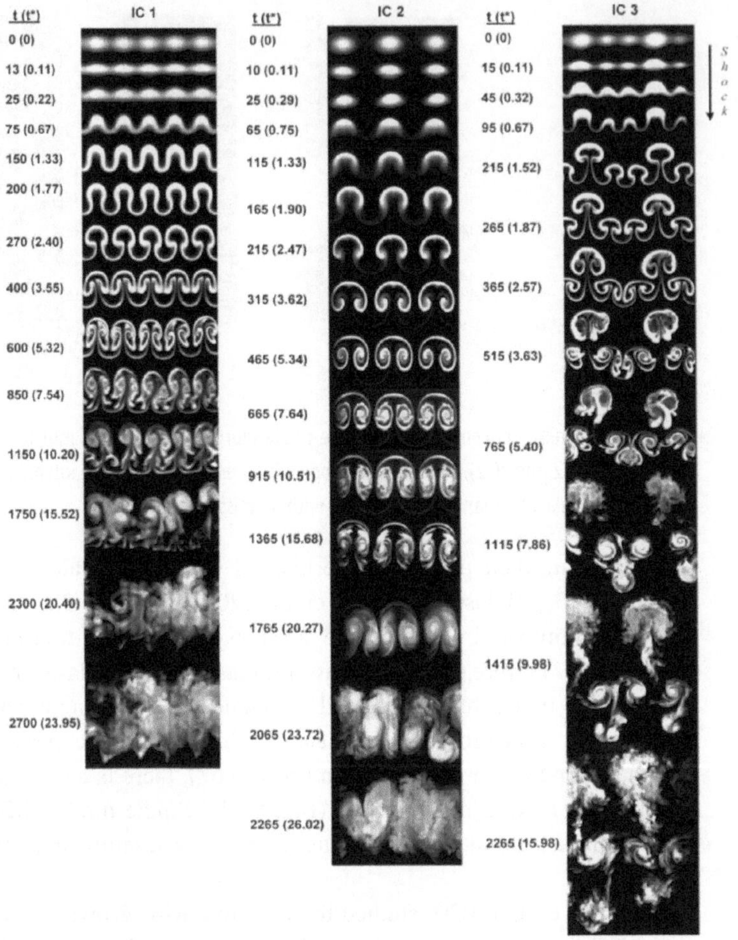

Fig. 9.11 Time evolution of density fields for three different initial conditions. Far left: single-mode short wavelength (IC 1); center: single-mode long wavelength (IC 2); and far right: multi-mode (IC 3). White indicates SF_6, and black indicates air. The time $t = 0\ \mu s$ corresponds to the initial condition. In all the sets of images, t is the absolute time, and t^* is the normalized time based on the initial linear growth. Fig. 5 of Balasubramanian et al. (2013), with permission.

obtained using a soap film technique, which has significantly lower mixing-width growth. Note, however, the significant differences of the Atwood number, 0.67 of Luo et al. and 0.15 previously used by Jacobs and co-workers (Zhou, 2017a). Motivated in part by this observation, Liang et al. (2021c) obtained an exact mathematical expression of a 3D-interface perturbation using a spectral analysis. The initial perturbation profile was found to consist of a superposition of transverse 2D single-mode and 3D multi-mode components. Such information allowed the authors to investigate the influence of mode-coupling on bubble evolution and to observe the resulting different behaviors for the bubble and spike width.

For 2D multi-mode systems, Figs. 9.14 and 9.15 illustrate Schlieren images of the shocked interface for both large and small initial amplitudes. The interface morphologies

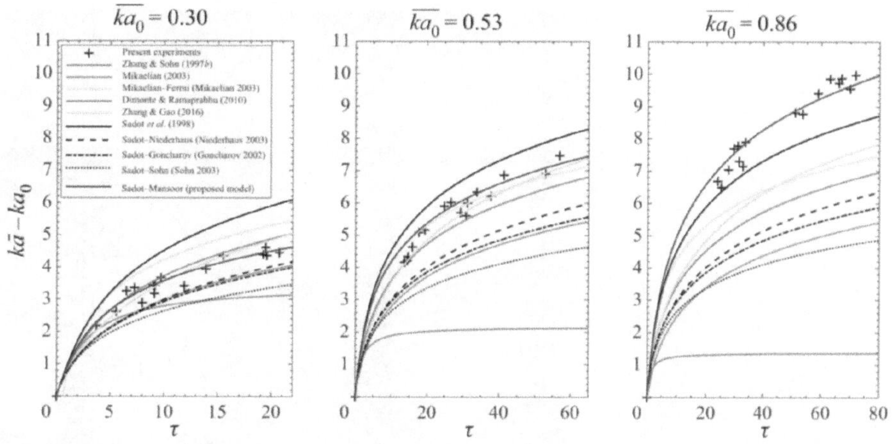

Fig. 9.12 Evolution of dimensionless perturbation amplitude versus dimensionless time. The data represents modern experiments and historied nonlinear models based on heuristic interpolation methods (Sadot et al., 1998; Mikaelian, 2003), Padé approximation approach (Zhang and Sohn, 1997b) and simulations (Dimonte and Ramaprabhu, 2010). Fig. 7 of Mansoor et al. (2020).

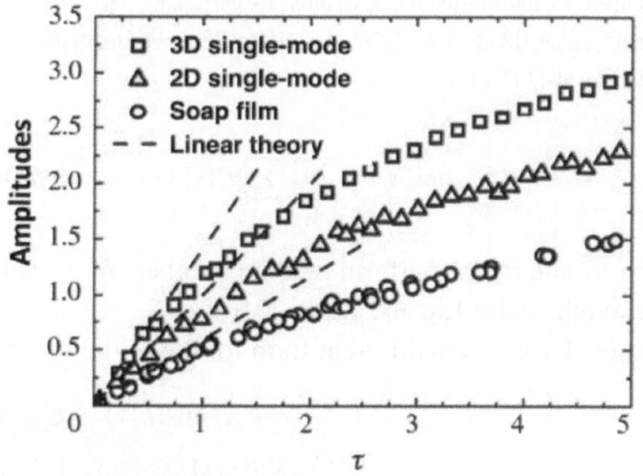

Fig. 9.13 Dimensionless growth of the perturbation amplitude in the symmetry plane is compared with the experimental results of the 2D single-mode (Niederhaus and Jacobs, 2003) and 3D single-mode (Chapman and Jacobs, 2006). Fig. 5a of Luo et al. (2013).

are qualitatively similar among the eight cases. However, if the initial amplitude is large, the vortices appear earlier, due to greater baroclinic vorticity production (Liang et al., 2021b). Additionally, the phases of the constituent modes greatly affect both the initial interface shape and the later interface evolution.

For many numerical studies, the multi-mode surface interface is initialized in Fourier space as a power spectrum of the form in Eq. 9.2. The random interface perturbation, $\hat{a}(x, y)$, can be obtained from $P(k)$ as follows:

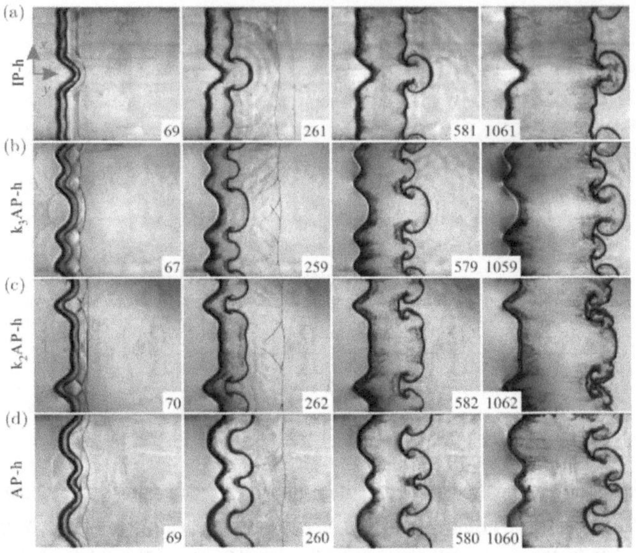

Fig. 9.14 Experimental Schlieren images of multi-mode interface evolution for systems with large initial amplitude. The extended soap-film technique generates a periodic multi-mode interface with a controllable initial shape separating SF_6 from air. TS, transmitted shock; w, interface perturbation width; LS, large spike; LB, large bubble; SS, small spike; SB, small bubble; UP, upstream point of interface; DP, downstream point of interface. Numbers denote time in μs. Fig. 3 of Liang et al. (2021b).

$$\hat{a}(x, y) = \sum_{k_x} \sum_{k_y} P(k) \cos(k_x x + \pi_x) \sin(k_y y + \pi_y), \tag{9.3}$$

where π_x and π_y are uniform random numbers (e.g., Thornber and Zhou, 2012; Olson and Greenough, 2014; Liu and Xiao, 2016).

Others have used a different form for the initial random perturbation

$$\begin{aligned} \hat{a}(x, y) = {} & C_{g1} \cos(k_x x) \cos(k_y y) + C_{g2} \cos(k_x x) \sin(k_y y) \\ & + C_{g3} \sin(k_x x) \cos(k_y y) + C_{g4} \sin(k_x x) \sin(k_y y) \end{aligned} \tag{9.4}$$

where C_{gi}, $i = 1, 2, 3, 4$ are set to Gaussian random variables having the same standard deviation. (e.g., Dimonte et al., 2007; Youngs, 2013; Thornber et al., 2017). For the periodic case it is important that the random perturbation is statistically homogeneous in both the x and y directions (for the perturbations quoted above this is true).

It is interesting to consider the differences between ICs characterized by spectral ranges that scale as $m = -1, -2$, or -3 (Groom and Thornber, 2020), which correspond to the self similar cases with $\theta = 2/(m + 5)$, Eq. 4.19. At the time of saturation of the largest initial wavelength, the smallest wavelengths at the end of the $m = -3$ simulation (Fig. 9.16) are still in a relatively early stage of nonlinear development, since the interface is still mostly connected. However, the smallest wavelengths have had more time to become nonlinear in the $m = -1$ simulation and there is correspondingly more fine-scale structure and multiple regions in the flow in which blobs of fluid have separated from the main interface.

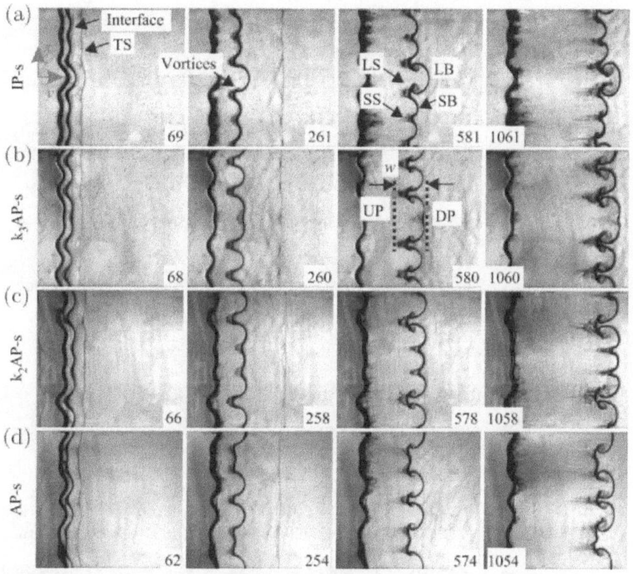

Fig. 9.15 Experimental Schlieren images of multi-mode interface evolution for systems with small initial amplitude. The extended soap-film technique generates a periodic multi-mode interface with a controllable initial shape separating SF_6 from air. TS, transmitted shock; w, interface perturbation width; LS, large spike; LB, large bubble; SS, small spike; SB, small bubble; UP, upstream point of interface; DP, downstream point of interface. Numbers denote time in μs. Fig. 2 of Liang et al. (2021b).

Figure 9.17 shows the evolution of the integral mixing width for three multi-mode simulations. Specifically, these runs correspond to three distinctive ICs, in which (k_x, k_y) in Eq. 9.3 take the values of (4,8), (16,20), and (28,32), respectively. While the compression rates for the three cases are almost identical, the late-time multi-mode growth rates clearly increase with the scale of the initial perturbations (Liu and Xiao, 2016).

Large-scale ICs are preserved in the velocity spectrum, but not in the concentration and concentration flux spectra (Soulard et al., 2018). Moreover, large scales retain their initial anisotropy so that the turbulent mixing layer only partially returns to isotropy.

9.3 Nonuniform initial flow fields

9.3.1 RTI

We next briefly consider the situation in which the initial flow field is non-uniform. Using DNS of the Boussinesq system in a generic 3D case, this illustrative problem can be investigated by assuming the initial unstable vertical temperature distribution:

$$T_0(z) = -\frac{\Delta T}{2}\mathrm{sgn}(z)\left(\frac{|z|}{L}\right)^{\tilde{c}}. \tag{9.5}$$

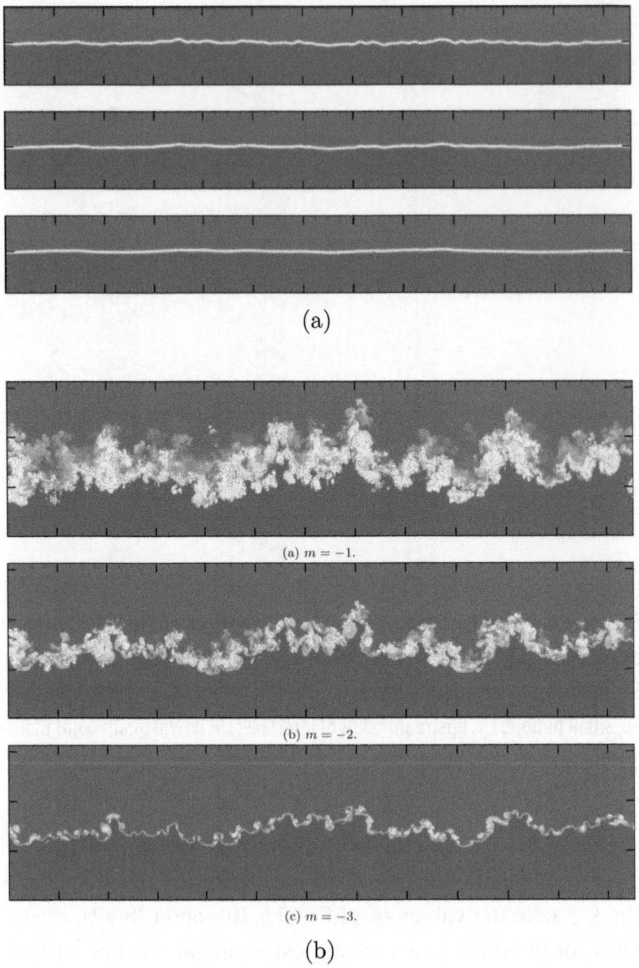

Fig. 9.16 Initial conditions (a) and late-time (b) of volume faction for three broadband cases, with initial conditions defined by Eq. 9.2, with $m = -1, -2$, and -3. Figs. 1 and 5, Groom and Thornber (2020).

Here L is a characteristic length scale, $-L \leq z \leq L$, and the constant driving force is proportional to $g\mathcal{A}$, where ΔT is the temperature jump between the two layers, β_B is the thermal expansion coefficient, and $\mathcal{A} = \beta_B \Delta T/2$ (Biferale et al., 2018). The exponent of the singularity lies in the interval $-1 < \tilde{c} < 1$. The value $\tilde{c} = 0$ recovers the standard RT configuration, the upper limit corresponds to a smooth profile and the lower limit ensures that the potential energy density of miscible fluids does not diverge near the interface (Fig. 9.18).

A dimensional consideration of Eqs. 2.50a and 9.5 results in

$$u(t)^2/h(t) \simeq \beta_B g \Delta T \Big(h(t)/L\Big)^{\tilde{c}}, \tag{9.6}$$

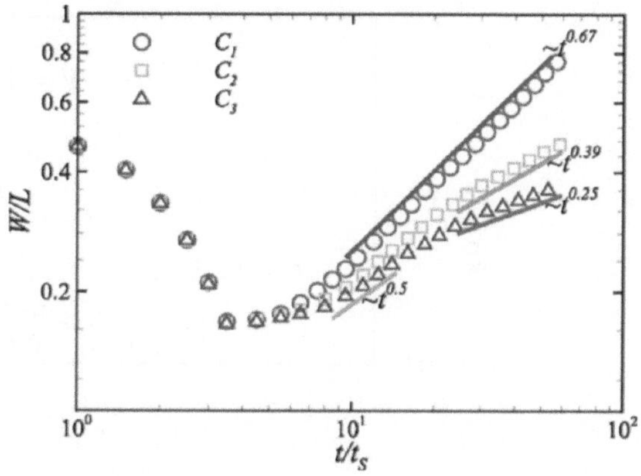

Fig. 9.17 Mixing width versus time for three simulation cases denoted as C_1, C_2, and C_3. The $(k_x, k_y)(L/2\pi)$ of Eq. 9.3 takes the values of (4,8), (16,20), and (28,32). Fig. 5 of Liu and Xiao (2016).

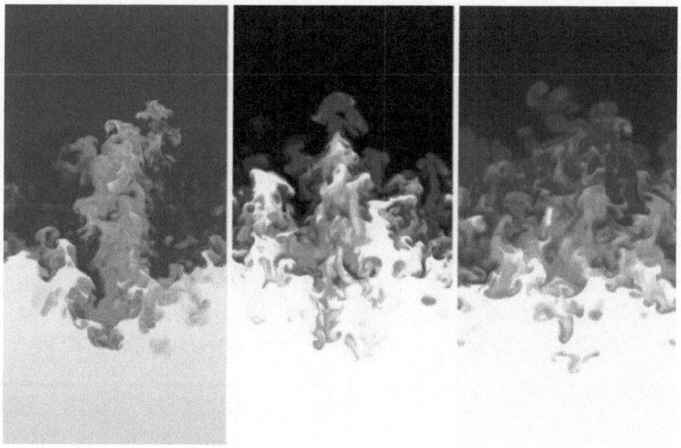

Fig. 9.18 Snapshots of the vertical section of the temperature field T for three simulations of RT turbulence with power-law initial condition with $\tilde{c} = -.25$ (left), $\tilde{c} = 0$ (center), and $\tilde{c} = 0.25$ (right) at three different times corresponding to the same mixing length $h(t)/L_z \simeq 0.4$. Fig. 1 of Biferale et al. (2018).

and

$$h(t) \simeq L\left(\frac{t}{t_*}\right)^{\tilde{\gamma}}, \tag{9.7}$$

where $\tilde{\gamma} = 2/(1-\tilde{c})$ and $t_* = \sqrt{L/(\beta_B g \Delta T)}$.

To force the late-time scaling of the mixing layer into its familiar quadratic form, the redefined Atwood number, $\mathcal{A}_{\tilde{c}}$, is given by $\mathcal{A}_{\tilde{c}} = (\beta_B \Delta T)^{1/(1-\tilde{c})}(gt^2/L)^{\tilde{c}/(1-\tilde{c})}$:

$$h(t) = \alpha_{\tilde{c}} \mathcal{A}_{\tilde{c}} g t^2. \tag{9.8}$$

Meanwhile, $\alpha_{\tilde{c}}$ is now the generalized α coefficient.

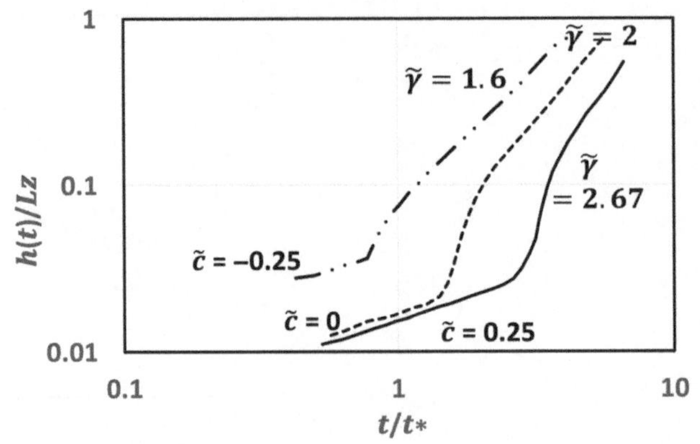

Fig. 9.19 Temporal evolution of the mixing layer. From left to right: $\tilde{c} = -0.25$, $\tilde{c} = 0$, and $\tilde{c} = 0.25$. The three lines represent the power law predicted by the formula with $\tilde{\gamma} = 1/(1 - \tilde{c})$. Fig. 2 of Biferale et al. (2018).

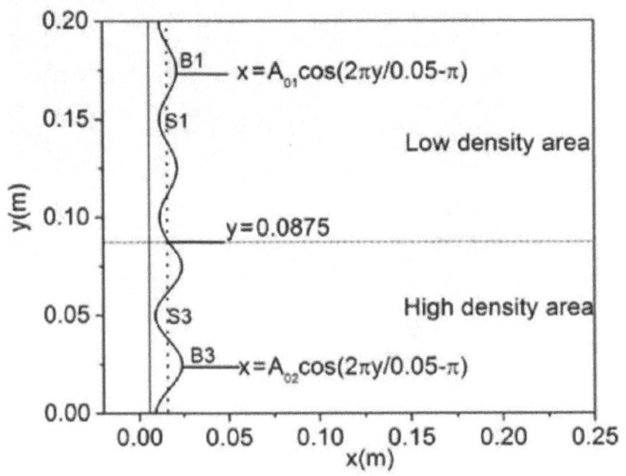

Fig. 9.20 Initial structure diagram of a non-uniform RMI flow. Fig. 1 of Xiao et al. (2016).

Figure 9.19 shows that the evolution of $h(t)$ provides insight into situations where the mixing proceeds over a nonuniform background, for example, in a thermally stratified atmosphere. The mixing layer develops under the driving of the time-dependent Atwood number, which in turn is dependent on the exponent of the singularity. As a result, the long-time growth is still self-similar, but no longer proportional to t^2, and the power-law exponents computed for the three different values are in reasonable agreement with Eq. 9.7 or Eq. 9.8.

9.3.2 RMI

As one might expect, the initial flows of an RMI application are often non-uniform. This has been investigated by several experimental and numerical studies (Bai et al., 2010b, 2012b; Liu et al., 2012b). Figure 9.20 displays the structure diagram for studying the initial

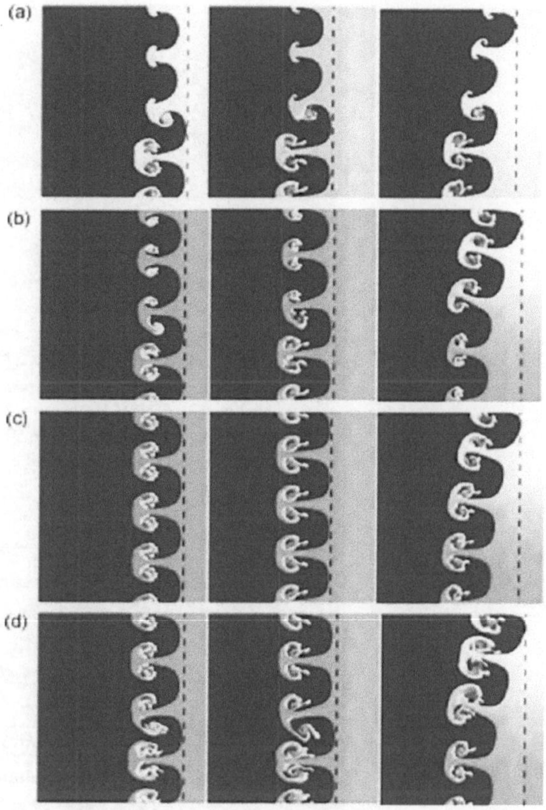

Fig. 9.21 Density contour images of the numerical simulation result at $t = 1.8$ ms under different groups of initial amplitudes. (Left) low density uniform flows; (middle) high-density uniform flows; (right) nonuniform Gaussian function flows. Fig. 5 of Xiao et al. (2016).

RMI perturbations. Numerical simulations used multi-mode cosine interfaces with various amplitudes, under both uniform and nonuniform flows (Xiao et al., 2016).

Visualizations such as the one in Fig. 9.21 could shed additional new light. The left, middle, and right column images correspond with simulated results of the low-density uniform flow, high-density uniform flow, and nonuniform flow with Gaussian function distribution, respectively. Each row corresponds to one group of initial amplitudes (mm): (a) [2.5, 7.5], (b) [5.0, 7.5], (c) [7.5, 7.5], and (d) [7.5, 10.0] (Xiao et al., 2016).

9.4 Realistic initial interface profiles

Unfortunately, only three existing methods are capable of providing precise initial interface perturbations: dedicated high energy density experiments with a solid-state target, numerical simulations, and magnetic fluids. The discussion of the first case will be deferred to Chapter 18. For the second case, it is a relatively trivial task once the code is written.

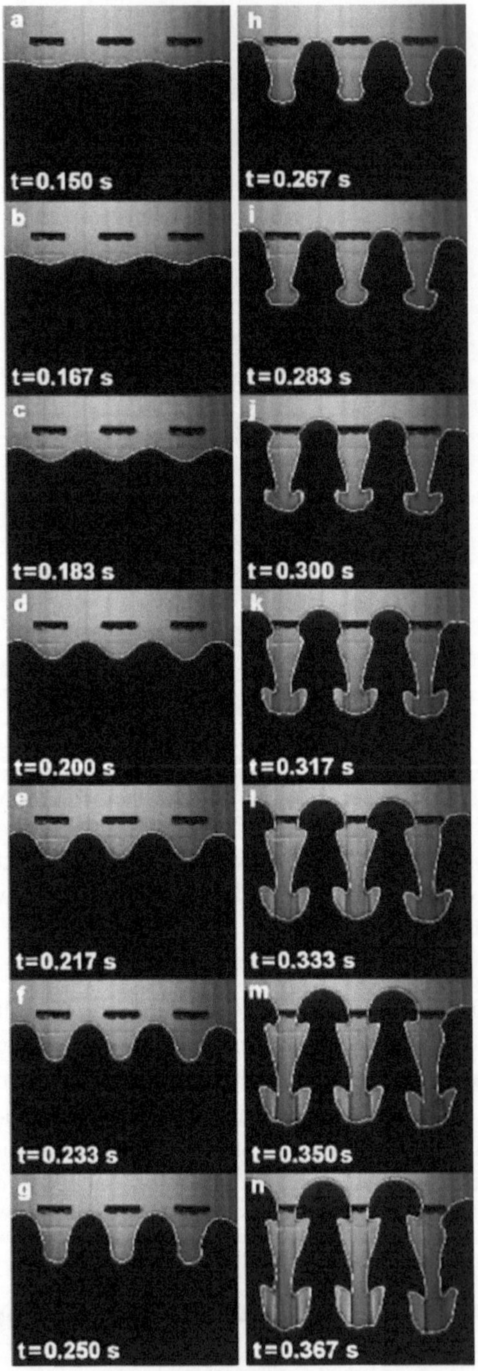

Fig. 9.22 Time evolution of RTI for an initial deformation of 12.5 mm wavelength. Fluid densities are 1398 and 773 $\mathrm{kg/m^3}$, respectively, and surface tension is 7.7 $\mathrm{mN/m}$. The dark horizontal segments are magnetically permeable wires placed outside the cell, forcing the initial single-mode interface deformation. Fig. 1 of Renoult et al. (2013).

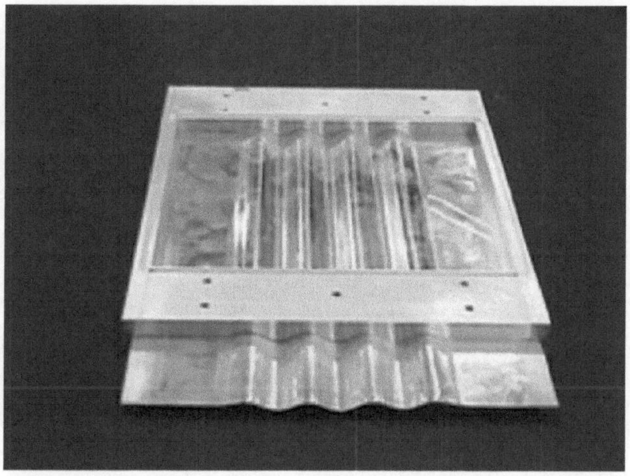

Fig. 9.23 Sinusoidally shaped copper plate and support frame. Fig. 1 of Puranik et al. (2004).

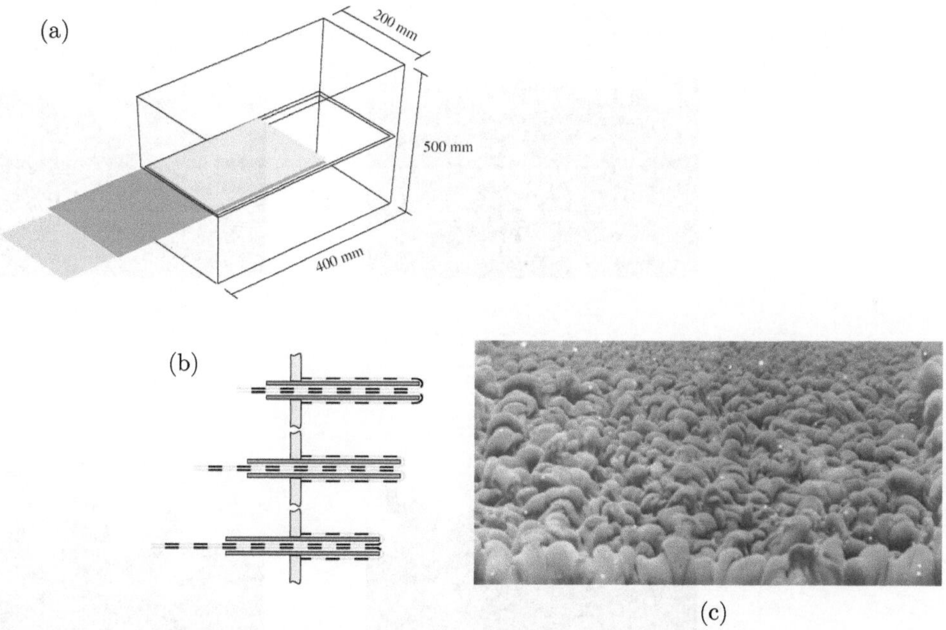

Fig. 9.24 (a) Sketch of experimental apparatus. The hollow stainless-steel barrier is shown as dark gray and the nylon fabric as light gray. The experiments were carried out in a Perspex tank 500 mm deep, 400 mm long, and 200 mm wide. The tank has a removable aluminum sheet 1.5 mm in thickness separating heavy and light fluids (Linden and Redondo, 1991). Figure 1 of Dalziel et al. (1999). (b) Illustration of process by which fabric (dashed) is removed through the gap between the plates (Andrews and Dalziel, 2010). (c) Early 3D development of RTI as the composite barrier (white band near the top of the picture) is withdrawn. Photograph courtesy of Dr. A.G.W. Lawrie, University of Bristol, UK.

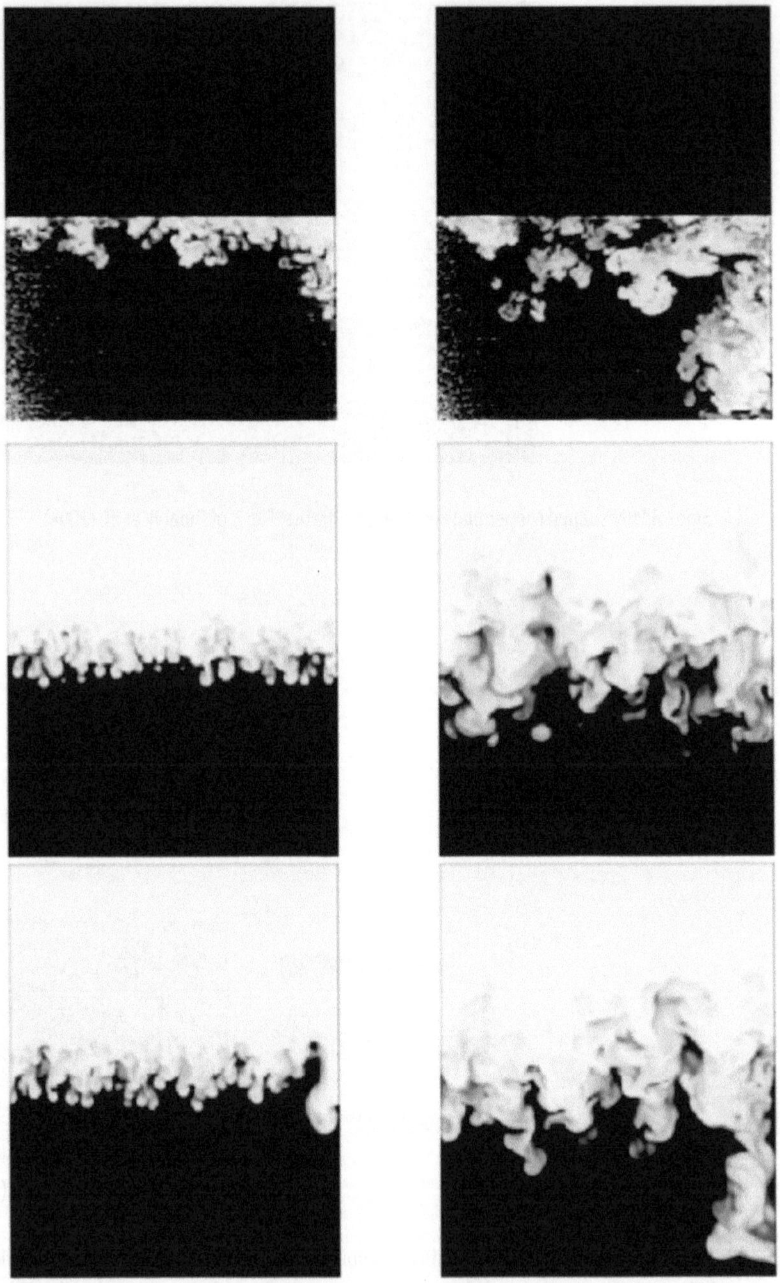

Fig. 9.25 Comparison between a typical experiment (top) and simulations using idealized initial conditions (middle) and initial conditions measured from experiments (bottom). The flows are shown for (a) $t = 5$ s and (b) $t = 10$ s. Fig. 6 of Dalziel et al. (1999).

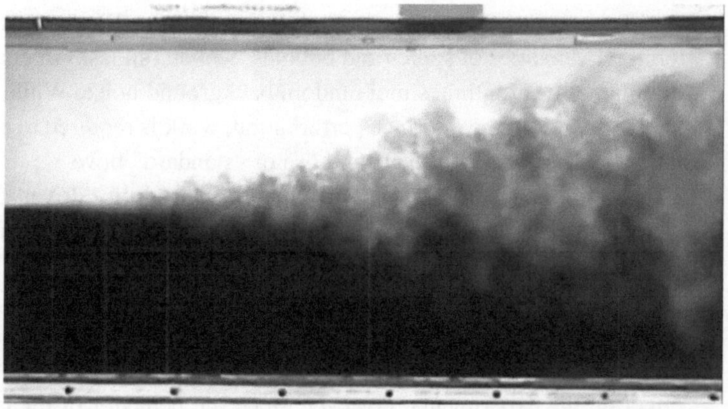

Fig. 9.26 The Texas A&M University water tunnel RT experiment. The experimental configuration was described by Snider and Andrews (1994), Wilson et al. (1999), and Andrews and Dalziel (2010). Photograph courtesy of Dr. A. Banerjee, Lehigh University.

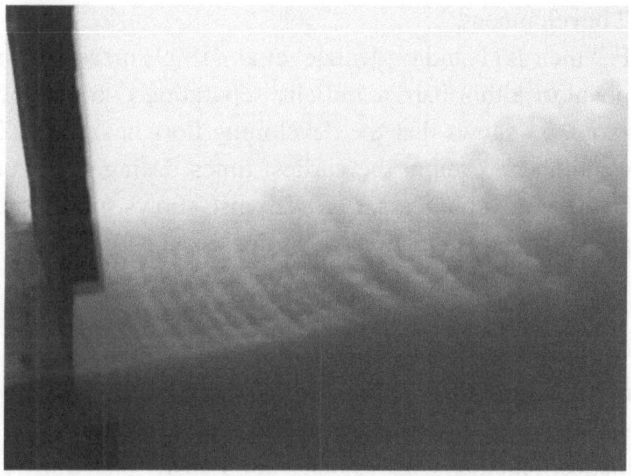

Fig. 9.27 Photograph of the initial development of a Rayleigh–Taylor mixing layer in the water channel, where a small amount of milk has been added to the bottom stream for visualization purposes. The flow is from left to right and corresponds to a dimensionless time interval [0,0.8] in the experiment. Fig. 6 of Mueschke and Schilling (2009a), with the permission of AIP Publishing.

In this subsection, we will first discuss the use of magnetic fields to precisely tune the interfacial perturbations. The utilization of magnetic levitation allows researchers to hold a heavy paramagnetic mixture on top of a light fluid (Adkins et al., 2017; Carlès et al., 2006; Huang et al., 2007; Pacitto et al., 2000; Renoult et al., 2013; Tsiklashvili et al., 2012; White et al., 2010). Switching off the stabilizing magnetic field initiates the RTI.

The unique properties of magnetic fluids have been exploited to create arbitrarily chosen and well-controlled initial perturbations. Indeed, the late time flow behavior is similar to that obtained in the standard manner (see Fig. 9.22). In other examples, White et al.

(2010) examined the nonlinear growth phase of the 2D single-mode RTI and obtained the terminal velocities of spikes and bubbles, whilst Tsiklashvili et al. (2012) investigated RTI originating from infinitesimal random background noise. While magnetic fluids allow one to precisely craft the interface perturbation, work is required to elucidate how the magnetic fluid properties may differ from those of "standard" flows.

The importance of ICs on hydrodynamic instability driven flows has been manifested in several ways. The self-similar asymptotic values of the concentration and velocity correlations depend on the properties of the large scales. Figure 9.23 shows a picture of a thin, sinusoidal plate that initially separates the driven gas (CO_2) from the test gas (air) in an RMI shock tube experiment. According to Puranik et al. (2004), manufacturing limits confine the plate shaping to the central 70% of the plate width. This has no adverse effect on the experiments, however, since the behavior of the central (sinusoidal) section is unaffected by the neighboring fluid. After the plate is withdrawn from the shock tube, a shock wave accelerates the interface into the test section where it is imaged. However, the retraction of the plate drags a small amount of fluid and generates a wake behind it. Furthermore, Puranik et al. found that as the plate begins to retract, the two gases come into contact between the shock tube window and the edge of the plate. This leads to immediate RTI development.

For their RTI studies, Dalziel et al. (1999) measured the velocity field generated by the removal of a thin barrier initially separating two fluids of different densities (Fig. 9.24). Figure 9.25 shows that the developing flow has a clear 3D character growing from high wavenumbers, even at the earliest times. Using the 2D components of the experimental ICs at low wavenumbers, Fig. 9.25 also shows substantially improved agreement between experiment and simulations using idealized ICs.

Using the Texas A&M University water tunnel (Fig. 9.26), Mueschke et al. (2006) quantified their initial streamwise and spanwise velocity and density perturbations. These developed into a small Atwood-number, incompressible, miscible RTI mixing layer, measured using high-resolution thermocouples and PLIF. It was shown that the measured ICs describe an anisotropic state (Fig. 9.27). The evolution of various velocity and density statistics, together with velocity and density spectra, were measured. They were used to investigate the early-time evolution and onset of strongly nonlinear, transitional dynamics within the mixing layer. The early-time evolution of the density and vertical velocity spectra indicated that velocity fluctuations are the dominant driving mechanism. A DNS was subsequently carried out (Mueschke and Schilling, 2009a,b; Schilling and Mueschke, 2010) using the initial conditions, geometry, and physical parameters of this experiment. Large-scale quantities such as the bubble front penetration and the mixing layer growth parameter and higher-order statistics such as velocity variances, the molecular mixing parameter, vertical velocity spectra, and density variance spectra were all in favorable agreement with the experiment. The DNS slightly underestimated the growth of the bubble front, but correctly predicted the RTI growth factor $\alpha = 0.07$ at the latest time.

In a related development, Ramaprabhu and Andrews (2004) found that numerical simulations of RTI mixing agree with experimental self-similar growth rates when the computations are initialized with experimental data.

10 Flow properties

10.1 Surface tension

The study of surface tension in fluids dates back more than two centuries to the celebrated works by P.S. Laplace (Laplace, 1805) and T. Young (Young, 1805). In our present context, surface tension modifies the evolution of the RT and RM fluid instabilities by acting to suppress their growth. It also influences the energy spectrum and scaling behavior in RT and RM turbulence, reducing the effective mixing rate. The RTI flow asymmetry at high $\mathcal{A}$, which sharply increases the spike growth factor α_s (see experimental LEM data in Fig. 4.12), can be attributed to the meniscus surface tension (Dimonte and Schneider, 2000).

It is fascinating to note that surface tension is intimately connected to a curious phenomenon that has been observed in wine glasses for centuries. Anyone who has enjoyed a glass of wine has undoubtedly noticed the regular pattern of liquid beads that fall along the inside of the glass, or "tears of wine." In the right setting, one can observe a thin layer of water-ethanol mixture that travels up inclined surfaces against gravity and proceeds to fall in beads resembling tears (Fig. 10.1). The phenomenon of wine tears was first analyzed qualitatively by James Thomson (Thomson, 1855), the older brother of Lord Kelvin. His step-by-step experiments to reveal the nature of its physical process can also be found in one of the books published by famous physicist James Clerk Maxwell (Maxwell, 1891). Contemporary studies of this topic include, among others, those by Fournier and Cazabat (1992), Vuilleumier et al. (1995), Venerus and Simavilla (2015), Nikolov et al. (2018), and Dukler et al. (2020).

The pressure difference on the interface is balanced by the interfacial force due to surface tension,

$$p_2 - p_1 = \sigma \varkappa, \tag{10.1}$$

where $\varkappa$ and σ denote the curvature of the interface and the surface tension, respectively (Laplace, 1805; Young, 1805),

The starting point of our investigation is the second-order ordinary differential equation (ODE) of Lord Rayleigh, but now with surface tension effects included (Chandrasekhar, 1961; Mikaelian, 1990b)

$$\frac{d}{dz}\left\{\rho \frac{du_3}{dz}\right\} = k^2 u_3 \left\{\rho - \frac{g}{\gamma^2}\frac{d\rho}{dz}\right\} + \frac{k^4}{\gamma^2}\sum_i \sigma_i \delta(z - z_i) u_3, \tag{10.2}$$

where u_3 is the perturbed velocity of the fluid having an arbitrary density profile $\rho(z)$, and σ_i is the surface tension at interface i located at z_i.

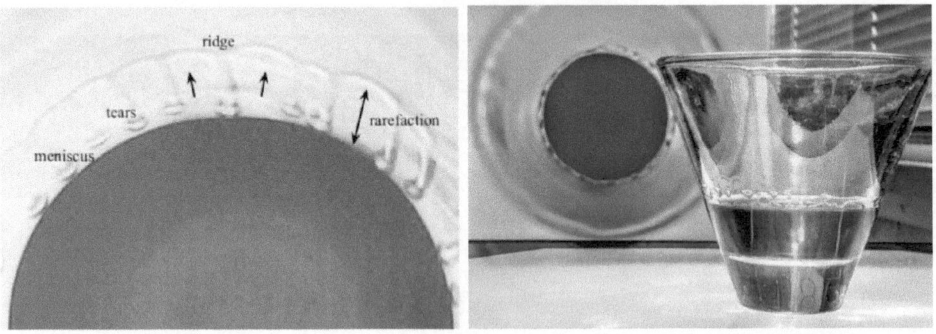

(a) (b)

Fig. 10.1 The "tears of wine" has been attributed to the physics of surface tension since 1855. Top view (left) and side view and projection (right) of a stemless martini glass with inclination angle of 65 degrees, using 18% ABV port wine. Swirling the wine around the glass creates a front that forms out of the meniscus. The draining film advances up the glass and destabilizes into wine tears. This figure is reproduced from Fig. 11 of a recent experiment published by Dukler et al. (2020).

The immiscible, viscous, incompressible two-fluid system may be formulated as a one-fluid system with variations of density and viscosity only in the neighborhood of the interface. The governing forced Navier–Stokes equation is given by

$$\frac{\partial \rho \mathbf{u}}{\partial t} + \nabla \cdot (\rho \mathbf{u}\mathbf{u} + p\mathsf{I}) = \nabla \cdot \boldsymbol{\tau} + \mathcal{F}, \tag{10.3}$$

where $\mathcal{F} = \rho \mathbf{g} + \sigma \kappa \delta(d)\mathbf{n}$, $\mathbf{g}$ is the gravitational field, $\mathbf{n}$ is the outward unit normal vector at the free surface, and $\delta(d)$ is the Dirac delta function, whose argument is the distance normal to the surface, d.

The derivation of the potential-flow equations with surface tension is essentially similar to the case without surface tension. The primary effect of surface tension is to introduce a cutoff wave number k_c (or equivalently, a cutoff wavelength λ_c) beyond which perturbations are stable (e.g., Bellman and Pennington, 1954; Mikaelian, 1990b)

$$k_c = \sqrt{\frac{(\rho_2 - \rho_1)g}{\sigma}} = \frac{2\pi}{\lambda_c}. \tag{10.4}$$

The RTI linear growth rate is then given by (Chandrasekhar, 1961; Young and Ham, 2006)

$$\gamma = \sqrt{\mathcal{A}gk\left[1 - \left(\frac{k}{k_c}\right)^2\right]} = \sqrt{\mathcal{A}gk - \frac{k^3\sigma}{(\rho_2 + \rho_1)}}. \tag{10.5}$$

With numerical calculations, Daly (1969) found that the variation of the linear growth rate with the surface tension coefficient, and (for a fixed coefficient) with the wavenumber of the perturbation, is in good agreement with Chandrasekhar's (1961) analytic prediction. Later, Emmons et al. (1960) arrived at the same conclusion after carrying out a weakly nonlinear asymptotic analysis of the 2D RT instability. It should be noted that Nayfeh (1969) corrected an error in the analysis of Emmons et al., where they failed to account for the singular nature of the perturbation problem. Also, Jacobs and Catton (1988a) derived a pair of weakly nonlinear asymptotic solutions for 3D RTI.

Similar to Goncharov's analysis without surface tension, the 2D and axisymmetric RT bubble velocities approach their late time asymptotic values (Young and Ham, 2006),

$$u_b = \sqrt{\frac{2\mathcal{A}}{3(1+\mathcal{A})}\frac{g}{k} - \frac{k}{9}\frac{\sigma}{\rho_2}}, \tag{10.6}$$

and

$$u_b = \sqrt{\frac{2\mathcal{A}}{(1+\mathcal{A})}\frac{g}{k} - \frac{3k}{16}\frac{\sigma}{\rho_2}}, \tag{10.7}$$

respectively. In both expressions, the terminal bubble velocities are reduced by surface tension. The model results agree well with numerical simulations of a viscous, single-mode RT instability.

To capture the nonlinear saturation of the single-mode interface modulation growth, it is necessary and sufficient to compute the third-order nonlinearity. Guo et al. (2017) considered the weakly nonlinear RMI in incompressible fluids with surface tension. Analytically investigating the evolution of a perturbed interface through the third-order solution, the authors found that the surface tension has a stabilizing effect that depends on $\mathcal{A}$.

A limitation of the Layzer model is that it fails to yield a quantitatively correct prediction for the bubble curvature. An alternative source-flow model, which describes the bubble as a potential with a point source (Zufiria, 1988), may provide a more satisfying prediction for the bubble motion (Sohn, 2004b). This potential-flow model produces the same functional form for the terminal bubble velocity as Eq. 10.6, but with the constants replaced with functions of the curvature. Nevertheless, the model produces a similar bubble Froude number to the Layzer model (Sohn and Baek, 2017).

Figure 10.2 shows the ratio of simulated and modeled single-mode v_b versus h_b, for two different Atwood numbers. In both 2D and 3D flows with surface tension, the bubble velocity overshoots in the $\mathcal{A} = 0.1$ case, but uniformly approaches the analytical results for $\mathcal{A} = 0.99$.

The role of surface tension in the multi-mode RTI is of significant interest (e.g., Cherfils and Mikaelian, 1996; Chertkov et al., 2005; Sohn, 2009). Garnier et al. (2003) derived a weakly nonlinear model for studying the RTI in the presence of surface tension. This model addresses the case of an initial multi-mode perturbation and uses statistical analysis.

Fixing $\mathcal{A} = 0.3$, Young and Ham (2006) varied the surface tension to investigate its effect on RT mixing. Figure 10.3 shows the time evolution of the bubble amplitude h_b for four different values of σ. The solid lines are from simulation data and the dashed lines represent linear growth. Here, the dimensionless surface tension, σ, is given by $\sigma = \sigma_0/(g\rho_h L^2)$, where σ_0 is the dimensional surface tension coefficient, $L = 30$ cm is the container horizontal size, and gravity is $g = 980$ cm/s^2. The time is reported in units of $\sqrt{L/g}$. When the surface tension is small, one would expect the flow to be close to that of miscible RT mixing with finite density contrast. For $\sigma = 2 \times 10^{-6}$ the linear growth starts after an initial transient, and the nonlinearity becomes "important" at $t_{nl} \approx 2.7$, when the bubble height $h_{nl} \approx 0.10$. It is clear that surface tension reduces the effective RT mixing rate in numerical simulations.

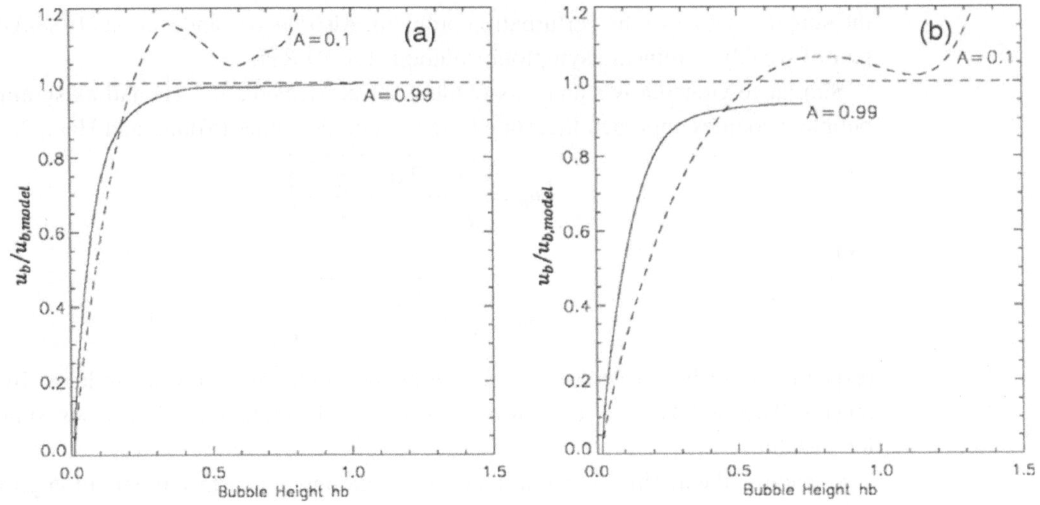

Fig. 10.2 Simulated bubble velocity for (a) 2D and (b) 3D single-mode RTI with surface tension. Equations 10.6–10.7 are used for normalization. Fig. 4 of Young and Ham (2006).

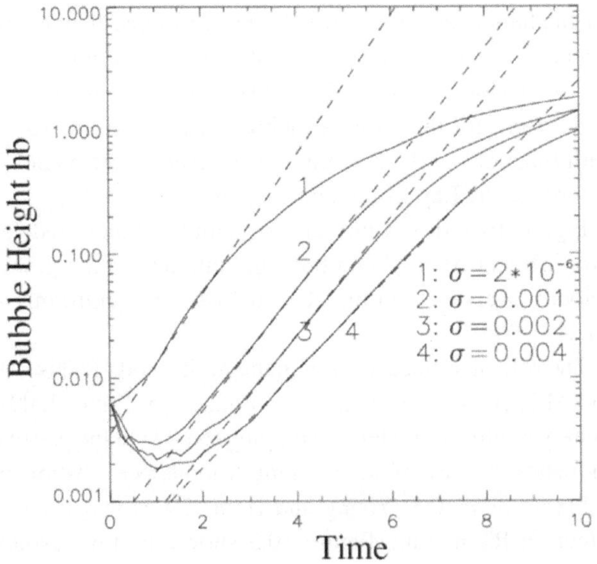

Fig. 10.3 The effect of surface tension on the simulated incompressible Rayleigh–Taylor mixing flow. Bubble height versus time on a log-linear plot, shown for four values of σ and $\mathcal{A} = 0.3$. Fig. 5 of Young and Ham (2006).

As σ increases, the random perturbations first decay and undergo a long transient, after which they begin to grow exponentially at a reduced growth rate (reduced due to the surface tension). Hence, for large surface tension, nonlinearity becomes important later in time (for $\sigma = 1 \times 10^{-3}, t_{nl} \sim 5.0, h_{nl} \sim 0.15$) – see Fig. 10.4.

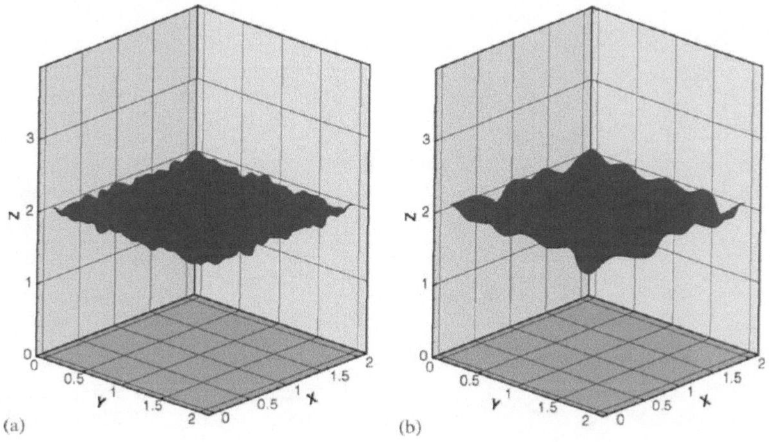

Fig. 10.4 Simulated RTI fluid interface at the onset of nonlinear evolution, $t = t_{nl}$, for (a) $\mathcal{A} = 0.3$ and $\sigma = 2 \times 10^{-6}$, (b) $\sigma = 1 \times 10^{-3}$. Fig. 6 of Young and Ham (2006).

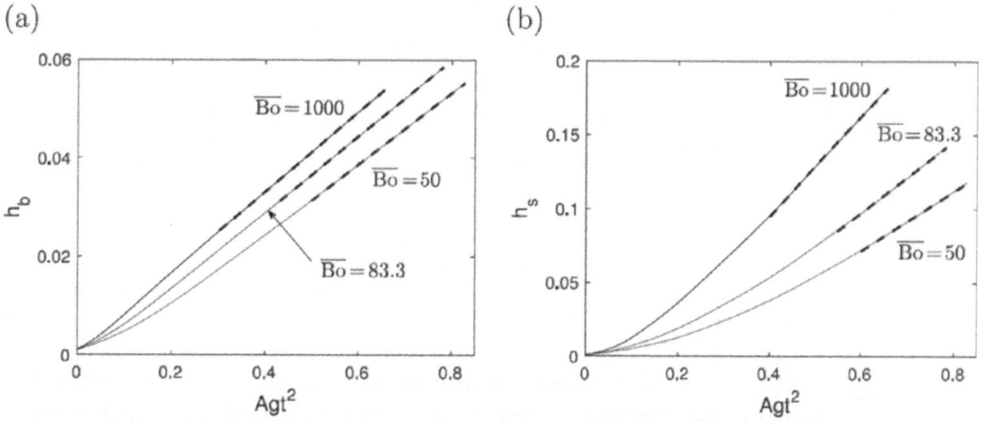

Fig. 10.5 Growth of the simulated multi-mode RTI mixing front vs. Agt^2 for three Bond numbers: 50, 83.3, and 1000. Panels show (a) bubble height and (b) spike height. Linear fits are shown by dashed lines. Fig. 11 of Shin et al. (2021). Reproduced with permission from Elsevier.

Shin et al. (2022) investigated the multi-mode RTI and focused on the case $\mathcal{A} = 1$, in which the numerical method of the vortex sheet model is most advanced. The authors show the competition and merging of fingers, as well as pinching and elongation. Moreover, the bubble front grows with the scaling law even in the presence of surface tension, while the spike front follows the scaling law weakly (Fig. 10.5). Finally, the bubble and spike growth rates decrease with surface tension, but the decrease for spikes is much larger than the decrease for bubbles (Figs. 10.6 and 10.7). Here, growth coefficients of bubbles and spikes are plotted against the inverse Bond number. The Bond number, also known as the Eötvös number, is a nondimensional parameter characterizing the relative importance of

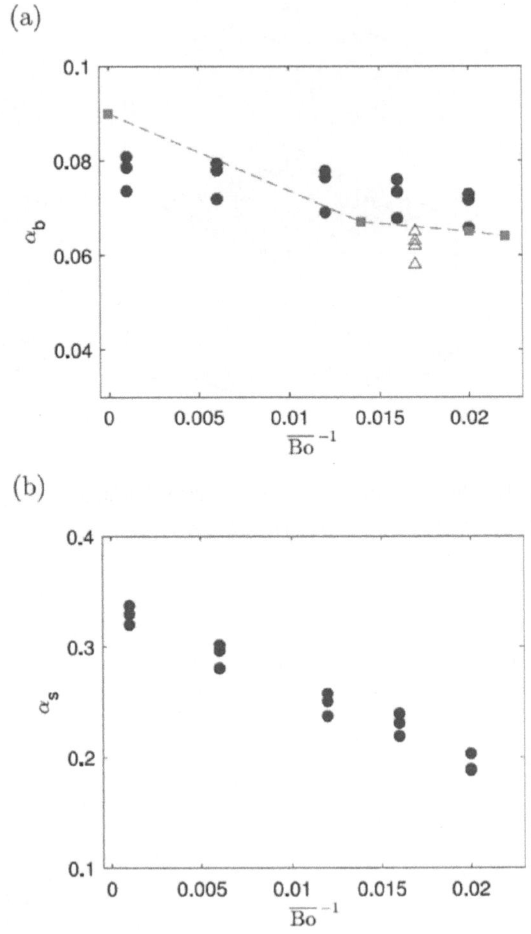

Fig. 10.6 Growth coefficients of (a) bubbles and (b) spikes versus the inverse Bond number for a simulated multi-mode RTI. The square represents the result of the front-tracking simulation (George et al., 2006), and the triangle represents the experimental data in Read (1984). Fig. 12 of Shin et al. (2021). Reproduced with permission from Elsevier.

gravitational forces compared to surface tension forces (Clift et al., 1978; De Gennes et al., 2013; Sohn and Baek, 2017)

$$Bo = Eo = \frac{(\Delta\rho)gL^2}{\sigma}. \tag{10.8}$$

For the multi-mode cases, we introduced a rescaled Bond number, $\overline{Bo} = Bo/N_b^2$, where N_b is the average number of initial bubbles.

By applying the boundary integral method, Matsuoka (2009) investigated the motion of a planar interface in incompressible RMI and RTI with surface tension. The surface tension term is expressed by the derivative of the interface curvature. Therefore, this term contains the highest spatial derivative in the governing equations. Together with the convective term, it governs the interfacial motion at large wavenumbers, which, in turn, causes vortex motion. When the surface tension term is sufficiently large, it may suppress vortex

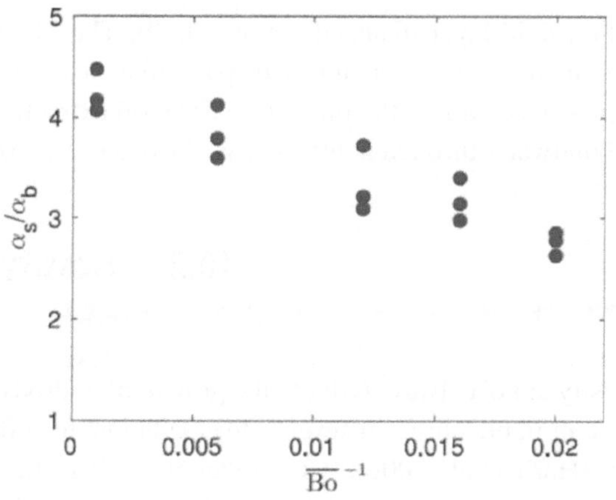

Fig. 10.7 Ratio of the bubble and spike growth rates for a simulated multi-mode RT. Fig. 13 of Shin et al. (2021). Reproduced with permission from Elsevier.

motions such as the rollup of the interface (Matsuoka et al., 2003; Matsuoka and Nishihara, 2006c, 2020).

At high Reynolds numbers, relatively high surface tension is found to have a noteworthy impact on the RT evolution. Hamzehloo et al. (2021) found that the shape of the bubble region and the stem of the spike are similar to those with the same Atwood number but lower Reynolds numbers. This is because a high surface tension force countermands the shear forces and consequently prohibits KHI formation in the bubble and the stem of the spike. Using an improved phase-field LB method, Liang et al. (2019) performed high-resolution DNS of multi-mode immiscible RTI with a low Atwood number ($\mathcal{A} = 0.1$). They focused on the effects of the Reynolds number, using initial conditions of varying perturbation wavelength and amplitude. Huang et al. (2021), Liang et al. (2021a), Liu et al. (2022), Ma et al. (2022), and Tavares et al. (2021b) have explored some other aspects of immiscible RTI with the LB method.

Tang et al. (2021b) have presented results of nonlinear and compressible numerical simulations of the RMI with surface tension in the linear, nonlinear and transitional regimes. These simulation results agree with the Mikaelian (1990d) model (with an appropriate modification based on the theory of Vandenboomgaerde et al. (1998) to accommodate Atwood numbers of either sign) in the linear regime and indicate that the model of Sohn (2009) correctly predicts the nonlinear evolution.

George et al. (2006) reported on a set of 3D RT chaotic mixing simulations based on an improved tracking algorithm in the front tracking code FronTier. It was clear that accurate tracking to control numerical mass diffusion and accurate modeling of physical scale-breaking phenomena (such as surface tension) were the critical steps to obtain simulations in agreement with the experiments of Read (1984) and Smeeton and Youngs (1987).

There is also considerable interest in studying the motion of an underlying incompressible fluid in a porous medium under the effects of gravity and surface tension. The main aspects of this subject, known in the literature as the Muskat problem, are the surface tension stabilization of the RTI and the possible formation of drops or fingering phenomena in a thin fluid layer (Gancedo et al., 2020). The Muskat problem has received sustained attention in a variety of fields. In particular, it is used to model flows in oil reservoirs (water is injected into the oil well to drive oil extraction) and in hydrology to model flows of groundwater through aquifers (e.g., Jacobs et al., 2021).

10.2 Viscosity

Viscosity almost always reduces the growth of hydrodynamic instabilities and its effect has important implications in several diverse fields: ferrofluids (Rosensweig et al., 2008), tectonics (Harig et al., 2008), ICF (Weber et al., 2014b), exploding foils (Willey et al., 2016), aerobreakup (Theofanous, 2011), and astrophysics (Blinova et al., 2016). The viscosity's influence is complex; a fourth order differential equation must be solved in order to find the growth rate in viscous incompressible fluids (Chandrasekhar, 1961).

Harrison (1908) was the first to consider the linear RTI of viscous fluids. Bellman and Pennington (1954) developed a linear theory and found that viscosity decreases the growth rate at low wavenumber and causes damping of oscillatory solutions at high wavenumbers. A useful approximation is the upper bound given by,

$$\gamma = \sqrt{k\mathcal{A}g + \tilde{\nu}^2 k^4} - \tilde{\nu}k^2, \tag{10.9}$$

where

$$\tilde{\nu} = \frac{\mu_1 + \mu_2}{\rho_1 + \rho_2}. \tag{10.10}$$

The formula implies that γ obtains a maximum value for

$$\lambda_{\max} = 4\pi \left(\frac{\tilde{\nu}^2}{\mathcal{A}g} \right)^{1/3}. \tag{10.11}$$

The first approximate "dispersion relation" was proposed by Hide (1955a)

$$\gamma^2 + 2\tilde{\nu}k^2\gamma - gk\mathcal{A} = 0, \tag{10.12}$$

the largest root of which, γ^+, is already given in Eq. 10.9 (Mikaelian, 1993a). Hide's derivation, based on an original suggestion by Chandrasekhar, was followed up and discussed further by Reid (1961), Menikoff et al. (1977), and Mikaelian (2013). More recent simulation and modeling efforts in this area include those from Tryggvason and Unverdi (1990), Cao et al. (2011), Statsenko et al. (2013), Movahed and Johnsen (2013), Zhang (2013), Doludenko and Fortova (2015), Chen et al. (2016), and Scase and Sengupta (2021).

Carlès and Popinet (2001, 2002) performed an asymptotic analysis of the Navier–Stokes equations using singular perturbation techniques and found that the weakly nonlinear law of interface deformation for the RMI can be written as

$$\frac{a}{a_0} = (1 + \mathcal{A}\Delta Vkt)\cos(kx) - \frac{16}{3\sqrt{\pi}}\frac{\sqrt{\rho_1\mu_1}\sqrt{\rho_2\mu_2}}{(\rho_1+\rho_2)(\sqrt{\rho_1\mu_1}+\sqrt{\rho_2\mu_2})}\mathcal{A}\Delta Vk^2t^{3/2}\cos(kx) - \frac{\sigma}{2(\rho_1+\rho_2)}k^3t^2\left(1+\frac{1}{3}\mathcal{A}\Delta Vkt\right)\cos(kx) - \frac{1}{2}a_0\mathcal{A}^3(\Delta V)^2k^3t^2\cos(2kx) \tag{10.13}$$

In above equation, each distinctive effect (viscosity, surface tension and nonlinearity) appears under the form of a specific term that attenuates the linear growth rate.

Restricting our attention to the viscosity term, Eq. 10.13 reduces to

$$\dot{a} = ka_0\mathcal{A}\Delta V\left[1 - \frac{8}{\sqrt{\pi}}k\sqrt{\nu_{CP}t}\right]\cos(kx), \tag{10.14}$$

where the density weighted kinematic viscosity is given by

$$\nu_{CP} = \frac{(\rho_1\mu_1)(\rho_2\mu_2)}{\left[(\rho_1+\rho_2)\left(\sqrt{\rho_1\mu_1}+\sqrt{\rho_2\mu_2}\right)\right]^2}. \tag{10.15}$$

This expression has been compared against the model of Mikaelian (1993a),

$$\frac{a}{a_0} = \left[1 + \mathcal{A}\Delta V\frac{1-\exp(-2k^2\tilde{\nu}t)}{2k\tilde{\nu}}\right]\cos(kx), \tag{10.16}$$

the expansion of which model for small times yields (Carlès and Popinet, 2002)

$$\frac{a}{a_0} = \left[1 + \mathcal{A}\Delta Vkt\right]\cos(kx) - \mathcal{A}\Delta Vk^3\tilde{\nu}t^2\cos(kx). \tag{10.17}$$

Using DNS results as a benchmark, Carlès and Popinet (2001, 2002) further argued that their model is superior to the Mikaelian model (Fig. 10.8). However, Mikaelian (2014b) commented on the Carlès and Popinet (2002) work. Since the fluid viscosities appear as a product in this model, it predicts that there are no viscous effects if one of the two fluids has zero viscosity. But, as pointed out by Mikaelian (2014b), this is not correct because the viscosity of the other fluid is sufficient to influence the flow, and the viscosities should appear as a sum rather than as a product.

Robey (2004) noted a distinctive feature of these two models regarding the RMI growth rate: the Mikaelian growth rate asymptotes to zero for $k \to \infty$, whereas the Carlès and Popinet method yields a finite wavenumber cutoff due to viscosity. Nevertheless, he concluded that both methods give similar magnitudes for the growth rate reduction. Walchli and Thornber (2017) revisited this issue and found that the discrepancies are by no means insignificant (Fig. 10.9(a)), but they do not differ by several orders of magnitude, as claimed by Carles and Popinet (2001).

Sohn (2009) studied solutions for RTI and RMI bubbles by generalizing the Layzer model to incorporate viscosity. The solutions are in good agreement with 2D numerical results (Sohn, 2009), but less satisfying than those from 3D calculations at $Re = 128$ (Walchli and Thornber, 2017). In particular, the simulated 3D behavior at $Re = 16$ agrees with the model at late times; however, results are inconsistent at early times (Fig. 10.9(b)), which may be attributed to a shortcoming in Sohn's linear model (Mikaelian, 2013).

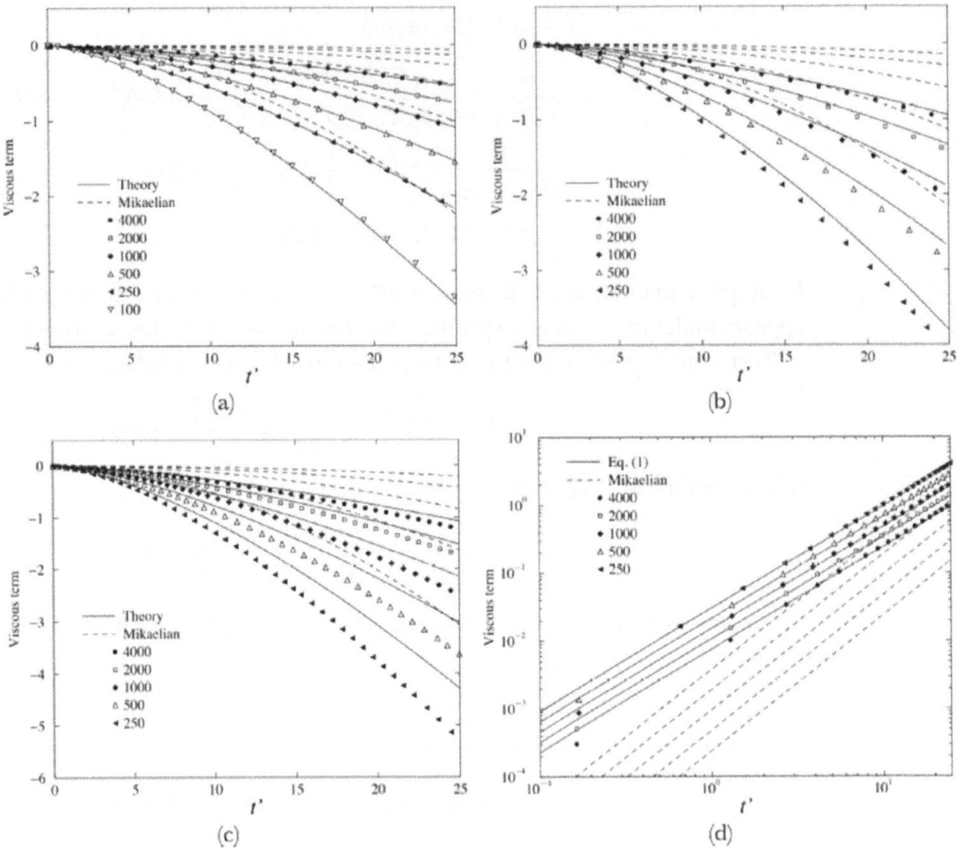

Fig. 10.8 Figure 3 of Carlés and Popinet (2002), with permission. (a)–(c) Viscous correction to the linear growth based on the authors' and Mikaelian's models, shown along with numerical simulations for several Reynolds numbers (from 100 to 4000, calculated for the heavier fluid) and several density ratios: 2, 4, and 8, respectively. (d) Repeat of panel (b) with a log–log plot. Carlès and Popinet (2002), with the permission of Elsevier.

As discussed previously, the vorticity near the bubble head provides a valuable method for predicting v_b. In this regard, the role of viscosity is directly relevant to the late-time behavior of 2D single-mode RTI at intermediate *Re* (Hu et al., 2019). Using statistical analysis and numerical simulations, Xie et al. (2017b) reported that the viscous RT mixing growth can be further retarded by superimposing an optimized suppression mode on its random initial perturbations.

Recall that circulation deposition is a dominant physical process in the early-time RMI evolution (e.g., Zabusky, 1999). Liu et al. (2020) found a hidden positive contribution of the viscous source to the circulation deposition in RMI, especially at high Mach number (up to 11% of total circulation rate).

Just to show how the problem can become extremely complex when we include viscosity, a very recent paper of Mikaelian (2019) focused on four special cases:

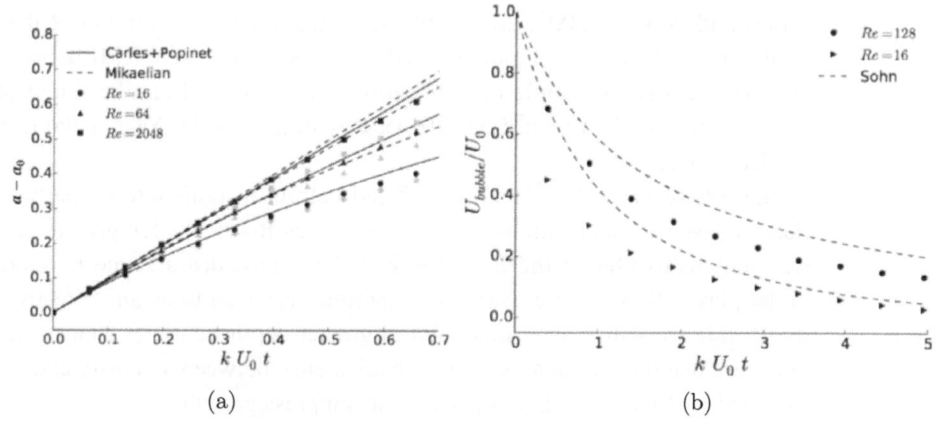

Fig. 10.9 (a) Comparison of simulated RMI layer amplitude (black) and bubble (gray) amplitude against the theoretical models of Mikaelian (1993a) and Carlès and Popinet (2001). (b) Bubble velocity comparison with the theoretical model of Sohn (2009). Figs. 8 and 7 of Walchli and Thornber (2017).

$$\text{Case A:} \quad \mathcal{A}_\mu = 1, \quad \mathcal{A} = 1 \quad \text{(single fluid)}, \tag{10.18a}$$
$$\text{Case B:} \quad \mathcal{A}_\mu = \mathcal{A}, \quad \mathcal{A} \text{ arbitrary}, \tag{10.18b}$$
$$\text{Case C:} \quad \mathcal{A}_\mu = 1, \quad \mathcal{A} \text{ arbitrary}, \tag{10.18c}$$
$$\text{Case D:} \quad \mathcal{A}_\mu = -1, \quad \mathcal{A} \text{ arbitrary}. \tag{10.18d}$$

Here, a "viscous Atwood number," owing to the use of the Boussinesq approximation, is defined by

$$\mathcal{A}_\mu = \frac{\mu_2 - \mu_1}{\mu_2 + \mu_1}. \tag{10.19}$$

Mikaelian (2019) offered detailed considerations from the perspectives of both eigenvalue (EV) and initial-value (IV) solutions. This work cleared up some earlier approximations and inaccuracies in Carles and Popinet (2001, 2002) and Sohn (2009).

Using a diffuse-interface model that combines the Cahn–Hilliard equation (which governs the volume fraction evolution of one fluid) with the Boussinesq Navier–Stokes equations, Zanella et al. (2020) investigated the effect of viscosity and perturbation wavelength on the mass transport at low Atwood numbers.

For many applications, surface tension and viscosity must be considered simultaneously. After linearization with respect to the perturbation amplitude the RT growth rate is (e.g., Sohn, 2009)

$$\gamma = \sqrt{k\mathcal{A}g + \tilde{\nu}^2 k^4 - \frac{k^3\sigma}{(\rho_2 + \rho_1)}} - \tilde{\nu}k^2. \tag{10.20}$$

Sohn (2009) obtained both the linear and nonlinear asymptotic solutions and carried out numerical studies for unstable interfaces with arbitrary surface tension and viscosity, both of which decreased the asymptotic RTI bubble velocity, v_b. For RMI, the model suggests that the decay of v_b depends on the relative importance of viscosity and surface tension.

Carlès and Popinet (2001, 2002) obtained a quantitative prediction of the effects of viscosity and surface tension on the weakly nonlinear impulsive RMI between two fluids of arbitrary density. A sample of publications in this regard can be found in Reid (1961), Gupta et al. (2012), Li and Luo (2014), Roy et al. (2014), Yakovenko (2014), and Yang and Cao (2021).

The role of viscosity in RTI has offered helpful insights into properties of solid materials in practical applications. Higher viscosities lower the RT growth rate, with shorter wavelengths exhibiting the largest effects. This provides a segue to solid-state strength in the plastic flow regime, which is sometimes referred to as an "effective lattice viscosity." While we will discuss this point in greater depth in a later chapter, it is sufficient to note here that there is an approximate equivalence between viscosity and material strength (denoted by Y), both being properties that suppress growth:

$$\frac{Y}{\mu} = \frac{2\sqrt{gk\mathcal{A}}}{3}, \tag{10.21}$$

and

$$\frac{Y}{\mu} = \frac{2|\dot{a}_0|k}{3}, \tag{10.22}$$

for RT and RM, respectively.

As discussed in Mikaelian (2013, 2019), this is only a crude, albeit useful, equivalence – see López Ortega et al. (2014). An example of such equivalence appears to hold up even in nonlinear jet formation, as described in He et al. (2018).

10.3 Diffuse interface

The reduction in the RTI growth rate caused by a continuous density variation was investigated in three classical publications (LeLevier et al., 1955; Hide, 1955b; Case, 1960). In particular, Hide (1955b) and Case (1960) considered an exponential (i.e., diffusive) density distribution, with Hide using approximation techniques and Case finding exact solutions.

Chandrasekhar (1961) derived a fourth-order ordinary differential equation (ODE) for the amplitude of the z-velocity component by applying normal mode analysis to the linearized Navier–Stokes equations (Chapter 10, Equation (22) in Chandrasekhar (1961)):

$$\begin{aligned}\frac{d}{dz}\left\{\left[\rho - \frac{\mu}{\gamma}\left(\frac{d^2}{dz^2} - k^2\right)\right]\frac{du_3}{dz} - \frac{1}{\gamma}\frac{d\mu}{dz}\left(\frac{d^2}{dz^2} + k^2\right)u_3\right\} \\ = k^2\left\{-\frac{g}{\gamma^2}\frac{d\rho}{dz}u_3 + \left[\rho - \frac{\mu}{\gamma}\left(\frac{d^2}{dz^2} - k^2\right)\right]u_3 - \frac{2}{\gamma}\frac{d\mu}{dz}\frac{du_3}{dz}\right\},\end{aligned} \tag{10.23}$$

where ρ is the horizontally averaged density, u_3 is the perturbed velocity of the fluid in the z-direction, and μ is the dynamic viscosity (Morgan et al., 2016). For a given wave number k, the determination of the admissible growth rates γ becomes an eigenvalue problem (Lord Rayleigh, 1883), corresponding to eigenfunctions u_3 vanishing at $z \to \pm\infty$.

Assuming a system of two pure fluids separated by a discontinuous interface, Chandrasekhar found the eigenvalues with an appropriate ODE matching condition and finally obtained the dispersion relation

$$\left[\frac{k\mathcal{A}g}{\gamma^2}+1\right](\alpha_2 q_1+\alpha_1 q_2-k)-4k\alpha_1\alpha_2+\frac{4k^2}{\gamma}(\alpha_1\nu_1-\alpha_2\nu_2)[(\alpha_2 q_1-\alpha_1 q_2)-k\mathcal{A}]$$
$$+\frac{4k^3}{\gamma^2}(\alpha_1\nu_1-\alpha_2\nu_2)^2(q_1-k)(q_2-k)=0, \tag{10.24}$$

where

$$\alpha_i=\frac{\rho_i}{\rho_1+\rho_2}, \qquad (i=1,2) \tag{10.25}$$

and

$$q_i=\sqrt{k^2+\frac{\gamma}{\nu_i}}. \tag{10.26}$$

Naturally, the $\nu_i=\mu_i/\rho_i$ are the kinematic viscosities of the two fluids (Morgan et al., 2016).

Equation 10.23 shows the important role of the viscosity, as it is responsible for raising the ODE is to a higher order. Neglecting it for now, we arrive at the second-order ODE first obtained by Lord Rayleigh

$$\frac{d}{dz}\left\{\rho\frac{du_3}{dz}\right\}=k^2\left\{\rho-\frac{g}{\gamma^2}\frac{d\rho}{dz}\right\}u_3. \tag{10.27}$$

This equation was solved analytically by Lord Rayleigh by assuming an exponential density profile (Cherfils and Lafitte, 2000).

Since only a limited number of density profiles admit exact analytic solutions, approximation techniques are often used. The first such technique was introduced by LeLevier et al. (1955), who considered the effect of a density gradient on RTI when the transition between the low and high density fluids is very smooth. In this report, the authors extended Taylor's derivation of an accelerated density discontinuity to the case of a smooth density transition. The authors derived a modification to the classical RTI growth rate,

$$\gamma_{LeL}=\sqrt{\frac{k\mathcal{A}g}{1+k\mathcal{A}L_0}}, \tag{10.28}$$

where L_0 is the width of the diffuse layer. The diffuse interface prevents $\gamma_{LeL}\to\infty$ at high wavenumber (the paper credits Dr. Edward Teller with suggesting this behavior).

Using Eq. 10.27, Cherfils and Lafitte (2000) obtained the exact analytical expressions for the growth rate with a linear density profile. In particular, the authors obtained a transcendental dispersion relation, which yields a family of eigenvalues. We will not provide the detailed results here, except to note that the formula reproduces the short- and long-wavelength limits

$$\lim_{k\to 0}\gamma(k)=\sqrt{kg\mathcal{A}},$$
$$\lim_{k\to\infty}\gamma(k)=\sqrt{g/L_0}. \tag{10.29}$$

These values are consistent with those from the widely used estimate in the ICF community (Lindl, 1995) – see Chapter 19 on ablative RT instabilities.

In his pioneering work, Lord Rayleigh (1883) considered the stabilizing effect of a continuous density gradient. The exact equation, sometimes called the "Rayleigh equation," can be cast in the form of Schrödinger's equation (Mikaelian, 1996). The RTI growth rate can be rewritten as $\gamma_{RT} = \sqrt{g}\,\Gamma$, where Γ is the solution to Rayleigh's eigenvalue equation with the density gradient. The so-called stabilization factor is given by $\Gamma/\Gamma_{\text{classical}}$, where $\Gamma_{\text{classical}} = (k\mathcal{A})^{1/2}$ corresponds to a sharp density profile (i.e., no density-gradients).

Duff et al. (1962) integrated Chandrasekhar's inviscid eigenvalue equation with an error function density profile and obtained the dynamic diffusion model,

$$\gamma_D = \left(\frac{k\mathcal{A}g}{\psi_D} + \tilde{\nu}^2 k^4\right)^{1/2} - (\tilde{\nu} + \mathcal{D})k^2, \tag{10.30}$$

where $\tilde{\nu}$ is the average kinematic viscosity of the two fluids, and $\mathcal{D}$ is the mass diffusion coefficient between the two fluids (Morgan et al., 2016). It should be noted that only some of the referenced studies consider the case of time-evolving diffuseness (e.g., Duff et al., 1962). With increasing interface thickness, the growth reduction factor ψ_D increases.

Morgan et al. (2016) investigated the RTI initiated with a diffuse interface. Experimental data were obtained for three different wavenumbers, with a diffuse interface between two gases of different density made unstable by rarefaction wave acceleration. The full linear stability formulation of Chandrasekhar, Eq. 10.23, is solved numerically with an error function diffusion profile using the Riccati method.

Figure 10.10 demonstrates that the inviscid model of Duff et al. (1962) agrees well with the complete viscous solution for these relatively inviscid experiments. The Duff et al. (1962) model, however, disagrees with the complete viscous diffuse-interface linear stability theory in the large-wavenumber regime, where the complete solution predicts larger growth rates and a larger most-unstable wavenumber. This is shown in Fig. 10.11.

Xie et al. (2017a) studied viscous RTI flows with and without diffusion by inspecting the classical Bellman and Pennington (1954) and Duff, Harlow and Hirt (1962) models. They found that the Duff et al. model inherits the maximum error ($\sim$12%) from the approximate solution of Bellman and Pennington (1954) (Hide, 1955b; Reid, 1961; Menikoff et al., 1977; Mikaelian, 1993b). The authors suggested an explicit solution of the implicit dispersion relation, which improved the accuracy of the models, but at the expense of added complexity.

Beyond the linear (e.g., Ruev et al., 2006; Mikaelian, 1986; Cherfils and Lafitte, 2000) and nonlinear (e.g., Dong et al., 2019; Fan and Dong, 2020; Han et al., 2021) regimes, RTI-induced mixing with an initial diffuse, or premixed, interface must be solved numerically, an example of which may be found in Tian et al. (2011), Song et al. (2021a,b), and Wang et al. (2022b).

Experimentally, Morgan et al. (2018c) found that at moderate Atwood numbers and in the nonlinear regime, the bubble and spike velocities approach larger values than those predicted by Goncharov's model. These late-time velocity trends are predicted well by numerical simulations using the LES, and by the model of Mikaelian (2009) that extends Layzer type models to variable acceleration and density.

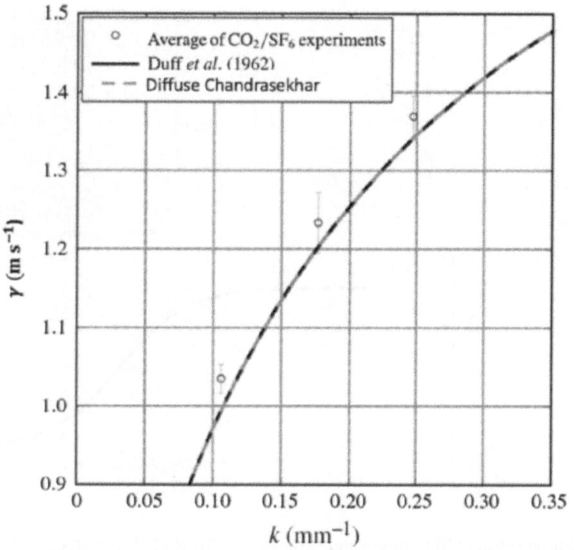

Fig. 10.10 The growth rates from diffuse-interface RT linear stability theory (diffuse Chandrasekhar in the legend) and the inviscid model of Duff et al. (1962) are compared again experimental measurements. The growth rates are averaged over four experiments using CO_2/SF_6, and the error bars represent the 95% confidence limits. Fig. 15 of Morgan et al. (2016).

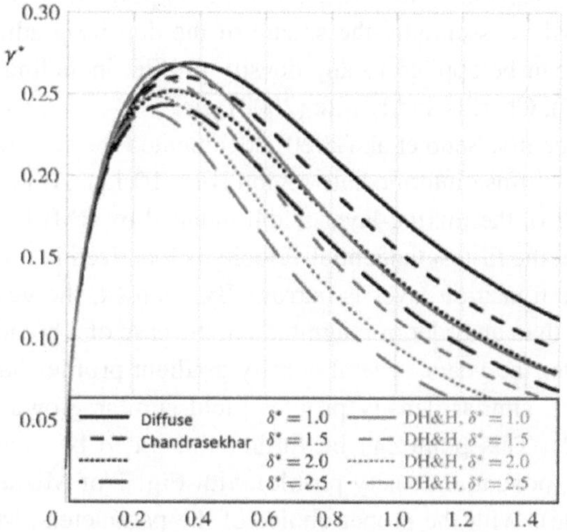

Fig. 10.11 A comparison of the dispersion curves generated using the diffuse-interface full linear stability solution (black lines), and the dynamic diffusion model. Here the Atwood number is 0.5 and δ^* denotes the characteristic length scale of the diffusion layer in the error function. Fig. 1 of Morgan et al. (2016).

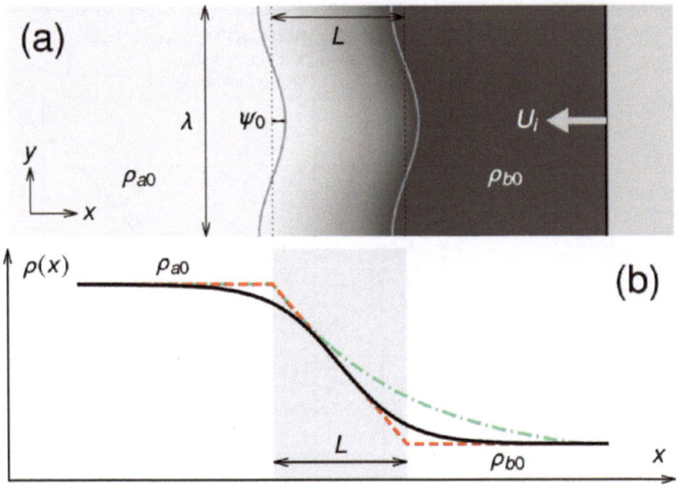

Fig. 10.12 (a) Initial setup of the single-mode analysis for the RMI. A planar shock hits a corrugated interface between the heavy and light fluids. (b) Density distributions of the transition layer adopted in the analysis, which are the tanh-type (solid line), linear-type (dash line), and exp-type (dash-dot line). Fig. 1 of Sano et al. (2020).

Mikaelian (1991) turned his attention to density gradients in the RMI. Using the profiles considered by Saffman and Meiron (1989) (hyperbolic-tangent) and Duff, Harlow, and Hirt (1962) (error-function), he treated the shock as an instantaneous acceleration of incompressible fluids. Mikaelian (1991) pointed out that the density-gradient-stabilization in RMI is essentially the square of the density-gradient-stabilization in RTI, $\Gamma / \Gamma_{\text{classical}}$. This can be applied to any density profile, including those considered by Lelevier et al. (1955), Cherfils and Lafitte (2000), etc.

Recently, Sano et al. (2020) documented the sensitivity of RMI growth to the thickness of the diffuse intermediate region (Fig. 10.12). At the time of the snapshot shown in Fig. 10.13(a), the mixing-layer width induced by RMI is still growing. Even at this nonlinear phase, the fastest tangential velocity is localized in the mushroom's roll-up region, so long as the transition layer is narrow. By contrast, the generation of velocity is weakened by more than an order of magnitude in the case of a broad transition layer – see Fig. 10.13(b).

Over the years, several density gradient profiles have been investigated. As one would expect, similar density profiles yield similar growth rates (see selected profiles in Fig. 10.12). This point can be further illustrated by comparing Fig. 1 of Mikaelian (1986) (an exponential density profile) with Fig. 1 of Mikaelian (1991) (the hyperbolic tangent profile). With the proper choice of the parameters, Mikaelian (1991) determined that the error-function and hyperbolic tangent profiles (Eq. 16 of his article, not shown) coincide within 3.5%. Furthermore, Fig. 5 of Mikaelian (2015) shows a similar profile when an error-function is employed.

Brouillette and Sturtevant (1989, 1994) as well as Motl et al. (2009) reported on experiments to measure the shock-induced growth of sinusoidal perturbations on thick interfaces separating two gases of different densities. Their results showed that the growth rates are reduced as the interface thickness is increased.

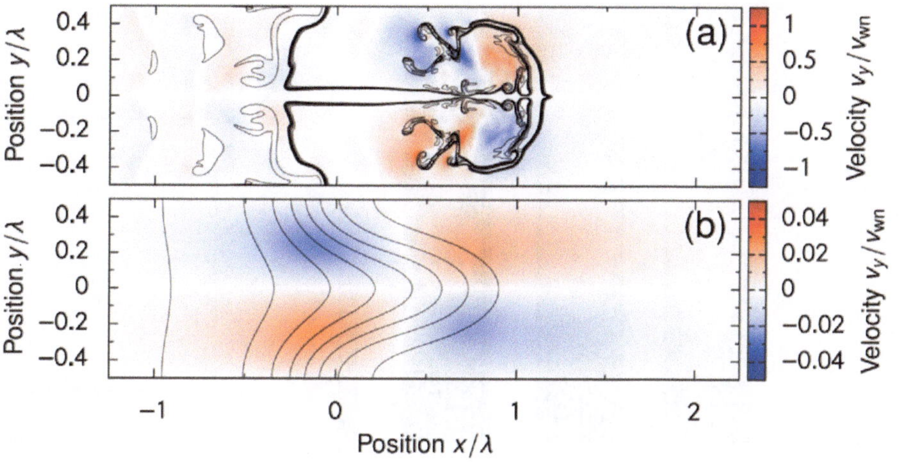

Fig. 10.13 Spatial distributions of the density and the tangential velocity in the nonlinear RMI regime. Fig. 2 of Sano et al. (2020).

To improve the agreement of the fitted curves with the numerical solutions in high Atwood number regimes, Dong et al. (2019) developed an asymptotic solution for a fluid interface with smoothly varying density based on the Wentzel–Kramers–Brillouin approximation for high wavenumbers tangential to the interface. A further study of the multiplicity of unstable eigenmodes focused on the excitation and nonlinear evolution of compressible RTI flows (Fan and Dong, 2020).

Finally, Olsthoorn et al. (2019) considered the role of the nonlinear equation of state (EOS) on the growth rate of the linear instability. This results in asymmetric mixing of the density profile. Gorodnichev et al. (2024) investigated the interface instability development induced by the bulk density perturbations in accelerated media described by the Mie–Grüneisen equation of state. The reader is referred to the original papers and references therein for the full details.

10.4 Stratification

Lafay et al. (2007) demonstrated that increasing stratification always stabilizes the Rayleigh–Taylor flow. Subsequently, Gauthier (2013) argued that the influence of the stratification may be classified into various stages. During the first stage of RTI growth, stratification does not affect the evolution and the flow is essentially that of a classical RT system. In the second stage, the mixing layer begins to smooth the density discontinuity, and the effective Atwood number, based on the mean density profile, decreases. Moreover, this smoothing of the density profile leads to decreased turbulence production. The baroclinic source term in the vorticity equation also begins to decrease, leading to saturation of the mixing layer thickness. The larger the stratification, the sooner the flow saturates (Fig. 10.14).

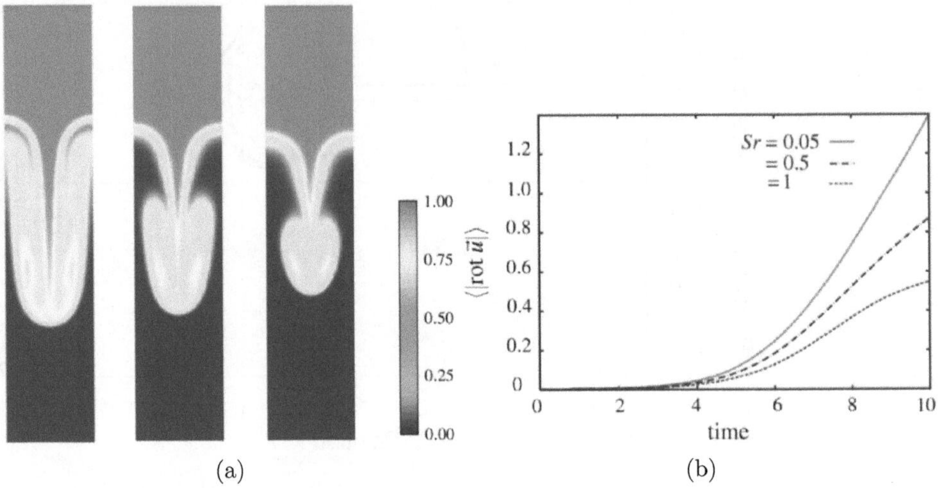

Fig. 10.14 (a) Stratification effects: concentration fields at dimensionless time $t = 10$ for three values of the stratification parameter: $Sr = 0.05$, $Sr = 0.50$ and $Sr = 1.00$ (from left to right). The Reynolds number is $Re = 1,000$. (b) Mean value of vorticity vs. time, for three values of the stratification parameter ($Sr = 0.05$, 0.50 and 1.00). Figs. 7 and 9 of Lafay et al. (2007).

For geophysical flows, Jacobs and Dalziel (2005) recognized that the RT instability will often develop as the result of breaking internal gravity waves, producing initial conditions characterized by both shear and complex stratifications that contain both stable and unstable portions. As demonstrated by their experiments, in the classical configuration, the RTI mixing layer will grow until it reaches the upper and lower bounds of the domain. However, if the instability develops at an unstable interface between stably stratified layers, the subsequent mixing layer will be inhibited by the stable stratification and may not ultimately extend to the domain boundaries (Dalziel and Mouet, 2021).

Lawrie and Dalziel (2011b) as well as Davies Wykes and Dalziel (2014) further investigated the mixing efficiency of stratification in confined RTI for a variety of initial density profiles. As an example, Fig. 10.15 shows that the initial behavior of the instability is similar to the classical case. As it becomes confined by stratification, the mixing zone growth slows down and then stops. The motion of the flow fields eventually dies away, except for internal waves present on the upper surface of the mixing region.

Recently, Dalziel and Mouet (2021) presented experiments for which one of the two layers initially has a stable density stratification, while the density of the other fluid is uniform. In this case, mixing occurs throughout the region of initially uniform density. Depending on the stratification strength compared with the density jump across the initial unstable interface, the mixing may be arrested before it reaches the lower domain boundary (Fig. 10.16).

Motivated by ICF applications, Mikaelian generalized the results of the classical RTI to a system of an arbitrary number N of stratified fluids (Mikaelian, 1982, 1983). Subsequently, he derived the corresponding equations for the RMI (Mikaelian, 1985). These problems

(a)
(b)
(c)
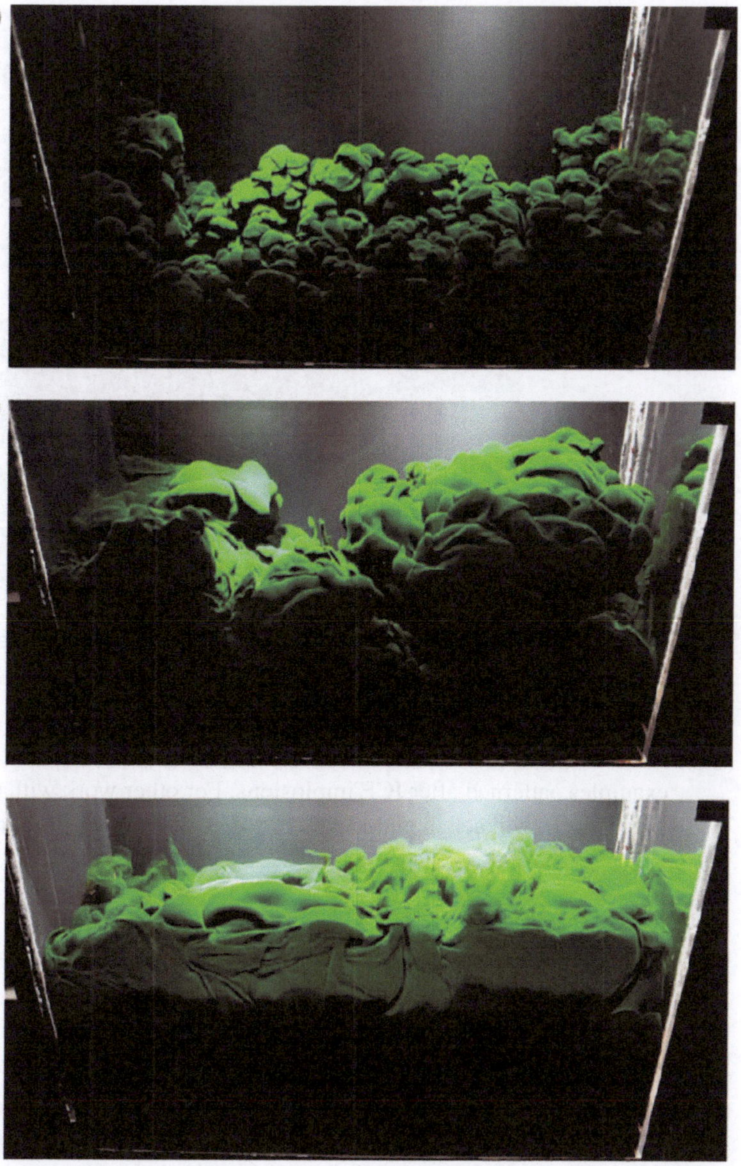

Fig. 10.15 Photographs showing the mixing region being confined by the stable density stratification for three dimensionless times, τ: (a) $\tau = 4.14$, large-scale structures extend upwards into the mixing region from the lower layer; (b) $\tau = 9.07$, the stratification has arrested further vertical growth of the mixing region; (c) $\tau = 13.37$, motion has decayed, except for internal waves present on the upper surface of the mixing region. Here $\mathcal{A} = 2 \times 10^{-3}$, initial stratification was quadratic with increasing density gradient. This view was taken with the camera looking slightly down at the upper layer. Fig. 7 of Davies Wykes and Dalziel (2014).

were addressed by solving Chandrasekhar's incompressible eigenvalue problems based on variational principles. He illustrated the cases of $N = 2$ (classical results) and $N = 3$ by

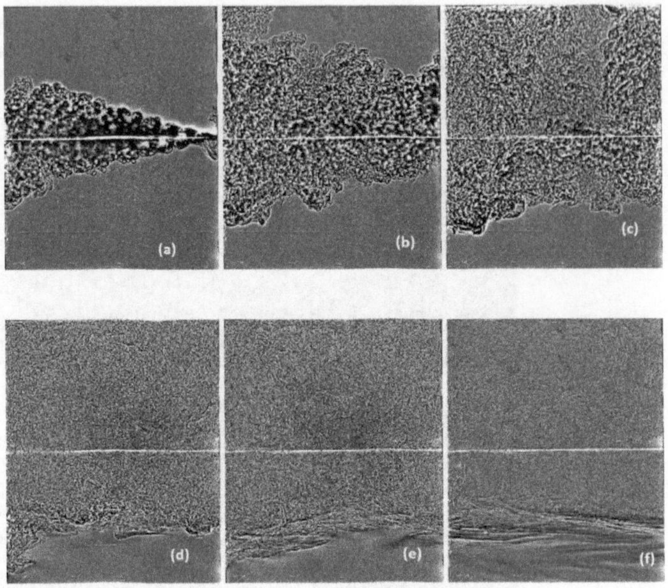

Fig. 10.16 Sequence of enhanced shadowgraph images showing the development of an experiment with an Atwood number of 0.0071. The times are (a) $t = 0$ s, (b) $t = 4.1$ s, (c) $t = 8.1$ s, (d) $t = 16.4$ s, (e) $t = 24.5$ s, and (f) $t = 65.3$ s. Fig. 6 of Dalziel and Mouet (2021).

examples patterned after ICF implosions. For other work with the same methodology, see, for example, Mikaelian and Lindl (1984), Mikaelian (1986), and Munro (1988).

10.5 Compressibility

10.5.1 Compressibility for RT

Two types of compressibility, known as the "static" (flow) and "dynamic" (fluid) compressibilities (Gauthier, 2017), may affect RTI growth. These quantities are related to the thermodynamic state and the stratification of background density and pressure fields (Livescu, 2004). Flow compressibility is associated with quantities that are independent of fluid properties (e.g., the velocity or thermodynamic state). On the other hand, fluid compressibility is related to material properties (e.g., the equation of state, differences in the specific heat ratio between the two fluids) that can only be changed by changing the fluid itself (Wieland et al., 2019).

Compressibility effects have been studied in LES by decomposing density fluctuations into an entropic part, due to variation of composition, and an acoustic part, due to intrinsic compressibility (Mellado et al., 2005). The latter is found to be less than 10% of the total density fluctuation, indicating that intrinsic compressibility effects are indeed small.

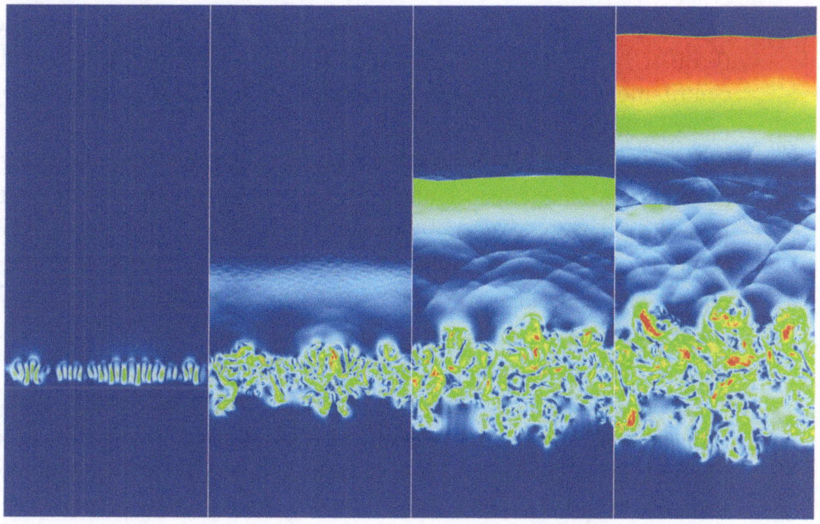

Fig. 10.17 Local Mach number from a simulation of the compressible RT instability. The upper fluid is xenon and the lower fluid is neon. Simulation times from left to right are: $t = 25, 50, 75$, and 100 s. Fig. 1 of Olson and Cook (2007).

Consequently, key features such as the quadratic time evolution of the mixing depth, the anisotropy of the Reynolds stresses, and the value of the mixing parameters compare well with those observed in the incompressible cases reported in the literature.

When the flow domain is expanded beyond that of Mellado et al. (2005), it can be demonstrated that RTI does exhibit strong compressibility effects, as evidenced by the formation of shock waves in the upper fluid. A series of large-eddy simulations illustrated that the rising bubbles of light fluid act like pistons, compressing the heavy fluid ahead of the fronts and generating shocklets (Fig. 10.17). These shocklets coalesce in multidimensional fashion into a strong normal shock, which increases in strength as it propagates upwards (Olson and Cook, 2007).

Single mode

The bubble growth in compressible RTI and RMI can also be studied using a modified Layzer-type potential flow theory (i.e., Goncharov, 2002). In addition to the familiar expression for the velocity potential near the bubble tip (Zhao et al., 2020a),

$$\phi^h = \tilde{a}_1(t)\cos(kx)\exp[-k(y-a_0)], \tag{10.31a}$$

$$\phi^L = \tilde{b}_1(t)\cos(kx)\exp[k(y-a_0)] + \tilde{b}_2(t)y, \tag{10.31b}$$

the dynamic densities near the bubble are also expanded in a similar fashion (Gupta et al., 2009):

$$\rho^h = \rho_0^h + \rho_1^h(t)\cos(kx)\exp[-k(y-a_0)], \tag{10.32a}$$

$$\rho^L = \rho_0^L + \rho_1^L(t)\cos(kx)\exp[k(y-a_0)]. \tag{10.32b}$$

By expanding the interface amplitude a to second order near the bubble tip, $(x, y) = (0, a(0, t))$, the bubble velocities for RTI and RMI can be obtained. The compressibility contributions that must now be included are the density, dynamic pressure, and adiabatic index of the heavy and light fluids.

The results of RTI potential flow model calculations suggest that the upper fluid density has a destabilizing effect on the compressible RT instability (Zhao et al., 2020a), while the lower fluid density has a stabilizing effect (Alon et al., 1995; Goncharov, 2002; Sohn, 2004; White et al., 2010). Similarly, the adiabatic index of the upper fluid has a destabilizing effect, while the adiabatic index of the lower fluid has a stabilizing effect. The former result regarding the fluid density is consistent with that of incompressible fluids, while the latter result is in agreement with the conclusion of He et al. (2008).

In Section 3.7.2, the dependence of 2D single-model RTI growth on M has already been documented, in the general form of plotting time series of v_b and v_s (Fig. 3.16). In particular, both velocities approach zero for larger Mach numbers, indicative of the complete suppression of RTI for relatively strong compressibility.

More recently, DNS of 2D single-mode RTI carried out to extreme late-times (Reckinger et al., 2016), for various Mach numbers and Atwood number $\mathcal{A} = 0.1$, confirmed that flow compressibility reduces the RTI growth, see Fig. 10.18. This provides support to the result deduced from linear stability analysis (Livescu, 2004). The authors suggested that

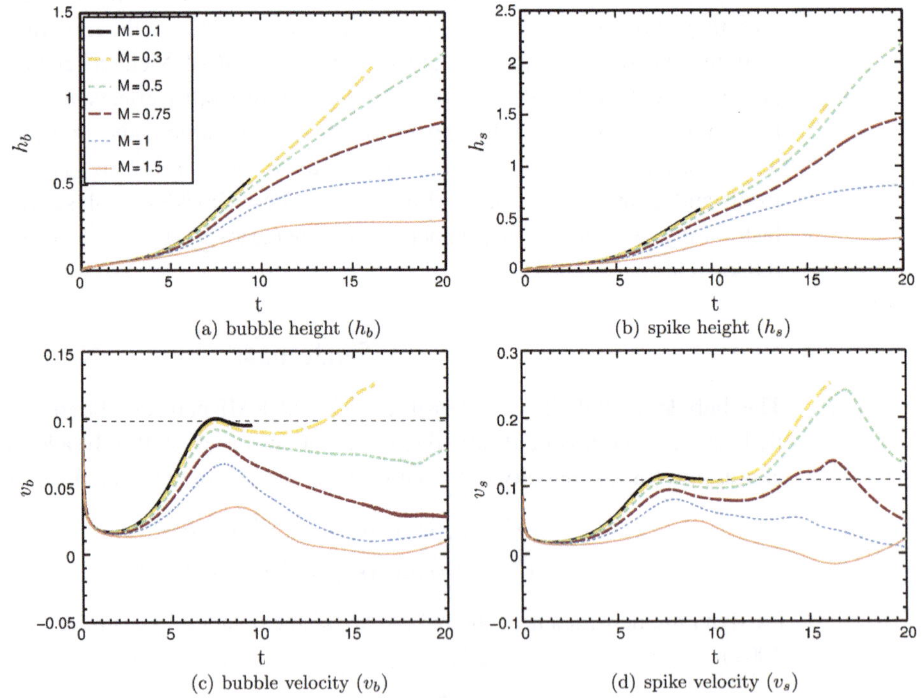

Fig. 10.18 Evolutions of bubble height (h_b), spike height (h_s), bubble velocity (v_b), and spike velocity (v_s). The dotted line represents the asymptotic velocity values from potential flow theory. Fig. 16 of Reckinger et al. (2016).

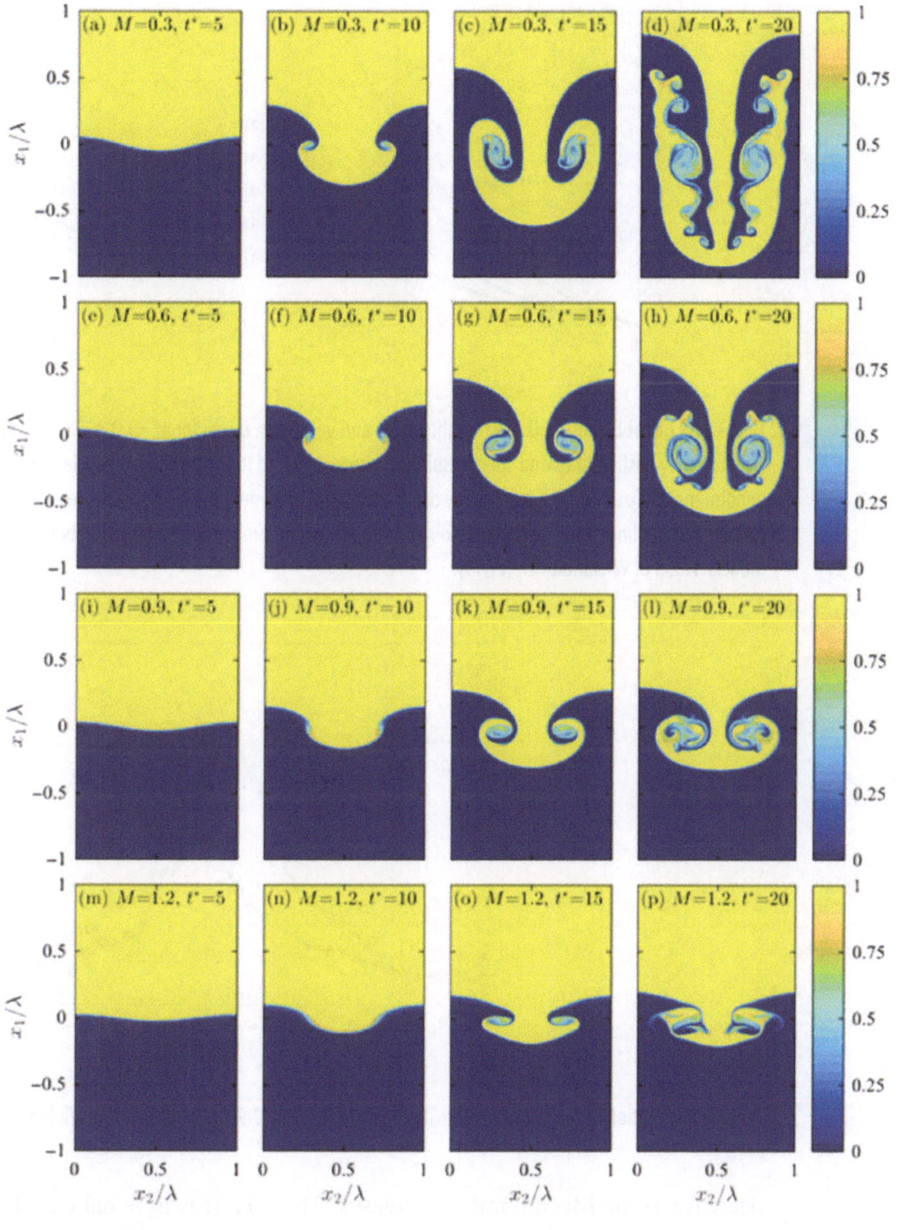

Fig. 10.19 Instantaneous fields of the heavier species mass fraction in the RTI, as a function of nondimensional time $t^* = t\sqrt{g/\lambda}$, for stratification strengths $M = 0.3, 0.6, 0.9$, and 1.2 (increasing from top to bottom). Fig. 2 of Wieland et al. (2019).

for compressible flow, the perturbation might spread more slowly than for incompressible flow, which results in the energy accumulation near the interface.

Figure 10.19 shows RTI growth as a function of time for each of the Mach numbers under consideration: 0.3, 0.6, 0.9, and 1.2. The corresponding bubble and spike growths

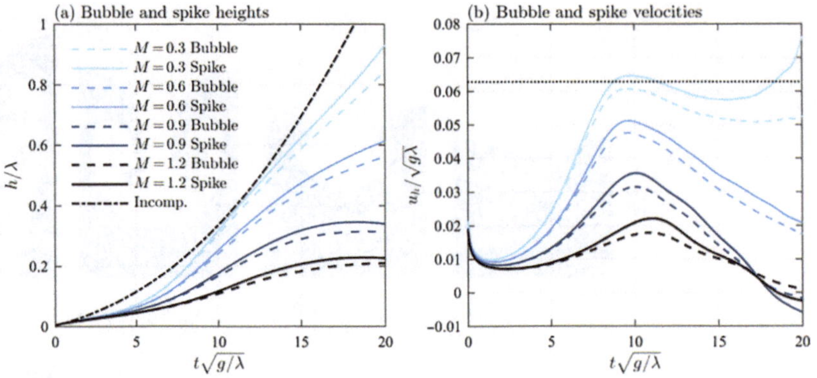

Fig. 10.20 Time series of bubble and spike tip heights, h (a) and velocities, u_h (b) for $M = 0.3, 0.6, 0.9$, and 1.2. Bubble results are shown by dashed lines and spike results are shown by solid lines. Heights, velocities, and times have all been nondimensionalized using λ and g. The dash-dotted line in panel (a) shows incompressible results from Wei and Livescu and the horizontal dotted line in panel (b) shows the predicted bubble velocity from drag and potential flow models. Fig. 3 of Wieland et al. (2019).

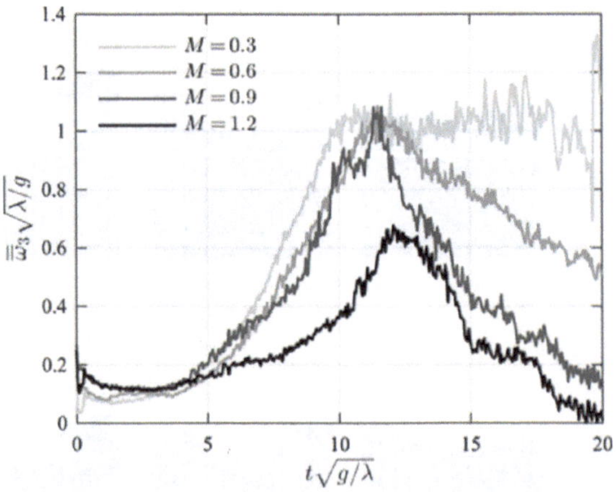

Fig. 10.21 Temporal evolution of the average vorticity for $M = 0.3, 0.6, 0.9$, and 1.2. Fig. 7 of Wieland et al. (2019).

decrease as the Mach number increases. The RTI growth is halted relatively early for the strongest Mach number, while secondary vortices are most prominent for the weakest Mach number. Figure 10.20 shows the RTI suppression compared to the incompressible case. For the two highest Mach numbers (0.9 and 1.2), the bubble and spike each reach maximum heights and then stop growing (Wieland et al., 2019).

An examination of the vorticity transport equation reveals that incompressible baroclinicity drives RTI growth for small and moderate Mach number, but increasing the compressibility leads to the suppression of vorticity production and RTI growth (see Fig. 10.21). The temporal evolution of the nondimensional vorticity ω_3 (the only nonzero component of vorticity in 2D) is shown in Fig. 10.22. The overall spatial extent of vorticity

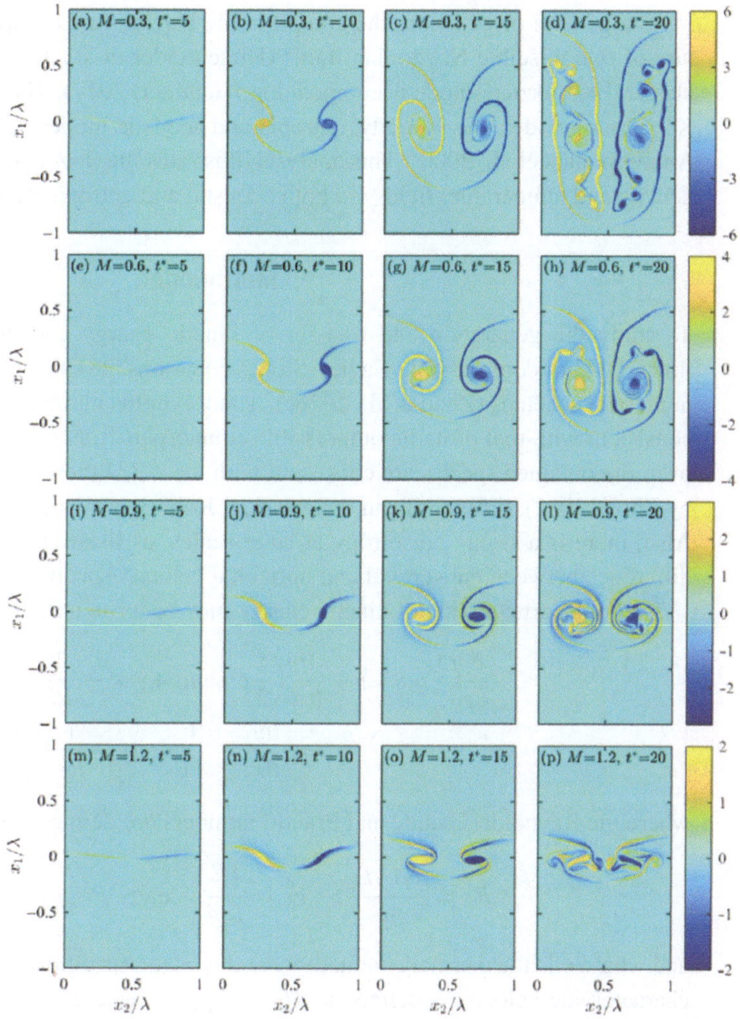

Fig. 10.22 Instantaneous fields of the nondimensional vorticity, as a function of nondimensional time, for Mach numbers increasing from top to bottom. The progression in time is shown in columns from left to right. Fig. 6 of Wieland et al. (2019).

production is greatest for low Mach numbers. The overall vorticity magnitude is also seen to decrease with increasing Mach number. In each case, the vorticity field initially develops as a vortex pair, one with generally positive vorticity and the other with negative vorticity. These vortex pairs evolve by moving downward slowly in the domain, while the KHI on the sides of the bubbles and spikes sheds further vortex pairs.

Le Creurer and Gauthier (2008) numerically simulated a single-mode RTI between compressible miscible fluids. Four regimes appeared: linear, nonlinear steady bubble rise, return toward equilibrium, and acoustic waves. The final regime is damped by both physical viscosity and thermal conduction.

As a general proposition, the compressible Navier–Stokes equations for a binary mixture of two miscible Newtonian fluids (Hirschfelder et al., 1964) may be analyzed using classical Kovàsznay-mode decomposition (Gauthier, 2017). The system contains the three Kovásznay modes, the vorticity, entropic and acoustic modes (Chu and Kovásznay, 1958; Andreopoulos et al., 2000). This analysis illustrates the flow compressibility effects on the density and temperature fields via both acoustic and entropic modes.

Multimode

To study the generation and transfer of kinetic energy and enstrophy in RTI induced flows, DNS of compressible Navier–Stokes equations have been performed with high order numerical schemes (Zhao et al., 2020c). The Reynolds number increases with a t^3 scaling, consistent with that of its incompressible counterpart from Cook and Zhou (2002). Moreover, the obtained spectra are consistent with the $-7/4$ inertial range power law predicted by Zhou (2001), different from the classical Kolmogorov law value of $-5/3$ (Fig. 10.23). Also, there is a strong anisotropy at large scales, as illustrated by an order of magnitude difference between the vertical and horizonal velocity spectra.

The transport equation of kinetic energy may be written in the form

$$\frac{\partial}{\partial t}\left(\frac{1}{2}\rho u_i u_i\right) + \frac{\partial}{\partial x_j}\left(\frac{1}{2}\rho u_i u_i u_j + p u_j - \frac{1}{Re}u_i\tau_{ij}\right) = p\frac{\partial u_j}{\partial x_j} - \frac{1}{Fr}\rho g u_3 - \frac{1}{Re}\tau_{ij}\frac{\partial u_i}{\partial x_j}, \tag{10.33}$$

where the Reynolds, Froude and Prandtl numbers are defined as

$$Re = \frac{\rho_I U_I L_Z}{\mu_I}, \quad Fr = \frac{U_I^2}{L_z g}, \quad \text{and} \quad Pr = c_p \frac{\mu_I}{\kappa}, \tag{10.34}$$

and where κ is the thermal conductivity and c_p is the specific heat at constant pressure. The characteristic velocity is defined as $U_I = \sqrt{p_I/\rho_I}$, where ρ_I and p_I are the initial density and pressure at the interface, and the vertical length L_z of the flow domain is selected as the characteristic length. It is worth noting that the external acceleration and viscosity are now included via the Froude and Reynolds numbers. Integrating Eq. 10.33 over the domain volume gives the global kinetic energy budget (Zhao et al., 2020c)

$$\frac{dK}{dt} = \mathcal{W} - \varepsilon - \frac{d\mathcal{P}}{dt}, \tag{10.35}$$

where $\mathcal{P} = (1/Fr)\int \rho z dV$ is the potential energy, and $\varepsilon = (1/Re)\int \tau_{ij}(\partial u_i/\partial x_j)dV$ is the energy dissipation rate. The pressure work due to compression in RT flows is $\mathcal{W} = \int p(\partial u_k/\partial x_k)dV$. This pressure-dilatation power term is responsible for the local conversion between kinetic and internal energies.

Figure 10.24 shows the development of the energy budget terms, Eq. 10.35. Here, the term $d\mathcal{P}/dt$ is always negative, converting the potential energy into kinetic energy. Such conversion acts as a major mechanism for the generation and growth of kinetic energy. The dissipation ε increases continuously to account for the kinetic energy dissipation.

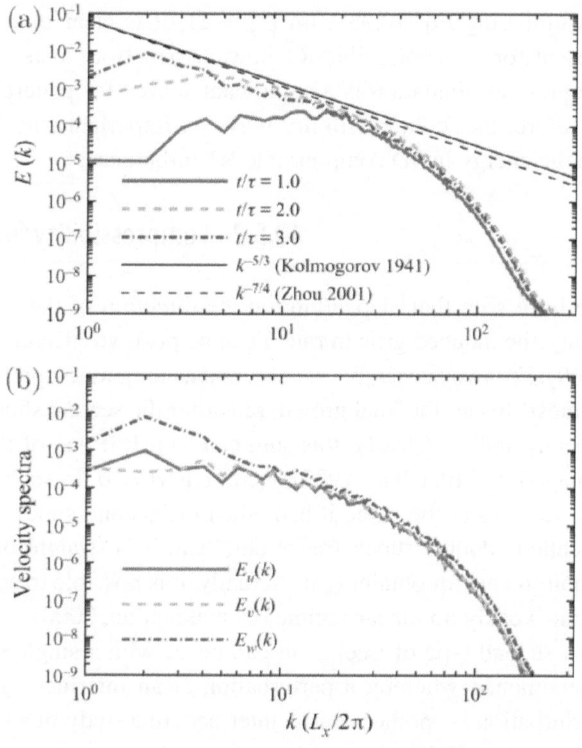

Fig. 10.23 Kinetic energy spectra, $E(k)$, (a) at nondimensionalized time $t/\tau = 1.0$, 2.0 and 3.0, and velocity spectra. The obtained spectra are consistent with the $-7/4$ inertial range power law predicted by Zhou (2001). (b) at $t/\tau = 3.0$ in the horizontal plane at $z = 0$. Here, the horizontal wavenumber is $k = \sqrt{k_x^2 + k_y^2}$. Fig. 2 of Zhao et al. (2020c).

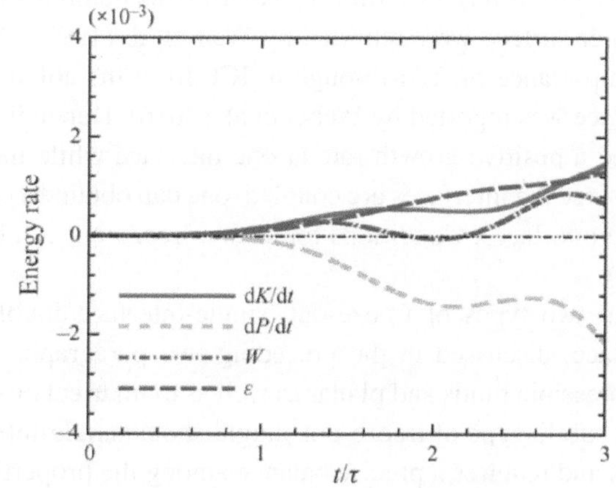

Fig. 10.24 Evolution of the kinetic energy terms in Eq. 10.35. Fig. 4 of Zhao et al. (2020c).

Comparing Eq. 10.35 with Eq. 6.21, it is clear that the pressure-dilation term is not present for incompressible RT flows. Apart from a short period of time around $t/\tau \approx 2$, the pressure-dilatation is a significant source for generating kinetic energy. It is evident, therefore, that compressibility plays an important role in the generation and transfer of kinetic energy in 3D compressible RT turbulence.

10.5.2 Compressibility for RM

We have seen that, depending on the direction of the shock (light-to-heavy or heavy-to-light), the induced growth rate $\dot{a}$ can be positive (Richtmyer, 1960) or negative (Meshkov, 1969). For two successive shocks, and as long as the perturbation remains single-scale and (almost) linear, the final growth rate after the second shock is the algebraic sum of the two rates $\dot{a}_1$ and $\dot{a}_2$. Clearly, this sum can vanish if one of the terms is positive and the other is negative (Mikaelian, 1985), leading to $\dot{a} = 0$. In such a case, the amplitude is "frozen" and remains at the value it had when the second shock arrived. For obvious reasons, this is called "double-shock freeze-out," and it is straightforward to express analytically the requirements for obtaining it. Actually, it is possible to achieve this type of freeze-out even in the weakly nonlinear regime (Charakhch'an, 2000).

A second type of freeze-out can occur with a single shock via feed-through, a general phenomenon whereby a perturbation at an interface interacts with or is coupled to the perturbation at another nearby interface. In a study of a single fluid with two free surfaces, G.I. Taylor (1950) showed how the evolution of a perturbation at one of the interfaces depends on the other. Generalizing to an arbitrary number of stratified fluids Mikaelian (1983) derived the coupling of an i-th interface to the two interfaces $i - 1$ and $i + 1$ on either side. In this manner, a perturbation at any interface is coupled to any other interface via intermediate fluid layers - even when a fluid starts with no perturbation, it acquires one by feed-through from a nearby interface. Feed-through occurs in both RTI and RMI (Mikaelian, 1985). Gelatin ring experiments demonstrating feed-through from an inner to an outer surface were reported by Weir et al. (1998). A recent experiment demonstrating the importance of feed-through in ICF from the ablating surface to the inner fuel/ablator interface was reported by Weber et al. (2016). Depending on the stratification, a shock can induce a positive growth rate at one interface while inducing a negative rate at the other and, since the interfaces are coupled, one can obtain freeze-out ($\dot{a} = 0$) at one (but not both) of the interfaces. Conditions for such a freeze-out can be derived analytically (Mikaelian, 1995).

The two types of freeze-out, single-interface/double-shock and single-shock/double-interface, discussed in the preceding two paragraphs occur in both incompressible and compressible fluids and planar as well as cylindrical or spherical geometry. A third, and far more subtle, type of freeze-out, single-shock/single-interface, occurs only in compressible fluids and requires a precise balance among the properties of the fluids (Mikaelian, 1993b, 1994a). Of course, this is the classical configuration of a single shock passing through a single interface considered by Richtmyer (1960) and Meshkov (1969), and where a perturbation at the interface (almost) always grows with time. In special cases, however, the perturbation can "freeze" and not grow at all. To understand its physical origin, recall that after the shock, the interface continues to interact with both the transmitted and the reflected shocks as they move away (Richtmyer, 1960). Under certain conditions they combine to suppress the growth resulting in freeze-out: the interface amplitude is compressed by the shock but it stays at that compressed value, that is, $\dot{a} = 0$. The simplest such case is when the compressibilities of the fluids differ by a factor of four – see Mikaelian (1993b, 1994a). This type of freeze-out was subsequently confirmed by Yang et al. (1994b). For weak shocks, Fraley (1986) gave an analytic expression for the fully compressible growth rate and an analysis of that expression reveals if and when it can vanish. Like Richtmyer, Fraley studied only the case when a shock is reflected. To date there are no analytic expressions for the case when a rarefaction is reflected, even for weak shocks. However, numerical solutions reveal (Wouchuk and Sano, 2015) that freeze-out can also occur in this case. Nor is this type of freeze-out limited to weak shocks – one can find it with strong shocks as well, but they must be found by numerical methods (Wouchuk and Nishihara, 1997). An important distinction between this type of freeze-out, versus the above two types, is that the conditions for double-shock or double-interface freeze-out depend on the wavelength of the perturbation, whilst the condition for single-shock/single-interface freeze-out is independent of wavelength. While the assumption of linearity is necessary to derive analytic conditions for any type of freeze-out, it is expected that nonlinear perturbations will also exhibit the same phenomena, perhaps at a reduced scale. That is, $\dot{a}$ will not be strictly zero, but it will be small compared to the "classical" growth rate given by Richtmyer (1960).

The challenging nature of single-shock/single-interface freeze-out can be seen by noting that none of the well-known prescriptions can predict it: Eq. 2.26 (Richtmyer, 1960), Eq. 2.27 (Meyer and Blewett, 1972), and Eq. 2.28 (Vandenboomgaerde et al., 1998) all predict a finite growth rate, when it should be zero. One should remember that these are heuristic modifications that perform well in the majority of cases, but there are exceptions. In fact, Eqs. 2.27 and 2.28 predict that the growth rate vanishes when $\mathcal{A}^+ = 0$, but there are cases with finite growth rates even when $\mathcal{A}^+ = 0$ (Mikaelian, 1993b, 1994a).

10.6 Porous media

Turbulent flows in porous media have been observed in a broad range of scientific and engineering applications (Wood et al., 2020), from practical industrial catalysis in packed beds and heat exchange in nuclear reactor vessels, to highly exotic cases such as the flow of superfluid helium past superconducting elements in high-energy particle accelerators (Soulaine et al., 2017).

Recently, RT mixing in porous media has received renewed interest due to the potential application in mitigating the effects of emissions on climate change, specifically of the CO_2 sequestration process in saline aquifers (Huppert and Neufeld, 2014; Boffetta et al., 2020).

With the Boussinesq approximation, RT turbulence in porous media has been numerically modeled (Fig. 10.25) using a disordered distribution of spheres with different void

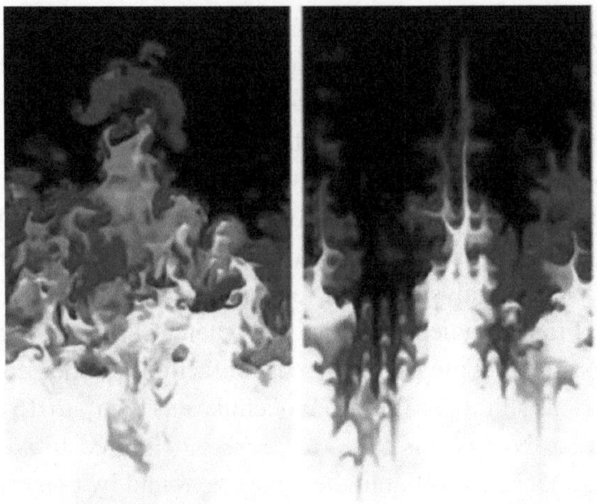

Fig. 10.25 Vertical sections of the temperature field for simulated Rayleigh–Taylor turbulence. Left: standard RT turbulence in homogeneous fluid with porosity coefficient $\varphi = 1$. Right: fully resolved simulation of porous RT turbulence with $\varphi = 0.6$. Fig. 1 of Sardina et al. (2018).

fraction values. The porosity of the domain, φ, is defined as the ratio of the void volume to the total volume (i.e., $\varphi = 1$ corresponds to the pure fluid case). Figure 10.26(a) shows the time evolution of the mixing layer for different values of the porosity (Sardina et al., 2018). For $\varphi = 0.8$, the mixing layer at late times has already departed from the standard t^2 scaling of the pure fluid case, and the scaling exponent is further modified to become linear for the smallest values of porosity. Figure 10.26(b) also shows the growth of the mixed mass (Eq. 5.19) for these porosity values. For sufficiently low values of φ, it displays a clear linear behavior instead of the t^2 scaling.

Under the Boussinesq approximation, the so-called Darcy–Rayleigh–Taylor (DRT) equation is often used to describe the motion of a single-phase fluid in porous media (Muskat, 1937; Nield and Bejan, 2006; Borgnino et al., 2021):

$$\partial_t \rho + \mathbf{u} \cdot \nabla \rho - D \nabla^2 \rho = 0, \tag{10.36a}$$

$$\mathbf{u} - \frac{\wp}{\mu \varphi}(-\nabla p + \rho g) = 0, \tag{10.36b}$$

$$\nabla \cdot \mathbf{u} = 0, \tag{10.36c}$$

where $\wp$ is the isotropic permeability and the Darcy flow rate is φu. The pressure p is given by the Poisson equation $\nabla^2 p = \rho \mathbf{g} \wp / (\mu \varphi)$.

Through 2D numerical simulations of porous media flows, Sabet et al. (2021) has shown that large viscosity contrasts profoundly affect the RTI, as described by the Darcy–Rayleigh–Taylor (DRT) equation. Their 2D domain had an isotropic porous medium containing a denser and more viscous fluid atop another miscible, less dense, and less

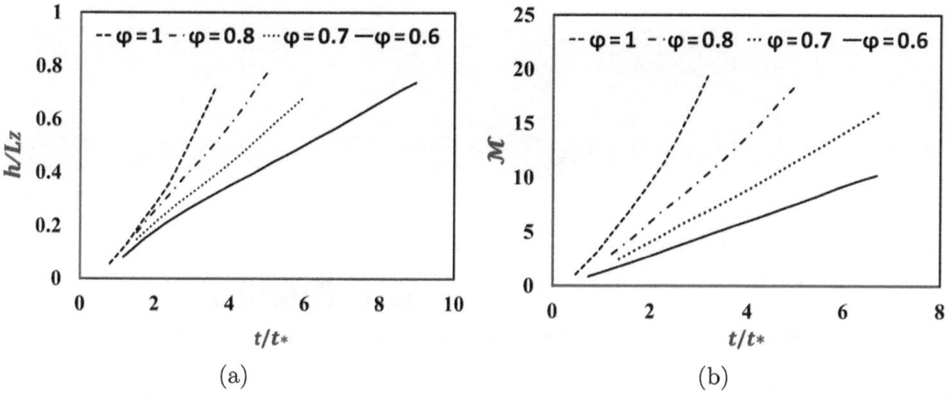

Fig. 10.26 (a) Temporal evolution of the mixing width h in four simulations of RT turbulence with different porosity. Dashed lines represent t^2 behavior. These quantities are shown in a lin-lin plot to emphasize the linear growth at later times for the two smallest porosity cases. (b) Temporal evolution of the mixed mass for different values of porosity. Figs. 2 and 3 of Sardina et al. (2018).

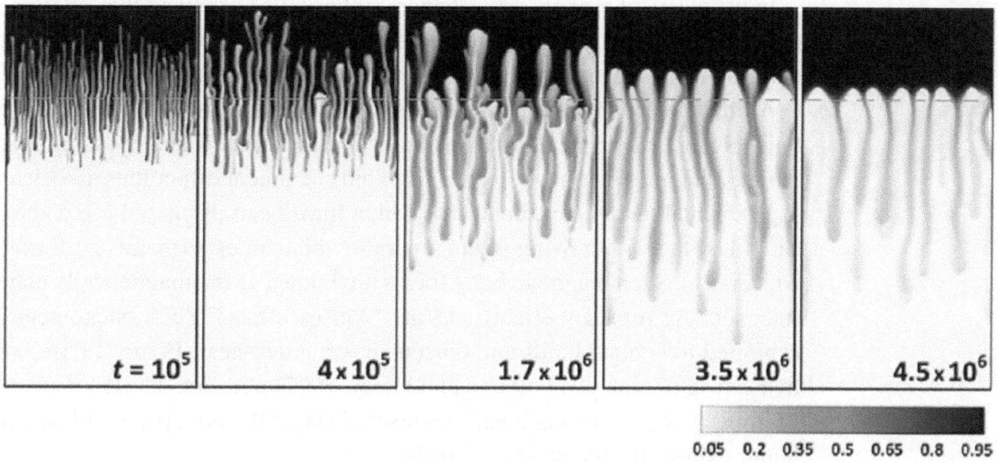

Fig. 10.27 Concentration field at the time shown on the panel at different log-viscosity ratios. As the ratio increases, the up-down symmetry of fingering breaks. The dashed line shows the initial interface location. Fig. 1 of Sabet et al. (2021).

viscous fluid. Figure 10.27 shows snapshots of the concentration field for different values of log-viscosity ratio, ranging from 0 to 8. For small values of the log-viscosity ratio, upward and downward fingers grow symmetrically at the same rate. This symmetry breaks, however, at greater viscosity ratios, with the downward fingers becoming more prominent as the density ratio increases. At the extremely large log-viscosity ratio of 8, the fingering becomes almost one-sided, as the upward fingers grow at a much smaller rate than the downward fingers.

11 Rotation and time-dependent acceleration

11.1 Rotation

Rotating RTI has found several important applications, including among others, rotating stars (García-Senz et al., 2018; Maeder et al., 2013), deep convection events in the ocean (Davey and Whitehead Jr, 1981), and complex dusty plasma (Dolai and Prajapati, 2018; Ma et al., 2006). The influence of rotation on the RTI was first studied by Hide (1956a,b) for a viscous incompressible fluid, which rotates uniformly about an axis with an angle relative to a gravitational field.

In his classical monograph, Chandrasekhar (1961) showed that rotation suppresses RTI growth. To focus on the essential elements of the problem, he took the axis of rotation along an angle relative to a gravitational field and neglected viscous effects.

A challenge of RTI experiments is to allow the heavy fluid to be placed atop a light fluid. Recent experimental studies have overcome this issue by using magnetic forces to initiate RTI in a rotating system. This avoids technical difficulties associated with standard experimental techniques, details of which have been discussed elsewhere (e.g., the wake caused by the withdrawing of the separating plate in experiments performed at Cambridge University). Such magnetic body forces are known as the magnetically enhanced buoyancy (magneto-Archimedes effect) and the "Moses effect." Such phenomena have also been exploited to levitate fluids and objects in weightlessness. Figure 11.1(a) shows a levitated living frog (Geim, 2011; Berry and Geim, 1997) and Fig. 11.1(b) displays a gold coin, a £1 coin, a silicon crystal, a gallium arsenide crystal, and a piece of lead, all floating in the liquid oxygen (Catherall et al., 2003).

Experiments have studied the RTI with a rotation axis parallel to gravity (Baldwin et al., 2015; Scase et al., 2017, 2020; Scase and Sengupta, 2021). Figure 11.2 shows the development of the instability, with "bubbles" of the paramagnetic layer intruding into the diamagnetic layer below it. Here, the RTI is caused by the magnetic attraction of the weakly paramagnetic upper layer to a superconducting magnet beneath the tank (not visible in these images) and by the magnetic repulsion of the colorless diamagnetic layer. In the upper image, the cylindrical container is not rotating; in the lower image, it rotates at approximately 24 revolutions per minute. The images illustrate how rotation restricts the size of the structures that form at the interface. In other words, compared to the nonrotating system, the dominant unstable scale is reduced (Fig. 11.3).

To isolate the effect of rotation on RTI growth, Carnevale et al. (2002) performed a series of numerical simulations of low Atwood number flows (the Boussinesq approximation). The rotation axis was parallel to the acceleration and the flow was in a regime with

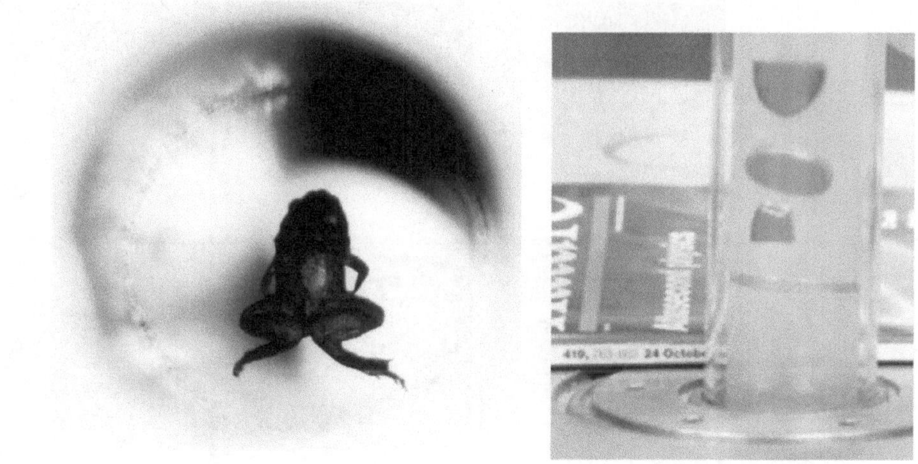

(a) (b)

Fig. 11.1 Rotating RTI experiments utilized an established method with the magnetically enhanced buoyancy (magneto-Archimedes effect) and the "Moses effect." Previously, the scheme demonstrated levitation of (a) a frog (Berry and Geim, 1997) and (b) dense objects floating in liquid oxygen in a glass Dewar vessel positioned just above the top of the magnet bore. Top to bottom: a gallium arsenide crystal, a £1 coin, a piece of lead, and a gold coin (Catherall et al., 2003). The celebrated "Of flying frogs and levitrons" experiment earned Michael Berry and Andre K. Geim the 2000 Ig Nobel Prize. Subsequently, Sir Andre was awarded the 2010 Nobel Prize in Physics for his groundbreaking work on graphene (Geim, 2011), becoming the only recipient of both Nobel and Ig Nobel Prizes (www.nobelprize.org/prizes/physics/2010/geim/facts/).

mass diffusion comparable to momentum diffusion (order one Schmidt number). For cases with random initial perturbations, the RTI can be delayed and its growth rate diminished by rotation (Fig. 11.4). Further, bubble formation is completely suppressed at sufficiently high rotation rates. Interestingly, rotation effects occur even in the earliest phase of the instability, even before the establishment of a mixing zone. This suggests that it should be possible to understand the effect of rotation, at least in the early phases, in terms of bubble and vortex ring formation.

Direct numerical simulations within the Boussinesq approximation show two snapshots of the temperature field, demonstrating that rotation produces clear qualitative effects (Fig. 11.5, Boffetta et al. (2016)). Figure 11.6(a) shows that rotation reduced the intensity of turbulent velocity fluctuations. By increasing the rotation rate, the vertical velocities are reduced. Eventually, for the highest rotation rate considered, the velocities near saturation. This should be compared with the non-rotating case, in which the vertical velocity fluctuations are observed to grow linearly in time, in agreement with the dimensional prediction.

As shown previously, vertical velocity fluctuations are responsible for the conversion of potential energy into kinetic energy. The efficiency of kinetic energy production, Σ, takes the value of 0.5 in the non-rotating case: half of the potential energy is converted

Fig. 11.2 Rayleigh–Taylor instability developing at the interface between a layer of 0.06 M manganese chloride solution (dyed fluorescent green) lying above a more dense 0.4 M sodium chloride solution (colorless) in a 90-mm diameter cylindrical tank. The instability is caused by the magnetic attraction of the weakly paramagnetic upper layer to a superconducting magnet situated beneath the tank (not visible in these images) and by the magnetic repulsion of the colorless diamagnetic layer from the magnet. In the upper row of images, the cylindrical container is not rotating; in the lower row, the container is rotating at 2.52 rad/s about its vertical axis. The images illustrate how rotation restricts the size of the structures that form at the interface. Fig. 2 of Scase et al. (2020). This figure is available under the terms of the Creative Commons Attribution 4.0 International license.

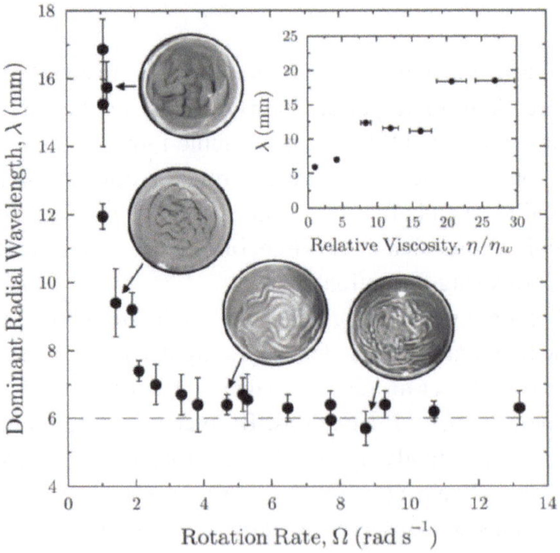

Fig. 11.3 The dominant scales of perturbation after the onset of instability. Error bars are associated with goodness-of-fit from the spatial autocorrelation algorithm. The solid line is an empirical best fit. (Baldwin et al., 2015). Open access article distributed under the terms of the Creative Commons CC BY license.

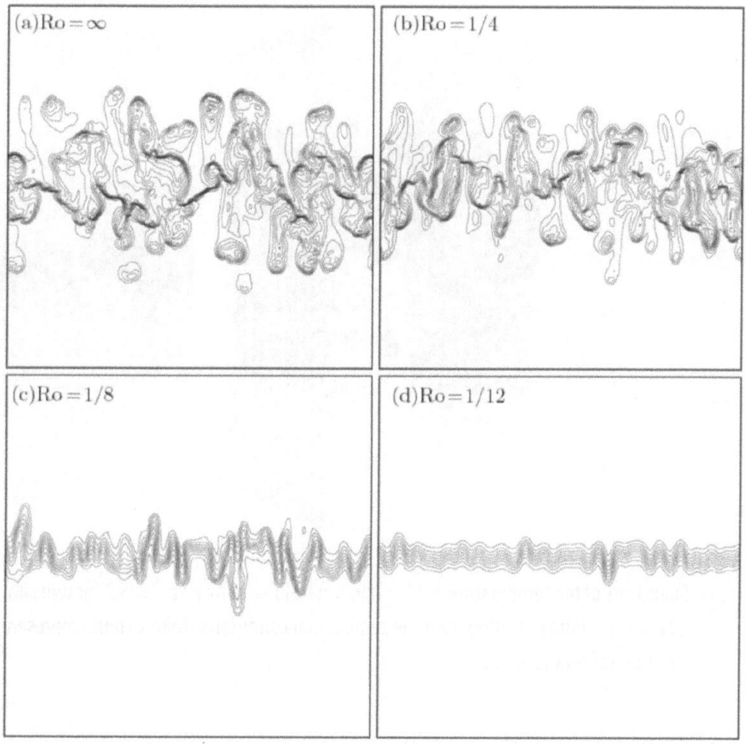

Fig. 11.4 The effect of rotation on RTI growth, measured with the nondimensional Rossby number (R_o), which is inversely proportional to the rotation rate. (a) $R_o = \infty$, (b) $R_o = 1/4$, (c) $R_o = 1/8$, (d) $R_o = 1/12$. Fig. 1 of Carnevale et al. (2002).

into large-scale kinetic energy, while the other half feeds the turbulent cascade, eventually being dissipated by viscosity. On the other hand, in the case of rapid rotation, Σ is reduced to 0.25–0.3 (Fig. 11.6(b)).

Unsurprisingly, as the rotation rate increases, the structure functions for velocity and temperature fluctuations reflect the same changes. The very short RTI "inertial range" scaling deteriorates rapidly as the rotation rate increases (Fig. 11.7).

Theoretical and numerical studies of centrifuged fluids have shown that rotation about an axis perpendicular to the acceleration inhibits RTI (Tao et al., 2013). Dávalos-Orozco (1996), as well as Dávalos-Orozco and Aguilar-Rosas (1989a,b), investigated the effect of a general rotation field on the RTI of a continuously stratified fluid and clearly demonstrated the important contribution of the horizontal rotation component on the stability in different directions. Duan et al. (2015) applied a general method of studying a family of modified variational problems (Guo and Tice, 2010) and illustrated that rotation diminishes the growth of instability.

These results suggest that it may be possible to slow or even eliminate RTI by the application of strong rotation. It is worth noting that a similar effect exists regarding flow

Fig. 11.5 Snapshots of the temperature field at nondimensional time $t/\tau = 5.2$ for two simulations with $\Omega\tau = 0$ (left) and $\Omega\tau = 10$ (right), starting from the same initial conditions. Yellow (blue) represents hot (cold), light (heavy) fluid. Fig. 1 of Boffetta et al. (2016).

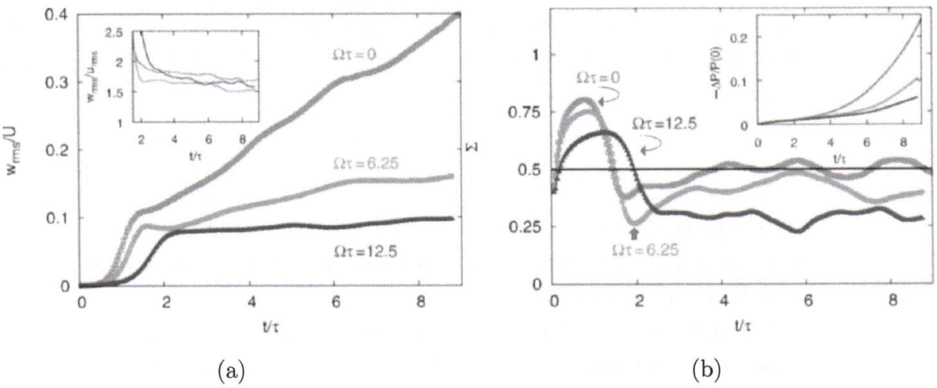

Fig. 11.6 (a) Evolution of vertical velocity fluctuations in the mixing layer for three different rotations. Inset: ratio of vertical to horizontal velocity fluctuations as a function of time, for the same values of rotation. (b) Efficiency of kinetic energy production as a function of time for three different rotations. Inset: potential energy loss for the same three values of rotations (from top to bottom). Figs. 2 and 5 of Boffetta et al. (2016).

curvature: Sen and Storer (1997) found that flow curvature can reduce the RTI growth. Also, the linear and the nonlinear RMI perturbation growth is affected by the interfacial shapes and curvatures (Guan et al., 2017).

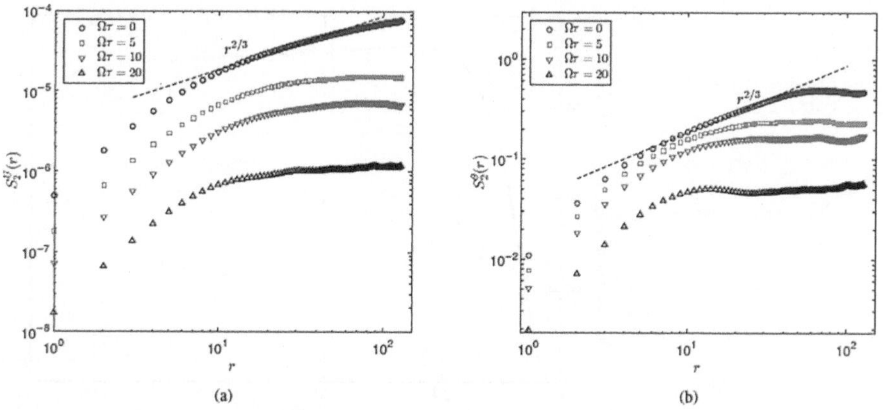

Fig. 11.7 Second order structure functions of the velocity and temperature from the lattice Boltzmann method for simulated rotating RTI. Fig. 15 of Wei et al. (2022).

11.2 Time-dependent acceleration/deceleration

Thus far, we have focused on the standard case of constant acceleration. However, there are important applications in which the acceleration changes in time and even reverses sign, including pulsating stars, supernovae explosions, and ICF (e.g., Colgate and White, 1966; Atzeni and Meyer-ter Vehn, 2004; Lindl, 1998; Betti et al., 2007).

While the AWE rocket-rig experiments were designed principally for constant RTI accelerations, time-dependent effects were also observed (Read and Youngs, 1983; Smeeton and Youngs, 1987). Indeed, the tank acceleration for experiment 110 ramped up during the initial motion before attaining a roughly constant value after 12 ms (Fig. 11.8(a)). RT flows with more complex acceleration histories have also been studied experimentally (Dimonte and Schneider, 1996, 2000; Dimonte et al., 2007), analytically (Mikaelian, 2009a, 2010), and numerically (Ramaprabhu et al., 2013, 2016; Aslangil et al., 2016, 2020, 2022; Boffetta et al., 2019; Livescu et al., 2021). The evolution of RT mixing with alternating acceleration and an additional localized perturbation was numerically investigated in Statsenko et al. (2013).

The LEM experiments have been performed with several deliberately chosen time-dependent acceleration profiles. Figure 11.8(b) shows the measured $g(t)$ history, scaled to Earth's gravity (Dimonte and Schneider, 1996, 2000). Dimonte et al. (2007) noted that two RT-unstable acceleration stages are generated by independently triggered capacitor banks at different times. The intermediate RT-stable deceleration stage is produced with a mechanical leaf-spring brake. The transitions are gradual due to the rise and fall times of the circuits and the gradual engagement and disengagement of the brake by the cell.

Additionally, RT unstable interfaces may be generated by using the acceleration provided by a rarefaction wave (Morgan et al., 2016, 2018c; Morgan and Jacobs, 2020; Wang et al., 2022b). The rarefaction wave generates a large but nonconstant acceleration, and

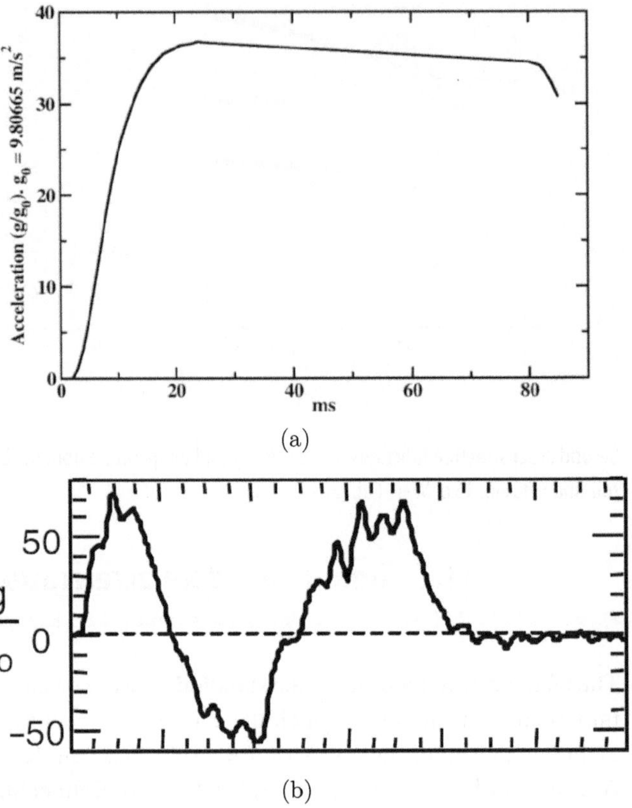

Fig. 11.8 (a) Smeeton and Youngs (1987), with permission. UK Ministry of Defence © Crown Owned Copyright 2022/AWE. (b) Scaled LEM acceleration vs time. Fig. 1 of Dimonte et al. (2007).

results in time varying densities (and Atwood numbers). As described by Morgan et al. (2016), the interaction of the rarefaction wave with the interface leads to a sustained large-magnitude acceleration. Figure 11.9(a) is an $z - t$ diagram generated using the method of characteristics, showing the refraction of the rarefaction wave, initially shown as a rarefaction centered at the diaphragm location. The refraction process results in the generation of a transmitted rarefaction wave, a reflected compression wave and the displacement of the interface down the rarefaction tube. Figure 11.9(b) shows the acceleration history produced in flat-interface experiments for the CO_2 /SF_6 systems. The induced acceleration is maximum when the rarefaction wave first impacts the interface, after which it decays with time.

As will be shown in Chapter 22, so-called Reynolds-averaged Navier–Stokes equations (RANS) models are frequently employed for routine engineering and ICF design calculations. In the context of flows induced by hydrodynamic instabilities, these RANS models are known as "mix models," because the model constants are calibrated using a suite of RTI, RMI, and KHI experimental or numerical data, as discussed in this book. As noted by several authors (e.g., Llor, 2005; Dimonte et al., 2007). a full evaluation of mix models

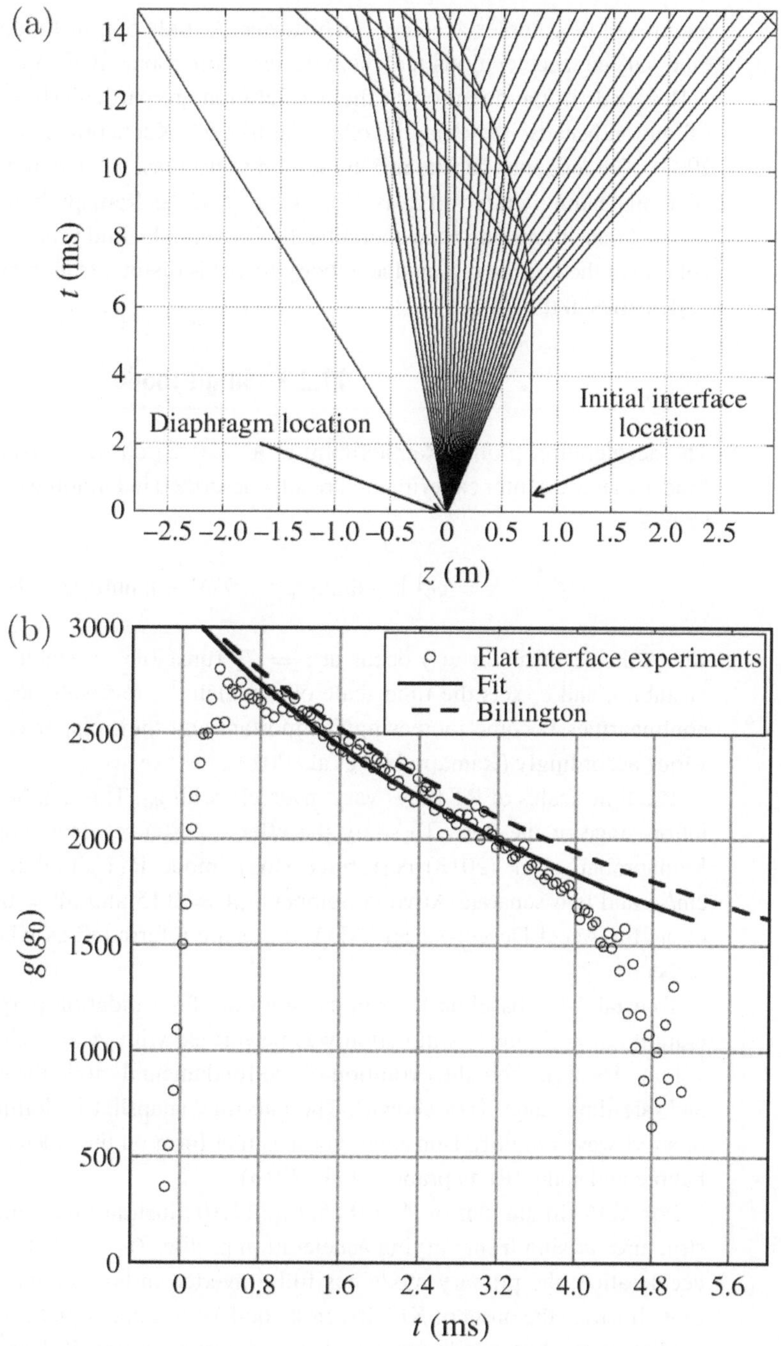

Fig. 11.9 (a) The wave and interface interactions in the z–t plane of a shock tube for CO_2/SF_6. The refracted rarefaction wave (fanned lines) generates a sustained acceleration on the interface, causing it to travel downwards. (b) The acceleration profile extracted using a moving parabolic fit to two flat-interface, CO_2/SF_6 experiments. This profile is compared with the theory of Billington (1956) and a decaying power-law fit, where g_0 is Earth's acceleration due to gravity. Morgan et al. (2016).

requires more complex acceleration histories. A study of how fluid interfaces evolve during time-varying accelerations may help design more robust ICF capsules (Llor, 2003).

Reversals in the direction of the acceleration are particularly useful for providing data validation because they may involve "demixing" (Kucherenko et al., 1997; Zaytsev et al., 2003). The tendency for the mixing region to decrease in size under deceleration is somewhat misleadingly referred to as "demixing," since as Ramaprabhu et al. (2013) has shown, variable density mixing is an energetically irreversible fluid process and compression of the volume of the mixing region due to body forces is distinct from rearranging the distribution of densities in the fluid.

11.2.1 Single mode

The acceleration profile Ramaprabhu et al. (2013) chose to examine is inspired by the Linear Electric Motor experimental results described in Dimonte et al., and is approximated here as

$$g = g_0\left[1 - \tanh[c_g(t - T_1)] - \tanh[c_g(t - T_2)]\right], \tag{11.1}$$

where the reversals in $g(t)$ occur at $t = T_1$ (unstable $\rightarrow$ stable) and $t = T_2$ (stable $\rightarrow$ unstable), and c_g sets the time scale over which the reversals occur. As RT flows evolve, nonlinearities become increasingly important and their response to acceleration reversal varies accordingly (Ramaprabhu et al., 2013).

The time scales of RT flows vary inversely with g_0. This can be used to control the evolution stage of the flow. To study the effect of $g(t)$ reversals applied at different times, Ramaprabhu et al. (2013) performed single-mode RTI simulations with $g_0 = \{2, 4, 8\}$ cm/s^2 and two separate Atwood numbers, $\mathcal{A} = 0.15$ and $\mathcal{A} = 0.90$. Based on the value of g_0, the Accel-Deccel-Accel (ADA) cases are referred to as ADA2, ADA4, and ADA8, respectively.

To establish a baseline for comparison (and for validation purposes), a constant acceleration ($g_0 = 2$ cm/s^2) simulation was performed with $\mathcal{A} = 0.15$. Vertical planes of fluid volume fraction show the evolution of the fundamental mode through the linear, nonlinear and late-time stages (not shown). The late-time snapshot is dominated by the maturation of short wavelength KH-unstable vortices that form on the sides of the fundamental mode bubble and spike (Ramaprabhu et al., 2013).

For ADA simulations of $\mathcal{A} = 0.15$, Fig. 11.10 illustrates the complexity of the secondary structures arising from varying acceleration profiles ($g_0 = \{2, 4\}$ cm/s^2). At the end of the deceleration, the primary mode has fully inverted in both cases. The images at the latest time illustrate the onset of KH-driven secondary structures. In the case of $g_0 = 4$ cm/s^2, the nonlinear development is further advanced, where similar KH-unstable vortical structures appear with smaller length-scale features.

When $\mathcal{A}$ is increased to 0.9, the skinny spikes penetrate the light fluid shortly after the first acceleration reversal (Ramaprabhu et al., 2013). Moreover, the secondary KHI at the spike-bubble boundary is inertially suppressed. When g_0 is further increased to 4 cm/s^2, the advanced nonlinearity of the primary waveform creates more convoluted nonlinear

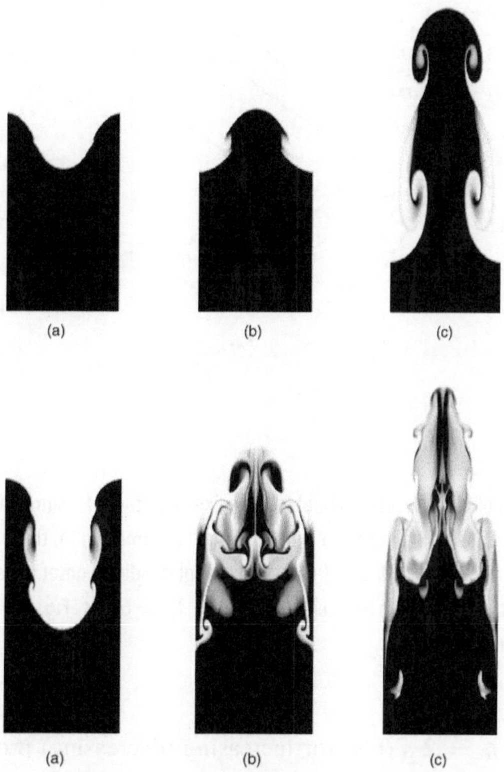

Fig. 11.10 Diagonal slices of volume fraction contours in simulations of single-mode RTI with accel-decel-accel histories, showing bubble and spike structures. Acceleration histories, (Top panel) $g_0 = 2$ cm/s^2, (Bottom panel) $g_0 = 4$ cm/s^2. (a) $t = 3.33$ s (T_1), (b) 6.66 s (T_2), and (c) 10 s (T_{end}). The simulations were performed with $\mathcal{A} = 0.15$. Figs. 4 and 5 of Ramaprabhu et al. (2013), with the permission of AIP Publishing.

structure and mixing than its counterpart with $g_0 = 2$ cm/s^2. However, mixing is still reduced relative to the corresponding $\mathcal{A} = 0.15$ simulations in Fig. 11.10 (not shown).

Figure 11.11 shows a decrease of the mixed widths during the deceleration period $T_1 < t < T_2$. In particular, the reduction of the mixing region volume is demonstrated to be a direct result of the phase reversal. For the $g_0 = 8$ cm/s^2 (ADA8) calculation, an asymmetry between h_b and h_s develops even at $\mathcal{A} = 0.15$. Increasing $\mathcal{A}$ only exacerbates the situation (Ramaprabhu et al., 2013).

Mikaelian (2009a, 2010) extended the potential flow models to study nonlinear RTI and RMI driven by variable accelerations in 3D. These simple potential flow models are only valid for low $\mathcal{A}$. In particular, he examined two idealized $g(t)$ profiles inspired by the linear electric motor experiments of Dimonte and Schneider (2000),

$$g(t) = g_0 \frac{t}{t_0}, \qquad t \le t_0, \tag{11.2}$$

$$= g_0 \left(1 - \frac{t - t_0}{t_L - t_0}\right), \qquad t > t_0, \tag{11.3}$$

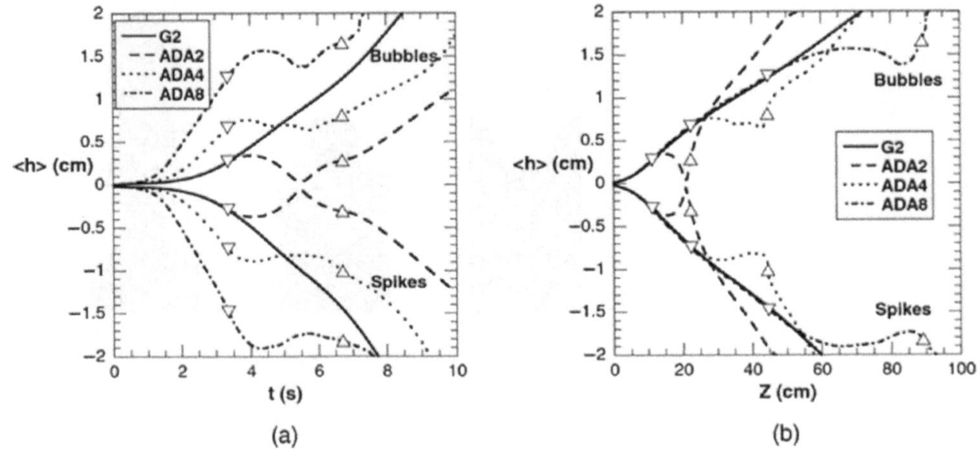

Fig. 11.11 Evolution of planar-averaged bubble and spike amplitudes for simulations with acceleration history of Eq. 11.1 and $\mathcal{A} = 0.15$. The amplitudes are plotted against (a) time and (b) the interface displacement Z, $Z(t) = \int_0^t \int_0^{t''} g(t')dt'dt''$. Downward triangles indicate onset of deceleration at $t = T_1 = 3.33$ s, while upward triangles indicate return to acceleration at $t = T_2 = 6.66$ s. Fig. 7 of Ramaprabhu et al. (2013), with the permission of AIP Publishing.

where $t_0 = 2$ s (6 s) for increasing (decreasing) profiles and the time of the end of the "experiment" $t_L = 8$ s for both cases (Ramaprabhu et al., 2016).

Ramaprabhu et al. (2016) revisited single-mode RTI driven by time-varying acceleration profiles, and $g(t)$ profiles that are suggestive of blast waves. Here, numerical simulations are compared with a potential flow model similar to Mikaelian's, as well as the buoyancy-drag model (see Section 3.2.2).

11.2.2 Multimode

We now turn to consider the case of multi-mode RTI. The first example is RTI induced flows in the presence of an alternating acceleration (Boffetta et al., 2019). Simulations of four mixing layers, obtained using DNS of the Boussinesq equations, are initiated with different periods of the alternating acceleration. The authors perturbed the initial condition by adding 5% of white noise to the initial temperature profile.

Mikaelian (2014c) considered another acceleration profile, $g(t) \sim t^{\tilde{n}}$. Previously, Llor (2003) considered this case in great detail for both positive and negative $\tilde{n}$ (theory, 3D simulation, buoyancy drag model). It should also be pointed out that this gives an important class of self-similar cases.

The initial conditions (ICs) may also affect the late-time evolution toward turbulence under acceleration reversal. This is a further elaboration of Chapter 9's discussion of ICs and classical RT-unstable flows, which has received considerable attention in the literature. Table 11.1 summarizes a study of four different ICs, but with the same Atwood number of 0.5 and an identical variable acceleration profile (Aslangil et al., 2016),

Table 11.1 ICs for all simulation cases reported in Aslangil et al. (2016).

Case	Wave-number range	Spectral index	Acceleration history
AS(0)-CG	32–64	0	constant gravity
AS(0)	32–64	0	ADA
AS(-2)	32–64	−2	ADA
BB	4–64	0	ADA

$$g(t) = g_0[1 - 2H(t-2) + 2H(t-4)] \tag{11.4}$$

where $g_0 = 4$ cm/s^2 and the Heaviside function, $H(x)$, represents the time

$$H(x) = \begin{cases} 1, & \text{for } x > 0 \\ 0, & \text{for } x < 0. \end{cases} \tag{11.5}$$

These ICs are imposed on the RT problem to investigate the effects of (a) spectral bandwidth and (b) spectral power distribution on the subsequent evolution (Aslangil et al. 2016). The AS(0) case is the starting point of the assessment. The spectral index of AS(0) is set to 0 to serve as a benchmark for comparison with the constant gravity α-group study (denoted as case AS(0)-CG). The case AS(0) can be compared with that of AS(−2), which has a spectral index of −2. Finally, the case AS(0) can be contrasted to a broadband IC case but now with the same zero spectral indexes.

Figure 11.12 displays volume fraction contours of the density field for both constant gravity and the ADA AS(0) case, with vertical slices taken along the domain center. The large coherent structures from the accelerating flow rapidly disintegrate during deceleration, leaving only smaller scales in the flow and a smaller range of densities. However, the situation is reversed quickly once the acceleration is restored.

Several statistics have been collected in Fig. 11.13 to shed light on the mechanism responsible for this observation. The vertical, dashed lines mark the times of acceleration reversal. Evolutions of h_s, h_b, and normalized vertical velocity variance are plotted as a function of the length scale

$$Z(t) = \int_0^t \int_0^{t''} g(t')dt'dt'', \tag{11.6}$$

where the intermediate variables t' and t'' are essentially arbitrary. Despite averaged quantities appearing independent of the ICs throughout the deceleration phase, some memory of the ICs persists into the second acceleration phase, and the trajectories of these three statistics diverge.

During deceleration, mixed widths were reduced for all three cases (Fig. 11.13(a)). The mass flux values during this period indicate no net conversion from potential to kinetic energy (Fig. 9 of Aslangil et al. (2016), not shown); the residual kinetic energy in the flow undergoes decay due to viscous dissipation. As a result, all cases observed a reduction in the normalized vertical velocity variance during the deceleration phase (Fig. 11.13(b)).

The RTI and RMI configurations differ, of course, in their initial acceleration mechanisms. While long-duration buoyancy forces generate RTI vorticity on the interface, an

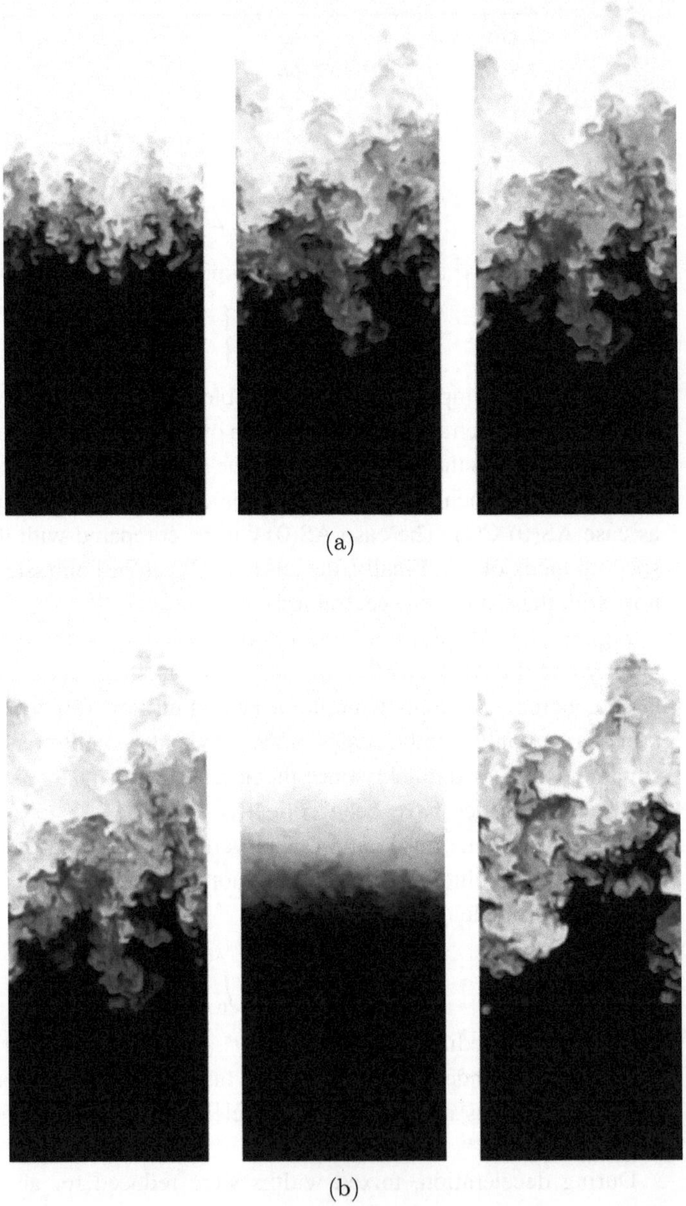

Fig. 11.12 Density contours of the RT mixing layer ($x-z$ planar slices) taken along the center of the domain; the initial conditions of AS(0)-CG (top panel) and AS(0) ADA (bottom panel) cases are defined in Table 11.1. Fig. 4 of Aslangil et al. (2016).

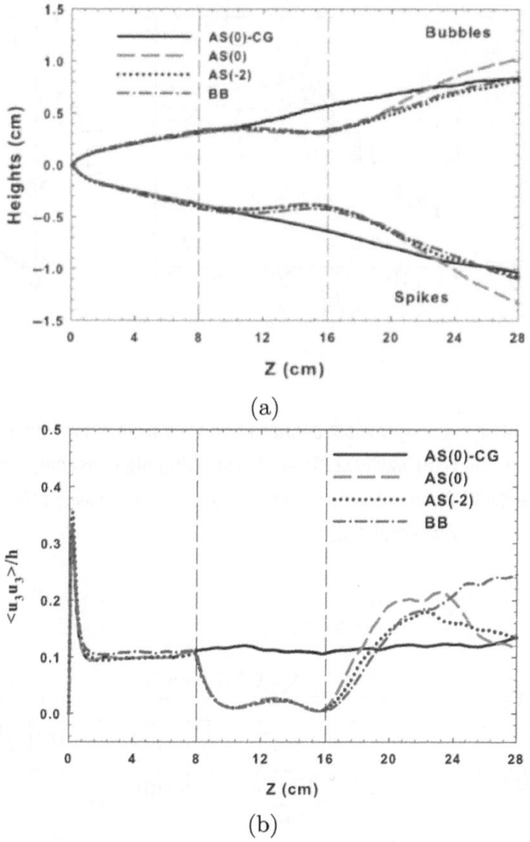

Fig. 11.13 (a) Evolution of a number of key measurements vs interface displacement (Z). (a) the bubble and spike heights and (b) normalized vertical velocity variance. Figs. 5 and 10 of Aslangil et al. (2016).

RMI mixing layer evolves with no external forcing after the initial shock-induced vorticity injection. An interesting case, which is closely related to the RMI, is one in which the applied acceleration is completely withdrawn at some point during the evolution (Aslangil et al., 2020; Livescu et al., 2021).

Figure 11.14 presents mixed-width results from DNS of RTI at Atwood numbers up to 0.9. After the layer width had developed substantially, additional branched simulations have been run under reversed and zero gravity conditions. Here, RTI+, RTI−, and RTI0 denote the forward ($g > 0$), reversed ($g < 0$), and zero ($g = 0$) gravity cases, respectively (Livescu et al., 2021).

Whereas the anisotropy tensor oscillates in the RT- configuration, in the RTI0 case, there is simply a progressive decay of anisotropy (Fig. 11.15(b)), which scales self-similarly with kinetic energy, once again consistent with RMI. The turbulent Reynolds number follows

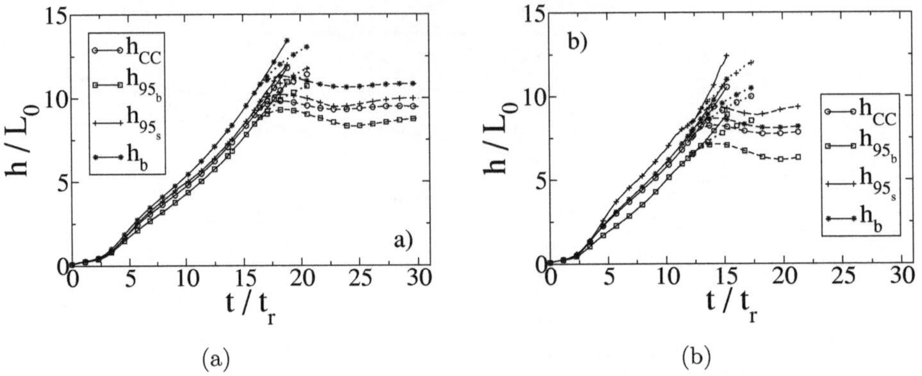

Fig. 11.14 The evolution of the layer width defined using several definitions detailed in Livescu et al. (2021). The main simulations (forward gravity or RTI+), shown with continuous lines, are branched into additional simulations with zero gravity (RTI0), shown with dotted lines, and reversed gravity (RTI−), shown with dashed lines. (a) $\mathcal{A} = 0.5$ and (b) $\mathcal{A} = 0.9$. Livescu et al. (2021).

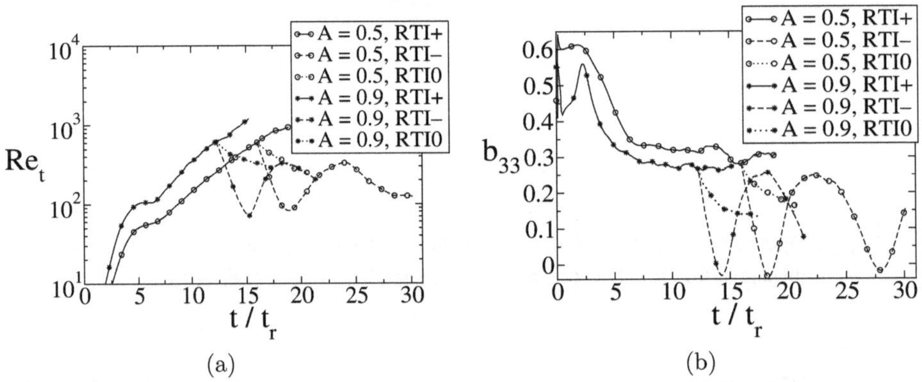

Fig. 11.15 (a) The evolution of the turbulent Reynolds number at the location where TKE reaches its maximum value, for RTI+, RTI−, and RTI0 cases with different Atwood numbers. (b) Evolution of the anisotropy tensor for time-dependent RTI cases with different Atwood numbers. The main simulations (forward gravity or RTI+), shown with continuous lines, are branched into additional simulations with zero gravity (RTI0), shown with dotted lines, and reversed gravity (RTI−), shown with dashed lines. Livescu et al. (2021).

the expected decay for the RTI0 cases but seems to oscillate without significant overall decay for the RTI− cases (Fig. 11.15(a)). This indicates that the flow undergoes significant transitory changes after the gravity reversal.

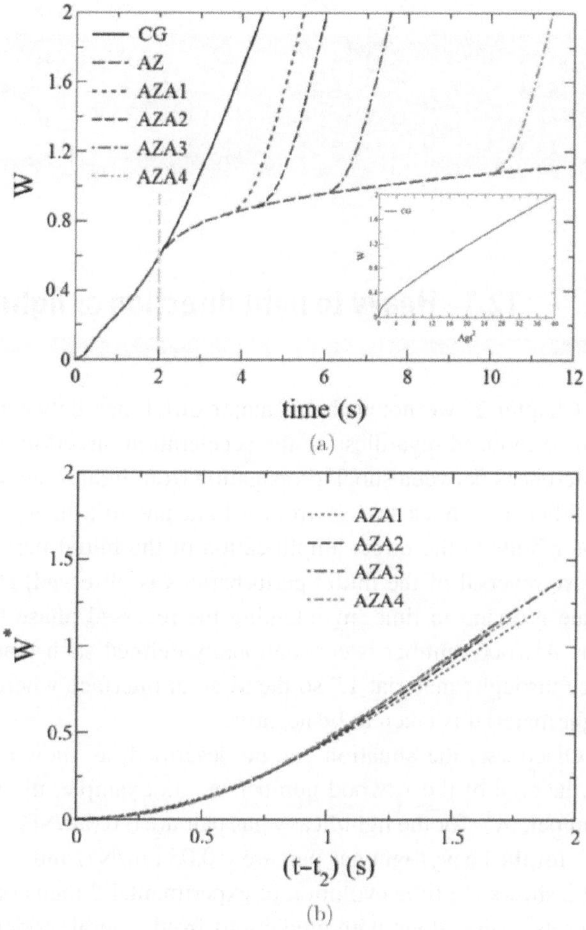

Fig. 11.16 The time evolution of (a) the mixing layer width of the RT instability for all cases considered. Three different acceleration profiles are examined in this study: (i) constant-gravity (CG): classical RTI, in which the acceleration remains constant during the whole flow evolution, (ii) AZ: configuration in which the RTI begins as in case (i) until the applied acceleration is removed, and (iii) AZA: a sequence of configurations that follow the acceleration profile of (ii) until the acceleration is re-applied [Note that the cases CG (solid line) and AZ (long dashed line) correspond to RT+ and RT0 of Figs. 11.14 and 11.15.] (b) the normalized mixing layer width from the time acceleration is restored. Aslangil et al. (2020).

Furthermore, Aslangil et al. (2020) varied the intermediate duration over which acceleration is not applied. They found that as growth resumes, the re-accelerated mixing layer is statistically indistinguishable from classical self-similar RTI (Fig. 11.16).

12 Direction, strength, and shape of incident shock waves

12.1 Heavy to light direction or light to heavy direction

In Chapter 2, we noted that a major difference between RTI and RMI is that the latter can be induced regardless of the acceleration direction. In this chapter, we illuminate the differences between shock propagation from light to heavy gas or vice versa.

When the shock travels from a light gas to a heavy gas, the shock-interface interaction results in the direct amplification of the initial perturbations. In the opposite case, a phase reversal of the initial perturbations is observed, after which the perturbations continue growing in time, maintaining the reversed phase (e.g., Jourdan and Houas, 2005). The Atwood number is conventionally defined such that the shock approaches the interface through "material 1," so the $\mathcal{A}$ at an interface where the shock passes from heavy to light material is taken to be negative.

Of course, the situation can be described, as shown in Fig. 12.1, using the language of the sign of the Atwood number. As an example, the values of the post-shock Atwood number, $\mathcal{A}^+$, for the light/heavy gas pair are 0.6 (air/SF_6) and 0.19 (air/CO_2). The values of $\mathcal{A}^+$ for the heavy/light gas pair are -0.05 (air/N_2) and -0.77 (air/He), respectively. Figure 12.2 shows the time evolution of experimental dimensionless bubble amplitudes for these four gas pairs, along with predictions from several models.

After further refining their experimental methods, Mariani et al. (2008) carried out another set of experiments for both the heavy–light and light–heavy sinusoidal interfaces. In particular, stereolithography has been used to design the membrane supports that initially materialize the gaseous interface, setting up highly accurate initial conditions. Two grids have been built for this study: The Grid 1 perturbation represents a single period sinusoidal bump, whereas the Grid 2 perturbation has a two-period sinusoidal shape. Figure 12.3 shows laser sheet images of both grids for two gas configurations, air/SF_6 and air/He. On each picture, the perturbation amplitude can be deduced by measuring between the extreme points of the interface contour (Fig. 12.4). Experiments are in very good agreement with both theory and simulations for the heavy–light case, but quickly deviate from them in the light–heavy configuration (Mariani et al., 2008), probably due to membrane effects.

Continuing their previous shock tube experimental studies (e.g., Zhai et al., 2011; Si et al., 2012) Luo et al. (2013) developed a method to create a discontinuous gaseous interface with a minimum-surface feature using the soap film technique. This interface has a zero-mean curvature (Isenberg, 1992), which differs from the interface used by Yosef-Hai et al. (2003), Chapman and Jacobs (2006), and Long et al. (2009). The initial condition

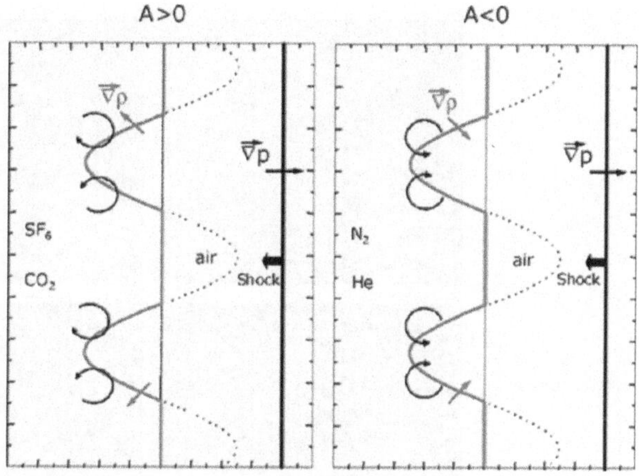

Fig. 12.1 Schematic of the initial perturbations, showing the deposition of vorticity in light/heavy and heavy/light cases (—— effective, ... imaginary). Fig. 1 of Jourdan and Houas (2005).

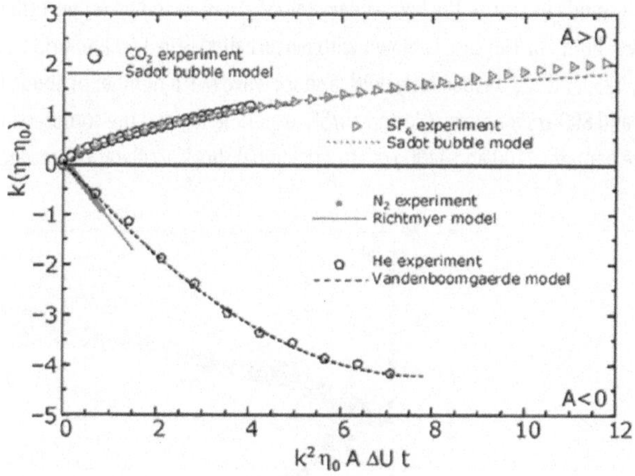

Fig. 12.2 Dimensionless bubble amplitude (before the reshock) as a function of dimensionless time for both positive and negative Atwood number cases. Fig. 3 of Jourdan and Houas (2005).

can be controlled, since the interface formed is free of a supporting mesh or pins. Five interfaces with different amplitudes are realized in the shock-tube experiments. The initial amplitude of the boundary varies from 3 to 7 mm (with a fixed wavelength of 35 mm). Figure 12.5 shows Schlieren sequences of the light–heavy, single-mode air(left)/SF_6 interface evolution in the experiments.

The instability evolution in the opposite SF_6/air configuration has also been investigated with the soap film technique and appears quite different (Fig. 12.6). After the shock impact, a phase inversion occurs, and a bubble-like structure (denoted as "S1") is observed behind the interface, which may be caused by the density variation. As the interface amplitude

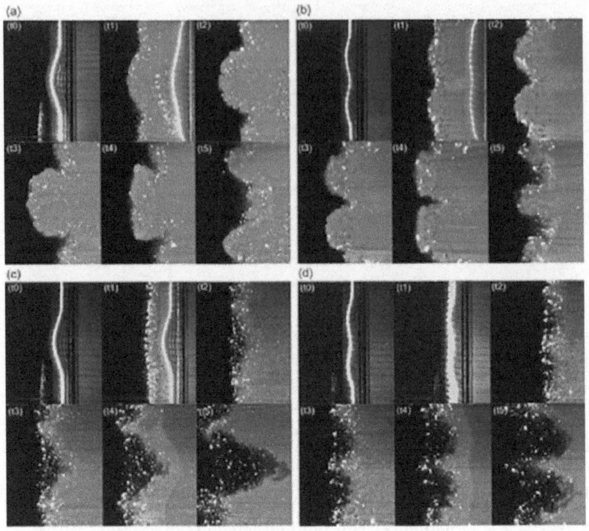

Fig. 12.3 Laser sheet pictures showing the interaction of a shock wave (moving from right to left) with a stereolithographed single-mode interface. The light–heavy (air/SF_6) case is shown for perturbation Grid 1 (a) and Grid 2 (b), and the heavy–light (air/He) case is shown with perturbation Grid 1 (c) and Grid 2 (d). (a), (b), (c), and (d) are from runs 126, 120, 125, and 123, respectively, with a shock wave Mach number of about 1.4. Time between frames is approximately 300 and 800 μs for the air/He and air/SF_6 experiments, and the compression of the interface by the reflected shock wave happens 1100 and 3200 μs after the initial shock acceleration, respectively. Fig. 1 of Mariani et al. (2008).

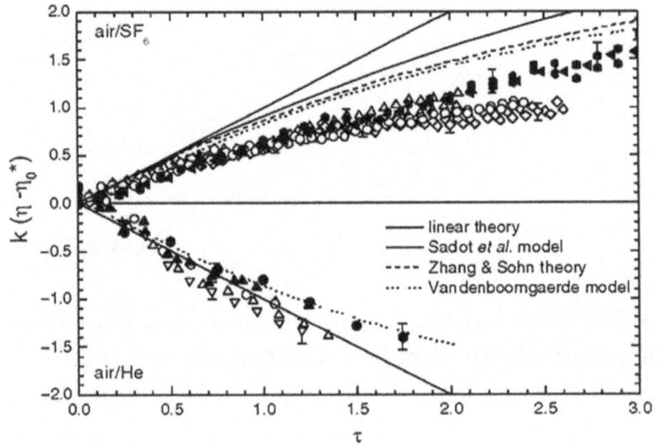

Fig. 12.4 Dimensionless-time evolutions of the amplitude for light–heavy (up) and heavy–light (down) sinusoidal interfaces. Black and white symbols represent experimental results at high and low Mach numbers, respectively. Fig. 2 of Mariani et al. (2008).

grows, S1 becomes more obvious. Due to the stronger shock–shock interaction (Luo et al., 2016c), it also lasts longer.

Using their edge contrast diagnostic, Probyn et al. (2021) show plots for two cases of 2D single-mode RMI. They have post-shock conditions $a^+/\lambda = 0.01$ and $\mathcal{A}^+ = \pm 0.5$,

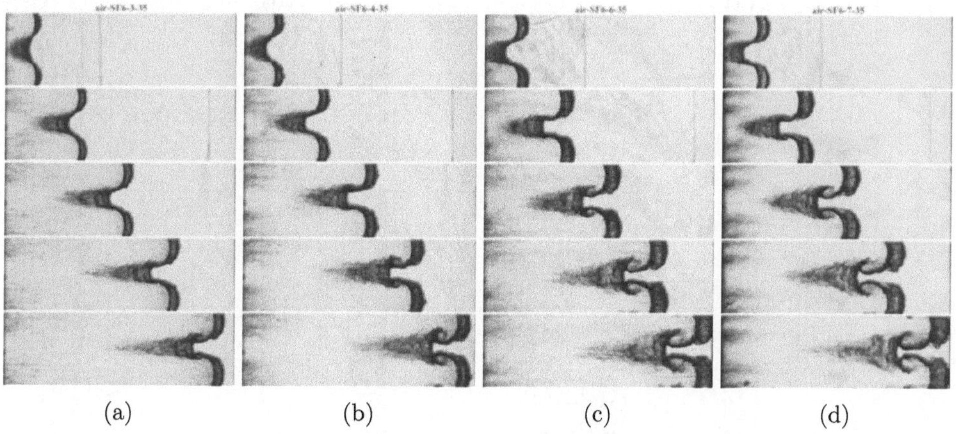

Fig. 12.5 Schlieren sequences showing the evolution of shocked air(left)/SF_6(right) interface for initial amplitudes varying from 3 to 7 mm. The shock propagates from left to right. Inserted numbers indicate the time in ms. Fig. 2 of Luo et al. (2013).

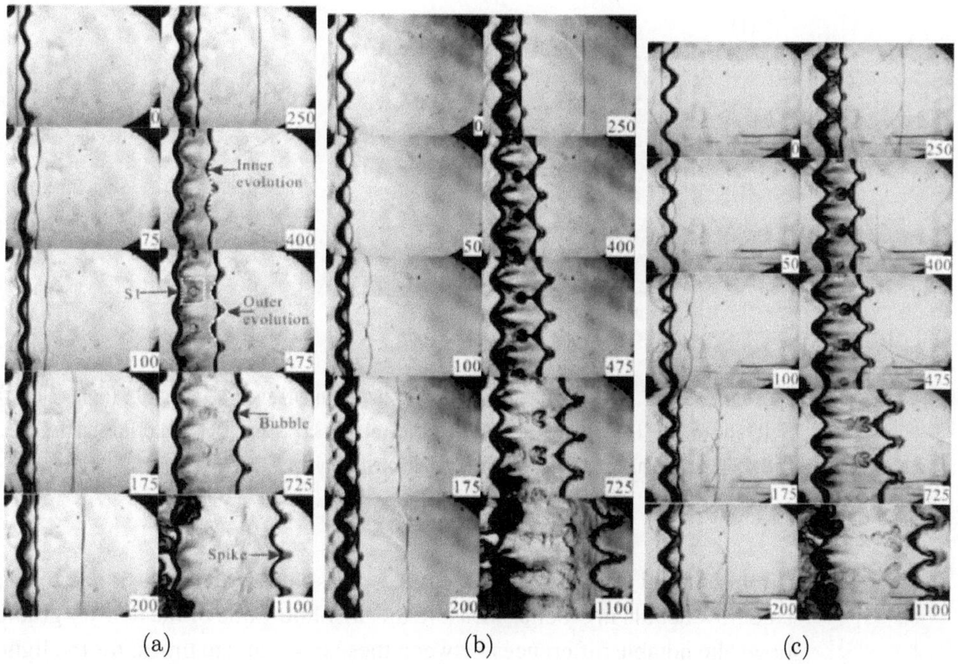

Fig. 12.6 Schlieren sequences showing the evolution of the shocked SF_6/air interface with three initial amplitudes: (left) $a_0 = 3$ mm, (middle) $a_0 = 5$ mm, and (right) $a_0 = 7$ mm. Here the numbers indicate time in μs. Figs. 4–6 of Luo et al. (2016c).

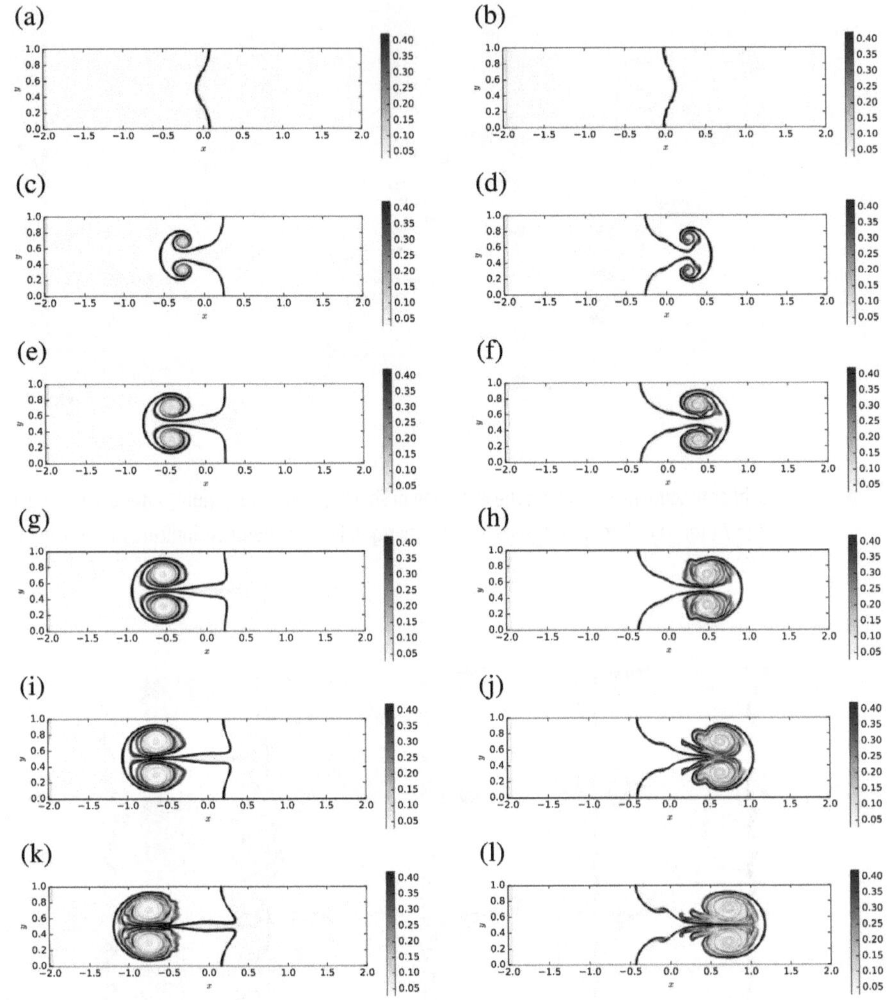

Fig. 12.7 Edge contrast for $a^+/\lambda = 0.01$ with $\mathcal{A}^+ = 0.5$ (left column) and $\mathcal{A}^+ = -0.5$ (right column). (a) and (b) $t = 0.1$, (c) and (d) $t = 1.0$, (e) and (f) $t = 2.0$, (g) and (h) $t = 3.0$, (i) and (j) $t = 4.0$, (k) and (l) $t = 5.0$. Fig. 12 of Probyn et al. (2021). UK Ministry of Defence © Crown Owned Copyright 2022/AWE.

with low initial amplitudes. The edge contrast technique is designed to draw out the fine-scale flow details in order to measure the absolute value of the density gradient. Figure 12.7 shows the notable differences between these cases at late times: for the light-to-heavy case, the growing bubble flattens and inverts, while for the heavy-to-light case, the vortex roll-up appears to be less smooth and regular (Probyn et al., 2021).

Further light can be shed on the problem by analyzing it from the vorticity viewpoint. Figure 12.8 illustrates the out-of-plane vorticity component. The vorticity distribution in the heavy fluid is particularly important in determining the shape of the bubble. In both

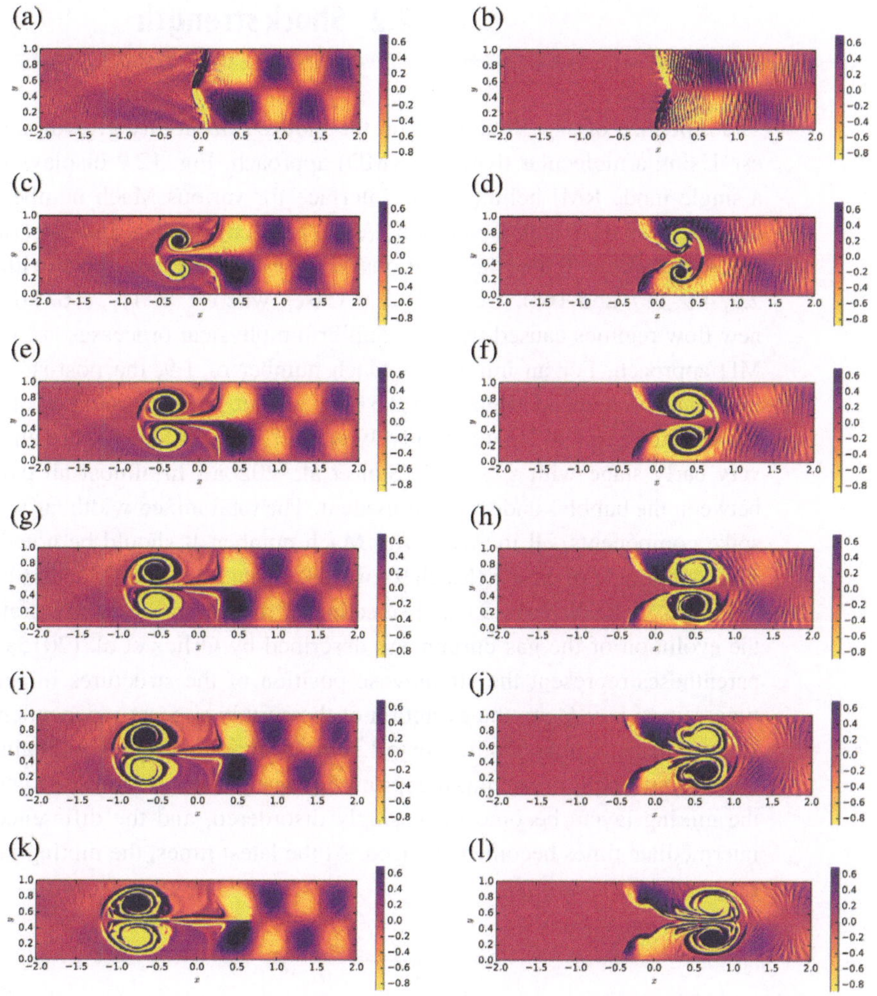

Fig. 12.8 Vorticity for $a^+/\lambda = 0.01$ with $\mathcal{A}^+ = 0.5$ (left column) and $\mathcal{A}^+ = -0.5$ (right column). (a) and (b) t = 0.1, (c) and (d) t = 1.0, (e) and (f) t = 2.0, (g) and (h) t = 3.0, (i) and (j) t = 4.0, (k) and (l) t = 5.0. Fig. 14 of Probyn et al. (2021). UK Ministry of Defence © Crown Owned Copyright 2022/AWE.

cases, the strongest bulk vorticity deposition is by the transmitted shock, although in the light-to-heavy case, there is also vorticity deposition by the reflected shock. In the light-to-heavy case, the vorticity deposited at the initial interface is progressively dragged with the spike material. In the heavy-to-light case, the reflected rarefaction deposits minimal vorticity, so the shape of the bubble will be determined by an irrotational velocity field (Probyn et al., 2021).

Finally, it should be noted that Velikovich et al. (2014) considered the change from $\mathcal{A}$ to $-\mathcal{A}$ in the weakly nonlinear limit.

12.2 Shock strength

The influence of shock intensity on the induced instability growth is of significant interest. Using a molecular dynamics (MD) approach, Fig. 12.9 displays the development of a single-mode RMI helium/argon interface for various Mach numbers. The densities of the gases are 0.24 g/cm^3 and 1.26 g/cm^3, respectively. The initial perturbation amplitude is low at $a_0/\lambda = 0.05$. These various Mach number experiments, with values of 1.9, 2.4, 2.9, 4.9, 9.0, and 16.0, are marked as Case$_i$, where $i = 1, \ldots, 6$. For microscopic RMI, new flow regimes caused by non-equilibrium physical processes are well captured by the MD approach. For an initial low Mach number of 1.9, the post-shock Atwood number is 0.59; it is reduced slightly to 0.51 for an initial Mach number of 16. The amplitude growths for cases 1–3, with relatively weaker incident shocks, are almost identical in the very early stage with $\tau < 0.4$ (Sun et al., 2020a). In almost all cases, the asymmetry between the bubbles and spikes is evident. The total mixed width, as well as its bubble and spike components, all increase with Mach number. It should be noted, however, that the spike amplitude of case 5 (Mach 9) overtakes that of case 6 (Mach 16).

Figure 12.10 displays a small selection of measured SF_6 concentration maps, showing the evolution of the gas curtain. As described by Orlicz et al. (2015a,b), the numbers in parentheses represent the streamwise position of the structures in centimeters. At early times, up to $x \approx 6$ cm, the evolution of the curtain appears independent of Mach number. At intermediate times, from 6 cm to 13 cm, the fluid mixing layers begin to grow differently and exhibit smaller scale mixing, primarily near the vortex cores. At late times, $x > 13$ cm, the mixing layers become increasingly disordered, and the differences first observed at intermediate times become enhanced. At the latest times, the mixing layers exhibit a wide range of scales.

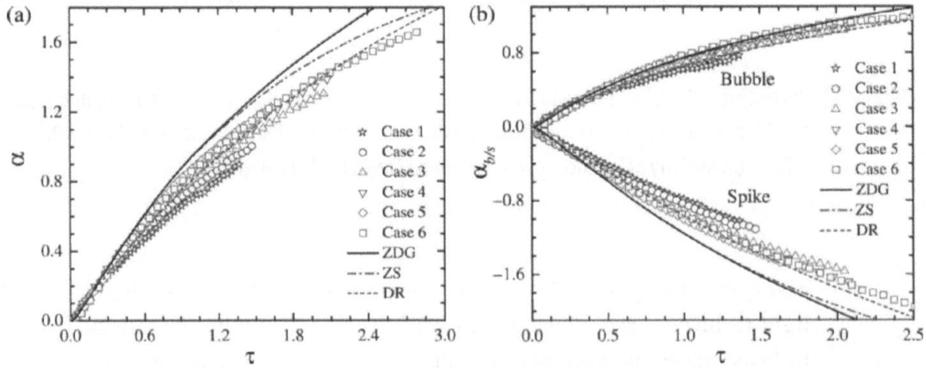

Fig. 12.9 (a) The full perturbation amplitude and (b) the amplitudes of bubbles and spikes. Symbols stand for the MD results and lines for the model predictions. The Mach numbers of the simulations, marked as Case i, i = 1,...6, are $M = 1.9$, 2.4, 2.9, 4.9, 9.0 and 16.0. Nonlinear models plotted here are those from Dimonte and Ramaprabhu (2010) (DR), Zhang and Sohn (1997) (ZS), and Zhang et al. (2018b) (ZDG). Fig. 3, Sun et al. (2020a), with the permission of AIP Publishing.

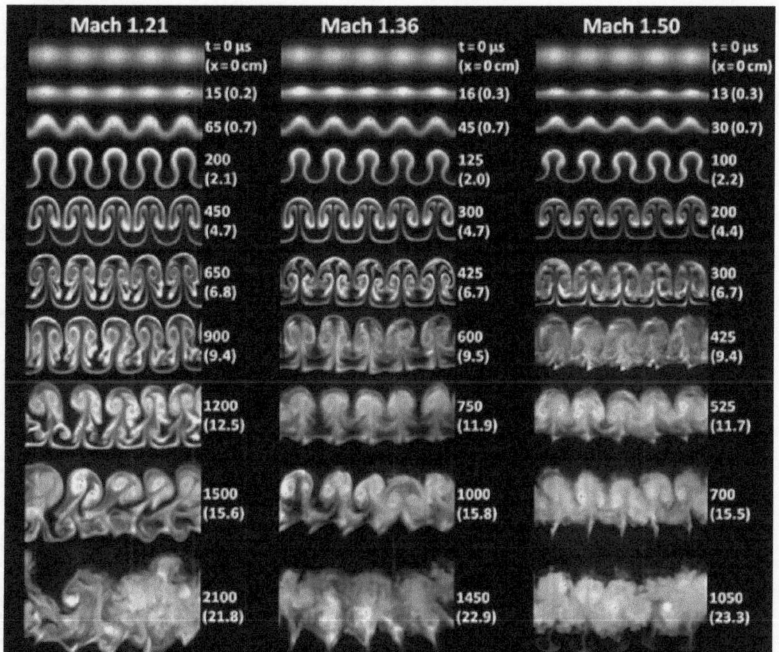

Fig. 12.10 A selection of SF_6 volume fraction maps from each Mach number experiment. Images in each row were acquired at roughly the same location as indicated in the figure. Fig. 1 of Orlicz et al. (2015a).

The time evolution of the 2D vorticity fields, measured simultaneously with the concentrations fields, is shown in Fig. 12.11. As expected, values for vorticity are higher for higher Mach numbers. where the primary vortices also dominate a greater region of the flow (Orlicz et al., 2013). In each case, as time progresses, the alternating blobs of positive and negative vorticity begin to lose their symmetry. This precipitates a transition of the vorticity field to a more disordered state and a breaking down of the primary vortices into smaller vortices.

Figure 12.12 shows that similarities between the two Mach number images are remarkable given the $\sim 2x$ difference in interface velocity. The greater compression of the $M = 2.2$ flow results in a thinner mixing layer at the same window location (Weber et al., 2014a).

Figure 12.13(a) shows the mixedness measurement, Θ, for three values of the Mach number (Orlicz et al., 2013). These values are significantly higher than those reported in numerical RMI simulations at five different Mach numbers (Lombardini et al., 2012), where $\Theta = 0.85$ was found to be independent of Mach number (Fig. 12.13(b)).

More recently, Reese et al. (2018) also experimentally investigated RMI in a vertical shock tube. A broadband initial condition was imposed on an interface between a helium–acetone mixture and argon ($\mathcal{A} \approx 0.7$). In this work, the density interface is accelerated by either a Mach 1.6 or 2.2 planar shock wave and a shear layer is introduced at the interface to serve as a statistically repeatable, broadband initial condition. Measurements

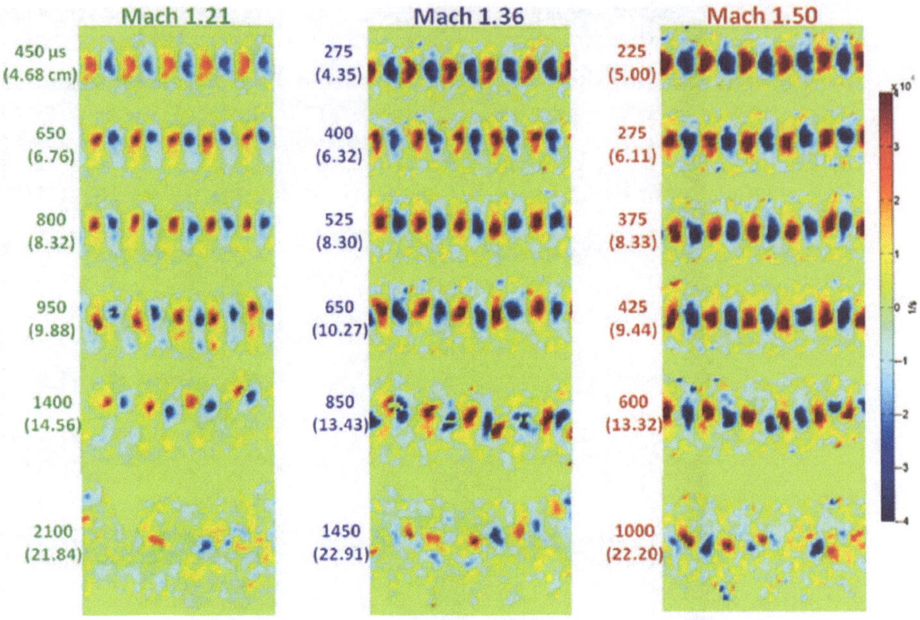

Fig. 12.11 Comparison of the vorticity evolution from each Mach number experiment shows the breaking down of the primary vortices over time. The color scale is held constant for all maps. Fig. 14, Orlicz et al. (2013), with the permission of AIP Publishing.

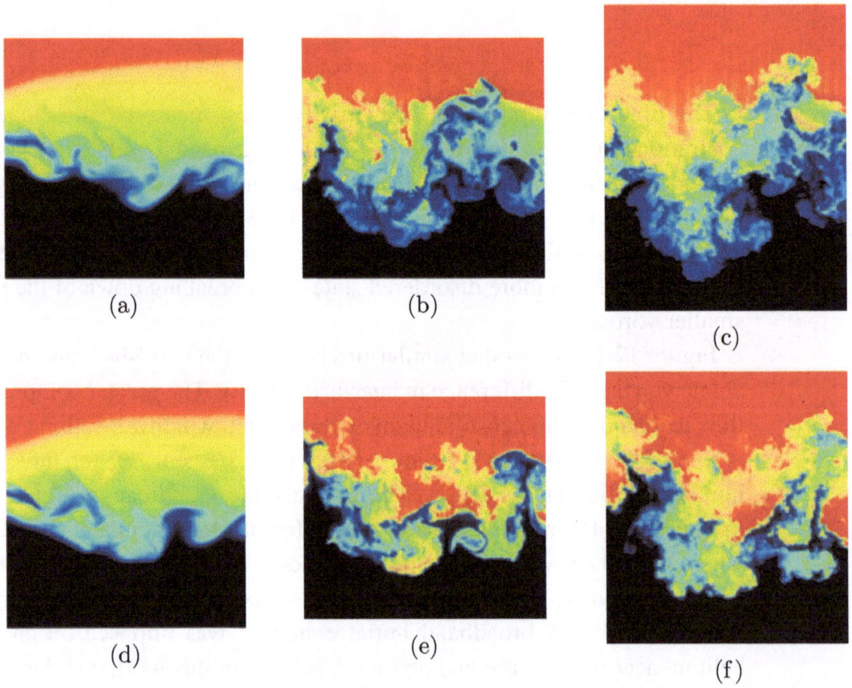

Fig. 12.12 Sequence of PLIF images for a He/Ar interface. Top row: Mach 1.6; Bottom row: Mach 2.2. Times are from initial shock interaction. Figs. 3 and 4 of Weber et al. (2014a).

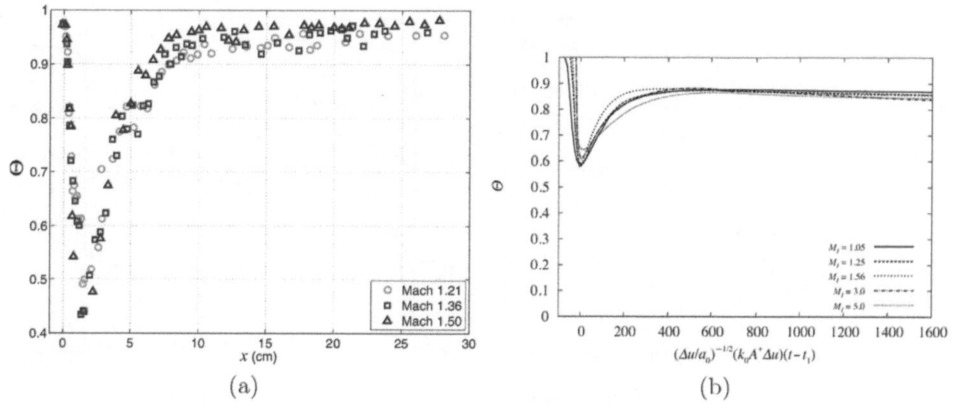

Fig. 12.13 Mixedness, Θ, measured numerically. (a) Fig. 24 of Orlicz et al. (2013), with the permission of AIP Publishing. (b) Fig. 17 of Lombardini et al. (2012).

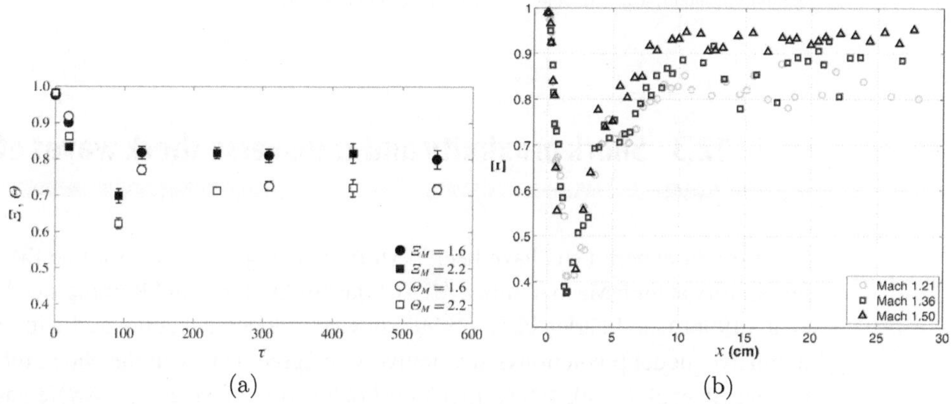

Fig. 12.14 Mixedness, Σ and Θ, measured experimentally. (a) Fig. 10 of Reese et al. (2018). (b) Fig. 23 of Orlicz et al. (2013), with the permission of AIP Publishing.

show that for both values of Mach number, Θ reaches an even lower final value, slightly above 0.7 (Fig. 12.14(a)).

As for the second mixedness measurement, Ξ, the asymptotic values obtained by Reese et al. (2018) are near 0.8 for both Mach cases (Fig. 12.14(a)). This is fairly close to the value reported for Mach 1.21 in Orlicz et al. (2013), but much lower than those corresponding to Mach 1.36 and 1.50 (Fig. 12.14(b)).

Figure 12.15 shows the simulated mixed width, the integral length scale, the Taylor microscale, and the Kolmogorov length scale for the three shock Mach numbers, 1.05, 1.2, and 1.5. It verifies the increasing separation of scales with increasing shock Mach number (Tritschler et al., 2014a). In the experimental investigation of Weber et al. (2013), however, the separation of scales appears not to depend on the shock Mach number.

Further research, perhaps on a single platform, is needed to understand the discrepancy among these measured mixedness and length parameters. The corresponding values of the more recently proposed mixed mass would also improve our understanding of this issue.

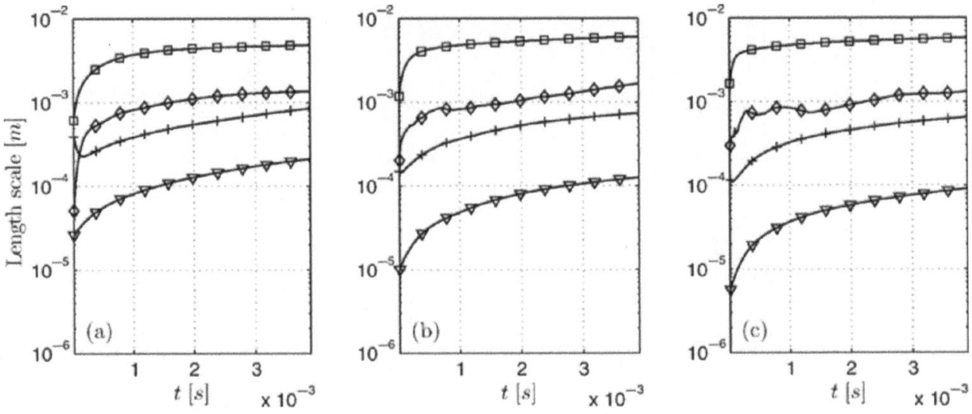

Fig. 12.15 Summary of simulated length scales for M = 1.05 (a), M = 1.2 (b), and M = 1.5 (c). The solid lines with symbols represent the mixing zone width (squares), the integral length scale (diamonds), the Taylor microscale (plus), and the Kolmogorov length scale (triangles). Fig. 8 of Tritschler et al. (2014a).

12.3 Shock proximity and transverse shock waves effects

A variety of experiments have been performed using lasers to examine the effect of compressibility at high Mach number, M. The data of Dimonte and Remington (1993), Dimonte et al. (1996b), and Farley et al. (1999) showed reduced v_0 (compared with incompressible impulsive model predictions), qualitatively in agreement with the shock tube experiments of Aleshin et al. (1990, 1997) (incident shock waves of Mach 2.5, Ar/Xe gas pair (Aleshin et al., 1990) and Mach 3.5, He/Xe gas pair (Aleshin et al., 1997)). Importantly, these experiments used large initial amplitudes, ($ka_0 > 1$, and as large as $ka_0 = 4$) to achieve high initial v_0. While it is natural to suppose that experimental deviations from the predictions of small-amplitude ($ka_0 < 1$) models is mainly related to the high M compressibility effects (Aleshin et al., 1997; Dimonte et al., 1996b; Holmes et al., 1999), it has also been suggested by Velikovich and Dimonte (1996), among others, that high ka_0 reduces the growth velocity (see also, Wang et al., 2023a, b).

In order to better distinguish between the compressibility and amplitude effects, Fig. 12.16 shows the measured initial growth velocity of Aleshin et al. (1990, 1997) and Dimonte et al. (1996b) relative to the linear extrapolation of the small-amplitude calculation. Also included is the data for (low) Mach 1.2 shock-tube experiments carried out by Rikanati et al. (2003) with a similar range of high initial amplitudes. The solid and dashed lines are drawn to guide the eye on the two classes of behavior.

Figure 12.16 shows that most experiments (class "A") follow the same trend in their reduction of the growth velocity. The qualitatively similar behavior, observed for a wide range of Mach numbers, suggests that in these experiments the large initial amplitude has a strong effect.

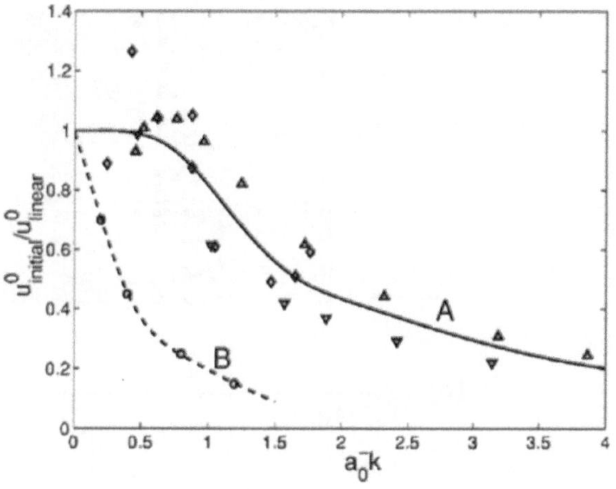

Fig. 12.16 Measured initial growth velocity relative to the linear extrapolation of the small-amplitude calculation. Fig. 1 of Rikanati et al. (2003).

Schematically marked as class "B," the Mach 3.5 He/Xe experiment of Aleshin et al. (1997) appears to exhibit a different behavior from all other datasets, suggesting the existence of a distinctive mechanism reducing the initial growth velocity. In this experiment, ka_0 is smaller than unity, so that the effect of the large initial amplitude is almost negligible (Rikanati et al., 2003). Experiments conducted on the Omega laser also showed reduced v_0 in a strongly shocked system with initial amplitudes $a_0k = 0.29$ and 0.92 (Glendinning et al., 2003).

The mechanism behind the observed decrease of the growth velocity v_0 relative to the small-amplitude calculation is often referred to as *shock proximity effect*. When a strong shock reaches an interface between two densities, the high compression will result in a transmitted shock that recedes slowly from the interface. Indeed, the shock will represent an impenetrable boundary, reducing v_0 (e.g., Holmes et al., 1999; Hurricane et al., 2000; Glendinning et al., 2003).

Schlieren photographs from a Mach 2, $a_0k = 0.72$ air/SF_6 experiment are shown in Fig. 12.17(a). It is seen clearly that the shock wave lingers in the vicinity of the perturbed interface and the shock reverberation is traveling in a direction tangential to the interface propagation. Due to the small initial conditions, the perturbation initially evolves in accordance with the nonlinear classical model (Fig. 12.17(b)). The deceleration and negative velocity effect flatten the tip of the bubble and increase the bubble radius of curvature, as is evident in the insert of Fig. 12.17(c).

As described in Guo et al. (2020), when the incident shock transmits through the high amplitude interface, the reflected shock travels across the interface and transverse shock waves (TRS) are formed (Fig. 12.18). The interaction of the transmitted shock waves from the upper and lower half-planes generates additional transverse shock waves (R1). These transverse shock waves then create regions of high pressure in front of the spikes and bubbles, thereby prohibiting (promoting) the bubble (spike) growth. The continuous interaction of the transverse shock waves with the interface leads to interface deformation.

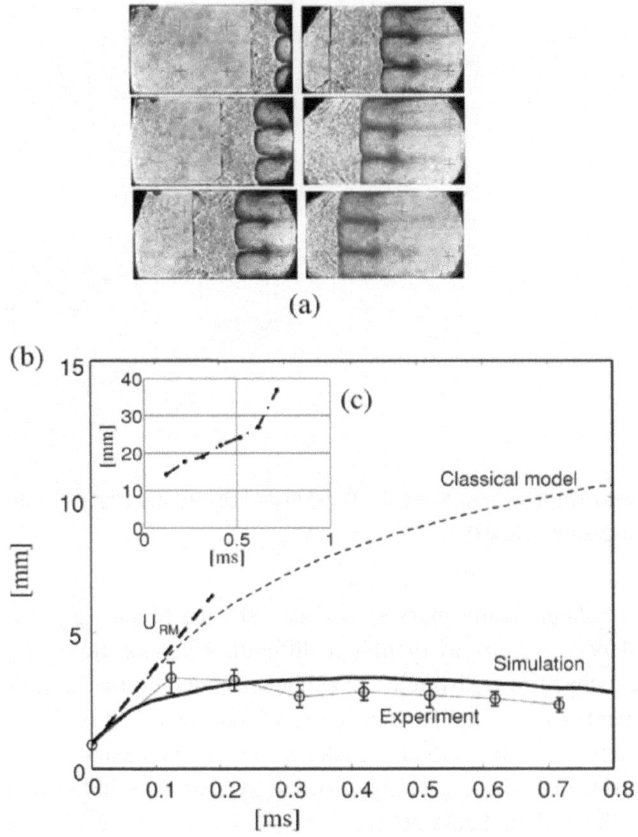

Fig. 12.17 (a) Schlieren photographs from a M = 2 air/SF_6 shock-tube experiment. (b) Comparison between the nonlinear model – dashed line (Sadot et al., 1998) – and the high Mach number experiment together with full 2D numerical simulation – full heavy line – and the linear theory – heavy dashed line (Richtmyer, 1960). The reduction from the nonlinear model at late time is clear. (c) The bubble radius of curvature versus time. Fig. 4 of Sadot et al. (2003).

12.4 Nonstandard RM with shaped shock waves

Nonstandard RMI flows have also been investigated by considering the stability of a uniform contact surface when a nonuniform shock wave passes through the surface (Ishizaki et al., 1996). The nonuniform shock is generated by a rippled piston moving with constant velocity. The initial amplitude and width can be extracted from the Schlieren images (Liu et al., 2014a). The total height and extent of the rippled shock can be defined in the same fashion as a perturbed single-mode interface in standard RMI (Table 12.1) (Zou et al., 2017).

The quantitative shape of the nonplanar incident shock front depends on how it was constructed. It can be characterized by a nondimensional parameter: the ratio of the spacing from the cylinder to the interface to the cylinder diameter, Figs. 12.19 and 12.20. Figure 12.21 illustrates that as this ratio increases, the penetration depth of the cavity decreases,

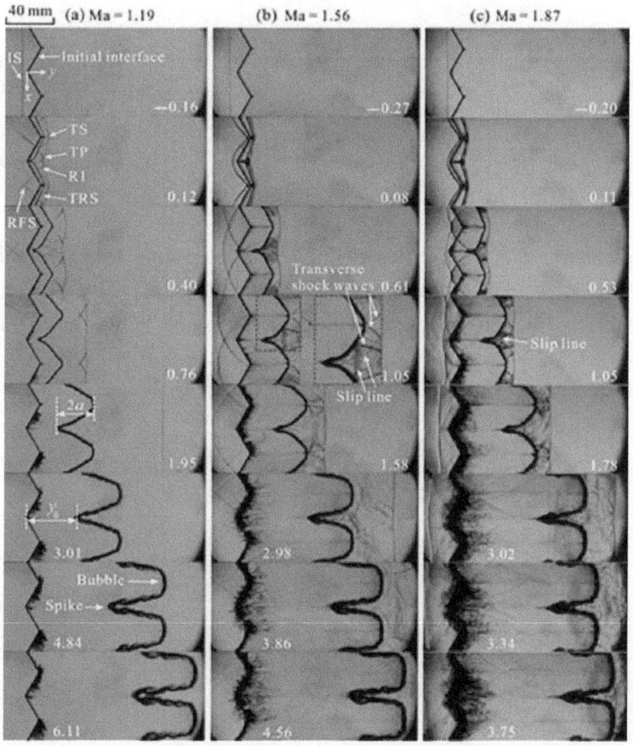

Fig. 12.18 Schlieren images of the air/SF_6 interfaces for the nonlinear cases impacted by a planar shock with Mach 1.19 (a), 1.56 (b), and 1.87 (c). IS, incident shock; TS, transmitted shock waves; RFS, reflected shock waves from the interface; TRS, transverse shock waves of the RFS refracting through the interface; and R1, transverse shock waves generated by the interaction of the TS from two half planes. Numerical labels represent dimensionless times. Fig. 2 of Guo et al. (2020).

Table 12.1 Experimental parameters of a non-planar shock and a comparison of the measured interface amplitude growth velocity with Richtmyer's theory prediction. The values of the Mach number and velocity jumps are approximately (1.22, 1.21, 1.23), (76.1 m/s, 73.0 m/s, 79.3 m/s), respectively. The values of the Atwood numbers are 0.68 for all cases. This table is based on the data provided in Table II and Table III of Zou et al. (2017).

Ratio between spacing from cylinder to interface to cylinder diameter	Amplitude of nonplanar incidental shock (mm)	Wavelength of nonplanar incidental shock (mm)	Amplitude/ wavelength	Standard RMI growth rate (theory) m/s	Nonstandard RMI growth rate (measured) m/s
2.0	2.8	52.8	0.05	16.25	4.68
3.3	1.6	39.3	0.04	12.47	2.94
4.0	1.5	79.6	0.02	6.77	2.25

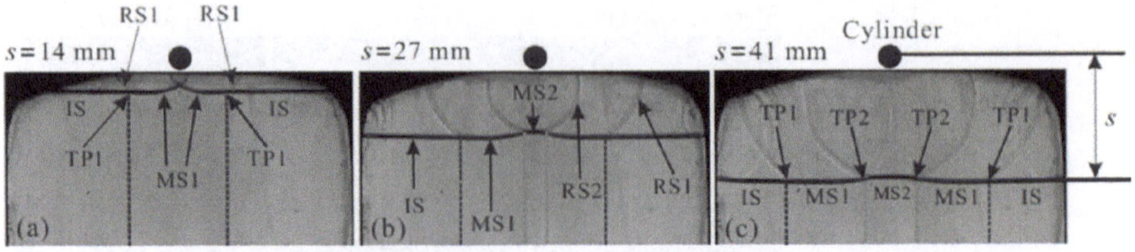

Fig. 12.19 Schlieren pictures displaying the wave pattern after the planar shock wave diffracts around a 10-mm-diameter rigid cylinder. The initial shock propagates from top to bottom, and *s* denotes the distance from the cylinder to the planar incident shock front. IS, incident shock; RS, reflected shock; TP, triple point; MS, Mach stem. Fig. 3 of Zou et al. (2017).

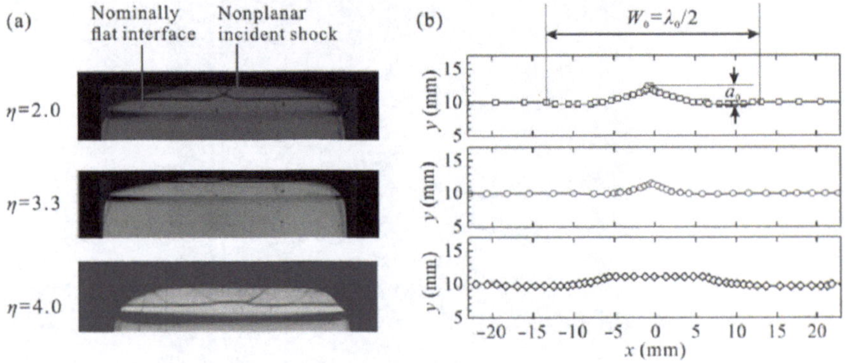

Fig. 12.20 (a) Schlieren pictures showing the shapes of the rippled incident shock and the interface. (b) Quantitative shape of the incident shock front extracted from the Schlieren images. a_0, initial amplitude of rippled shock, initial width of rippled shock, initial wavelength of rippled shock, and Fig. 4 of Zou et al. (2017).

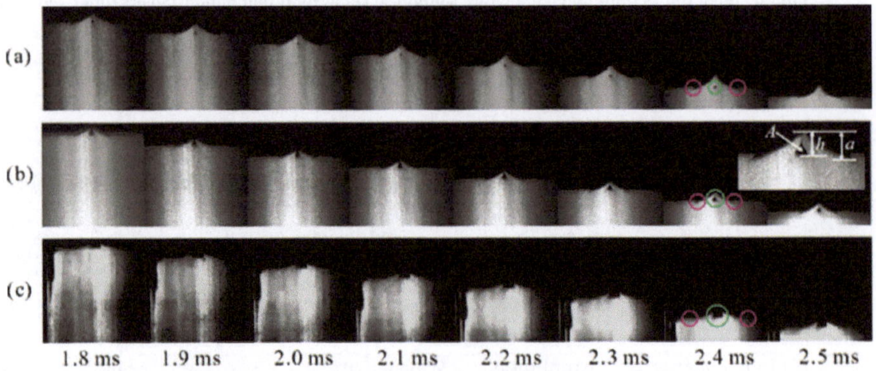

Fig. 12.21 Fog visualization sequences showing the evolution of a planar interface impacted by a rippled shock for three ratios of spacing from cylinder to interface to cylinder diameter. Fig. 5 of Zou et al. (2017).

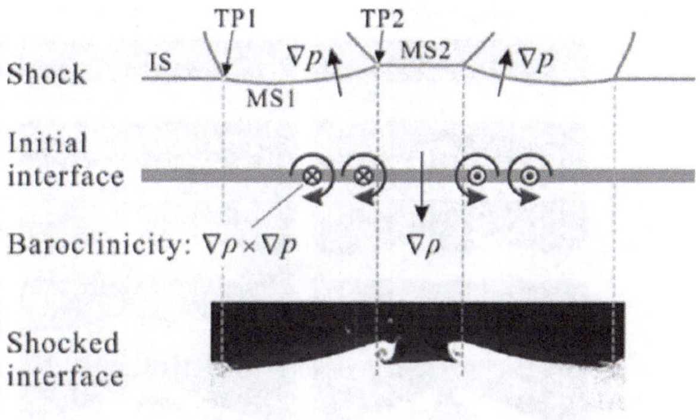

Fig. 12.22 Schematic of the baroclinic vorticity deposition on the uniform interface. The perturbed shock propagates downward toward the uniform interface and applies a pressure gradient ∇p across the density gradient $\nabla \rho$, which locally deposits baroclinic vorticity on the interface. Fig. 8 of Liao et al. (2019).

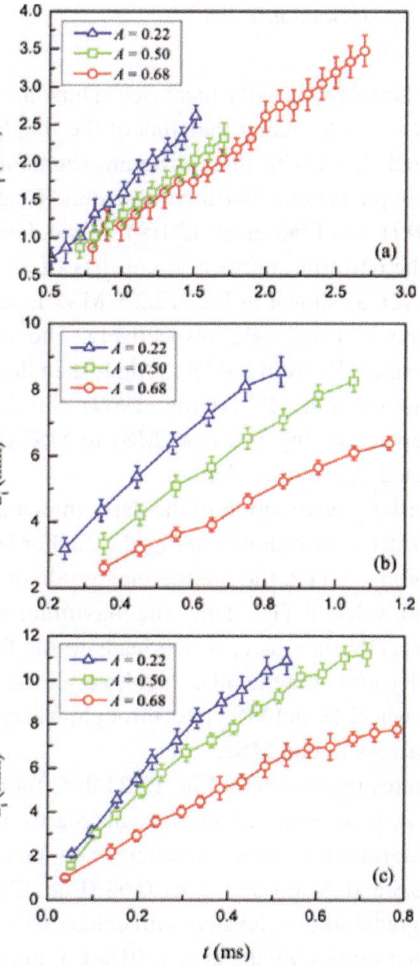

Fig. 12.23 Time evolution of the shocked interface amplitude for different initial Atwood numbers at Mach number (a) 1.2, (b) 1.5, and (c) 1.8. Fig. 9 of Liao et al. (2019).

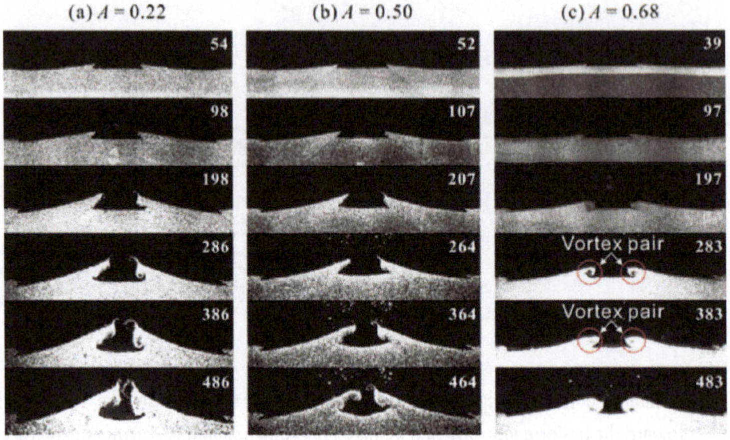

Fig. 12.24 Mie scattering sequences showing the evolutions of shocked interfaces of different initial Atwood numbers for Mach number 1.8. Fig. 7 of Liao et al. (2019).

while the area of the cavity increases. Thus, the growth velocity of this nonstandard RMI instability is much smaller than that of the standard RMI.

A rippled shock with three different segments separated by triple points is formed by the incident planar shock diffracting around a rigid cylinder. As described in detail by Zou et al. (2017) and Liao et al. (2019), immediately before the diffracted perturbed shock impacts the uniform interface, it consists of an incident shock (IS) and two Mach stems (MS1, MS2), as shown in Fig. 12.22. MS1 is separated from the IS and MS2 by the two triple points TP1 and TP2, respectively. The shock strength experiences a sudden jump upon crossing TP1 from the IS to MS1, then decreases continuously along MS1, reaching a local minimum at TP2 (Heilig, 1969). Similarly, there is a sudden jump in the shock strength upon crossing TP2 from MS1 to MS2 (Luo et al., 2017). Of course, there are also two reflected shocks (not shown).

The vorticity distribution in the early interaction between a perturbed shock and a uniform interface is also shown in Fig. 12.22. For MS1, the maximum misalignment between $\nabla\rho$ and ∇p is at TP2, though the magnitude of ∇p decreases continuously along MS1 to a minimum value at TP2. Thus, the maximum vorticity is deposited at the location where TP2 impacts the interface, which leads to the formation of a counter-rotating vortex pair near this location. On the other hand, very little vorticity is deposited on the two interface segments where IS and MS2 pass through, where $\nabla\rho$ across the interface is nearly collinear with ∇p across IS and MS2.

It is interesting to note in Fig. 12.23 that, for all the three Mach numbers, the amplitude growth rate of the shocked interface decreases with increasing Atwood number. An apparent counter-rotating vortex pair emerges on the top of the interface for the case of the Mach number 1.8 and Atwood number 0.68 (Fig. 12.24). This is because both the density and pressure gradients are elevated with enhanced Atwood and Mach numbers.

In related studies, Zhang et al. (2018c) simulated a planar interface subjected to a rippled shock of varying Mach number. Experimentally, a flat interface interacting with rippled shock waves produced by two or three cylinders can be found in Zhai et al. (2018).

13 Reshock and multi-shocks

Reshock occurs frequently in ICF and astrophysical applications. In these cases, multiple shockwaves pass through a mixing layer, for instance, due to reflections from the center of a spherical capsule, thereby causing further enhanced mixing and destruction of the ordered velocity field. Indeed, reshock deposits additional energy into a mixing layer formed by the RM instability. The reshocked mixing layer width deviates significantly from the single-shocked case, and the growth rate rises substantially (see, e.g., Andronov et al., 1976, 1982; Bonazza and Sturtevant, 1996; Houas and Chemouni, 1996; Jourdan and Houas, 1996; Erez et al., 2000; Nikiforov et al., 1995).

To isolate the physics most relevant to reshock, we will focus here on reshock events in a planar geometry, sidestepping additional complexities arising from a convergent geometry, a detailed description of which will be undertaken in Chapter 16. In shock tube experiments, reshock is typically due to the reflection of the first shock from the tube end-wall. As an example, membrane-less reshock experiments with Los Alamos' horizontal shock tube investigated a thin SF_6 acetone gas curtain suspended in air, a light–heavy–light configuration instead of the traditional light–heavy or heavy–light configuration (Balakumar et al., 2008, 2012; Balasubramanian et al., 2012; Tomkins et al., 2013).

The simultaneous PIV–PLIF snapshots shown in Fig. 13.1(a,b) illustrate concentration contours and fluctuating velocity vectors after the first shock and after reshock. Meanwhile, fluctuating velocity vector fields superimposed on vorticity contours show the redistribution of vorticity caused by the reshock (Fig. 13.1(c,d)). Before the reshock, the highest SF_6 concentration is present in the roll-ups around the vortex cores. A row of counter-rotating vortices is visible, each with vorticity concentrated at the vortex core. The ordered flow field is dramatically disturbed by the reshock wave, as evident in flow visualizations. The baroclinic deposition of smaller scale vorticity occurs throughout the structure, as illustrated by the complex distribution of density gradients. The ordered vorticity is redistributed throughout the reshocked structure and the initial spanwise symmetry is largely destroyed, suggesting an enhanced and evolving state of mixing after reshock.

The change in the flow structure due to reshock can be further demonstrated by a close examination of the vorticity fields. By looking at the PIV–PLIF fields of the evolving curtain immediately after reshock, Balakumar et al. (2012) provided insights into how the enhanced post-reshock mixing can be understood. These experiments were conducted on a varicose SF_6 fluid layer (heavy fluid) interposed in the air (light fluid) with an incident Mach number of 1.21 and a reflected reshock Mach number of 1.14. Unlike the pre-reshock case, which is driven by a row of approximately equispaced counter-rotating vortices, the post-reshock flow is driven by two rows of counter-rotating vortices (as shown in the

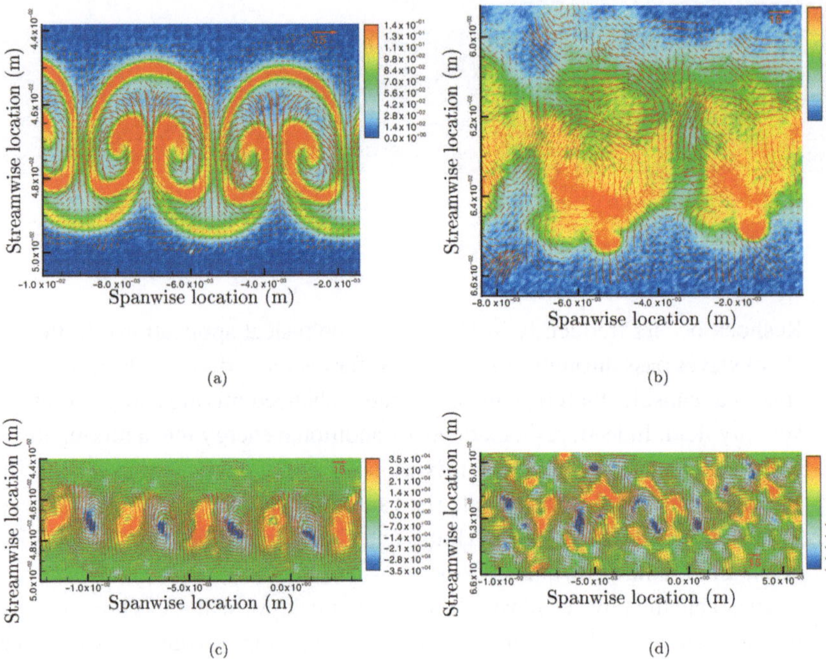

Fig. 13.1 (a) and (b) Simultaneous PIV–PLIF snapshots showing concentration contours and fluctuating velocity vectors (a) at $t = 465\ \mu s$ after first shock and (b) after reshock at $t = 665\ \mu s$. (c) and (d) Fluctuating velocity vector fields superimposed on vorticity contours showing the redistribution of vorticity after reshock. Vectors represent the fluctuating velocity. The solid line represents one primary wavelength. Source: Fig. 12 of Balakumar et al. (2008), with the permission of AIP Publishing.

sketches of Fig. 13.2). This latter arrangement is unstable, and therefore small perturbations after reshock will result in vortex dislocations. Thus, post-reshock mixing is probably driven both by vortex-induced fluid motion and by the motion of the vortices themselves, resulting in a faster transition to a turbulent state when compared to the single shocked case (Balakumar et al., 2012). The energy transfer in a high-resolution DNS of these reshocked heavy gas curtain experiments reveals the backscatter of kinetic energy transfer in the earlier stage, followed by the dominant forward scatter in the late stage (Zeng et al., 2018b).

13.1 Mixed width measurements

Understanding the reshock process requires both high-quality repeatable experimental data and validation by robust numerical algorithms (Shankar and Lele, 2014) capable of simulating the highly non-linear flow. Over the years, several prominent data sets have been developed by Vetter and Sturtevant (1995), Poggi (1997), Poggi et al. (1998), and Leinov et al. (2009), which are almost inevitably included in the experimental testing suite of any

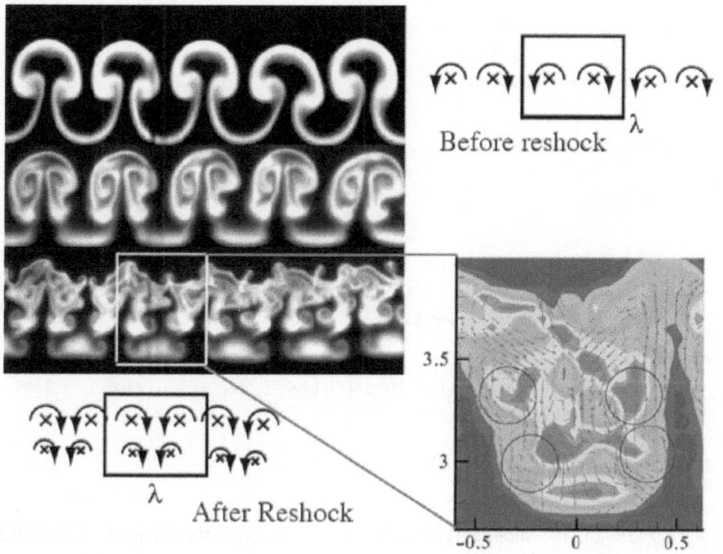

Fig. 13.2 Simultaneous PIV–PLIF field showing multiple interacting vortices immediately after reshock. Source: Fig. 4 of Balakumar et al. (2012).

given mix model developer. These mix models, often referred to as the Reynolds-averaged Navier–Stokes (RANS) equations in the literature, are very useful for routine engineering design calculations (see Chapter 22). Frequently, explicit and implicit LES are also invoked to compute turbulent statistics and structures of the mixing layers. One can take a step further to say that any reputable mix or subgrid model cannot be considered validated without first being "stress tested" against reshock experiments, such as those mentioned above.

In a seminal paper, Vetter and Sturtevant (1995) of the California Institute of Technology (Caltech) performed a series of shock-tube experiments with various combinations of high- and low-density gas, orientation of the shock direction, positions of the membrane, and positions of the wire mesh. As shown in Fig. 13.3(a), the previous LES results (e.g., Hill et al., 2006; Schilling and Latini, 2010) for the pre-reshock mixing layer are not satisfactory due to a suspected dependence on the initial conditions. However, a recently proposed model, based on a combined LES and RANS approach, has shown some promise, since it can be made to agree fairly well with experimental measurements (Bin et al., 2021).

In experiments performed in the vertical shock-tube of Commissariat à l̆ Énergie Atomique (CEA), a discontinuous heavy–light interface between sulfur hexafluoride (SF_6) and air was impulsively accelerated by a Mach 1.45 shock wave (shock strength 0.54) and decelerated by several shock waves reflected from the end wall. Each reflected shock had decreasing strength (0.38 for the first, 0.15 for the second) (Mügler and Gauthier, 2000). Here, the shock strength is defined as the pressure jump across the shock front, normalized by the pressure behind the shock (Rotty, 1962). Figure 13.3(b) shows the temporal evolution of the mixed widths (Mügler and Gauthier, 2000) from the CEA experiments documented in Poggi (1997) and Poggi et al. (1998). The error bars of these visual measurements are approximately $\pm 10\%$.

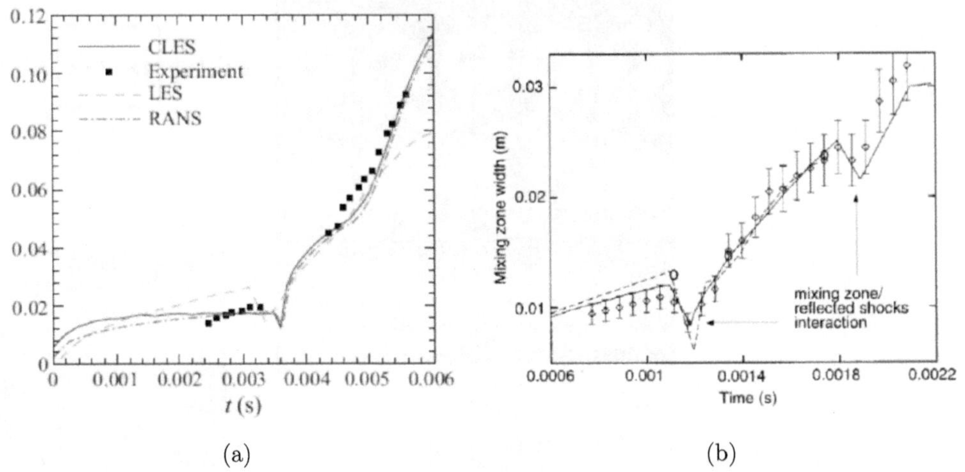

Fig. 13.3 Mixed-width measurements from two key reshock experiments: (a) The Caltech (California Institute of Technology) data by Vetter and Sturtevant (1995). Also shown are calculations from a LES and a RANS model, as well as a newly proposed model based on a combined LES and RANS approach (labeled CLES). Fig. 1a of Bin et al. (2021). (b). The CEA (Commissariat à l'Energie Atomique experiments) data (Poggi, 1997; Poggi et al., 1998). Evolution of the mixing-layer width versus time. Diamonds correspond to the experimental width measured from Schlieren pictures. The error bars of this visual measurement are $\pm 10\%$. The three small circles correspond to microdensitometry measurements of some Schlieren pictures. Full and dotted lines correspond to numerical widths obtained from the coarse and fine grids, respectively. Fig. 2 of Mügler and Gauthier (2000), with the permission of AIP Publishing.

13.2 Spectra enhancement and compressibility

From a different perspective, statistical measurements can be used to reveal the profound effects of reshock on the mixing layer evolution. Thornber et al. (2010) investigate two specific initial conditions through numerical simulations. The first is an initial perturbation characterized by a constant power spectrum over a narrow band of high wavenumbers. The second initial condition consists of a broadband linear combination of modes satisfying a k^{-2} power spectrum. The former is utilized to examine the growth of the RM instability at late times, while the latter has practical importance as it is representative of the measured surface finish of an inertial confinement fusion capsule (Barnes et al. 2002). Figure 13.4 illustrates the RMI turbulent kinetic energy spectra for the narrowband and broadband perturbations before reshock, just after reshock, and at a much later time. It is clear that the reshock induces a very strong amplification of greater than one order of magnitude in kinetic energy for all modes (Thornber et al., 2011). The amplification for the narrowband case, however, is more pronounced than in the broadband case. Also, immediately following reshock (where the shock has moved fully clear of the layer) there is a Kolmogorov $k^{-5/3}$ range present in the spectrum, while at the latest time when self-similarity is achieved, the spectrum has returned to the $k^{-3/2}$ scaling of Zhou (2001).

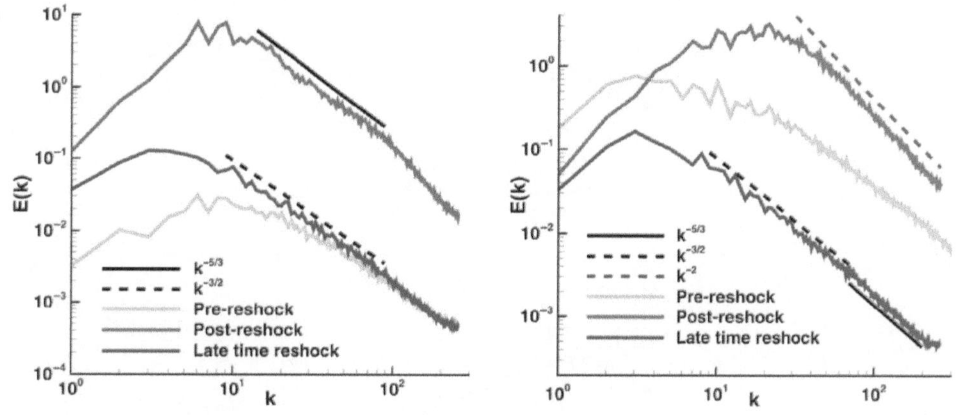

Fig. 13.4 Evolution of the kinetic energy spectra for the narrowband (left) and broadband case (right) during reshock and at late time. Fig. 15 of Thornber et al. (2011), with the permission of AIP Publishing.

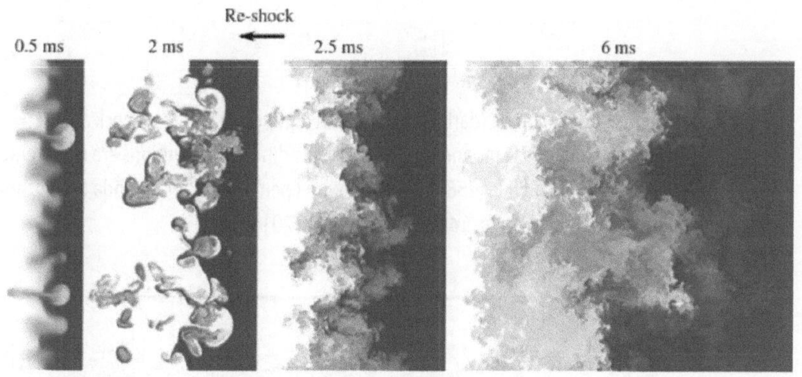

Fig. 13.5 Grayscale plots of species mass fraction of SF_6 gas from the Miranda code at various times. Data are taken from the finest grid. The contours range from 0.05 (white) to 0.75 (dark gray). Fig. 5 of Tritschler et al. (2014b).

The same phenomenon has been investigated numerically, from which the mass fraction of SF_6 gas at various times is shown in Fig. 13.5. The authors performed well-resolved LES using two separate independent numerical methods that fundamentally differ with respect to the employed subgrid-scale regularization. In agreement with the results of Thornber et al. (2011), Fig. 13.6 shows the energy spectra before and after reshock, exhibiting the $k^{-3/2}$ scaling of Zhou's self-similar theory. The destruction of the vortical structures initiated by the reshock leads to the formation of small scales (Tritschler et al., 2014b). Meanwhile, the evolution at large scales is independent of the initial condition (IC).

An important counter example can be found in a Los Alamos National Laboratory (LANL) experiment – strong evidence has been presented that the imprint of ICs persists to the end of their experiments (see Fig. 13.7). The instantaneous streamwise velocity contours (superimposed on the velocity vector field) show the presence of IC imprints after reshock (Balakumar et al., 2012).

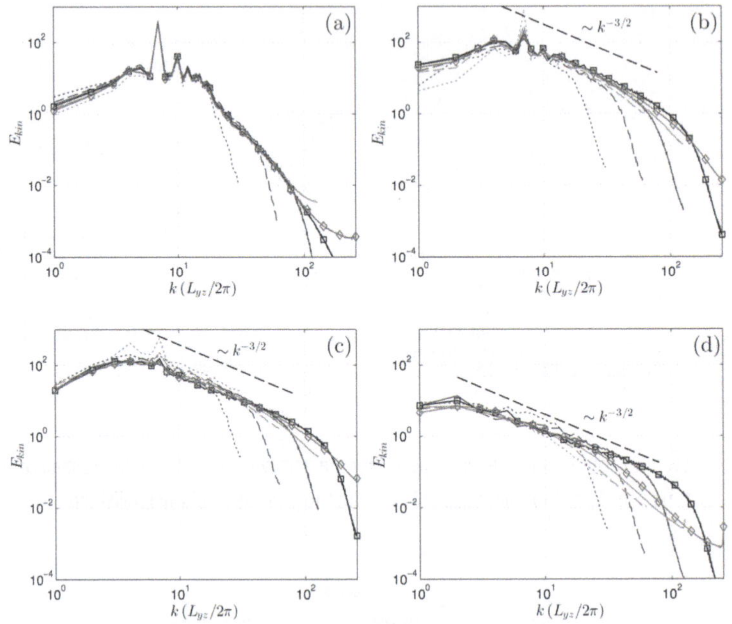

Fig. 13.6 Spectra of TKE from Miranda (dark gray) and INCA (light gray) before reshock at (a) t = 0.5 ms and (b) t = 2 ms and after reshock at (c) t = 2.5 ms and (d) t = 6 ms. The different resolutions are represented as a dotted line (64^3), dashed line (128^3), solid line (256^3), solid line with open squares for Miranda (512^3), and solid line with open diamonds for INCA (512^3). Fig. 14 of Tritschler et al., 2014.

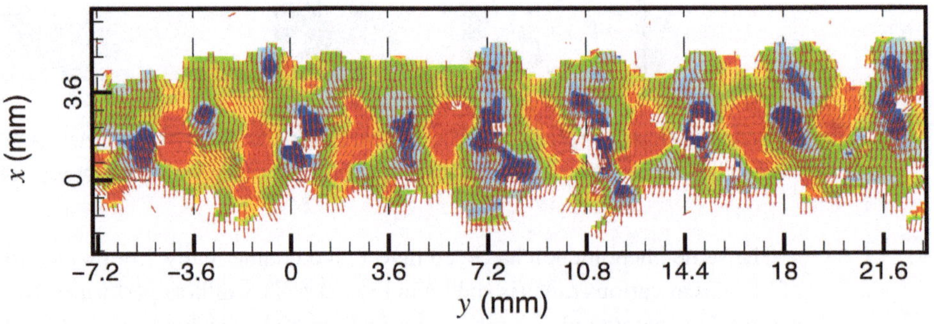

Fig. 13.7 Instantaneous streamwise velocity contours superimposed on the velocity vector field. This shows the presence of IC imprints after reshock, at a nondimensional time $t = 22.9$. Red and blue represent high and low streamwise velocity regions, respectively. Fig. 12 of Balakumar et al. (2012).

In addition to the phenomena already described above, many practical applications suffer additional complications from the presence of rarefaction/compression waves. Li et al. (2019a) performed an RMI simulation with an air/SF_6 gas combination subject to reshock and rarefaction/compression waves. Figure 13.8 indicates similar reshock effects to those already reported in Figs. 13.4 and 13.6, namely, the amplification all kinetic energy modes.

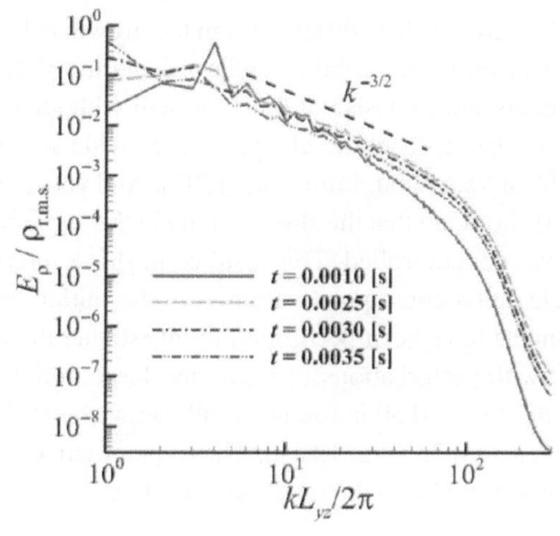

Fig. 13.8 Radial energy spectrum of density. Fig. 13 of Li et al. (2019a), with the permission of AIP Publishing.

Interestingly, the above mentioned $k^{-3/2}$ scaling is again observed. It remains quite robust, despite the arrival of the rarefaction and compression waves.

13.3 Initial condition when the incident shock arrives

Early reshock experiments were performed with a membrane to maintain the initial gas separation. Andronov et al. (1976) conducted one of the earliest of these, using a heavy/light gas configuration (air/He), in which the width of a mixing layer was measured over time using the Schlieren diagnostic technique. Weber et al. (2012) carried out reshock experiments with six different initial amplitudes, but only followed the system for a relatively short time after the reshock.

As already hinted above, ICs are an important factor in the reshock experiments. Vetter and Sturtevant (1995) experimentally studied the evolution of (light/heavy) air–SF_6 mixing layers in shock tubes with 3D perturbations, finding that the thin membrane has a non-negligible influence on the initial growth. However, the measured growth rates after the first reflected shock are independent of the membrane configuration. Consistent with this finding, Erez et al. (2000) found that the initial membrane thickness affects the growth rate after the incident shock, but does not influence the post-reshock growth rate.

Jacobs et al. (2013) performed reshock experiments using the membrane-less shock tube design pioneered by Jones and Jacobs (1997) and Collins and Jacobs (2002). The experiments were performed with an air/SF_6 gas combination with a Mach 1.2 incident shock wave. Perturbations were introduced by vertically oscillating the gas column to produce Faraday waves on the interface. The ICs were introduced in two distinct ways. The measurements of the mixed width before reshock suggested two growth behaviors, both having power-law growth. The differences in the growth behavior are most likely caused by differences in the initial perturbation at the arrival time of the first shock. Following reshock, both experiments show consistent linear growth with identical growth rates $dh/dt = 9.26$ m/s.

At first glance, Jacobs et al. appear to provide another data point in support of the observation from Vetter and Sturtevant (1995). And yet, an updated work, documented in Sewell et al. (2021), noted that the shock timing in the authors' previous experiments (Jacobs et al., 2013) was not controlled. This resulted in shock interaction at a random phase of the forcing cycle and a consequent variation in the initial amplitudes. New air/SF_6 RMI reshock experiments have been performed to investigate the evolution of the instability when initialized with perturbations of small and large initial amplitudes, see Fig. 13.9. A growth exponent of $\theta \approx 0.50$ is found for all cases except the high-amplitude reshocked regime, in which $\theta \approx 0.33$ (Fig. 13.10). These quantitative measurements demonstrate the lasting effect of ICs well into the reshocked regime.

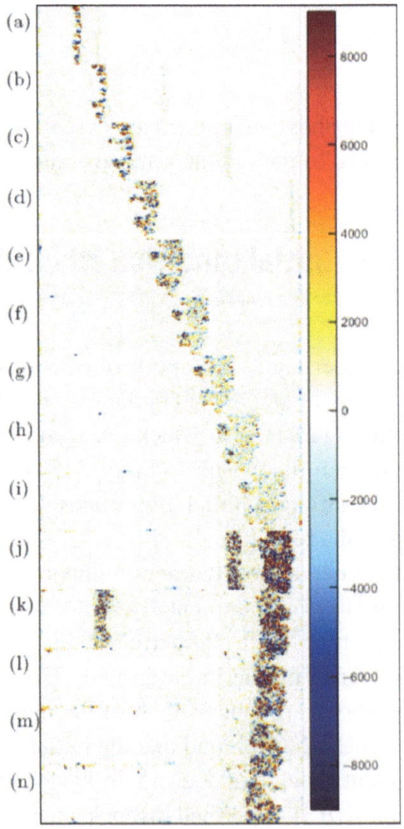

Fig. 13.9 A sequence of vorticity fields obtained from PIV processing of a typical (high amplitude) experiment at (a) t = 0.27 ms, (b) t = 0.77 ms, (c) t = 1.27 ms, (d) t = 1.77 ms, (e) t = 2.27 ms, (f) t = 2.77 ms, (g) t = 3.27 ms, (h) t = 3.77 ms, (i) t = 4.27 ms, (j) t = 4.77 ms, (k) t = 5.27 ms, (l) t = 5.77 ms, (m) t = 6.27 ms, (n) t = 6.77 ms. Reshock occurs at ≈4.6 ms. Sewell et al. (2021).

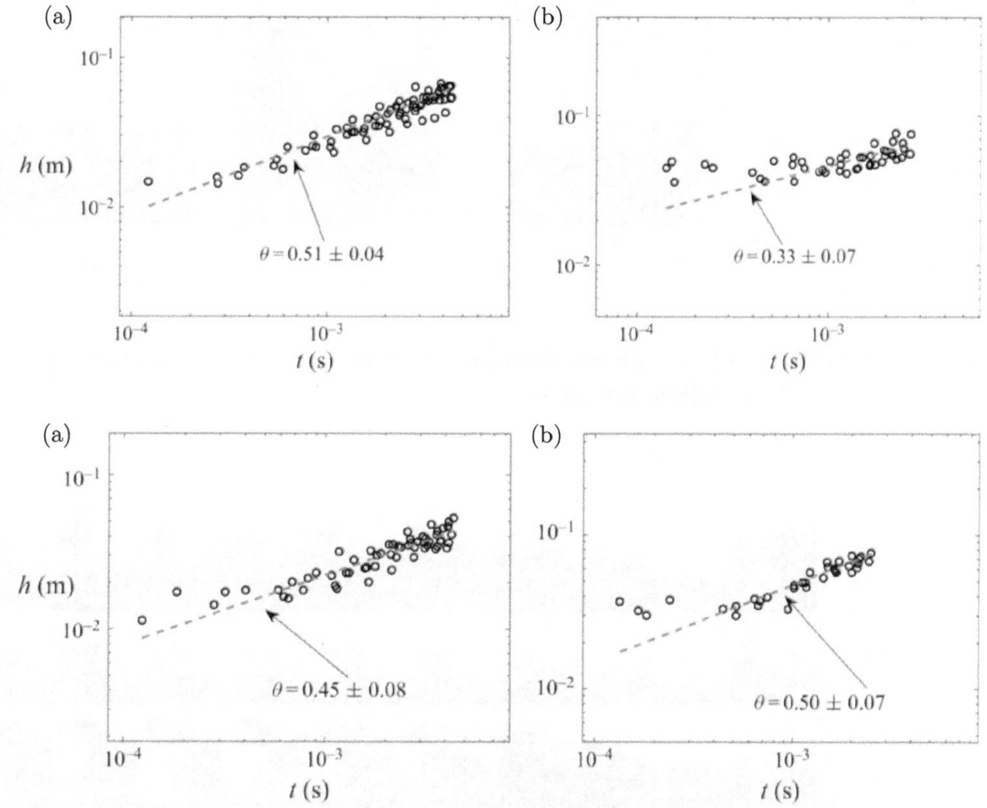

Fig. 13.10 Logarithmic plots of mix-width (h) versus time for the high- (top panel) and low-amplitude (bottom panel) experiments following the incident shock interaction (a) and following reshock (b). The dashed lines represent curve fits to the ensemble data. The growth exponents were computed from the mean of individual experimental runs. Also shown are the 95% confidence intervals of the mean. Sewell et al. (2021).

In the Wisconsin Shock Tube Laboratory (Anderson et al., 2000; Bonazza, 2022), the gas interface is accelerated by a shock of nominal strength, and is then accelerated again by the transmitted shock that reflects off the tube end wall (Noble et al., 2020a,b). The IC is a shear layer with broadband diffuse perturbations at the interface between a helium-acetone mixture and argon. The IC for each experiment is shown in Fig. 13.11. HS4 and HS8 have similar structures while the structure of HS6 is markedly different. This will allow an assessment of the influence of ICs on moment evolution. Figure 13.12 shows a subset of the resulting light gas mole fraction fields. Each row is a time series in which the first image is the reshock IC and the second image is after the reflected shock traverses the interface.

In their ILES studies of Vetter and Sturtevant (1995) experiments, Grinstein et al. (2011) found that velocity fluctuations in the mixing layer are more isotropic when the initial interface perturbations also include short wavelengths, in which case the flows nearly recovered full isotropy. When the authors used only long-wavelength perturbations, the mixing layer remained anisotropic except for a narrow range on the heavy-gas side.

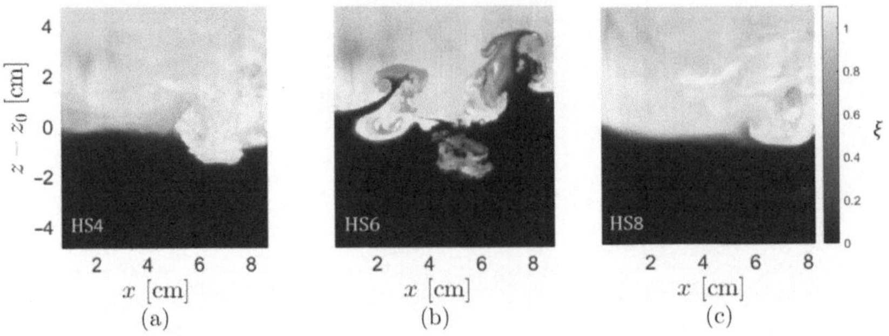

Fig. 13.11 Light gas mole-fraction of the initial conditions of the three experiments analyzed in further detail. (a) HS4, (b) HS6, (c) HS8. Fig. 6 of Noble et al. (2020a).

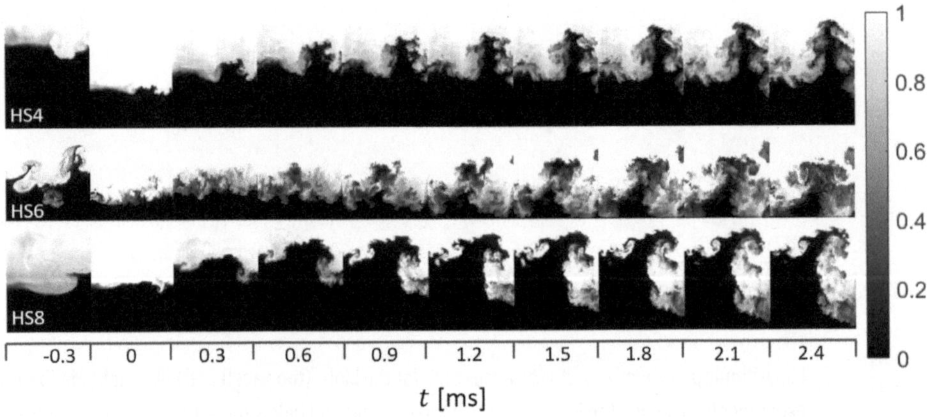

Fig. 13.12 Example mole fraction fields evolving with time from three experiments. Fig. 4 of Noble et al. (2020a).

In a related inquiry, Nelson and Grinstein (2015) focused on the number of spectral modes present in a given wavenumber band of the initial perturbations. When the initial spectrum of perturbations is dense, the flow is much more readily adapted to provide the needed turbulent cascade to small scales. If the initial spectrum is sparse, however, the turbulent cascade needs additional time to populate the spectral modes for the inertial range.

Wang et al. (2019) numerically compared the interface morphology, mixed width, integrated vorticity, circulation, and TKE of the RMI flow fields before and after reshock. They focused on the effects of non-uniform flows from a sinusoidal incident shock wave interacting with a flat air/SF_6 interface. The evolution of the resulting mixing layer was strongly dependent on the flow non-uniformity in the linear and weak nonlinear regime.

More recently, Guo et al. (2022b) performed experimental studies of reshocked RMI heavy/light interfaces, considering quasi-single-mode interfaces with sawtooth and trapezoid shapes in comparison with the standard single-mode interface. The development of interface morphologies shows a strong memory of initial shapes. The authors also delineated large-amplitude effects on interface perturbation growth in RM flows with reshock (Guo et al., 2022a).

13.4 Reshock timing

One topic of particular interest is the sensitivity of the post-reshock mixing-layer growth to the state of the flow when the reshock arrives. In a typical shock tube facility, the reshock interacts with the evolving interface after bouncing from the end wall. As such, the reshock timing is dictated by the Mach number of the incident shock. Such experiments are not appropriate for assessing the topic at hand, because of a confounding factor: the previously shocked layer also depends on M. Fortunately, the timing can be independently altered by either shortening or lengthening the distance of the wall from the interface.

In an early study, Nikiforov et al. (1995) utilized the experimental data of Andronov et al. (1982) and Zaitsev et al. (1985) for comparison with their numerical calculations for different end wall distances.

In a fairly comprehensive paper, Leinov et al. (2009) examined air–SF_6 mixing layers, showing that post-reshock growth rates were relatively insensitive to the arrival time of the reshock at the previously shocked layer (Fig. 13.13). Indeed, the "universal" post-reshock growth rate, as Leinov points out, agrees with Mikaelian's proposal of $2\alpha\mathcal{A}\Delta V$ (Mikaelian, 1989). To make their numerical simulation concur with the experimental results, Malamud et al. (2014) inspected two distinctive initial perturbation profiles (Fig. 13.14). For

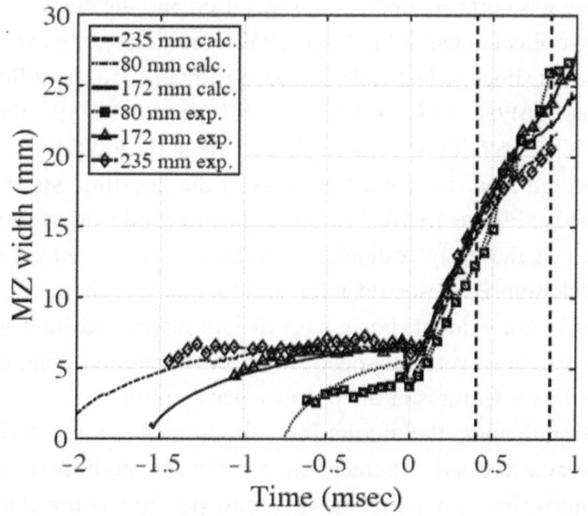

Fig. 13.13 Experimental mixing layer width as a function of shifted time. The end wall is at 80, 172 and 235 mm for −□−, −△−, −◇−, respectively. Lines show corresponding multi-mode DNS calculations (dotted, solid and dash-dot lines, respectively), Malamud et al. (2008). Vertical dashed lines represent the rarefaction wave arrival time in the 80 and 172 mm cases (Leinov et al., 2009).

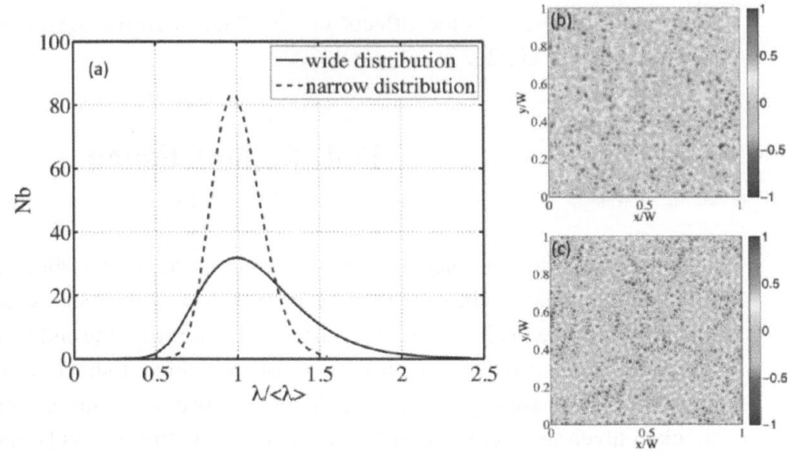

Fig. 13.14 The initial number of bubbles vs. normalized wavelength (a), and the initial interface between the air and SF_6 in arbitrary units, for the wide (b) and narrow (c) initial distributions used in the numerical simulations. Fig. 5 of Malamud et al. (2014), with the permission of AIP Publishing.

all end-wall positions and shock Mach numbers, they found this agreement could only be achieved when the initial perturbation was of a multi-mode form.

Balasubramanian et al. (2012) demonstrated that reshocking an interface at late time (representing steeper density gradients) causes the late-time structures to mix and generate smaller scales at a faster rate (Fig. 13.15). Here, a single-mode, membrane-free varicose heavy gas curtain (air-SF_6-air) was used and the density (concentration) and velocity fields were collected using PLIF and PIV. Specifically, the evolving interface induced by the first incident shock (Mach 1.2) had been reshocked at the following four reshock times (denoted as t_{RS}): (a) $t_{RS} = 90\ \mu$s, (b) $t_{RS} = 170\ \mu$s, (c) 280 μs, and (d) $t_{RS} = 385\ \mu$s.

This observation becomes more clear by looking at the corresponding power spectra collected from the center of mass of the late-time structures, approximately 210 μs after reshock. Figure 13.16 shows the spectra produced at four different reshock times. For the cases of the early reshocked interface at $t_{RS} = 90\ \mu$s and $t_{RS} = 170\ \mu$s, the dominant wavelength is preserved with some small length scales, and a transition to a well-mixed state is not evident. For a case of late reshocked interface with $t_{RS} = 280\ \mu$s and $t_{RS} = 385\ \mu$s, however, the spectrum shows increased smaller scales at the expense of losing significant features at the primary wavelength.

Significantly, the mixing layer might evolve fairly differently when the flow field is in a nonlinear instead of a linear state at the arrival of the reshock.

Innovative shock tube designs now permit a controllable time between the arrival of the first shock, which initiates the RMI, and the second shock, which ushers in the reshock regime. The Dual-Driver Vertical Shock Tube (DDVST) of the University of Arizona consists of two traditional single-driver shock tubes, each consisting of a single driver and driven section, oriented vertically opposed on opposite sides of a test section. Thus, the dual driver configuration permits the strength of the first and second shocks to be independently

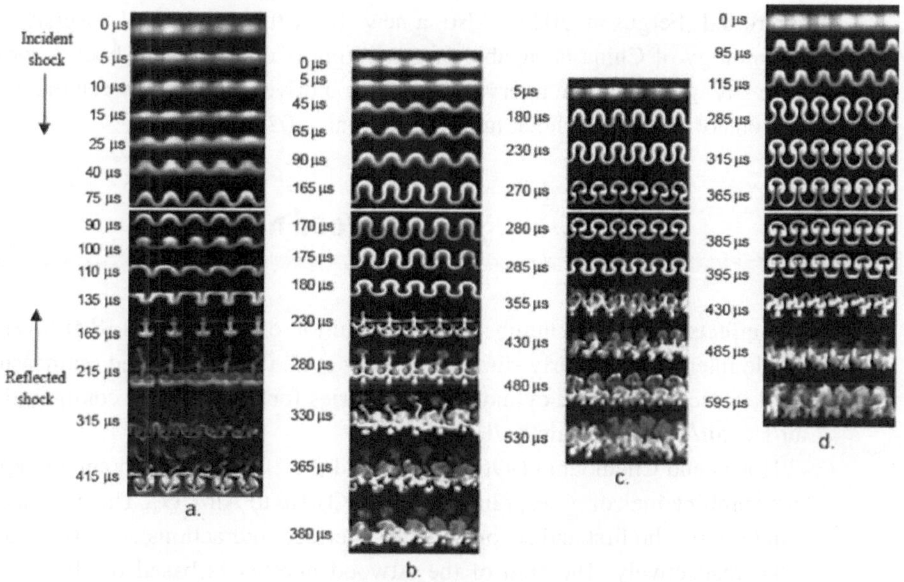

Fig. 13.15 Four sets of experiments, each beginning with the varicose gas curtain (top), shocked with a Mach 1.2 shock that travels downward. Each experiment is then reshocked at a different time, as indicated below the white bar crossing the time series. Reshock occurs at times (a) 90 μs, (b) 170 μs, (c) 280 μs, and (d) 385 μs. At late times after reshock, approximately 210 μs, there is a large variation in mixing for each of the three cases. Fig. 3 of Prestridge et al. (2012), with the permission from Springer.

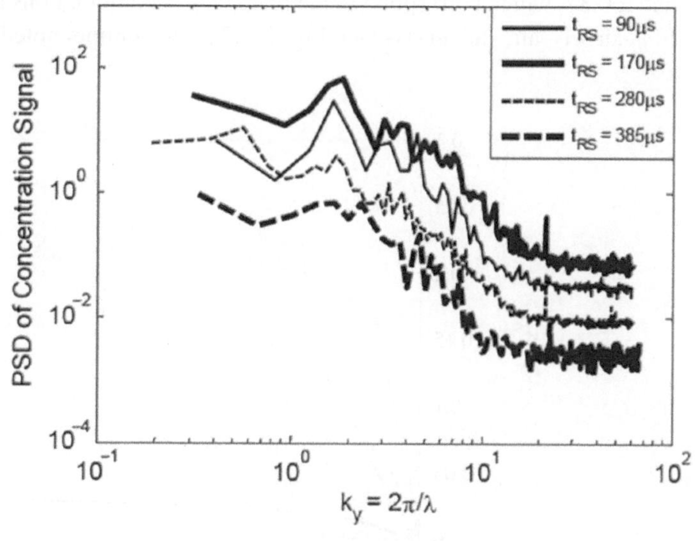

Fig. 13.16 Power spectral density of the concentration signal for late-time flow structure 200–210 μs after reshock, for four experiments with different reshock times. Reshock occurs at times (a) 90 μs, (b) 170 μs, (c) 280 μs, and (d) 385 μs. Fig. 5 of Prestridge et al. (2012), with the permission from Springer.

controlled (Ferguson, 2022). Also, a new shock tube from the University of Science and Technology of China is capable of generating two successive shock waves with controllable strengths and time intervals by adding a driver section between the driver and driven sections of a standard shock tube (Wang et al., 2022a).

13.5 Atwood number

The effects of Atwood number on the evolution of a reshocked RMI layer are of considerable interest. In an early study, Andronov et al. (1982) reported on measurements of the RMI in rectangular and cylindrical geometries for different gas configurations, including: air/He, air/He/air, and air/Xe/He.

Houas and Chemouni (1996) investigated the effect of membrane thickness using various combinations of gases, ranging from CO_2/He to Air/CO_2. The corresponding Atwood numbers for the first and second shock–interface interactions are -0.83, 0.17 and -0.10, 0.85, respectively. The sign of the Atwood number is based on the direction in which the density increases and changes upon the passage of the reshock. Following an assessment of the diagnostic techniques, the authors discussed the main experimental difficulties, including the presence of membrane fragments, and the disturbances induced by the wall boundary layers.

Lombardini et al. (2011) studied the Atwood-number dependence of RM flows under reshock conditions using large-eddy simulations. Separate cases had $\mathcal{A} = \pm 0.21, \pm 0.67$, and ± 0.87, which correspond to the realistic gas combinations air–CO_2, CO_2–air, air–SF_6, SF_6–air, H_2–air, and air–H_2 (see Fig. 13.17). The authors noted that there are fundamental

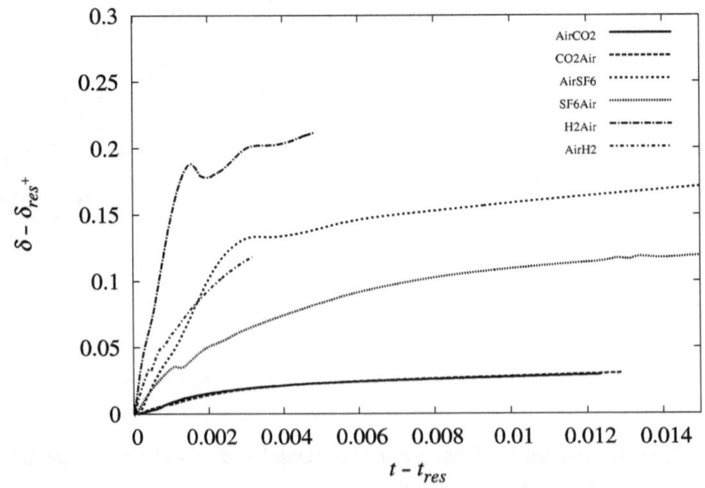

Fig. 13.17 Simulated post-reshock mixing layer width, versus, $t - t_{res}$ for various gas combinations. Note that the reshock event occurs at $t = t_{res}$. Fig. 6 of Lombardini et al. (2011).

differences between the heavy–light and light–heavy mixing layer growth. In heavy–light configurations, the primary reshock greatly accelerates the growth initiated by the initial shock interaction, while subsequent wave interactions have a minor effect on the growth amplitude. The light–heavy configurations can be described by two post-reshock phases: a post-reshock growth approximately linear in time, followed by a saturation phase occurring after 2–3 times the reshock time. The passage of the expansion and compression waves following the primary reshock have a stronger signature on the layer width.

13.6 Mach number

In the canonical geometry of a rectangular shock-tube with square cross-section, Hill et al. (2006) performed LES with the configuration of the gases, the geometry of the domain, and the strength of the incident shock all chosen to match the experiments of Vetter and Sturtevant (1995). Figure 13.18 shows the time evolution of the mixing layer for the plane discontinuous interface between air and SF_6, for incident shock waves with Mach numbers 1.24, 1.50, and 1.98.

Leinov et al. (2009) discussed the available methods for controlling the reshock strength experimentally. Of course, varying the incident Mach number will change the strength of the reflected shock wave (Fig. 13.19(a)). However, a comparison between such experiments is not appropriate because the perturbations will have evolved for different lengths of time prior to the reshock arrival. This is essentially the same argument articulated in Section

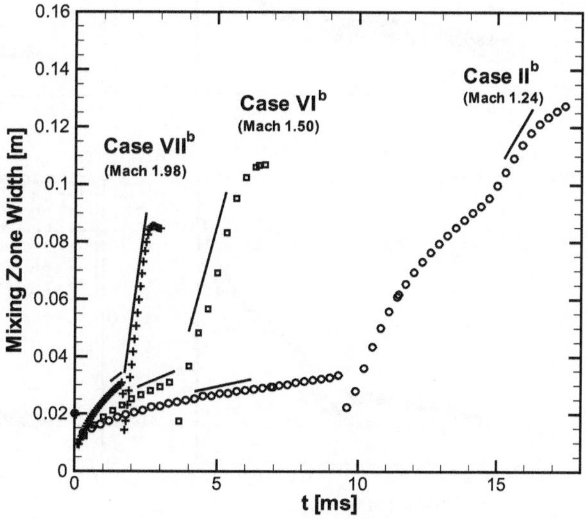

Fig. 13.18 Evolution of the reshocked-RMI mixing-layer width. The growth rates inferred from the air/SF_6 experiments are indicated as straight lines with the appropriate slopes, drawn in the approximate times over which they were observed. Fig. 4 of Hill et al. (2006).

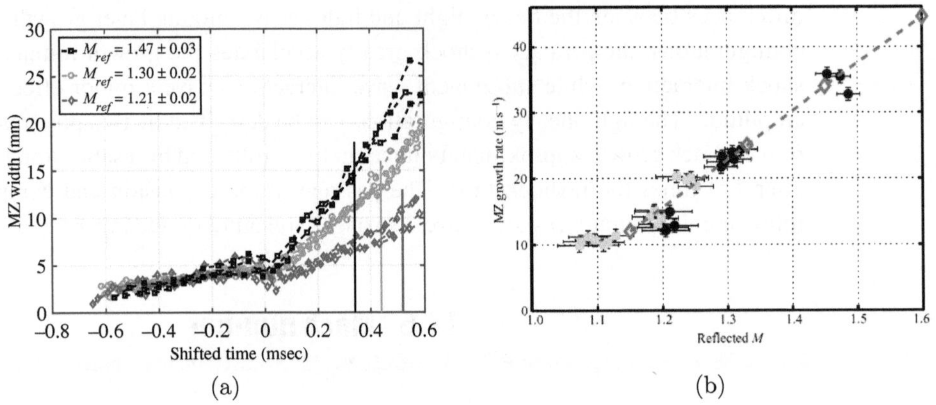

Fig. 13.19 (a) Mixing width measurements as a function of shifted time for three different incident shock-wave Mach numbers (the endwall is fixed at 80 mm). Fig. 13 of Leinov et al. (2009). (b) Mixing layer growth rate as a function of reshock Mach number for six different experimental sets. Symbols: ● rigid endwall, □ elastomeric foam endwall, and ◇ multi-mode DNS calculations. The endwalls are at 80 mm. Gray dashed line: linear trend line intended to guide the eye. Fig. 17 of Leinov et al. (2009).

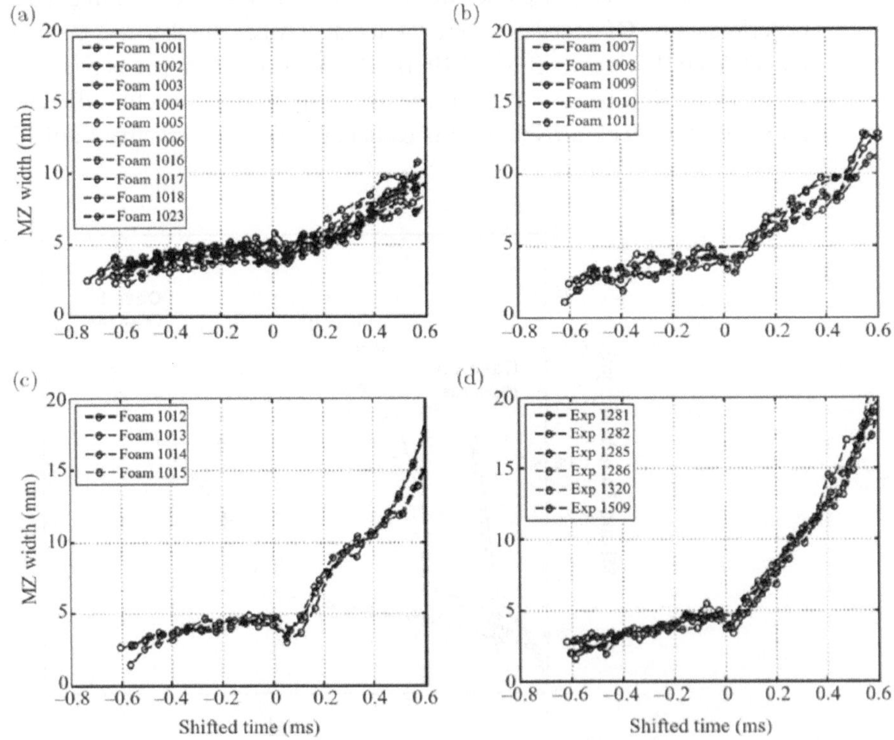

Fig. 13.20 Mixing width measurements as a function of shifted time for four different endwall rigidities. All endwalls are at 80 mm: (a) elastomeric foam endwall, (b) elastomeric foam and cardboard endwall, (c) elastomeric foam and aluminum plate endwall and (d) rigid endwall. Fig. 16 of Leinov et al. (2009).

Table 13.1 Reshock Mach numbers, overall velocity jumps and the growth rates of the mixing layer following reshock for four experimental sets with different endwall rigidity. The endwall is at 80 mm. Table 3 of Leinov et al. (2009).

Endwall feature	Reshock Mach number	Overall velocity jump (m/s)	Growth rate (m/s)
Elastomeric foam	1.09 ± 0.03	36.3 ± 6.0	10.5 ± 0.8
Elastomeric foam & cardboard	1.19 ± 0.01	45.4 ± 5.0	14.3 ± 1.0
Elastomeric foam & aluminum plate	1.25 ± 0.02	73.0 ± 6.0	19.7 ± 1.0
Rigid	1.30 ± 0.02	96.4 ± 5.0	23.1 ± 1.0

13.4 against using the incident shock strength for the purpose of studying the effect of differing reshock arrival times. A resolution in that case is obtained by shifting the end wall of the shock tube, as described previously.

For the present consideration, Leinov et al. modified the end wall's rigidity in order to achieve the objective. With the end wall fixed at 80 mm, Fig. 13.20 shows the growth rates of the mixing layer following reshock for six experimental sets. Figure 13.20 displays the mixing width measurements from the three different configurations of elastomeric foams and from the rigid endwall configuration at 80 mm. As the Mach number of the reshock increases in Fig. 13.19(b), the subsequent mixing layer growth rate also increases (see also, Table 13.1).

13.7 Mix measurements

Recall that an important feature of the mixed mass ($\mathcal{M}$) – a measure of how much mixed product exists in the flow – is that it does not decrease with the passage of a reshock. This stands in contrast to the mixing layer width, which decreases due to the compression of the reshock wave.

Figure 13.21(a) shows that the growth rate of the mixed mass is higher than that of the mix width at late time, as the reshock has changed the efficiency of mass mixing. Note also that the value of the normalized mixed mass, Ψ, is lower than its mixedness counterparts, Ξ and Θ (Fig. 13.21(b)).

Mohaghar et al. (2019) investigated both the single- and multi-mode inclined interfaces created via shear and buoyancy superposed on the dominant inclined perturbation. Analysis of mixing width, mixedness, mixed mass, and mixed-mass thickness is carried out to demonstrate the effect of ICs on the development of this RMI flow at different scales. Figure 13.22(a) shows the evolution of δ, defined in a similar fashion to the integrated mixing widths discussed in Section 5.1.1,

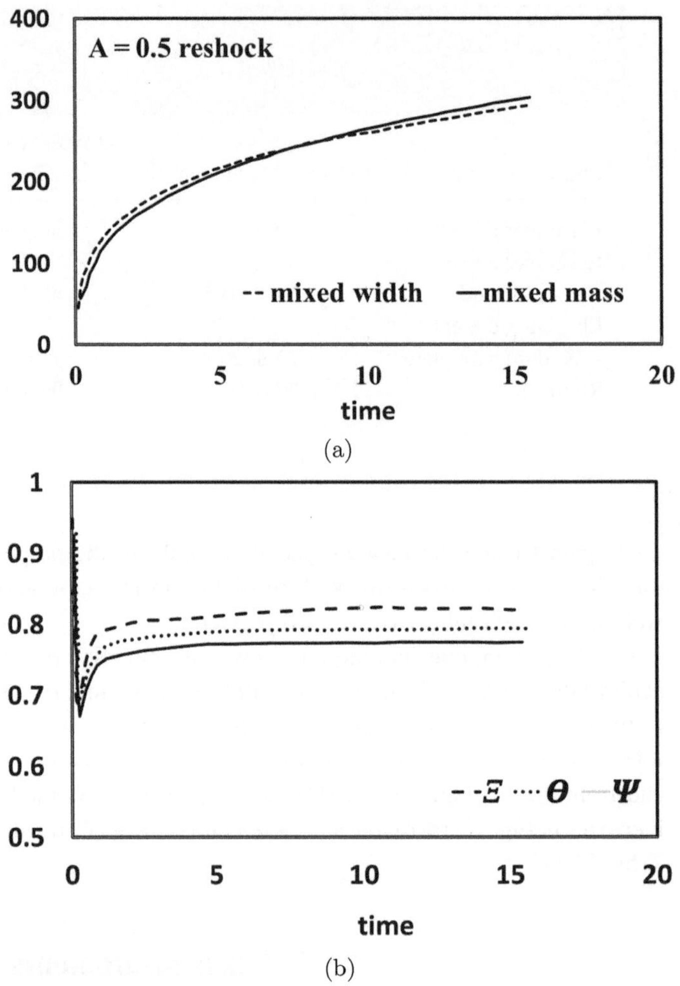

Fig. 13.21 (a) Temporal evolution of the mixed mass for the reshock case with $\mathcal{A} = 0.5$. Also, the corresponding results for the mixed width (dashed lines), multiplied by arbitrary constants. (b). Temporal evolution of the mixedness parameters and normalized mixed mass for the reshock case with the same Atwood number. All curves are plotted from the time of reshock. Zhou, Cabot, and Thornber (2016).

$$\delta = \frac{\iint 4Y_{N_2} Y_{CO_2} dxdy}{\int dy}. \tag{13.1}$$

The nondimensionalized time is defined so that the reshock arrival time will be zero for both the Mach 1.9 and 1.55 cases. Here, the multi-mode interface consistently grows faster than the single-mode case before reshock. After reshock, this length scale becomes insensitive to the additional modal content of the multi-mode interface, as all cases continue to evolve similarly (Mohaghar et al., 2019).

As illustrated in Figure 13.22(b), a linear time-dependence of $\mathcal{M}$ is observed for all cases before reshock. The multi-mode case has a higher growth rate of mixed mass compared to

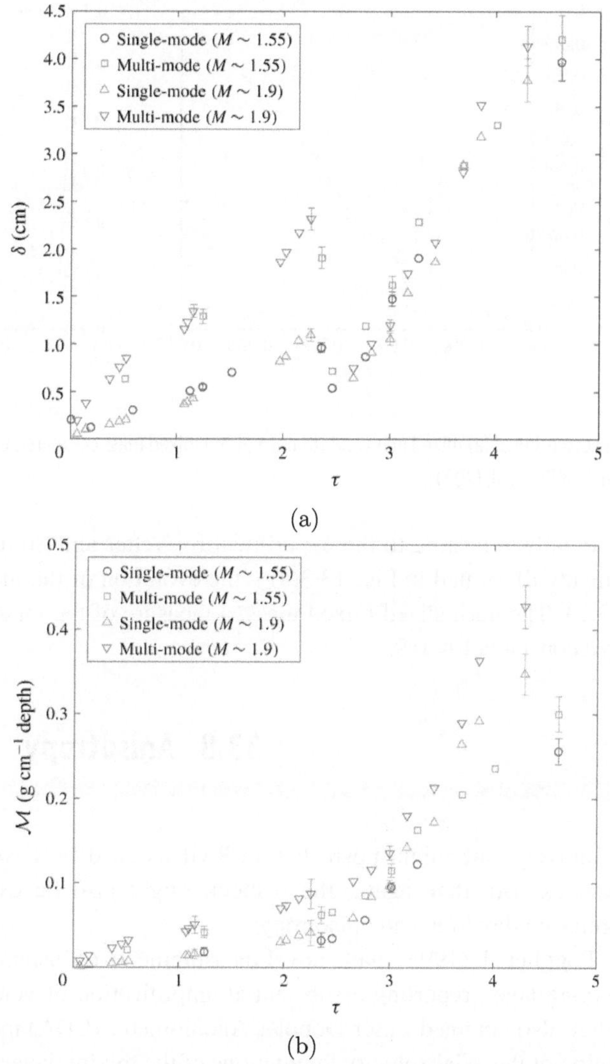

Fig. 13.22 (a). Evolution of integral mixed-mass thickness, δ, in dimensionless time, τ, for all cases. (b). Evolution of integral mixed mass, $\mathcal{M}$, in dimensionless time, τ, for all cases (Mohaghar et al., 2019).

the single-mode case, which is similar to the mixed width. Furthermore, the single- and multi-mode cases at high Mach have slightly higher mixed mass compared to low Mach number cases before reshock.

Unlike the behavior of δ, the post-reshock mixed mass is noticeably higher for higher Mach number, due to the larger density jump, particularly after reshock. This observation reveals that the rate of entrainment is similar for the two Mach numbers, while the rate of molecular mixing is higher for high Mach number (Mohaghar et al., 2019).

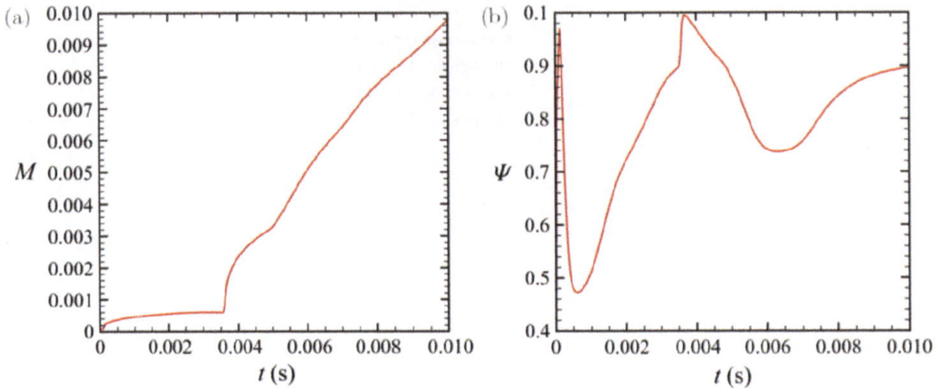

Fig. 13.23 The evolution of (a) mixed mass and (b) normalized mixed mass calculated using a combined LES and RANS approach. Fig. 2 of Bin et al. (2021).

Finally, returning to the experiment by Vetter and Sturtevant (1995) – its mixed-width already illustrated in Fig. 13.3(a) – the evolution of the mixed mass, $\mathcal{M}$, is shown in Fig. 13.23. The normalized mixed mass, a measure of the mixed product, appears to converge to a constant of ≈ 0.9.

13.8 Anisotropy

Intuitively, one might posit that an RMI-induced flow would be return to isotropy after reshock. But then again, the reshock might also be expected to introduce additional preferred direction and anisotropy.

Poggi et al. (1998) made novel measurements of instantaneous velocities in an SF_6 –air mixing layer, reporting a substantial amplification of velocity fluctuations after reshock. They also obtained Laser Doppler Anemometry (LDA) measurements, including the evolution of the axial velocity fluctuations of the mixing layer, before and after its interactions with reshocks.

Figure 13.24 illustrates the components of measured Reynolds stress tensor at the 161, 169, and 178.5 mm abscissas. The total error on the experimental data is estimated to be $\pm 15\%$ at these measurement locations. One can clearly see the enhancement of the turbulent energy induced by the first reshock (between the 125.5 and 161 mm abscissas). The interaction of the second reshock with the turbulent mixing layer is observed by LDA at the 178.5 mm abscissa. Once the waves have left the mixing layer, turbulence measurements are again available around the 178.5 mm abscissa measurement.

Figure 13.25 illustrates the anisotropy of the Reynolds stress tensor at the 169 mm abscissa, that is, approximately 0.3 ms after the interaction with the first reflected shock wave. At the profiles' maxima the anisotropy ratio $(\overline{R_{11}}/\overline{R_{22}})$ reaches a value of about 3 in the experiments. This strongly anisotropic state can be explained by resorting to the expected phenomenology of the shock–turbulence interaction, in which production of the Reynolds stress takes place mainly in the longitudinal direction.

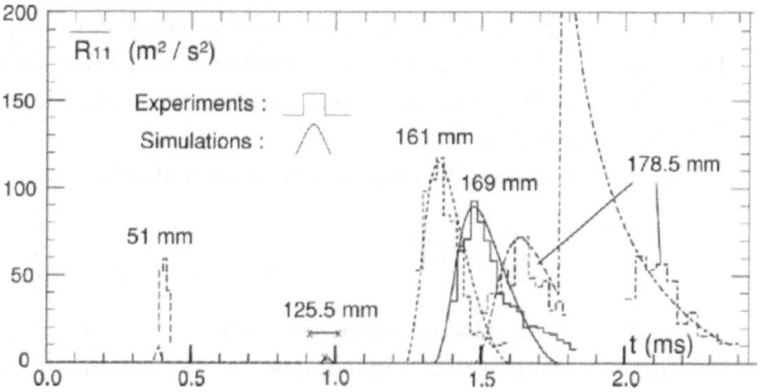

Fig. 13.24 CEA shock tube experiment results for $\overline{R}_{11}$. Broken lines represent the experimental data (see [Poggi et al., 1998], figure 5, and [Poggi, 1997] for data after 2.0 ms) while curves are the results of mix model calculation. To facilitate a comparison, the calculation results have been shifted forward by 0.03, 0.055 and 0.075 ms for the 161, 169 and 178.5 mm abscissa, respectively, so that the abscissa where the profiles reach their maxima coincide. $\overline{R}_{11}$ at the measurement locations, 51 (– – –), 125.5 (× — ×), 161 (- - - -), 169 (———), and 178.5 (— · —·) mm above the interface's initial location. Fig. 4 of Grégoire et al. (2005), with permission.

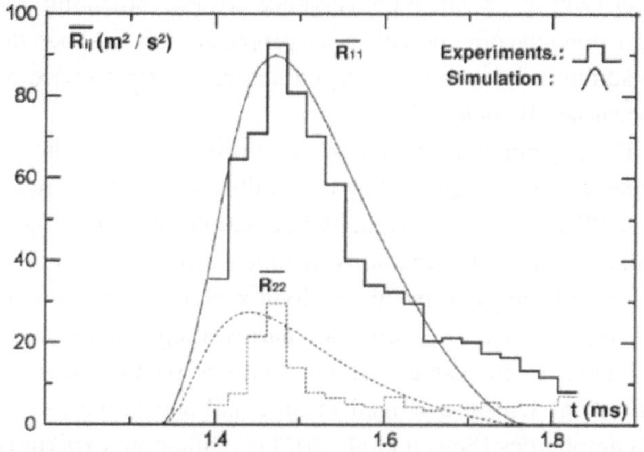

Fig. 13.25 CEA shock tube experiment: $\overline{R}_{11}$ (———), $\overline{R}_{22}$ (- - - -), at the abscissa 169 mm above the initial location of the interface. Broken lines represent experimental data while curves reflect turbulence model calculation results. As in Fig. 13.24, calculation results have been shifted forward by 0.055 ms. Fig. 5 of Grégoire et al. (2005), with permission.

As mentioned previously, the reshock experiments reviewed here have frequently been used as a verification and validation (V&V) testbed for mix models. The RANS calculations have been included in Figs. 13.24 and 13.25, but detailed model equations will be deferred to Chapter 22.

Numerically, Thornber et al. (2011) reported that anisotropy is increased, rather than decreased, after reshock. The final stage of the RMI-induced flow is the decay of all components of the turbulent kinetic energy as a power law in time, while the level of anisotropy (measured by TKX/TKY)[1] remains approximately constant. For the narrowband case, TKX/TKY $= 1.7 \pm 0.1$, and for the broadband case, TKX/TKY $= 1.8 \pm 0.2$. These values are higher than the previous results for a single-shocked layer (where TKX/TKY ≈ 1.6 and 1.5, respectively).

Tritschler et al. (2014b) evaluated the degree of anisotropy after reshock by analyzing

$$a(x, y, z, t) = \frac{|u_1'|}{|u_1'| + |u_2'| + |u_3'|} - \frac{1}{3}, \tag{13.2}$$

for the streamwise velocity component. Here, $a(x, y, z, t) = 2/3$ corresponds to having all TKE in the streamwise velocity component. The volume-averaged anisotropy in the inner mixing layer approaches an asymptotic limit of 0.04, implying no full recovery of isotropy, even after reshock.

In another simulation study, Ristorcelli et al. (2013) found that just before reshock the value of $\langle b_{ij} \rangle$ is about 0.28, which corresponds to three times as much energy in the axial direction than in either of the transverse directions. At the end of the simulation, after reshock, the anisotropy is much lower, $\langle b_{ij} \rangle \sim 0.1$, corresponding to about 30% more energy in the axial component than in either of the transverse components.

From their experimental data, Balakumar et al. (2012) found, however, the streamwise and spanwise velocity variances are approximately equal in magnitude throughout the fluid layer (within statistical convergence errors), indicating that any anisotropy in the velocity field along the direction of shock propagation is less than the statistical measurement errors. In addition, in an area-averaged sense, the streamwise and spanwise velocity variances appear nearly identical.

The examination of anisotropy of reshocked RMI flows may be complicated by its ICs dependence, as suggested by Gowardhan et al. (2011) and Thornber et al. (2012). Grinstein et al. (2011) noted that velocity fluctuations in the mixing layer are more isotropic when the initial interface perturbations include short wavelengths, in which case the authors nearly recovered full isotropy. When long-wavelength perturbations are used, the mixing layer remains anisotropic except for a narrow range on the heavy-gas side.

Figure 13.26 displays recent findings on the mean of the diagonal elements of the anisotropy tensor, averaged in time and across the mixing layer, with high and low initial amplitudes (Sewell et al., 2021). While both experiments are brought to similar levels of anisotropy following reshock, the behaviors of the flow field during the progression to late time are initial-value-dependent. Anisotropy is initially high following shock interaction, with high amplitude experiments exhibiting a slightly larger average anisotropy due to the larger interpenetration speeds of the growing layer at the earliest times. Given the higher overall energy of the flow fields, the anisotropy quickly falls to the asymptotic value since mode coupling occurs more quickly.

Youngs (private communication, 2017) pointed out that the differences in the results described above can be attributed to the influence of initial conditions and to the difficulty in achieving sufficient time to establish self-similarity. He further noted that for the narrowband case the reshock is just an additional (albeit nonlinear) narrowband perturbation. In

[1] In this work, the direction of shock propagation is x.

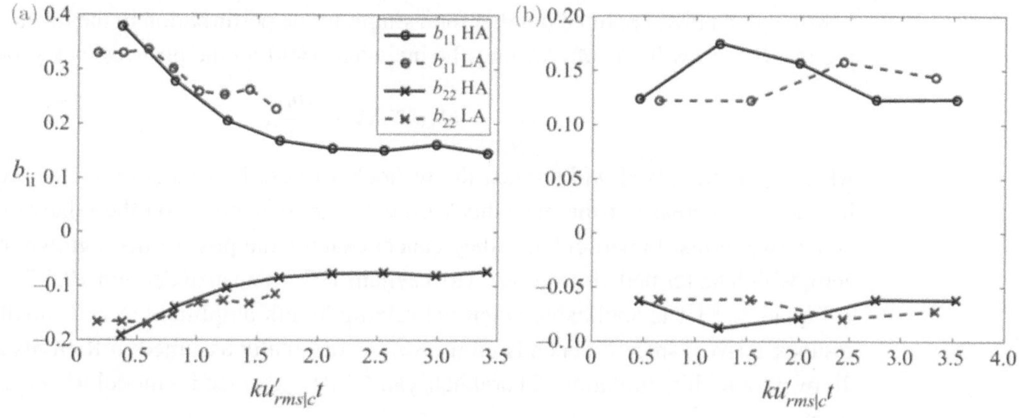

Fig. 13.26 Ensemble mean of the diagonal elements of the anisotropy tensor across the mixing layer versus dimensionless time. Solid lines indicate high amplitude experiments, while dashed lines indicate low amplitude experiments. Anisotropy following the incident shock (a) falls more rapidly in the high amplitude experiments than in the low amplitude experiments. After the second shock interaction (b) both groups of experiments have achieved similar levels of isotropy. Fig. 15 of Sewell et al. (2021).

this case, therefore, self-similar anisotropy should be expected to be the same for the single-shock and the reshock cases. For the self-similar broadband case with initial perturbation, spectrum $P(k) \sim k^m$, it has been argued that θ should be different for the single-shock and reshock cases, and this is likely to give rise to different levels of anisotropy. Experiments are likely to have broadband perturbations, but not necessarily of the self-similar form. This could lead to near-self similar behavior during the experiment but with further differences in anisotropy.

13.9 Reshock linear and nonlinear models

13.9.1 Single mode

To extend Richtmyer's model to account for reshock, Brouillette and Sturtevant (1989, 1994) generalized Richtmyer's work to model the growth of the total width of a single-mode interface subject to $N + 1$ impulsive accelerations, as observed in their experiments,

$$\left(\frac{dh}{dt}\right)_N = 2k \sum_{i=0}^{N} \frac{a_i^+ \mathcal{A}_i^+ \Delta V_i}{\psi_{D,i}}, \tag{13.3}$$

where k is the initial perturbation wave number. As one might expect, for the i-th shock wave, a_i^+, ΔV_i, $\mathcal{A}_i^+$, and $\psi_{D,i}$ are the post-shock amplitude, velocity jump, Atwood number, and the post-shock growth reduction factor (Duff et al., 1962). Of course, $N = 2$ for the "single reshock" case. Similarly to the assumption in Richtmyer's original model, Eq. 13.3 is applicable when the mixing length amplitude is very small at the time of reshock arrival.

Mikaelian (1985) pointed out that for a single-mode perturbation in the linear regime the pre-reshock growth rate dh_1/dt must be included, yielding the post-reshock growth rate

$$\frac{dh_2}{dt} = kh_r\mathcal{A}^+\Delta V - \frac{dh_1}{dt}, \tag{13.4}$$

where h_r is the mixed-width when the reshock arrives. This model may lead to either an increase or decrease of the post-shock growth rate, depending on the relative sign of the above two terms. In particular, if they cancel exactly, the post-reshock growth rate can be zero, which he termed "freeze-out" (Mikaelian, 1985; See also Section 10.5.2).

Equation 13.4 is applicable when the mixing length amplitude is very small when the reshock arrives, since it depends on the kh_r, as originally assumed in Richtmyer's model. To overcome this limitation, Charakhch'yan (2001) advanced a model with the following form

$$\frac{dh_2}{dt} = C_c\mathcal{A}^+\Delta V - \frac{dh_1}{dt}, \tag{13.5}$$

where dh_1/dt is the growth rate immediately before the reshock and C_c is an empirical constant ≈ 1.25 (see also, Charakhch'an, 2000). With well-defined initial heavy/light 2D interfaces, Guo et al. (2022b) found that the value of C_c obtained from experiments is smaller than the model value (0.66 ± 0.07 for the single-mode case; 0.47 ± 0.06 for the trapezoid case; and 0.96 ± 0.07 for the sawtooth case). This model is no longer a function of h_r and k, based on an assumption that the growth rate is only a weak function of the initial conditions at the reshock.

Schilling et al. (2007) analyzed the post-reshock growth rates of 2D single-mode RMI simulations. The initial conditions and computational domain were adapted from the Mach 1.21 air (acetone)/SF_6 shock tube experiment of Collins and Jacobs (2002). For later times, the simulated data suggested that the Brouillette-Sturtevant model (1989) underestimates the growth at all times following reshock, while the models of Mikaelian and Charakhch'yan overestimate growth.

Note that dh_2/dt and dh_1/dt have opposite signs, as they correspond to perturbation growths in opposite directions. If dh_1/dt is small, Charakhch'an's model and Mikaelian's reshock model are governed by the same equation, differing only in the values of the coefficient (Ukai et al., 2011).

13.9.2 Multimode

For multi-mode 3D perturbation, Mikaelian's reshock model (Mikaelian, 1989) employed the following linear correlation,

$$h = C_M\mathcal{A}^+\Delta Vt, \tag{13.6}$$

where C_M was estimated as 0.28, extrapolated from constant-acceleration RTI experiments (Read, 1984) applied to shock tube experiments.

The Mikaelian (1989) model achieved good agreement with the simulation of reshocked multi-mode RMI flows using the WENO scheme (e.g., Zhang et al., 2003, 2006), with a two-mode initial perturbation with superposed random noise to model the Mach 1.5 air/SF_6

Vetter–Sturtevant shock tube experiment (Schilling and Latini, 2010). The simulation indicated a velocity (dh/dt) obtained from the model about 10% lower than the data before reshock and around 3% lower than the data after reshock.

From their shock tube experiments, Leinov et al. (2009) obtained a growth coefficient of 0.28–0.39, as later studies by Dimonte and Schneider (2000) and Oron et al. (2001) revealed that there is an asymmetry between the fronts of the bubbles and spikes of the mixing layer. From their membraneless reshock experiment, Jacobs et al. (2013) obtained a value of 0.26 for this coefficient. Changing the length of the shock tube (Griffond et al., 2017), Bouzgarrou et al. (2014) determined that values of C_M can be constrained between 0.38 and 0.43.

More recently, Mikaelian (2011) pointed out that his 1989 reshock model should only describe the early stages of a shock or reshock. He proposed that the subsequent growth rate begins to decay and the mix width transitions from the early $h \sim t$ behavior to $h \sim t^{\theta}$. The transition from t to t^{θ} was argued to be a result of the mixing width forgetting the direction of the shock, that is, turbulence becoming isotropic around $t = t^*$. Thus, the extended shock or reshock model takes the form,

$$\begin{aligned} h &= h_0 + 2\alpha \mathcal{A} \Delta V t, && t \leq t^*, \\ h &= h^* \left(1 + \frac{\dot{h}^*}{\theta h^*}(t - t^*)\right)^{\theta}, && t \geq t^*, \end{aligned} \tag{13.7}$$

where $\alpha = \alpha_b$ or α_s, $h = h_b$ or h_s, and h^* and $\dot{h}^*$ are the mix width and its growth rate at $t = t^*$, ensuring continuity of both measurements. The performance of the model obviously depends on how $t = t^*$ is determined, with Mikaelian originally proposing $t^* = (h_0^{b+s}/\Delta V)\beta_M$ and $\beta_M \sim 6$.

The most recent reshock RMI model is due to Mikaelian (2015), who has refined his previous procedure and suggested the following expression:

$$t^* = \frac{(1 + 2\alpha\beta_M \mathcal{A})h_{\text{before}} - h_{\text{after}}}{2\alpha \mathcal{A} \Delta V}, \tag{13.8}$$

where h_{before} and h_{after} are the total mixing width before and after reshock. It is worth reporting that this model agrees with the experiments of Vetter and Sturtevant (1995), Leinov et al. (2009), and Jacobs et al. (2013), as well as the simulations of Thornber et al. (2017).

Mikaelian and Olson (2020) emphasized the separation of mix widths into bubbles and spikes because they play different roles in applications. Figure 13.27(a) shows the total mix widths, $h = h_b + h_s$, and compares the model's performance against the shock-tube experiment of Jacobs et al. Figure 13.27(b) exhibits total mix widths for the theta-group problem calculated by the MIRANDA code and the model.

Mikaelian and Olson (2020) noted that the abrupt change at the transition time cannot be physical – nothing special is occurring at t^*. If this transition signals a loss of memory of shock direction, it cannot occur instantaneously. Clearly, such a two-stage description is only a simplified approximation to what must be a fully continuous evolution, a possibility they explore in their article.

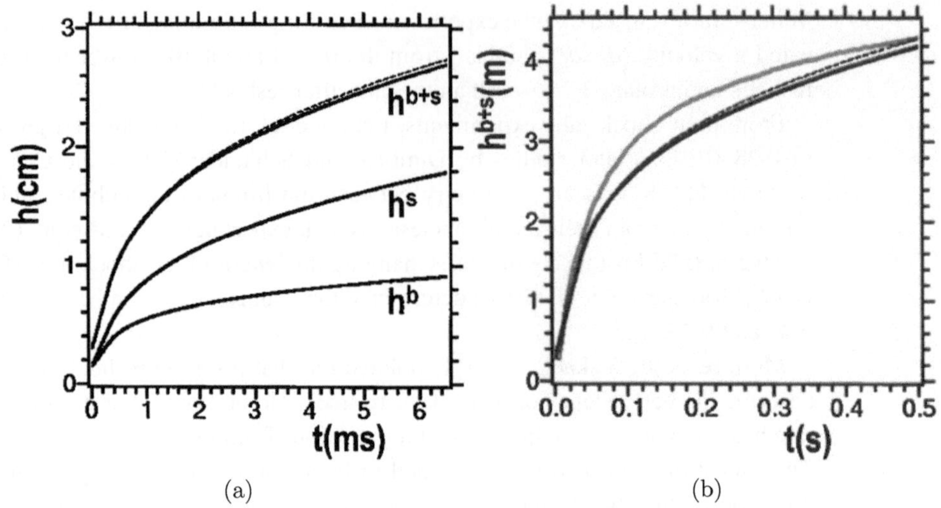

Fig. 13.27 Total mix widths for (a) a shock-tube experiment of Jacobs et al. (2013) and (b) the theta-group problem of Thornber et al. (2017) calculated by MIRANDA code (Upper gray curve). The solid black curve is the sum $h^b + h^s$ calculated separately with $\theta^b = 0.25$ and $\theta^s = 0.31$, while the dashed black curve shows h^{b+s}, calculated with a single $\theta = 0.29$. Figs. 1 and 3 of Mikaelian and Olson (2020).

13.10 Multi-shocks

Figure 13.28(a–c) shows the mass fraction isosurface, enstrophy, and density cross-section from a multi-mode, multiply-reshocked RMI simulation with an air-SF_6 system of Vetter and Sturtevant (1995). Following the passage of the incident shock, spikes and bubbles form at $t = 0.25$ ms. The spikes begin to roll-up, which can barely be seen prior to $t = 3$ ms. Reshock occurs at $t \approx 3.25$ ms, at which point the mixing layer is compressed. At $t \approx 5$ ms, the reflected rarefaction wave induced by the shock's refraction at the interface, interacts with the evolving interface. Following reshock, the instability develops a complex structure and the enstrophy field transforms from elongated tubular structures into short, disordered tubular structures with random orientations. The arrival of the reflected rarefaction represents the last important wave–interface interaction. Consequently, the flow nearly becomes a purely decaying flow after $t = 6$ ms (Schilling and Latini, 2010).

Figure 13.29 presents a sequence of laser sheet images showing the evolution of the instability in the air/SF_6, air/CO_2, air/N_2, and air/He cases. With an Atwood number close to -1, the air/He interface (Fig. 13.29(d)) overturns immediately following the incident shock. Later, we observe a quick distortion of the interface into mushroom structures, which are enhanced by the successive reshocks (Jourdan and Houas, 2005).

Lapushkina et al. (2019) and Azarova et al. (2020) produced mixing by a much higher Mach number (Mach 5–6) (see also, Azarova and Gvozdeva, 2018). Multiple generations of the shear layer and RM instabilities have been obtained as the result of the interaction of a shock wave with thermal strata, which is the area formed by ionization instability

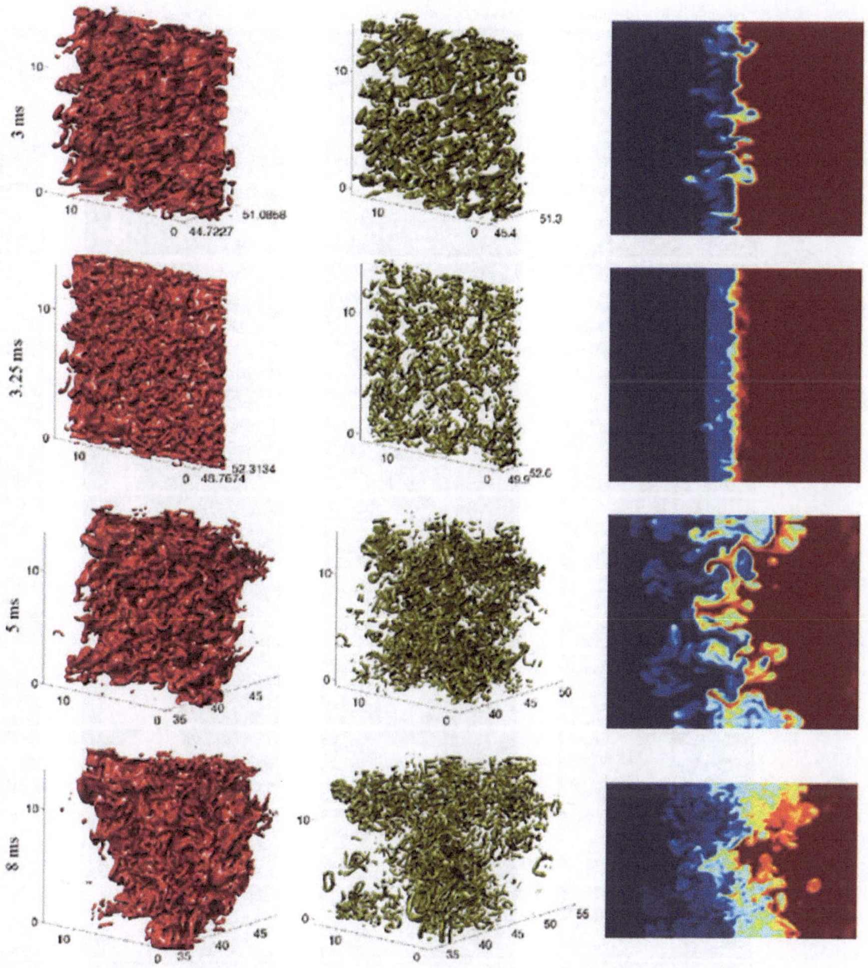

Fig. 13.28 Visualization of the mass fraction isosurface (left column), enstrophy isosurface (middle column), and density (x, z)-cross section (right column), where blue represents air and red represents SF_6 at 3, 3.25, 5, and 8 ms. The height of each density cross section is 27 cm. Fig. 8 of Schilling and Latini (2010).

in gas discharge plasma at a pressure of 3–7 Torr and gas-discharge current of 100–400 mA. Numerical simulations provide further insight by allowing one to analyze not only the density and pressure profiles, but also the internal and kinetic energy, as well as other statistics difficult to obtain experimentally (Fig. 13.30). It should be noted that the effect of thermal conductivity is essential in the considered phenomena. The instabilities formed in a viscous flow are numerous and highly pointed (Fig. 13.31(a)), while thermal conductivity "cuts off" the amplitudes of these peaks and generally smooths out the solution (Fig. 13.31(b)).

While the efficacy of reshock has been demonstrated in enhancing mixing and speeding up the transition to turbulence, Fig. 13.32, taken from Li et al. (2019a), suggests that the

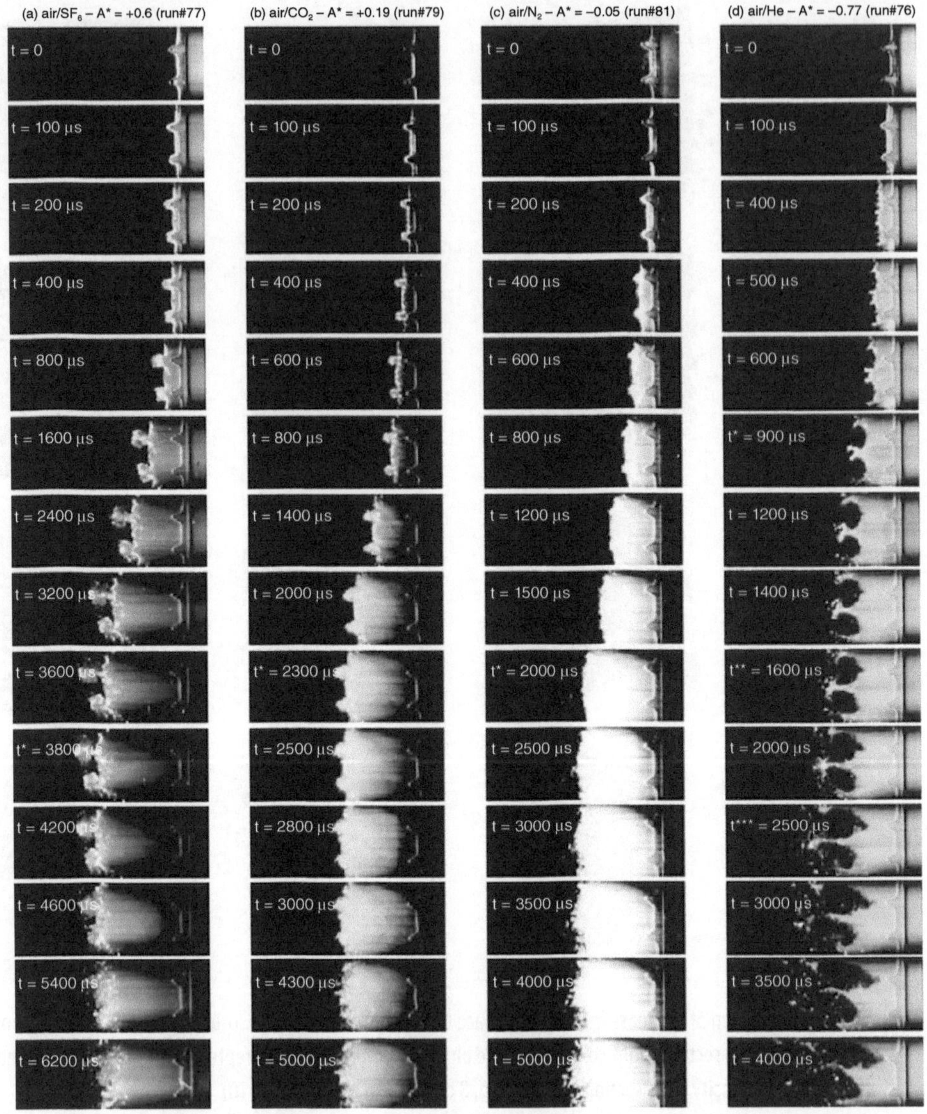

Fig. 13.29 Laser sheet time sequence showing the interaction of a shock wave (moving from right to left) with two high-amplitude single-mode perturbations for (a) air/SF_6, (b) air/CO_2, (c) air/N_2, and (d) air/He gas interfaces. t^*, t^{**}, and t^{***} correspond approximatively to the 1st, 2nd, and 3rd reshock/interface interaction times, respectively. Fig. 2 of Jourdan and Houas (2005).

rarefaction and compression waves have reduced the growth rate of the mixed width from that obtained during the time interval of the reshock, between 0.0021 s and 0.0027 s. For this multi-mode case, the value of the mixedness, Θ, increased very modestly due to the rarefaction/compression waves.

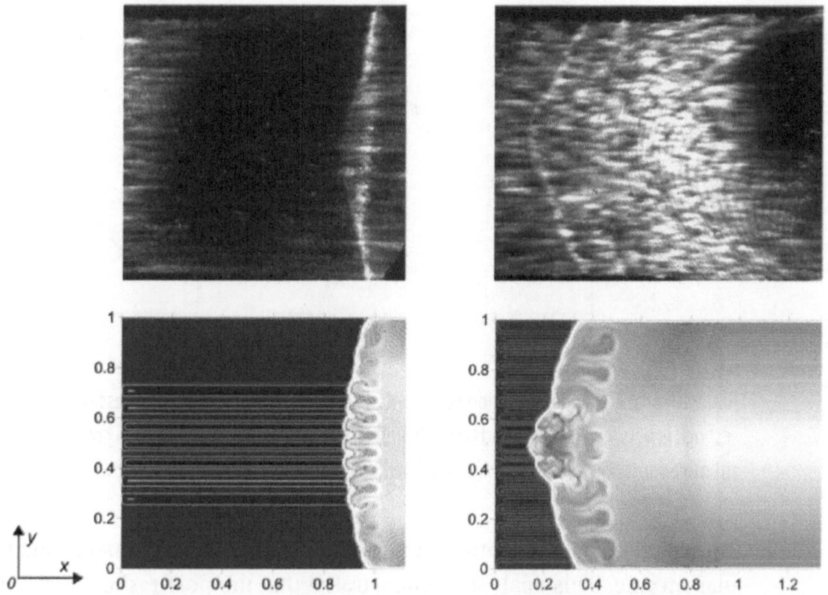

Fig. 13.30 Entrance of an initially-plane shock wave to a region formed by the ionization of an unstable plasma (left column) and distortion of its front during the passage through the region (right column): top panel – experiment, Schlieren images; bottom panel – calculations, density (isochors). Fig. 12 of Lapushkina et al. (2019), with permission from Aerospace.

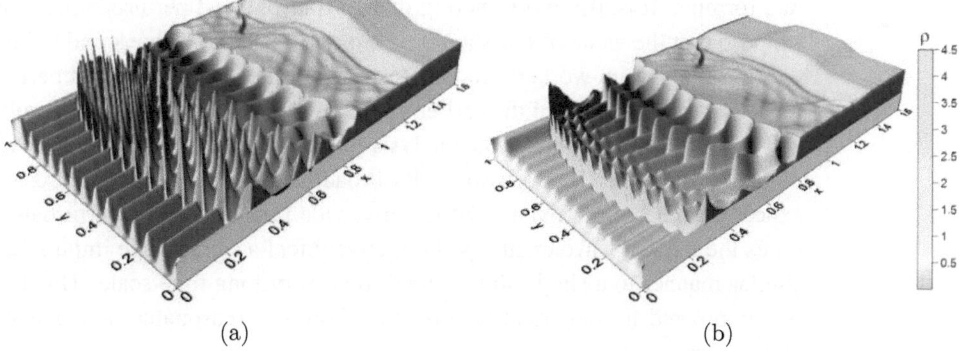

Fig. 13.31 Effect of heat conductivity on the density field. The coefficient of thermal conductivity is zero (left figure) and non-zero (right figure). Fig. 11 of Azarova et al. (2020). Figure is used under the terms of the Creative Commons Attribution 3.0 license.

It appears that the RM scaling parameter (θ) is likely to depend on the initial conditions, as suggested in two reshock models proposed by Thornber et al. (2011). The first model is a reshocked self-similar mixing layer initiated from a narrowband perturbation, valid for early times. This model was developed through consideration of a local Atwood number,

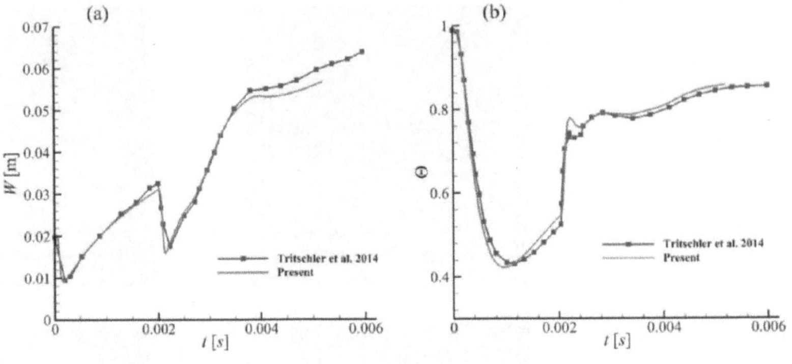

Fig. 13.32 RMI with air/SF_6 gas combination subject to reshock and rarefaction/compression waves. The time duration of the reshock is between 0.0021s and 0.0027s before the arrival of rarefaction and compression waves. (a) Mixed width and (b) mixedness. Fig. 4 of Li et al. (2019a), with the permission of AIP Publishing.

generated by a combination of a spatial mean-density gradient (due to mixing) and molecular mixing. This analysis demonstrated that the post-reshock growth rate is a function of the variable, $\sqrt{1-\Theta}$, where Θ is the molecular mix prior to reshock. As this molecular mix depends on the initial conditions, there is a sustained impact of these initial conditions on the post-reshock growth. This model suggests an initial growth that is linear in time, and a comparison with Mikaelian's model (1989) suggests that the coefficient C_M will include dependence on pre-reshock molecular mixing. Comparison with numerical simulations indicates that the sensitivity of the model's free parameter is substantially reduced compared to Mikaelian's model, due to the integration of molecular mix in the growth rate formula. It is also worth noting that for narrowband perturbations, it is reasonable to suppose that the value of C_M should be similar for the first shock and the reshock.

Thornber and co-workers' model describes reshock of broadband perturbations, which represents an extension of the earlier model of Youngs, presented in Thornber et al. (2010), and follows a similar line to the analysis of Haan for the RT instability (Haan, 1989). The model predicts that for a substantially broadband perturbation, each reshock reduces the expected value of θ. Given a sufficiently wide range of initial perturbations, at very late times the longest wavelengths will effectively feel a single large impulse and respond in a similar manner to a single shock, albeit on a very long time scale. This broadband model was compared to numerical results, and showed a reasonable agreement, particularly at early times.

Because of the heuristic nature of their construction, the reshock models reviewed here are designed to fit, but not to predict, the growth rate after the passage of the reshock.

14 Combined instabilities

14.1 RTI and RMI

The persistent, well-collimated, relativistic jet is a universal structure observed in astrophysical compact object accretion disks (e.g., Begelman et al., 1984; Ferrari, 1998; Mirabel and Rodriguez, 1999). The RTI and RMI between the jet and the external medium can destroy cylindrical jet configuration as a result of spontaneously induced radial oscillating motion (Matsumoto and Masada, 2013).

Balick and Frank (2002) discuss the kinematics of ejecta in astrophysical planetary nebulae. These ejecta result from stellar wind interactions late in the life of intermediate-mass stars, when the stars rapidly lose mass and evolve through alternate phases where the stellar winds are slow, cool, and dense, or fast, hot, and diffuse. The internal collisions between these successive outflows give rise to shocks, and induce RT and RM instability within the ejecta. The perturbed shape of the supernova remnant may also be attributed to simultaneous RTI and RMI actions (Aschenbach et al., 1995).

Non-premixed combustion is an application in which both chemically reacting RT and RM instabilities occur. As noted recently in Attal and Ramaprabhu (2015), in the absence of an incident shock, burning results in the formation of so-called combustion waves, which spontaneously trigger RM- and RT-like instability growth of the interface. When an incident shock is also present, the aforementioned combustion waves and shock waves interact, creating RTI and RMI that significantly distort them. When the unstable interface is reshocked, the mixing, combustion efficiency, and combustion product formation are all significantly enhanced.

ICF implosions are subject to RT and RM instabilities during several phases of the implosion. As the shell is accelerated to its peak implosion velocity, RM instabilities at the various interfaces are followed by ablation-front RT growth as the low-density blowoff plasma accelerates the dense shell of solid ablator and fuel. During the later acceleration phase, RTI is also possible between the ablator shell and DT fuel if compression sufficiently increases the Atwood number (Hammel et al., 2010; Clark et al., 2011). Finally, at stagnation, RTI acts on the decelerating shell, and RMI is stimulated by the multiple shocks reflected from the implosion center (Haan et al., 1995; Lobatchev and Betti, 2000; Herrmann et al., 2001). Both the RT and RM instabilities might be involved simultaneously in the deceleration phase of ICF implosions, which has been investigated, for example, by Bradley (2014) with a hydrodynamics code to examine the relative amounts of shock-induced RMI and RTI mixing.

To untangle the physics of multiple instabilities, a planar geometry is usually the preferred starting point. As mentioned already, Brouillette and Sturtevant (1989) generalized Richtmyer's linear model to describe the growth of the total width of a single-mode interface subject to multiple impulsive accelerations observed from their experiments. A nonlinear model incorporating both the RT and RM instabilities can be constructed for all density ratios, based on a simple potential-flow approach (Sohn, 2003). Numerically, Chen et al. (2018) addressed the collaboration and competition between the single-mode RT and RM instabilities. The growth rate of the disturbance increases in such coexisting systems. This is different from the pure RM case, in which the growth rate decays due to the gradual increase in nonlinear effects.

At the Institute of Technical Physics (RFNC–VNIITF), Kucherenko et al. (2010) observed the statistical and spectral characteristics of the turbulent mixing layer formed under consecutive RM and RT instabilities in a multi-functional shock tube (Kucherenko et al., 2003). Meshkov and his colleagues investigated the features of turbulent mixing layer development at an interface accelerated by a non-stationary shock wave (Fig. 14.1). The authors found that coupled RT and RM instabilities suppress each other in some situations (Meshkov, 2013).

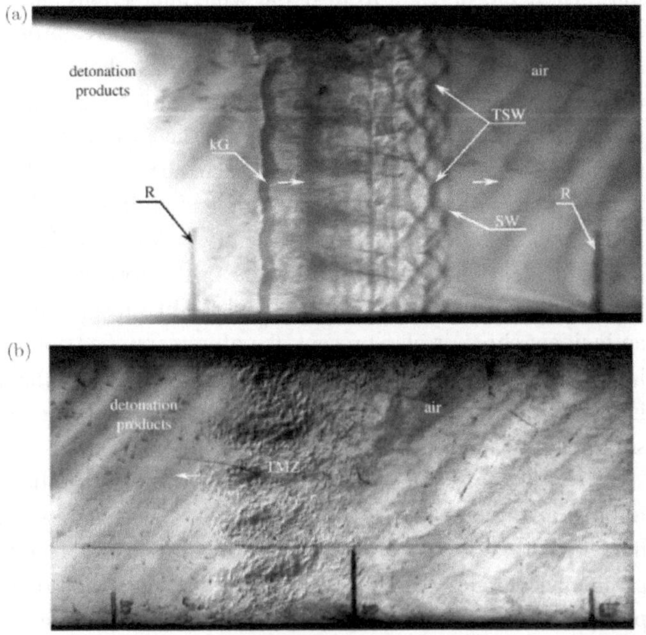

Fig. 14.1 The flow pattern in an acetylene shock tube for two times (μs): (a) 100 (b) 700. The time is counted from the time of initiation of the acetylene–oxygen mix layer; SW, shockwave; TSW, transverse shockwave system; kG, detonation products–air interface; TMZ, turbulent mixing zone; R, reference. Arrows show flow direction. The top figure illustrates a complicated and regular flow structure behind the shock wave front at the initial times in the shock tube. The observed pattern represents the flow behind the shock wave front with a 3D periodic perturbation. In panel (b), the turbulent mixing layer develops quickly after the wave is reflected. Fig. 11 of Meshkov (2013), with permission.

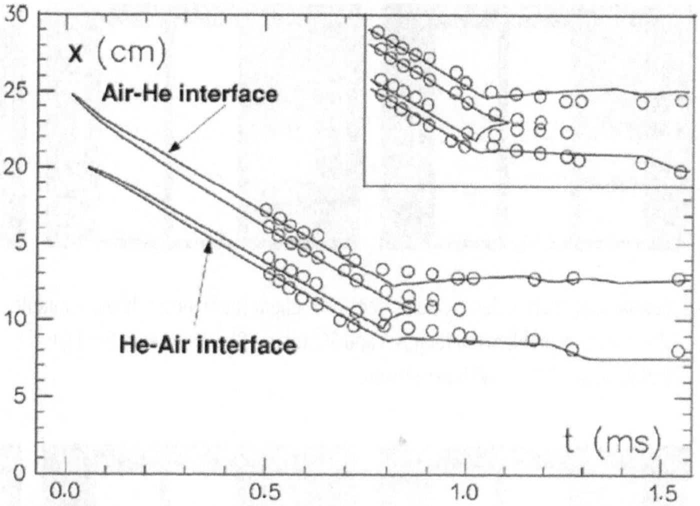

Fig. 14.2 Three-layer VNIIEF shock tube experiment: location of the boundaries of the turbulent mixing zones obtained from the experiments of Andronov et al. (1982) (◯) and mix model of Grégoire et al. (2005) (——). In the insert, the experimental data have been shifted by 45 μs and by −0.5 cm. Fig. 8 of Grégoire et al. (2005), with permission.

Andronov et al. (1982) described their three-layer VNIIEF shock tube experiments. The gases are air, helium, and air initially separated by membranes (Bliznetsov et al., 1997), where the intermediate layer has an initial thickness of 4.8 cm. As shown in Fig. 14.2, two turbulent mixing zones develop simultaneously, and due to their proximity, the number of interactions increases notably. This flow undergoes two RMI phases of opposite Atwood numbers, since the Mach 1.3 shock wave propagates from air to helium at one interface and from helium to air at the other.

Using the discontinuous Galerkin method, Henry de Frahan et al. (2015c) solved the 2D multi-fluid Euler equations to investigate the interaction of a shock wave with successive perturbed interfaces, and specifically RMI (due to shocks) and RTI (due to rarefactions) growth. The thickness of the intermediate layer is the same as the wavelength of the single-mode perturbation profiles. The single-mode initial perturbation of the two interfaces is in phase. At each interface, the incoming shock generates reflected and transmitted shocks and rarefactions, which further interact with the interfaces. Particular attention is paid to shock interaction with two successive interfaces, specifically the dependence of mixing on the density of the third gas. If the third gas is heavier, the reflected shock decreases the growth and tends to reverse the perturbation growth as the thickness of the second gas increases (Fig. 14.3). If the third gas is lighter than the second gas, however, the reflected rarefaction at the second interface amplifies the growth at the first interface. This has two main reasons: the reflected rarefaction deposits vorticity in the same direction as the incoming shock, and the perturbation amplitude has grown before its interaction with the rarefaction (Fig. 14.4). Another area of study is the perturbation growth at both interfaces as a function of the dimensionless interface spacing.

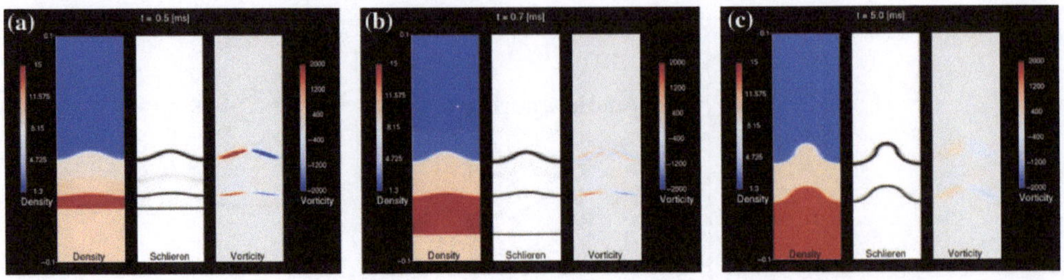

Fig. 14.3 Several snapshots of density (left), density gradient (numerical Schlieren, middle), and vorticity (right) fields for the three-layered problem (air for gas A and SF_6 for gas B) with a heavy third gas (heavier than SF_6). Fig. 10 of Henry de Frahan et al. (2015c), with permission.

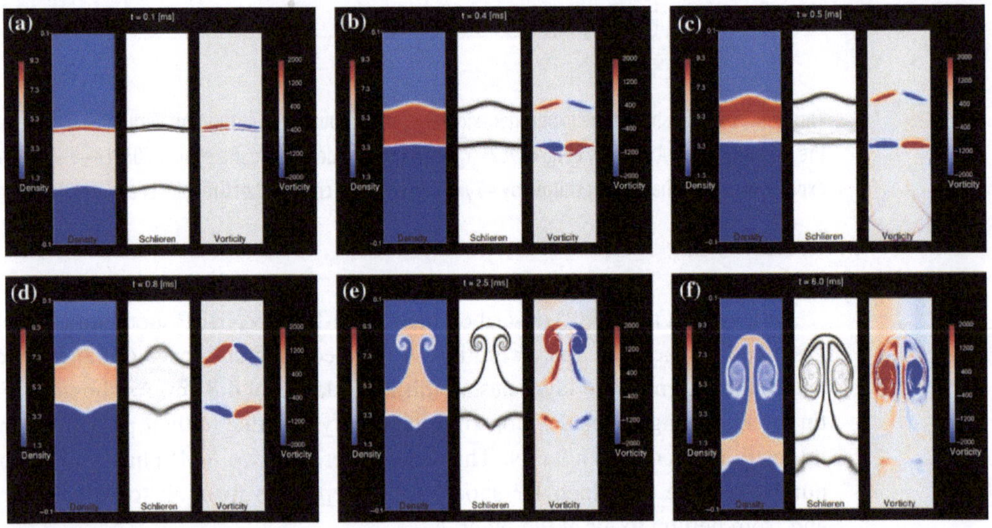

Fig. 14.4 Several snapshots of density (left), density gradient (numerical Schlieren, middle), and vorticity (right) fields for the three-layered problem (air for gas A and SF_6 for gas B) with a light third gas (helium). Fig. 5 of Henry de Frahan et al. (2015c), with permission.

For multi-mode initial conditions, Li et al. (2021a) simulated a shock tube with an end wall. This created a complex mixing process from successive waves, such as shock, reshock, rarefaction, and compression waves. At the end of the simulations, the mixed-width reached a value that is notably higher for the light-heavy (air-SF_6) than the heavy-light gas (SF_6-air) configuration (Fig. 14.5, top panel). In the meantime, after the passage of the first shock wave, the corresponding values for the mixedness, Θ, increase quickly, arriving at 0.86 and 0.82, just before the arrival of the reshock. After the reshock, the curves for Θ decrease to a minimum of approximately 0.71 and 0.74, before recovering to around 0.8 and 0.78 by the experiment's end (Fig. 14.5, bottom panel). During the passage of rarefaction and compression waves, the RM, RT, stretching and compression effects are all important. Additionally, a combined numerical and modeling study has been

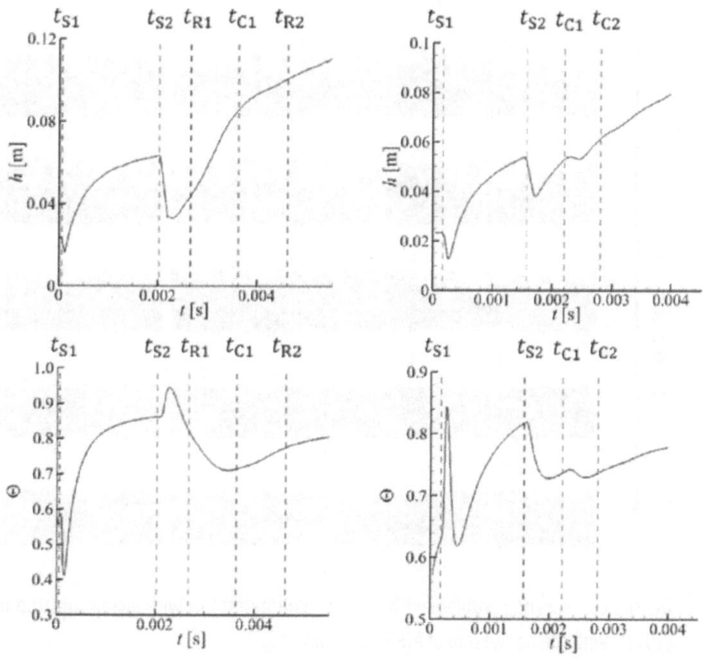

Fig. 14.5 The mixing width (h, top panel) and the molecular mixing fraction (Θ bottom panel). The results are from air/SF_6 (left column) and SF_6/air (right column). For air/SF_6, the dashed lines mark the time when the waves (the first shock wave, second- or reshock, first rarefaction wave, first compression wave, and second rarefaction wave) meet the mixing layer. For SF_6/air, the dashed lines mark the time when the waves (the first shock wave, second or reshock wave, first compression wave, and second compression wave) meet the mixing layer. Fig. 6 of Li et al. (2021a), with permission.

carried out to further elucidate the relative contributions from RT and RM in the growth of the mixing layer.

The 3D RT and RM instabilities can be induced by a laser pulse on a metal surface in a "covered target configuration." The environment evolves into complex fluid flow with wave-vortex phenomena and turbulent mixing. Such an environment is generated in a semi-confined configuration, which causes the extended vapor/plasma lifetime and multiple reshocks that strike the density interface (Lugomer, 2019, 2020).

14.2 RTI and/or RMI with KHI

Combined KHI and RTI (KHRT) has been measured using a network of temperature sensors in ocean half a kilometer deep, as mentioned briefly in Chapter 1. As an another example, in planetary impacts, additional vorticity can emerge due to RT and KH instabilities, depending on the impact conditions and geometry (Ravid et al., 2021).

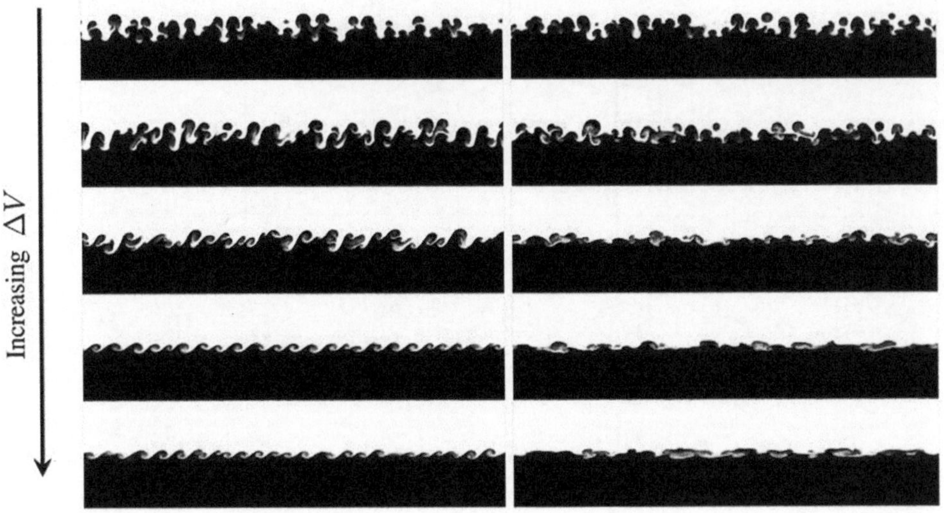

Fig. 14.6 Density contours of a problem with fixed RT parameters but with increasing amounts of shear. Fig. 14 of Olson et al. (2011), with the permission of AIP Publishing.

Recently, Vadivukkarasan and Panchagnula (2017) considered the 3D destabilization characteristics of an annular liquid sheet when subjected to the combined RTI and KHI mechanisms. As discussed previously, the KHI is the origin of the mushroom structures during the nonlinear RTI and RMI (Yabe et al., 1991; Zabusky, 1999), and it plays a critical role in the time-dependent transition to turbulence in these flows (Zhou et al., 2007, 2003a,b).

In the combined KHRT nonlinear regime, Olson et al. (2011) found numerically that adding a small amount of shear decreases the growth rate and changes the flow structure such that less energy is channeled into vertical mixing. At the point of minimum peak mixing rate, the structure has changed to largely resemble a pure KH instability (Fig. 14.6). Continued increase of the shear increases the amount of energy available and therefore enhances the mixing rate.

Akula et al. (2013) and Akula and Ranjan (2016) carried out an experimental investigation of coupled KHRT in the multi-layer gas tunnel facility (Fig. 14.7), an improved version of the two-layer gas tunnel used by Banerjee and Andrews (2006). To properly model their experiments, one must modify the classical RTI scaling, as the mixing layer growth is initially driven by the KHI. Akula et al. (2017) have attempted to quantify the switch from KH to RT dynamics in an unstably stratified free shear flow using a transitional Richardson number

$$Ri = -\frac{g\left(\frac{\partial \rho}{\partial z}\right)}{\rho\left(\frac{\partial u}{\partial z}\right)^2} = -\frac{2gh\mathcal{A}}{\Delta_u^2}, \tag{14.1}$$

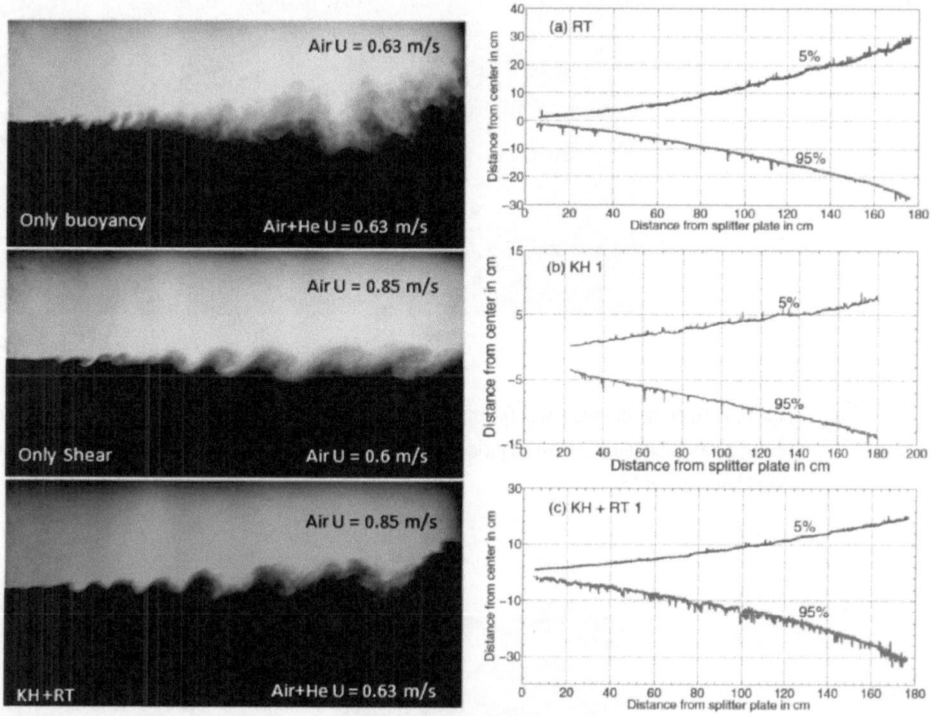

Fig. 14.7 Sample flow images with $\mathcal{A} = 0.032$ and (a) only buoyancy, (b) only shear, and (c) combined buoyancy and shear. Lines show volume fraction contours of the bottom stream fluid (0.05 and 0.95) (ensemble average of 200 images taken over 200 s). Fig. 3 of Akula et al. (2013), with permission.

where Δ_u is the velocity difference, and ρ and h are the measured mean density and the mixing width, respectively. The Richardson number values are positive for stably stratified flows and negative for unstably stratified flows.

Broadly speaking, oblique shocks generate all three instabilities: RT, RM, and KH (Mikaelian, 1994b). An interesting experimental study considered the evolution of an unstable elliptically shaped inhomogeneity embedded in ambient air. Separate experiments aligned it normally and at an angle of 24° to an incident plane blast wave with $M = 2.15$ (Sembian et al., 2018). While no KHI is observed in the former case, a vortex train similar to the KHI was observed for the non-normal case. The transmitted shock was seen to adjust itself to travel normally to the inclined inhomogeneity and to spawn uneven precursor shock regions (Fig. 14.8).

The interaction of a planar shock wave with a triangular SF_6 interface surrounded by air is a related case (Fig. 14.9). This configuration can be constructed using the soap film method (Wang et al., 2013c; Luo et al., 2015a). For such a triangle, the RMI will deposit a large number of small-scale baroclinic vorticities on the gas interface (Zeng et al., 2018a). These vorticities then accumulate and redistribute to form the primary KH billows. Along the gas interface, many small-scale vorticities are subsequently developed, resulting in significant mixing layer growth. Other numerical simulations have been carried out to study

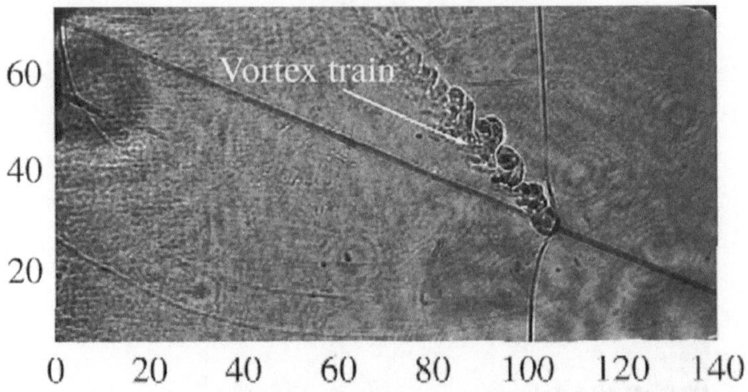

Fig. 14.8 Single shot experimental shadowgraph of an unstable elliptically shaped inhomogeneity that was embedded in ambient air and aligned at 24° to an incident $M = 2.15$ plane blast wave. The vortex train structure is visible. Fig. 13 of Sembian et al. (2018).

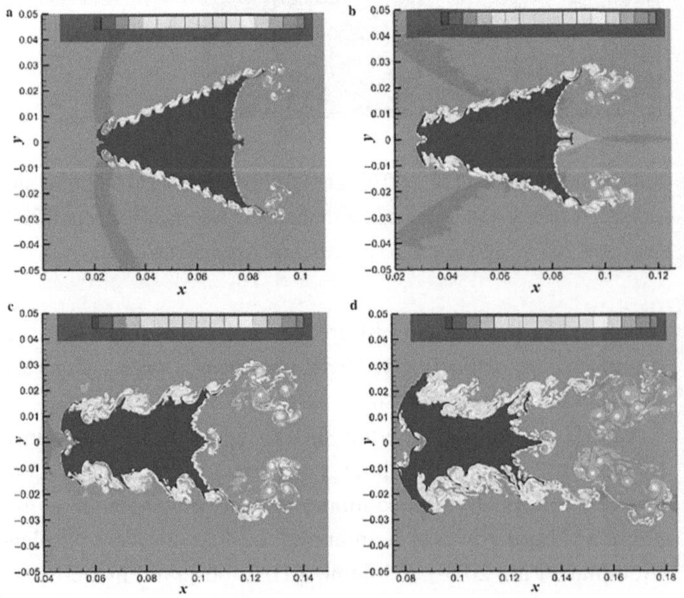

Fig. 14.9 Contours of air density at various late stages of fluids mixing (ms): (a) 0.5; (b) 0.65; (c) 1; and (d) 1.5. Fig. 11 of Zeng et al. (2018a), with permission.

interactions between a shock wave and an object in rectangular, elliptical, diamond-like, and other triangular shapes (Fan et al., 2012).

Within the linear growth regime, Ye et al. (2011) analytically evaluated the RTI and KHI competition for arbitrary $\mathcal{A}$ and continuous density and velocity profiles. They showed that this competition depends on the Froude and Atwood numbers, and the combined linear growth rate increases with the acceleration, but decreases with the width of the transition layer (Wang et al., 2010). Using Layzer's model, Mandal et al. (2011) studied the nonlinear evolution of two fluid interfacial structures arising due to the combined action of RTI and KHI. They found that the growth rate of the RTI-induced nonlinear interfacial structures is further enhanced by the KHI.

Banerjee et al. (2012) studied the development of interfacial structures such as RTI/KHI or RMI/KHI bubbles in viscous fluids, both analytically and numerically. It was found that the growth and normal velocity of the structures are dependent on the relative velocity shear and the coefficient of kinematic viscosity. For instance, the growth rate and the bubble growth are decreased significantly for fluids of high viscosity coefficient and small velocity shear difference. The 3D destabilization characteristics of an annular liquid sheet and an inviscid cylindrical interface were also examined under combined RTI and KHI (Vadivukkarasan, 2021; Vadivukkarasan and Panchagnula, 2017).

15 Geometrical configurations

15.1 Initial interfacial perturbation shapes

As revealed in Section 14.2, a shock wave interacting with an inclined perturbed interface introduces enhanced KHI features compared to its single- and multi-mode planar counterparts. We now will expound on this observation.

A series of shock tube experiments conducted at AWE included two experiments with initial interfacial configurations of standard and inverted chevron (Smith et al., 2001; Holder and Barton, 2004). The use of a fine wire grid to support the membrane initially used to separate the gases allowed for the modification of the initial interface geometry. The inverse chevron experiment involves a block of dense sulfur-hexafluoride (SF_6) gas, seeded with olive oil droplets, encased in the air within a tube. Figure 15.1(b) contrasts the initial flow field for the inverse-chevron test case with that of the standard double planar setting (Fig. 15.1(a)).

We note that the inverse chevron experiment provides a test case which is 2D on average, and which may thus be used for the validation of 2D Reynolds-averaged Navier–Stokes equations (RANS) models. The richness of the physics involved in such experiments leads to its frequent use as a "stress test" for validating simulations or mix models, similar to those from reshock experiments.

The model used to describe initial perturbations is an important issue, as it has a significant influence on the eventual results of the simulations. In order to produce a realistic material interface, Fig. 15.2 compares the experimental observations against the numerical results from surface irregularities approximated by a power spectrum, $P(k)$, proportional to wavenumber k or to an inverse-square law k^{-2}.

While it is somewhat difficult to visually distinguish the impact of the initial conditions on the flow development from the photographs, the evolution of the total turbulent kinetic energy sheds considerable light on the situation. Specifically, the power spectrum $P(k) \sim k$ generates more total mixing in the early stages (Fig. 15.3(b)), but cannot maintain the turbulent mixing rate at late times due to the lack of long-living large energetic scales (Fig. 15.3(a)). This is an example in which experimental measurements can be supplemented by simulations in order to achieve a fuller picture of the physical processes.

Visualizing vortex-core isosurfaces by volume fraction (white for pure air, black for pure SF_6) offers insight into the results of numerical simulations. Figure 15.4(a) shows the results of a recent simulation after 2.22 ms, in which the majority of the vortex structures are primarily composed of the lighter material. The wall-normal velocity and shear forces that develop at the inclined interface slightly increase the vortex-structure composition of

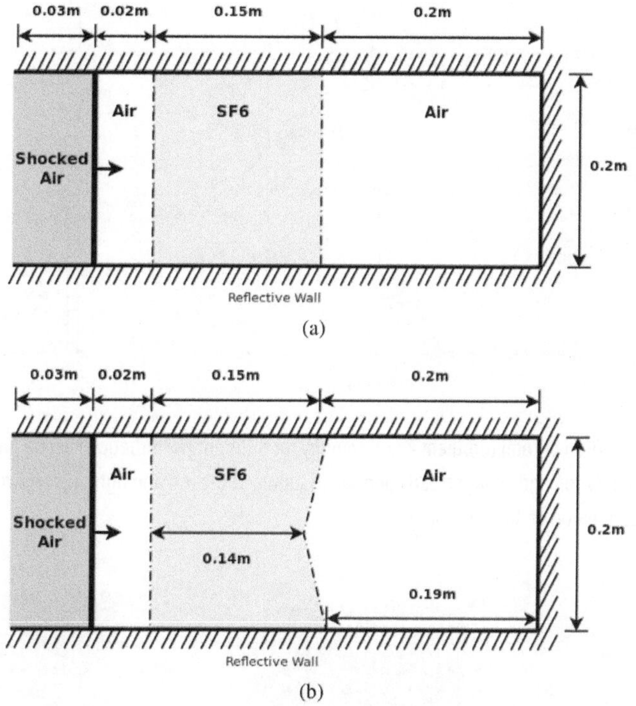

Fig. 15.1 Schematic illustration of initial conditions used for (a) double-planar and (b) inverse-chevron experiments and simulations. Fig. 1 of Kokkinakis et al. (2020).

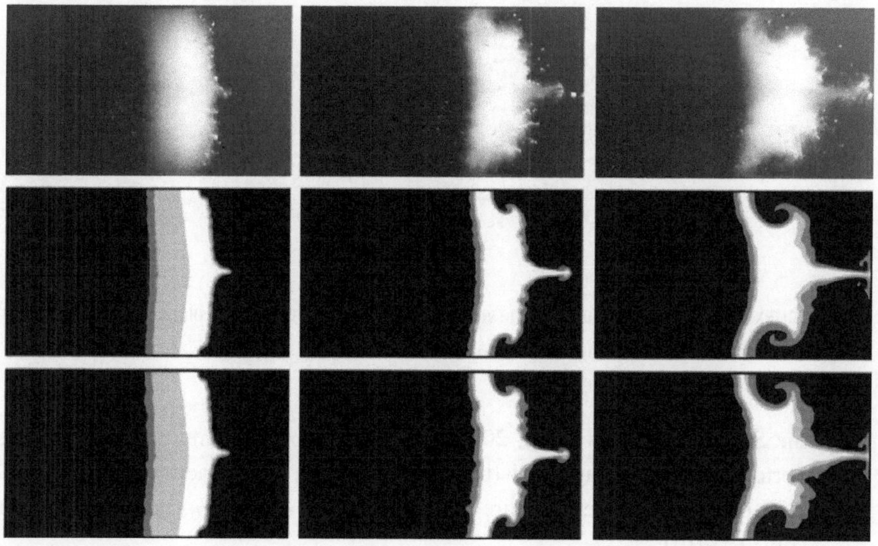

Fig. 15.2 Comparison of the flow development between experiments and simulations. Left to right: snapshots at times 1.9, 2.7, and 3.3 ms; top to bottom: experiment, k-spectrum, and k^{-2}-spectrum. Fig. 5 of Hahn et al. (2011), with the permission of AIP Publishing.

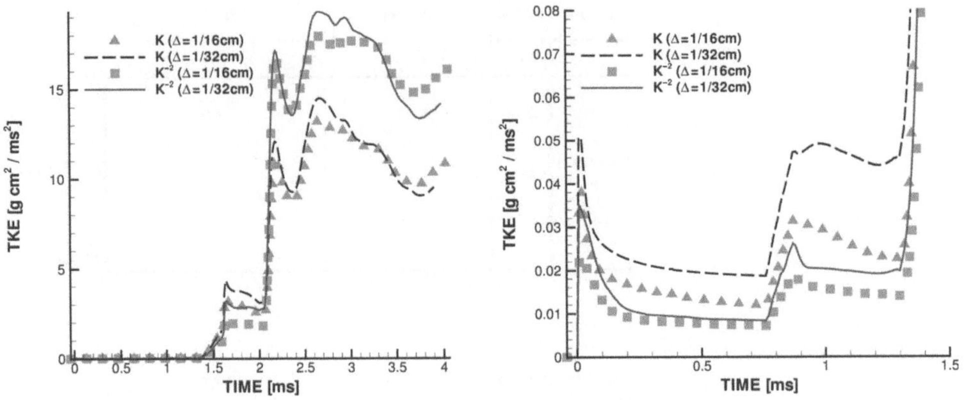

Fig. 15.3 Evolution of total turbulent kinetic energy for different perturbation spectra and grid resolutions. The entire simulation and enlarged early period are shown at the left and right, respectively. Fig. 9 of Hahn et al. (2011), with the permission of AIP Publishing.

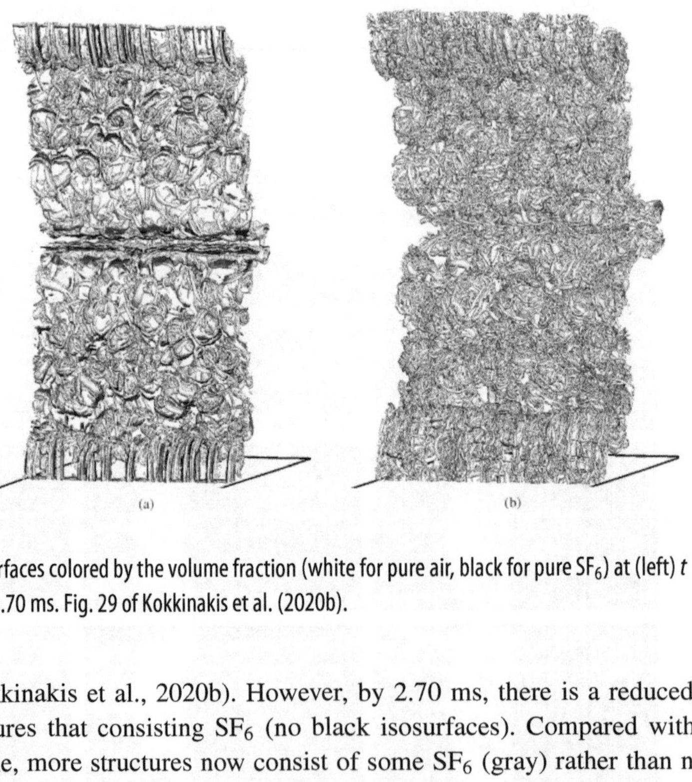

Fig. 15.4 Vortex-core isosurfaces colored by the volume fraction (white for pure air, black for pure SF_6) at (left) $t = 2.22$ ms and (right) $t = 2.70$ ms. Fig. 29 of Kokkinakis et al. (2020b).

the SF_6 (Kokkinakis et al., 2020b). However, by 2.70 ms, there is a reduced fraction of vortex structures that consisting SF_6 (no black isosurfaces). Compared with the earlier 2.22 ms frame, more structures now consist of some SF_6 (gray) rather than none (white) (Fig. 15.4(b)).

More complex initial profiles could be constructed using the soap film technique to experimentally investigate the early stages of the bubble merger process (Guo et al., 2019). Four initial interface arrangements – the nominal case and three cases with various major and minor inverse-chevron shapes – are shown in the first row of Fig. 15.5. For these latter

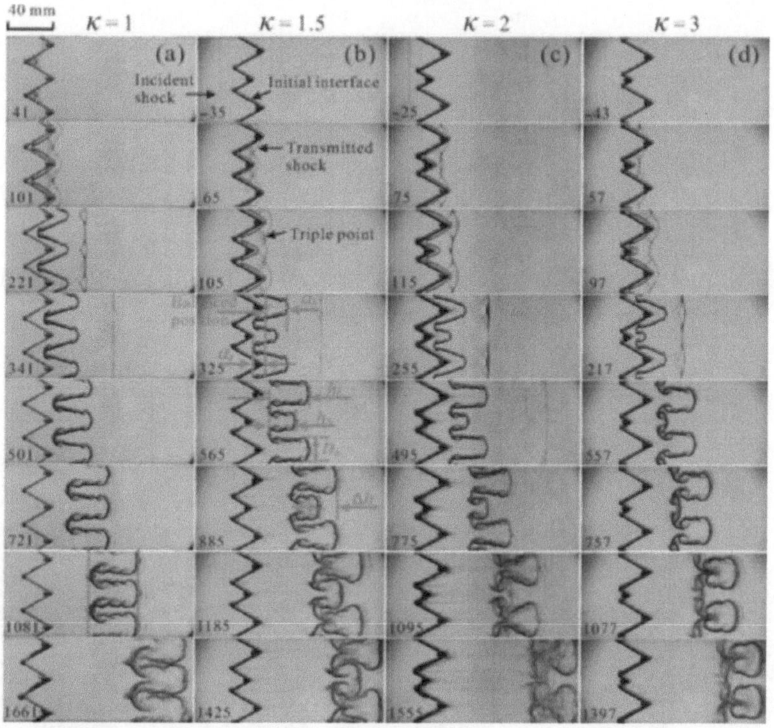

Fig. 15.5 Schlieren images of the air-SF_6 interfaces impacted by a planar shock. Numbers in the lower left corners represent times (in μs). The columns correspond to the four initial interface arrangements of the major and minor inverse-chevron shapes, where the wavelength ratios (as well as the initial amplitude ratios) are equal to 1, 1.5, 2, and 3, respectively. Fig. 3 of Guo et al. (2019).

cases, the values of the wavelength ratio (as well as the initial amplitude ratio) are equal to 1.5, 2, and 3, respectively. The control case with equal chevron sizes is shown in Fig. 15.5(a). For the smallest ratio, shown in Fig. 15.5(b), the major bubble front expands and occupies the downstream space of the neighboring minor bubble fronts at a later time. While the minor bubbles survived, the growth of their fronts was seriously inhibited. Such behavior should be compared to that of the two large ratio cases. As time elapses, Fig. 15.5(c,d) reveals that the minor bubbles were "swallowed" by the neighboring major ones and essentially disappeared near the end of the data collection.

Many distinct initial interfacial profiles may be setup for experimental or numerical RMI studies. A popular profile is the enlarged double-bump (Holder et al., 2003b), "V" (Mikaelian, 2005b; Luo et al., 2016b; Zhai et al., 2016; Li et al., 2016; Wu-Wang et al., 2024), "W" (Zhou et al., 2018), and sawtooth (Sadot et al., 1998; Rikanati et al., 2003; Zhai et al., 2020).

To demonstrate how quickly things can become complicated, Fig. 15.6 displays schematics of four distinctive initial interfacial profiles of a SF_6 gas layer, which is surrounded by air. A baseline case (far left plot) is also included, in which no perturbation is imposed on either side of the layer (Liang et al., 2020). For each perturbed interface,

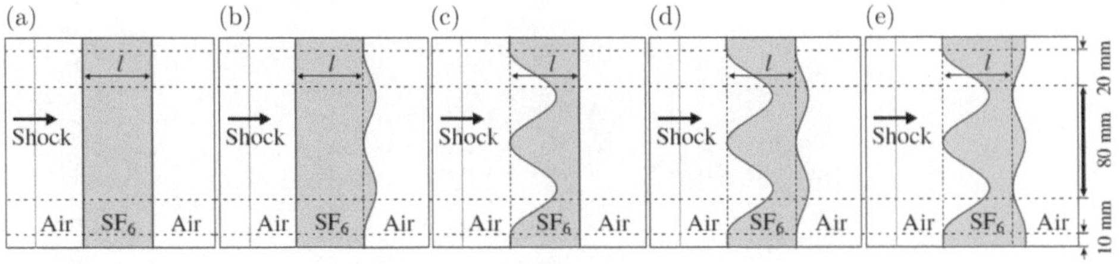

Fig. 15.6 Schematics of the initial gas layer configurations: (a) 1-D, (b) DS, (c) US, (d) IP and (e) AP. The incident shock propagates from left to right. Fig. 1 of Liang et al. (2020).

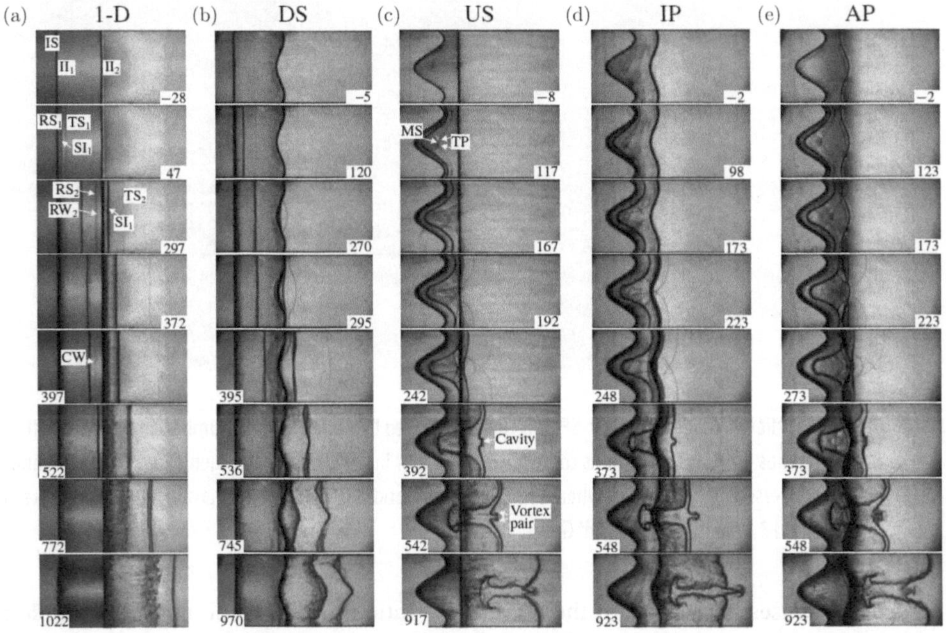

Fig. 15.7 Schlieren pictures of the gas layer evolution in different cases. IS, incident shock; II$_1$ and II$_2$, the initial first and second interfaces, respectively; TS$_1$ and TS$_2$, the transmitted shocks through the first and second interfaces, respectively; SI$_1$ and SI$_2$, the shocked first and second interfaces, respectively; RS$_1$, the shock reflected from the first interface; RS$_2$, the shock reflected from the filaments; RW$_2$ and RW$_1$, the rarefaction waves reflected from the second and first interface, respectively; CW, compression wave reflected from the first interface; MS, Mach stem; TP, triple point. Numbers denote the time in μs. Fig. 3 of Liang et al. (2020).

the initial single-mode perturbation amplitude on the first and second interface measured 15 mm and 4 mm, respectively, and the wavelength and wavenumber were fixed at 60 mm and 104.72 m^{-1}.

Figure 15.7 exhibits the alteration to RMI growth due to coupling between the first and second interfaces (borrowing the language of ICF, "feed-through"). Figure 15.8 shows the evolution of the dimensionless interface amplitudes as a function of dimensionless time. RMI growth can be observed even for the initially unperturbed interface, albeit at

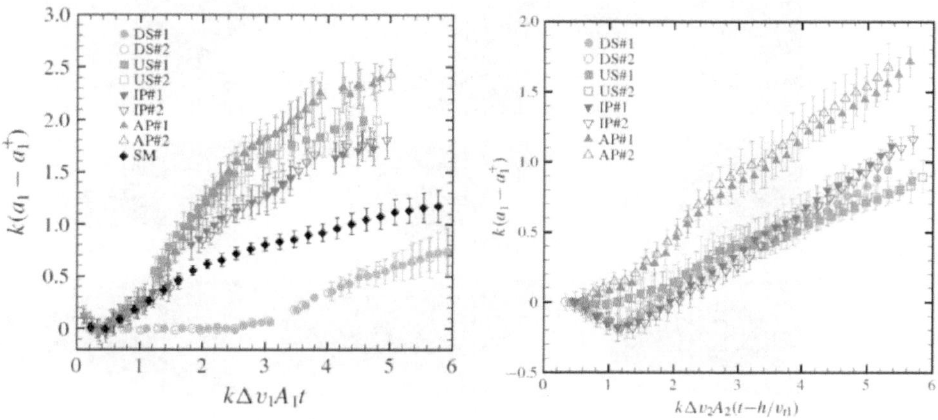

Fig. 15.8 (Left) the dimensionless amplitude of the first interface. (Right) comparison of the dimensionless amplitude of the second interface. Figs. 6a and 8 of Liang et al. (2020).

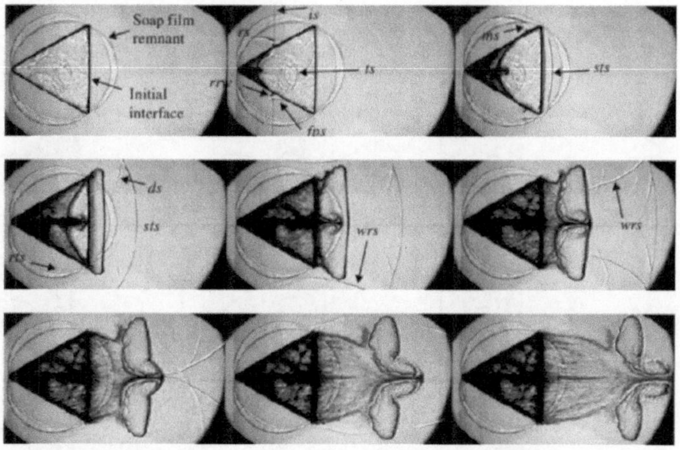

Fig. 15.9 Experimental Schlieren images of a triangular N_2 volume surrounded by SF_6 impacted by a planar shock wave (Mach 1.29). is, incident shock; ts, transmitted shock; fps, free precursor shock; rs, reflected shock; tp, triple point; ms, Mach stem; sts, secondary transmitted shock; ds, diffracted shock; rts, reflected transmitted shock; wrs, reflected shock from the wall; rrw, reflected rarefaction waves. Fig. 4 of Zhai et al. (2014).

a later time. More recently, the evolution of SF_6 fluid layers is investigated experimentally and theoretically to reveal the mechanism of instabilities and motions of waves under reshock conditions (Cong et al., 2022). Other related articles can be found in Liang and Luo (2021a,b, 2022, 2023); Liang (2022); Matsuoka (2020).

A separate study has examined the effect of different initial cylinder shapes (specifically circular and elliptic cylinders) on the RMI mixing layer development (e.g., Bai et al., 2010a). More recently, Zhai et al. (2014) studied RMI mixing for a polygonal N_2 volume surrounded by SF_6. For a Mach 1.29 ± 0.01 incident shock, they considered triangle, diamond, and square interface configurations (Figs. 15.9–15.11). The evidence suggests that

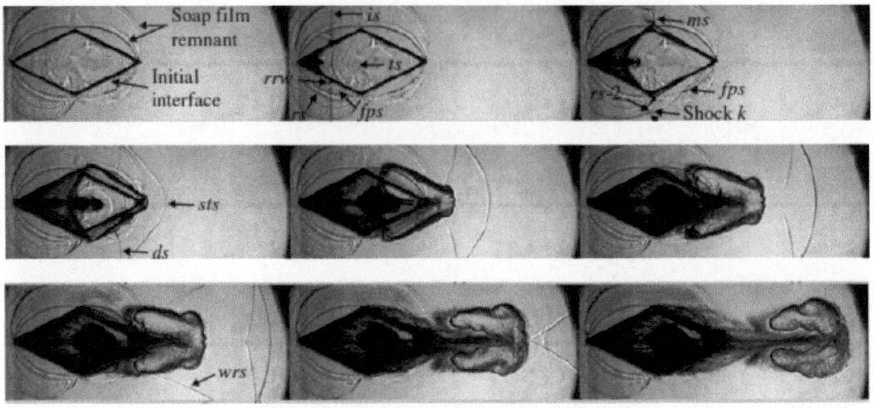

Fig. 15.10 Experimental (upper) and synthetic numerical (lower) Schlieren images of a diamond N_2 volume surrounded by SF_6 impacted by a planar shock wave (Mach 1.30). rs-1, reflected shock 1; rs-2, reflected shock 2, the other symbols are the same as those indicated in Fig. 15.9. Fig. 5 of Zhai et al. (2014).

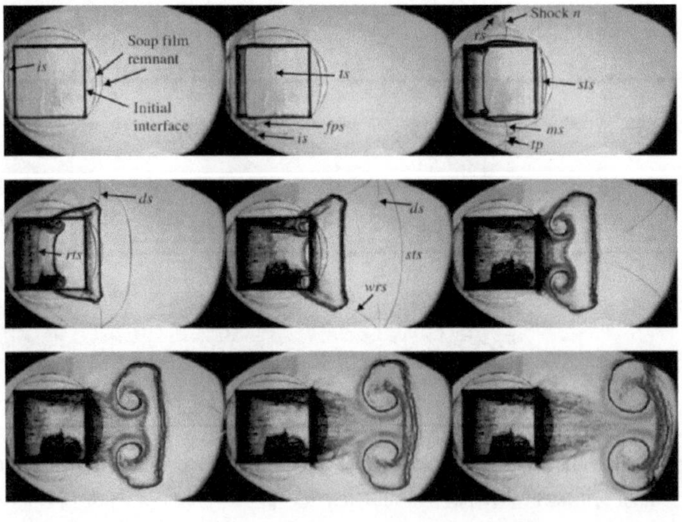

Fig. 15.11 Experimental (upper) and synthetic numerical (lower) Schlieren images of a square N_2 volume surrounded by SF_6 impacted by a planar shock wave (Mach 1.28). The other symbols are the same as those indicated in Fig. 15.9. Fig. 3 of Zhai et al. (2014).

the evolution of the polygonal interfaces depends on their initial shapes. In the triangular and diamond cases, the small "SF_6 jet" is the most striking feature, which grows constantly with time. As time proceeds, the distance between the jet head and the downstream boundary becomes shorter. Eventually, the jet head penetrates the downstream boundary, forming two independent vortices. In the square interface, however, two vortices are generated shortly after the shock impacts the left corner, and they grow with time to dominate the flow field.

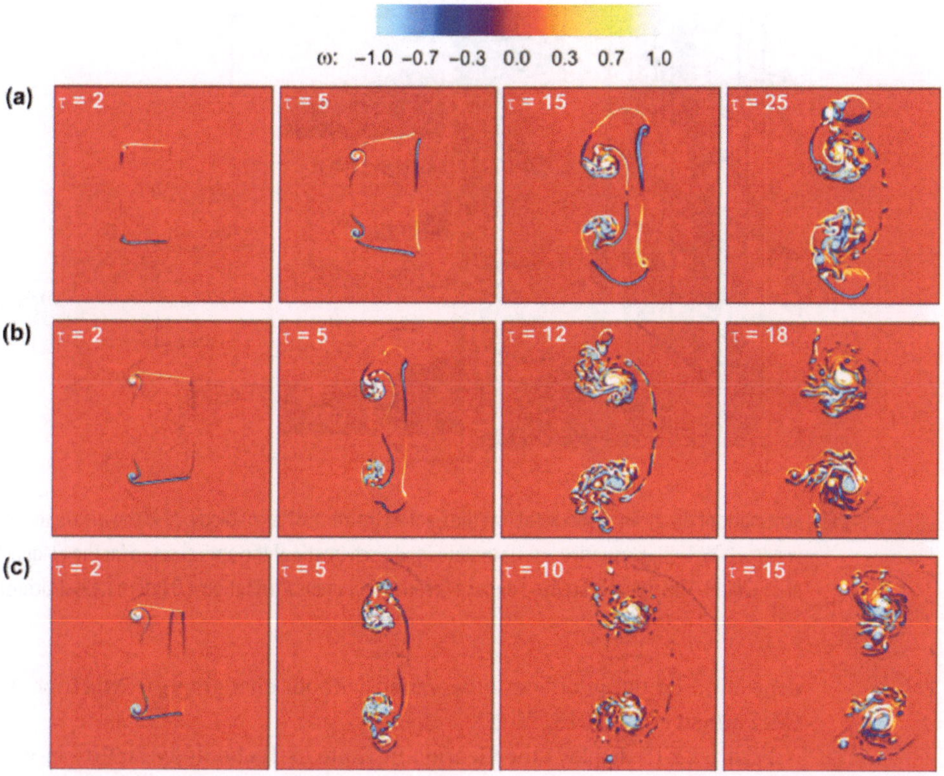

Fig. 15.12 Effect of Mach number on a shock-accelerated square helium bubble: Contours of vorticity profile for Mach (a) 1.21, (b) 1.7, and (c) 2.1 at different times. Fig. 9 of Singh (2021).

Figure 15.12 exhibits the effects of different incident shock strengths on the vorticity distribution, focusing on the interaction of a planar shock wave with a heavy square polygon. The vorticity magnitude is maximal at the top and bottom of the bubble, where the density and pressure gradients are orthogonal. As a result, a significant quantity of positive (negative) vorticity is generated on the upper (lower) horizontal side of the bubble interface. The vorticity, which is generated around the rolled-up vortices on the bubble interface, becomes more pronounced for high Mach numbers (Singh, 2021).

15.2 Inclined shock tube

This method of perturbing the fluid interface is created by inclining the shock tube with respect to the gravitational field. The interface remains stratified by gravity, while the incident shock is at an angle to the interface. This simple method provides a highly repeatable perturbation capable of directly controlling the cross-product of the pressure and density gradients in the vorticity equation (see Eqs. 2.30 and 2.31), thereby allowing the energy

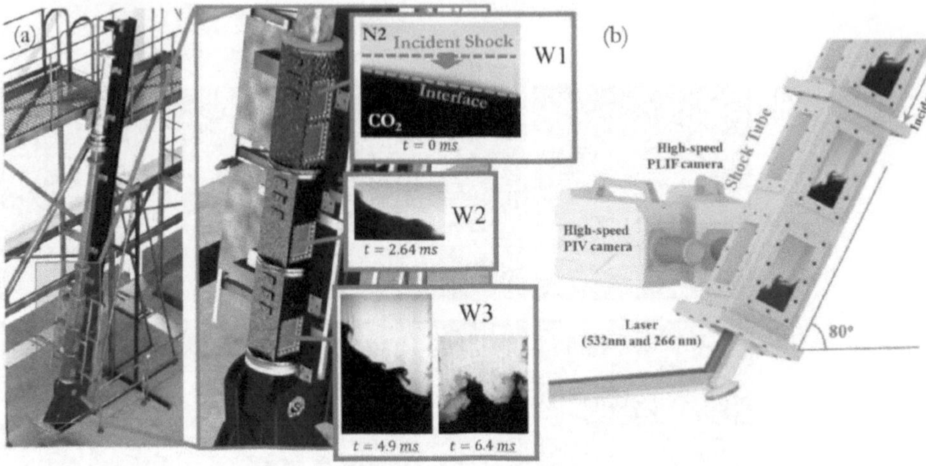

Fig. 15.13 Schematic of the inclined shock-tube at the Georgia Institute of Technology. Also shown are the measurement windows (W 1–3), illustrative interface states and a schematic of the optical setup for the diagnostics. Fig. 1 of Carter et al. (2019). This figure is licensed under a Creative Commons Attribution 4.0 International License.

deposition on the interface to be varied without changing the Mach or Atwood numbers (McFarland et al., 2013, 2015).

Figure 15.13 displays the Georgia Institute of Technology facility in which the shock is inclined at 10° to the interface (80° with respect to the horizontal). The tube can be pivoted about its base from horizontal to fully vertical (Carter et al., 2019; McFarland et al., 2014).

Exact solutions for the baroclinic vorticity deposition of an oblique shock onto an interface for polytropic gases have been published previously (Samtaney and Zabusky, 1994; Samtaney et al., 1998). The Birkhoff–Rott equation (Birkhoff, 1962; Rott, 1956) is appropriate for describing the evolution of both the KH (Krasny, 1986) and RM/RT (Sohn, 2004b) and takes into account the calculated vorticity distribution, allowing the velocity induced by the vorticity to itself transport the vorticity distribution (Rasmus et al., 2019).

As always, it is interesting to compare and contrast single and multi-mode initial conditions (Mohaghar et al., 2022; Pathikonda et al., 2022). Two sets of concentration field measurements are illustrated in Fig. 15.14 for single-mode and Fig. 15.15 for multi-mode cases, respectively. Here, the evolution of the N_2/CO_2 interface is driven by an incident shock strength of Mach 1.9. The reshock occurs 4 ms after the incident shock reflects off the bottom wall of the tube, traveling upwards and appearing as a sharp gradient in the PLIF image (Mohaghar et al., 2019).

For the single-mode case, the complex interface topology is evident, with two well-developed roll-up features clearly visible before reshock (Fig. 15.14). These form the initial condition when the reshock arrives. The interface is compressed by reshock and some small-amplitude, short-wavelength features can be clearly observed immediately after its traversal. Between images at 3.9 and 4.8 ms, the interface inverts, and the uppermost extent of the heavy fluid before reshock becomes the lowest extent of the lighter fluid. Additionally, the two prominent roll-up features, where the heavy fluid penetrates the light fluid

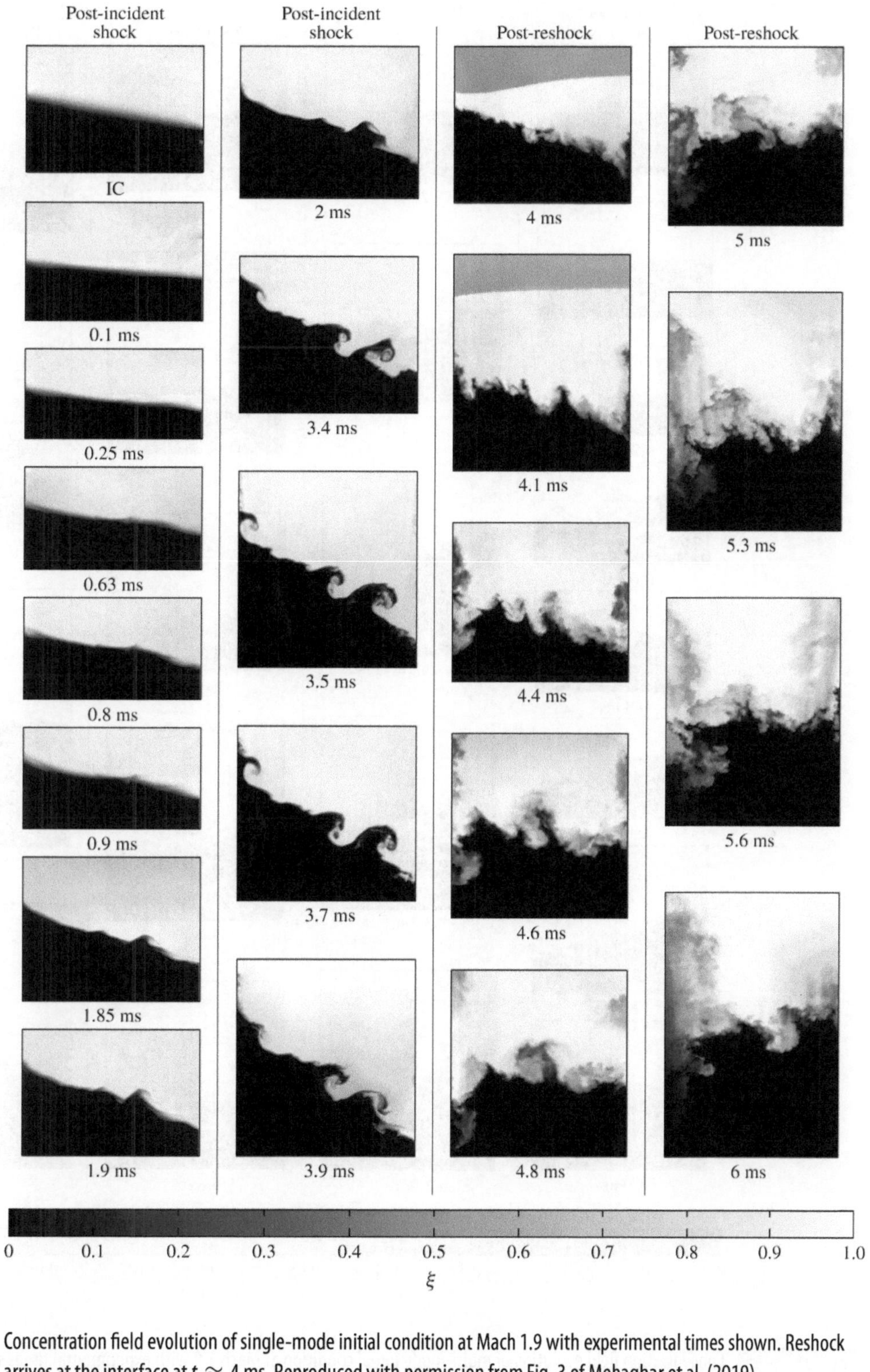

Fig. 15.14 Concentration field evolution of single-mode initial condition at Mach 1.9 with experimental times shown. Reshock arrives at the interface at $t \approx 4$ ms. Reproduced with permission from Fig. 3 of Mohaghar et al. (2019).

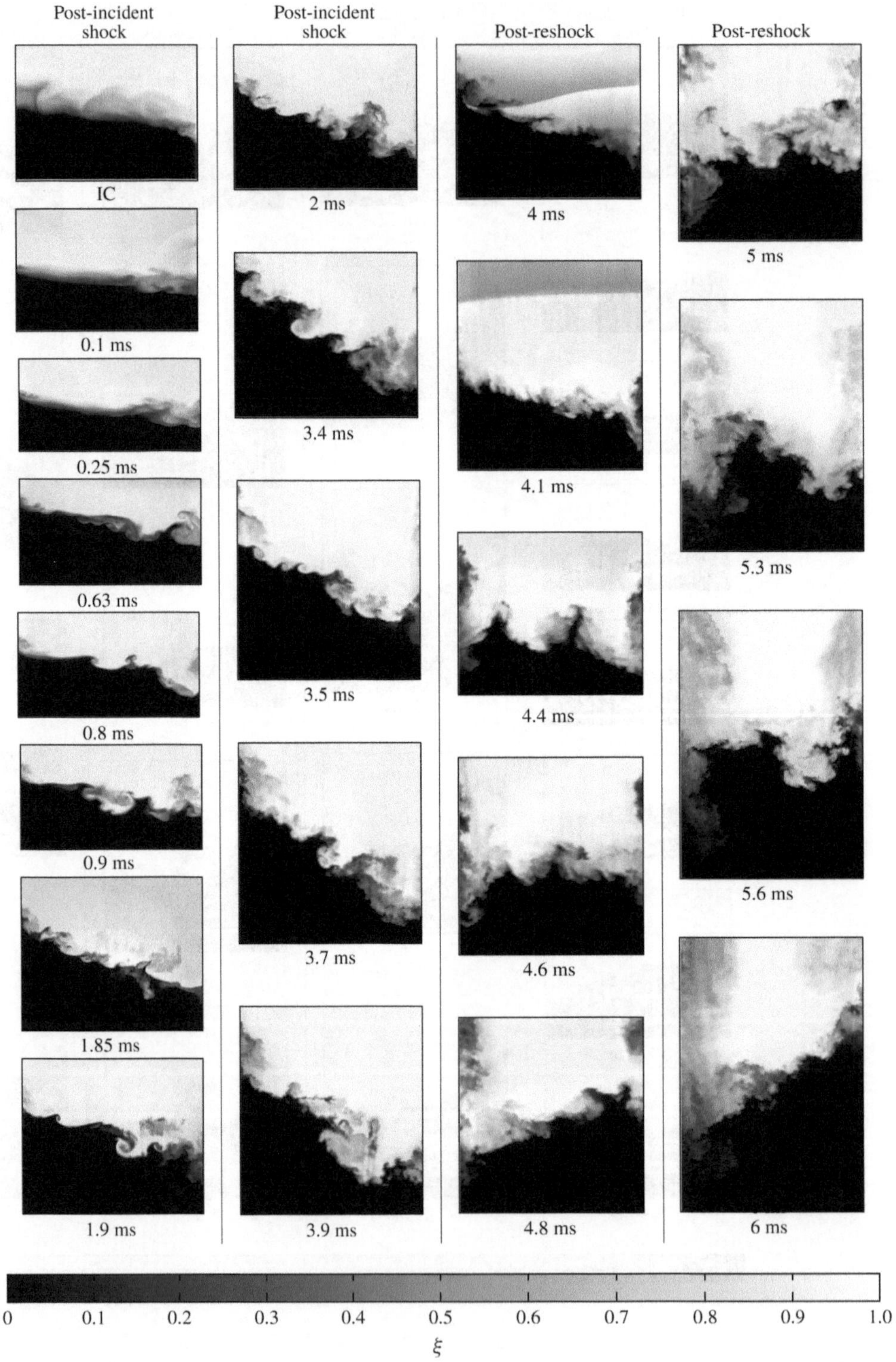

Fig. 15.15 Concentration field evolution of multi-mode initial condition at Mach 1.9 with experimental times shown. Reshock arrives at the interface at $t \approx 4$ ms. Reproduced with permission from Fig. 4 of Mohaghar et al. (2019).

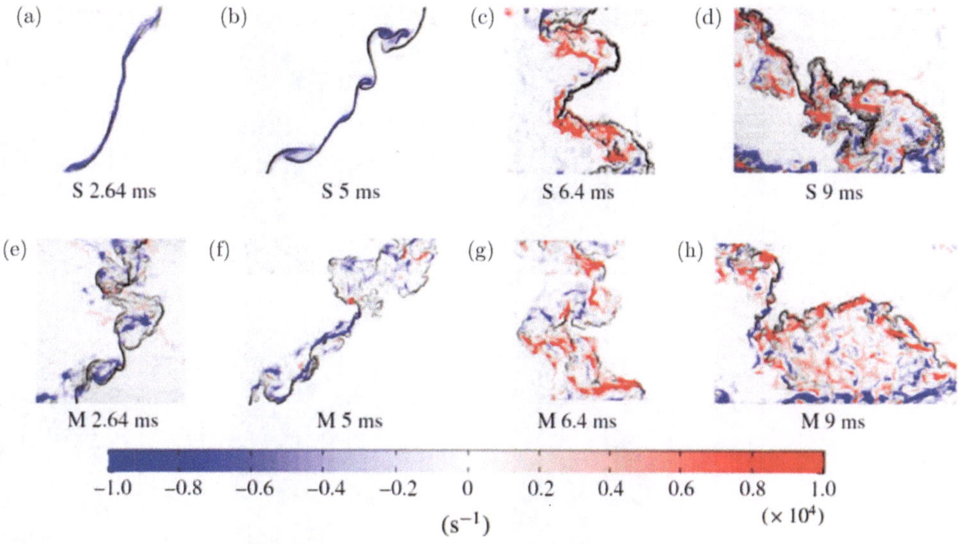

Fig. 15.16 Vorticity evolution overlaid on simulated Schlieren images for reshocked RMI from Wisconsin shock tube experiments. S refers to single-mode case and M refers to multi-mode case. Time of reshock: ∼ 6 ms for both cases. Reproduced with permission from Fig. 10 of Mohaghar et al. (2017).

before reshock, are observed to be inverted into two regions of light fluid penetrating heavy fluid. After 4.8 ms these features become indistinguishable, and the interface looks increasingly chaotic.

A similar sequence showing the RMI evolution from a multi-mode initial condition is presented in Fig. 15.15. Before reshock, the multi-mode case shows additional small-scale features that were not present in the single-mode case. Also, the interface seems to have fewer large-scale features immediately after reshock. These observations were anticipated by Mohaghar et al. (2019), owing to the greater modal content in the initial condition and in the multi-mode features at late times before reshock. These results are similar to those previously observed for low Mach number (1.5) (Mohaghar et al., 2017).

The vorticity evolution is shown in Fig. 15.16, where red and blue represent positive and negative vorticity, respectively. For the single-mode case, the incident shock wave passing from light to heavy gas creates strong negative vorticity (Fig. 15.16(a)). After the shock, the vorticity deposition causes the initial perturbations to grow and form larger vortical structures (Fig. 15.16(b)). As the reshock wave passes through the interface, it deposits such a large amount of vorticity that the majority of the field becomes positive. The interface itself begins to rotate in the opposite direction immediately after reshock (Fig. 15.16(c)). The presence of positive and negative vorticities in a larger area of the reshocked flow, shown in Fig. 15.16(d), is a result of vorticity deposition on an interface with a much less unidirectional gradient.

For the multi-mode case, shown as a progression in Fig. 15.16(e–h), the general physics is similar to the single-mode case described above. A key difference highlighted by Mohaghar et al. (2017), however, is that the multi-mode vorticity deposition results in

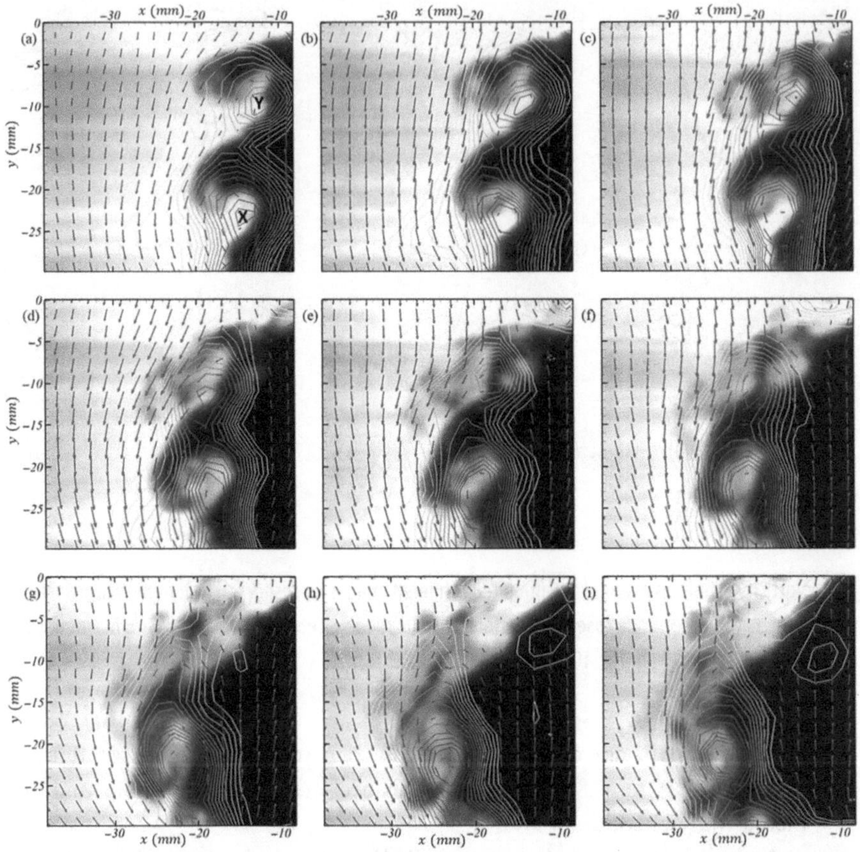

Fig. 15.17 An illustrative temporal evolution of merging vortices [X and Y in (a)], and the associated scalar evolution shown in snapshots 67 μs apart, starting at 5.550 ms after incident shock in (a), 5.617 ms in (b), 5.684 ms in (c), 5.751 ms in (d), 5.818 ms in (e), 5.885 ms in (f), 5.952 ms in (g), 6.019 ms in (h) and 5.086 ms in (i). The vectors are velocities in relevant frames of reference. Fig. 9 of Carter et al. (2019). This figure is licensed under a Creative Commons Attribution 4.0 International License.

higher-amplitude positive vorticity post-reshock than in the single-mode case. This is due to the presence of positive vorticity pre-reshock. Although the amount of vorticity deposition is equivalent, its distribution is quite different. In the single-mode case, the vorticity deposition occurs along the interface, as it closely follows its shape. In the multi-mode case, the distribution of deposited vorticity is broader and more diffuse.

The utility of simultaneous measurements of velocity and vorticity fields lies in the ability to track the temporal evolution of small-scale phenomena, which represent the primary mixing mechanism in the early stages of vortex-dominated flows. Figure 15.17 shows an illustrative temporal sequence highlighting these small-scale dynamics. The vorticity contours overlaid on PLIF images identify two vortices (marked as X and Y in Fig. 15.17(a)). The velocities are marked in the frame of reference of the vortex X, showing the swirling motion around the vortex. Both vortices are formed from the reshock-initiated vortex sheet, according to Carter et al. (2019), and subsequent roll-up is due to its unstable nature. Under their mutual influence, the symmetric vortices approach each other and eventually merge. This "mode-merging" of vortices results not only in backward scatter of turbulent energy to smaller wavenumbers (larger scales), but also in a local spike in scalar mixing, as can be seen in Fig. 15.17(f–i).

15.3 Tilted tank

Tilted tank experiments consist of a tank filled with light fluid above heavy, which is then tilted to one side of the apparatus. The tilt gives rise to an "angled interface" in relation to an acceleration history, similar to the inverse chevron RM experiment. The tilting of the initial interface relative to the sides of the tank leads to a large-scale overturning motion, in addition to the growth of a turbulent mixing layer. The flow is, on average, 2D.

The "Tilted-Rig" is part of the series of important AWE (The Atomic Weapons Establishment) experiments carried out using the Rocket-rig facility previously shown in Fig. 9.7. In the experiment, the acceleration from the rocket motors varied in time, with a magnitude averaging approximately 35 times Earth's gravity. The rig requires approximately 20 ms to rise to a near-constant plateau. Figure 15.18 shows the schematic of the system and a sequence of experimental images at times 45.3, 59.8, and 71.1 ms, from case 110 of the experiment (Youngs, 2017). The experimental rig was inclined at an angle 5°56′ to the vertical. The associated Atwood number is 0.48, given that the heavy and light fluids utilized are a NaI solution and hexane, respectively.

The Tilted-Rig is a unique 2D RT mixing experiment for which experimental data, DNS data (Andrews et al., 2014), and ILES and DNS results all exist (Youngs, 2017). The distributions of volume fraction, turbulent kinetic energy (K), and kinetic energy dissipation ($\mathcal{E}$),

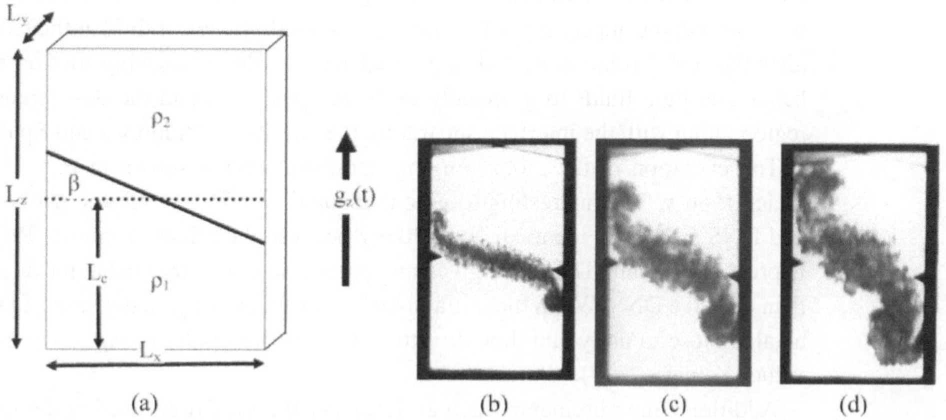

Fig. 15.18 Schematic of the "Tilted-Rig" and the sequence of experimental images at times 45.3, 59.8, 71.1 ms, from case 110 of the experiment Fig. 11 of Youngs (2017). UK Ministry of Defence © Crown Owned Copyright 2022/AWE.

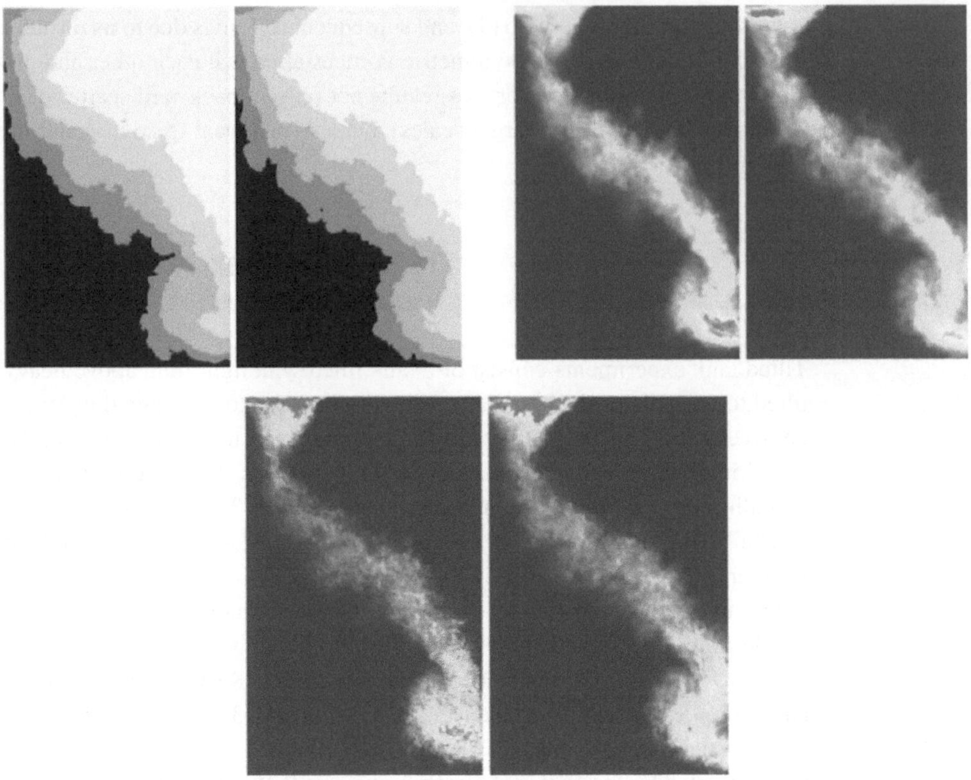

Fig. 15.19 "Tilted-Rig" calculations from ILES and DNS: (top left) volume fraction, (top right) kinetic energy, and (bottom) dissipation rate. Figs. 12(a,b,c,d) and 13(a,b) of Youngs (2017). UK Ministry of Defence © Crown Owned Copyright 2022/AWE.

obtained from the DNS, are shown in Fig. 15.19 (Youngs, 2017). Three phenomena may be observed: the appearance of oppositely directed plumes of fluid at the side walls shortly after the initial rotation period, a gradual large-scale overturning motion that causes the heavy and light fluids to ultimately exchange positions, and the development of a mixed region along with the interface separating the two fluids (Andrews and Spalding, 1990).

The evolution of the global mixing parameter (Θ) is shown in Fig. 15.20 versus the scaled time τ. It is interesting to note that for $\tau > 1.75$, the values of $\Theta \sim 0.7$ for ILES and DNS are nearly identical, most likely because the effect of infinite Reynolds number is presumed small. The ILES gives an approximation to this behavior at an earlier stage than does the DNS. When the initial interface diffuses away, in the case of DNS, Θ reaches a value close to unity and then dips to ~ 0.5, before finally rising to the asymptotic-time value (Youngs, 2017).

Additional measurements, such as density-self correlations used in the BHR mix model (described in Chapter 22), have also been computed by Andrews et al. (2014). The "Tilted-Rig" problem has been formally developed into a standard test problem with its main purpose to provide a benchmark problem for validating engineering models. The combined

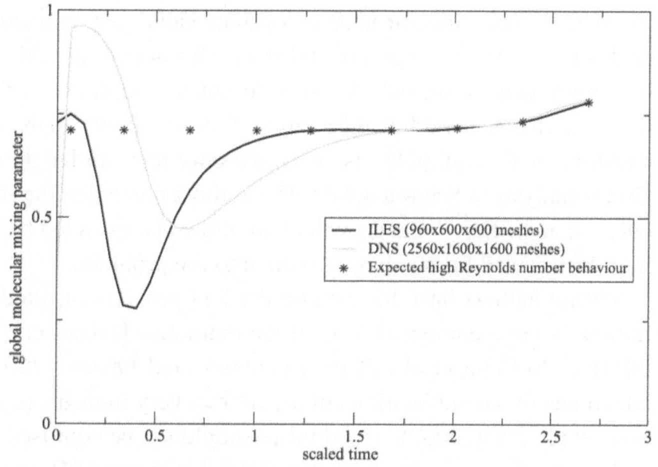

Fig. 15.20 Global mixing parameter, Θ, versus scaled time, τ. 2017 Crown copyright. Reproduced with the permission of Queen's Printer for Scotland and University of Strathclyde. Fig. 14 of Youngs (2017), UK Ministry of Defence © Crown Owned Copyright 2022/AWE.

experimental and numerical datasets can be very helpful for validating 2D mix (RANS) models (Andrews et al., 2014; Denissen et al., 2014; Kokkinakis et al., 2019; Xiao et al., 2021; Xie et al., 2021a; Ferguson et al., 2023). A detailed discussion of this point is deferred to Chapter 22.

Three subsequent tilted RTI experiments use much lower acceleration than those discussed previously. First, the experiment of Andrews and Spalding (1990) places a heavy fluid above a light one by inverting a narrow tank mounted on a horizontal axle. The tank sits within a rigid frame that may be tilted by placing wedges under one side to superimpose a 2D motion on the mixing process. Second, the experiments of Holford et al. (2003) were carried out in the tank previously shown in Fig. 9.24 (Dalziel et al., 1999), in which a barrier separating saline and freshwater was removed to initiate the instability. In that case, the whole tank was tilted at an angle (up to 10°) to the horizontal. Finally, Jiang et al. (2018a) carried out tilted RT experiments driven by non-constant accelerations – a further development of a previous study reported in Liu et al. (2012a). The interfacial obliquity increases with time parabolically at first, and as a result of a competition between the mixing induced by RTI and KHI, the fluid interface turns over in the later stage of evolution.

15.4 Confinement

Kadau et al. (2004, 2007) evaluated the RTI mixing process among atomistic simulations, continuum simulations, and experiments. Whereas the continuum calculations exhibit self-similar growth, atomistic simulations revealed a regime change from t^2 behavior to linear-t behavior in the penetration depth. After an interval in which the velocity increases linearly in time, it then varies around a constant value. As stressed by Kadau et al. (2007), an analysis of previous experimental data (Dimonte et al., 2004) showed that even the best of experiments is not yet able to rule out the possibility of such a regime change. This behavior change is excluded for any self-similar flow – in the solution of the Navier–Stokes equations with negligible viscosity, the only relevant length scale is gt^2 (see also, the Lie Group analysis in Section 4.4.3). The authors noted that the self-similarity could be broken only when the horizontal growth of structures is restricted by the system width, which then introduces an additional length scale into the problem.

Several authors have focused on the RTI problem of tilted tubes in which the region of mixing is large compared to the tube diameter (Dalziel et al., 2008; Lawrie and Dalziel, 2011b,c; Boffetta et al., 2019; Inogamov and Oparin, 2003), the geometrical shape of which has found applications in the oil recovery industry (e.g., Debacq et al., 2001, 2003; Séon et al., 2004, 2007). Tilted tubes should not be confused with the tilted tank discussed in the last subsection. In particular, Dalziel et al. used PIV to show that the flow eddies arising from the RTI remain approximately isotropic, despite only being confined in the lateral direction. Instead of the quadratic RT growth of the layer width, a $h(t) \sim t^{2/5}$ scaling emerges, displaying a sub-diffusive growth.

From its exposition in Dalziel et al. (2008), Lawrie and Dalziel (2011b) obtained a similar solution to the 1D diffusion equation, extended it to finite domains via numerical integration, and compared it with results from 3D ILES simulations. However, use of initial conditions with one stably-stratified layer introduces an additional length scale to the problem, breaking the global self-similarity. The development of this modified instability can be explored utilizing the methodology of Lawrie and Dalziel (2011c).

Furthering this line of investigation, Horne and Lawrie (2020) employed a thin tank with dimensions (1,000 $\times$ 250 mm) $\times$ $\tilde{z}$ mm, where $\tilde{z}$ is the thickness dimension, which was varied between experiments, from $\tilde{z} \approx 1$ to 12 mm, by adding a sequence of acrylic spacer plates. For $\tilde{z} = 6$ and 12 mm, Fig. 15.21 shows an increase, followed by a reduction, in large structures toward the end of the observation period. This development can be characterized in three phases (Horne and Lawrie, 2020). First, the classic unconstrained 3D RT growth, in which the instability height is small compared to any of the lateral dimensions. When the growth is of a comparable length scale to the smallest lateral dimension, the original form of self-similar behavior is lost. The subsequent growth depends on a fixed length scale as well as time, so length- and time-scales can no longer increase in proportion. The final phase is one in which vortex-stretching becomes sufficiently strong that diffusive processes rapidly reduce the baroclinic engine that drives further kinematic activity.

The volume fraction's temporal evolution for $\tilde{z} \approx 1$ mm is markedly different, in that the development of the large structures is unabated (Fig. 15.22(a)). The corresponding mixing layer possesses a self-similar quadratic scaling with $\alpha \approx 0.012$. One should also note the non-self-similar adjustments that strongly perturb the growth profiles away from the quadratic ideal for the 6 and 12 mm cases (not shown). Moreover, an abrupt change from k^{-4} (6 and 12 mm) to k^{-3} (1 mm) spectral slopes for energy is just a foretaste of what will happen if a much higher aspect ratio is enacted. Figure 15.22(b) suggests that the RT mixing layer spectrum may be consistent with expectations from Kraichnan for 2D flows (Kraichnan, 1967).

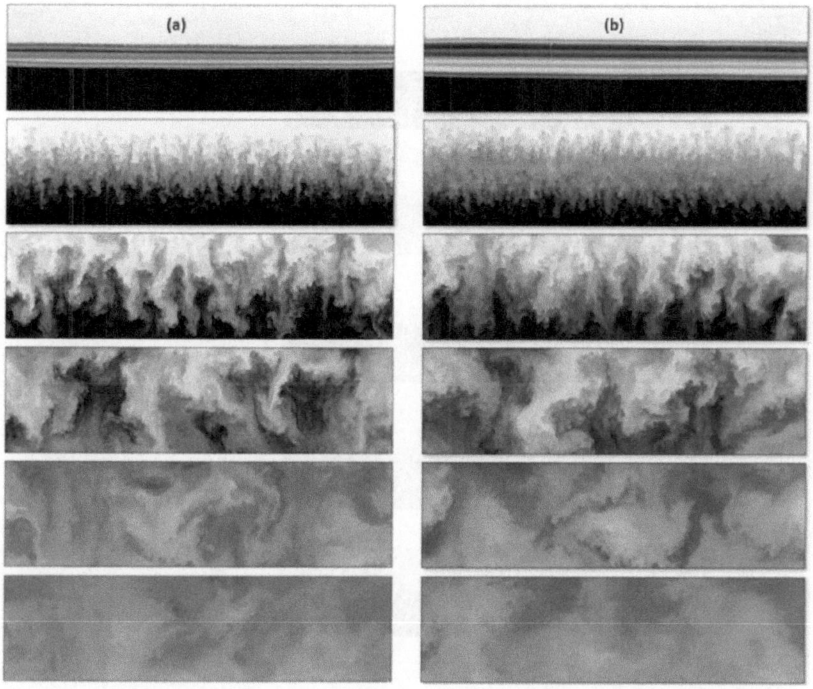

Fig. 15.21 Volume fraction spectrum development with time for (a) 6 mm thick experiment and (b) 12 mm thick experiment. The horizontal bands at zero time arise due to line-of-sight averaging of shear on the interface during overturning. Fig. 14 of Horne and Lawrie (2020).

Additionally, Lawrie and Dalziel (2011a) and Wykes et al. (2015) considered vertical confinement by buoyancy. They focused their attention on the distribution of irreversible energy conversions (mixing efficiency) between quiescent initial and final states in a miscible RT-driven system. The authors contended that the physical mechanisms intrinsic to RTI provide optimal conditions for mixing.

Within the Boussinesq approximation, Boffetta et al. (2012) performed high-resolution numerical simulations of miscible RTI induced flows at low Atwood numbers and high aspect ratio, in which one transverse side is 32 times smaller than the other. Figure 15.23(a) plots the ratio between the energy growth rate (dE/dt) and the viscous dissipation rate ($\mathcal{E}$) to ascertain if a transition from 3D to 2D turbulent behavior could be detected. Since both quantities grow linearly in time when the flow is in the 3D regime, their ratio becomes constant after an initial transient. Once the transition occurs, the inverse cascade sets in and the mixing layer asymptotes to quadratic growth, with $\alpha \approx 0.035$ (Figure 15.23(b)). What distinguishes this work from the previously discussed experiments and simulations of Horne and Lawrie (2020) is that the energy spectrum follows the $k^{-11/5}$ scaling of Bolgiano Jr (1959), instead of the typical 2D k^{-3} spectrum from Kraichnan.

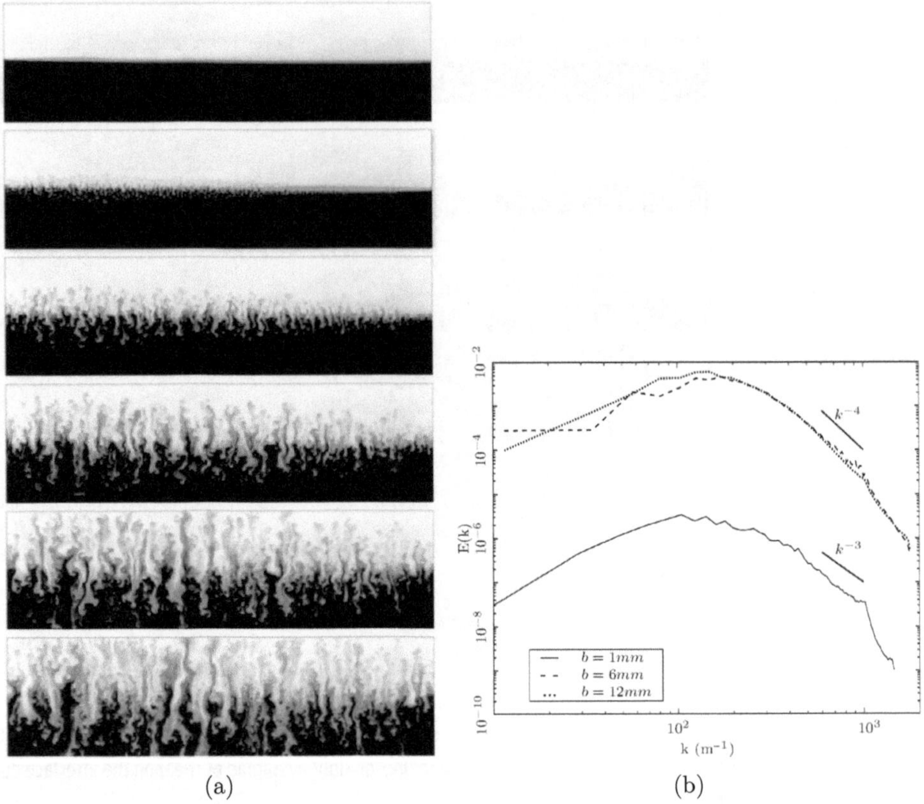

Fig. 15.22 (a) Volume fraction development with time for $\tilde{z} \approx 1$ mm experiment. (b) The experimental measured energy spectra for $\tilde{z} \approx 1, 6,$ and 12 mm. Figs. 8 and 15 of Horne and Lawrie (2020).

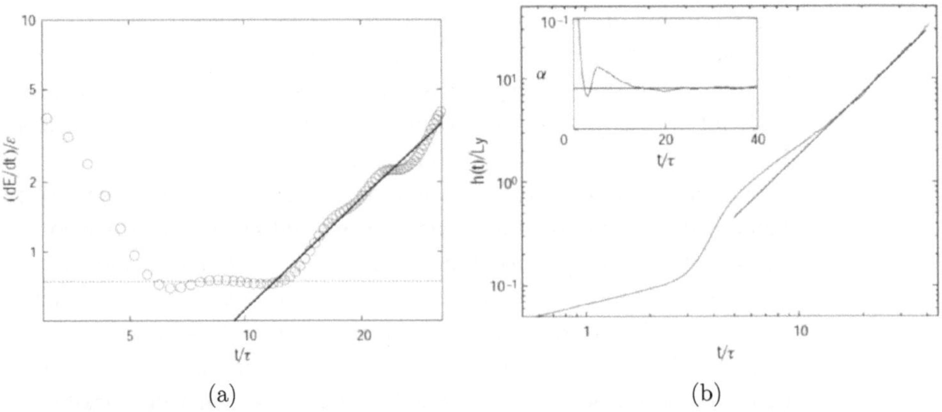

Fig. 15.23 (a) The ratio of the kinetic energy growth rate to viscous energy dissipation rate as a function of time. The solid straight line represents the behavior $t^{8/5}$. (b) Temporal evolution of the mixing layer width $h(t)$. The straight line represents the quadratic law with $\alpha = 0.035$, obtained from the compensated plot of $h(t)/\mathcal{A}gt^2$ shown in the inset. Boffetta et al. (2012).

Utilizing extensive direct numerical simulations of the Darcy–Rayleigh–Taylor system, Borgnino et al. (2021) found that a sharp transition from 3D to 2D regimes occurs when the transverse scale is of the order of the typical horizontal width of the fingers. This should be compared to the similar configurations of turbulent flows in general (Alexakis and Biferale, 2018), and RTI induced flow in particular (Boffetta et al., 2012), in which there is a continuous transition and coexistence of both the 2D and 3D regimes.

16 Convergent/divergent geometry

For analytical simplicity, most research to date on RT and RM instabilities has focused on planar geometries. Such a simplified design is very helpful in easing the diagnostic requirements for laboratory or HEDP experiments. However, in the limited observations of Chapter 15, we have already witnessed that other geometric configurations may alter the mixing layer growth significantly.

In a variety of important applications, one must deal with imploding/exploding flows, the prime examples of which are ICF implosions (convergent geometry) and supernova explosions (divergent geometry). In these configurations, the flows are radially accelerated/decelerated. In contrast to planar geometry, where only RM growth is expected to occur, converging/diverging shock-accelerated interfaces can be RT unstable as they geometrically contract or expand (e.g., Lombardini et al., 2014a,b).

An important distinction between the works of Binnie (1953), Chandrasekhar (1955), Gupta and Lawande (1986), and Terrones and Carrara (2015), on the one hand, and the works of Bell (1951) and Plesset (1954), on the other, is that the former papers deal with the stability of a static interface under a gravitational force, and hence constant radius, while the latter papers treat the dynamic problem of a moving interface in which $R = R(t)$. In other words, they study the spherical analog of the two problems treated in a planar geometry by Lord Rayleigh (static interface in a gravitational field) and subsequently by G.I. Taylor (accelerating interface). While the results of Rayleigh and Taylor are identical in a planar geometry, in a spherical geometry there are differences between the two approaches, which are generally known as "Bell-Plesset effects (BP)."

Indeed, Bell and Plesset realized that when considering perturbed surfaces in a curved geometry, additional effects appear even without the presence of acceleration. Thus, it is fair to state that there is no analog to BP effects in a planar geometry.

More specifically, Plesset considered two incompressible fluids with an arbitrary $\mathcal{A}$. Bell restricted his considerations to the case $\mathcal{A} = 1$, but did allow for compressibility under the assumption that compression was uniform. Remarkably, Penney and Price (1942) derived the Bell–Plesset model for $\mathcal{A} = 1$ several years before the G.I. Taylor paper. Penney and Price (1942) showed that the surface of an underwater explosion bubble is similarly unstable. Unfortunately, their British Report SW-27 has become somewhat obscure after being discussed in Cole's book (Cole, 1948) and Birkhoff's Los Alamos report (Birkhoff, 1954).

16.1 Degrees of convergence

The governing equations for RTI evolution in planar, cylindrical, and spherical geometries were developed by the classical work of Rayleigh (1883), Taylor (1950), Bell (1951), and Plesset (1954). As a baseline, recall that the equation of motion for the planar RT interface perturbation amplitude $a(t)$ can be written as (Eq. 2.23),

$$\frac{d^2a}{dt^2} - kg\mathcal{A}a = 0. \tag{16.1}$$

The so-called Bell's equation describes the RT instability in a cylindrical geometry (though as noted by Mikaelian (2005a), the RM instability was not yet discovered at that time),

$$\frac{d^2a}{dt^2} + 2\frac{\dot{\mathcal{R}}}{\mathcal{R}}\frac{da}{dt} - [n\mathcal{A} - 1]\frac{\ddot{\mathcal{R}}}{\mathcal{R}}a = 0, \tag{16.2}$$

where n is the mode number of the perturbation and $\mathcal{R}$ is the radial coordinate of the cylindrical geometry. Bell derived the governing equations for two specific cases, $\mathcal{A} = \pm 1$.

As for the spherical geometry, there were some early papers on perturbations of spherical interfaces (Elliott, 1965a,b), the governing equation of which is

$$\frac{d^2a}{dt^2} + 3\frac{\dot{\mathcal{R}}}{\mathcal{R}}\frac{da}{dt} - nA(n)\frac{\ddot{\mathcal{R}}}{\mathcal{R}}a = 0. \tag{16.3}$$

Here, $\mathcal{R}$ is the radial coordinate for the spherical geometry and

$$nA(n) = \frac{n(n-1)\rho_2 - (n+1)(n+2)\rho_1}{n\rho_2 + (n+1)\rho_1} = \frac{2n(n+1)\mathcal{A}}{2n+1-\mathcal{A}} - 2. \tag{16.4}$$

The equations above are given in the order of increasing convergence effects by Mikaelian (1990b,c, 2005a). As one moves from cylindrical geometry to the more convergent spherical geometry, the coefficient of $\dot{R}\dot{a}/R$ increases from 2 to 3. Note that there is no corresponding term in a planar geometry, Eq. 16.1.

As first accomplished by Richtmyer for the planar geometry, the same RTI equations can be generalized to the RMI by assuming an impulsive acceleration, $g = \ddot{\mathcal{R}} = \Delta V\delta(t)$. The RM results can be written as (Mikaelian, 2005a)

$$a(t) = a(0)(1 + k\mathcal{A}\Delta Vt), \tag{16.5}$$

$$a(t) = a(0)[1 + (n\mathcal{A} - 1)(1 - \mathcal{R}_0/\mathcal{R})], \tag{16.6}$$

$$a(t) = a(0)\left\{1 + \frac{nA(n)}{2}\left[1 - (\frac{\mathcal{R}_0}{\mathcal{R}})^2\right]\right\}, \tag{16.7}$$

where $\mathcal{R} = \mathcal{R}_0 + \Delta Vt$.

The RTI solutions are not so simple in the cylindrical and spherical cases because they involve hypergeometric functions – see the appendices in Mikaelian's papers on the cylindrical (Mikaelian, 2005a) and spherical (Mikaelian, 1990a) geometries, respectively.

16.2 Bell–Plesset (BP) effects

While the term "accelerationless growth" has also been used (Hsing et al., 1997), Epstein (2004) pointed out that "undriven growth" would be more precise to denote convergence and compression effects in the absence of the buoyant force that drives the RT instability. The reason is that the term "accelerationless" suggests the more restricted case of zero acceleration. As yet, the analysis of Epstein indicates that BP effects depend on the RT growth rate, and in general, the net perturbation growth does not separate easily into acceleration-driven and undriven contributions.

After the previous chapters, it should not be surprising to find that the same potential flow methodology that proved so useful in describing planar geometries can also be profitably extended to curved configurations. As a reminder, the velocity potential, consisting of a background flow plus a perturbation, ϕ_ℓ, describes irrotational flow with uniform mass density and is governed by Bernoulli's theorem. The treatment in this section follows closely that from Epstein (2004).

To begin, the perturbation is assumed to be incompressible,

$$\nabla^2 \phi_\ell = 0. \tag{16.8}$$

For the same order of increased convergence effects, the subscript ℓ denotes the harmonic order of the perturbation mode corresponding to $\cos(2\pi\ell y/L)$ (planar), $\cos(\ell\theta)$ (cylindrical), and $Y_\ell^m(\theta, \emptyset)$ (spherical) geometries.

The solutions for each side of the interface for planar, cylindrical, and spherical geometries, respectively, are:

$$\phi_{\ell\pm} = b_{\ell\pm}(t)e^{\pm 2\pi\ell x/L}\cos(2\pi\ell y/L), \tag{16.9}$$

$$\phi_{\ell\pm} = b_{\ell\pm}(t)r^{\pm\ell}\cos(\ell\theta), \tag{16.10}$$

and

$$\phi_{\ell\pm} = b_{\ell\pm}(t)r^{[\ell,-(\ell+1)]}Y_\ell^m(\theta, \emptyset), \tag{16.11}$$

after imposing vanishing boundary conditions at large distances from the interface, and the z-dependent perturbation is disregarded for planar and cylindrical geometries.

For planar, cylindrical, and spherical geometries, the positions of the perturbed interface are

$$r_s(t, y) = \mathcal{R}(t) + a_\ell(t)\cos(2\pi\ell y/L), \tag{16.12}$$

$$r_s(t, \theta) = \mathcal{R}(t) + a_\ell(t)\cos(\ell\theta), \tag{16.13}$$

and

$$r_s(t, \theta, \emptyset) = \mathcal{R}(t) + a_\ell(t)Y_\ell^m(\theta, \emptyset), \tag{16.14}$$

respectively. The interface displacement, a_ℓ, is the spatial amplitude of the perturbation.

All quantities are evaluated at $r = r_s$ by expanding to first order in the perturbation about the unperturbed interface position $r = R$ (Epstein, 2004),

$$\left(-\gamma_\rho + \frac{d}{dt}\right)\frac{d}{dt}(a_\ell \rho) = \gamma_0^2 (a_\ell \rho), \tag{16.15}$$

$$\left(-\gamma_\rho + \frac{d}{dt}\right)\frac{d}{dt}(a_\ell \rho \mathcal{R}) = \gamma_0^2 (a_\ell \rho \mathcal{R}), \tag{16.16}$$

and

$$\left(-\gamma_\rho - \gamma_\mathcal{R} + \frac{d}{dt}\right)\frac{d}{dt}(a_\ell \rho \mathcal{R}^2) = \gamma_0^2 (a_\ell \rho \mathcal{R}^2), \tag{16.17}$$

for planar, cylindrical, and spherical geometries, respectively. Note that $\rho = (\rho_1 + \rho_2)/2$. This rate of compression is assumed to apply everywhere,

$$\gamma_\rho = \frac{\dot{\rho}_1}{\rho_1} = \frac{\dot{\rho}_2}{\rho_2}, \tag{16.18}$$

and the convergence rate is defined as

$$\gamma_\mathcal{R} = \frac{\dot{\mathcal{R}}}{\mathcal{R}}. \tag{16.19}$$

Equations 16.15–16.17 show a physically meaningful isolation of the RT effect. For the three geometries, the "driving" terms are proportional to the RT growth rates (γ_0)

$$\gamma_0^2 = k\mathcal{A}g_p, \tag{16.20}$$

$$\gamma_0^2 = \frac{\ell}{\mathcal{R}}\mathcal{A}g_p, \tag{16.21}$$

and

$$\gamma_0^2 = \frac{\ell(\ell+1)}{\mathcal{R}}\frac{(\rho_2+\rho_1)}{[\ell\rho_2 + (\ell+1)\rho_1]}\mathcal{A}g_p, \tag{16.22}$$

where $g_p = -(1/\rho)[\partial p(\mathcal{R}, t)]/\partial x$. Equation 16.20 is the familiar "classical" RT growth rate for incompressible planar flow. The spherical results (Eqs. 16.17 and 16.22) are equivalent to the result of Plesset (1954) for incompressible flow ($\gamma_\rho = 0$). For the special case of a free surface (either $\rho_1 = 0$ or $\rho_2 = 0$), Eqs. 16.15–16.22 reduce to Bell's final result (Bell, 1951).

Unfortunately, BP effects from the undriven limit generally cannot be combined with classical RT growth in a simple way to obtain a correct result. The first-derivative terms in Eqs. 16.15–16.17 represent "dynamic" BP effects (Ramshaw, 2000) that alter the mass-perturbation growth rates from their pure RT values given by Eqs. 16.20–16.22.

There are two limiting cases of particular interest. First, while rapid RT growth might amount to several e-foldings in, for example, an ICF implosion, BP effects can be much smaller and still amount to a significant effect. Second, Eqs. 16.15–16.17 are easily solved over time intervals in which the compression, convergence, and RT growth rates are constant. The resulting solution pairs are exponential in time with constant growth rates:

$$\gamma_\pm = \frac{1}{2}\gamma_\rho \pm \sqrt{\gamma_0^2 + \frac{1}{4}\gamma_\rho^2} \tag{16.23}$$

for both planar and cylindrical geometries and

$$\gamma_{\pm} = \frac{1}{2}(\gamma_\rho + \gamma_{\mathcal{R}}) \pm \sqrt{\gamma_0^2 + \frac{1}{4}(\gamma_\rho + \gamma_{\mathcal{R}})^2} \tag{16.24}$$

for spherical geometry.

As Epstein (2004) pointed out, this presentation of a unified body of results for several geometries and flow characteristics hopefully has conveyed a clearer sense of the origin, nature, and diversity of BP effects and their tendency to defy any simple characterization applicable over a wide range of circumstances. For related work, see Ge et al. (2022), Gou et al. (2022) and Cao et al. (2024).

16.3 Cylindrical geometry

Although ICF capsules are spherical, we will see that all convergence effects are captured in cylindrical geometries, albeit at reduced levels (Mikaelian, 2005a).

There are many efforts to develop weakly nonlinear or nonlinear incompressible RTI growth at cylindrically convergent interfaces. Guo et al. (2014), Liu et al. (2014b,c, 2015a), and Wang et al. (2013b, 2015a) have looked at various aspects of such models. For RMI, Matsuoka and Nishihara (2006a,b) performed the weakly nonlinear analysis up to third order. The analytical growth rates agree well with numerical vortex sheet calculations until the neighborhood of the numerical computation's breakdown. Liu et al. (2012c) found that the interface expression consists of two parts: a planar-type result and a modification due to cylindrical effects. Based on both a Padé approximation technique and a perturbation expansion performed directly on the perturbed interface, rather than the unperturbed interface, Liu et al. (2014d) obtained a nonlinear model in good agreement with previous simulations of Matsuoka and Nishihara.

Utilizing a vertical coaxial diaphragmless shock tube (Si et al., 2015), Lei et al. (2017) performed investigations of a sinusoidal interface interacting with a cylindrically converging shock (Fig. 16.1). Four single-mode air/SF_6 interfaces with different initial amplitudes or wavenumbers were generated using an extended soap film technique (Wang et al., 2015d). The shape of the interface took the form of $r(\theta) = R_0 + a_0 \cos(n\theta)$, where a_0 is the initial amplitude, n is the wavenumber, and R_0 is the radius. In all cases, the perturbation amplitude initially decreased due to the shock compression, after which a phase of rapid growth followed. Finally, the perturbation began to decrease quickly, due to the RTI stabilization caused by the interface deceleration present in the converging geometry (Fig. 16.2). Quantitatively, the authors have shown that the RMI, the BP, and the RT effects govern the whole process in the converging RMI.

Figure 16.3(a), taken from Wu et al. (2021), shows that the temporal positions of the interface and shock wave of their 2D single-mode simulations agree well with experiments (Lei et al., 2017). Figure 16.3(b) shows the amplitude and corresponding growth rate can be divided into three stages, as detailed by Wu et al. (2021): (i) the incident shock (IS) compresses and accelerates the interface; (ii) the perturbations first grow with increasing

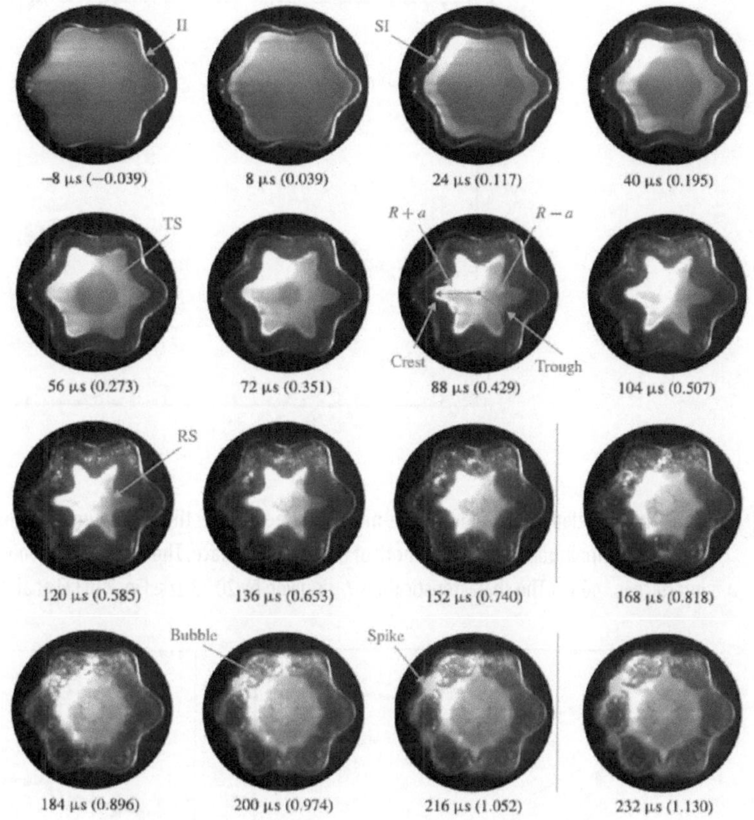

Fig. 16.1 Typical evolution of a sinusoidal air/SF6 interface ($a_0 = 2.0$ mm and $n = 6$) subjected to a converging shock. II: initial interface, SI: shocked interface, TS: transmitted shock, RS: reflected shock. The time interval between consecutive frames is 16 μs. The time when the converging shock meets the interface is defined as 0 μs. Numbers inside brackets are dimensionless times, based on a characteristic time of 205.3 μs. Fig. 5 of Lei et al. (2017).

growth rate due to the RM instability and BP effects. After the transmitted shock (TS) is reflected from the center, however, the inward moving interface slows down gradually and induces the RT stabilizing effect that suppresses the growth of the perturbations; (iii) after reshock, the perturbations undergo shock compression again and a simultaneous phase reversal is observed.

To delineate the individual contributions to mixing layer development, the classical formulation for a cylindrical geometry can be instructive. Bell (1951) derived the linear model for the perturbation growth rate of the planar RT instability (Taylor, 1950) into a cylindrical geometry. Under a small-perturbation assumption, the simplified compressible model in the early stage takes the form (Wu et al., 2021)

$$\dot{a} = \frac{\mathcal{R}_0^2}{\mathcal{R}^2(t)}\dot{a}_0 + \frac{n\mathcal{A} - 1}{\mathcal{R}^2(t)}\int_{t_0}^{t} \mathcal{R}(t')\ddot{\mathcal{R}}(t')a(t')dt' - \frac{\gamma_\rho}{\mathcal{R}^2(t)}\left[\int_{t_0}^{t} \mathcal{R}(t')\dot{\mathcal{R}}(t')a(t')dt' + \int_{t_0}^{t} \mathcal{R}^2(t')\dot{a}(t')dt'\right] \tag{16.25}$$

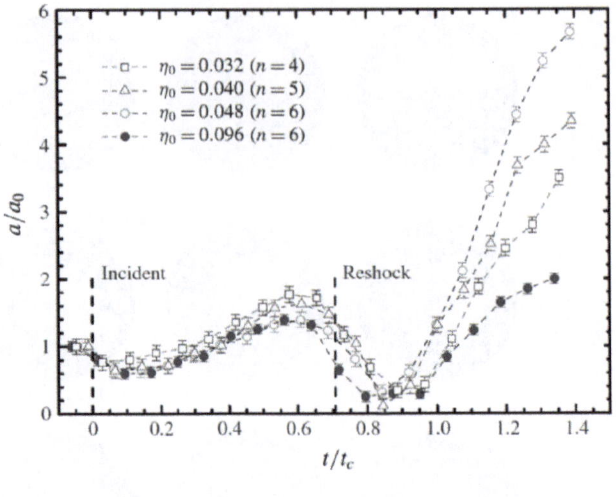

Fig. 16.2 Amplitude development for four single-mode interfaces under the converging RMI. Here, $\eta_0 = na_0/(2\pi\mathcal{R}_0)$ is the ratio of the amplitude to the wavelength of the initial interface. The time of the reshock interacting with the evolving interface is 146 μs.The characteristic time, t_c, is given by 205.3 μs. Fig. 7 of Lei et al. (2017).

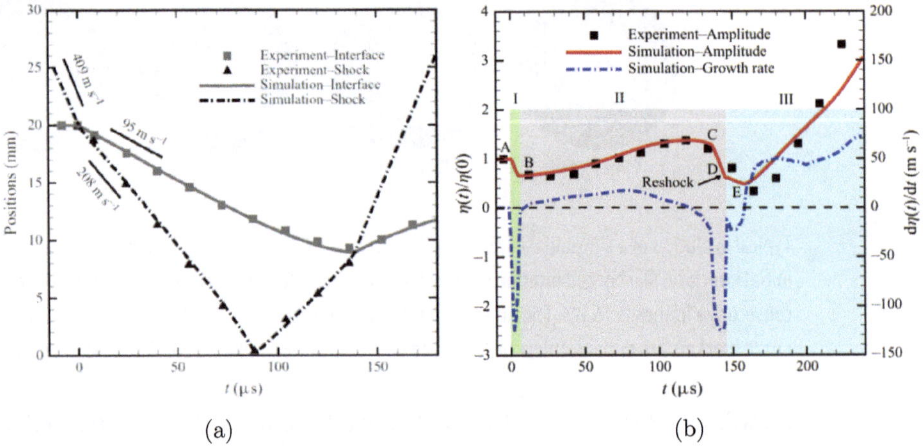

Fig. 16.3 (a) r-t diagram of the interface and shock wave for a cylindrically converging shock wave impinging on an initially unperturbed Air/SF6 interface. The lines and symbols represent the simulation and experimental data (Lei et al., 2017), respectively. (b) Evolution of the amplitude (solid line) and corresponding growth rate (dot-dashed line) for the single-mode cylindrical RM instability in comparison with experimental data (black squares). Figs. 5 and 8 of Wu et al. (2021).

where n is the wavenumber. The terms on the right-hand side of Eq. 16.25 represent the contributions of the RMI, RT stability/instability, and the compressibility effect, respectively. The first two terms appear as an incompressible model, which cannot correctly describe the evolution of the perturbations in stage II, due to ignorance of the compressibility effect, as discussed by Ding et al. (2017).

While the various effects mentioned above can be readily identified with individual terms of Eq. 16.25, the BP effect is coupled into each term and is difficult to isolate from the others. Thus, the total measured amplitude can be decomposed as

$$a = a_{RM} + a_{RT} + a_C + a_{up}. \quad (16.26)$$

The last term in this equation acknowledges the importance of taking into account the premixed width of the initially-generated interface, which is either due to unavoidable gas diffusion (in experiments) or artificially added to simulations to ensure numerical stability. For an initial shock of Mach number 1.25, Fig. 16.4 also shows that such premixed width must be taken into account when computing the model equation, Eq. 16.25, where the initial condition must be specified to achieve a good agreement between the model and simulated data (Wu et al., 2021).

Figure 16.4 demonstrates that the effect of the RMI dominates for the first three-fourths of phase II. The coupled RT stabilizing and BP effects are small and positive during this period, then increase rapidly to suppress the perturbation growth caused by the deceleration of the interface. As a result, there is a substantial cancellation between the RMI and RTI influences during the last quarter of the second phase. The coupled compressibility and BP effect are not significant factors in this case.

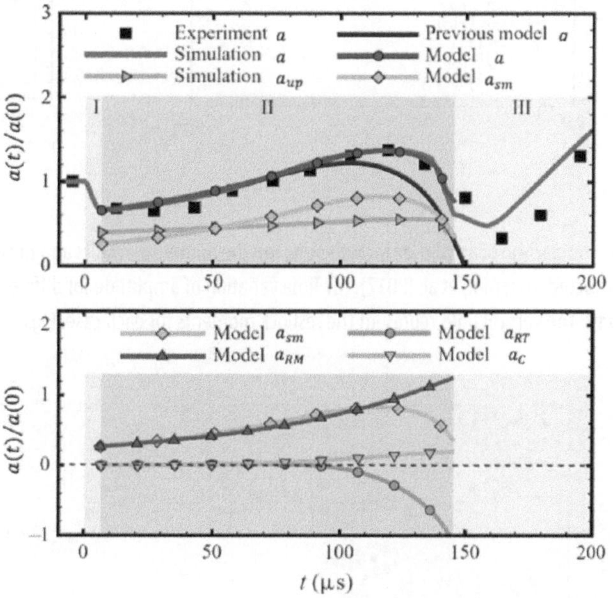

Fig. 16.4 (Top) Total amplitude (solid line with circles) and all contribution terms calculated by the modified compressible model. Here, $a_{sm} = a_{RT} + a_{RM} + a_C$ and $a = a_{sm} + a_{up}$. The data from the simulation (upper solid line) and experiment (squares) (Lei, 2017) are added for comparison. The curve of the predicted total amplitude by Bell's compressible model, also shown in lower solid line, deviates from the simulation and experimental results in late stage II. (Bottom) The terms represent the contributions of the RMI, RT stability/instability, and the compressibility effect, respectively. Fig. 9 of Wu et al. (2021).

Simulations of 2D single-mode light/heavy interfaces driven by convergent shock waves have been used to investigate the effect of the Atwood number on convergent RMI (Tang et al., 2021a). Figure 16.5(a) shows a good agreement with experimental measurements, including in the reshock stage (Ding et al., 2017). The authors considered five values of $\mathcal{A}$, ranging from 0.21 to 0.67. As the Atwood number was reduced, the perturbation grew more slowly because the density gradient was less prominent and less baroclinic vorticity was deposited at the interface (Figure 16.5(b)).

Sun et al. (2020b) performed an experimental study for the convergent RMI of an SF_6 layer with a uniform outer surface and a sinusoidal inner surface surrounded by air in a semiannular convergent shock tube (Luo et al., 2015b). Figure 16.6(a) displays the temporal evolution of the various perturbation amplitudes of the inner interface upon impact of the first transmitted shock. The initial ratios of a_0/λ are 0.03 (Case 4), 0.06 (Case

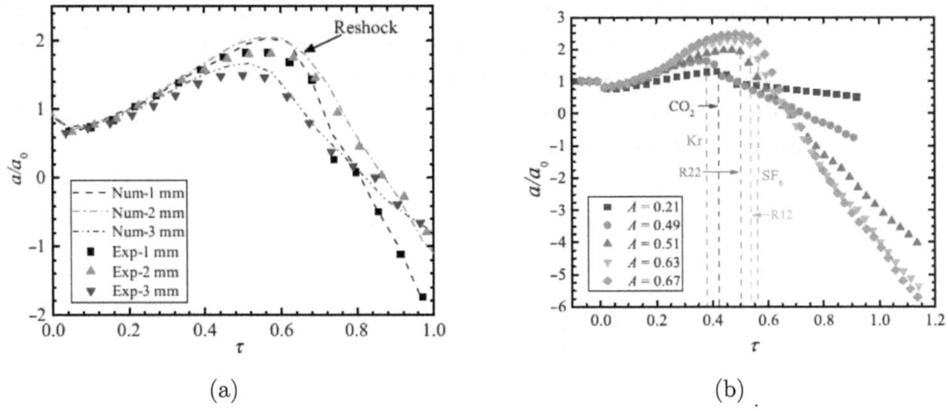

Fig. 16.5 (a) Comparison of amplitude growth between the numerical results and the experiments. The experimental results are extracted from Ding et al. (2017). (b) Time variation of amplitude for different Atwood number cases in dimensionless form. The vertical lines represent the reshock moments for each case. Figs. 2 and 7 of Tang et al. (2021a).

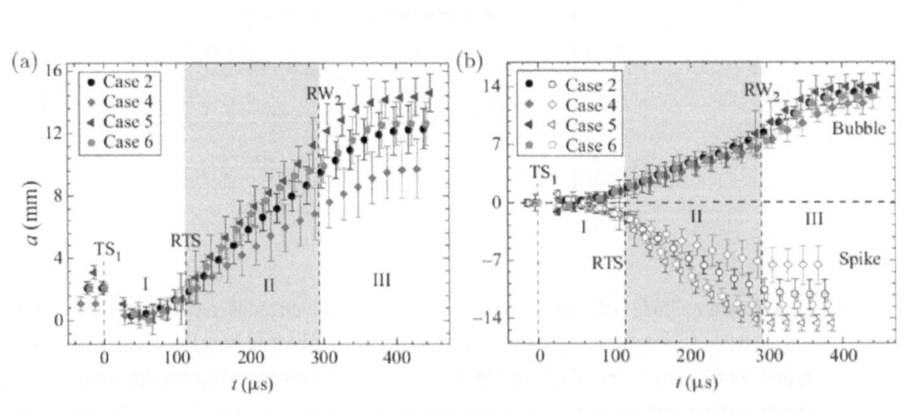

Fig. 16.6 The initial ratios of a_0/λ are, in increasing order, 0.03 (Case 4), 0.06 (Case 2), 0.10 (Case 5), and 0.11 (Case 6). Fig. 7 of Sun et al. (2020b).

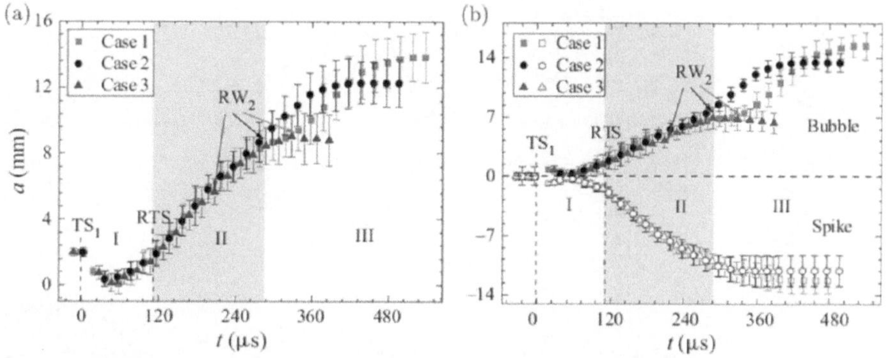

Fig. 16.7 Temporal variation of (a) the overall perturbation amplitude and (b) the individual spike and bubble amplitudes of the inner interface. The labels I, II, and III stand for the three stages of the amplitude growth process for Case 2. Fig. 5 of Sun et al. (2020b).

2), 0.10 (Case 5), and 0.11 (Case 6). After the phase reversal, in which the perturbation amplitude decreases gradually to zero, the mixing layer increases continuously. The perturbation amplitude subsequently experiences a long period of growth that is quasi-linear in time after reshock.

While the bubble growth has only a weak dependence on the initial conditions, Fig. 16.6(b) demonstrates that the spike growth's dependence on the initial amplitude-to-wavelength ratio is quite strong. The spike dynamics at late stages is dominated by two mechanisms: 1) the driving force produced by baroclinic vorticity, whose strength is positively correlated with the initial amplitude-to-wavelength ratio, and 2) the drag force caused by geometric convergence, which is independent of the initial perturbation. Essentially, the spike amplitude can grow to a higher saturation value for cases with a larger initial amplitude-to-wavelength ratio, because stronger baroclinic vorticity is deposited on the interface, resulting in a greater driving force acting on the growing spike (Sun et al., 2020b).

Figure 16.7(a) shows the overall temporal evolution of SF_6 layers with thicknesses of 35 mm (Case 1), 25 mm (Case 2) and 15 mm (Case 3), while the individual spike and bubble amplitudes of the inner interface are shown in Fig. 16.7(b). Sun et al. (2020b) found that the RT effect and the interface coupling, which dominate the late-stage development of the inner and outer interfaces, respectively, depend heavily on the layer thickness.

The dynamics of cylindrical converging RMI on a dual-mode interface can also be examined, where the initial configurations of different dual-mode interfaces is described by $r(\theta) = R_0 + a_1 \cos(k_1\theta) + a_2 \cos(k_2\theta)$. In particular, Zhou et al. (2020b) considered the evolution of a dual-mode interface subjected to a cylindrical shock for cases with distinctive features, labeled D-3, D-5, and D-6 by the authors. The result of the single-mode case (S-3) is given for comparison. For all dual-mode interfaces, the parameters for the first sinusoidal wave remain the same: $k_1 = 20$, $a_1 = 1.5$. Whilst the second initial amplitude, $a_2 = 0.2$, for both D-3 and D-5, their values of k_2 are 40 and 60, respectively. The cases of D-5 and D-6 both have $k_2 = 60$, but have initial amplitudes of opposite signs ($a_2 = 0.2$

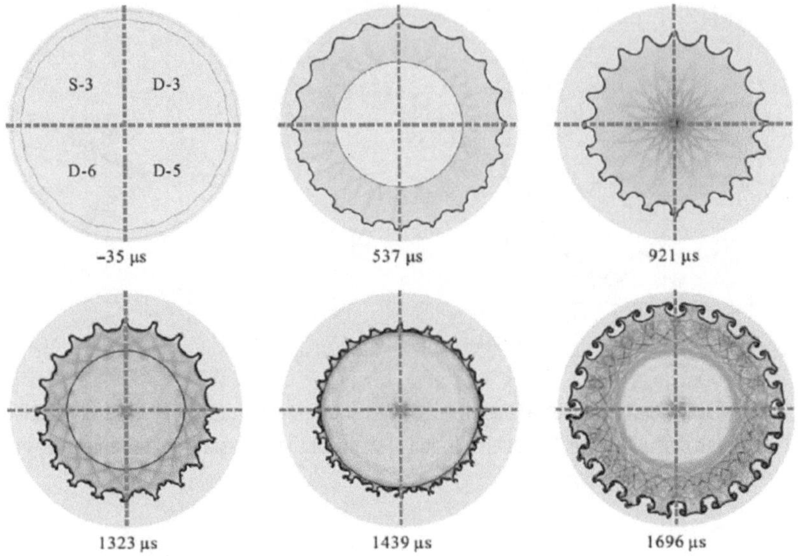

Fig. 16.8 Evolution of a dual-mode interface subjected to a cylindrical shock for cases D-3, D-5, and D-6. The result of single-mode case (S-3) is given for comparison. Note that the discrepancy among the four cases diminishes significantly after the reshock (1468 μs). Fig. 7 of Zhou et al. (2020b).

and -0.2) in order to study the influence of the phase difference between two basic waves on the instability development.

The result of single-mode case (S-3) is given in Fig. 16.8 for comparison, where $k = 20$, $a_0 = 1.5$, and $a_0/\lambda = 0.0245$. Three interesting observations can be made: (i) the evolution of D-3 has a strong resemblance with that of S-3, the single-mode case; (ii) the nonlinear bubble (spike) structures of D-5 and D-6 display quite different morphologies; (iii) after the reshock (1468 μs), the discrepancy among the four cases diminishes significantly.

Recently, Liang et al. (2023) carried out shock-tube experiments on the convergent RMI of a dual-mode interface. It is substantiated that the mode-coupling mechanism is closely related to the initial spectrum, including initial amplitudes of the constituent modes, relative phases, and azimuthal wavenumbers.

For the multi-mode cases in cylindrical geometry, the initial perturbation profiles may be described in a similar manner to those used in the planar geometry:

$$P(k) = \begin{cases} C_p k^m & \text{if } k_{min} < k < k_{max} \\ 0 & \text{otherwise,} \end{cases} \tag{16.27}$$

where $k = \sqrt{k_x^2 + k_y^2}$ is the wavenumber of the perturbation (Youngs, 2013; Dimonte et al., 2004; Ramaprabhu et al., 2005). Here, C_p is a constant chosen to ensure the initial perturbations are linear at the highest wavenumber (El Rafei and Thornber, 2020).

Ge et al. (2020) performed high-resolution numerical simulations of RMI at a complete cylindrical interface, with the imploding shock wave initially passing from the heavy fluid

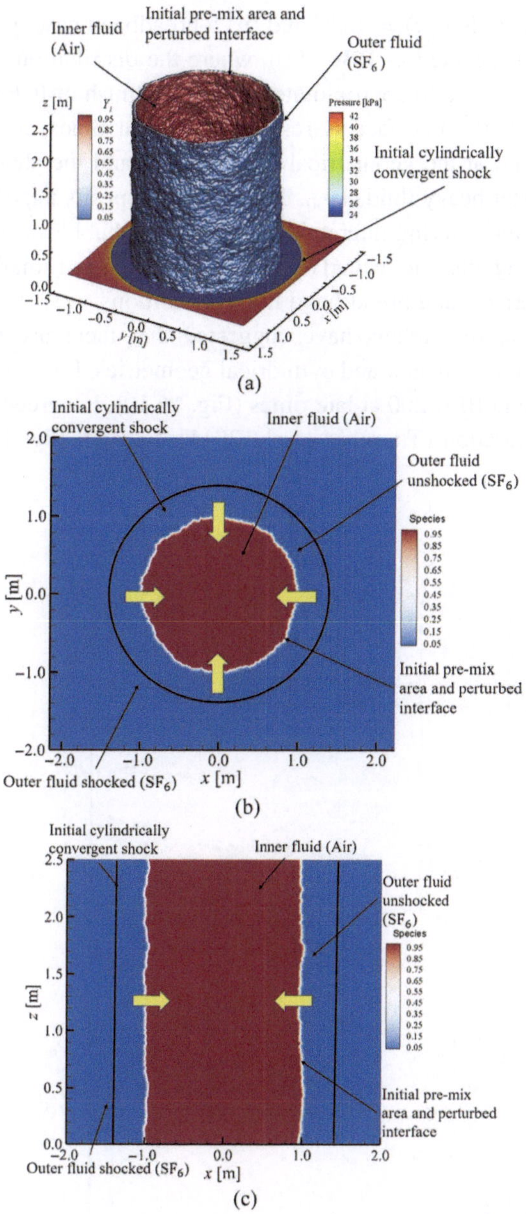

Fig. 16.9 3D isosurface of the simulated mass fraction of the light fluid and [(b) and (c)] its 2D mass species slices, where blue and red represent the heavy and light fluid, respectively. Fig. 1 of Ge et al. (2020), with the permission of AIP Publishing.

to the light fluid (at t_{10}). Figure 16.9 shows the simulation setup from different perspectives. Hence, two types of initial perturbations are applied: (i) a narrow-band case with the wavenumber modes ranging from 16 to 64 is named NB16-64, for which the distribution of energy is $P(k) = C_p k^0$ and (ii) a broadband case with the wavenumber ranging from 6 to 200 (denoted as BB6-200), where the distribution of energy is $P(k) = C_p k^{-2}$, a form which is used to approximate the surface finish of ICF capsules (Dimonte, 2004).

Here, the interface is reshocked several times, as the initial incident shock wave is reflected at the geometrical center at t_{20}, and the shock wave is reflected once again by the outer heavy fluid at t_{30}. The reshock deposits a large amount of vorticity in the mixing zone, intensifying interactions among both bubbles and spikes. At the end of calculations, and long after the second reshock, many large and small structures have emerged from both the narrow- and broad-band initial conditions.

Some researchers have suggested that there are noticeable common characteristics between the planar and cylindrical geometries. First, NB16-64 has a higher mixing parameter than BB6-200 at late times (Fig. 16.10), in agreement with observations of the planar configuration (Thornber et al., 2011). Second, Fig. 16.11 demonstrates that the density

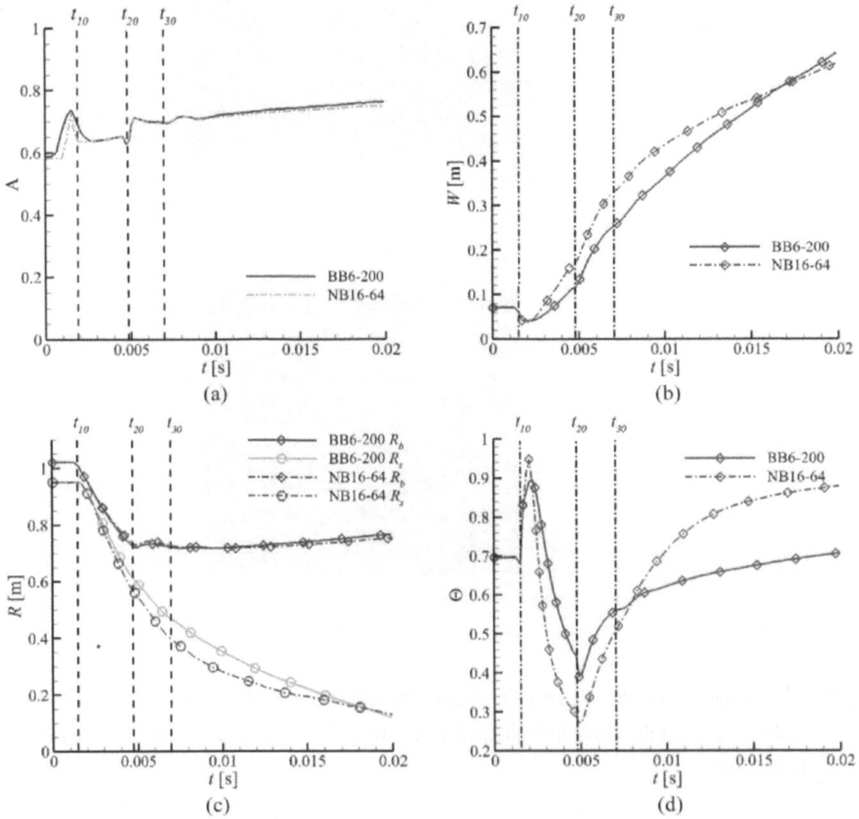

Fig. 16.10 The basic mixing parameters for multi-mode RMI in cylindrical geometry, including (a) the Atwood number, (b) the mixing width, (c) the bubble and spike fronts, (d) the mixing width, and (d) the molecular mixing fraction. Fig. 5 of Ge et al. (2020), with the permission of AIP Publishing.

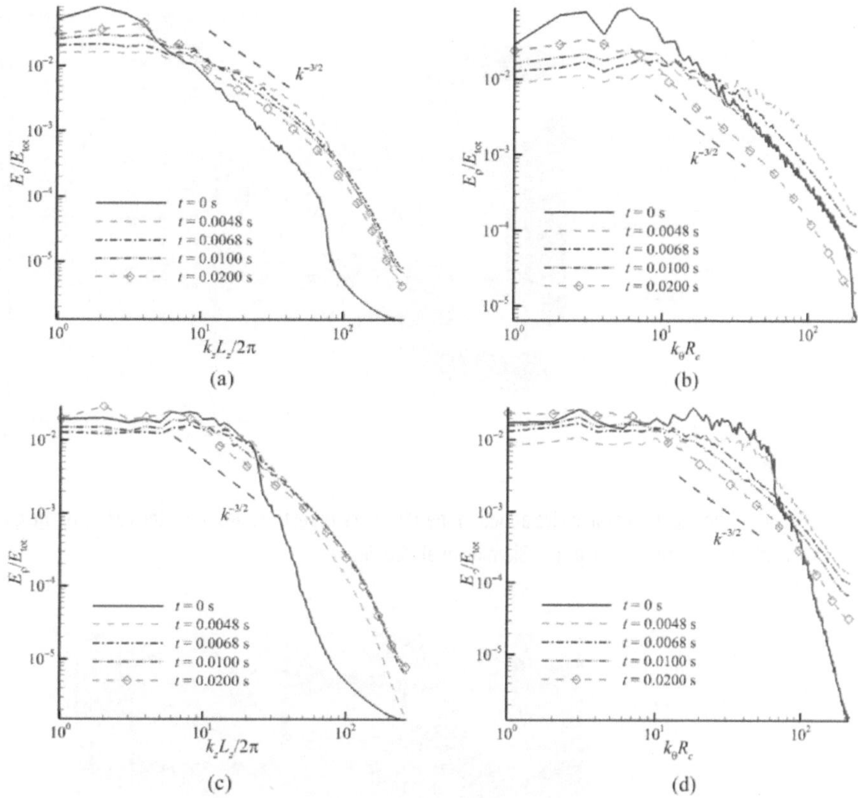

Fig. 16.11 Evolution of the density energy spectra for multi-mode RMI in cylindrical geometry in [(a) and (c)] the axial direction and [(b) and (d)] azimuthal direction for [(a) and (b)] the case BB6-200 and [(c) and (d)] the case NB16-64. In both cases, a wide range of the inertial subrange appears at the end of the calculations with a $k^{-3/2}$ scaling as previous predicted theoretically (Zhou, 2001). Fig. 4 of Ge et al. (2020), with the permission of AIP Publishing.

energy spectra for both the axial and azimuthal directions scale as $k^{-3/2}$ of Zhou (2001), as shown previously for the planar RMI cases (Figs. 7.6, 7.7, 7.8, 13.4, 13.6, and 13.8).

Dimotakis and Samtaney (2006) suggested a gas lensing technique capable of smoothly focusing a plane shock into a segment of a circular cylinder in a 2D wedge geometry. Vandenboomgaerde and Aymard (2011) generalized this method by proposing to use a suitable interface between two gases. In order to experimentally check the feasibility of this approach, Biamino et al. (2014) implemented the gas lens technique on the T80 conventional shock tube of IUSTI (Jourdan et al., 2004), where the convergent experimental chamber is illustrated in Figs. 16.12–16.13. Moreover, the flow field generated by a planar shock wave (Mach 1.15 in air) propagating through the three zones (air/SF6/air) is shown in Fig. 16.14.

In a laboratory setting, Weir et al. (1998) measured RT growth in a 2D cylindrically convergent geometry by imploding gelatin cylinders with high-pressure gases. This experiment featured both the precise control of initial perturbations and high-resolution

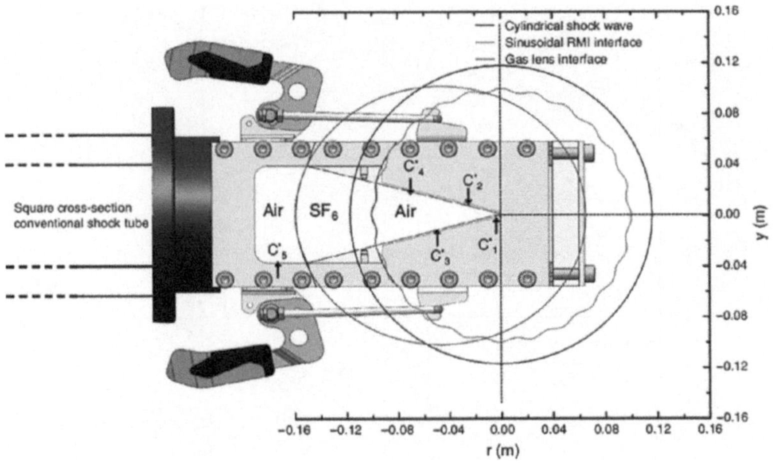

Fig. 16.12 Details of the experimental device added to the T80 shock tube; the lines show the curves of the gas lens and the perturbed RMI interfaces. Fig. 1 of Biamino et al. (2015).

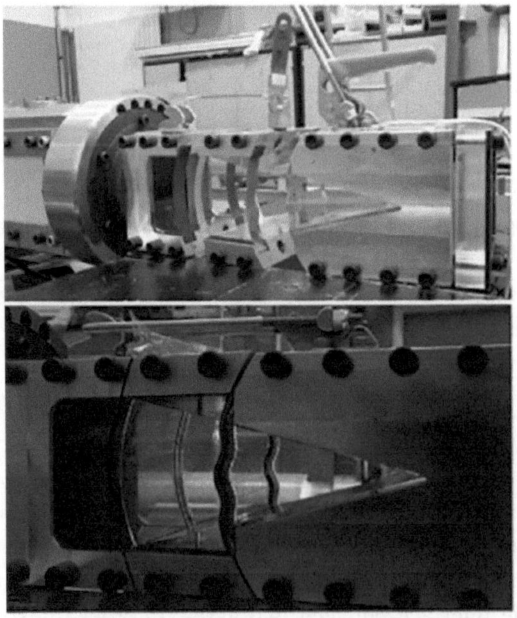

Fig. 16.13 Views of the convergent experimental chamber. Fig. 2 of Biamino et al. (2015).

imaging of the RT and feedthrough growth. Holder et al. (2003a) described a high-convergence experiment on the AWE convergent shock tube and comparison with 3D TURMOIL simulations.

HEDP cylindrical implosion experiments (Hsing and Hoffman, 1997; Hsing et al., 1997; Tubbs et al., 1999; Barnes et al., 2002; Lanier et al., 2003) have inspected compression

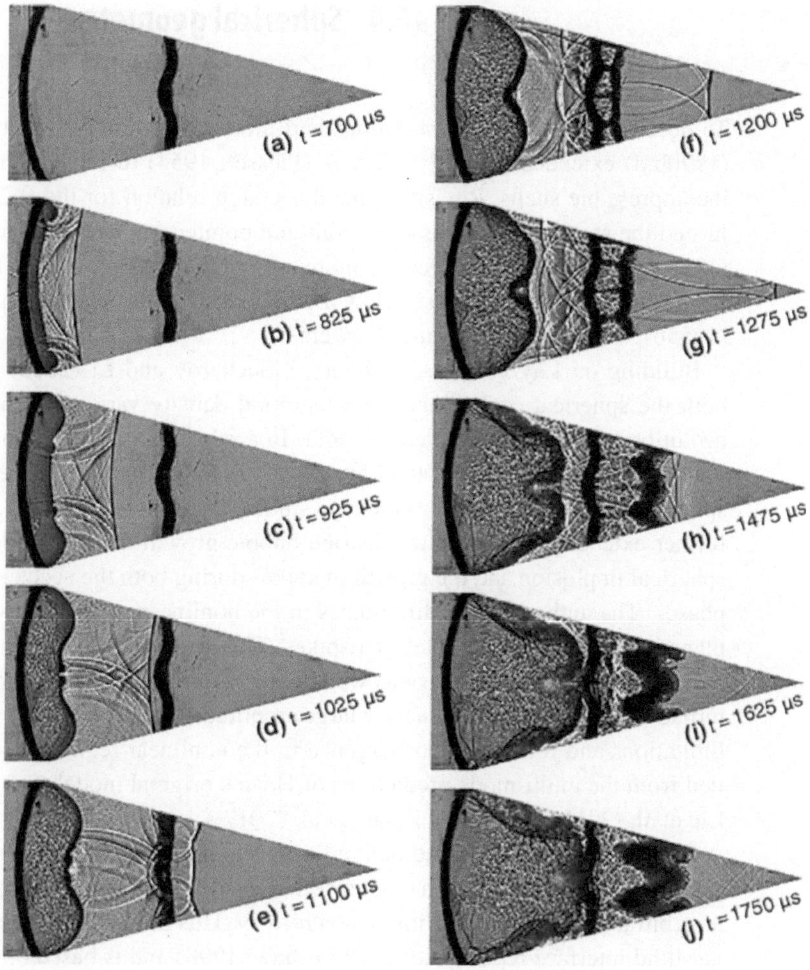

Fig. 16.14 Schlieren pictures showing the evolution of the converging Richtmyer–Meshkov instability issued from a Mach 1.15 incident plane shock wave leaving a conventional shock tube and refracted through an air/SF6 gas lens. Fig. 3 of Biamino et al. (2015).

and convergence effects, both as effects modifying perturbation growth and as the cause of perturbation growth. Two papers focusing on turbulent transport and mesh convergence in 2D RM in a cylindrical geometry highlight the different physics expected in 2D inhomogeneous mixing (Melvin et al., 2013, 2014).

Other interesting papers relevant to strongly shocked cylindrically convergent implosions include Marinak et al. (2001), Parker et al. (2004), Fincke et al. (2004, 2005), Perez et al. (2009), Ramis (2013), Luo et al. (2019), Ding et al. (2021), Sun et al. (2021), Li et al. (2022a), Piriz et al. (2022), and Zhang et al. (2024). Some relevant numerical simulations can be found in López Ortega et al. (2015), Tian et al. (2006), Zheng et al. (2008a), Zhai et al. (2019), Zeng et al. (2020), Yan et al. (2022), and Boffetta and Musacchio (2022).

16.4 Spherical geometry

To describe RT and RM instabilities and mixing in stratified spherical shells, Mikaelian (1990b,d) extended Plesset's analysis (Plesset, 1954) to include an arbitrary number of incompressible shells. Revisiting the dispersion relation for the RTI growth rate, he confirmed the soundness of Plesset's result and pointed out its difference from that of Binnie (1953) (Mikaelian, 1987). For the more complex matter of viscous fluids in planar geometry, some well-known classical work includes Bellman and Pennington (1954), Hide (1955b), and Reid (1961), among others.

Building on Layzer's (1955) theory, Goncharov and Li (2005) studied the effects of both the spherical convergence and temporal density variations on the nonlinear bubble evolution of single-mode, classical RTI. In a related study, Clark and Tabak (2005b) also obtained a generalization of the Layzer (1955) nonlinear bubble rise rate for a self-similar spherically converging flow of the type studied by Kidder (1976). Clark and Tabak (2005a) further extended the model to describe bubble growth during the deceleration phase of a spherical implosion and the growth of spikes during both the acceleration and deceleration phases. The authors found differences in the nonlinear growth rates when compared with planar results for both bubbles and spikes. The model is limited in assuming a self-similar background implosion history as well as only addressing growth from a single-mode perturbation of an already nonlinearly large amplitude. Clark and Tabak (2006) remedied these limitations and found that convergence in the nonlinear regime is not substantially modified from the multi-mode predictions of Haan's original model (See also related papers by Liu et al. (2015b, 2017) and Zhang et al. (2017)).

Ramshaw (1999) proposed both a linear and nonlinear model in spherical geometry for predicting the time evolution of an incompressible spherical fluid mixing layer subjected to an arbitrary time-dependent implosion history. This model is analogous to his unstable planar fluid interface representation (Ramshaw, 1998) and is based on a heuristic expression for the kinetic energy of the system of a linearly perturbed interface, but with a dynamically renormalized effective wavelength.

In a study of supernova dynamics, Duffell (2016) presented a method for approximating the multidimensional effects of RTI as a modification of the 1D spherical hydrodynamic equations. Following Gull and Longair (1973) and Shakura and Sunyaev (1973), an additional scalar quantity, representing the magnitude of turbulent RT fluctuations, is evolved along with the usual hydrodynamic variables.

Youngs and Williams (2008) performed accurate 3D TURMOIL hydro-code numerical mixing simulations in a simple spherical implosion, a simplified version of an inertial confinement fusion implosion. They found that the code has good low Mach number behavior, a key requirement for the calculation of turbulent flow, and the kinetic energy dissipation could be accurately quantified.

Lombardini et al. (2014a) performed large-eddy simulations of turbulent mixing at a perturbed, spherical interface that was impacted by a spherically imploding shock wave with Mach number 1.2. The authors built an initially isotropic perturbation at a spherical density interface of radius R_0 (i.e., no preferred directions when moving along the spherical surface). The heavy-fluid mass fraction field initially takes the form of a hyperbolic tangent profile with characteristic thickness $0.002R_0$.

Figure 16.15 shows the temporal evolution of the isosurfaces of the heavy-fluid mass fraction. The late-time light–heavy and heavy–light mixing layers resemble each other for the air-to-SF_6 and SF_6-to-air configurations. The authors also note that a phase reversal, characteristic of heavy-to-light interactions, has occurred: the initial spikes (which refer to fingering structures of heavy fluid penetrating light fluid) become bubbles (i.e., pockets of light fluid penetrating heavy fluid), and initial bubbles turn into spikes.

Lombardini et al. (2014b) also obtained a significant number of turbulence statistics, including the mixedness and effective Atwood number, A_e, which offers a better understanding of the physics and provides data for engineering modeling. Here, the effective Atwood number is given by (e.g., Cook et al., 2004)

$$\mathcal{A}_e = \left\langle \frac{\sqrt{\overline{\rho'^2}}}{\bar{\rho}} \right\rangle, \tag{16.28}$$

a parameter also known as the r.m.s. density fluctuation ratio (Grégoire et al., 2005). While the early behaviors of Θ and A_e are different, due to the phase reversals mentioned above, both of them ultimately reach the late-time values 0.8 and 0.6 for the light-heavy and heavy–light configurations (Fig. 16.16).

Flaig et al. (2018) compared results from four different codes simulating the growth of a single spherical harmonic mode perturbation on the inner surface of a dense collapsing shell. The problem can be considered as a highly idealized representation of an inertial confinement fusion implosion. The initial interface perturbations consists of a single spherical harmonic mode with mode number l and amplitude a_0,

$$\delta R(\theta) = a_0 Y_{l0}(\theta). \tag{16.29}$$

Heidt et al. (2021) additionally investigated the effect of initial amplitude and convergence ratio on instability development and deposited fluctuating kinetic energy in the single-mode RMI in spherical implosions.

For a spherical geometry, El Rafei and Thornber (2020, 2024) also studied initial perturbations of the form of Eq. 9.2, for both a narrow-band (NB) case where the distribution of energy is $P(k) = C_p k^0$ and a broad-band (BB) case with $P(k) = C_p k^{-2}$. To compare the behavior of the NB and BB layers, the same initial standard deviation was adopted for both cases. The interface between the light and heavy fluids is hit by the incident shock wave at 0.5 ns, which initiates the interface's motion. The shock wave reflects at the core of the domain and passes through the interface again around 1.9 ns (first reshock) and 2.4 ns (second reshock), before the interface stagnates at 2.5 ns.

The integral width of the NB and BB layers begin with the same amplitude and when the incident shock wave hits the interface, the NB layer grows at a faster rate than the BB layer (Fig. 16.17). The longer wavelength perturbations that exist in the BB case grow slowly in

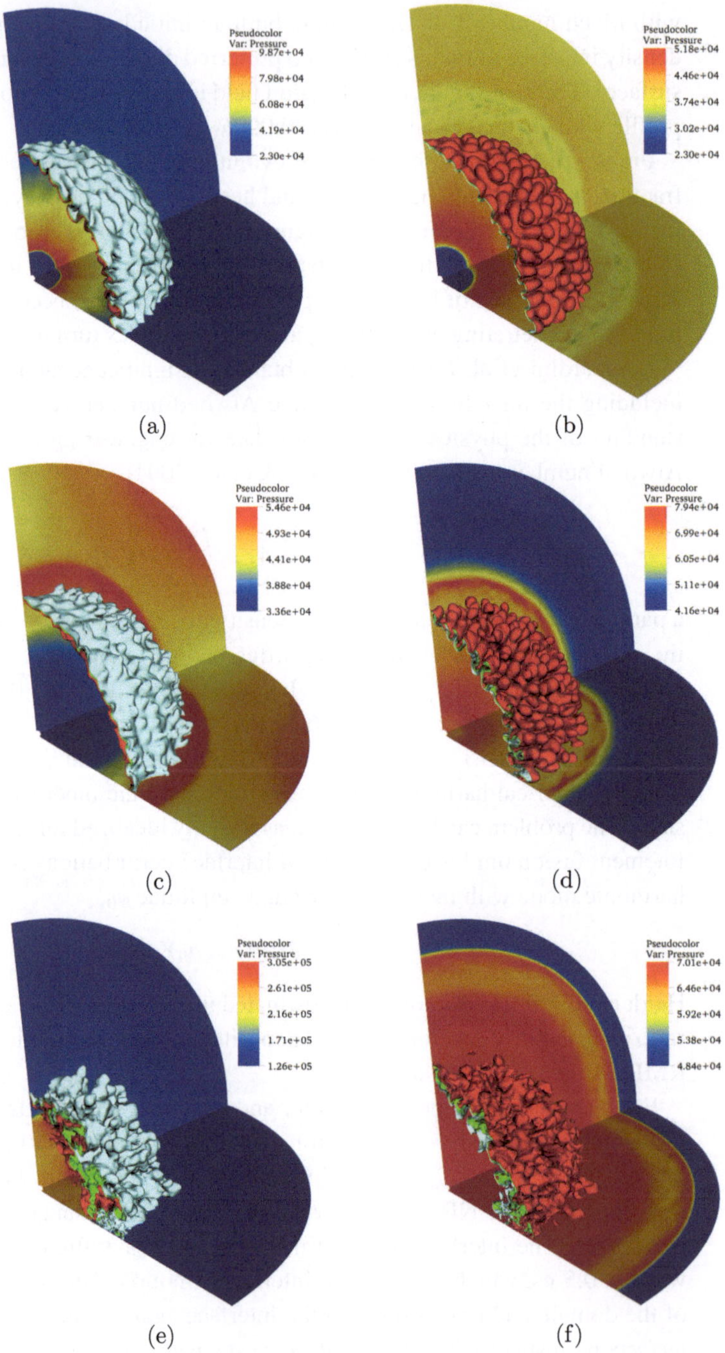

Fig. 16.15 Isosurfaces of the heavy-fluid mass fraction and contours of pressure, at three different times, nondimensionalized by the reshock time: 0.5 (top), 1.5 (middle), and 3 (bottom). The left and right panels display light–heavy and heavy–light configurations: air $\longrightarrow$ SF_6 (a, c, e) and SF_6 $\longrightarrow$ air (b, d, f), respectively. Fig. 5 of Lombardini et al. (2014a).

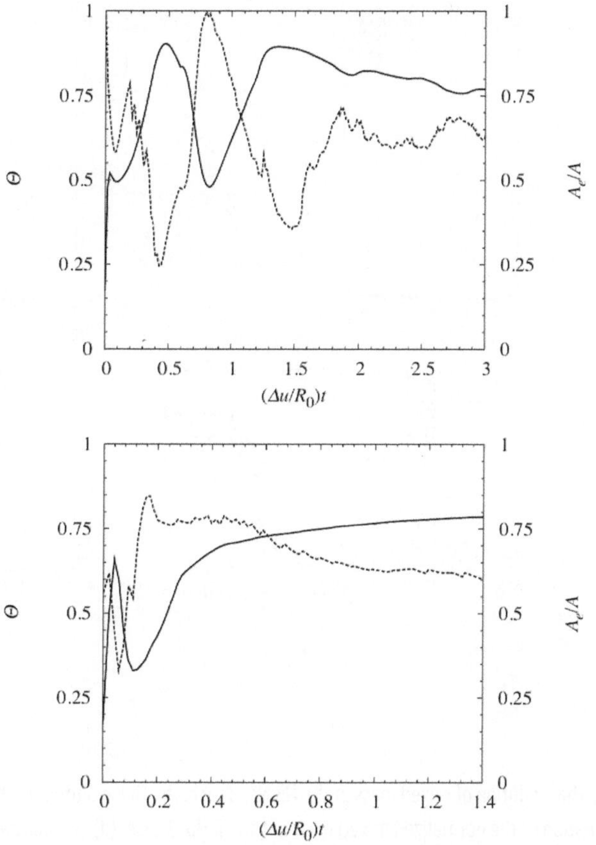

Fig. 16.16 Ratio of mixed to entrained fluid Θ (solid line) and effective Atwood number ratio (dashed) versus t: (a) air ⟶ SF_6 and (b) SF_6 ⟶ air. The nondimensionalized time of the first reshock is 0.6. Fig. 2 of Lombardini et al. (2014b).

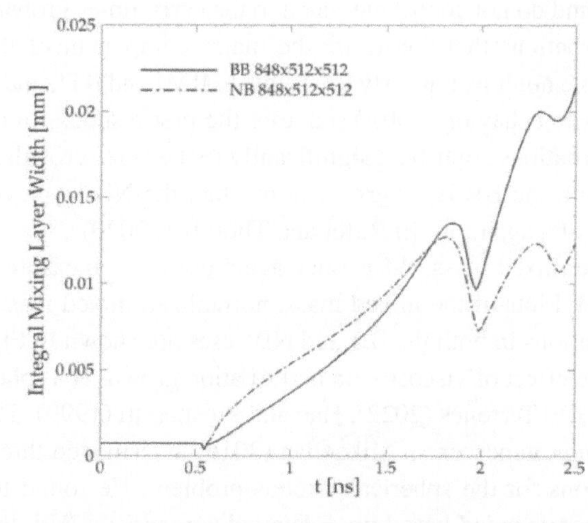

Fig. 16.17 Simulated integral mix layer width for narrowband and broadband cases at the highest grid resolution in 3D spherical implosions. Fig. 6 of El Rafei and Thornber (2020), with the permission of AIP Publishing.

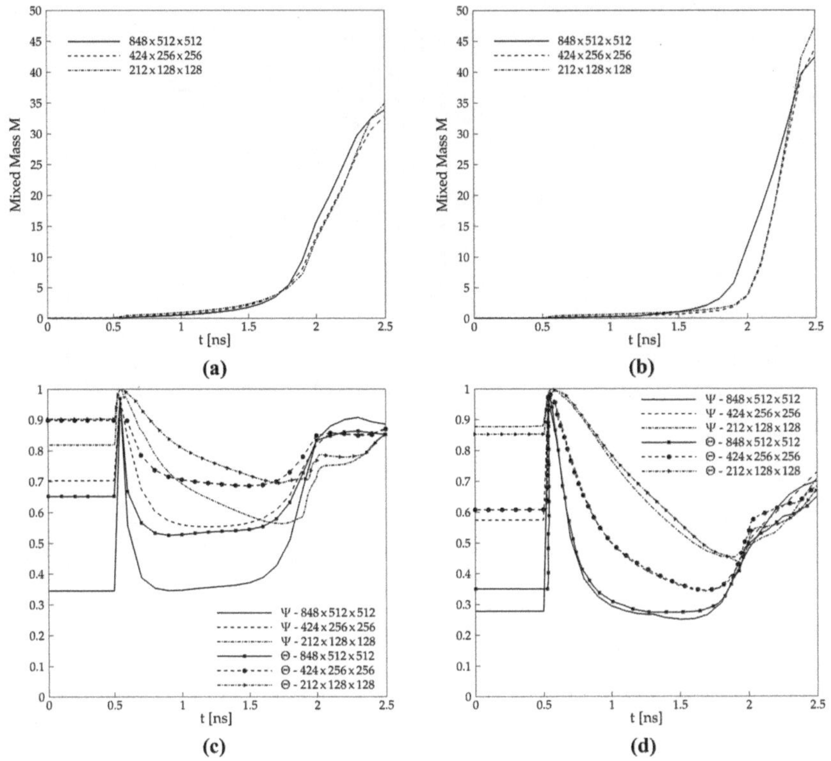

Fig. 16.18 (a) The evolution of mixed mass in the NB No. 3 case. (b) The evolution of the mixed mass in the BB No. 3 case. (c) The variation of the normalized mixed mass in the NB No. 3 case. (d) The normalized mixed mass in the BB No. 3 case. Fig. 12 of El Rafei and Thornber (2020), with the permission of AIP Publishing.

time and do not contribute much to the early-times growth. However, the short-wavelength perturbations that constitute the major component of the NB layer grow quickly. They become nonlinear at early times due to RMI and RTI, and are saturated before reshock. The opposite behavior is observed after the first reshock. In the BB case, the long-wavelength perturbations contribute significantly to the layer growth at late times after the reshock. As a result, the BB layer grows faster than the NB layer, reaching higher amplitudes at the point of stagnation (El Rafei and Thornber, 2020).

The mixed mass, $\mathcal{M}$ is once again used as a measure of how effectively two fluids are mixed. Plots of the mixed mass, normalized mixed mass, and mixedness at different grid resolutions in both the BB and NB cases are shown in Fig. 16.18.

The effect of viscosity on perturbation growth at a spherical interface was considered by Oren and Terrones (2022), Hao and Prosperetti (1999). The latest paper notes the relevance to sonoluminescence. Mikaelian (2016) investigated three successively simpler dispersion relations for the spherical viscous problem. He found that the simplest form, an ansatz, is recommended based on a detailed comparison with the exact results of Chandrasekhar (1955) as well as those of Terrones and Carrara (2015).

16.5 Cylindrical versus spherical

Joggerst et al. (2014) present simulations of the implosion of a dense shell in 2D spherical and cylindrical geometries, performed with four different compressible, Eulerian codes: RAGE, FLASH, CASTRO, and PPM. The authors find that perturbation growth is not dependent on the choice of grid geometry (Cartesian or polar meshes), but is influenced by the numerical choices employed in the different codes (Figs. 16.19–16.20).

More recently, Fu et al. (2020) performed DNS of RMI in spherical and cylindrical converging geometries with a Mach number of approximately 1.5. The shock wave converges from the heavy fluid (SF_6) into the light fluid (N_2). Figure 16.21(a) shows the temporal development of the main shock wave positions, which are determined by the absolute maximum of velocity divergence. As expected, the amplitudes of interface disturbances are first flattened by the incident shock waves, then grow after the phase inversion.

Figure 16.21(b) shows the evolution of the mixing layer widths for cylindrical and spherical converging RM instabilities. Additionally, temporal variations of inner and outer radii are included (Fu et al., 2020). Because the spherical interface converges in three dimensions to a center point, while the cylindrical interface converges only in two dimensions to a centerline, the pressure and light fluid density at the spherical interface are greater than those at the cylindrical interface, which restrains the outer radius of the spherical interface from converging to the center point.

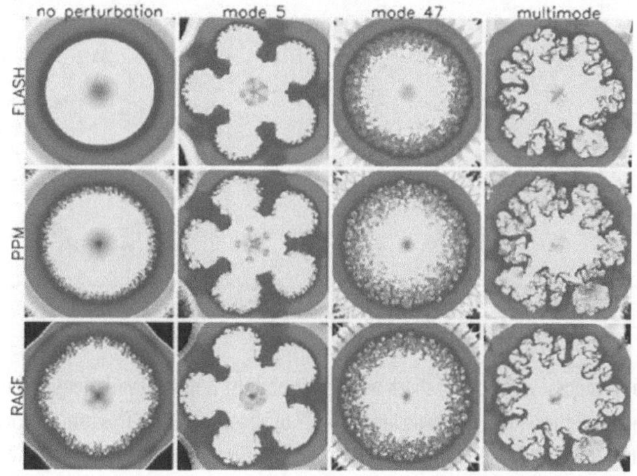

Fig. 16.19 Snapshots of fluid density in the cylindrical calculations at 2.5 s, near maximum compression, for FLASH (top row), PPM (middle row), and RAGE (bottom row). The inner cylindrical interface is preserved for the FLASH simulation (which uses a cylindrical mesh), but small perturbations seeded by the Cartesian grid have grown in the PPM and RAGE calculations, as expected. The unperturbed case, shown in the first column, shows the greatest difference between the Cartesian- and cylindrical-mesh simulations. The simulations for which perturbations were imposed appear similar between codes overall, with differences for a particular perturbation varying more greatly between Cartesian codes (PPM and RAGE) than between the Cartesian codes and the simulations performed on a cylindrical mesh (FLASH). Fig. 5 of Joggerst et al. (2014).

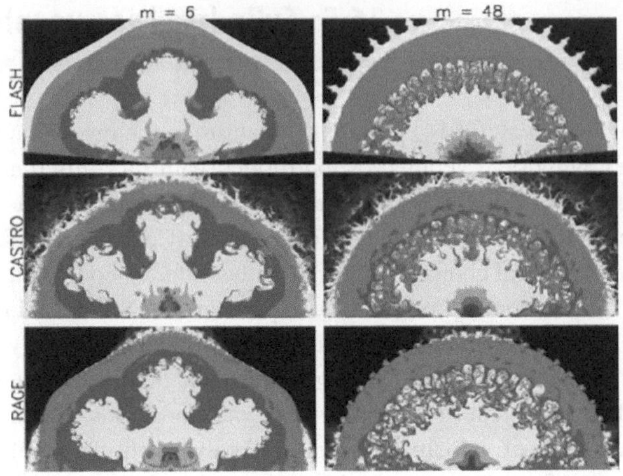

Fig. 16.20 Snapshots of fluid density at 2.5 s in spherical simulations, after maximum compression, for FLASH (top row), CASTRO (middle row), and RAGE (bottom row). The growth of higher-mode instabilities in the low-mode perturbation simulations (left panel) differs between the Cartesian codes RAGE and CASTRO, with CASTRO calculations showing more dramatic growth of secondary perturbations and a spectrum tilted toward lower mode numbers than RAGE. Fig. 10 of Joggerst et al. (2015).

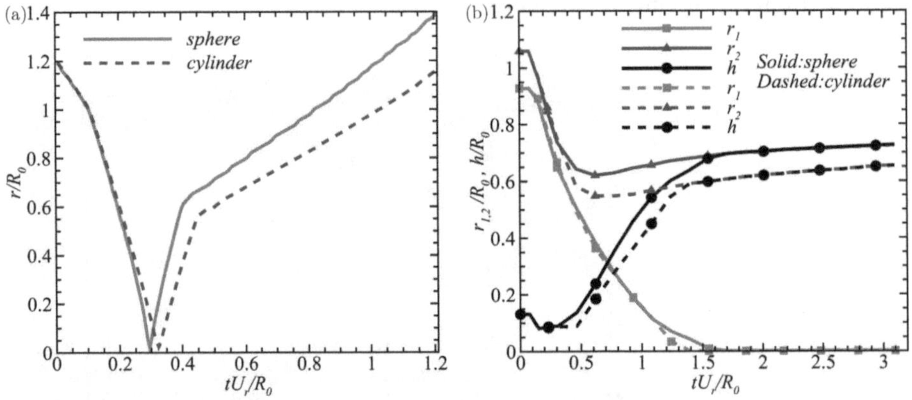

Fig. 16.21 (a) Temporal variations of shock wave positions. (b) Temporal variations of the spherical/cylindrical inner and outer radii and the widths of mixing layers. Fig. 7 of Fu et al. (2020). This article is distributed under a Creative Commons Attribution (CC BY) license.

For both cylindrical and spherical configurations, Figs. 16.22–16.23 demonstrate that small scale features are fairly quickly generated from large-scale structures upon the passage of the reshock waves through the material interface. The reshock-induced enhanced mixing is of course analogous to that previous discussed for the planar geometry. Figures 16.24–16.25 display the density fields, viewed from three perspectives, of the middle sections of the cylindrical and spherical converging RM instability at the end of the simulations.

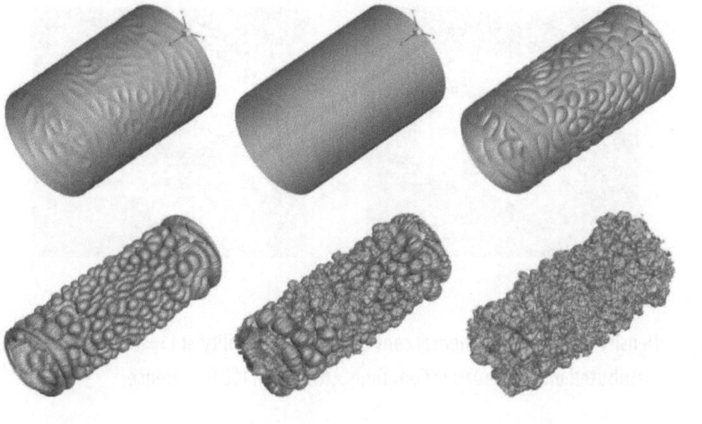

Fig. 16.22 Isosurfaces of $Y_{N_2} = 0.1$ for the cylindrical converging RM instability at t = 0 ms, 0.01 ms, 0.02 ms, 0.04 ms, 0.08 ms, and 0.2 ms. Fig. 4 of Fu et al. (2020). This article is distributed under a Creative Commons Attribution (CC BY) license.

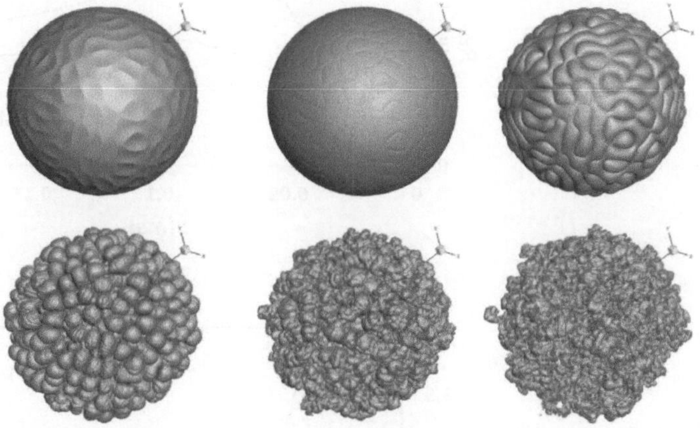

Fig. 16.23 Isosurfaces of $Y_{N_2} = 0.1$ for the spherical converging RM instability at t = 0 ms, 0.01 ms, 0.02 ms, 0.04 ms, 0.08 ms, and 0.2 ms. Fig. 3 of Fu et al. (2020). This article is distributed under a Creative Commons Attribution (CC BY) license.

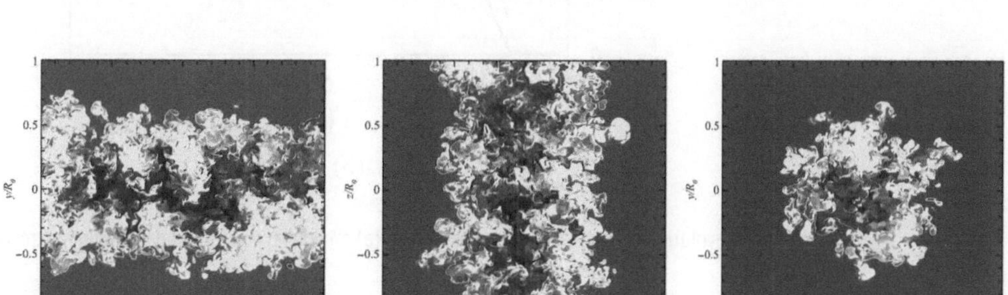

Fig. 16.24 Density fields for the cylindrical converging RM instability at t = 0.2 ms. Fig. 6 of Fu et al. (2020). This article is distributed under a Creative Commons Attribution (CC BY) license.

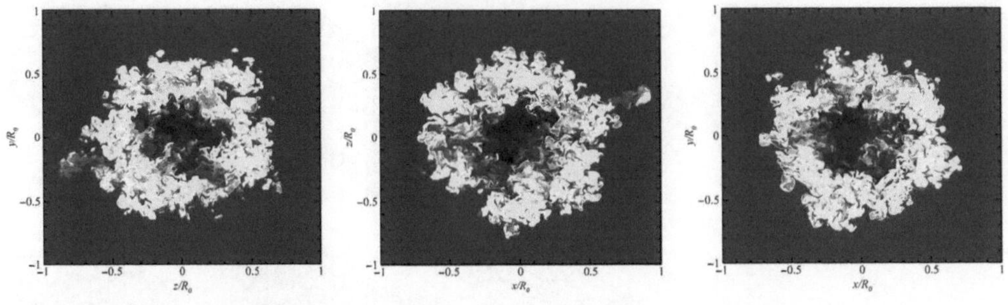

Fig. 16.25 Density fields for the spherical converging RM instability at t = 0.2 ms. Fig. 5 of Fu et al. (2020).

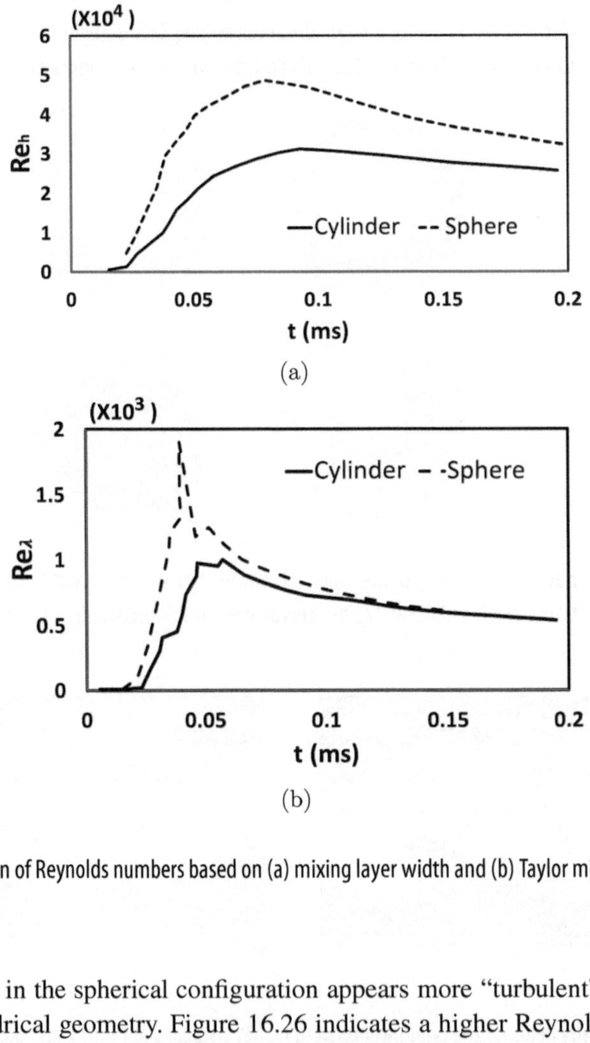

(b)

Fig. 16.26 The temporal evolution of Reynolds numbers based on (a) mixing layer width and (b) Taylor microscale. Fig. 10 of Li et al. (2021b).

The flow field in the spherical configuration appears more "turbulent" than its counterpart in the cylindrical geometry. Figure 16.26 indicates a higher Reynolds number for the former case than latter one, based on their radial mixing layer widths (Re_h) and Taylor microscales (Re_λ).

In addition, the analysis of the turbulence kinetic energy transport equation shows that the mechanisms of turbulent production and dissipation in the mixing layers of the spherical and cylindrical converging RM instabilities are the same, and the turbulent diffusion and dissipation are the main mechanisms of turbulence decaying in the later stage (Fu et al., 2020).

16.6 Divergent versus convergent

One of the most important physical quantities in studying single-mode RMI in a cylindrical geometry is the growth rate of the fingers at an unstable material interface in the nonlinear regime. Zhang and Graham (1998) and Graham and Zhang (2000) studied four classes of RM-unstable interfacial fluid mixing systems in curved geometries: a shock wave exploding (imploding) from light fluid to heavy and a shock exploding (imploding) from heavy fluid to light fluid. The presence of the reshock, which is unavoidable for the cylindrical geometry, significantly complicated the dynamics of the RMI. Zhang and Graham (1997) derived the scaling laws for a RM-unstable interface driven by strong shocks on the overall growth rate, the amplitudes, and the shape of the unstable interface. Numerical simulations confirmed these scaling laws are indeed satisfied once the Mach number is larger than 10. The authors suggested that these useful relationships might be applicable to spherical geometries as well.

Wang et al. (2021) performed experiments on a divergent RTI at well-controlled single-mode cylindrical interfaces between air and viscous liquid. Figure 16.27 illustrates the method of prefabricating a single-mode perturbation at the internal surface of a cylindrical

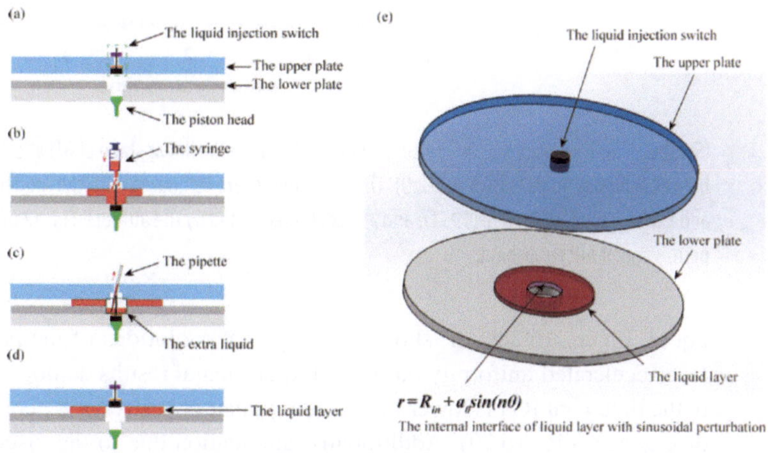

Fig. 16.27 (a)–(d) Procedure for prefabricating a single-mode perturbation at the internal surface of a cylindrical liquid. (e) Sketch of the prefabrication of a single-mode perturbation at the interface. Fig. 3 of Wang et al. (2021), with the permission of AIP Publishing.

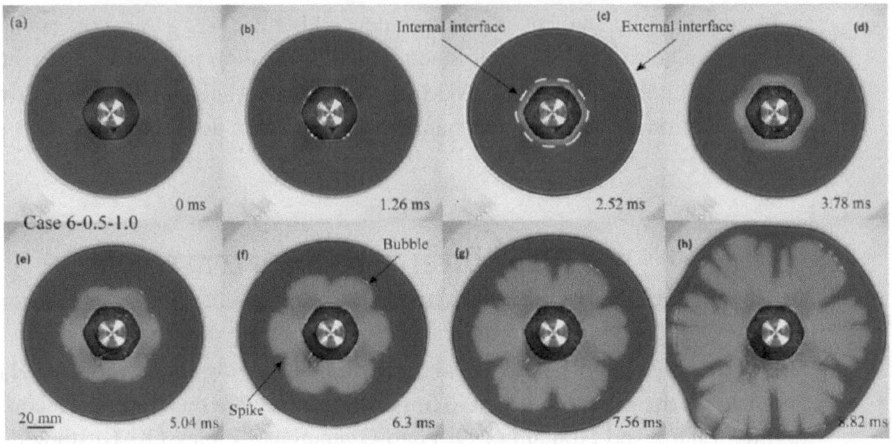

Fig. 16.28 Temporal evolution of a sinusoidal perturbation at an air/fluid interface subjected to a diverging impact (case 6-0.5-1.0). Fig. 8 of Wang et al. (2021), with the permission of AIP Publishing.

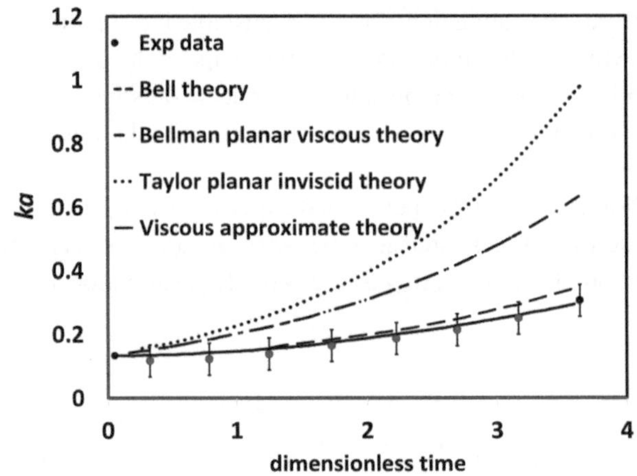

Fig. 16.29 (Symbols) Time variations of the perturbation amplitude in the linear stage of RTI at cylindrically divergent interfaces between viscous fluids (case 6-1.0-2.0). The experiments are compared with the models from Taylor (1950), Bellman and Pennington (1954), Bell (1951), and a model constructed by the authors. Fig. 12 of Wang et al. (2021), with the permission of AIP Publishing.

liquid. Figure 16.28(a–h) shows eight typical evolutions of the perturbed air/liquid interface accelerated uniformly outward. Experimental results demonstrate that the growth rate in the divergent RTI is lower than that in planar or convergent geometries due to geometric divergence (Fig. 16.29). Additionally, attenuation due to the viscosity is also reduced by divergent BP effects compared to the planar case. The experimental data can be reproduced well with the Bell theory and its modified version, which account for viscosity.

Li et al. (2020a) carried out a set of experiments for five single-mode interfaces with different amplitudes and wavelengths, with a_0/λ ranging from 0.058 to 0.117 (Fig. 16.30).

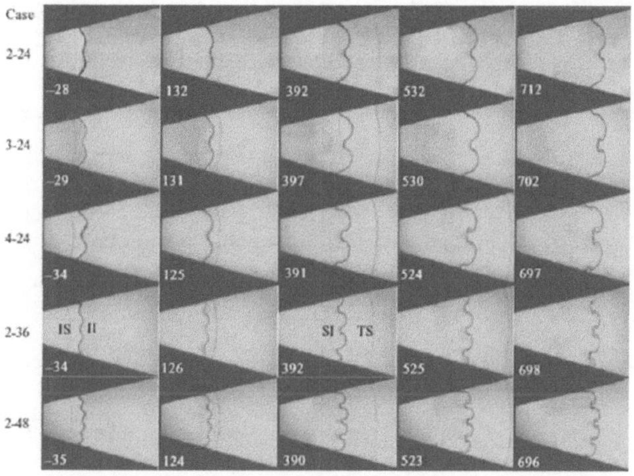

Fig. 16.30 Evolution of a single-mode air–SF_6 interface subjected to a cylindrically exploding shock for several initial conditions. The numbers have units μs. Fig. 3 of Li et al. (2020a).

The Mach and Atwood number numbers are approximately 1.25–1.28 and 0.58–0.60, respectively. Due to geometric divergence, the growth rate was found to be far lower than its counterpart in the planar or convergent settings. In particular, the nonlinearity is much weaker than that in planar or convergent RMI and has a negligible influence on the overall amplitude growth. Thus, the linear theory of Bell, which represents the effects of the geometric divergence, along with a term representing the RT stabilization caused by flow deceleration, can be utilized to match the experimental results from early to late stages (Fig. 16.31). See Zhang et al. (2023b) for a follow-up study.

Due to the effects discussed above, the divergent RMI cannot be treated simply as the opposite of convergent RMI. That is, it cannot be understood simply by replacing opposite geometric effects in the convergent RMI because the flow features of the former, as well as the behaviors of nonlinearity and compressibility, may be completely different from the latter (e.g., Li et al., 2020a).

The same cautious approach is also required for RTI applications, where the direction of acceleration has been shown to have a significant influence on the growth of the mixing layer. For example, Yu and Livescu (2008) found that the RTI grows faster in a 3D cylindrically convergent geometry (gravity acting inward) than in a Cartesian coordinate system but slower in the cylindrically divergent (gravity acting outward) configuration.

For inviscid incompressible fluids, the perturbation of the cylindrical RTI in the linear regime also exhibits exponential growth (Wang et al., 2013b; Guo et al., 2018), with a growth factor

$$\gamma = \sqrt{Ang/r_0}, \tag{16.30}$$

in either the convergent or divergent cases, where n is the number of perturbation waves and r_0 the radius of the unperturbed interface.

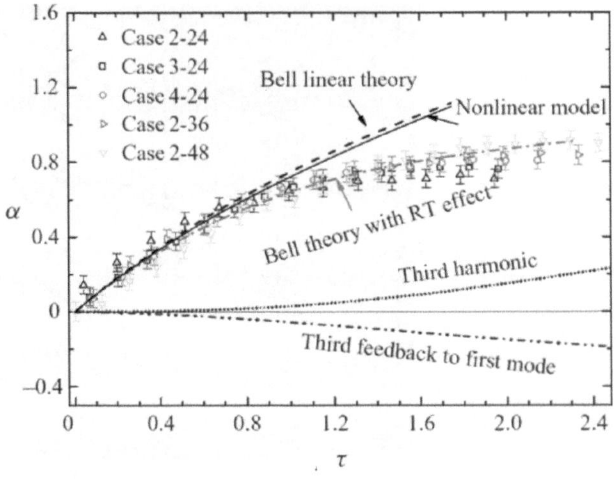

Fig. 16.31 Normalized variations of the overall perturbation amplitude. The dashed (dash–dot) line denotes the linear theory of Bell (1951) without (with) the RT effect, and the solid line denotes the nonlinear theory of Wang et al. (2015). The dotted line refers to the growth of the third harmonic and the dash–dot–dot line to the third-order feedback to the basic mode. Fig. 4 of Li et al. (2020a).

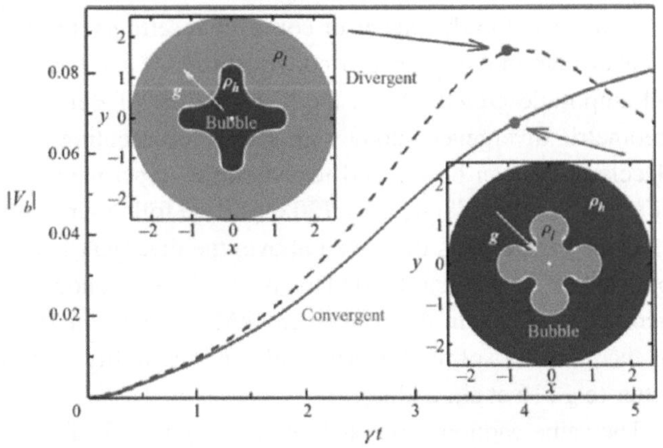

Fig. 16.32 Sketch of single-mode convergent and divergent configurations. Also shown is the bubble velocity growth of cylindrical RTI at Atwood number 0.6 obtained by DNS. Fig. 4 of Zhao et al. (2020b).

Figure 16.32 provides a visualization of the system's geometry, along with a set of DNS results for Atwood number 0.6 (also computed, but not shown in this figure, $\mathcal{A} = 0.3$). Following Layzer's treatment (Layzer, 1955), Zhao et al. (2020b) derived an analytical model of nonlinear evolution for 2D single-mode RTI in cylindrical geometry for inviscid, irrotational, and incompressible fluids.

Unlike in a planar geometry, the bubble velocity $v_b = \dot{a}_0$ cannot saturate to a constant value in a cylindrical geometry. It is especially interesting that $\dot{a}_0/\sqrt{a_0 + r_0}$ approaches a constant as $t \to \infty$,

$$\frac{\dot{a}_0}{\sqrt{a_0+r_0}} \to \left(\frac{4n(3n+1)^2\mathcal{A}g}{54(\mathcal{A}+1)n^4+9(15\mathcal{A}+17)n^3+3(35\mathcal{A}+53)n^2+24(\mathcal{A}+3)n+12}\right)^{1/2}, \tag{16.31}$$

$$\frac{\dot{a}_0}{\sqrt{a_0+r_0}} \to -\left(\frac{4n(3n-1)^2\mathcal{A}g}{54(\mathcal{A}+1)n^4-9(15\mathcal{A}+17)n^3+3(35\mathcal{A}+53)n^2-24(\mathcal{A}+3)n+12}\right)^{1/2}, \tag{16.32}$$

for both the convergent and divergent geometries (Zhao et al., 2020b).

The full bubble evolution can reproduce the earlier exponential growth, Eq. 16.30, as well as the nonlinear regime saturating with a quadratic-in-time ($r_0 + a_0 \sim t^2$) growth (Fig. 16.33).

When the cylindrical RTI moves from the linear into the nonlinear regime ($\gamma t \approx 4.0$), the bubble evolution changes from exponential to quadratic growth. We could call such a transition "nonlinear saturation" for the cylindrical RTI; it means that the bubble acceleration, $|V_b|/\sqrt{r_0 + a_0}$, tends toward a constant value (Fig. 16.34).

For multi-mode initial perturbations, Zhao et al. (2021) studied the nonlinear evolution of cylindrical RT turbulent mixing layers, driven by an acceleration from heavy to light fluid. Here, two classes of RT unstable systems in cylindrical geometries exist (Fig. 16.35),

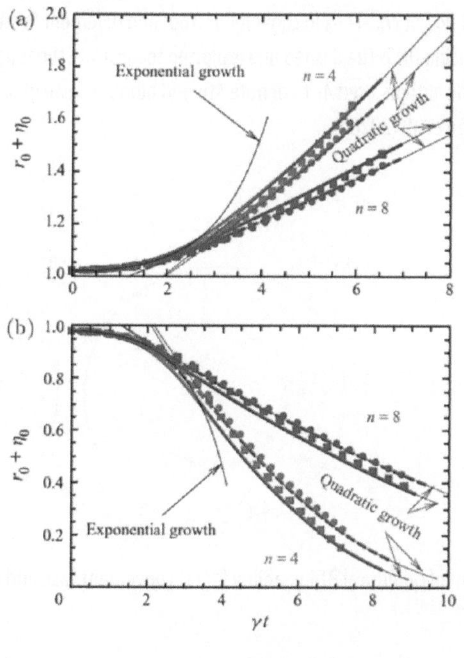

Fig. 16.33 Radial position of the bubble tip for the convergent (a) and divergent (b) cases at Atwood number 0.3 (solid lines and squares) and 0.6 (dashed lines and circles) obtained by the analytical models (solid and dashed lines) and DNS (squares and circles). For comparison, the profiles are also plotted for the exponential growth (solid thin line) in the linear regime given by Wang et al. (2013b), and for quadratic-in-time growth (thin lines) with $r_0 + a_0 \sim 0.5a_bt^2$ in the nonlinear regime. Fig. 5 of Zhao et al. (2020b).

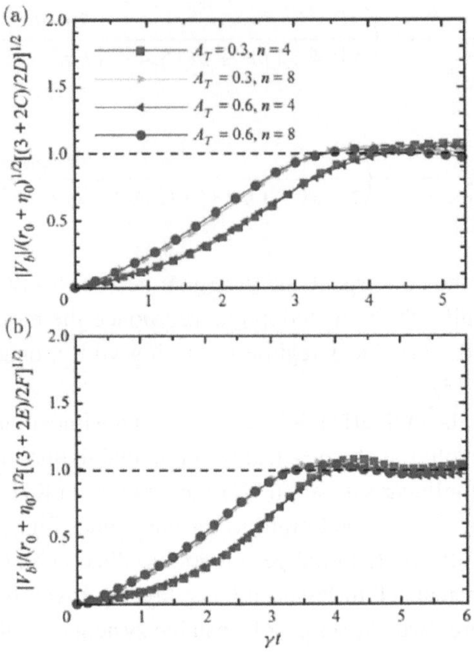

Fig. 16.34 Normalized $|v_b|/\sqrt{r_0 + a_0}$ in convergent (top) and divergent (bottom) cases at Atwood number 0.3 and 0.6, calculated using DNS. The dashed line is plotted to illustrate the tendency toward the asymptotic analytical solution. Note that the authors used A_T to denote Atwood number, instead of the $\mathcal{A}$ extensively employed in this manuscript. Fig. 6 of Zhao et al. (2020b).

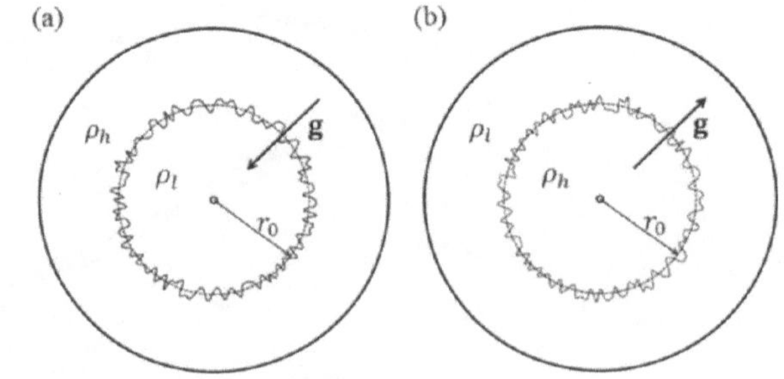

Fig. 16.35 The diagram of cylindrical RT instability for (a) convergent case and (b) divergent case. Fig. 1 of Zhao et al. (2021).

one comprises the heavy fluid surrounding the light fluid (convergent case) and the other comprises the light fluid surrounding the heavy fluid (divergent case). Six 2D simulations were performed for convergent (a)–(c) and divergent (e)–(g) cases, respectively. Another two calculations were carried out for 3D convergent (d) and divergent (h) cases. The dimensionless parameters, the Reynolds and Froude numbers, are given by

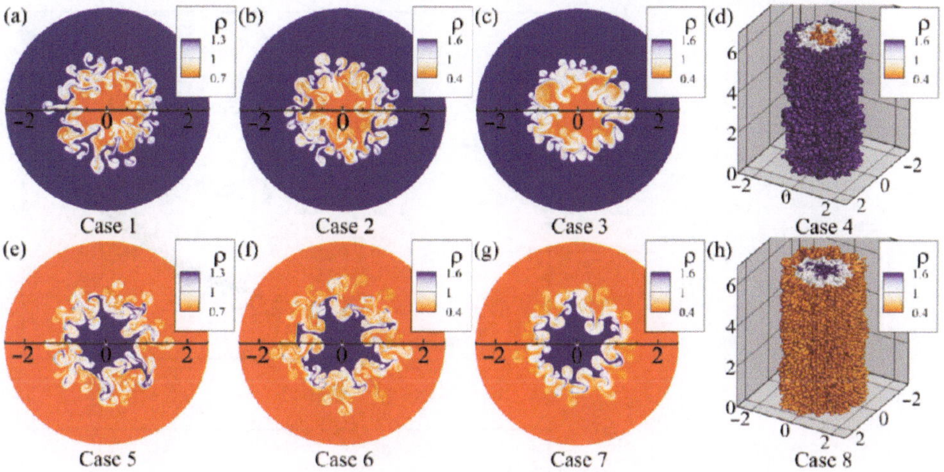

Fig. 16.36 Snapshots of instantaneous density fields for 2D convergent (1)–(3) and divergent (5)–(7) cases at the rescaled time of 3. Isosurfaces of the density for 3D convergent (4) and divergent (8) cases at the same rescaled time. Fig. 2 of Zhao et al. (2021).

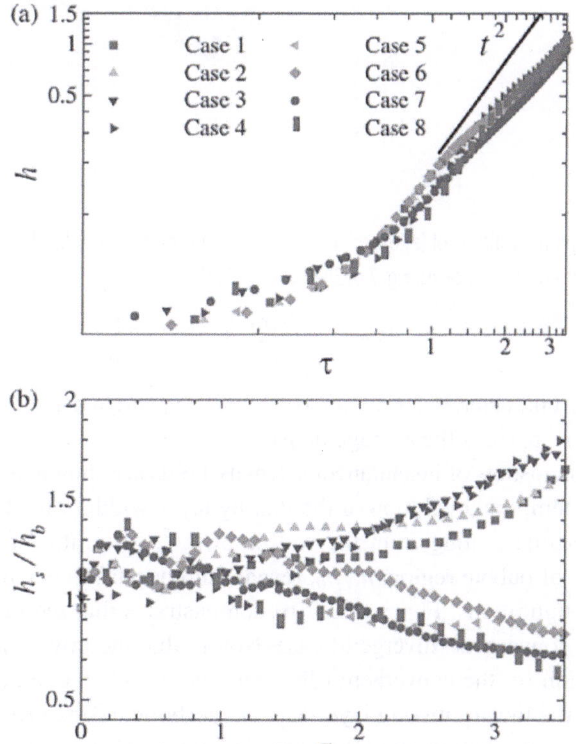

Fig. 16.37 Snapshots of instantaneous density fields for convergent (1)–(4) and divergent (5)–(8) cases at the temporal evolution of (a) the width of mixing layer h and (b) the ratio of width of spike region to that of bubble region, h_s/h_b. Fig. 6 of Zhao et al. (2021).

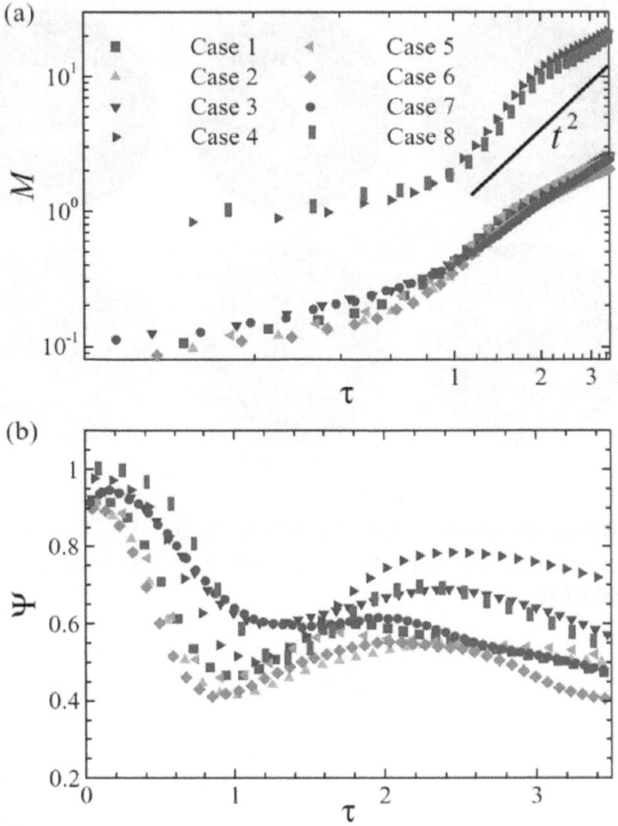

Fig. 16.38 Temporal evolution of (a) the mixed mass, $\mathcal{M}$, and (b) the normalized mixed mass, Ψ, for convergent (1)–(4) and divergent (5)–(8) cases. Fig. 7 of Zhao et al. (2021).

$$Re = \frac{\rho_I U_I r_0}{\mu_I}, \quad Fr = \frac{U_I^2}{r_0 g}, \quad Re_H = \frac{\rho_I \dot{h} h}{\mu_I}. \tag{16.33}$$

The characteristic velocity is $U_I = \sqrt{p_I/\rho_I}$, where p_I is the initial pressure and $\rho_I = (\rho_h + \rho_l)/2$ is the average density.

Snapshots of instantaneous density fields are shown in Fig. 16.36. Figure 16.37(a) shows the temporal evolution of the mixing layer width, h in all cases, there is a departure of the late-time scaling from the well-known t^2. The ratio of the width of the spike region to that of bubble region, h_s/h_b, reveals another difference between the planar and cylindrical configurations. Figure 16.37(b) demonstrates that the ratio h_s/h_b increases (decreases) in the convergent (divergent) case. Noting that the inward mixing layer is the spike (bubble) region for the convergent (divergent) cases; Zhao et al. (2021) suggested a cosine growth for the inward mixing layer and a hyperbolic cosine growth for the outward mixing layer.

The mixed mass and the normalized mixed mass are shown in Fig. 16.38. The scaling of $\mathcal{M}$ departs from t^2 in the planar geometry. Furthermore, Ψ can discriminate results based on the direction of the external acceleration in a cylindrical geometry. It should be noted, however, these curves have not saturated by the end of the calculations.

17 Magnetohydrodynamic fluid instabilities

17.1 Magnetohydrodynamics: Governing models

Magnetic fields can affect the flow of a fluid in ways that fundamentally change the fluid instabilities. Magnetohydrodynamics (MHD) is a fluid-mechanical formulation in which the fluid medium is assumed to be electrically conductive, so that it can interact with magnetic fields. In particular, the electromagnetic Lorentz force appears as a body force which acts on a fluid parcel, in addition to body forces such as gravity, while the magnetic field itself can change according to Faraday's law of induction, due to the motion of the fluid.

In deriving any of the usual formulations of MHD, several essential assumptions are made about the medium, which can be viewed as a continuum of separated positive and negative electrical charges. First, the fluid is electrically quasi-neutral. In a given region of the fluid, it may transpire that a group of positive charges will separate from a group of negative charges, so that one part of the region has a net positive charge and the other has a net negative charge. This separation causes an electrical field that will tend to force the charges back toward each other, so that large charge separations cannot be sustained in the plasma. Despite this, however, local thermal effects may cause charge imbalances on smaller scales which are characterized by the *Debye* length. By assuming a quasi-neutral plasma with some appropriate flow length scale, we are implicitly assuming the Debye length scale is sufficiently small, relative to the flow scales, to be neglected. We assume the same with respect to the Larmor radius, which is the radius of the circular trajectory a charged particle travels in a uniform magnetic field. Finally, if we treat the electrons and ions (which are really separate species) as a single unified fluid, we arrive at MHD (Goedbloed and Keppens, 2004; Goedbloed et al., 2010; Zhou et al., 2004).

There are several different formulations of MHD, but here we will focus primarily on *ideal* MHD. This is the most idealized model – it neglects viscosity, thermal conduction and electrical resistivity in the fluid. The resulting governing equations are:

$$\frac{\partial \rho}{\partial t} + \nabla \cdot (\rho \mathbf{u}) = 0, \tag{17.1}$$

$$\rho \left(\frac{\partial \mathbf{u}}{\partial t} + \mathbf{u} \cdot \nabla \mathbf{u} \right) + \nabla p - \mathbf{J} \times \mathbf{B} - \rho \mathbf{g} = 0, \tag{17.2}$$

$$\frac{\partial p}{\partial t} + \mathbf{u} \cdot \nabla p + \gamma p \nabla \cdot \mathbf{u} = 0, \tag{17.3}$$

$$\frac{\partial \mathbf{B}}{\partial t} - \nabla \times (\mathbf{u} \times \mathbf{B}) = 0, \tag{17.4}$$

$$\nabla \cdot \mathbf{B} = 0, \tag{17.5}$$

where Eq. 17.3 is obtained from an ideal gas equation of state with specific heat ratio γ, **J** is the electric current density, **B** is the magnetic field (formally, the flux density) and **g** is the gravitational field.

Equation 17.4 is obtained from the induction equation, coupled with Ohm's law for a perfect conductor. For a perfectly conducting fluid, electric field $\mathbf{E} = 0$ in the fluid's rest frame. The relativistic transformation of the electromagnetic fields to the laboratory frame then yields $\mathbf{E} = -\mathbf{u} \times \mathbf{B}$, where **u** is the Lorentz transformation velocity, equal to the fluid velocity. Inserting this expression into Faraday's law yields Eq. 17.4. Furthermore, taking the divergence of Eq. 17.4 gives $\partial \nabla \cdot \mathbf{B}/\partial t = 0$, rendering Equation 17.5 superfluous. If it holds at time zero, it must continue to hold for all later times.

The electric current density, **J**, is given by

$$\mathbf{J} = \frac{1}{\mu_0} \nabla \times \mathbf{B}, \tag{17.6}$$

where μ_0 is the magnetic permeability in a vacuum. This relation is obtained from Ampere's law (in the limit where relativistic effects are unimportant, allowing the displacement current to be neglected).

The key aspect to note about the ideal MHD equations is that the magnetic field influences the velocity field evolution in Eq. 17.2, and vice versa in Eq. 17.4; thus the magnetic field and velocity fields are nonlinearly coupled. This aspect leads to an important property of the magnetic field in ideal MHD: the field moves locally as if it were a line element with the same orientation embedded in the flow; the magnetic field lines are "frozen" into the fluid and simply follow the motion of the plasma. Field lines are carried along by fluid motion perpendicular to the field direction, whereas the field is left unaffected by fluid motion along the field line (since $\mathbf{u} \times \mathbf{B} = 0$ in Eq. 17.4).

The Lorentz force, taking the form $\mathbf{J} \times \mathbf{B}$ in Eq. 17.2, gives a force on the fluid perpendicular to both the magnetic field vector and the current vector. This force can be decomposed into two components:

$$\mathbf{J} \times \mathbf{B} = \left(\frac{\mathbf{B}}{\mu_0} \cdot \nabla \right) \mathbf{B} - \nabla \left(\frac{B^2}{2\mu_0} \right). \tag{17.7}$$

The first of these components is called the magnetic tension force, while the second is called the magnetic pressure force. A key parameter for measuring the importance of the magnetic field in an MHD fluid is the "plasma β," defined as the ratio of gas to magnetic pressure:

$$\beta = \frac{p}{B^2/(2\mu_0)}. \tag{17.8}$$

When $\beta \gg 1$, in general, the hydrodynamic forces dominate the system, while when $\beta \ll 1$, the magnetic forces dominate. It is important to note that because gravity scales differently with length than the other forces, even in a low β system, gravity can be important. The magnetic RT instability is just such a system.

When magnetic diffusion (η_D) is included, the magnetic field and the flow are no longer frozen in lockstep. Equation 17.4 then becomes

$$\frac{\partial \mathbf{B}}{\partial t} - \nabla \times (\mathbf{u} \times \mathbf{B} - \mu_0 \eta_D \mathbf{J}) = 0, \tag{17.9}$$

which causes diffusion of the magnetic field profile. Including diffusion in the induction equation also results in the addition of an Ohmic heating term of the form $\mu_0 \eta_D |\mathbf{J}|^2$ to the RHS of Eq. 17.3.

Since the diffusion term in Eq. 17.9 is in effect a second-order differential operator on **B**, it is more effective at smaller scales. The importance of diffusion at a given scale can be understood by calculating the magnetic Reynolds number (the ratio of the diffusion time to the advection time) given by

$$R_m = \frac{VL}{\eta_D}, \tag{17.10}$$

where V is a characteristic velocity of the system, and L is a characteristic length scale (e.g., Bera et al., 2022). In a collisional plasma, the magnitude of η_D is related to the degree to which the current is reduced as a result of the current carriers (normally electrons) colliding with other particles.

Using an ideal MHD system (no viscosity or diffusivity) and taking possibly the simplest situation of an isotropic, homogeneous plasma at rest, with density ρ, that is permeated by a uniform, unidirectional magnetic field in the absence of gravity, we can determine the wave spectra of this system. In this idealized system, there are three kinds of waves that exist, unlike (non-magnetic) hydrodynamics, which admit only sound waves. First, there is the Alfvén wave. These waves are incompressible and propagate along the magnetic field at phase speed $V_A = B_{x,0}/\sqrt{\rho}$. There are also two compressible waves, the slow- and fast-mode magneto-acoustic waves. More information on the MHD wave spectra can be found in Goedbloed and Keppens (2004), for example.

17.2 The magnetic Rayleigh–Taylor instability

Section 17.1 contained a discussion of the coupling of a magnetic field, through the framework of MHD, to the hydrodynamic body forces. This leads to an additional body force, namely the Lorentz force (e.g., Briard et al., 2022, 2024; Gréa and Briard, 2023). In this section, we will explore the effect of this additional force on the Rayleigh–Taylor instability.

The extension to MHD of the RT instability was first investigated by Kruskal and Schwarzschild (1954). Here, we will focus on cases where the magnetic field vector is perpendicular to the direction of gravity. The case with the magnetic field parallel to gravity, and other interesting examples, are presented in Chandrasekhar (1961).

17.2.1 Studying the magnetic Rayleigh–Taylor instability in an incompressible 2D model

Our main investigation will focus on a simplified setting of the linear stability of a discontinuous density inversion with a uniform, horizontal magnetic field under gravity in 2D. The initial conditions we study in this section are

$$\rho_0 = \begin{cases} \rho_1 & \text{if } z < 0 \\ \rho_2 & \text{if } z > 0 \end{cases}, \tag{17.11}$$

$$u_{x,0} = 0, \quad u_{z,0} = 0, \tag{17.12}$$

$$B_{x,0} = B_x, \quad B_{z,0} = 0, \tag{17.13}$$

$$p_0(z) = p_0(0) - \rho_0 g z. \tag{17.14}$$

We also assume the system is inviscid, ideal, and incompressible (i.e., we solve the evolution using Eqs. 17.1, 17.2, 17.4, 17.5, and $\nabla \cdot \mathbf{u} = 0$, the last of which removes the requirement to solve for the evolution of the gas pressure).

These assumptions and initial conditions result in the following linear equations for the perturbed quantities $\delta\rho$, $\delta\mathbf{u}$, and $\delta\mathbf{B}$:

$$\frac{\partial \delta\rho}{\partial t} = -\delta\mathbf{u} \cdot \nabla\rho_0, \tag{17.15}$$

$$\rho_0 \frac{\partial \delta\mathbf{u}}{\partial t} = -\nabla\delta p + (\nabla \times \delta\mathbf{B}) \times \mathbf{B}_0 + \delta\rho\mathbf{g}, \tag{17.16}$$

$$\frac{\partial \delta\mathbf{B}}{\partial t} = \nabla \times (\delta\mathbf{u} \times \mathbf{B}_0), \tag{17.17}$$

$$\nabla \cdot \delta\mathbf{u} = \nabla \cdot \delta\mathbf{B} = 0. \tag{17.18}$$

There are several ways in which the above equations can be manipulated to allow for further investigation of the ideal MHD instabilities. Here, we will apply the Lagrangian displacement vector field, since it is an effective way of reducing the equations to an eloquent form.

The Lagrangian displacement, denoted by the vector $\boldsymbol{\xi}$, describes the difference in position of a fluid element at a time t in two different flows, where the fluid element in each flow is at the same position $\mathbf{a}$ at time $t = 0$ (see Fig. 17.1). Taking the two flows to be the initial flow $\mathbf{u} = \mathbf{u}_0 = \mathbf{0}$ and the perturbed flow $\mathbf{u} = \mathbf{u}_0 + \delta\mathbf{u} = \delta\mathbf{u}$, $\boldsymbol{\xi}$ then describes the manner in which the instability displaces fluid elements. The change of $\boldsymbol{\xi}$ with time must therefore be

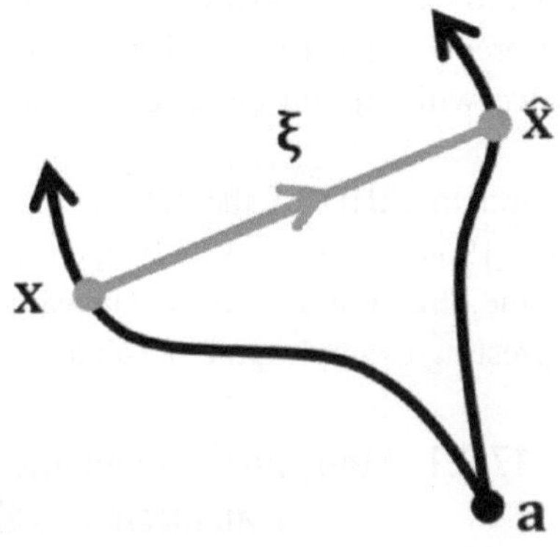

Fig. 17.1 The schematic diagram of the Lagrangian displacement $\boldsymbol{\xi}$ for two flows; one which takes the fluid parcel from position $\mathbf{a}$ to position $\mathbf{x}$ at time t and one which takes it to $\hat{\mathbf{x}}$. Fig. 41 of Zhou et al. (2021).

$$\delta \mathbf{u} = \frac{\partial \boldsymbol{\xi}}{\partial t}. \tag{17.19}$$

For a more detailed explanation of Lagrangian displacements see, for example, Hillier (2019).

Inserting Eq. 17.19 into Eqs. 17.15 and 17.17, and with the appropriate simplifications and choice of integration constants, we find

$$\delta \rho = -\boldsymbol{\xi} \cdot \nabla \rho_0, \tag{17.20}$$

$$\delta \mathbf{B} = \nabla \times (\boldsymbol{\xi} \times \mathbf{B}_0). \tag{17.21}$$

Thus, our linear equations become

$$\rho_0 \frac{\partial^2 \boldsymbol{\xi}}{\partial t^2} = -\nabla \delta p + (\nabla \times \nabla \times (\boldsymbol{\xi} \times \mathbf{B}_0)) \times \mathbf{B}_0 - (\boldsymbol{\xi} \cdot \nabla \rho_0)\mathbf{g}, \tag{17.22}$$

$$\nabla \cdot \boldsymbol{\xi} = 0, \tag{17.23}$$

as we also note that due to the form of Eq. 17.21, the solenoidal condition on the magnetic field is now automatically satisfied. Later, we will use Eq. 17.23 to remove the pressure from Eq. 17.22. Once this is done, we can re-cast Eq. 17.22 in the form

$$\rho_0 \frac{\partial^2 \boldsymbol{\xi}}{\partial t^2} = \mathbf{F}(\boldsymbol{\xi}). \tag{17.24}$$

Consequently, the second derivative of $\boldsymbol{\xi}$ with respect to time (the acceleration) relates to a force operator $\mathbf{F}$ that only depends on $\boldsymbol{\xi}$. This provides a quite general tool to investigate instabilities (though not one we will apply here). If, for a given $\boldsymbol{\xi}$, we have $\boldsymbol{\xi} \cdot \mathbf{F}(\boldsymbol{\xi})$ greater than zero somewhere in the fluid, then the displacement $\boldsymbol{\xi}$ has induced a force in the system which is working to increase the displacement (i.e., we have instability). Otherwise, the system is stable.

Derivation of the dispersion relation

Because of the initial conditions prescribed above, there is no variation in the equilibrium in the horizontal direction. This means that in these directions, a normal mode decomposition can be used, so that perturbations (both δp and the components of $\boldsymbol{\xi}$) are taken to be of the form

$$\delta f(\mathbf{x}, t) = \tilde{f}(z) \exp(i\mathbf{k} \cdot \mathbf{x} - i\omega t). \tag{17.25}$$

This results in the Lagrangian displacement, Eq. 17.19, giving

$$-i\omega \boldsymbol{\xi} = \delta \mathbf{u}. \tag{17.26}$$

In addition, when looking at the 2D plane defined by the gravity and magnetic field vectors, we obtain the following set of linear equations

$$-\rho_0\omega^2\xi_x = -ik\delta p, \tag{17.27}$$

$$-\rho_0\omega^2\xi_z = -\frac{d}{dz}\delta p + B_{x,0}\left(\frac{d^2}{dz^2} - k^2\right)\xi_z + \xi_z\frac{d}{dz}\rho_0 g. \tag{17.28}$$

We also have the incompressible condition

$$ik\xi_x = -\frac{d}{dz}\xi_z. \tag{17.29}$$

Taking the x derivative of Eq. 17.27 allows ξ_x to be eliminated from the equations,

$$-\rho_0\omega^2(ik\xi_x) = \rho_0\omega^2\frac{d}{dz}\xi_z = k^2\delta p. \tag{17.30}$$

Meanwhile, rearranging Eq. 17.30 and differentiating with respect to z gives

$$\frac{d}{dz}\delta p = \frac{1}{k^2}\frac{d}{dz}\left(\rho_0\omega^2\frac{d}{dz}\xi_z\right). \tag{17.31}$$

On substitution into Eq. 17.28, we obtain

$$\rho_0\omega^2\xi_z = \frac{1}{k^2}\frac{d}{dz}\left(\left(\rho_0\omega^2 - (kB_{x,0})^2\right)\frac{d}{dz}\xi_z\right) + (kB_{x,0})^2\xi_z - \xi_z g\frac{d}{dz}\rho_0. \tag{17.32}$$

Using this equation, we can now determine the variation of ξ_z with z and the dispersion relation.

Away from $z = 0$, Eq. 17.32 simplifies to

$$\left(\rho_0\omega^2 - (kB_{x,0})^2\right)\left(\frac{d^2}{dz^2} - k^2\right)\xi_z = 0. \tag{17.33}$$

In the following, we will assume the non-trivial case: $\rho_0\omega^2 - (kB_{x,0})^2 \neq 0$. Therefore, Eq. 17.33 reduces to

$$\left(\frac{d^2}{dz^2} - k^2\right)\xi_z = 0, \tag{17.34}$$

to which the solutions are

$$\xi_z(z) = C_0\exp(-kz) + C_1\exp(kz). \tag{17.35}$$

To maintain a finite energy in the system we require that as $z \to \pm\infty$, $\xi_z(z) \to 0$. Combining this condition with the continuity of ξ_z at $z = 0$ leads to the solution

$$\xi_z(z) = C_0\begin{cases}\exp(kz) & \text{if } z < 0,\\ \exp(-kz) & \text{if } z > 0\end{cases}. \tag{17.36}$$

With the full spatial distribution of ξ_z, ω can be determined. To do so, we integrate Eq. 17.32 from $-\epsilon$ to ϵ and take the limit $\epsilon \to 0$. Any terms of this equation that are continuous or form a step function at $z = 0$ will not contribute to the integration in the limit $\epsilon \to 0$. Therefore, we are only interested in terms containing the derivatives of step functions (i.e., the first and third terms of the right-hand-side of Eq. 17.32). Formally, this process is equivalent to demanding continuity of force at the discontinuity. Upon integration (and evaluation at $z = 0^{\pm}$) we obtain

$$\omega^2 k(\rho_2 + \rho_1) - 2k^3B_0^2 = -(\rho_2 - \rho_1)gk^2. \tag{17.37}$$

Finally, solving for ω gives

$$\begin{aligned}\omega &= \left[\frac{2k^2B_0^2}{(\rho_2+\rho_1)} - gk\mathcal{A}\right]^{1/2} \\ &= \left[\frac{1-\mathcal{A}}{2}k^2V_{\mathrm{A1}}^2 + \frac{\mathcal{A}+1}{2}k^2V_{\mathrm{A2}}^2 - gk\mathcal{A}\right]^{1/2},\end{aligned} \tag{17.38}$$

where V_{A1} and V_{A2} are the Alfvén speeds in the regions below and above the discontinuity, respectively.

The term inside the square root can either be positive or negative. The term that can drive the instability is the same term that drives the hydrodynamic RT and can be simply understood as the square of the inverse of the free-fall time over a distance of $1/k$, with a modified gravity of magnitude $\mathcal{A}g$. The positive term represents the suppression of instability by magnetic tension and is the square of the frequency of a surface Alfvén wave.

If we allow perturbations that have a component perpendicular to the magnetic field in the plane perpendicular to gravity (i.e., the y-direction), following the same procedure as above, we find the following dispersion relation:

$$\omega = \left[\frac{2{k_x}^2B_0^2}{(\rho_2+\rho_1)} - gk\mathcal{A}\right]^{1/2}. \tag{17.39}$$

Here k_x is the component of the wave vector aligned with the magnetic field. This equation highlights the fact that there are two fundamental perturbations to the system: the undular mode, where $\mathbf{k}$ is parallel to $\mathbf{B}_0$, and the interchange mode, where $\mathbf{k}\cdot\mathbf{B}_0 = 0$. For the interchange mode, the growth rate reduces to that of the hydrodynamic system. For the undular mode, instabilities at large k are suppressed by the magnetic field. Naturally, we often observe in nature a mixture of the two.

Variations in the direction of the magnetic field at the interface

A well-known issue with the magnetic RT instability is that certain perturbations (e.g., perturbations with $\mathbf{k}\cdot\mathbf{B} = 0$) have a growth rate which is unbounded with k. Simply put, for certain perturbations, as k goes to infinity, so does the growth rate. This issue also exists for hydrodynamic systems, where the problem is often regularized by the introduction of a region of finite size over which the density transitions, or the introduction of viscosity or surface tension.

Similarly, MHD terms can also be used to regularize the problem. This can be achieved by allowing for the magnetic field vector to be pointing in different directions above and below the density jump. Assuming a uniform magnetic field strength, the growth-rate of the instability is given by (Ruderman et al., 2014; Hillier, 2016):

$$\omega = \left[\frac{2(k_xB_x)^2 + 2(k_yB_y)^2}{\rho_2+\rho_1} - g\mathcal{A}k\right]^{1/2}, \tag{17.40}$$

where it has been assumed, without loss of generality, that the magnetic field is of the form $\mathbf{B} = [B_x, B_y, 0]$ above the discontinuity and $\mathbf{B} = [B_x, -B_y, 0]$ below (i.e., the y component

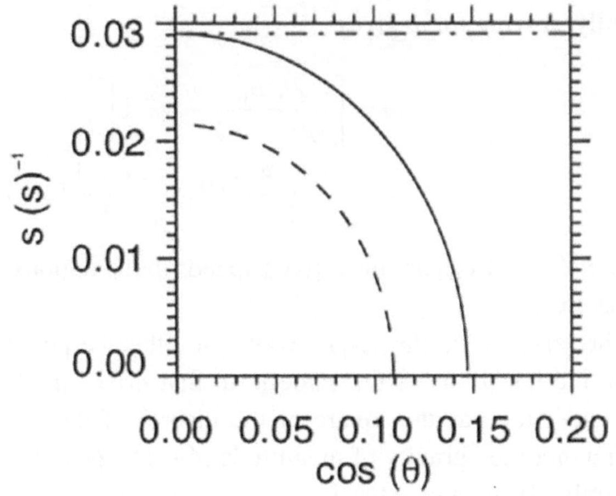

Fig. 17.2 Growth rate $s = i\omega$ of the uniform field (in the x direction) magnetic Rayleigh–Taylor instability (solid line), the sheared field case (dashed line) for the same field strength and the hydrodynamic case (dot-dashed line). For this calculation, we have $\rho_1 = 10^{-12}$kg/m^3 and $\rho_2 = 10^{-10}$kg/m^3. We use a wavelength of 2×10^6 m and θ is the angle between the wavevector and the x direction. In both magnetic cases, $|\mathbf{B}| = 5 \times 10^{-4}$ T, for the sheared field case $B_y = 0.1B_x$. For the uniform field case, only modes close to the interchange mode grow. The inclusion of the sheared field, in this case only a small amount of shear, noticeably reduces the growth rate. Fig. 42 of Zhou et al. (2021).

of the magnetic field changes sign, but not magnitude, across the discontinuity, whereas the x component is the same on both sides).

It is clear from Eq. 17.40 that there is no longer any wavevector for which the influence of the magnetic field vanishes. Therefore, we know that when there is magnetic shear at the density jump, magnetic tension will always work to suppress the instability, with higher wavenumbers experiencing greater suppression. Figure 17.2 shows the variation of the instability growth rate due to magnetic shear, highlighting the reduction in the growth rate as a result of magnetic shear and the anisotropic influence of the magnetic field overall.

The most unstable mode may be calculated by maximizing the gravitational term which drives the instability while minimizing the suppressive terms from the magnetic field. This can be done simply by determining which of B_x^2 or B_y^2 is smaller. Without loss of generality, we can assume that B_x^2 is the smaller. Therefore, to minimize the second term it is necessary to set k_y to be zero. Therefore, the k_x that satisfies

$$g\mathcal{A} - \frac{4k_x B_x^2}{\rho_2 + \rho_1} = 0, \tag{17.41}$$

is the k_x that gives the fastest growing mode in this system (Hillier, 2016). This result is the same as that of the fastest growing undular mode for a uniform magnetic field of strength $|B_x|$.

17.2.2 The nonlinear phase of the magnetic Rayleigh–Taylor instability

Having covered some key aspects of the linear stability of the magnetic RTI, we now turn our attention to the effect of the magnetic field on the nonlinear regime of magnetic RTI.

Nonlinearities in hydrodynamic systems can result in the occurrence of large shear flows that allow parasitic instabilities, for example, the Kelvin–Helmholtz instability, to develop. In MHD systems, these saturation mechanisms are still present. If we balance the temporal derivative of the velocity with the advective derivative, we have the following scaling,

$$\frac{\partial^2 \xi}{\partial t^2} = s^2 \xi \propto s^2 \boldsymbol{\xi} \cdot \nabla \xi \sim s^2 k \xi^2 \tag{17.42}$$

$$\Longrightarrow \xi \sim \frac{1}{k} \tag{17.43}$$

Therefore, it can be predicted that velocity shear created by the instability growth will saturate the linear stage of the instability when the boundary has been perturbed through a distance of approximately $1/k$.

Alternatively, the role of magnetic forces in halting the further evolution of the instability can be given by

$$s^2 \xi = \frac{\partial^2 \xi}{\partial t^2} = \frac{1}{\rho_0} \nabla \times \delta \mathbf{B} \times \delta \mathbf{B} \sim \frac{k_x^2 B_0^2}{\rho_0} \xi^2, \tag{17.44}$$

$$\Longrightarrow \xi \sim \frac{1}{k} \frac{s^2}{\omega_{\mathrm{A}}^2}, \tag{17.45}$$

where ω_{A} is the frequency of a surface Alfvén wave. Therefore, when $|\omega_{\mathrm{A}}| \gtrapprox |s|$ we would expect that magnetic forces will result in nonlinearity developing at smaller plume heights.

Tension of a magnetic field along the interface can fully suppress RT growth along the field direction. Modes perpendicular to this direction are also slightly suppressed, since their growth is forced into a 2D rather than 3D character (Walsh, 2022). Inclusion of the magnetic field diffusion due to Ohmic resistance notably changed the magnitude of the magnetic RT suppression in two separate numerical studies (Manuel et al., 2021; Walsh, 2022).

17.3 The magnetic Richtmyer–Meshkov instability

17.3.1 Shock waves

Like gases, plasmas support shock waves. The conditions on the shock waves in a plasma can be determined in a manner similar to the corresponding gas dynamic case.

While the character of MHD shock waves is derived in a similar way to that of their gas dynamic counterparts, the situation in the former case is much richer. This is because in non-conducting gases, information is transmitted throughout the fluid only through sound waves, and the shock Mach number (defined in terms of the sound speed) characterizes the shock jump conditions. That is, the jumps are controlled by only one parameter, which

Table 17.1 Types of MHD shocks.

Type	Upstream	Downstream
F(1−2)	$M_s > M_I > M_f > 1$	$M_s > M_I > 1 > M_f$
I(1−3)	$M_s > M_I > M_f > 1$	$M_s > 1 > M_I > M_f$
I(1−4)	$M_s > M_I > M_f > 1$	$1 > M_s > M_I > M_f$
I(2−3)	$M_s > M_I > 1 > M_f$	$M_s > 1 > M_I > M_f$
I(2−4)	$M_s > M_I > 1 > M_f$	$1 > M_s > M_I > M_f$
S(3−4)	$M_s > 1 > M_I > M_f$	$1 > M_s > M_I > M_f$

we usually define as the Mach number. By contrast, there are three types of waves that can travel through MHD fluids: the Alfvén wave, and the fast and slow magnetosonic waves (see Section 17.1 for more details), each traveling at a different characteristic speed; thus there are three parameters that control the plasma shock jump conditions. Hence, we can define three different Mach numbers for the MHD shock in terms of each speed, say M_f, M_I, M_s, respectively, the I representing the *intermediate* wave speed. We thus encounter a situation where, in the shock frame, six different configurations are permitted by the jump conditions. Assuming the characteristic speeds are distinct, these configurations are shown in Table 17.1, where F, S, and I designate the *fast* magnetosonic, *slow* magnetosonic, and *intermediate* shock types, respectively, and the numbering indicates where in the ordering of characteristic speeds the upstream and downstream velocities in the shock frame are placed.

Each of these shocks has its own unique character. The intermediate shocks in particular are controversial in the literature, and it is a subject of debate whether they can develop physically (Wu, 1987, 1990; Wheatley et al., 2005a; Falle and Komissarov, 2001). Bypassing this important, but rather complicated, discussion for the moment, it will suffice for us to discuss the first type (fast shocks). As can be seen from the first line of Table 17.1, the shock jump for this type of shock crosses only the fast magnetosonic speed (and is generally uncontroversial). For the other five shock types, they do not cross the fast magnetosonic speed, but may cross one or both of the other characteristic speeds.

Regardless of the shock type, the presence of a magnetic field will generally require a jump in both the tangential field components, $\mathbf{B}_t$, and the tangential velocity components, $\mathbf{u}_t$. These jumps tend to occur together. Unlike gas dynamic shocks, MHD shocks can support tangential velocity jumps (hence vorticity).

In the case of zero mass flux, we again obtain density interfaces, as in gas dynamics. Arbitrary jumps in tangential velocity, however, are only supported when the magnetic field does not penetrate the interface, that is, $B_n = 0$. In this case, a pressure jump is supported, provided the total pressure $p + B_t^2/2$ remains continuous. This is the MHD tangential discontinuity. However, with a finite B_n, all quantities except density must remain continuous on the interface, which is classified as an MHD contact discontinuity. Said another way, MHD contact discontinuities do not support vorticity.

The resulting situation is in general exactly the opposite of the gas dynamic case: while gas dynamic shocks cannot carry vorticity and tangential discontinuities can, MHD shocks can carry vorticity and contact discontinuities cannot. This has a fundamental impact on the RT and RM instabilities in MHD.

17.3.2 Suppression of the RM instability

As with the RTI, the RMI can also be suppressed by the presence of a magnetic field in MHD. This was in fact shown for RT by Chandrasekhar (1961) in his monograph on hydrodynamic stability, using a linear stability analysis. It was only much more recently that RM suppression was discovered by Samtaney (2003), who showed numerically, under the vorticity paradigm, that an oblique interface with an applied magnetic field in MHD did not exhibit the instability.

How then is the RMI suppressed? In the gas dynamic case, the incident shock hits the interface and produces, via shock refraction, a transmitted shock and a reflected shock. Neither of these shocks carries vorticity, unlike the interface, which is accordingly unstable as discussed previously. In MHD, the interaction produces four waves: transmitted fast and sub-fast slow shocks, and reflected fast and sub-fast shocks. Each of these shocks can carry vorticity, but the interface itself cannot. Hence, even though the incident shock deposits baroclinic vorticity on the interface, this vorticity is immediately transported away from the interface by the transmitted and reflected MHD waves. With the vorticity removed from the interface, the growth agent has disappeared, and the interface is stable (Zhou et al., 2021).

As the Mach number in one sense determines the severity of the RMI, so the strength of the magnetic field influences the extent of suppression observed in MHD. This is usually quantified using the parameter $\beta = 2p_0/B_0^2$ (see Eq. 17.8), which measures the reference magnetic pressure (with the permeability constant scaled out) compared to the thermodynamic pressure. Thus, gas dynamic behavior is reproduced in the limit $\beta \to \infty$, while small values of β correspond to strong magnetic fields. The MHD shocks arising from the interaction, particularly the sub-fast waves, tend to travel with higher velocities in relatively stronger fields (lower β), since the corresponding characteristic speeds are faster.

The extent of the instability suppression was also quantified via a solution to the *incompressible* linear problem with impulsive acceleration, in the form of an initial value problem solved with Laplace transform techniques (i.e., without using the method of normal modes), by Wheatley et al. (2005b). In the incompressible model, only the Alfvén waves arise, and these are the agents by which vorticity is transported away from the interface.

Generally, the stabilization mechanism requires that the vorticity is transported away from the interface more quickly than the interface grows; Sano et al. (2013) point out that this constitutes a critical condition for the instability to be suppressed. Figure 17.3 illustrates the transport of vorticity away from the interface. If v_{lin} indicates the linear perturbation amplitude growth rate in the hydrodynamic case, then the authors introduce the condition that the Alfvén velocity must be greater than some factor of this speed,

$$V_{A2} \geq C_\alpha v_{\text{lin}}, \tag{17.46}$$

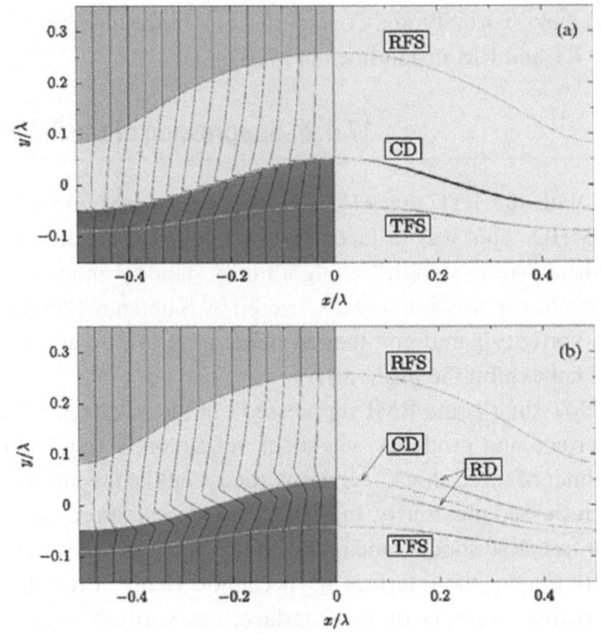

Fig. 17.3 Numerical results of (Sano et al., 2013) showing the importance of critical field strength for RM suppression, after passage of the shock. Contours show density distribution on the left and vorticity on the right. Magnetic fields lines are shown on the left. The shock refraction process produces a transmitted (TFS) and reflected (RFS) fast shock, and a contact discontinuity (CD). (a) For a weak magnetic field, the vorticity remains on the CD as the RM growth rate outstrips the Alfvén speed, so that RM is not suppressed. (b) For magnetic field above the critical strength, for the parameters chosen in the study, the vorticity is transported off the interface by rotational discontinuities (RD), which are a type of intermediate shock, suppressing RM. Note the deflection of the field lines at the vortex sheets. Fig. 2 of Sano et al. (2013), with permission.

where they estimate $C_\alpha \simeq 0.1$ based on numerical results. Since the Alfvén speed is determined by the field strength, the critical magnetic field strength is

$$B_{crit} = \sqrt{\rho_2^+} C_\alpha v_{\text{lin}}, \tag{17.47}$$

where ρ_2^+ is the post-shock density in the dense fluid.

More recently, Sano (2021) showed that the above criterion is applicable universally for both heavy-to-light and light-to-heavy configurations. In the situation in which the field is transverse to the interface, the vorticity never leaves the interface, but self-interferes in a way that causes the interface amplitude to oscillate (Wheatley et al., 2013).

MHD suppression of RT and RM is important because it suggests that inertial fusion techniques, such as ICF or pinch techniques, which feature cylindrical and spherical geometries, could be stabilized by applying a magnetic field (see Fig. 6 of Perkins et al. (2017) that indicates suppression of RT instability in ICF with an imposed magnetic field).

There are many additional aspects of the MHD RMI in cylindrical and spherical geometries that deserve attention, and subsequent studies have investigated some of these. Mostert

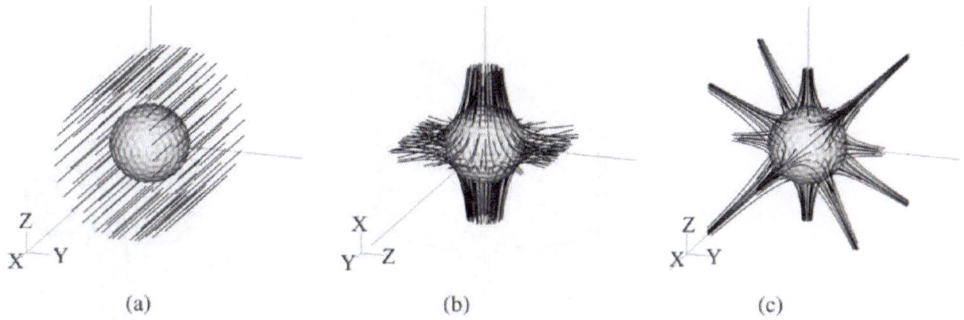

Fig. 17.4 Example field configurations for converging flows showing density interfaces and magnetic field lines in three dimensions from Mostert et al. (2017). (a) Three-dimensional uniform unidirectional field; (b) 3D "two-loop" field, notionally generated by two current loops placed above and below the target; (c) 3D "six-loop" field with octahedral symmetry, notionally generated by six current loops placed on the principal axes around the target. Fig. 1 of Mostert et al. (2017).

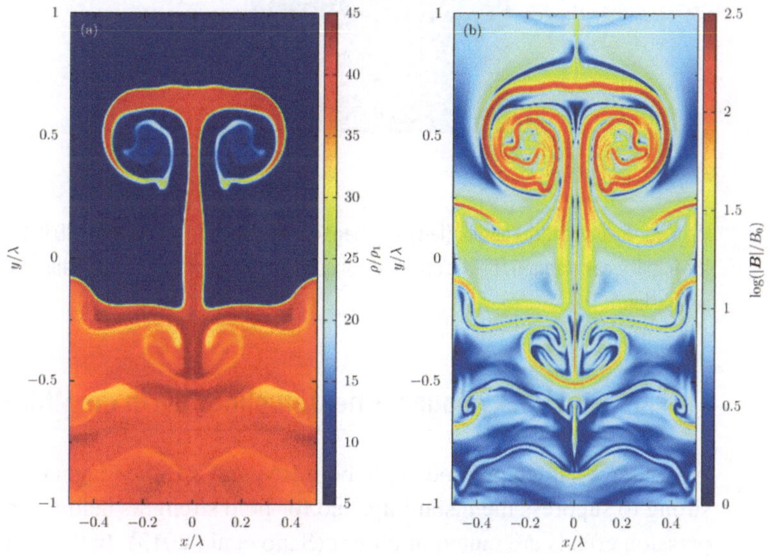

Fig. 17.5 Density (left) and magnetic field (right) contours of the MHD RM instability (Sano et al., 2012). The initial weak magnetic field is oriented horizontally, but is swept up in the nonlinear evolution of the RMI and amplifies in filamentary structures. Fig. 2 of Sano et al. (2012). Reproduced with permission.

et al. (2017) found that a field configuration with an octahedral symmetry group better maintains symmetry in the base flow, without compromising the extent of RM suppression (Fig. 17.4). Also, Li et al. (2018b) found that using a double density interface also shows suppression, but that the separation distance between the interfaces affects the quality of the suppression.

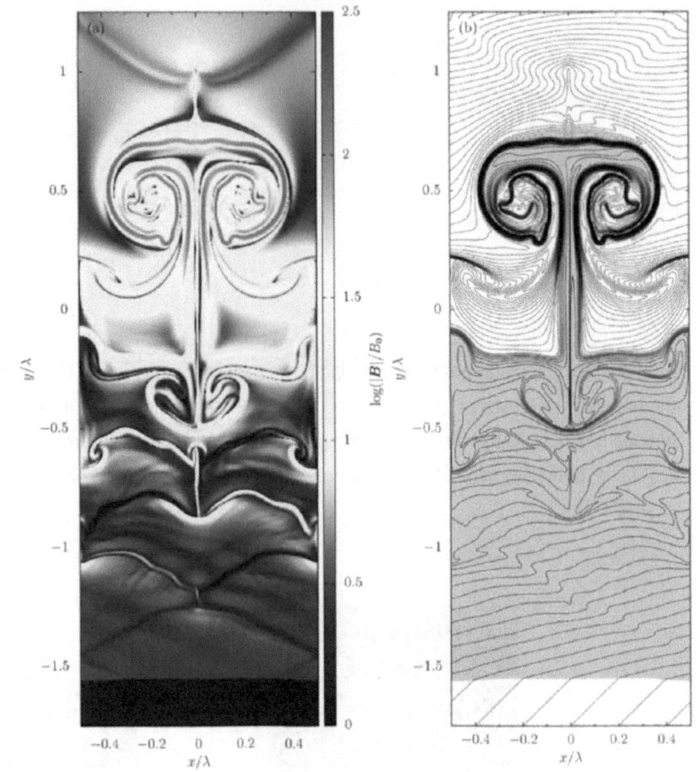

Fig. 17.6 Magnetic field strength contour (left) and magnetic field lines (right) of the MHD RM instability (Sano et al., 2012). The initial weak magnetic field is oriented obliquely to the interface. Fig. 6 of Sano et al. (2012). Reproduced with permission.

17.3.3 Magnetic field amplification in the MHD RM and RT

As we have shown previously, in both RM and RT, the magnetic field must be sufficiently strong to suppress the instability, and for field strengths below some critical value the suppression effects are muted or absent (Sano et al., 2013). In this circumstance, the magnetic field can be *amplified* by RT- or RM-induced mixing. Moving away from the inertial fusion context, this phenomenon has been used to explain some large magnetic fields in supernova remnants (SNRs). The essential mechanism for magnetic field amplification for both RM and RT follows from the same property of magnetic flux conservation in MHD and is related to the magnetic dynamo effect (Kulsrud, 2005).

Figures 17.5 and 17.6 show the development of the RMI when the magnetic field is oriented horizontally (Fig. 17.5) and obliquely (Fig. 17.6) (Sano et al., 2012). The shock travels vertically from top to bottom. After the shock passage, the unstable interface develops into its characteristic mushroom shape and eventually leads to turbulent mixing. As this development occurs, magnetic field lines that are embedded in the fluid close to the interface on one side are dragged along with the interface. As the complexity of the interface grows, the field lines are stretched and compacted next to each other, increasing the flux density locally. Clusters of field lines thus take on a filamentary character, which is strikingly shown in the right of Fig. 17.6. The magnetic fields can be magnified by orders of magnitude, up to the point where the (initially very weak) magnetic pressure balances the thermal pressure. The field strength may be locally greater than the critical value identified by Sano et al. (2013), in which case further development of the secondary Kelvin–Helmholtz instability may be inhibited. Sano et al. (2012) note that in the presence of plasma resistivity, the amplification cannot continue indefinitely, as magnetic dissipation will counteract the very high field gradients.

Generally, MHD turbulence tends to lead to local field amplification by the same mechanism described above (Giacalone and Jokipii, 2007; Guo et al., 2012), particularly in the presence of shocks (Mizuno et al., 2010, 2014). The particular role played by RT and RM instabilities in these processes is also an active research topic and their field amplification properties have been used in subsequent studies concerned especially with SNRs.

17.4 Beyond ideal MHD

While the ideal MHD results discussed thus far are promising with regard to understanding the RM and RT instabilities in plasmas, they are limited in their physical application by the assumptions underlying the MHD model, particularly the assumption that the various plasma length scales are much smaller than the smallest lengths of interest in the MHD problem. However, at least one ICF experiment has featured a plasma with at least one of these length scales *larger* than the hot spot radius at the implosion center (Hohenberger et al., 2012). Moreover, self-generated magnetic fields, which are not accounted for by ideal MHD models, can arise in inertial implosions (Manuel et al., 2012). Observations like these lead to more sophisticated MHD models, such as Hall MHD (Srinivasan and Tang, 2012) and especially *two-fluid* models, in which the ions and electrons are considered as separate fluid species, which interact both with each other and with the electromagnetic environment. This increased complexity brings with it significant theoretical and numerical challenges.

One key feature of two-fluid models is that the charge separation scales (e.g., between ion and electron species) are modeled explicitly. This is accomplished by the introduction of finite values for the Larmor and Debye lengths d_L and d_D, which are the reference plasma length scales, the vanishing limit of which reduces to the single-fluid MHD equations. In two-fluid models, since the electron sound speed is considerably higher than the ion sound speed, electron shocks will typically outrun ion shocks.

In a study of the two-fluid RMI without an applied magnetic field, Bond et al. (2017) shows numerically that this effect leads to rapid charge separation between species, along with a corresponding counteracting Lorentz-force acceleration on both species. This acceleration leads to oscillatory motion of the species and can result in an RT instability at ion and electron interfaces, including the shocked interface. The complex nonlinear

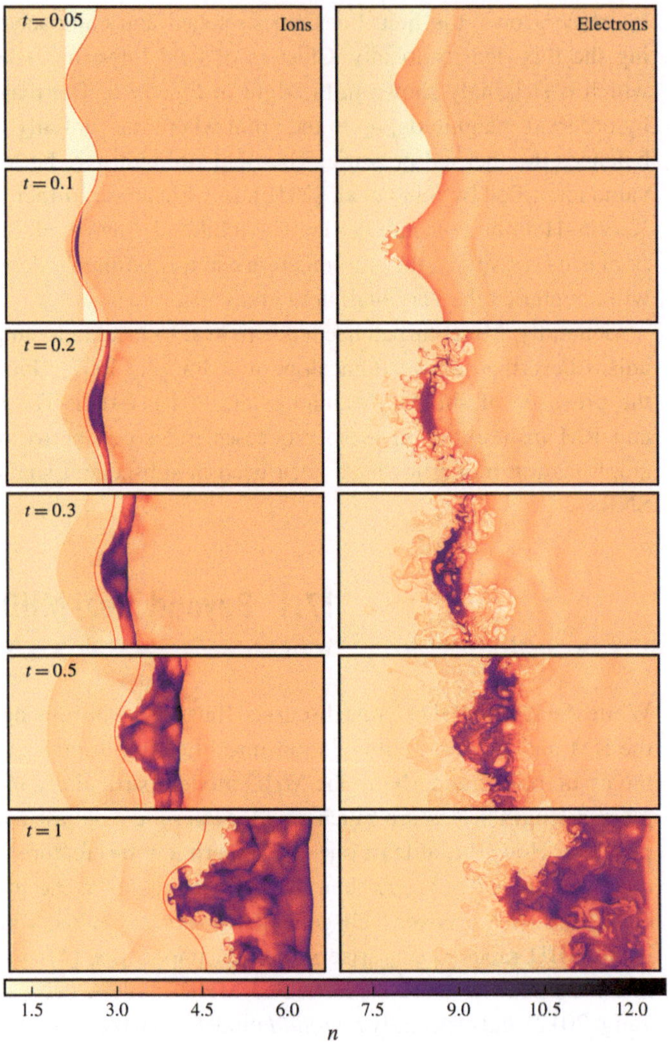

Fig. 17.7 Ion and electron number densities in two-fluid RM problem without an applied magnetic field, with hydrodynamic interface position overlaid, for reference. The Larmor and Debye lengths are $d_L = d_D = 0.1$, relative to the flow length scales (Bond et al., 2017).

interplay between electron and ion species can even destabilize the (ion) shock surfaces themselves. The overall effect can be seen in Fig. 17.7, noting particularly the faceted appearance of the ion shock at late times and the fine structure induced in the ion and electron number densities. The complexity of the physics and the mathematical structure of the model make the two-fluid RMI a challenging research area. Nevertheless, for now, it appears that RM suppression is retained in the presence of an applied magnetic field.

As articulated by Tapinou et al. (2023), such multi-fluid plasma (MFP) model effects fundamentally alter the evolution and severity of the generic plasma RMI (Bond et al., 2017) and produce a different RMI evolution for the isotope, species, and thermal density interface cases that are near indistinguishable in the ideal MHD and other single fluid models (Tapinou et al., 2022).

Another effect not observable in ideal or resistive MHD models is the appearance of self-generated magnetic fields via the Biermann battery effect (e.g., García-Rubio et al., 2021; Walsh and Clark, 2023), which appears through the inclusion of electron physics (Srinivasan and Tang, 2012). The generated field arises from density and pressure (or temperature) gradient misalignment in the flow (Evans, 1986; Srinivasan and Tang, 2012; Bond et al., 2017), typically entering into Hall-MHD or two-fluid formulations. This is a baroclinic generation term and can thus be linked to vorticity generation in the flow (Evans, 1986), and has been seen to arise in plasmas featuring the RTI (Manuel et al., 2012; Srinivasan and Tang, 2012). The extent of effects beyond ideal MHD, for example the Hall effect and the Biermann battery, in astrophysical flows is also of ongoing interest (Lebedev et al., 2019).

Magnetic fields have possible implications with respect to electron thermal conductivity inhibition, which has been observed in magnetized inertial flows (Hohenberger et al., 2012), and may affect transport processes in Z-pinches (Ryutov et al., 2000b). Even if the magnetic field pressure and tension are negligible (i.e., $\beta \gg 1$), magnetic fields can still have an indirect effect via the heat flux and other transport processes. These extended-MHD effects are especially important in high energy density plasmas, which usually have $\beta \gg 1$. If the plasma temperature is high enough, heat flux can become a dominant term in the hydrodynamic equations, tending to smooth out temperature gradients and pressure perturbations. This reduces the RT instability growth rate in a process known as ablative stabilization (discussed in Chapter 19). In effect, the heat flux across the fluid interface causes ablation of material into the hotter region. If part of the interface is perturbed towards the hotter side, the temperature gradient and ablation increase. This perturbed material is then ablated faster than other regions, reducing the amplitude of the perturbation. This effect can completely stabilize the RT instability for shorter wavelengths (Takabe et al., 1985). Growth is fastest for intermediate wavelengths, with a least stable mode set by the relative rates of ablation and interface acceleration. Ablation is a key mechanism that stabilizes laser driven plasma flows.

Heat flux is usually described by plasma fluid closure models that perturb the electron distribution away from a Maxwell–Boltzmann distribution. This perturbation can be driven by temperature gradients or currents in the plasma. The resulting anisotropy induces a flow of faster electrons down the temperature gradient, carrying energy with them. However, magnetic fields can also grow around temperature gradients due to the Biermann battery effect. Even in initially unmagnetized plasmas, the field strength can become so strong that the gyrofrequency becomes similar to the Coulomb collisional timescale. As explored by Braginskii (1965), this invalidates many of the assumptions in the usual heat flux models. Circular gyromotion of the electrons in the magnetic field can localize them perpendicular to the field direction, preventing heat conduction. This will reduce both the ablation and the ablative RT stabilization. However, motion of electrons along the magnetic field is unaffected. This leads to anisotropic heat transport and viscosity effects. Modeling of these effects requires inclusion of MHD and magnetized heat transport into the radiation hydrodynamics. This is a regime known as extended-MHD (XMHD). For example, recent 3D simulations using the Gorgon XMHD code (Walsh, 2022) demonstrated increased bubble-spike height when thermal conduction is suppressed by an external magnetic field. A similar phenomenon affects the Kelvin–Helmholtz instability (Sadler et al., 2022a). Application of a magnetic field was recently measured to increase the growth rate of the ablative RT instability (Matsuo et al., 2021) by 50% or more. The magnetic field insulates the heat flux and reduces ablative stabilization. This occurs so long as the electron magnetic gyrofrequency is similar to the Coulomb collision frequency, requiring field strengths of $\simeq 100\,\mathrm{T}$ in high power laser driven plasmas. Self-generated magnetic fields are expected (Walsh et al., 2017; Sadler et al., 2020) to reach $5\,\mathrm{kT}$ near RT spikes at the edge of the fusion hot-spot in current inertial confinement fusion experiments. This is strong enough to change the heat loss profile at the edge of the fuel, affecting the Lawson ignition threshold for thermonuclear ignition (Sadler et al., 2022b).

PART 3

FROM THE MICROSCOPIC TO COSMIC SCALES

[W]e realize that trying to understand the infinite wonders of nature with our limited human intellect is a story that has no ending.

Tsung-Dao Lee, Nobel laureate

18 High-energy-density physics

Intense lasers are now being used to probe the physics of fluids dynamics in the high energy density physics (HEDP) regime, a term roughly referring to thermodynamic pressures greater than 1 Mbar (Drake, 2018). These laser platforms are also employed to recreate aspects of astrophysical phenomena in the laboratory (Remington et al., 1999, 2000, 2006), a specialized research area frequently referred to as laboratory astrophysics (Fig. 18.1).

Fundamentally, the type of physics that can be studied using high-intensity lasers is completely determined by the platforms' power and diagnostics. Thus, we are led to ask, as the capabilities of the HEDP facilities continue to improve (see Fig. 18.2), might it be possible to design experiments on these platforms capable of observing the transition to fully developed turbulence? While at the moment the answer to that question is a "not yet," it is clear that many fascinating HEDP studies have already been carried out.

Smaller high energy density (HED) platforms (Nike, Obenschain et al. (1996) and Orion, Hopps et al. (2013, 2015)) have only a very limited energy (3–5 kJ) available for studying the linear regime of hydrodynamic instabilities. An example of such a study of RMI is that of Aglitskiy et al. (2002, 2010). With increased capacity (10–30 kJ), the Gekko XII (Yamanaka et al., 1981; Mima et al., 1996) and Omega (Boehly et al., 1997) facilities offered an opportunity to study the weakly nonlinear regime of the RTI (Smalyuk et al., 2005), RMI (Glendinning et al., 2003), and KHI (Harding et al., 2009; Hurricane et al., 2012). More recently, with the enhanced energies available at the National Ignition Facility (NIF) (Miller et al., 2004; Moses and Wuest, 2003) and the French megajoule laser project (LMJ) (André, 1999; Cavailler, 2005; Casner et al., 2018) it is possible to obtain much higher resolution and to do so at significantly later times. As a result, researchers have been able to push into deeply nonlinear regimes of RTI (Huntington et al., 2018; Casner et al., 2019; Martinez et al., 2015), reshocked RMI (Nagel et al., 2017, 2022; Bender et al., 2021), and KHI (Flippo et al., 2016, 2018). Accurate and robust transition criteria will be critical for guiding future design activities for creating fully developed turbulent flows on HEDP platforms.

18.1 RTI

In Chapter 4, we discussed the NIF experiments of Martinez et al. (2015) on ablative RTI in the transition from weakly- to highly-nonlinear regimes. There, our focus was on how the multi-mode modulation amplitudes, which are found in the nonlinear regime, grow beyond the Haan multi-mode saturation limit.

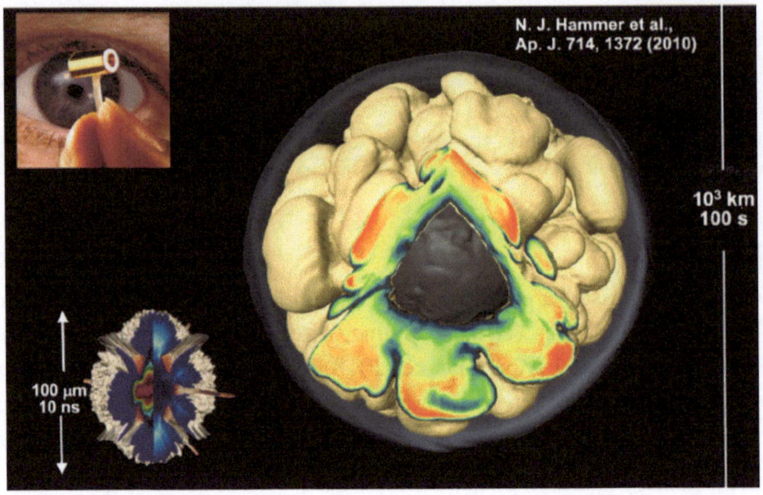

Fig. 18.1 The length- and time-scales of astrophysical flows and those generated in the HEDP platform. The SNe visualization is adopted from Hammer et al. Reproduced with permission from Hammer et al. (2010).

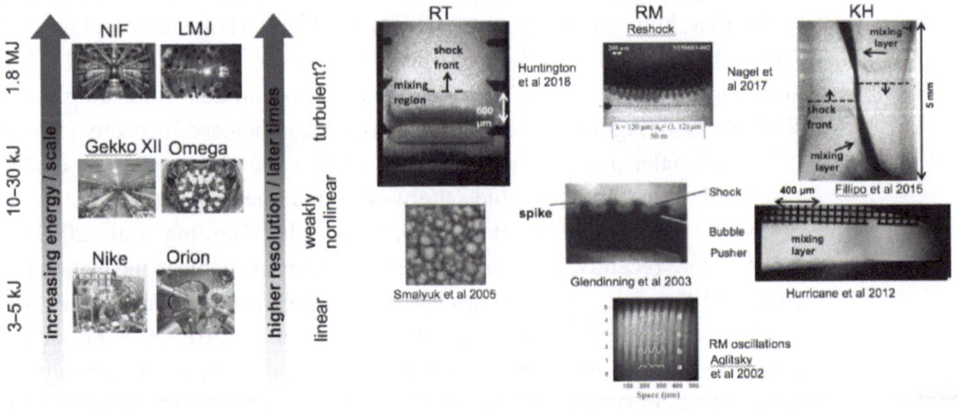

Fig. 18.2 HEDP facilities, showing the progression toward higher energy, increased longer-time drive, and enhanced resolution. Zhou et al. (2019).

The work of Martinez et al. (2015) shows a radiographic comparison of two samples with different amplitude, covering the 1 mm broadband region at 15.5 ns for a 2.8 μm (Fig. 18.3(a)) and 0.5 μm (Fig. 18.3(b)) initial rms amplitude. Line-outs are shown in Fig. 18.3(c). The size of the RTI features increases, while larger bubbles overtake smaller bubbles. Meanwhile, the mixing layers evolve toward the longer wavelengths and show insensitivity to the initial conditions.

Recently, Rigon et al. (2021b) presented micron-scale phenomena observed from an HEDP experiment (Fig. 18.4) at the SACLA facility in Japan, where a laser-produced RT unstable plasma flow evolved toward a possibly turbulent state. The inclusion of surface

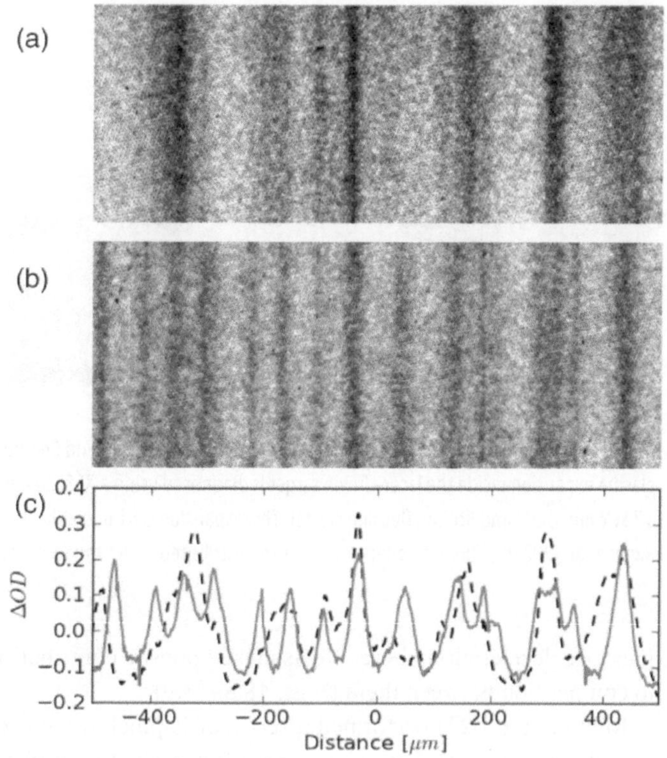

Fig. 18.3 (a) Processed NIF RTI data images at 15.5 ns for an iodine doped CH foil with an initial rms amplitude of 2.8 μm, where black represents an increase in the optical density (OD). (b) Data with an initial rms amplitude of 0.5 μm. (c) Line-outs from (a) dashed line and (b) solid line. Fig. 2 of Martinez et al. (2015).

modulation on the pusher layer favors the growth of a given Fourier mode via instability, allowing precise control of the experimental conditions. Both mono-mode (40 μm wavelength) and bi-mode modulations (15 and 40 μm wavelengths) were used to provide variations of the initial conditions. At late times, the expanding plasma appears streaked and chaotic and the RTI structures are lost. The loss of the initial conditions occurs between 50 ns and 60 ns for the mono-mode targets (Fig. 18.5). Moreover, Fig. 18.6 shows that the transition happens at earlier times for bi-mode targets (around 40 ns). The radiographs from the evolution of an RT unstable system allowed for the extraction of the spatial power spectrum, which the authors assess to be consistent with existing turbulence theory (Fig. 18.7).

In related work, Rigon et al. (2021a) report on a parametric study of the RTI in a decelerating phase. The authors performed experiments at the LULI2000 high-power laser facility, adopting a direct-drive approach to put into motion the modulated rear side of a solid target. The expansion of this target into a foam – used as a deceleration medium – is RT unstable. In these HED experiments, the change of the foam density resulted in both a modification of initial Atwood number across the interface and an alteration of

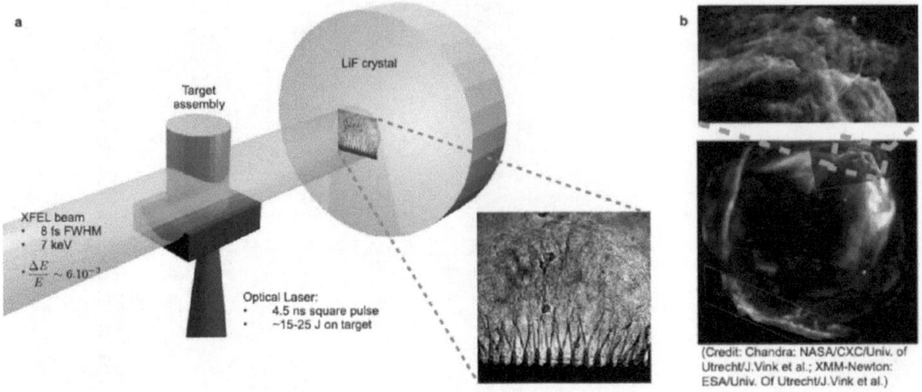

Fig. 18.4 RTI experimental setup. Diagram of the overall experiment setup (a) and Chandra's observation of RCW 86 (b). The plasma expansion inside the laser-driven target is diagnosed using a XFEL beam (>1 mm diameter, 8 fs, centered on a 7 keV emission) and lithium fluorine crystal. This expansion is RT unstable, as in the supernovae remnant. Fig 1 of Rigon et al. (2021b). This is an open access article distributed under the terms of the Creative Commons CC BY license.

interface deceleration; these effects make opposite contributions to the RTI growth, leading to competition between them (Figs. 18.8–18.9).

Ruiz et al. (2022) performed a series of Z-pinch driven[1] experiments investigating the growth and interaction of two magneto-Rayleigh–Taylor instability (MRTI) modes, k_1 and k_2, initially seeded on the outer surface of an imploding shell. Four images of the experimental radiographs captured the temporal evolution of the MRTI (Fig. 18.10). Figure 18.11 illustrates the harmonic generation of new daughter modes. Significantly, the amplitudes of several daughter modes are larger than the initial amplitude of the mother modes. In particular, the k_2-k_1 mode has the largest amplitude of the daughter modes and provides experimental evidence for the onset of energy transfer from small scales to large scales, that is, an inverse cascade occurring from the nonlinear coupling of discrete initial MRTI modes.

Sauppe et al. (2020) investigated RTI growth in laser-driven cylindrical implosion experiments, using two distinctive cylindrical targets to directly measure instability growth on the 30 kJ Omega and 1.8 MJ NIF laser facilities, respectively. The NIF target had three times the radius of the Omega target and took three times longer to implode, but the two experiments exhibited nearly identical hydrodynamic behavior in the high-energy-density regime. Figure 18.12 shows the same scale-invariant RTI growth in the fuel during the deceleration phase, even though the experiments operated at very different spatial and temporal scales.

A study of laboratory astrophysics illustrated how high energy fluxes may affect RTI growth in young supernova remnants. Kuranz et al. (2018) developed a radiative shock

[1] See also, Chapter 19 for further discussion.

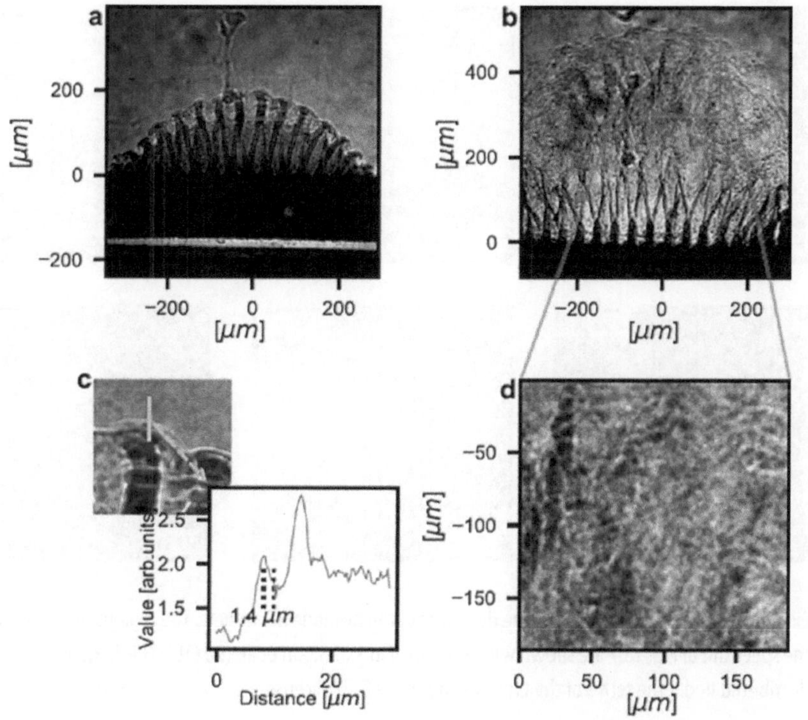

Fig. 18.5 Experimental radiographs of mono-mode targets of RTI. Experimental radiographs of mono-mode targets obtained before (40 ns (a) and (c)) and after (60 ns (b) and (d)) the transition to turbulence occurs between 50 and 60 ns. Following the initial Rayleigh–Taylor growth and pusher expansion (curved modulated surface of the two first radiograph), the system becomes turbulent. The structure becomes blurry and the power spectrum nearly isotropic, as can be observed in the zoom of the 60 ns radiograph. A precise study was made possible thanks to the high resolution of the radiography, which is illustrated on the zoom and line-out of the 40 ns radiograph (c). Fig 2 of Rigon et al. (2021b).

stabilized planar RT platform in indirect drive with drive radiation temperatures $T_R = 200$–350 eV on NIF (Fig. 18.13). These planar RT experiments used a strong shock launched into a doped plastic ablator, which then entered a low-density SiO_2 aerogel foam tamper on the back side of the target. This created a high temperature and a strong source of radiation and electron thermal diffusivity in the foam, which in turn ablatively stabilized the perturbation growth at the plastic-foam interface. For low drives, the shocked foam is not hot enough to stabilize the RTI and the predicted RT growth at fixed interface displacement is 40% larger (Fig. 18.14). In comparing their results with the Type II supernova SN1993J, the authors found that the energy fluxes produced by heat conduction appear to be larger than those due to radiation, and large enough to have a significant effect on the RT growth, which could affect our understanding of young supernova remnants (Kuranz et al., 2018; Huntington et al., 2018; Remington et al., 2019; Sadler et al., 2023).

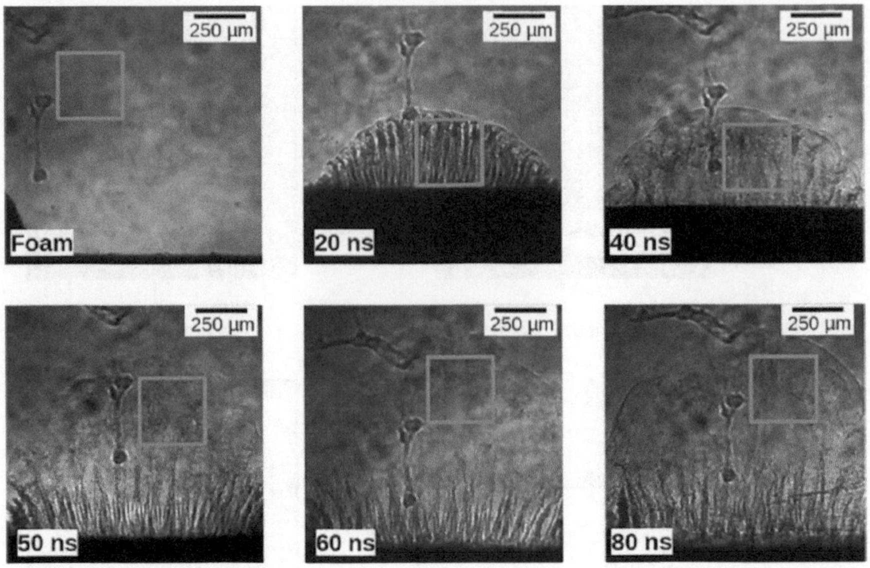

Fig. 18.6 Experimental radiograph showing the dynamics of the bi-mode RTI targets. The regions of interest taken to calculate the spectrum of Fig. 18.7 are shown with a square. Fig 3 of Rigon et al. (2021b). This is an open access article distributed under the terms of the Creative Commons CC BY license.

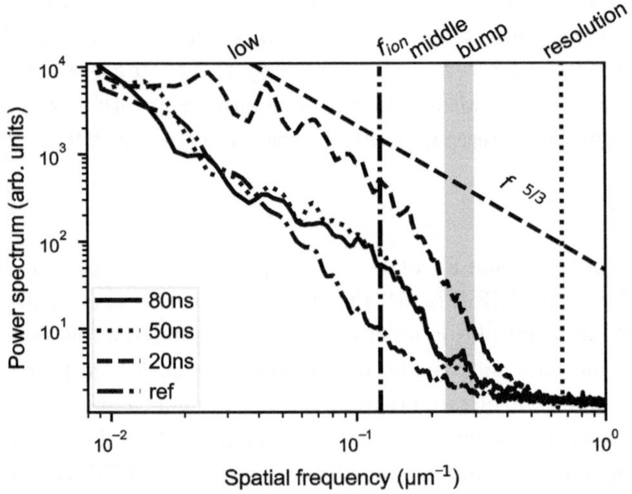

Fig. 18.7 Spatial power spectra of the RTI mixing layer, shown in Fig. 18.6. Fig 4 of Rigon et al. (2021b). This is an open access article distributed under the terms of the Creative Commons CC BY license.

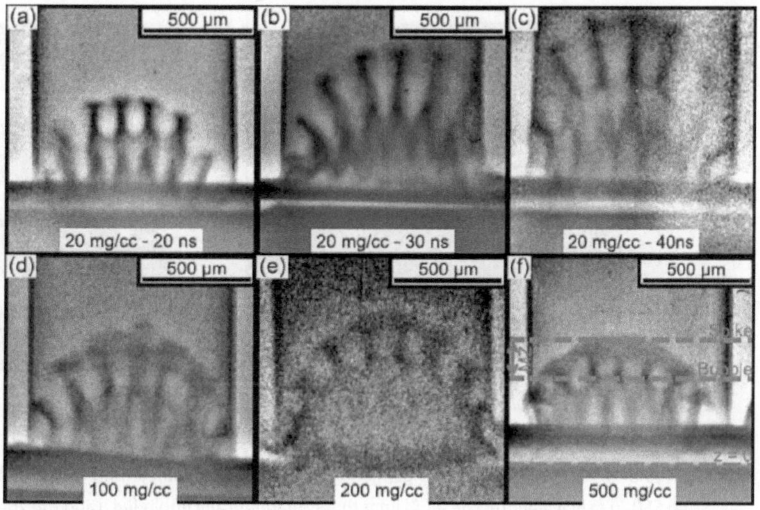

Fig. 18.8 Experimental radiographs (a, b, c) show the RTI evolution of a 20 mg/cm^3 foam. Similarly panels (d), (e), and (f) correspond to radiographs of 100, 200, and 500 mg/cm^3 foams taken 30 ns after the laser shot. The initial wavelength λ is 120 μm. Fig. 1 of Rigon et al. (2021a).

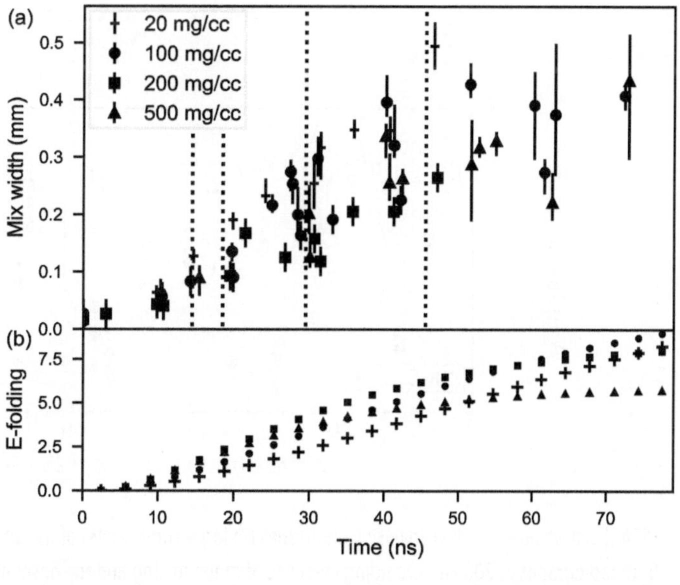

Fig. 18.9 (a) Evolution of the mixing width as a function of time for different foam densities: 20, 100, 200, and 500 mg/cm^3, corresponding to $\mathcal{A} = 0.97, 0.87, 0.75$, and 0.47. The vertical dotted lines indicate the time of maximum deceleration for each density: 45.5, 29.5, 18.5, and 14.5 ns, respectively. (b) Number of e-foldings for each foam. Fig. 4 of Rigon et al. (2021a).

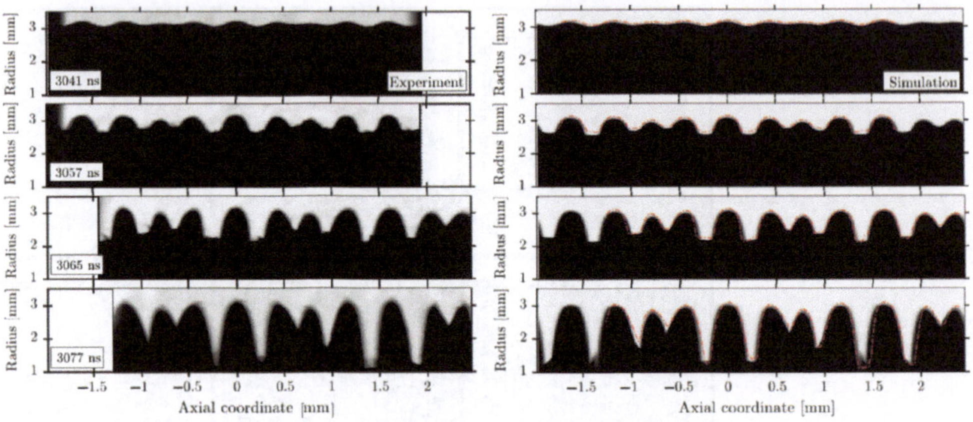

Fig. 18.10 Left column: enlarged images of experimental radiographs, obtained from two discrete preseeded axial modes (400- and 550-μm wavelengths) on an Al liner that is magnetically imploded using the 20-MA, 100-ns rise-time Z Machine at Sandia National Laboratories. The diagnostic has a 4-mm axial field of view. Right column: synthetic radiographs obtained from HYDRA code, using the initial two periodic perturbations. Both modes had the same 10-μm initial amplitude. Fig. 2 of Ruiz et al. (2022).

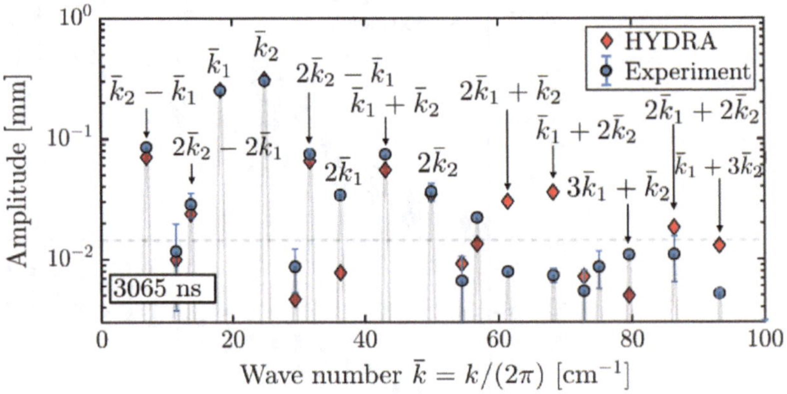

Fig. 18.11 LSSA (least-squares spectral analysis) amplitudes (in logarithmic scale) of the unfolded transmission contours inferred from radiographs at 3065 ns, providing evidence of mode mixing and the onset of an inverse-cascade process induced by the nonlinear coupling of two discrete preseeded axial modes. The error bars correspond to differences between the left- and right-hand sides of the experiment radiographs. Horizontal gray dashed line indicates the amplitude that can be measured to 30% uncertainty due to the fitting error. Fig. 3 of Ruiz et al. (2022).

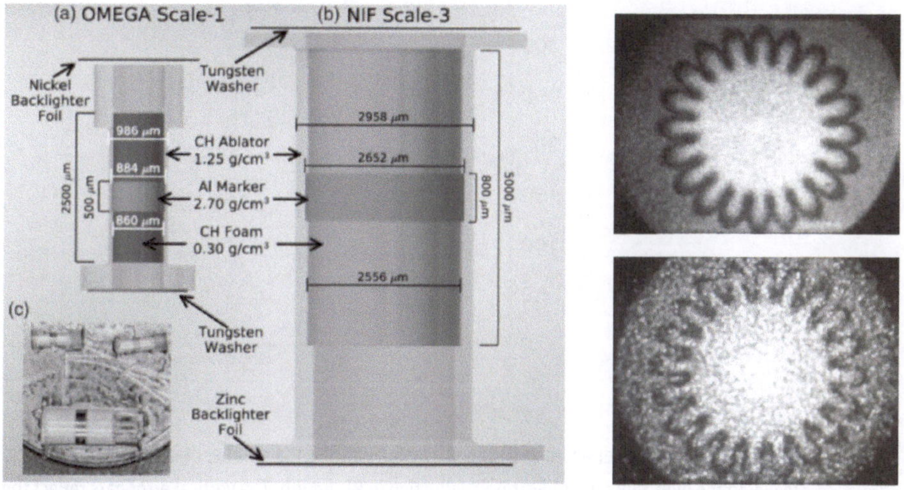

Fig. 18.12 Left: (a) Schematic of the scale-1 cylindrical target fielded at Omega and (b) the scale-3 target fielded at the NIF. (c) Two Omega targets (top) and one NIF target (bottom) on the surface of a penny. Right: RTI experimental radiographs from (top) Omega shot 93069 and (bottom) NIF shot N190212-003. Figs. 1 and 2 of Sauppe et al. (2020).

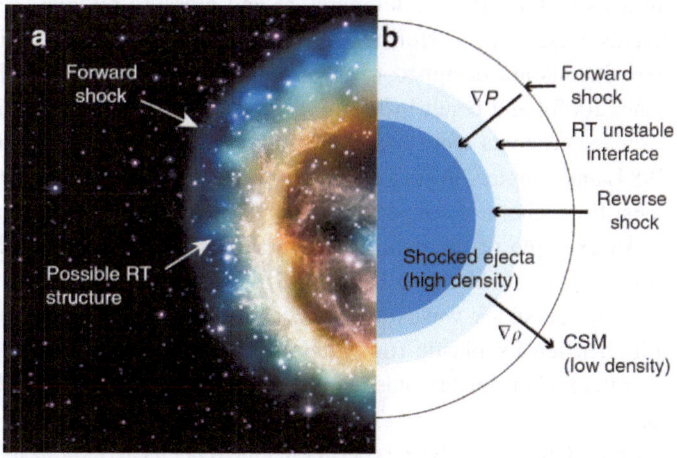

Fig. 18.13 Image of a supernova remnant. (a) False-color image of SNR E0102.2-72. This object is believed to result from a core-collapse supernova about 1000 years ago. One can see the edge of the forward shock. The modulated boundary within it might be structuring of the ejecta-CSM interface produced by RT. The brighter, inner colors are attributed to emission from the higher-Z, interior portions of the ejecta. Image credit: X-ray (NASA/CXC/MIT/D. Dewey et al. and NASA/CXC/SAO/J. DePasquale); Optical (NASA/STScI). (b) Schematic (size and shape not to scale) of inner structures of the supernova that creates the opposing density and pressure gradients to induce an RT unstable interface. Reproduced with permission from Kuranz et al. (2018), licensed under a Creative Commons Attribution (CC BY) license.

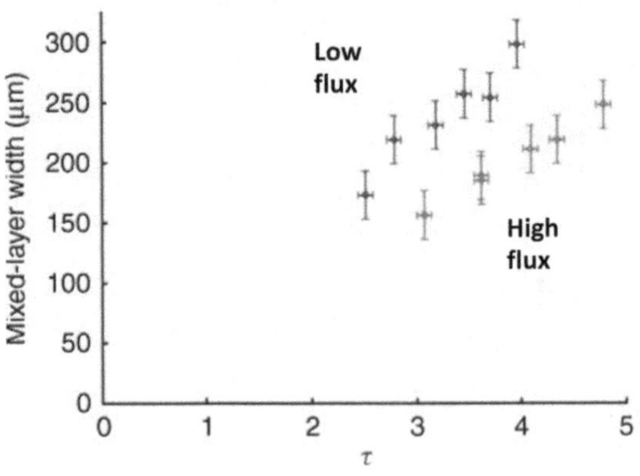

Fig. 18.14 Experimental mixed-layer width vs RT growth factor, which is the number of instability e-foldings. The mixed-layer is smaller in the high-flux case, indicating a reduction in RT growth. The vertical error bars reflect the uncertainty of the location of the RT spike tip and bubble head, and the horizontal error reflects the uncertainty in the experimental time. Reproduced with permission from Kuranz et al. (2018), licensed under a Creative Commons Attribution (CC BY) license.

18.2 KHI

To achieve relevant ICF temperature and pressure regimes, several KH instability experiments have been performed at high energy density facilities. In some cases, HEDP experiments are designed to have a certain aesthetic, as well as scientific value (Hurricane, 2008), and result in stunning images (Hurricane et al., 2009). The designs of these laser experiments were based on traditional (Brown and Roshko, 1974; Ho and Huerre, 1984) and counter-flowing (Papamoschou, 1995; Alvi et al., 1996; Strykowski et al., 1996) shear flow laboratory experiments performed in the liquid or gas phase.

An example of the evolution of traditional KHI-induced flow obtained at the Omega laser facility is shown in Fig. 18.15 (Hurricane et al., 2009; Raman et al., 2012). Here, laser-driven shock waves propagated through low-density plastic foam placed on top of a higher-density plastic foil. Behind the shock front, lower-density foam plasma flowed over the higher-density plastic plasma. The interface between the foam and plastic was KH unstable.

Other Omega and NIF experiments have been carried out using counter-flowing geometries, in which the flows enter the apparatus from opposite directions (e.g., Doss et al., 2013; Merritt et al., 2015) The counter-flowing geometry was used to further increase the velocity difference across the fluid interface. Flippo et al. (2018) varied the initial perturbation profiles by changing the surface roughness of the diagnostic tracer at the shear interface. Figure 18.16 demonstrates that the behavior of the mixing layer changes significantly by using similar Al foils of varying initial roughness. The mixing layer growth is shown in Fig. 18.17. Wang et al. (2015b) conducted a sensitivity study with 2D and 3D numerical simulations of counter-propagating NIF shear flow experiments and identified a 3D geometrical effect at late times if the energy drives from the two ends of the shock tube are asymmetric.

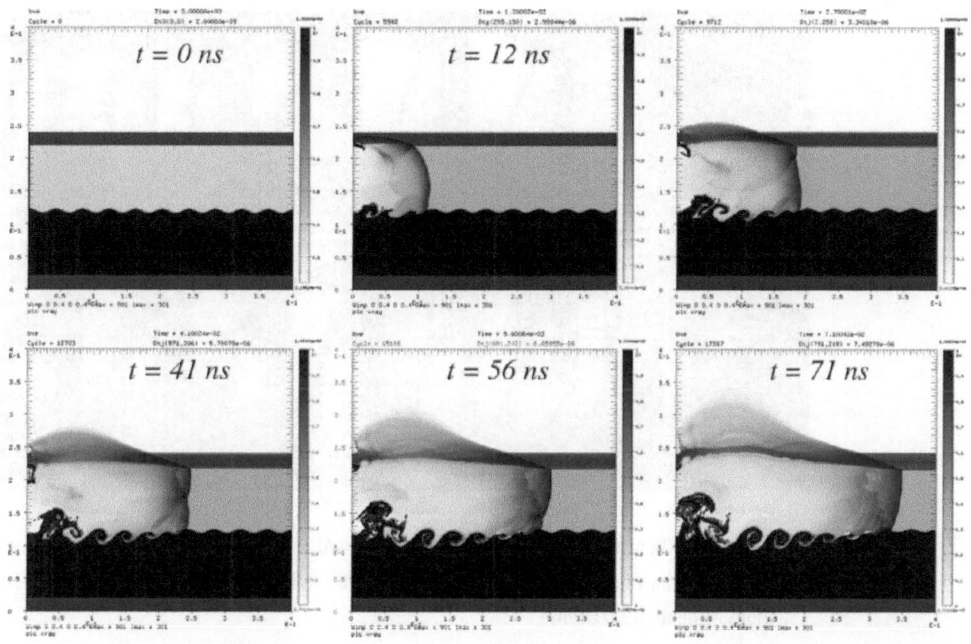

Fig. 18.15 Frames of 2D KHI simulations show the expected target behavior in a synthetic X-ray image as a function of time (Hurricane et al., 2009). The simulation shows that the region of target to the left of $x = 1$ mm contains debris from the gold shield on the laser drive-end of the target, so the region of hydrodynamic interest is $x > 1$ mm. As the primary shock moves from left to right in the low density foam, transmitted shocks move into the high density plastic part of the target (not visible) and reflected shocks are created in the post-shock flow region (visible in the upper half of the target). The small zero density voids visible above a few of the vortices in the late-time images are a numerical artifact of the synthetic X-ray software. While these 2D simulations did a good job of predicting the general experimental behavior, the best fidelity was obtained with 3D simulations performed many years later (see Raman et al., 2012). Hurricane et al. (2009), with the permission of AIP Publishing.

To build an intuition for the effect of a dominant initial single-mode on the KHI, it will be helpful to consider some experimental observations performed on the Omega-60 laser facility (Di Stefano et al., 2014). A target was designed and machined to have a multi-mode initial perturbation, but the radiographs show that the initial seed perturbations in the experiment have a multi-mode spectrum with a dominant single-mode of 16 μm wavelength. The late-time measured mixed-width has surpassed 100 μm, an order of magnitude larger than what one would expect from the saturated value from single-mode-dominated KHI flows. Shimony et al. (2016) inspected the vortex dynamics of such an initial spectrum, using both numerical simulations and a statistical model. They concluded that the KHI evolution in the experiment is dominated by multi-mode, vortex-merger dynamics, thereby overcoming the dominant initial mode.

In supersonic flow regimes, the shocked, flowing material creates a shear layer between two plasmas from well-characterized initial conditions (e.g., Malamud et al., 2013b). The Omega-EP facility was used to produce and sustain a steady shockwave over a foam-plastic

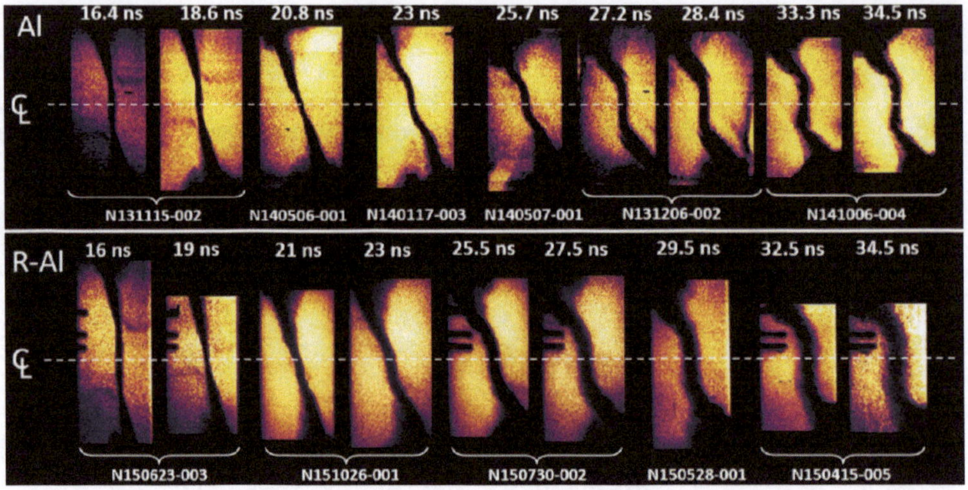

Fig. 18.16 Edge-view time KHI history for smooth Al (top) and rough Al, R-Al, (bottom) from 16 ns to 34.5 ns with shot numbers indicated under the images. Fig. 12 of Flippo et al. (2018). This figure is available under the terms of the Creative Commons Attribution 3.0 License.

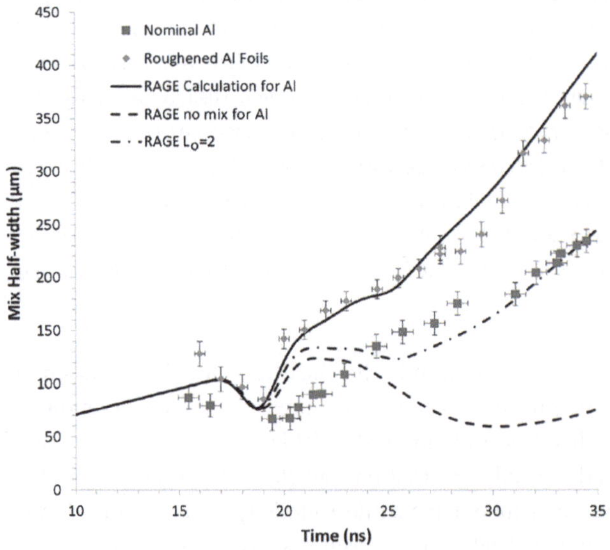

Fig. 18.17 Smooth and rough Al KHI half-widths taken from Fig. 18.16. The squares are smooth Al tracer half-widths and the diamonds are roughened Al tracer half-widths. The RAGE hydrocode was run to simulate the data with the BHR model in the following three cases: (1) turned-off (dashed line), that is, pure hydro; (2) with an initial scale-length of 2 μm (dot-dash line); and (3) an initial scale-length of 8 μm (solid line). Note that the BHR model will be discussed in a later chapter. Fig. 4 of Flippo et al. (2016). This figure is available under the terms of the Creative Commons Attribution 3.0 License.

interface. Wan et al. (2015, 2017) reported their observations of the evolution of the KHI from single- and dual-mode initial conditions. The latter case, with precision-machined seeded interface profiles, also provided measurements of the instability's vortex merger rate and data on the inhibition of the instability's growth rate in a compressible flow.

18.3 RMI

Some of the earliest HEDP RMI experiments were conducted by Dimonte and Remington (1993) using targets of single-mode surface corrugations at the Nova laser facility. The authors measured instability growth resulting from a single-shock impact on a perturbed interface between an ablator material (e.g., beryllium or plastic) and a lower-density foam. The shock – which converted the solid materials to plasmas – was driven by an X-ray bath, generated via laser irradiation of the interior of a cylindrical hohlraum. The authors reported observations of linear growth and saturation; this work was followed by Dimonte et al. (1996b) with additional experiments varying the Atwood number and shock strength. Peyser et al. (1995) and Farley et al. (1999) focused on observations of nonlinear growth of sawtooth perturbations. These works also leveraged radiation hydrodynamics simulations for experimental design and analysis.

Malamud et al. (2013a) and Di Stefano et al. (2015a,b) discussed RMI experiments involving multi-mode perturbations, along with supporting simulations. Examples of radiographs obtained from their experiments are shown in Fig. 18.18, for the two-mode initial perturbation, at 20 and 35 ns. In these images, the ablator, CRF (carbonized resorcinol formaldehyde), and plastic layers are clearly identifiable. At 35 ns, the shock has broken out of the tracer layer and is identifiable in the undoped plastic layer (Di Stefano et al., 2015a). The solid lines in Fig. 18.19 show the growth of the three most prominent modes: the two initial modes and their lowest coupled harmonic. The accompanying simulations clearly show that the growth rates begin to fall off earlier and at a smaller amplitude in

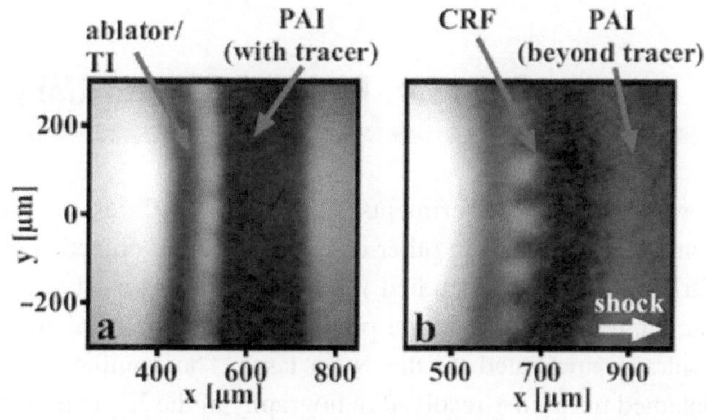

Fig. 18.18 (a) Experimental radiograph, taken at 20 ns; (b) experimental radiograph, taken at 35 ns. The RM growth of the interface structure during the elapsed time is evident from the two frames. Fig. 2 of Di Stefano et al. (2015a).

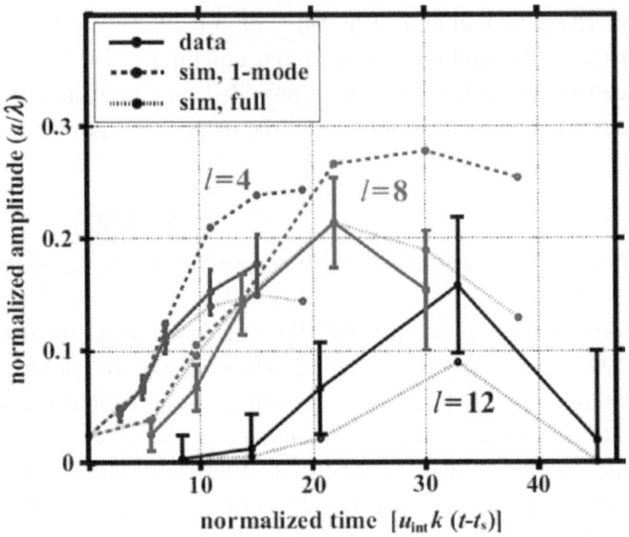

Fig. 18.19 Time evolution of the two fundamental RM modes ($l = 4, 8$) and their lowest coupled harmonic ($l = 12$), as measured using Fourier techniques for RM experiments. Here, the authors use the notation l for the experimental mode wavenumbers, which are normalized such that $l = 400/\lambda[\mu\text{m}]$). Thus, $l = 4$ and $l = 8$ correspond to the initial $\lambda = 100\ \mu$m and $\lambda = 50\ \mu$m modes, respectively, while $l = 4 + 8 = 12$ corresponds to their coupled harmonic with $\lambda = 33\ \mu$m. The solid lines correspond to the experimental radiographs, the finely-dotted lines to the full experimental simulation, and the heavily-dotted lines to the single-mode simulations. The five data points for each mode correspond to images taken at 20, 22.5, 25, 30, and 35 ns, respectively. Note that the time scale is normalized to the standard, circular wavenumber k. Fig. 6 of Di Stefano et al. (2015a).

each others' presence than would occur in isolation, and that the coupling effect is evident, beginning at an amplitude-to-wavelength ratio of roughly 0.1–0.15, as expected from a theoretical analysis. At late times, the bubble-merger absorption of the shorter-wavelength modes by the longer-wavelength ones can also be observed.

18.4 RMI with RTI and RMI with KHI

For many HEDP experiments, multiple instabilities cannot be avoided as an unfortunate complication, while in other cases they are the objective of the research. As an example, Farley et al. (1999) studied the growth of an unstable density interface from a nonlinear, single-mode, 2D initial perturbation using a miniature shock tube attached to a gold hohlraum irradiated by the Nova laser. The nonlinear growth of the mixing layer was obtained from time-resolved radiography of the X-ray transmission through the shock tube. A strong Mach 30 shock wave propagated across the perturbed density interface causing the onset of the RMI. The interface subsequently experienced a relatively weak RTI unstable deceleration.

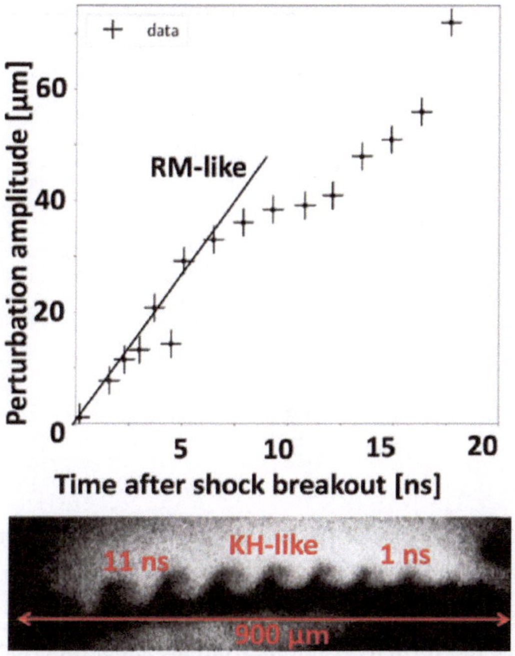

Fig. 18.20 Results from oblique shock experiments at Omega-EP. Top: the perturbation amplitude history. Bottom: an experimental radiograph showing perturbation morphology. Note that the perturbation amplitude history is a linear function of time at early times (as in RM), while the perturbation morphology is rolled over at late times (as in KH). Fig. 2 of Rasmus et al. (2019), with the permission of AIP Publishing.

Di Stefano et al. (2017) performed an Omega-EP experiment designed to observe coupled RM and RT growth. This results in a system in which, for a normal-incidence shock, the interface first experiences a period of pure RM growth, followed by a period of coupled RM and RT growth. This situation, in which an initial RMI is subsequently dominated by the RTI, is of practical importance, as most real-world shocks decay.

More recently, Rigon et al. (2022) used an experimental platform developed to create radiative shocks on high-energy laser facilities such as LULI2000 and GEKKO XII. The authors investigated the effects of a radiative shock on the morphology of jet-like objects subjected to RTI and RMI. The jets produced during their experiments can be compared to astrophysical radiative cooling jets (Balick et al., 2013).

The sustained shocks available on Omega EP have also been utilized in the oblique case, where a tilt was introduced between the shock and unstable interface to produce coupled RM and KH growth, seeded with a sinusoidal, single-mode perturbation (see Fig. 18.20 for the perturbation amplitude history and shock-interface interaction diagram). In such a case, the perturbation growth is dictated by the competition between the impulsive acceleration due to the normal component of the shock velocity, which drives the RMI, and the shear flow across the interface due to the tangential component of the shock velocity, which drives the KHI. Indeed, the experiments of Rasmus et al. (2018) showed that the growth was impulsive at early times, as in RM, but asymptoted to a KH-like state at late times.

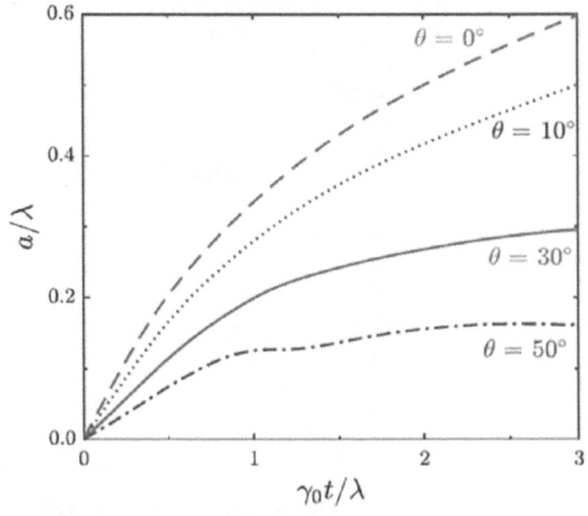

Fig. 18.21 Time evolution of the RM & KH perturbation amplitude for different tilt angles. Pellone et al. (2021), with the permission of AIP Publishing.

Since the vorticity resulting from the shock-interface interaction is confined to the interface, Rasmus et al. (2018, 2019) describe the perturbation growth using a 2D vortex-sheet model. More recently, Pellone et al. (2021) pointed out that two elements could contribute to the behavior observed in Fig. 18.20: the increased importance of shear (relative to impulsive acceleration) and saturation due to the onset of nonlinearities. This behavior is quantitatively illustrated in Fig. 18.21, which shows the time evolution of the perturbation amplitude with an increasing tilt angle, from 0° to 50°.

18.5 Reshock RMI

Welser-Sherrill et al. (2013) presented an experiment featuring two independent halfraums, one to drive a primary shock and one to drive a reshock. Haines et al. (2013) conducted 3D implicit large-eddy simulations of the reshock experiments of Welser-Sherrill et al.

Figure 18.22 shows image sequences of the instability growth obtained from several NIF shots performed with nominally identical targets and drives.[2] The initial interface perturbation is a sinusoidal ripple of wavelengths 120 μm and 200 μm, but with different initial amplitudes ($a_0 = 3$ μm and 12 μm) on the two sides of the target. The two initial side-by-side amplitudes allow us to continuously compare the "macroscopic" growth rate.

The subsequent data shows the reshock compressing the perturbation to a minimum, after which it grows at an accelerated rate. After the reshock, the mix width curves for the

[2] By using a split radiographic tracer technique, Huntington et al. (2020) were able to simultaneously measure the penetration of bubbles and spikes in a deeply nonlinear flow, yielding the full mix width. With the use of the Crystal Backlighter Imager, Do et al. (2022) were able to improve the resolution by a factor 3, down to 7 μm with a good 81% contrast imaging of hydrodynamic spikes, greatly improving the image quality.

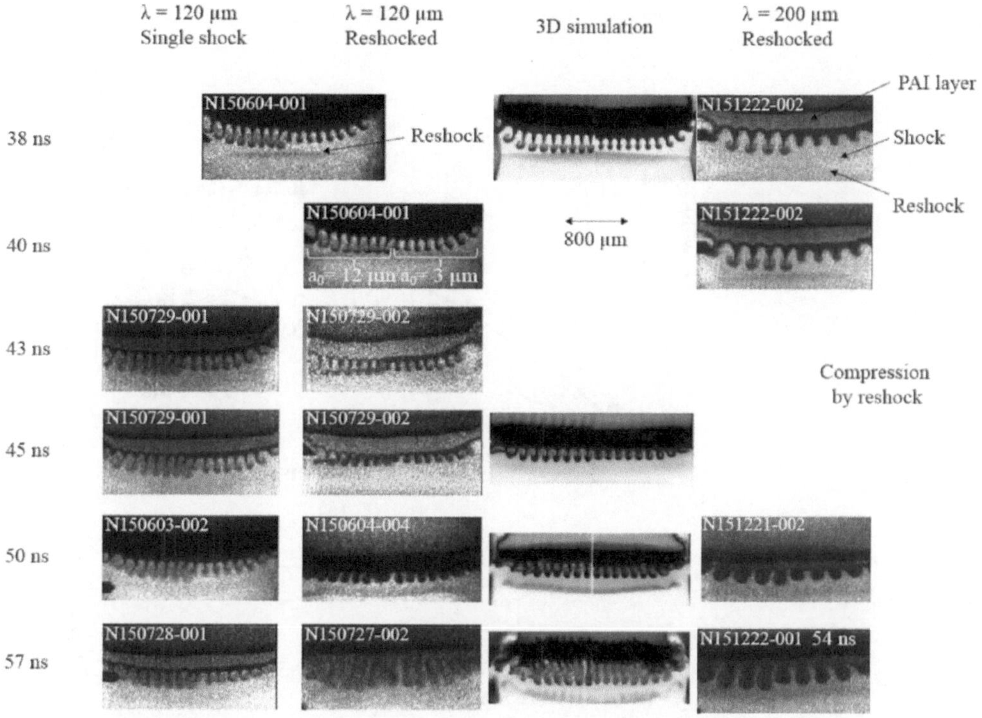

Fig. 18.22 Image sequences for different initial conditions of reshocked RMI. In each frame, the interface is driven from the top of the image and the (opposing) reshock propagates from bottom to top. Each target includes a region of smaller (right) and larger (left) initial amplitude ripples. The two initial amplitudes are 3 and 12 μm. The first image for the 120 μm wavelength (at 38 ns) is prior to reshock, and is a data point for both columns, single and reshocked. The 3rd column includes synthetic radiographs from 3D simulations of the 120 μm wavelength and two initial amplitudes of 3 and 12 μm. Nagel et al. (2022).

different initial conditions (shown in Fig. 18.23(a) and (b)) have a similar slope, indicating that the post-reshock dynamics are comparatively insensitive to the initial condition. This result is consistent with observations from many previous non-HED reshock experiments. The data demonstrate that the growth is faster after the reshock, a finding also captured by simulations.

Nagel et al. (2022) found that large-scale flow structures (comparable to the initial wavelengths, like the regular interval streaks we observe) can persist in HED experiments – if the initial perturbations have large enough wavelengths – even though there is well-developed mixing at smaller scales in regions between the bands of light material.

For a multi-mode initial perturbation, Fig. 18.24 shows six measurements of the mixing-layer width from NIF shots. This figure also plots time histories of the simulated mixing-layer width using a 1%–99% mass-fraction criterion. At $\sim$ 11 ns, the main shock arrives at the interface, initiating perturbation growth via the RM instability. At $\sim$ 31 ns, the reshock arrives at and compresses the developing mixing layer. At $\sim$ 33 ns, the rapid post-reshock

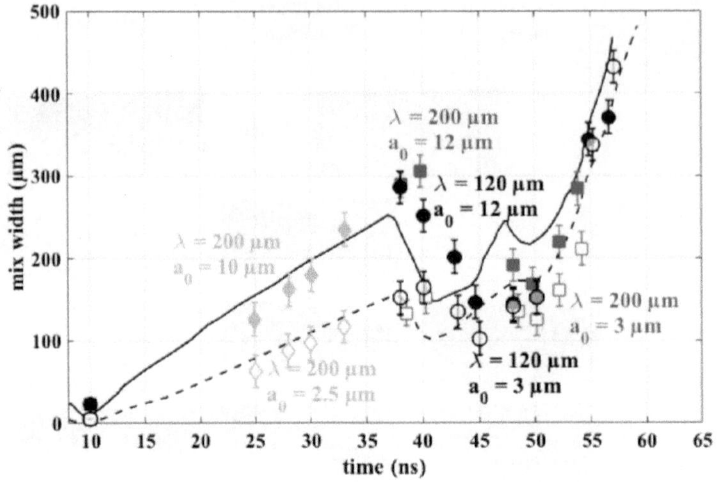

(a)

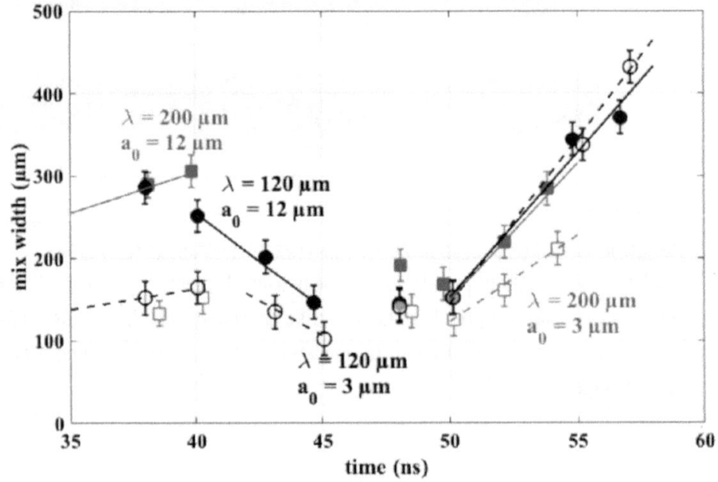

(b)

Fig. 18.23 (a) Mix width evolutions from measured NIF reshocked RMI data for 450 μm thick ablator and foam density of 119 $\pm$ 5mg/cc. The symbols represent the following: black and circles for $\lambda = 120\ \mu$m, gray and squares for $\lambda = 200\ \mu$m, and lighter gray and diamonds show data taken with a thinner shock tube, Nagel et al. (2017). The larger amplitudes ($a_0 = 12\ \mu$m) are solid symbols and solid lines while the smaller amplitudes ($a_0 = 3\ \mu$m) are open symbols and dashed lines. Note: The solid and dashed lines are mixed-width measurements from 3D simulations. (b) Focused on the reshock portion of mix width measurements shown in panel (a), experimental data only. Here the lines indicate fits to particular time intervals. Figs. 7 and 8 of Nagel et al. (2022).

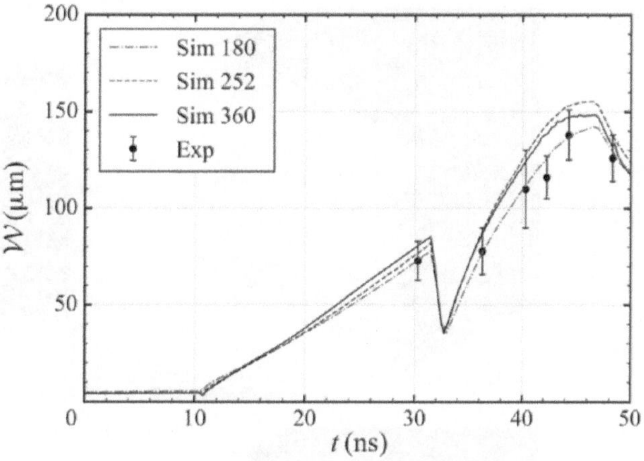

Fig. 18.24 Evolution of the reshocked RMI mixing-layer width ($\mathcal{W}$) from NIF experimental measurements (Exp; symbols) and high resolution simulations (Sim; lines) discussed in Bender et al. (2021).

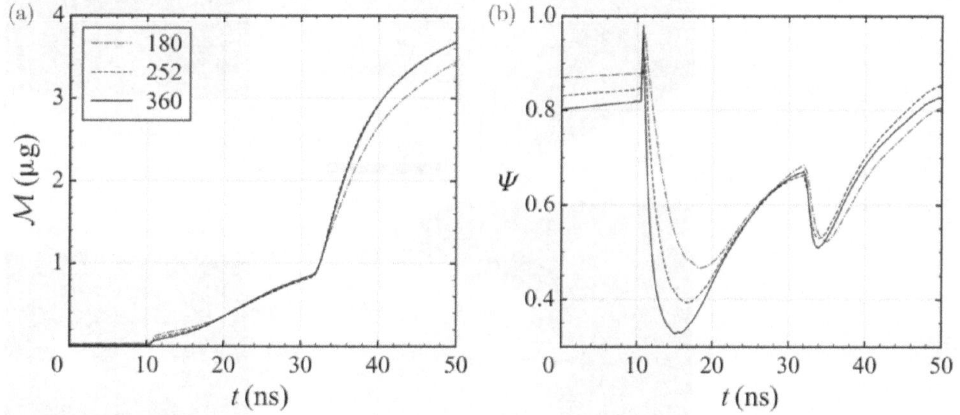

Fig. 18.25 Evolution of (a) the mixed mass and (b) the normalized mixed mass for reshocked RMI. Bender et al. (2021).

growth begins. The growth rate decreases at later times, with the maximum width reached at $\sim$ 45 ns.

As recalled in detail in Bender et al. (2021), the mixed mass, $\mathcal{M}$, is monotonically increasing and is notably not reduced by the reshock compression, illustrating its usefulness as a progress variable for mixing-layer development (Zhou et al., 2016). Growth of $\mathcal{M}$ is much more rapid after reshock than after the first shock, and a discernible increase in the magnitude of dM/dt occurs as soon as the reshock enters the mixing layer (see Fig. 18.25(a)). Figure 18.25(b) shows a related metric, Ψ, the normalized mixed mass. Before the arrival of the first shock, Ψ increases slightly due to mass diffusion. The first shock initiates perturbation growth, stretching coherent packets of the heavy and light plasmas across the interfacial region and sharply decreasing Ψ. Then, as smaller-scale structures

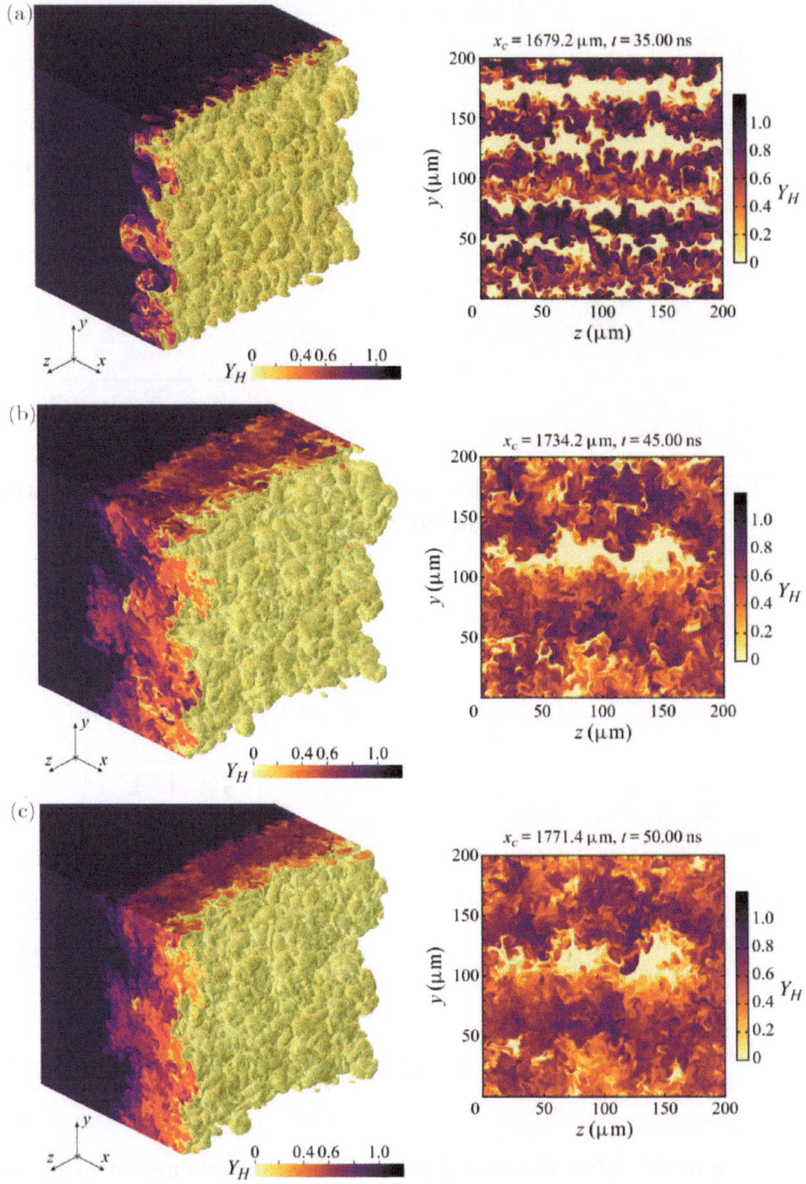

Fig. 18.26 Contours of the instantaneous mass fraction in the reshocked RMI mixing layer of the finest-resolution baseline simulation at three times after reshock: (a) 35 ns, (b) 45 ns and (c) 50 ns. Fig. 7 of Bender et al. (2021).

develop and physical and numerical dissipation ensues, Ψ rises toward values between approximately 0.80 and 0.85, with an interruption due to reshock.

The mixing-layer development can be visualized via contours of the instantaneous mass fraction. The first post-reshock figure, Fig. 18.26(a), shows the mixing layer just after reshock compression, with a significant interpenetration of the heavy and light fluids. The second post-reshock time is 45 ns (Fig. 18.26(b)) and the third post-reshock time, at the end of the simulation, is 50 ns (Fig. 18.26(c)).

In a related work, Di Stefano et al. (2020) considered multiple shocks in the laser platform. Also, Desjardins et al. (2019) presented a new experimental program to investigate mixing after three or more shocks.

18.6 Euler scaling and laboratory astrophysics

Euler scaling is an important tool that allows one to investigate astrophysical phenomena in the laboratory with intense lasers. As illustrated in Fig. 18.1, the objective of laboratory astrophysics is to study the behavior of long-time, large-spatial-scale astrophysics problems (e.g., Hammer et al., 2010) on an HEDP platform with a short-time, small-spatial-scale target. Striking similarities exist between hydrodynamic instabilities in ICF capsule and core-collapse supernova explosion simulations, as shown in Fig. 18.27.

Euler scaling relates the variables in the laboratory system to the astrophysical system. Specifically, the transformation takes the form:

$$\mathbf{r} = a\mathbf{r}_1; \ \rho = b\rho_1; \ p = cp_1; \ t = a\sqrt{\frac{b}{c}}t_1; \ \mathbf{v} = \sqrt{\frac{c}{b}}\mathbf{v}_1, \tag{18.1}$$

where the subscripted variables are those of the laboratory system and a, b, and c are scaling coefficients. A more complete discussion and applications of the Euler scaling can be found in Ryutov et al. (1999), Remington et al. (2000, 2006), and Rigon et al. (2019).

A key requirement for Euler scaling is that the Reynolds number must be high (Ryutov et al., 2000a, 2001). However, the required spatial separation for the "minimum state," discussed in Chapter 8, should be sufficient to satisfy this basic condition (Zhou, 2007).

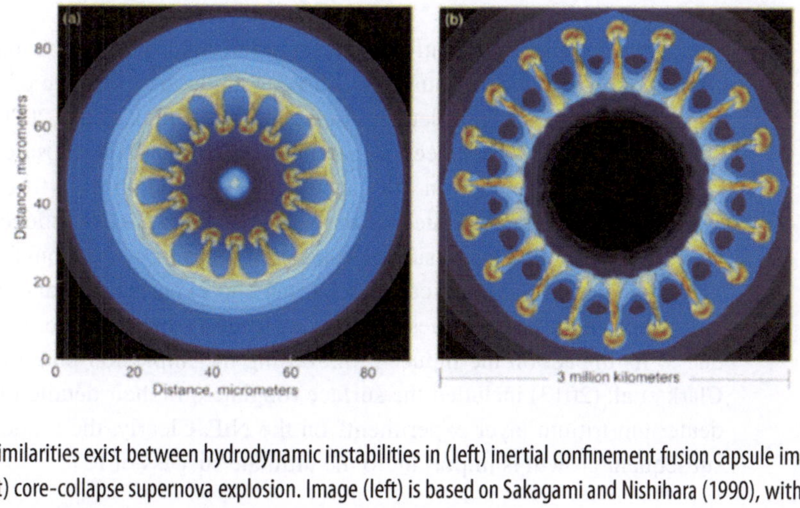

Fig. 18.27 Striking similarities exist between hydrodynamic instabilities in (left) inertial confinement fusion capsule implosions and (right) core-collapse supernova explosion. Image (left) is based on Sakagami and Nishihara (1990), with the permission of AIP Publishing; Image (right) is based on Hachisu et al. (1991).

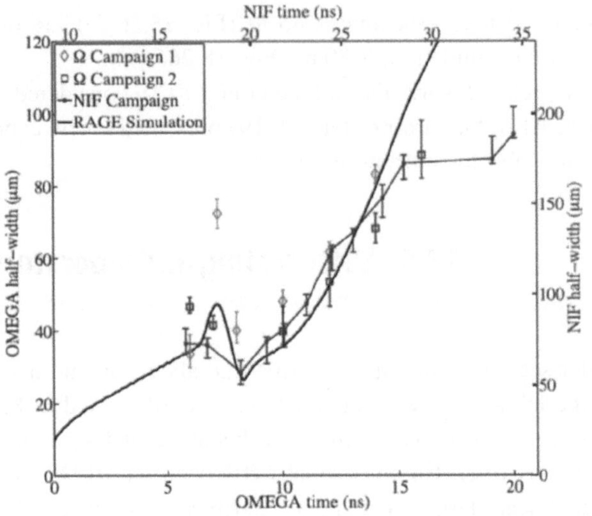

Fig. 18.28 KHI mixing layer widths as a function of time at Omega and NIF. Left and bottom axes refer to widths and times of data from Omega, while right and top axes show widths and times taken at NIF. The scaling between the two axes is such that the hydrodynamics are equivalent. Doss et al. (2015), with the permission of AIP Publishing.

Finally, Euler scaling can be utilized to relate experiments performed at HED facilities of different energetic limits. Figure 18.28 shows that the mixing layer width from the counter-flowing shear experiments taken from both Omega and NIF can be nicely collapsed (Doss et al., 2015).

18.7 Surface perturbation

One advantage of HEDP platforms is that initial conditions can be machined quite precisely on the solid-state target. Thus, one is able to study the influence of the initial perturbation on an ICF implosion in a detailed manner (e.g., Zulick et al., 2020a,b; Velikovich et al., 2020). Recent years have seen remarkable progress in NIF capsule fabrication, including significant improvements in the capsule surface smoothness (Cook et al., 2008; Nobile et al., 2006). This is illustrated by the lines in Fig. 18.29 (reproduced from Fig. 2 of Cook et al.), which correspond to surface roughness measurements from fabricated NIF capsules.

Significantly, 3D numerical simulations and experimental data show that the neutron yield of an ICF capsule varies with the surface roughness of the inner shell of the capsule, due to its impact on the induced mix during the implosion phase (Marinak et al., 1996). Clark et al. (2013) included the surface roughness in their detailed implosion modeling of deuterium-tritium layer experiments on the NIF. Clearly, the reduction of this seed or its subsequent growth is important to the ultimate success of ICF.

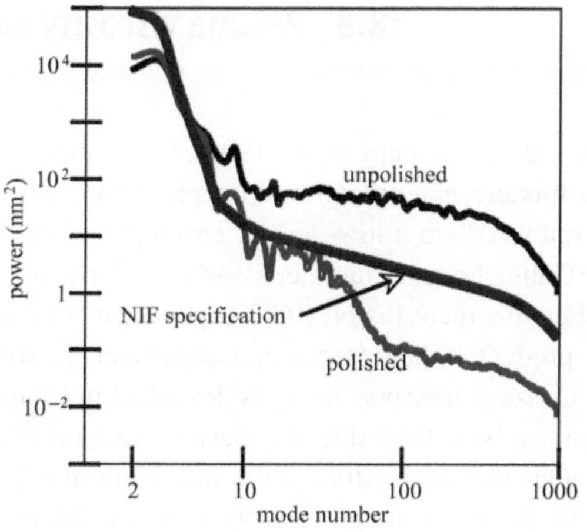

Fig. 18.29 Example of the effect of polishing on the outer surface power spectra. Fig. 2 of Cook et al. (2008).

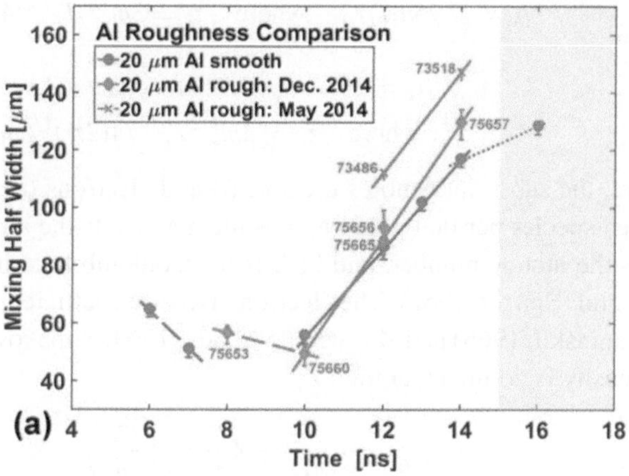

Fig. 18.30 Mix half-width for various aluminum foil surface roughnesses. The shot number for each data point is included in the plots. Fig. 8 of Merritt et al. (2015), with the permission of AIP Publishing.

As an example of the above observations, Merritt et al. (2015) used counter-propagating shear experiments carried out at the Omega Laser Facility to evaluate the effect of initial target conditions, the characteristics of a tracer foil located at the shear boundary, on Kelvin–Helmholtz mixing layer evolution. A series of experiments illustrated that increasing the surface roughness (surface perturbation scale spectrum) can increase the relative mixing in the system, see Fig. 18.30.

For single-mode reshock RMI experiments in a planar geometry, two distinctive a_0/λ values can induce rather different evolutions, as shown previously in Fig. 18.22.

18.8 Plasma viscosity and diffusivity

As stressed by Clérouin et al. (1998), the problem of estimating the viscosity of a dense plasma mixture is a crucial one for predicting astrophysical mixing of heavier core elements outward into a low-density envelope, or during the compression of a laser fusion pellet (Galmiche and Gauthier, 1996). Fuel-pusher mix is one of the major obstacles in inertial confinement fusion (ICF) experiments that has to date hindered realization of the energy production goal (see, e.g., Larroche et al., 2018; Yin et al., 2019).

The classical transport theories described in Chapman and Cowling (1990) and Spitzer (2006) must be extended to the plasma transport of a high-temperature, dense (collision-dominated) plasma mixture. Note that while most of the attention has been paid to the viscosity, the electron conductivity (Lee and More, 1984) and diffusion (Paquette et al., 1986) coefficients are also important for dense plasmas. In the absence of magnetic fields the relevant coefficients of viscosity are given by Braginskii (1965):

$$\mu_i = 0.96 n_i T_i \tau_i, \quad \text{where} \quad \tau_i = 3m_i^{1/2} T_i^{3/2} / 4\pi^{1/2} n_i Z^4 e^4 \ln \Lambda, \tag{18.2}$$

$$\begin{aligned} \mu_e &= (0.73 \to 1.81) n_e T_e \tau_e, \quad (Z = 1 \to \infty), \\ \text{where} \quad \tau_e &= 3m_e^{1/2} T_e^{3/2} / 4(2\pi)^{1/2} n_i Z^2 e^4 \ln \Lambda \end{aligned} \tag{18.3}$$

Here, the subscript denotes the ions (i) and electrons (e), n is the number of particles of a given species per unit volume, m is the mass, T is the absolute temperature of the plasma, Z is the atomic number, and $\ln \Lambda$ is the Coulomb logarithm, proportional to the collision integral (Spitzer, 1962). The electron viscosity coefficients have been determined for $Z = 1$ (Braginskii, 1965) and $Z \to \infty$ (Whitney, 1999; Simakov and Molvig, 2014). The electron viscosity is dominant at high Z,

$$\frac{\mu_e}{\mu_i} \sim Z \frac{\tau_e}{\tau_i} \sim \left(\frac{m_e}{m_i} \right)^{1/3} Z^3. \tag{18.4}$$

Equations 18.2–18.4 indicate there is an ion mass and charge dependence. They are applicable at high temperatures when the plasma resembles a gas, rather than a liquid, and are valid when the ionic coupling is small (Baus and Hansen, 1980; Ichimaru et al., 1987). The plasma coupling parameter

$$\Gamma_i = \frac{Z^2 e^2}{k_B T_i a_i} \quad \text{where} \quad a_i = \left(\frac{3}{4\pi n_i} \right)^{1/3}, \tag{18.5}$$

is the ratio of the energy due to Coulomb forces to the thermal energy.

Plasma viscosity increases in both the high- and low-temperature limits (Dufreche and Clerouin, 2000; Saigo and Hamaguchi, 2002). Murillo (2008) advanced a practical method of computing the viscosity based on mapping the system of interest onto the Yukawa model; he compared the model with quantum molecular dynamics results for compressed iron, relevant to the Earth's core, and to the experimental data for many liquid metals, as well as to simulation results for dense deuterium, relevant to inertial confinement fusion

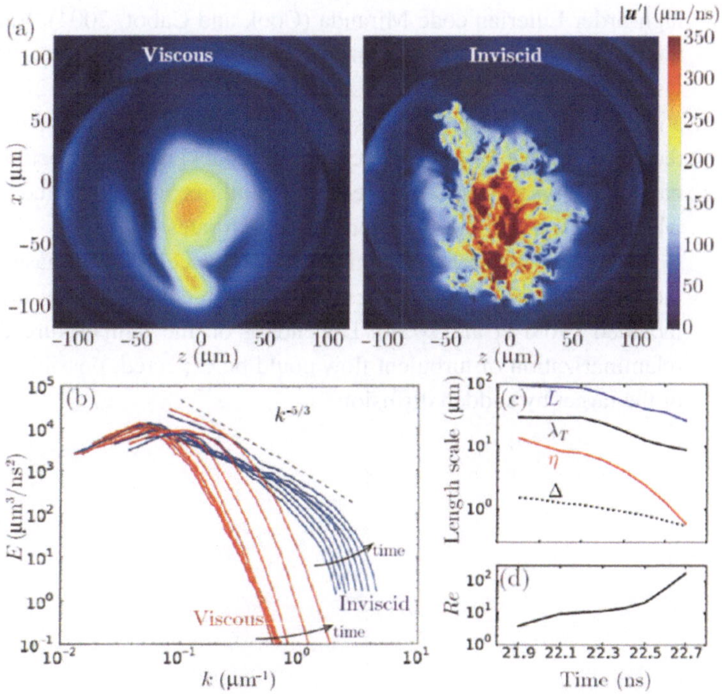

Fig. 18.31 Viscous effects on the fluctuating velocity. (a) Fluctuating velocity magnitude at 22.21 ns with (left) and without (right) viscosity. (b) Kinetic energy spectra in the hot spot from the viscous and inviscid simulations from 21.9 to 22.6 ns. (c) Turbulent length scales and (d) Reynolds number from the hot spot. Fig. 4 of Weber et al. (2014b), with permission.

experiments. An improved shear viscosity formulation for asymmetric plasma mixtures at thermodynamic conditions relevant to astrophysical and inertial confinement fusion plasmas has been achieved largely with assistance from large-scale molecular dynamics (MD) calculations.

Haxhimali et al. (2015) proposed a simple hybrid model that interpolates between the strong and weak coupling limits, with Murillo's Yukawa viscosity model and classical transport theory, respectively. In the latter, the collision integrals in the 1st approximation of Chapman and Cowling's formulation are tabulated by Paquette et al. (1986) or Stanton and Murillo (2016) for static screened Coulomb (Yukawa) potentials with weak to moderate coupling. Higher-level approximations for classical transport theory formulated by Simakov and Molvig (2016a,b) are only valid in the weakly coupled (Lorentz) limit. Ticknor et al. (2016) have developed similar hybridizations for viscosity and diffusivity in asymmetric binary plasma mixtures using one-component plasma (OCP) models and classical transport theory. Over a wide range of thermodynamic conditions, these hybrid models compare reasonably well with molecular dynamics simulation results.

In spherical geometry, Weber et al. (2014b, 2015) and Clark et al. (2015) carried out 3D simulations including radiation and viscosity, all of which act to suppress, but not eliminate the RT instability. Weber et al. (2014b) presented large-eddy simulations with the high-order Eulerian code Miranda (Cook and Cabot, 2004), linked from simulations with the ALE code HYDRA (Marinak et al. (1996, 1998)). Figure 18.31 illustrates the plasma viscosity's role in smoothing small scale fluctuations in a 2D ICF fuel region. Applying a normal mode analysis, Gerashchenko and Livescu (2016) studied the viscous effects on the compressible RTI in the presence of a background temperature gradient. Using a set of parameters taken from Weber et al., the authors also illustrated the importance of viscosity, which can significantly dampen the growth rates.

Recent ICF work has examined the RT deceleration phase instability. In this regime, the plasma transport properties of viscosity and diffusivity are important and should be included (Vold et al., 2021). Depending on the temperature dependence of viscosity, a relaminarization of turbulent flow could be expected, possibly inducing a massive mixing of the target by sudden diffusion.

19 Inertial confinement fusion implosion

Inertial confinement fusion (ICF) is an approach to fusion that relies on the inertia of an imploding fuel mass to provide confinement (Nuckolls et al., 1972; Lindl, 1998; Atzeni and Meyer-ter Vehn, 2004; Hurricane et al., 2023) and that has the goal of making fusion energy a viable alternative energy source. In a pioneering paper, Nuckolls et al. (1972) first noted the extreme densities and pressures achieved in astrophysical settings. Indeed, hydrogen is believed to exist at more than one thousand times liquid density and at a pressure greater than 10^{11} atmospheres (temperature $\sim 1-2$ keV) in the center of the Sun (Schwarzschild, 2015).

Nuckolls et al. (1972) pointed out that hydrogen may be compressed to more than 10,000 times liquid density by an implosion system energized by a high energy density laser. As detailed in Lindl (1998) and Atzeni and Meyer-Ter-Vehn (2004), ICF aims to achieve thermonuclear fusion by compressing deuterium-tritium (DT) fuel to extreme densities and temperatures. ICF requires both heating the fuel to sufficiently high temperatures, in order to initiate fusion burn (i.e., reach a sufficiently large DT fusion reaction cross section) and also to compress the fuel to high densities so that the inertia of the fuel itself can confine the burning plasma long enough to achieve significant fuel consumption.

As the shock travels through the DT mixture and reaches the center, it heats and compresses the mixture to hundreds of millions of Kelvin and billions of atmospheres, causing the fuel to ignite in a nuclear fusion reaction. The surrounding fuel, having been accelerated inward by the shock, is confined close to the ignition point by its own inertia – hence the name – and can thus maintain the fusion burn. However, since the density of the fuel mixture is not uniform, RT and RM instabilities are excited and can promote mixing between the different layers, disrupting the spherical symmetry of the system and inhibiting confinement and fusion burn (Zhou et al., 2025).

More specifically, ICF capsules are rapidly compressed to a small fraction of their initial radius so that both geometric compression and convergence effects play a pronounced role. The physics involved is complex, as the compression of an ICF capsule typically involves multiple shocks, each of which modifies the RMI growth rate (e.g., Mikaelian, 1985). As the implosion decelerates and the fuel core begins to slow the inward motion of its higher-density shell, the RTI will further amplify the perturbation and mixing. Both baroclinic instabilities can act to degrade the overall yield of the ICF target (Atzeni and Meyer-ter Vehn, 2004; Lindl, 1998).

In the U.S., a number of intense laser based platforms with increased capabilities have been built over the years (see Chapter 18), with the National Ignition Facility (NIF) recently achieving an igniting fusion fuel (thermonuclear instability). The results of this record experiment have been published in three papers (Abu-Shawareb et al., 2022; Kritcher et al., 2022; Zylstra et al., 2022).[1]

The December 2022 NIF experiment was the first to reach scientific energy breakeven in the laboratory by delivering 2.05 megajoules (MJ) of energy to the target, resulting in 3.15 MJ of fusion energy output (Abu-Shawareb et al., 2024; Hurricane et al., 2024; Kritcher et al., 2024; Pak et al., 2024).[2] The fusion reactions happened within just 0.1 billionths of a second, temporarily exceeding the global power usage of all humans.

When complete, the French Laser Mégajoule (LMJ), under construction near Bordeaux, France (André, 1999; Tassart, 2004; Fleurot et al., 2005; Bourgade et al., 2008; Casner et al., 2015; Liberatore et al., 2023), will be very similar to NIF. Very large facilities such as NIF and LMJ provide excellent opportunities in the effort to achieve ignition, a necessary step to make fusion energy a feasible alternative energy source (Edwards et al., 2013; Danson and Gizzi, 2023).

In the United Kingdom, the Vulcan Petawatt laser at the Rutherford Appleton Laboratory has been used to carry out interesting experimental work on RTI (Palmer et al., 2012) and developed turbulence and nonlinear amplification of magnetic fields (Meinecke et al., 2015; Gregori et al., 2015). Hopps et al. (2013, 2015) described in detail the Orion laser facility at the Atomic Weapons Establishment (AWE), which has been operational since April 2013, fielding experiments that require both its long- and short-pulse capability. The most advanced Russian ICF facility is ISKRA-6 at the Institute of Experimental Physics, Russian Federal Nuclear Center in Sarov (e.g., Galachov et al., 1999; Karlykhanov et al., 2004; Chizhkov et al., 2005). The ShenGuang-III (SG-III) laser facility was developed by the Laser Fusion Research Center, Chinese Academy of Engineering Physics, Mianyang, China for ICF studies in China (e.g., Jiang et al., 2018b; Pu et al., 2018; Zhang et al., 2005; Zheng et al., 2008b).

Separately, the Z-pinch configuration was recognized and analyzed before the development of such concepts as ICF (see, e.g., Bennett (1934) and Tonks (1937) for early analyses of it). Its potential for controlled fusion, by constricting the interior plasma to the required temperature and pressure, was developed in the 1950s (Bishop, 1958). The Z-pinch, the most powerful of which was built in Sandia National Laboratory (Spielman et al., 1998), is designed based on converting the stored electrical energy in a pulsed-power accelerator into X-rays (Matzen, 1997; Ryutov et al., 2000b; Sinars et al., 2020). Outside the U.S., Z-pinch experiments have been performed on the mega-ampere generator for plasma implosion experiments (MAGPIE) pulsed power facility at Imperial College, London (Lebedev et al., 2001, 2005) and at the Angara-5 facility in Russia (Al'bikov et al., 1990; Grabovskii et al., 2004). Deng et al. (2016) and Li et al. (2019b) provided a broad overview of major Z-pinch research carried out at the primary test stand (PTS) and several institutions in China, respectively. Recently, Wang (2019b) and Wu et al. (2019) also summarized relevant studies engaged in Tsinghua and Xi'an Jiaotong Universities.

[1] Laser Fusion Experiment Unleashes an Energetic Burst of Optimism, Kenneth Chang, *The New York Times*, Section D, Page 1, Aug. 24, 2021.

[2] A Blast of 192 Lasers Achieves a Breakthrough in Nuclear Fusion, Kenneth Chang, *The New York Times*, Section A, Page 1, Dec. 14, 2022.

In this chapter, we will illustrate the significant impediment hydrodynamic-instability-induced mix represents in the quest to achieve ignition. After a brief discussion of the relevant laser drives, the so-called central-spot ignition and several alternative designs will be discussed to illustrate the roles played by hydrodynamic instabilities in these platforms.

19.1 Direct and indirect laser drives and magnetic drive with pulsed power

The route toward inertial fusion has been investigated through both a laser-direct drive (LDD) (Craxton et al., 2015; Campbell et al., 2020a) and a laser-indirect (X-ray) drive (LID) (Lindl, 1995) scheme (Fig. 19.1). In the LDD, the spherical shell is directly irradiated by a laser incident on the capsule's outer surface. Direct drive (DD) exhibits higher conversion efficiency from laser energy to shell kinetic energy than does indirect drive (ID),

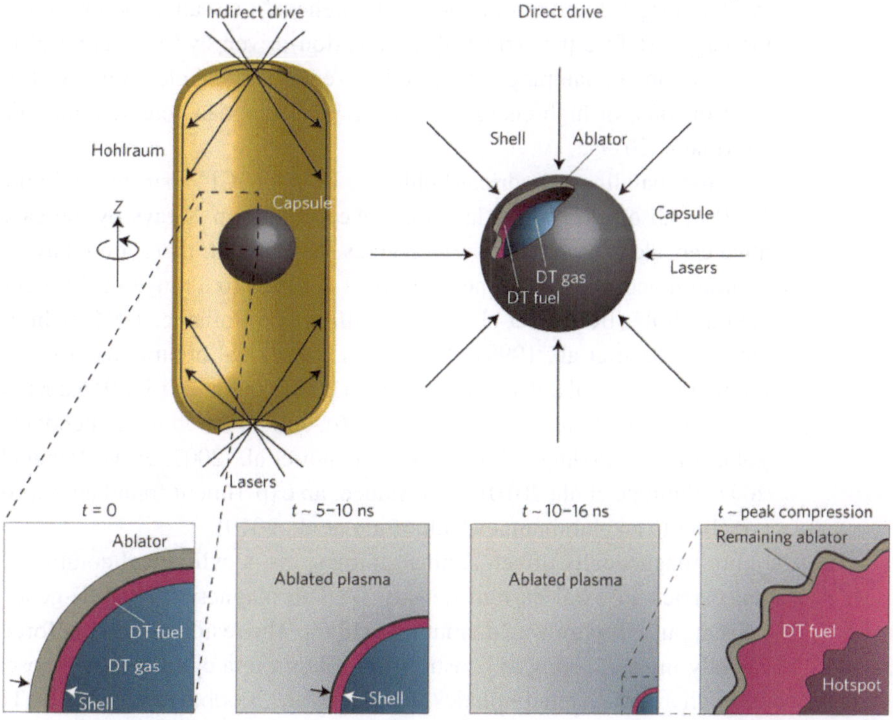

Fig. 19.1 Typical targets used in laser-driven ICF are indirectly driven (upper left) or directly driven (upper right). In either case, a spherical capsule is prepared at $t = 0$ with a layer of DT fuel on its inside surface. As the capsule surface absorbs energy and ablates, pressure accelerates the shell of remaining ablator and DT fuel inwards – an implosion. By the time the shell is at approximately one-fifth of its initial radius it is traveling at a speed of many hundreds of kilometers per second. By the time the implosion reaches minimum radius, a hot spot of DT has formed, surrounded by colder and denser DT fuel. Adopted from Betti and Hurricane (2016).

thereby allowing an implosion of greater DT fuel mass. A disadvantage of direct drive is a reduced uniformity of the illumination (Betti and Hurricane, 2016). Experiments in which the capsule is driven directly by laser beams (direct drive) are being studied, principally at the University of Rochester's Laboratory for Laser Energetics (LLE) and U.S. Naval Research Laboratory.

A LID target takes the form of a spherical cryogenic capsule consisting of an ablator material, DT solid fuel, and DT gas, see Fig. 19.1. The target is encased in a cylindrical high-Z enclosure or hohlraum with two laser-entrance holes (LEHs) at opposite ends (Lindl et al., 2004). At the NIF, the indirect drive approach to ICF is being pursued where NIF's 1.8 MJ of laser energy are first converted to soft X-rays inside of a high-Z hohlraum. The resulting near-Planckian bath of X-rays then ablates the outer surface of the spherical capsule suspended in the middle of the hohlraum, driving strong shocks into the capsule to implode and heat the fuel.

As laser energy is transported into the ablator, it heats up, either by electron conduction in the case of DD, or by X-rays in the case of ID. As the ablator ionizes in response to the heating, pressure is generated that drives flows, compression waves, or shock waves into the target. In addition, so-called "preheat" may alter the properties and geometry of the target. Before the arrival of the ablation-driven hydrodynamic disturbance, processes such as Landau damping can drive high-energy "hot" electrons (tens to hundreds of kilo-electronvolts) or high-energy X-rays (2 keV) that can penetrate the target (Hurricane and Herrmann, 2017).

In the mainline cylindrical-hohlraum LID for ICF, a strong limitation is the inefficient ($\leq$10%) absorption of the laser-produced hohlraum X-rays by the capsule, as set by relative capsule-to-hohlraum surface areas. As a result, there are many attempts to alter the hohlraum geometry from the traditional cylinder to a range of different shapes, including "tetrahedral" (Bennett et al., 2000; Phillion and Pollaine, 1994; Schnittman and Craxton, 1996; Wallace et al., 1999), "three-axis cylindrical hohlraums" (Kuang et al., 2016), "I-raum" (Robey et al., 2018), foam-lined (Zhang et al., 2015, 2016a; Moore et al., 2020a,b), "octahedral hohlraums" (Lan et al., 2016), and "frustraum" (Amendt et al., 2019). Using a gold, rugby-ball-shaped hohlraum (Amendt et al., 2007, 2008; Vandenboomgaerde et al., 2007; Philippe et al., 2010), for instance, an experiment found an improved $\sim$30% energy coupling to an aluminum capsule (Ping et al., 2019).

For magnetically driven Z-pinch implosions, a cylindrical column of conducting material carries an axial current density j_Z. The magnetic field self-generated by the rising current pulse is purely an azimuthal field B_θ. The resultant Lorentz force ($\mathbf{j} \times \mathbf{B}$) is directed radially inward, acting to constrict the column in a cylindrical collapse, and increasing in strength as the system implodes (Ryutov et al., 2000b; Haines, 2011). The earliest work on Z pinches was undertaken in Holland by Martinus van Marum in 1790 (Haines et al., 2000).

In experiments on the Z machine at Sandia National Laboratories, the Z-pinch plasma was formed from an initially solid metal tube commonly referred to as a liner. A rising current pulse is applied along the liner's axis (the z-axis) with a rise time of 100 ns and a peak current of about 20 MA, which creates a radially-inward force on the liner and turns it into plasma. Indeed, such implosions require multi-Mbar pressures surpassing the yield strength of the initially solid targets (Schmit et al., 2016). This results in

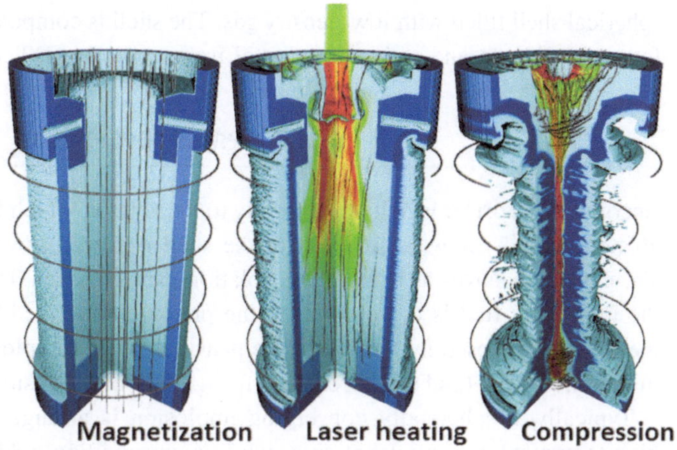

Fig. 19.2 A schematic representation of the three critical components of the MagLIF concept. An axial current creates a force that is used to implode a gas-filled, premagnetized, cylindrical target. Near the start of the implosion, the fuel is heated by the laser. The liner compresses and further heats the fuel to fusion-relevant conditions at stagnation. Gomez et al. (2014).

accelerating, convergent shells of electrically conducting fluid (namely, dense plasma) subject to magnetically enhanced RTI (Harris, 1962). The RTI may compromise the integrity of magnetically imploded liners (Awe et al., 2013; McBride et al., 2013; Yu et al., 2023a,b).

The recently proposed Magnetized Liner Inertial Fusion (MagLIF) concept (Fig. 19.2) involves using a pulsed electrical current to drive the implosion of an initially solid cylindrical metal tube (liner) filled with preheated and premagnetized fusion fuel (deuterium or DT) (e.g., Slutz et al., 2010; Slutz and Vesey, 2012; Cuneo et al., 2012; Gomez et al., 2014, 2020; Yager-Elorriaga et al., 2022). Figure 19.2 illustrates the three critical components of MagLIF: magnetization, laser heating, and compression. More specifically, the fuel is preheated to 100–300 eV to ease liner-convergence requirements, while the cylindrical liner and fuel are premagnetized with a slow-rising, fully-diffused axial field ($B_{z,0}$) of 10 to 30 T, which is then rapidly compressed by the metallic, flux-conserving liner to $B_{z,fuel} > 1000$ T (Awe et al., 2013, 2016). In the fuel, the ultra-high field reduces heat loss and enhances alpha particle energy deposition (Lindemuth and Kirkpatrick, 1983; Basko et al., 2000).

19.2 Hot spot ignition

In this section, a basic description of the so-called central-spot ignition scheme, which has dominated ICF for decades, is given. To achieve conditions under which inertial confinement is sufficient for efficient thermonuclear burn, high-gain ICF targets have features similar to those shown in Fig. 19.1 (Betti and Hurricane, 2016). A capsule is generally a spherical shell filled with low-density gas. The shell is composed of an outer region, which forms the ablator, and an inner region of frozen or liquid DT, which forms the main fuel.

19.2.1 Central-spot ignition

Energy from a driver is delivered rapidly to the ablator, which heats up and expands. As the ablator expands outward, the rest of the shell is forced inward to conserve momentum. The capsule behaves as a spherical, ablation-driven rocket. The work that can be done on the imploding fuel is the product of the pressure generated by the ablation process and the volume enclosed by the shell. The peak achievable implosion velocity determines the minimum energy (and mass) required for ignition of the fusion fuel in the shell.

Typically, a spherically converging implosion is arranged to heat a low-density, but high-temperature, central hot spot, while a surrounding colder fuel layer is compressed to high density, but at much lower temperatures, to achieve the necessary confinement (Rosen, 1999). This separation into a high-temperature hot spot ($\sim$10 keV) and a cold but dense main fuel (>500 g/cm^3) minimizes the overall energy investment necessary to reach ignition conditions. Ideally, if the central hot spot is large enough that the fusion product α-particles slow down within and self-heat the hot spot, run away fusion ignition can occur, propagating a detonation wave into the surrounding dense fuel. At a hot spot density of $\sim$100 g/cm^3, this requires a hot spot radius of $\sim$30 μm and an energy investment of $\sim$40 kJ. With $\sim$100 mg of surrounding DT fuel compressed to >500 g/cm^3, a burn duration of $\sim$50 ps is possible resulting in a burn-up fraction of $\sim$1/3 and a yield of $\sim$10 MJ. Assuming a total coupling efficiency of $\sim$10%, this idealized scaling suggests that substantial energy gain is possible with only $\sim$400 kJ of absorbed energy. Decades of experimental and theoretical ICF research have shown this scaling to be highly optimistic, however, in large part due to the impact of hydrodynamic instabilities during the implosion.

19.2.2 Ablative Rayleigh–Taylor instability

At the ablation front, the ablation of the outer capsule surface exerts a strong stabilizing effect on RT growth (Bodner, 1974; Takabe et al., 1985; Kilkenny et al., 1994). This stabilization is essentially due to the "polishing away" of the perturbations that occur under ablation: the steepened temperature gradients in the neighborhood of a perturbation maximum causes enhanced heat flux and elevated drive pressure near that maximum, both of which tend to suppress the perturbation maximum, reducing its growth. Sufficiently short wavelengths are, in fact, completely stabilized and only a finite spectrum of modes can grow in the presence of ablation, unlike in the classical case.

Here, a slightly enhanced discussion will be provided on the ablation-front Rayleigh–Taylor instability (ARTI) (Bodner, 1974; Kilkenny et al., 1994; Kull and Anisimov, 1986; Sanz, 1994; Betti et al., 1995; Goncharov et al., 1996a,b; Budil et al., 1996, 2001; Glendinning et al., 1997, 2000; Li et al., 2020b).

An expression for the linear growth-rate of ablation driven RTI, the Takabe formula, is given by (Takabe et al., 1985)

$$\gamma_{ARTI} = \alpha_A \sqrt{kg} - \beta_A k V_a, \tag{19.1}$$

where k is the perturbation wavenumber, V_a is the ablator velocity (speed at which the denser fluid interface is eroded away), and α_A and β_A are numerical factors fitted to the eigenvalue data (Takabe et al., 1985; Bodner, 1974). After a number of computational and experimental studies, the value of β_A is now recognized to be in the range 1–4, depending on the ablation scheme (Takabe, 2004).

The model has also been used in a modified form, the so-called modified Takabe formula (Betti et al., 1993; Kilkenny et al., 1994; Budil et al., 2001),

$$\gamma_{ARTI} = \alpha_A \sqrt{\frac{Akg}{1 + kL_\rho}} - \beta_A k V_a, \tag{19.2}$$

where L_ρ is the density gradient scale-length of the ablation front. Budil et al. (1996) found that this simple model does a reasonable job of reproducing the Nova laser data using a value of $\beta_A = 3$. The classical RT growth-rate, $\gamma_{RT} = \sqrt{Akg}$ is easily recognizable in Eq. 19.2. It is also worth noting that the $1 + kL_\rho$ factor comes from a paper by LeLevier et al. (1955) on density gradient stabilization.

19.2.3 Nonlinear stage of ablative RTI

Ablation also influences the nonlinear RTI stages. Betti and Sanz (2006) derived an ablation driven reacceleration stage, in which the bubble velocity exceeds the Goncharov terminal velocity. This increase is caused by the centrifugal force from swirling fluid inside the bubble, and is thus related to the bubble interior vorticity. The vortex is shown in the schematic (Fig. 19.3). The spike (gray) is ablated into the bubble by heat conduction and other energy fluxes, causing a counter-clockwise vortex. The other half of the bubble (not shown) has a clockwise rotating vortex. There is minimal vorticity in the dense material to counteract the additional pressure at the bubble vertex from the fluid rotation. This additional pressure increases the terminal bubble velocity to $\sqrt{v_0^2 + \rho_1 \zeta^2/(4k^2 \rho_2)}$, where ζ is the vorticity in the bubble[3] and v_0 is the classical terminal velocity. If the ablation speed V_a is known, Betti and Sanz reasoned through mass conservation that the vortex rotation rate must be $\zeta \simeq kV_a \rho_2/\rho_1$, with vortex radius $\simeq \lambda/2$.

The increased terminal velocity acts in contrast to ablation in the linear regime, which is stabilizing. In three dimensions, the situation is exacerbated even further beyond the Betti and Sanz model. In the 3D simulations of Yan et al. (2016), with heat flux in an accelerated reference frame, small diameter bubbles did not reach a terminal velocity. The vorticity accumulation and bubble acceleration continued indefinitely.

In conclusion, ablation may reduce the bubble growth or even stabilize RTI completely for large k and $ka < 1$. However, if a perturbation reaches $ka > 1$ (or starts that way) then ablation may drive vortices that actually increase the RTI growth above the classical rate for all k. This imposes obvious tolerances on the defect amplitudes in ICF capsules and the drive asymmetry.

[3] As noted in footnote 3 of Chapter 2, this book follows the notation convention of the vorticity from the atmospheric sciences community, ζ, instead of ω used by Betti and Sanz and many other authors.

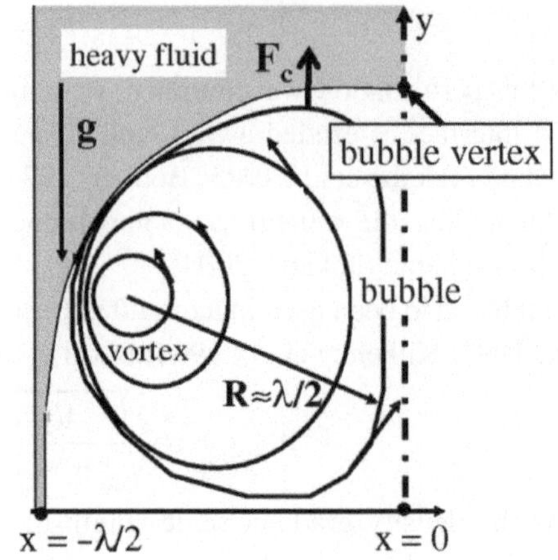

Fig. 19.3 Schematic of the bubble (white) and spike (gray) structure in the ablative RTI. Ablation of the spike tip causes a transverse mass flux from the spike into the bubble, which creates a vortex in the bubble fluid. The rotation velocity can exceed the bubble velocity. The vortex centrifugal force F_c forces the bubble front upwards at a velocity exceeding the classical value. Fig. 1 of Betti and Sanz (2006), with permission.

Recently, Xin et al. (2019) studied the coupling and evolution of two-mode ARTI in 2D geometry via numerical simulations. Furthermore, Zhang et al. (2020a) found that the nonlinear ARTI is dominated by bubble competition, which is dependent on both the initial perturbation amplitude and the vorticity effect. As a result, ablative stabilization of the nonlinear ARTI in ICF may not be as effective as expected if the initial perturbation is significant.

As for the spike velocity in the nonlinear stage, there is little consensus in the literature. However, as it is the bubbles that penetrate and break up the dense ICF shells, the bubble growth is the key quantity that must be limited in experiments. Ablation appears to erode away the spikes in both the linear and nonlinear stages. Wang et al. (2012) found in simulations that KHI is suppressed at spikes in ablation fronts, and the resulting mushroom caps do not appear. As thermal conductivity increases with temperature, ablation increases further as the spikes penetrate the hotter fluid. This causes the spikes to decelerate in the very late RTI stages (Ye et al., 2010). This is in contrast the case of RTI at large Atwood number without ablation, in which spikes never reach a terminal velocity.

There are two further remarks we would like to make. Liu et al. (2023b) studied the dynamic of shock-bubble interactions and nonlinear evolution of ablative hydrodynamic instabilities initialed by capsule interior isolated defects. Li et al. (2022b) and Zhang et al. (2023a), among others, considered the effects of high-intensity heat conduction on turbulence induced by the ARTI.

19.2.4 Various phases of hydrodynamic instabilities

The high compression necessary to achieve the relevant densities for ICF at laser energies in the MJ range inevitably implies a high velocity implosion (>350 km/s) and

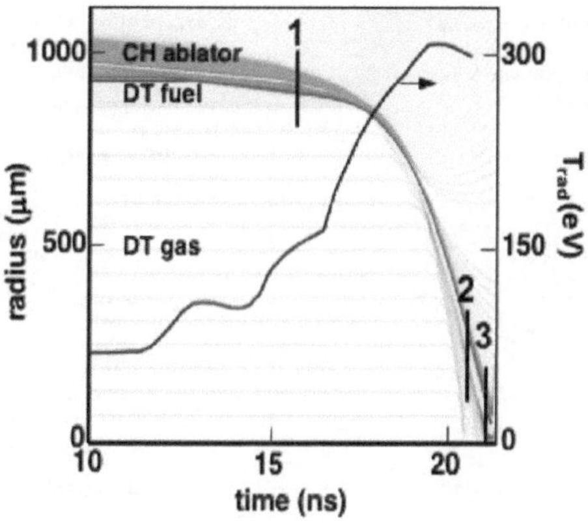

Fig. 19.4 r-t diagram and radiation temperature of a typical ICF implosion. Zhou et al. (2019), Phys. Plasmas.

high convergence. The imploding DT-containing shell should have a convergence ratio of $C_R \equiv R_0/R_{hs} > 30$ (Hurricane et al., 2014), where R_0 and R_{hs} are the initial shell and final hot spot radius, respectively. In turn, this high convergence and acceleration to high velocity provide ample opportunity for hydrodynamic instabilities to develop, predominantly the well-known RT and RM instabilities (Zhou et al., 2019).

Significant perturbation growth of the interface occurs throughout the acceleration phase due to RTI, and during the coasting phase, to Bell-Plesset (BP) related convergence effects (Bell, 1951; Plesset, 1954). After the convergent shock reflects from center and returns to the shell, several (typically two) impulsive deceleration (RM growth) phases follow before a continuous deceleration (RT growth) phase. BP convergent effects are magnified during the deceleration phase because of the large convergence ratio (Temporal et al., 2006).

Details of a typical ICF implosion are shown in Fig. 19.4. The figure shows radius vs. time of an imploding ICF capsule from a 1D Lagrangian simulation. The different traces show the different materials making up the shell: the interior volume is filled with low-density DT gas, the main fuel is a thin layer of DT ice, and the outer shell is composed of various layers of doped plastic (CH) that are ablated by the incident X-ray flux. The time dependence of the hohlraum radiation temperature (or X-ray flux) that drives the implosion is shown by the thick black curve with the scale on the right. Each step in the radiation temperature launches a shock into the capsule and ultimately accelerates the shell to a velocity of more than 350 km/s by the peak of the X-ray drive. Stagnation, and hopefully fusion ignition, occurs in the last few hundred picoseconds of the implosion.

Line-outs of the shell density from the three representative times marked in Fig. 19.4 are shown in Fig. 19.5. During the first time shown in Fig. 19.5, several strong shocks can be seen propagating through the DT fuel and plastic ablator shell. As the shocks cross the various interfaces in the capsule, including the ablation front (or outer surface of the imploding capsule), RM instabilities can develop. While the RM modulation amplitudes are generally quite small due to the very smooth surfaces of the initial capsule and grow only slowly with time, RM growth is important in setting the initial amplitudes (and phases) for subsequent RT growth (Peterson et al., 2014; Clark et al., 2014; Landen et al., 2016).

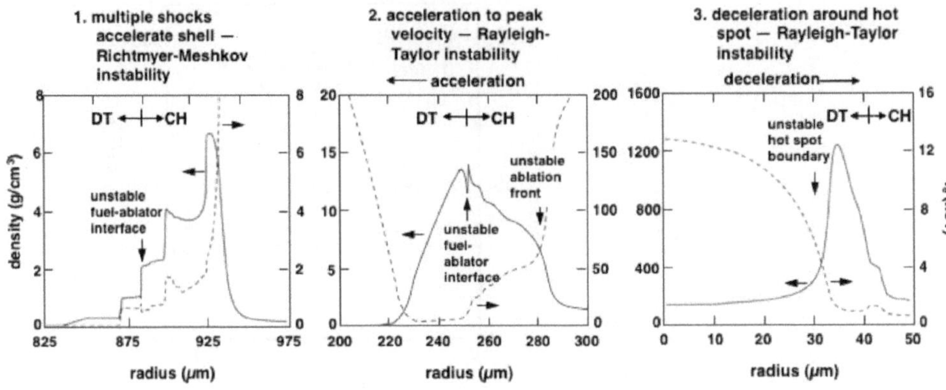

Fig. 19.5 Distinctive hydrodynamic instabilities during three phases of ICF implosion process. Note the change of the scales. Zhou et al. (2019), Phys. Plasmas.

After the shocks break out from the inside of the shell, the shell begins to accelerate, and a subsequent RT unstable phase of the implosion begins, as represented by the second time in the figure. This phase again involves instability growth at both the ablation front and at any interior interfaces with an unstable density gradient.

19.2.5 Instabilities at interior interfaces

At the interior interfaces, particularly the interface between the DT fuel and plastic ablator, there is no ablative stabilization and classical RT instability growth can occur (Hammel et al., 2010; Clark et al., 2011). The only means of controlling this growth is by managing the amplitude of the unstable Atwood number at this interface throughout the implosion. It is precisely for this reason that the dopant layers shown in Fig. 19.4 are added to the plastic ablator. By doping the plastic and increasing its opacity, the interior layers of the ablator are subject to less X-ray heating from the driving X-ray radiation. This causes the interior ablator layers to remain relatively cool, and hence dense, if they are compressed to the same pressure. In turn, this reduces the unstable Atwood number and subsequent instability growth between the DT fuel and ablator. This interior stabilization comes at a cost, however, in that adding dopant to the ablator also steepens the outside ablation front during the implosion, causing that surface to become more RT unstable. ICF implosion design, and indeed the success of ICF generally, depends on properly managing these competing instability demands.

Finally, in the last few hundred picoseconds of the implosion, the shell stagnates to form a low density but high temperature and high pressure hot spot. In this phase, labeled "3" in Figs. 19.4 and 19.5, the low density hot spot now decelerates the denser shell, and RM and RT instabilities can again develop at the interface between them. If the perturbation seeds are large enough or the instability growth fast enough during this phase, perturbations of the hot spot-main fuel boundary can grow to amplitudes large enough to quench the desired hot spot ignition.

Note that none of the various phases of instability growth shown in Figs. 19.4 and 19.5 occurs in isolation. Indeed, a principle complicating factor is that the ablation front instabilities can feed through the finite thickness shell (e.g., Weber et al., 2016) and seed instability growth during deceleration around the edge of the hot spot. Moreover, there are multiple initial imperfections in real ICF targets that can act as seeds for RM and RT growth. These include the roughnesses inherent to each surface in the capsule at the start of the implosion, both the external ablator surface and all internal surfaces between the ablator layers, the ablator and DT fuel, and finally the inner surface of the DT ice layer. Bulk inhomogeneities, either in density or opacity, can also seed instabilities, as can asymmetries in how the implosion is driven, either by X-rays or directly by lasers. Finally, all of the discrete features inherent in any practical target can also seed instability growth, including the 45 nm thick plastic membrane or "tent" used to support the capsule at shot time (Nagel et al., 2015; Tommasini et al., 2015; Hammel et al., 2016; Weber et al., 2017) and the micron-scale fill tube (Edwards et al., 2005; Hammel et al., 2010; MacPhee et al., 2017, 2018; Weber et al., 2020) used to pre-fill the capsule with DT.

The interactions of these effects, together with the weakly nonlinear instability evolution, make a detailed understanding of hydrodynamic instabilities in ICF extremely complicated. Naturally, heavy reliance must then be made on numerical simulation to obtain a quantitative understanding of the implications. Over the previous few decades, instability modeling in the ICF context has become quite sophisticated, relying on highly developed multi-physics codes that couple radiation transport, hydrodynamics, thermonuclear burn, and detailed material properties (equations of state and opacities) in realistic 3D geometries (Marinak et al., 2001; Gittings et al., 2008; Igumenshchev et al., 2016). Indeed, quantitative comparisons between simulation and experimental data from high-convergence NIF implosions have only proven feasible with these fully 3D, multi-effect simulations (Clark et al., 2013, 2015, 2016, 2017).

Controlling hydrodynamic instabilities is important to the success of ICF implosions. This is evident in the experimental data from NIF, where neutron yields have increased two orders of magnitude in progressing from highly unstable low-foot implosions to much more stable HDC ablator implosions. The progress is visually apparent in the simulation results of Fig. 19.6.

In inertial fusion, the fluids are in fact plasmas, which interact with magnetic fields, and there is the prospect of using this fact in some way to control or otherwise influence the RTI and RMI development. One concept is to use the implosion to amplify the magnetic field (e.g., Perkins et al., 2013), in a process called *flux compression*, to inhibit electron transport near the hot spot at the center, increasing its ability to retain heat. Using the Omega laser at the University of Rochester, it has been shown that embedding a strong magnetic field in an ICF-type capsule could yield an increase in ion temperature of 15% and neutron yield of 30%, a greater increase than even predicted by a parallel 1D numerical code (Chang et al., 2011; Hohenberger et al., 2012). This discrepancy was attributed to the inability of the code to capture higher-dimensional effects. It also worth noting that RT can generate magnetic fields as in Manuel et al. (2012), Srinivasan et al. (2012), and Walsh et al. (2017). The magnetic field may also affect performance in other ways. However, the reduced electron conductivity afforded by field line electron confinement may also be detrimental to

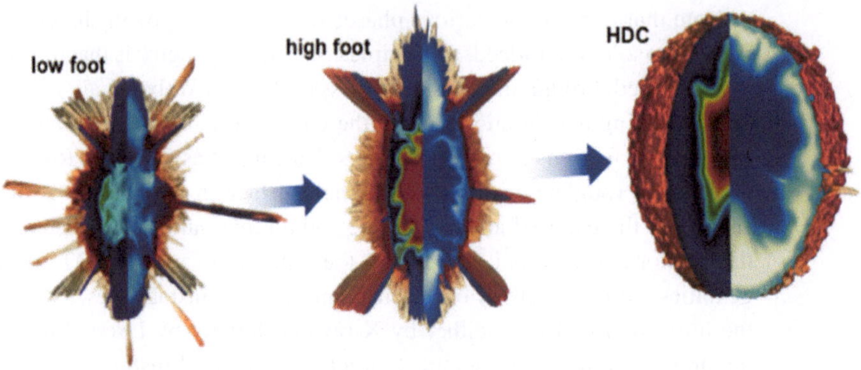

Fig. 19.6 3D simulations of NIF implosion. Zhou et al. (2019).

performance because of a reduced coupling between drive laser energy and the capsule (Manuel et al., 2012). Another line of research relevant to our discussion has focused on the hydrodynamic aspects of the plasma motion using magnetohydrodynamics (MHD); this idea is based on the observation that RM can be suppressed altogether in MHD (Samtaney, 2003), along with the long-known similar observation regarding RT (Chandrasekhar, 1961). As we have seen, fundamental numerical investigations in MHD along these lines have been promising for the possibility of instability suppression. Along with the property of flux compression (Chang et al., 2011; Hohenberger et al., 2012), RM and RT instabilities can also amplify the magnetic field during their development (Jun et al., 1995; Sano et al., 2012), which can further inhibit electron transport (Srinivasan and Tang, 2013).

19.3 Alternative capsule designs

19.3.1 Pushered single shells

The Pushered Single Shells (PSSs) shown in Fig. 19.7(c) concept is an example of innovative alternative capsule designs. The double shell in Fig. 19.7(b) is also a promising design, offering several advantages and disadvantages when compared with the more standard design in Fig. 19.7(a).

To study the resulting effects on pusher-gas mix as well as radiation trapping, Dewald et al. (2019) added a low concentration of Ge ($\approx$ 2.5%) to the pusher as a surrogate for higher-Z materials and used the method of separated nuclear reactants (Casey et al., 2014). The first PSS capsule used in the NIF contains a high-Z pusher surrogate which is colocated with the deuterated plastic inner layers and the capsule gas consists of a 75%/25% hydrogen-tritium (H/T) mixture.

Relying on such nuclear data to assess implosion performance, the 14-MeV neutron signal from DT fusion reactions characterizes the degree to which the pusher and fuel mix. In the mean time, neutrons from deuterium-deuterium (DD) reactions and 5–10 MeV neutrons

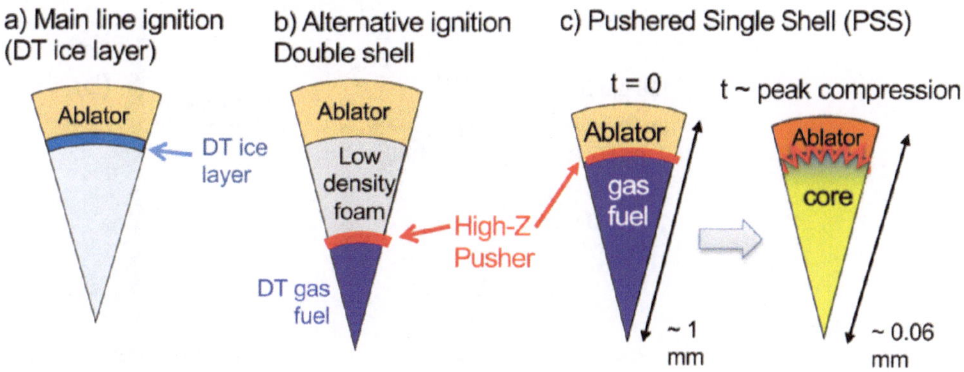

Fig. 19.7 Implosion capsule designs for (a) central-spot (mainline) ignition, (b) double shells, and (c) pushered single shells (PSSs). Dewald et al. (2019), Phys. Plasmas, with the permission of AIP Publishing.

generated from tritium-tritium (TT) reactions characterize the warm unmixed region of the implosion and the core, respectively. The mix yield shows dramatic changes when the Ge dopant is added: The DT yield is reduced by 75% and the DT ion temperature decreases by 25%. Capsule X-ray spectra validate simulations of implosion hydrodynamics and agree with nuclear measurements.

A novel evolution of the pushered single shell design uses a radial density gradient in the ablator shell to eliminate the otherwise classically unstable step in density between ablator and heavy pusher (MacLaren et al., 2021). The density gradient is achieved by blending chromium or molybdenum with the beryllium that serves as the ablator material, and the gradient profile is designed to optimize the tradeoff between performance and stability. Figure 19.8 illustrates the density and concentration profiles for an ablator shell from MacLaren et al. (2021) that is graded from pure beryllium to a mixture of 80% beryllium and 20% molybdenum. This pre-imposed density gradient in the ablator helps stabilize an otherwise classically unstable interface against RT growth, yet still results in a significant increase in stagnated mass and confinement time versus a conventional low-Z ablator. We note that the design in Fig. 19.8 incorporates a suspended cryogenic fuel layer as well as a thin layer of pure Be between pusher and fuel to mitigate mixing. Indeed, with the improved stability, this graded density PSS implosion achieves ignition via a hot spot similar to conventional ICF capsules. However, the enhanced confinement provided by the dense pusher permits accessing the ignition threshold at a lower hot spot temperature and therefore slower implosion velocity, while the anti-mix layer improves performance during burn by minimizing the migration of high-Z species into the fuel. Additionally, the opacity of the molybdenum in the stagnated shell reduces Bremsstrahlung radiation losses as the hot spot heats, further reducing the compression and temperature required to reach ignition. The first graded metal PSS capsule implosions on the NIF indicate that the implementation of the principle design feature, the graded density inner metal layer, has succeeded in producing a stable implosion with performance in agreement with predictions (Dewald et al., 2022).

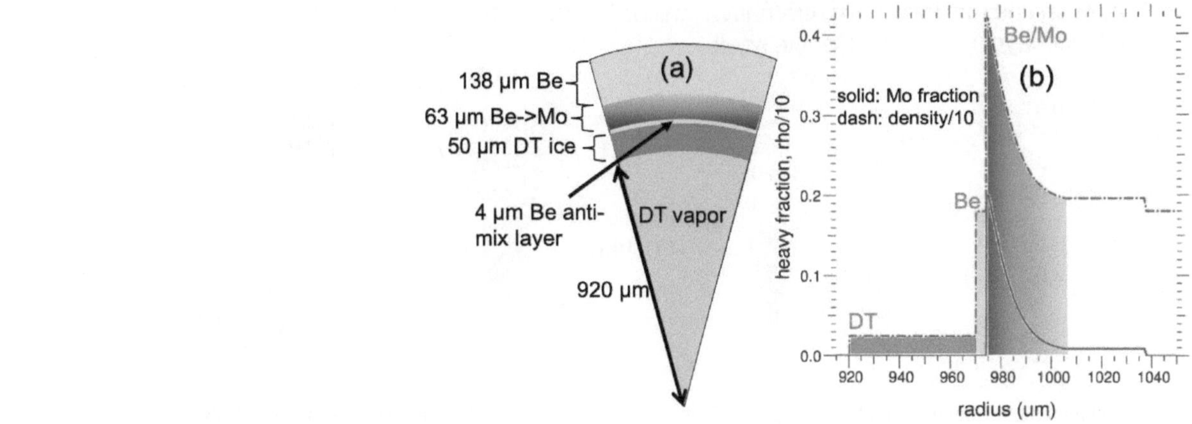

Fig. 19.8 (a) Schematic of the graded metal-shell PSS capsule design. (b) Detail of the Mo density and concentration gradient. Fig. 2 of MacLaren et al. (2021).

19.3.2 Multishell capsule designs

Volumetric ignition is an alternative approach to ICF which does not include a cold high density fuel layer and is designed to heat gaseous fuel volumetrically. Although the potential energy gain is limited, such approaches can be designed to enhance radiation trapping and lower core ignition requirements (Basko, 1990; Wilson et al., 2000).[4]

Here, the cryogenic DT ice layer (≈ 0.25 g/cm^3) may be referred to as the "pusher." This highlights its hydrodynamic function of initially shocking and compressing the DT gas. For central-spot (mainline) ignition designs, the DT pusher is in intimate contact and ablatively driven by a relatively hot, ablated plasma material. Clearly, there are several well known physics challenges and possible uncertainties in accelerating the DT ice pusher to the required 350–400 km/s velocities and high convergence ratios of $C_R > 30$, in order to ignite the DT gas (Lindl, 1995). Uncertainties in the ablator opacity, thermal conduction, radiation transport, and non-Planckian aspects of the radiation field (hohlraum Mband) are exacerbated by the required high convergence of an ablatively driven low-Z pusher (e.g., Montgomery et al., 2018). Furthermore, a DT pusher is very susceptible to hot electron preheat due to laser-plasma instabilities during the acceleration phase (Robey et al., 2014).

Several multi-shell concepts have been proposed, motived by a desire to mitigate these challenges. The designs considered, in order of increasing complexity, can be classified as either (a) double shells, or (b) the revolver. These schemes strive to separate the shell acceleration from the hot spot formation. However, such remedies can introduce extra tamping and render the capsule emission opaque to X-ray diagnostics.

The multi-shell configuration has additional interfaces to consider from the perspective of hydrodynamic instabilities. Each shell is accelerated from its outer surface by a high-pressure lighter fluid and decelerated during collision by the pressure on its inner surface

[4] A recent description of volume ignition without radiation trapping has been published (Amendt et al., 2020) and opens the door to volume ignition of a low-Z single shell by driving a high-adiabat implosion.

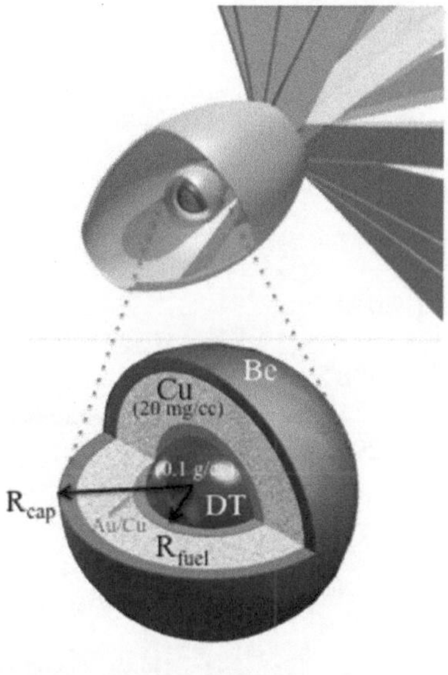

Fig. 19.9 Schematic of a double-shell target design in a hohlraum. The nominal DT fuel density is 0.1 g/cm^3 and the intra-shell Cu foam density is 20 mg/cm^3. Amendt et al. (2010), with the permission of International Atomic Energy Agency.

by a presumed lighter fluid, so that both surfaces of each shell are expected to induce RTI. The mixing layer between shells is exposed to multiple reflected shocks that generate a quasi-isobaric pressure to compress the subsequent shell (Molvig et al., 2016).

Double-shell targets

Double-shell targets (see Fig. 19.7(b)) have been suggested as an alternative design for demonstrating and exploring ignition on the NIF (Amendt et al., 2002, 2007, 2010; Canaud et al., 2011; Haines et al., 2021), as the volumetric ("PdV") ignition in double shells is fundamentally different from the conventional "hot spot" ignition approach. Double-shells consist of a low-Z ablator outer shell that impacts a high-Z inner shell filled with high-density DT gas at room temperature (Fig. 19.9), and therefore does not require cryogenic preparation below the triple point of DT ($\approx$ 19K). The outer shell acts as an efficient absorber of hohlraum-generated X-rays, magnifying and transferring the acquired energy density when it spherically converges and collides with the inner shell.

The main function of the high-Z inner shell is to limit radiative losses from the igniting fuel and to provide added inertia for delaying fuel disassembly. The pusher is hydrodynamically driven, initially by the main shock from ablation at the outer shell, then further driven by the increasing pressure in the foam cushion as the outer shell compresses it after converging only a factor of a few (Montgomery et al., 2018). Indeed, double-shell implosion relies on a series of energy transfer processes starting from X-ray absorption by the outer shell, followed by transfer of kinetic energy to an inner shell, and finally conversion of kinetic energy to fuel internal energy (e.g., Merritt et al., 2019). This process is illustrated in the numerical simulation shown in Fig. 19.10.

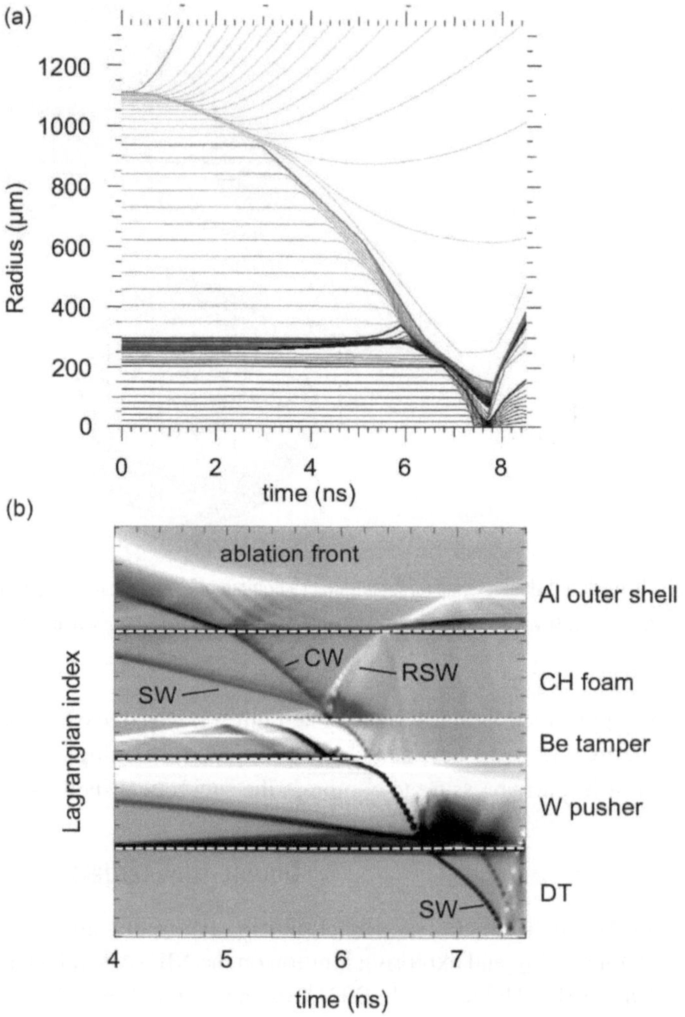

Fig. 19.10 Radius versus time plot for the fall-line optimized double-shell design shown in panel (a). Contours of the logarithmic derivative of material density versus Lagrangian index and time for the fall-line optimized design shown in panel (b), so-called shock plots. Some of the main shock waves (SW), compression waves (CW), and reflected shock waves (RSW) are labeled. Montgomery et al. (2018).

Thus, double-shell implosions are predicted to ignite at a lower convergence ratios, lower temperatures, and lower implosion velocities than single-shell designs. Yet, despite these desirable features, the utilization of numerous interfaces with significant density gradients renders double-shell designs especially vulnerable to interfacial and fill tube instabilities.

Amendt et al. (2002) stressed that an important concern for double-shell stability is the feed-through (e.g., Mikaelian, 1983) of outer-surface perturbations on the inner shell to the inner surface. At the time of shell collision, an impulsive-like acceleration will be delivered to the outer surface of the inner shell, promoting the RMI. The subsequently induced RT instabilities will demand careful considerations in such designs.

Finally, we note the work by Milovich et al. (2004) on using a graded density gradient to dramatically reduce RM and RTI on the outside of the inner shell. This provides at least some hope that instability growth might be controllable with target fabrication tricks such as this.

Triple-shell targets

The Revolver is a direct-drive triple-shell ICF concept that aims to turn a 6 ns laser drive pulse on the ablator into a 1.5 ns pressure pulse delivered to the liquid DT fuel by the innermost high-Z (Au or W) pusher shell (Molvig et al., 2016; Scheiner et al., 2019). The main features of this class of targets are low convergences (<3) throughout for each shell collision and low fuel convergence ≈ 9. In particular, the lower DT fuel convergence could reduce the mixing of material from the innermost shell into the fuel (Krasheninnikova et al., 2020). Recently, Scheiner et al. (2019) showed results of experiments on OMEGA laser, demonstrating the ablator-driver shell collision relevant to the outer two shells of the Revolver triple-shell inertial-confinement-fusion concept (Fig. 19.11).

With the additional energy available from a direct drive laser configuration, the massive pusher confined the fuel for an extended period of time and delayed the disassembly for 100 ps after initial stagnation. The main burn phase in a Revolver ignition capsule has fuel confined by the pusher, so that burn proceeds without fuel expansion for an extended period of time.

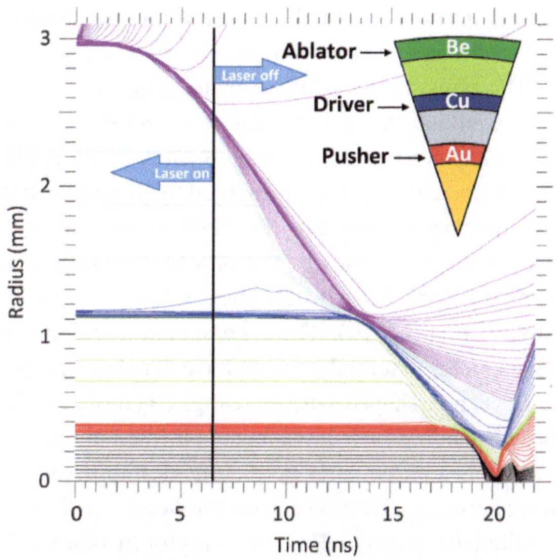

Fig. 19.11 A Lagrangian plot of the ignition-scale Revolver implosion including a cutaway pie diagram of the shell configuration. Reprinted with permission from Molvig et al. (2016). Copyright 2016 The American Physical Society.

The Revolver design was aimed at achieving the upstream ignition state with negligible RTI growth of the pusher-fuel interface. Using the late-time t^2 scaling of the mixing layer growth, Molvig et al. (2016) estimated that interfaces' shell penetration by the bubbles is small and the shell will remain intact throughout the acceleration phase. More recently, Molvig et al. (2018) reported from their simulation studies that the fuel-pusher interface is stable to RTI through the burn phase and for an additional 50 ps. This feature would add significant appeal to the three shell design.

19.4 Z-pinches

If a current is passed through a cylindrical column of conducting material, the resulting Lorentz force perpendicular to the current acts to constrict the column in a cylindrical collapse. This configuration, called a *Z-pinch*, was considered and analyzed before the development of such concepts as ICF (e.g., see Bennett (1934) and Tonks (1937) for early analyses of the Z-pinch). Its potential for controlled fusion, by constricting the interior plasma to the required temperature and pressure, was developed in the 1950s (Bishop, 1958).

Z-pinches are important in certain classes of fusion schemes and, as we will see below, in the reproduction of astrophysical phenomena in the laboratory. They do, however, feature an interesting array of challenges. The traditional Z-pinch obtains an equilibrium in which the plasma column is sustained by a balance between thermodynamic and magnetic pressures. This equilibrium is susceptible to various instabilities, most famously the *kink* and *sausage* instabilities (e.g., Dai et al., 2023). The kink instability, for example, arises from the amplification of helical perturbations, caused by an unbalanced increase of magnetic pressure on the perturbations and resulting in a runaway growth and disruption of the pinch. The growth rates of these instabilities are such that the use of Z-pinches for magnetically confined fusion is impractical, and this is a major reason for the relative lack of interest in Z-pinches until recently (Ryutov et al., 2000b).

In an alternate configuration enabled by more recent developments in pulsed-power technology, the current pulse may be rapid, so that an equilibrium of the above type is never reached. If the timescales of the process are sufficiently short, it may be able to run to completion before these instabilities, whose growth rates are known, can spoil the effort (Goedbloed et al., 2010). These are called *fast* Z-pinches (Ryutov et al., 2000b). Several variations on the essential concept have also been proposed, including quasi-spherical wire arrays or even X-shaped wire crossings (Ryutov, 2015).

Notwithstanding their advantages outlined above, fast Z-pinches are susceptible to the RTI, which is a particularly pernicious destabilizing influence on the cylindrical collapse, featuring a free acceleration across a nonuniform density plasma. In this context, it is sometimes called the *magnetic* Rayleigh–Taylor instability (MRTI), given the integral role of the magnetic field generated by the Z-pinch (Curzon et al., 1960; Harris, 1962). The so-called MRTI (Kruskal and Schwarzschild, 1954; Chandrasekhar, 1961) is similar to the classical RTI, but now in its frame of reference, the magnetic field – a "light fluid" or more

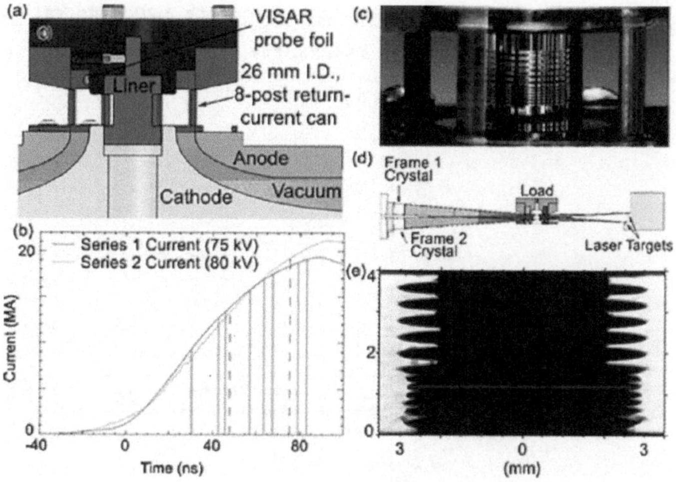

Fig. 19.12 Description of Magneto-RTI experiments. (a) Half-section diagram of the power-feed hardware surrounding the load. (b) Measured currents for the experiments with radiographic times overlaid as vertical bars. (c) Photo of a target inside the return-current can. (d) Schematic of the 2-frame backlighter illustrating the 3 degree angle above or below horizontal. (e) Sample radiographic data taken at 83 ns. Fig. 1 of Sinars et al. (2010). This article is available under the terms of the Creative Commons Attribution 3.0 License.

accurately "massless fluid" – supports the plasma – a "heavy fluid" – against the gravity or acceleration (Haines et al., 2000; Haines, 2011; Ryutov et al., 2000b). Thus, Z-pinches are quite similar (Douglas et al., 2001) to electromagnetic guns, rail guns, and other systems in which plasma or solid objects are electromagnetically accelerated. On the other hand, stabilizing mass ablation does not occur in magnetically driven liner implosions (Schmit et al., 2016). This is in contrast to those in laser driven spherical capsule implosions, in which the RTI is controlled via ablative stabilization.

The MRTI has been experimentally observed in foils (Peterson et al., 1996) and in gap puffs (Douglas et al., 1997; Osin et al., 2011; Narkis et al., 2022). Using multiframe 6.151 keV radiography, Sinars et al. (2010, 2011) performed solid liner implosions experiments to measure single-mode MRTI growth (Fig. 19.12). The single-mode growth was seeded by machined sinusoidal perturbations on the outer surface of the aluminum (Al) liner. The perturbation wavelengths were varied from 25 to 400 μm, while the peak-to-valley amplitudes were fixed at 5% of the wavelength. Figure 19.13 shows the MRTI amplitude (the difference between the spike radius and the bubble radius) versus time measured from the experiments plotted as black crosses, where the measurement error is indicated by the extent of the crosses. These experiments and 2D LASNEX simulations were shown to be in good agreement. It is interesting that the experiments of Sinars et al. (2011) (Figs. 8 and 9, not shown) showed only a factor of $\sim$2 difference in MRTI amplitude from a liner with 10–30 nm initial roughness when compared to one seeded with 2.5–5 μm perturbations. The radiographs showed observable differences between the 200 and 400 μm waves, such as the jet formation (Fig. 19.14).

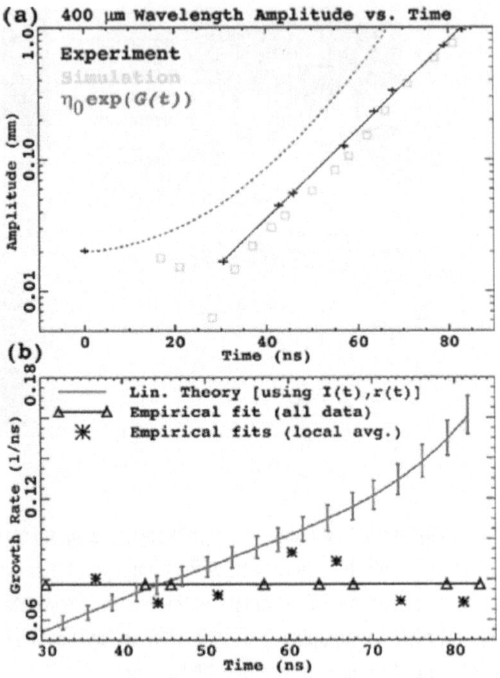

Fig. 19.13 (a) Peak-to-valley amplitude vs time for experiments, simulations, and linear theory. The extent of the crosses represents the measurement error. The line represents an exponential fit to the data. (b) The instantaneous growth rate is shown in the solid line. An exponential growth rate averaged over all the data is shown in a horizontal line, along with asterisks showing the local growth rate calculated for each pair of points in the data. Fig. 3 of Sinars et al. (2010). This article is available under the terms of the Creative Commons Attribution 3.0 License.

Equations 53 and 54 of Harris (1962) are linearized equations describing the growth of the MRTI amplitude. These two second-order differential equations have a total of four solutions, with one exponentially decaying, one exponentially growing, and two oscillatory solutions. Rearranging these equations, Sinars et al. (2011) obtained a neat equation that describe the exponentially growing and decreasing solutions,

$$\frac{d^2a}{dt^2} = \gamma^2_{MRTI}(t)a, \tag{19.3}$$

where $a(t)$ is the amplitude and $\gamma^2_{MRTI} = kg(t)$, where the time-dependent acceleration can be evaluated from Newton's second law. It is clear that the MRTI growth is still in the linear regime as the linearized solution, Eq. 19.3, agrees well with the measured data throughout the duration of the experiment.

Some ideas have been suggested to attempt to stabilize the RTI appearing in these pinches, including introducing a velocity shear, a rotation around the pinch axis, or an axial magnetic field in addition to the existing azimuthal field (Ryutov et al., 2000b). The latter idea in particular, of applying an additional magnetic field to a Z-pinch configuration, has led to the development of MagLIF (Slutz et al., 2010; Ruiz et al., 2023a,b) and magnetized target fusion (MTF) (Lindemuth and Kirkpatrick, 1983), although with the stated aim

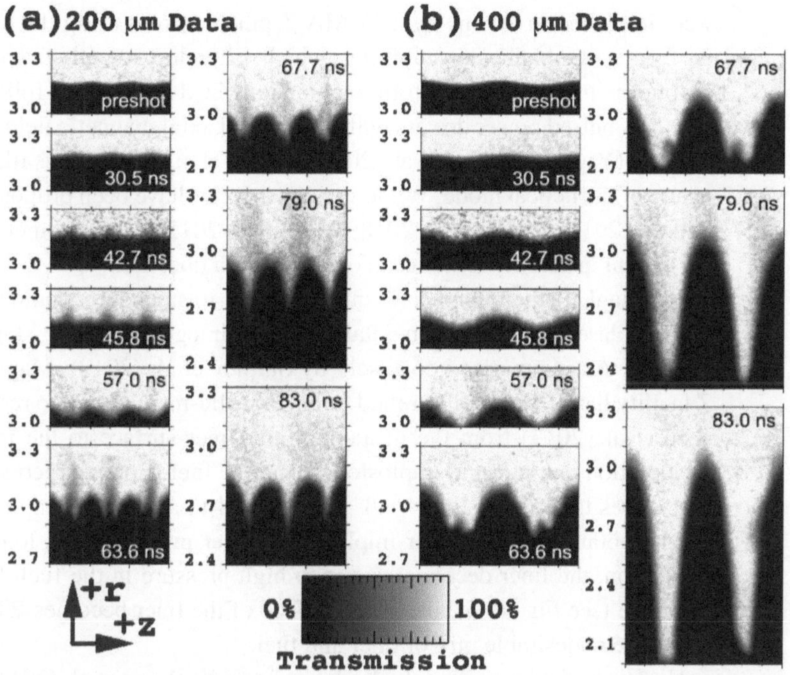

Fig. 19.14 Expanded views of portions of the radiographs. (a) Central three of the five $\lambda = 200\ \mu$m waves. (b) Two of the six $\lambda = 400\ \mu$m waves. Fig. 2 of Sinars et al. (2010). This article is available under the terms of the Creative Commons Attribution 3.0 License.

of inhibiting electron heat transport rather than hydrodynamic stabilization. However, the unstable flute modes associated with the MRTI were found in MagLIF studies to develop a surprising helical instability (Awe et al., 2013), which shows some evidence of reduced growth relative to the standard (i.e., zero axial magnetic field) case (Awe et al., 2014).

Another possible means of MRTI stabilization is the dynamic screw pinch (Schmit et al., 2016; Shipley et al., 2019; Campbell et al., 2020b), in which the magnetic field dynamically rotates.

For MagLIF, the loss of liner integrity prior to stagnation could cause the design to fail. To prevent this from happening, a thick liner with an aspect ratio (AR, defined as initial liner outer radius/initial liner wall thickness) of less than 10 is generally considered to be imperative.

McBride et al. (2012) performed Beryllium (Be) liner implosion experiments on the Z accelerator in preparation for MagLIF, using acquired penetrating radiographs to quantitatively characterize the MRTI. The authors found that the MRTI amplitude grows nearly linearly with the distance moved, which is consistent with results from classical RT experiments in the nonlinear regime. McBride et al. (2013) observed large-scale MRTI on imploding liners, which were initially very smooth.

Awe et al. (2013) studied experimentally the implosion dynamics of premagnetized ($B_{z,0} = 7$ or 10 T) MagLIF-relevant liners. In their study, premagnetized Beryllium liners were imploded in 100 ns by a 17 MA Z-pinch current using the Z facility. Data revealed the formation of helical instabilities, which differ dramatically from the mostly azimuthally symmetric instabilities that form on unmagnetized liners. Thin foil experiments performed on 1 MA pulsed power drivers with an external axial magnetic field also reproduced helical patterns (Yager-Elorriaga et al., 2016; Atoyan et al., 2016), thus illustrating the ubiquitous nature of the helical mode. While various theories have been proposed (Ryutov et al., 2014; Sefkow, 2016; Seyler et al., 2018; Weis et al., 2015; Woolstrum et al., 2020), the origin of the helical instability remains an open research question.

For MagLIF, the loss of liner integrity prior to stagnation could cause the design to fail. As with laser-driven ICF, maintaining liner integrity prior to stagnation is essential for MagLIF. As summarized concisely by Shipley et al. (2019), MagLIF experiments on the Z facility have traditionally relied on thick, solid-metal liners to reduce MRTI feedthrough (Lau et al., 2011) from the unstable, outer liner surface to the inner surface during the acceleration phase of the implosion. Since the fuel is premagnetized and thermal conduction losses to the cold liner wall are inhibited in a MagLIF implosion, fusion conditions can be obtained with slower implosions, lower pressures and longer burn times. During stagnation, the liner decelerates due to high pressure in the fuel. Similar to the ICF capsule case (see Fig. 19.5), the inner surface of the liner becomes RT unstable, which could introduce undesirable mix of liner and fuel.

Using a resistive magnetohydrodynamic code, Wang et al. (2019a) carried out a numerical study on MRTI for thin liner implosions with the PTS facility. As expected for the large-radius and aspect ratio liner implosions, both the long implosion distance and mode coupling contribute to the larger perturbation and wavelength of the dominant mode.

Pulsed-power facilities moving beyond the basic Z-pinch configuration have also proven useful for laboratory reproduction of astrophysical phenomena. The plasmas generated by Z-pinches can often be modeled by ideal MHD if care is taken in the setting of the various pinch parameters (Ryutov, 2015) and may also resemble the plasmas in these astrophysical contexts. By an appropriate arrangement of wire arrays in conical, spherical, or planar settings, it has been possible to investigate jets emitted from young stellar objects (a review is given by Reipurth and Bally (2001)), jet interactions with interstellar plasma "winds" (Ciardi et al., 2008), astrophysical shock waves, and even to some extent, the accretion disks which form around celestial objects (Pringle, 1981). A review of experimental efforts in this regard is given by Lebedev et al. (2019). Many of these astrophysical phenomena can be investigated in terms of resistive or ideal MHD and likewise feature hydrodynamic instabilities such as RTI. Examples of the RT instability in particular can be found in the linear, highly collimated plasma-jets mentioned above, and in arch-shaped jets strongly guided by magnetic field (Bellan et al., 2005). The experiments generated using pulsed-power facilities can in turn be used to validate numerical models of these phenomena, particularly in MHD, providing a powerful investigative tool for understanding astrophysical flows often in the absence of direct observational data (Lebedev et al., 2019).

20 Laboratory applications

20.1 Material strength

The classical presentation of RT and RM instabilities in Chapters 1–8 is indeed useful, as it allows one to develop an intuition regarding the fundamental physics at work. However, as is evident from the myriad of applications discussed in this manuscript, the RT and RM instabilities' "real world" manifestations are inevitably complicated by any number of "non-classical" considerations. At first glance, these complications can easily be viewed as an annoyance in which the fundamental physics is obscured by "non-fundamental" details that introduce deviations from the classical behavior. However, if the experimental or observational tools for probing a system are sufficiently precise, these deviations may provide a means of studying, and even quantifying, the complicating physics. One such example is found in the behavior of solid materials under extreme pressures where they exhibit hydrodynamic properties more typical of fluids.

The essential point suggesting the possibility of inferring material strength through RT or RM growth measurements in solids is that material strength tends to suppress this growth (e.g., Remington et al., 2004; Park et al., 2010; Vogler and Hudspeth, 2021; Grigoryev et al., 2022). Strength, meaning resistance to deviatoric (i.e., shear) deformation, moderates the growth rates of the instabilities (Prime et al., 2017). Thus, at one extreme is the case of classical RT or RM growth (no strength), while at the other extreme (infinite strength) the growth is suppressed entirely. Between these extremes lies a wide range in which the RT or RM instability will be more or less suppressed, and the degree of this suppression may be used as a window into the material's strength. Unfortunately, this window is more translucent than transparent, as material properties such as strength are decidedly non-constant as the material evolves from its initial ambient condition to the pressures and temperatures required to generate measurable RT or RM growth in solids.

Material strength is conventionally separated into two regimes, those of elastic and plastic deformation. Elastic deformation is assumed to be reversible up to a certain yield stress, described by the shear or bulk modulus. This assumption is reasonable under dynamic loading conditions, but does not take into account longer-term deformation mechanisms such as ductility, creep, brittle failure, or fatigue (Coble, 1963; Ellis and Stöckhert, 2004; Marrow et al., 2004; Renshaw and Schulson, 2001).

The RMI has been used as a means to "measure" material strength (Piriz et al., 2008, 2009; Dimonte et al., 2011; Jensen et al., 2015; Prime et al., 2017, 2019; Grigoryev et al., 2022) using the configuration illustrated in Fig. 20.1(a). From a modeling perspective,

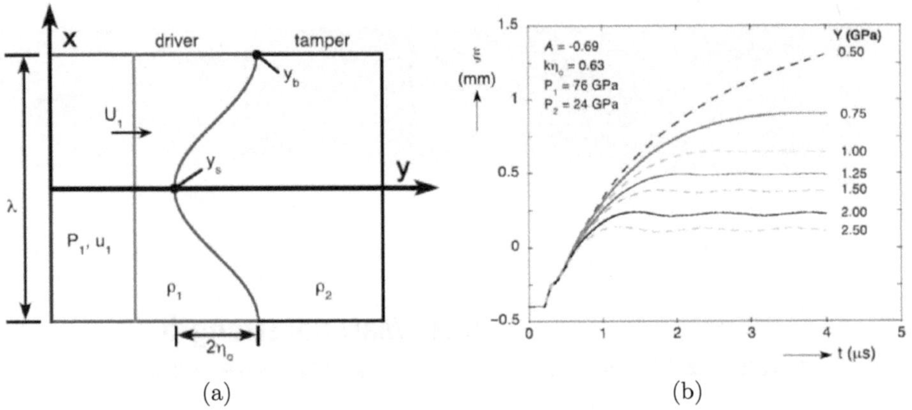

Fig. 20.1 (a) Schematic of the initial configuration for an RMI experiment. The shock propagates through the driver, across the interface with the sinusoidal perturbation, and into the tamper. (b) Spike amplitude evolution from simulations with varied strength. Figs. 1 and 3a of Vogler and Hudspeth (2021). This is distributed under a Creative Commons Attribution 4.0 International License.

Mikaelian (2013) considered the case of arbitrary $\mathcal{A}$ and found that the length of the arrested spikes is given by

$$k(a_\infty - a_0) = \frac{1}{3}\mathcal{A}^2(ka_0)^2\frac{\bar{\rho}u_2^2}{\bar{Y}}, \tag{20.1}$$

where $\bar{\rho}$ and $\bar{Y}$ are the effective density and strength for the two-material configuration, a_0 and a_∞ are the initial and asymptotic perturbation amplitudes, respectively, and the material 2 is shocked to velocity u_2. For the limiting case of $\mathcal{A} = -1$, Eq. 20.1 differs from that of Piriz et al. (2008, 2009) only in the constant prefactor. This relationship is similar to other related relationship of Dimonte et al. (2011) for the case of $\mathcal{A} \approx -1$, but is slightly different from that of Chen et al. (2019) for the case when material 1 is an elastic-plastic solid and material 2 is a fluid (Vogler and Hudspeth, 2021).

Figure 20.1(b) shows the results from simulations with the strength varied for a spike amplitude evolution for the case of Ta jetting into a high-density tamper fluid (Na_2WO_4 solution), the Atwood number for which configuration is -0.69. As described by Vogler and Hudspeth (2021), the Ta driver with an initial perturbation given by $ka_0 = 0.63$, impacted by a thick Ta impactor traveling at a velocity of 2 km/s. For simplicity, an elastic-perfectly plastic strength model with $Y = 0.75$ GPa is used for the Ta, while the tamper is treated hydrodynamically (i.e., no strength). Results obtained with this very simple model are qualitatively similar to those obtained from a more physically motivated strength model such as the Preston–Tonks–Wallace (PTW) model (Preston et al., 2003). The spike amplitude grows with a gradually decreasing rate and eventually arrests with an amplitude of about 0.91 mm.

Figure 20.2(a) shows the more experimentally accessible free surface ($\mathcal{A} = -1$) RMI configuration. The bottom figure is the initial condition and the top is post-shock. The perturbed surface of the sample is accelerated by a shock from an impact or HE detonation.

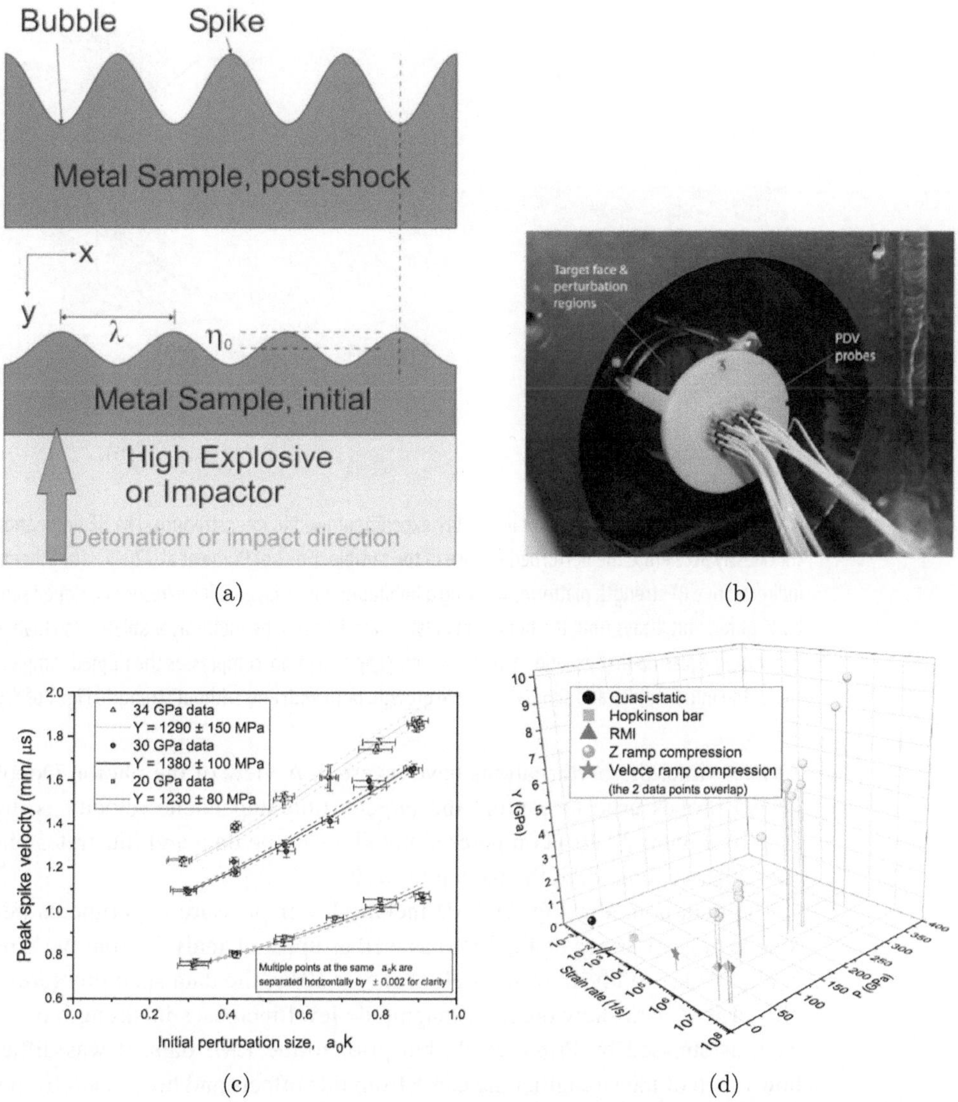

Fig. 20.2 (a) Schematic of a Richtmyer–Meshkov instability experiment for $\mathcal{A} = -1$. The perturbed surface of the sample is accelerated by a shock from an impact or HE detonation. At later time, the perturbations have inverted. The bottom figure is the initial condition and the top is post-shock. Perturbation amplitudes are exaggerated. Fig. 2 of Prime et al. (2017). (b) A view of the target free surface and the PDV probes prior to execution. (c) Model-based average strength estimates fit the measured data quite well. (d) The RMI data on tantalum shows modest rate hardening at rates up to 10^4/s compared to apparently more significant pressure hardening. Figs. 5, 12, and 15 of Prime et al. (2019). Published by the American Physical Society under the terms of the Creative Commons Attribution 4.0 International license.

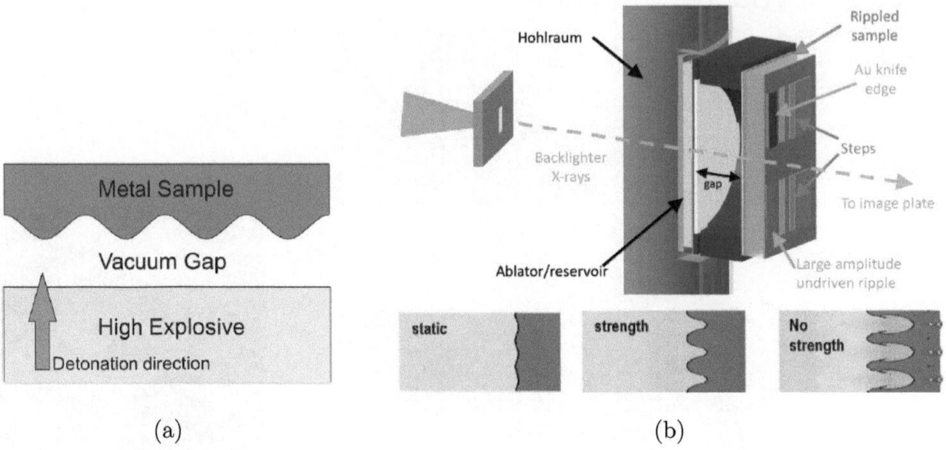

Fig. 20.3 (a) Schematic of a Rayleigh–Taylor instability experiment used to infer strength. The HE detonation products shocklessly accelerate the perturbed surface of the sample. Fig. 1 of Prime et al. (2017). (b) Schematic of the indirect-drive RT strength platform, showing a hohlraum, multi-layer ablator/reservoir, ripped sample target, and backlighter. Soft X-rays from the hohlraum ablate material from the multi-layer ablator and foam reservoir, generating a plasma that expands across a vacuum gap and ramp-compresses the rippled sample. The bottom inset shows the initial ripples and simulated ripple growth, both with and without strength. Fig. 1 of Krygier et al. (2019).

At later time, the perturbations have inverted. A view of the photon Doppler velocimetry (PDV) probes used to capture the target's time-dependent velocity is displayed in Fig. 20.2(b). Figure 20.2(c) compares simulations to the data and illustrates that the estimated strengths, in general, fit the data quite well.

The data points in Fig. 20.2(d) include lower pressure experiments performed on the Veloce pulser (Brown et al., 2013) as well as updated analysis from the Z machine (Brown et al., 2014a). With strain rates all near 5×10^5/s, the data span pressures from 15 to 350 GPa. The Z data show order-of-magnitude level increases in strength over lower pressure data, as stressed by Prime et al., but prior to the RMI data, it was difficult to establish how much of the strengthening came from rate effects and how much from pressure effects (Prime et al., 2017, 2019).

Scientists have long used RTI to infer material strength (Barnes et al., 1974; Colvin et al., 2003; Lebedev et al., 1996; Henry de Frahan et al., 2015a). Figure 20.3(a) shows a typical RT experiment, in which gaseous detonation products from a high explosive (HE) expand across a vacuum gap and shocklessly accelerate a sample with initial perturbations. The perturbation growth rates, measured experimentally via radiography, are then used to indirectly estimate the strength of the sample at very high strain rates (Prime et al., 2017).[1]

More recently, several related platforms have been developed to extract strength-relevant RT data from high-energy-density experiments. Here, we restrict our consideration to the indirect-drive platform which has been applied at both the NIF and Omega lasers. This method, as depicted in Fig. 20.3(b), builds upon the concept first suggested by Barnes

[1] Another interesting approach is to use carefully characterized soft solids (mostly food colloids) as the elastic-plastic material (see, e.g., Polavarapu et al. (2019) and references therein).

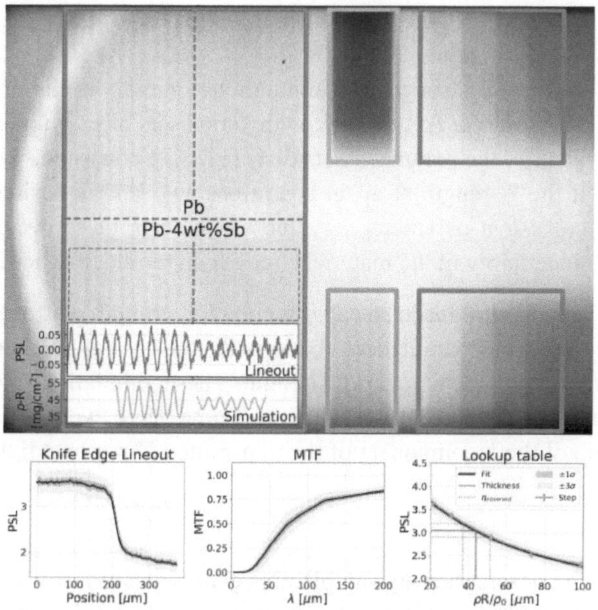

Fig. 20.4 Radiographic data and analysis plots from an RT strength experiment performed on lead at NIF with a $\sim$ 400 GPa peak pressure. The modulated X-ray intensity is clearly visible in the left half of the target, indicating the sinusoidally varying thickness of the rippled lead sample. The right half contains a knife-edge, undriven ripples and "steps" of various thickness, which provide a means of inferring ripple growth through establishing a *lookup table*, as shown at bottom right. Fig. 3 of Krygier et al. (2019).

et al. (1974) and the development of "indirect" hohlraum-based reservoir-gap ramp drives (Edwards et al., 2004). In this platform, laser beams enter a hohlraum, which generates an intense X-ray source that in turn is incident upon a multi-layer ablator and foam reservoir. The reservoir is thereby converted into a plasma, which expands across a gap and is incident upon a sample with preformed sinusoidal ripples, generating an RT instability in the solid rippled sample. These ripples grow (Zhou, 2017a,b) and are characterized at a later time via face-on X-ray radiography (Park et al., 2010, 2015).

The corresponding experimental data point can be evaluated from face-on radiographic data, a typical example of which is shown in Fig. 20.4 (Krygier et al., 2019). On the left side of the target, the sinusoidal modulation of the transmitted X-ray intensity is clearly visible. This intensity is correlated to the material thickness through the presence of both undriven ripples (i.e., ripples not subject to the laser-induced pressure drive) at bottom-center of the target, as well as the "steps" of various known thicknesses at bottom-right.

The task of an RT strength study in a laser platform might be described as follows: first, obtain some number of RT growth measurements with various "loading" (pressure) and temperature histories, then "invert" this data to obtain the strength function $Y(p, T, \ldots)$. Of course, this inversion is by no means unique, even for large data sets. Given the limited experimental time at the few facilities capable of reaching the necessary conditions, in practice the available data sets will be relatively small, perhaps $\sim$10 measurements. The path forward is thus:

1. postulate a functional form of $Y(p, T, \ldots)$ based either on phenomenology or a microscopic model;
2. perform hydrodynamic simulations of the various experiments using the postulated form of $Y(p, T, \ldots)$ along with varied sets of its unknown parameters;
3. identify the parameter sets of $Y(p, T, \ldots)$ consistent with the experimental data;
4. if the "strength at given conditions (p_*, T_*)" is desired, evaluate $Y(p_*, T_*, \ldots)$ for all consistent parameter sets; the variation in these evaluations gives an estimate of the uncertainty in the material strength under those conditions.

Having introduced an experimental RT strength platform, we now proceed to discuss the hydrodynamic simulations that provide the crucial link between experimental data and the hitherto unknown material strength. These simulations allow for the direct calculation of the RT growth factor as a function of time. Here, we define the growth factor, $GF(t)$, as the ratio of the k-component of the transmitted X-ray backlighter ($\tilde{I}_{\text{trans}}$) to its $t = 0$ value:

$$GF(t) = \frac{\tilde{I}_{\text{trans}}(k, t)}{\tilde{I}_{\text{trans}}(k, 0)}. \tag{20.2}$$

For a constant density sample, this expression reduces to the geometric growth of the rippled interface, while in the more general case of a density varying sample, such as is realized in RT strength experiments, it continues to represent the experimentally accessible data obtained via face-on radiography.

For a given set of material strength model parameters and one or more experimental data points, the question is then asked: *is the experimental data consistent (within errors) with the simulated growth factor?* If not, that portion of parameter space is excluded. As the number of experiments increases, ideally performed at different pressures and temperatures, the parameter space consistent with the growing data set is narrowed, reducing the uncertainty in the material model. Finally, the model space consistent with the experimental data can be evaluated for material strength, thereby inferring the strength of the material itself. This procedure can be repeated for several strength models, providing not only a specific value of the material strength at the specified pressure and temperature condition, but also an indication of its uncertainty, through the distribution of these evaluations.

In an illustrative example, Fig. 20.5 shows the measured growth factor as a function of time and compares the NIF data with the simulated results from the ARES hydrodynamic code (Darlington, 2001) with various constitutive flow stress models (Park et al., 2015). The predictions of the Livermore multiscale strength (LMS) model (Barton and Rhee, 2013; Barton et al., 2011) agree with the experimental data well, but the models of PTW, Steinberg–Guinan (SG) (Steinberg et al., 1980), Steinberg-Lund (SL) (Steinberg and Lund, 1989) all underpredict the Ta flow stress (strength) at these strain rates and pressures. Clearly, the high-pressure material strength significantly reduces the RTI growth.

Finally, the growth factor data in Fig. 20.6, measured at NIF with ripples ramp compressed to ≈400 GPa peak pressure, which is among the highest-pressure strength measurements ever reported on any platform, show no difference between Pb and Pb-4wt%Sb beyond the experimental error bars. The observed growth factor for Pb is described well by simulations using the SG flow stress model for high-pressure body-centered-cubic (bcc) Pb (Krygier et al., 2019).

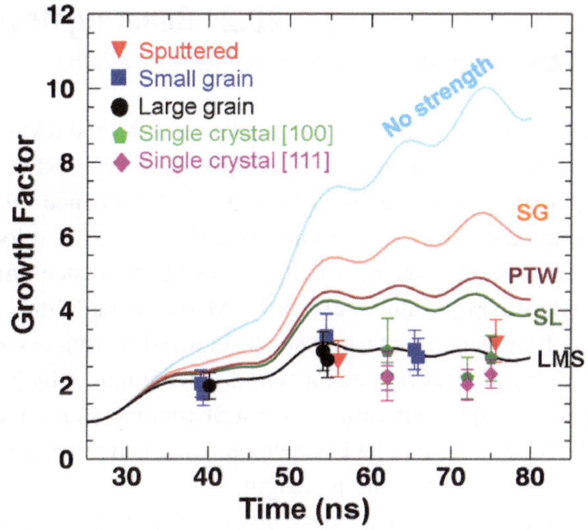

Fig. 20.5 Measured growth factor (points) compared with 2D simulations by using different flow stress (strength) models referenced in the text (curves). Fig. 5 of Park et al. (2015).

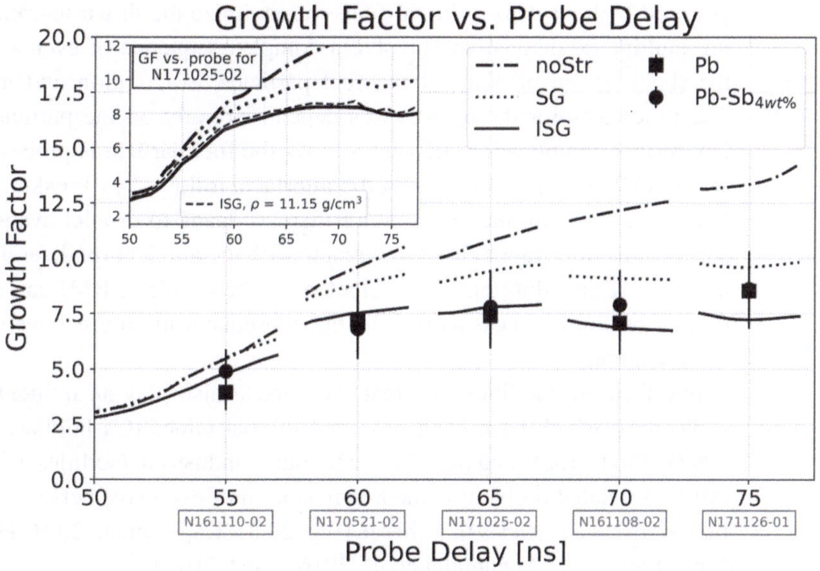

Fig. 20.6 Growth factor data for Pb (squares) and Pb-4wt%Sb (circles) with $\approx 0.4\ \mu$m initial ripple amplitude and hydrodynamic simulations using no strength, face-centered-cubic (fcc)-based Pb SG model, and the body-centered-cubic (bcc)-based Pb SG model (labeled as ISG). Discontinuities in the simulation curves are due to shot-to-shot variations. The inset shows full growth factor curves for 65 ns delay with an additional simulation curve using Pb SG with Pb-4wt%Sb initial density (11.15 g/cm^3). Fig. 5 of Krygier et al. (2019).

20.2 Reacting flows

In this section, we discuss the occurrence of RT and RM instabilities in flows with chemical reactions accompanied by heat release. Early interest in the interaction of a shock with a flame bubble was motivated by a desire to enhance mixing in scramjet applications. In hypersonic flight, in order to minimize losses and heat load, the combustion must occur in a supersonic gas stream, which reduces the residence times and requires greater and more efficient mixing (Yang et al., 1993; Marble et al., 1989, 1990; Waitz et al., 1993). Marble et al. (1989) suggested using shock-induced mixing associated with the RMI to accelerate the mixing process between a cylindrical column of hydrogen fuel and the ambient air. In a typical scramjet configuration (or a Shcramjet: shock-induced combustion ramjet (Veraar et al., 2009)), the fuel jet is supersonic and is intersected by an oblique stationary shock at every cross section along its length.

RMI-mediated mixing also plays a significant role in multiphase detonation, which occurs in propulsion applications with liquid fuels and in dust explosions. Pulse detonation engines (PDEs) or rotating detonation engines (RDEs) rely on detonative combustion, which is more efficient, and are expected to outperform engines designed around the Brayton cycle (Kailasanath, 2000). However, for practical applications these devices must be powered by high-density liquid fuels to maximize the thrust-to-weight ratios. The resulting multiphase detonations involve a complex sequence of events (Fig. 20.7), including the shock-driven breakup of the fuel particles, evaporation and mixing, and reactions. The timescales for these processes depend strongly on the particle sizes and can overlap, further complicating the analysis. As the fuel particle is processed by the detonation wave, RMI leads to stretching of the interface, followed by breakup into "child particles." Following breakup, the temperature increase leads to droplet evaporation, resulting in a supercritical fuel vapor cloud that reacts with the ambient oxidizer after a chemical induction time. Depending on the parameters of the problem, RMI can strongly influence the droplet breakup process as well as the subsequent mixing between the supercritical fuel and the ambient.

Interfacial instabilities in a reactive context also play an influential role in dust cloud explosions, which have been observed in coal mines (Cashdollar, 1996; Dastidar et al., 1997), food processing facilities and other industrial facilities (Eckhoff, 2005; Irving, 2013). A related problem is the augmentation of explosive performance through the addition of reactive particles to a charge (Fig. 20.8) (Kapila et al., 2001; Houim and Oran, 2016; Ling et al., 2012; Annamalai et al., 2016; Frost, 2018).

RT instabilities have also been identified as improving the performance of ultra-compact combustors (UCCs) in gas turbine systems (Cottle and Polanka, 2016; Zelina et al., 2004). UCC designs under consideration have the potential to greatly reduce the weight of gas turbine engines, thus increasing the thrust to weight ratio. In addition, their compact size allows for the inclusion of a reheat cycle between turbines, thus increasing the efficiency of the system.

Similarly, pressure-gain combustors using shock-flame interactions have been recently considered as a means to improving the efficiency and specific fuel consumption of gas

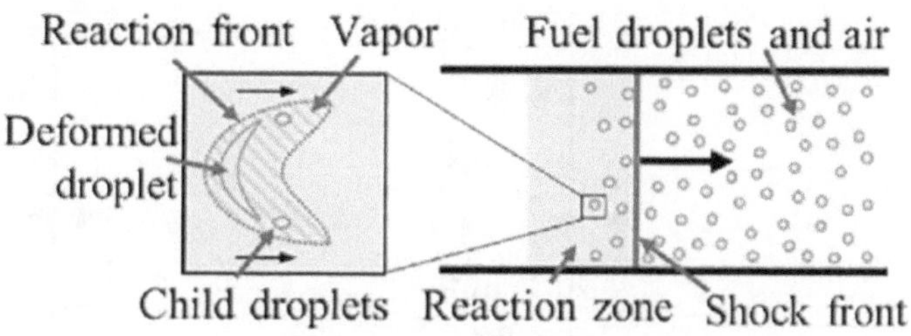

Fig. 20.7 Fuel droplets processed by a shock front in a multiphase detonation application. Inset shows details of the droplet deformation due to shock interaction, breakup of the parent droplet through shock-driven instabilities, and subsequent evaporation. Figure courtesy of Dr. Jacob McFarland, Texas A&M University.

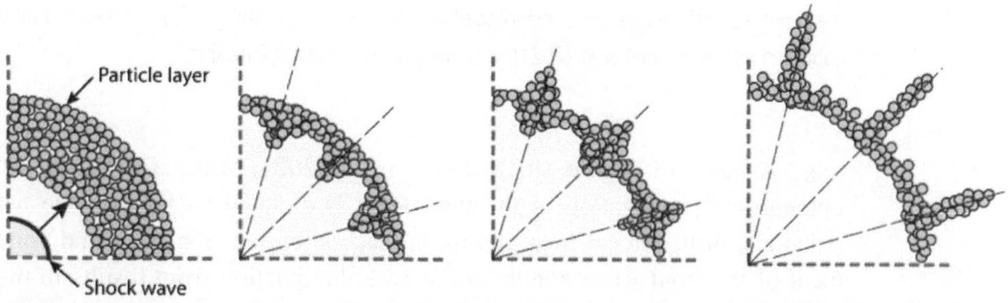

Fig. 20.8 Schematic of the instability that develops on the interface of the particle layer for a circular ring of particles dispersed by a weak shock wave (adapted from the schematic in Fig. 6 of Rodriguez et al. (2017), based on their experiments of the impulsive dispersal of fine particles contained within a 2D Hele-Shaw cell). Note that the perturbations initially develop on the inner boundary of the particle layer. The particle clusters catch up to and penetrate the outer portion of the powder layer to form the particle jets observed experimentally. Images reproduced from Frost (2018) with permission.

turbines. A promising approach is to allow shock waves to interact with the pre-mixed flame within the combustor, so that RM instabilities increase the flame surface area and hence the heat release rate (Lutoschkin, 2013).

20.2.1 Categories of reacting RM and RT flows

Typically, the chemical reaction time scales are orders of magnitude shorter than the flow timescales. This means the hydrodynamic timescales (set by the above interfacial instabilities) determine the rate and efficiency of the overall process. Reacting RM and RT instabilities occur in a wide array of applications including SCRAMJET combustion, inertial confinement fusion, type Ia supernovae, vapor cloud explosions, multiphase combustion in propulsion applications, and gas turbines. However, the flow in all the above applications can be analyzed in the framework of premixed or non-premixed combustion.

In *premixed* combustion, a homogeneous mixture of fuel and oxidizer is ignited resulting in a self-sustaining flamefront that then separates the burned material from the unburned

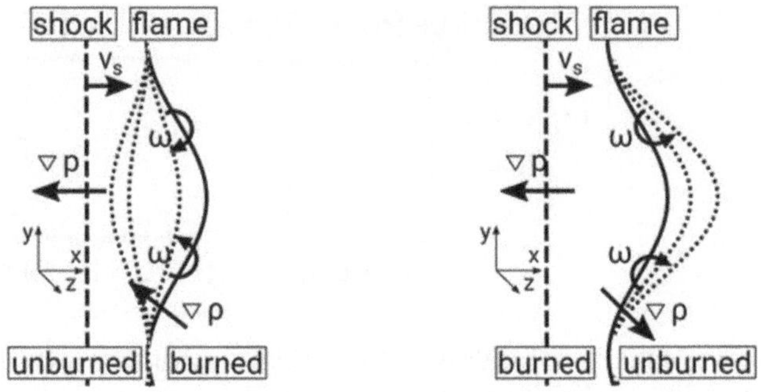

Fig. 20.9 Amplification of flame surface disturbance (solid line) and its temporal evolution (short dashed line) due to baroclinic torque caused by the misalignment of density and pressure gradient. Left: shock wave travels with velocity V_s from the unburned to the burned side, including phase reversal effect. Right: shock wave travels from the burned to the unburned side. Bambauer et al. (2021), with the permission of AIP Publishing.

(e.g., Aslani and Regele, 2018; Sykes et al., 2021; Wang and Dong, 2020). The development and propagation of the flamefront is dictated by the balance between chemical reactions, diffusion (of heat and mass) across the reaction front and convective development of the front. Heat conduction across the reaction front results in the combustion of additional unburned gases, so that the flame front propagates forward with a given flame speed. RT and RM instabilities often aid the transition of the flamefront to a turbulent state, which is accompanied by an increase in the surface area available for combustion. As a result, the burning velocity for the turbulent flame is higher than the laminar counterpart. Through a sequence of processes initiated by RM or RT, and the transition of the flame brush to turbulence, the flame deflagration may transition to a detonation front, a process termed DDT (deflagration-to-detonation transition). In this context, the RMI is caused by the interaction of a sharp pressure gradient, generated by a shock wave and the density gradient at a flame surface (Fig. 20.9). This interaction leads to the production of baroclinic torque at the flame surface, which causes flame wrinkling (Dorofeev, 2011; Bambauer et al., 2020, 2023). The influence of the RMI can be a decisive factor for DDT, as it causes heavy wrinkling of the flame brush region and subsequently a strong increase in the integral reaction rate (Bambauer et al., 2021). Turbulence generated from hydrodynamic instabilities are thus key to DDT, without which the run-up distances of the corresponding 1D flame would approach lengthscales of kilometers (Massa and Jha, 2012; Law, 2006).

When the fuel and oxidizer are separated by a diffuse interface, the resulting combustion is confined locally to the interface and is rate-limited by the diffusion. This scenario is referred as *non-premixed* combustion, and results in a flame that does not propagate in either the fuel or the oxidizer streams. As a result, non-premixed or diffusion flames are safer, but less efficient since the burning is dictated by the mixing process across the interface. Thus, hydrodynamic instabilities can significantly influence the outcomes in diffusion flames through local stretching of the flame surface as well as enhancing the mixing through the deposition of localized vorticity. In the extreme cases, hydrodynamic instability-driven turbulence can improve the mixing between fuel and oxidizer streams through increased entrainment, thereby resulting in higher burning efficiencies.

Based on the above discussion, we may classify reacting RM and RT flows into four distinct categories (i) non-premixed RM, (ii) premixed RM, (iii) non-premixed RT and (iv) premixed RT. In premixed RM/RT, the flame front separates burned and unburned mixtures, so that the density difference across the interface can lead to instability when a suitable acceleration is applied (impulsive or otherwise). In non-premixed RM/RT, a sharp or diffuse interface initially separates fuel from the oxidizer. When accelerated impulsively by a shock or through other means accompanied by heat addition, stirring at the interface brings the fuel and oxidizer streams in direct contact followed by flame formation. While reacting flows occurring in applications are complex and involve multiple physical processes, the above classification allows us to study each component process in detail by isolating it from other physics.

20.2.2 Flame physics

When chemical/nuclear reactions and heat release are present alongside instability-driven mixing, the resulting flow evolution can be very complex governed by multiple scales and parameters. In both premixed and non-premixed flames, ignition is accompanied by generation of pressure waves, the acceleration from which can further enhance (or impede) the growth of the underlying instability. In non-premixed flames, the formation of the reaction zone can sometimes alter the stability of the flow. Often, the initial perturbation seed for RT/RM in premixed flames is the growth of modes due to the Darrieus–Landau (DL) instability (Matalon, 2018; Darrieus, 1938; Landau, 1944). The DL instability occurs due to hydrodynamic effects induced by the thermal expansion of the burned gas, and is dependent on the density ratio between the unburned and burned gases, $\sigma_r = \rho_u/\rho_b$. The thermal expansion induces a jump in the tangential velocities across the perturbed interface, while the resulting vorticity induces perturbation growth at all wavenumbers according to linear stability analysis (Darrieus, 1938; Landau, 1944). However, Markstein (1951) argued based on observations, that small-scale perturbations must be stabilized due to curvature-dependent local stretching. This effect is captured by modifying the flame speed (S_f) according to Matalon (2018); Markstein (1951),

$$S_f = S_{f,\infty} - \kappa L_M, \tag{20.3}$$

where $S_{f,\infty}$ is the laminar flame speed of the unperturbed flame, L_M is the Markstein length that sets the cutoff for the critical wavelength, and $\kappa = \nabla \cdot \mathbf{n}$ is the interface curvature. The consequence of Eq. 20.3 is to differentially modify the flame speeds at the crests and troughs in such a way as to stabilize the DL instability (Matalon, 2018; Markstein, 1951) for wavelengths shorter than $\lambda_c = 4\pi\sigma_r L_M/(\sigma_r - 1)$. Spherically expanding flames, such as those occurring in the astrophysical applications (Akkerman et al., 2011; Akkerman and Law, 2013), approach the infinitesimally thin flame limit as their radius increases, and are thus particularly vulnerable to DL instabilities without the stabilizing effect of a diffusion thickness.

In addition to seeding perturbations, the DL instability could also potentially play a role in the self-acceleration of a spherically expanding flame. Akkerman et al. (2011) illustrate such a scenario comprised of the following sequence: an initially laminar, smooth and spherical flame is formed as a result of energy deposition; as the flame front expands, it is susceptible to the formation of cusps and creases due to the DL instability; the flame surface area increases as a result of the appearance of these features, thereby increasing the consumption rate of the fresh charge; the flame obtains a sustained self-acceleration.

An RT-unstable flame front may also exhibit self-regulating behavior (Gamezo et al., 2003; Hicks, 2015). The flame speed increases through turbulent wrinkles generated by RTI. However, a corrective to this trend is that as cusps form on the flame front they may be lost to burning, due to their high curvatures (Eq. 20.3). As a result, a measure of balance may be achieved between the production of flame surface through RTI and the corresponding losses due to cusp burning (Hicks, 2015). However, it is not clear if this balance will be maintained through late times, as the RT-unstable flame front is populated by low wavenumber modes with low curvatures and are thus likely to survive the burning process with their flame speeds largely unaffected.

20.2.3 RM flames

The earliest experimental studies of a shock interaction with a flame were undertaken by Markstein (Markstein, 1951; Markstein and Schwartz, 1955; Markstein, 1957a,b), who examined the passage of a planar shock through a n-butane-air stoichiometric flame. The experiments were conducted in a vertical shock tube, in which a premixture of fuel and air were admitted, while the shock was introduced by rupturing a diaphragm separating compressed and ambient air. From high-speed photography and pressure tracers, the authors diagnosed multiple interactions, first between the incident shock and the flame and followed by interactions between reflected shocks and expansion waves. The interactions resulted in funnels (spikes in the RMI context), of cold air penetrating the flamefront, corrugating the flame surface and enhancing burning. The funnel features were correctly attributed to RM (due to shock/flame interaction) and RT (due to subsequent expansion wave interactions with the flame).

Recent experiments include the study by Thomas et al. (2001) of a shock interaction with a spheroidal ethylene flame bubble. In the experiments, the flame was distorted by the passage of several reflected shocks, ultimately leading to a turbulent state as well as the transition to a detonation wave. Figure 20.10 shows high-speed Schlieren images from the experiment, including the appearance of RM-created spike strands at late times.

Rudinger (1958) presented a wave diagram approach for analytically computing the interaction of a planar (1D) shock with a flame surface approximated using small perturbation theory. Early simulation work was performed by Picone et al. (1985) who reported 2D simulations of a cylindrical flame-like region driven by a planar shock under conditions similar to Markstein's experiments (Markstein and Schwartz, 1955; Markstein, 1957a,b). In their simulations, however, the flame was represented without reactive effects as a region

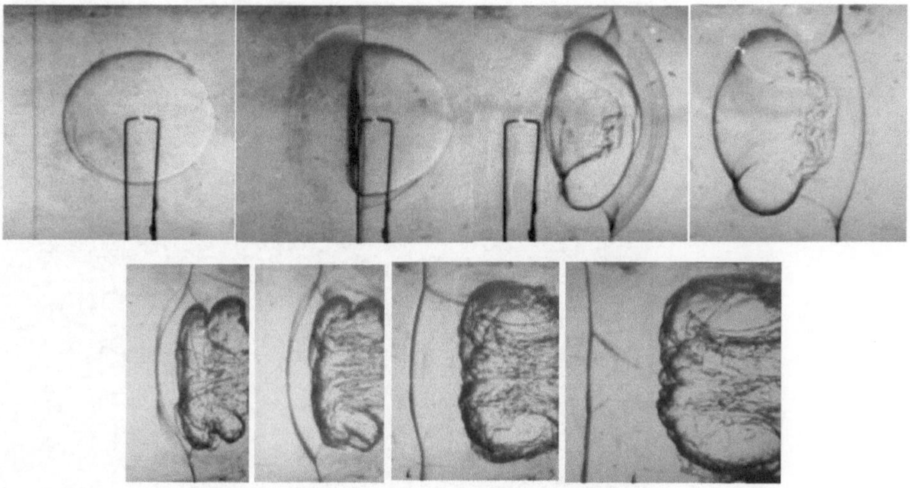

Fig. 20.10 Schlieren images of an ethylene flame bubble impacted by an incident shock (traveling rightward), and later by a reflected shock (traveling leftward), from experiments of Thomas et al. (2001). The incident shock Mach number was 1.7. Images are reproduced from Thomas et al. (2001).

of reduced density and increased temperature. A nonlinear theory of vorticity deposition was proposed to explain the simulation results.

Khokhlov et al. (1999) employed 2D and 3D numerical simulations to investigate the modification of a stoichiometric acetylene-air flame due to shock interaction. They found that the shock induces RMI at the sinusoidally perturbed flame site, and the associated vorticity deposition sustains the flame to late times against the stabilizing effects described in the Markstein model above. The formation of a "funnel" due to RMI is evident in their simulations, as it transports fresh unburned material toward the flame. At late times, the mushroom head detaches from the rest of the spike, and proceeds to burn out leaving behind a cusped spike. The simulations of Khokhlov et al. (1999) showed an increase in the energy release rate of a factor of 20–30 for the RM flame over the unshocked case. Additionally, the energy release rate per unit area was found to increase weakly with the shock Mach number, as the increased density at higher shock strengths leads to greater mass flow through the flame.

Using experiments in a vertical Hele-Shaw cell, La Fleche et al. (2017) investigated RMI on a cellular flame. The Hele-Shaw plates had a gap width of 5 mm filled with a hydrogen-air mixture at ambient pressure, ignited using spark electrodes. As cellular features develop on the flame (due to thermo-diffusive instabilities), an expanding shock parallel to the flame front is launched. The shock interacts with the flame, triggering RM seeded by the cellular structures. Both light-to-heavy and heavy-to-light interactions were studied, with the RM in the latter situation resulting in a phase inversion of the seeded perturbations. More recently, Yang and Radulescu (2021) presented a more complete work, whose Figs. 4, 7, or 9 show a similar sequence to the one just described for different incident shock strengths of Mach

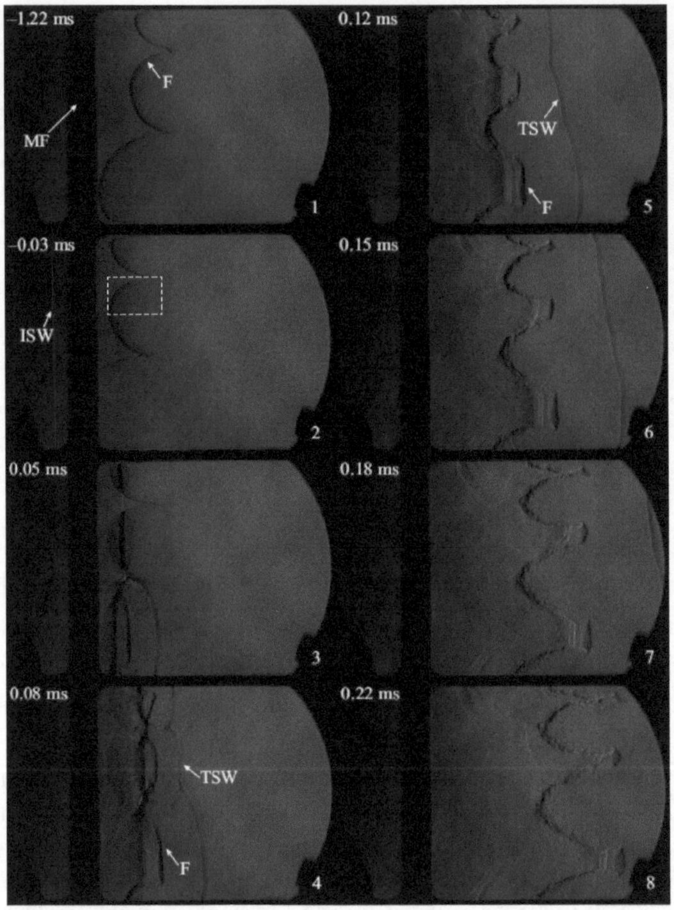

Fig. 20.11 Schlieren image sequence of the interaction of a Mach 1.9 incident shock wave with stoichiometric hydrogen-air flame. Here, TSW is the transmitted shock wave, and MF is the middle flange of the shock tube. Time 0 corresponds to the beginning of the interaction. Images reproduced from Fig. 4 of Yang and Radulescu (2021).

1.53, 1.75, and 1.9. Figure 20.11 displays Schlieren images of the last case, showing the interaction of a Mach 1.9 incident shock wave with stoichiometric hydrogen-air flame.

In contrast to the extensive studies of premixed RM summarized above, the corresponding non-premixed problem has received little attention, despite its significance to applications such as multiphase combustion. In Valentine and Wohletz (1989), results from a comparative study of premixed and non-premixed combustion were reported at different incident shock Mach numbers. For the conditions of their shock tube experiment, the RMI-induced mixing did not result in the combustion of a pure Hydrogen bubble in an oxygen environment. In contrast, premixed bubbles of H_2/O_2 (in O_2 or N_2 surroundings) underwent ignition when shocked at the same Mach numbers. Billet et al. (2008) and Attal et al. (2015) reported 2D results from simulations of a non-premixed shock-bubble flame interaction. In both studies, the effects of detailed chemistry and transport coefficients were included, while the problem setup involved an initially quiescent cylindrical H_2 bubble embedded in air and shocked by a Mach 2 planar shock. The development of the diffusion flame proceeded through a complex sequence of events and was fundamentally different from the premixed examples discussed earlier. Ignition was first observed in a thin rib connecting the kidney-shaped vortex features on either side. The thin ridge was flanked on either side by air, and thus constituted a double-diffusion flame surface. As the vortices developed more small-scale features, increased mixing was observed, leading to combustion within the vortex cores.

20.2.4 RT flames

In the experiments of Elyanov et al. (2018), a combustible mixture of hydrogen and air was pressurized in a cylindrical container and spark ignited. The diagnostics included high speed photography of the flame front, from which the radial trajectory of the expanding flame could be inferred. The authors in Elyanov et al. (2018) concluded from the radial trajectory that the flame was RT unstable, corroborated by the observation of flame bubbles ascending towards the unburned material. RTI was also found to be influential in the transition of so-called tulip flames (TF) to distorted tulip flames (DTF) in premixed combustion.

Hicks (2015) performed 3D simulations of a premixed RT flame front to investigate the accuracy of commonly used subgrid models for flame thickness and flame speeds in astrophysical calculations. The simulations were initialized with a multi-mode perturbation, while reactions were modeled using a bistable reaction term implemented in an advection-diffusion-reaction equation. The simulations revealed the flame widths were thinner than suggested by traditional turbulent combustion models, likely due to stretching by the RT modes.

The authors of Hicks (2015) attribute the failure of standard turbulence models to the formation of cusps on the flame surface, which are unaccounted for in the models. The likelihood of cusp formation and their prevalence scales with the RT forcing, so that at large values of $g\mathcal{L}$ (where $\mathcal{L}$ represents a dominant RT scale and g is the acceleration) such features are widely present. The cusps increase the flame surface area, as well as creating a focusing effect for the thermal flux (Hicks, 2015) enhancing the flame speed beyond the theoretical estimates.

Attal and Ramaprabhu (2020) investigated modifications to the single-mode RTI, in the presence of chemical combustion occurring in a non-premixed flame setting. Figure 20.12 shows scaled bubble and spike amplitudes and corresponding Froude numbers for reacting and non-reacting single-mode interfaces against scaled time. The authors also found that the properties of the flame, the mixing layer, and the stability of the system can depend on the global Atwood number. This problem has relevance to ultra-compact combustors, which rely on RT-driven mixing to minimize the residence times, and thereby the dimensions of the combustion chambers.

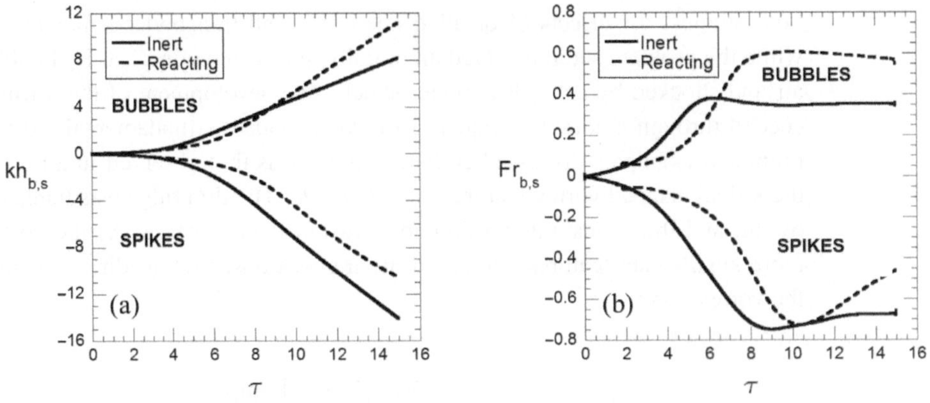

Fig. 20.12 (a) Scaled bubble and spike amplitudes and corresponding (b) Froude numbers for reacting and non-reacting single-mode interfaces against scaled time ($\tau = t\sqrt{\mathcal{A}gk}$). Fig. 6 of Attal and Ramaprabhu (2020).

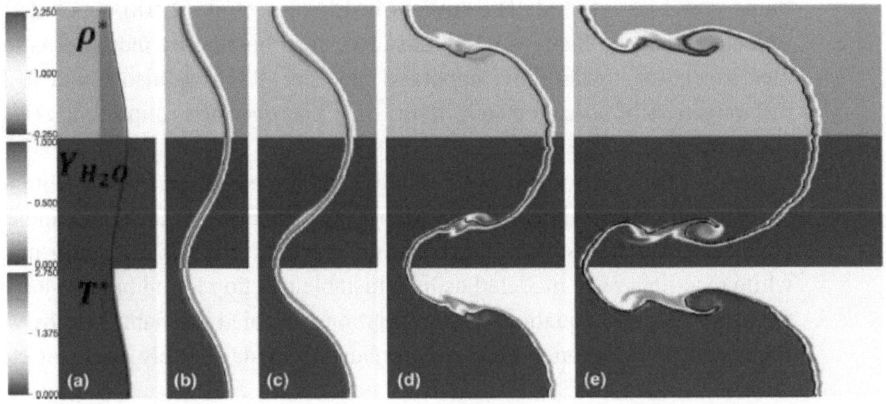

Fig. 20.13 Contours of scaled density (top), H_2O mass fraction (middle) and scaled temperature (bottom) from 2D numerical simulations of a sinusoidally perturbed, non-premixed H_2-O_2 flame shocked by a Mach 1.2 planar shock. Fig. 17 of Attal and Ramaprabhu (2015).

20.2.5 Reacting RT and RM

Non-premixed combustion is an application in which both the chemically reacting RT and RM instabilities occur. The sinusoidally perturbed, non-premixed RM problem was studied by Attal and Ramaprabhu (2015) using high-resolution numerical simulations with detailed chemistry (Fig. 20.13). In the absence of an incident shock, burning results in the formation of combustion waves, which spontaneously trigger RM- and RT-like instability growth of the interface. When an incident shock is also present, the aforementioned combustion waves and shock wave interact, resulting in significant non-planar distortions of each from combined RTI and RMI actions. When the unstable interface is reshocked, the mixing, combustion efficiency, and combustion product formation are all significantly enhanced.

Finally, the potential of water hammer events for igniting hydrogen–oxygen mixtures was examined in an experimental study (Coronel et al., 2021). A key finding in this study is that the interaction of the liquid pressure wave with the liquid-gas interface causes rapid growth of disturbances on the initially planar interface and dispersion of the water into the gas. The authors attribute the observed interfacial instability to a combination of the impulsive acceleration of the interface (RM mechanism) and the continuous deceleration of the interface (RT mechanism) occurring during the compression of the gas pocket.

20.3 Explosively expanding single and multiphase flows

This section offers a brief discussion of RT and RM instabilities in flows issued from and related to the detonation of an explosive charge. The study of these instabilities in this extreme regime is particularly difficult experimentally due to the challenges of obtaining detailed measurements, while simulations are complicated by the restrictive grid and time-stepping requirements.

Explosive flows could be classified into two broad categories: (i) homogeneous explosive (i.e., bare explosive charges) driven flows in which only an energetic reactive substance is involved and the resulting flow is entirely in the gaseous phase; (ii) heterogeneous explosive (i.e., particle-embedded or particle-surrounded explosive charges) driven flows in which the explosively driven flow is a mixture of gaseous products of the detonation process along with the particulate matter initially seeded. In both cases, the evolution of the pressure and density profiles in the flow is such that RT and RM instabilities occur multiple times within the typical period of interest. While for the homogeneous explosives, these instabilities occur at the material front (or contact discontinuity) between the explosive products and the surrounding medium, they also take place at the advancing front of the particle cloud for heterogeneous explosives.

20.3.1 Some examples of explosive flows

Figure 20.14 shows the surface of a rapidly expanding turbulent reactive material front resulting from a thermobaric test. The effect of RT and RM instabilities is clearly seen in the highly wrinkled nature of the expanding front. The early growth of these instabilities and their later development into a highly turbulent interface is often difficult to observe in such explosively driven flows, due to the rapid evolution of the process and the intense brightness associated with an explosion. The difficulties associated with describing the initial detonation process are avoided if we consider the problem of a sudden release of a highly pressurized spherical or cylindrical region of gas into an ambient of much lower pressure. These idealized configurations may be referred to as spherical or cylindrical shock tubes, as they are analogous to the classical planar shock tube problem, and the sudden release is equivalent to bursting of the diaphragm in a conventional shock tube. Figure 20.15 shows Schlieren images of the flow resulting from such a spherical shock tube, where a glass sphere plays the role of the diaphragm. The outermost density jump

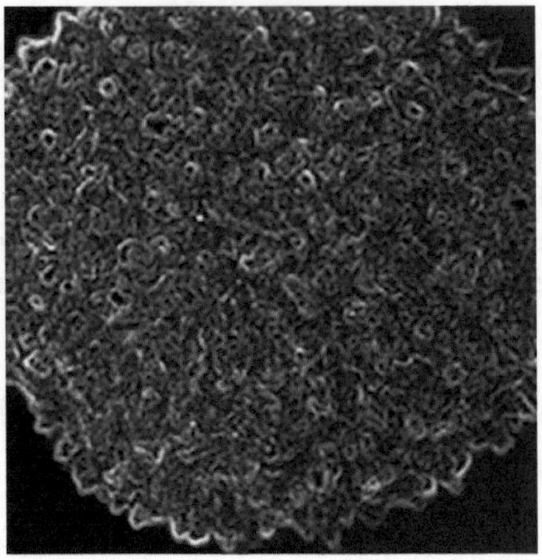

Fig. 20.14 Rapidly expanding turbulent reactive front resulting from a thermobaric test (Baer, 2022). Figure courtesy of Dr. Melvin R. Baer, Sandia National Laboratories.

seen in panel (b) is the outward propagating primary shock (PS). In contrast, the material interface (or the contact surface) between the gas initially contained within the glass sphere and the ambient undergoes instabilities, which later develops into the highly turbulent interface seen in panel (c).

The presence of particles (or droplets) in the rapidly expanding fluid greatly alters the nature of RT and RM instabilities. In addition to the density jump across the gas-gas interface between the initially pressurized and ambient fluids, we also have a strong density jump across the particle front. The particle front separates the high-density fluid-particle mixture from the lighter, particle-free fluid, and thus can undergo RT and RM instabilities. The time evolution of the interface in this scenario, showing the development of RT and RM instabilities, is shown in Fig. 20.16 for a spherical explosive charge of nitromethane embedded initially with zirconium particles.

Examples of experiments studying the particle-surrounded explosive charge configuration include those by Frost et al. (2012) and more recently, by Hughes et al. (2018). In a weak cylindrical casing, Frost et al. set up an annulus of packed glass particles sandwiched between two thick annuli of packed iron powder to enforce quasi-two-dimensionality of the glass particles dispersal. A PETN (pentaerythritol tetranitrate) chord ran through the length of the arrangement and was lit at the far end, opposite the camera. High-speed recordings of the post-detonation flow showed the formation of late-time aerodynamically-stable particle jets. These can be seen in the top row of Fig. 20.17. Interestingly, a substantial difference in the particle jet instabilities is observed in the top row between the dry powder (left) and the water-saturated wet condition (right). Hughes et al. (2018) furthered the study of this configuration through experiments that were instrumented to capture gas, particles, and gas-particle flow data. The middle and bottom row of Fig. 20.17 show axial and transverse views of a tungsten-particle shot (left) and a steel-particle shot

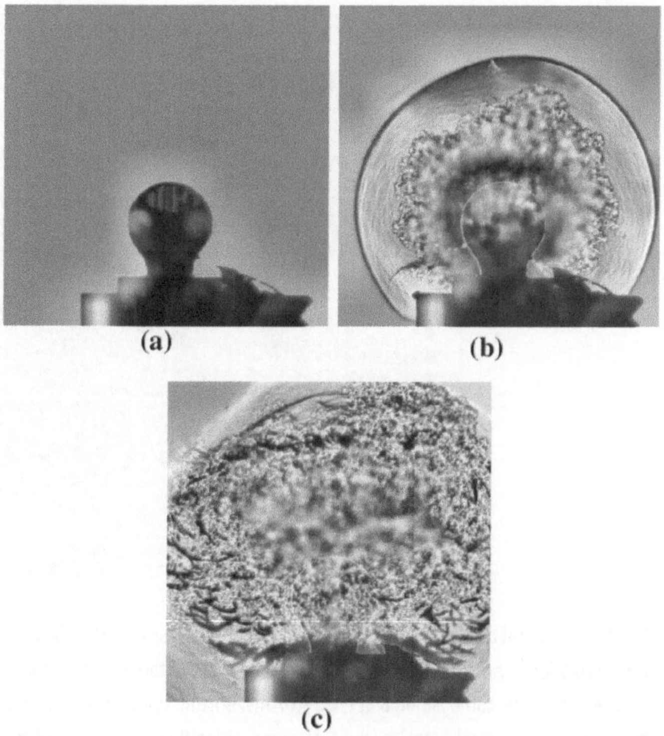

Fig. 20.15 Schlieren images of a pressurized glass sphere experiment: (a) before explosion; (b) strong blast wave phase; (c) end of reshock phase. The secondary shock is visible near the top left corner of the picture. Fig. 13 of Courtiaud et al. (2018).

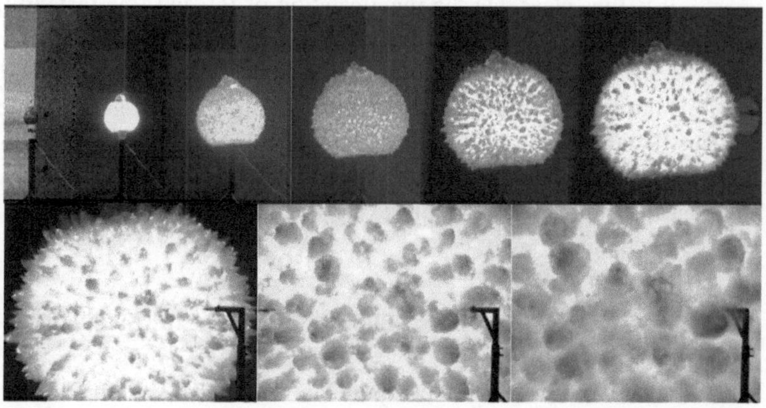

Fig. 20.16 Development of instabilities on fireball surface from detonation of a charge packed with zirconium particles saturated with nitromethane (Frost et al., 2005). Reproduced with permission.

Fig. 20.17 Top row: dispersal of a layer of dry (left) and wet (right) glass particles shown at $t = 12$ ms. (Frost et al., 2012), with the permission of AIP Publishing. Middle row: axial view of the dispersal of layers of tungsten particles (left) and steel particles (right) by composition B at $t = 3.67$ ms. Bottom row: transverse view of the dispersal of layers of tungsten particles (left) and steel particles (right) by composition B at $t = 3.60$ ms (Hughes et al., 2018). Reproduced with permission.

(right). Luminescence of the tungsten particles aside, one can clearly distinguish different shapes of instabilities between these shots.

It is thus clear that the manifestation of RT and RM instabilities can show substantial variation in terms of the observed evolution of material interfaces. The factors that seem to influence the evolution are (i) the geometry of the problem (spherical versus cylindrical), (ii) the instability of rapidly expanding single-phase gas-gas interface versus multiphase particulate front, (iii) the particles being initially embedded within the high-pressure, high-density gas versus emplaced outside, and (iv) other factors such as the initial pressure and density ratio between the inside and the ambient, size, density, and packing volume fraction of the particles, etc.

20.3.2 Phenomenology of explosively driven flows

Brode (1958) was first to compute the 1D laminar solution of the blast wave created by the detonation of a spherical TNT charge. In three dimensions, the flow becomes turbulent and rich in RTI and RMI mechanisms (e.g., Balakrishnan et al., 2010, Saurel et al., 2012). Figure 20.18 shows cross-sectional views of the evolution of the temperature field created by the detonation of a 1-g spherical TNT charge or aluminum powder in air in an unconfined space (Kuhl et al., 2013). The computations were performed in three dimensions, and the

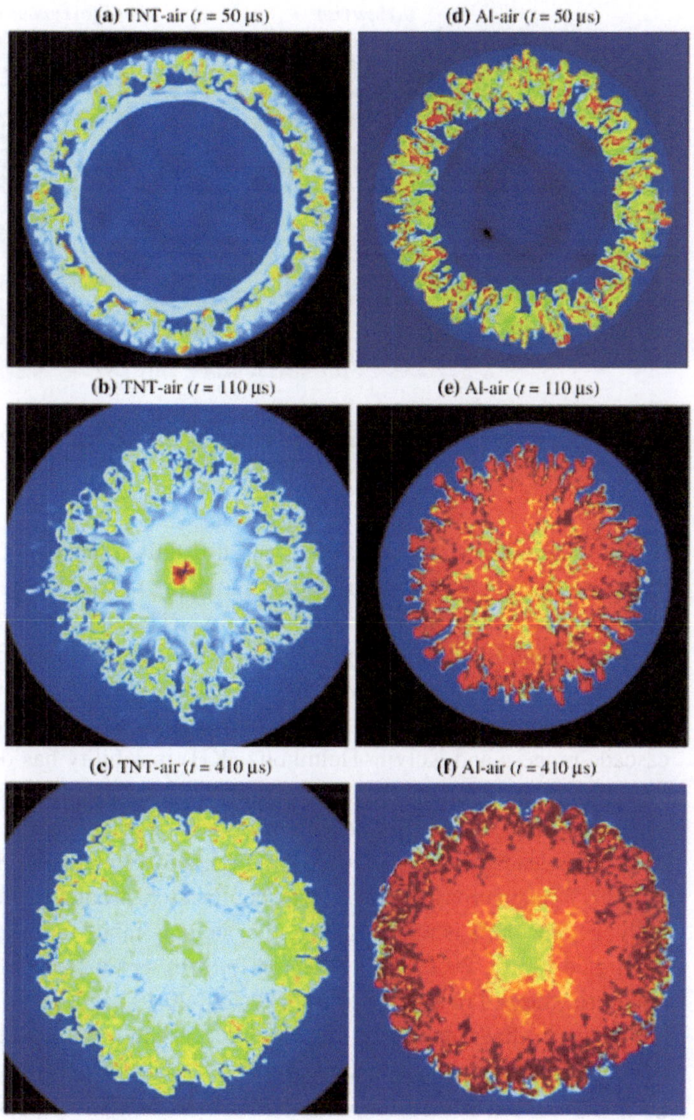

Fig. 20.18 Cross-sectional view of the temperature field of spherical combustion clouds: frames a–c illustrate TNT–air combustion, while frames d–f depict Al–air combustion (colors denote temperature – red $3,000 < T < 4,000$ K; yellow: T = 2,000 K; turquoise: T = 1,000 K). Fig. 2 of Kuhl et al. (2013), with permission from Shock Waves.

structure of the detonation wave was modeled by a similarity solution, using thermodynamic properties of TNT as specified in the Cheetah code (Kuhl et al., 2011; Kuhl, 2015). In the physical experiment, perturbations arise either from granular irregularities on the charge surface or from molecular fluctuations, whereas in the computation they originate from discretization errors (Boris, 1989). As detailed in Kuhl (1996), the interface is first subjected to impulsive accelerations that can lead to RMI. As the fireball expands, the

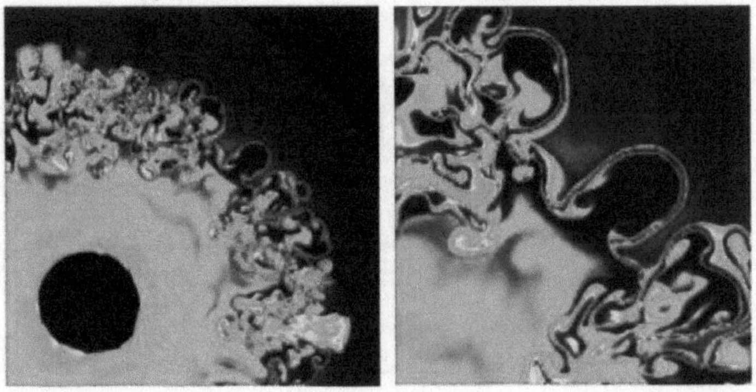

Fig. 20.19 Exothermic flame sheet in the TNT-air fireball ($t = 100\ \mu s$). Fig. 3 of Kuhl et al. (2013), with permission from Shock Waves.

shock-compressed air decelerates it, and it is thus subject to RTI. The resulting structures are drawn inward by a backward facing shock (first predicted by Brode) that implodes at the center and rebounds as a spherical blast wave. This deposits more vorticity, which intensifies the mixing. By 50 μs, the perturbations have evolved into mushroom-shaped structures characteristic of RM- and RT-driven mixing layers. These structures rapidly evolve into the nonlinear regime. By 100 μs, they have generated new fine structures through a nonlinear cascade process and Kelvin–Helmholtz (KH) instability has developed due to shear flow along the walls of the bubbles. Perturbations on the fireball interface cause a misalignment of pressure and density gradients that creates vorticity by the inviscid baroclinic mechanism. This leads to a self-induced production of vorticity in the mixing region. The fuel rich TNT detonation products form combustion products when they come into contact with air. Since this is a non-premixed system, combustion occurs in thin flame sheets (as illustrated in Fig. 20.19, left panel). A blowup cross section shows a mushroom-shaped structure wrapped in a thin flame sheet (Fig. 20.19, right panel).

The wave diagram in Fig. 20.20 represents the main events that follow the detonation of an explosive charge surrounded by a bed of particles in spherical or cylindrical geometry. At the end of the detonation, there exists a spherical or cylindrical region of very high-pressure and high-density gas surrounded by the ambient fluid. An outward propagating blast wave (BW) and an inward propagating expansion wave (EW) are generated. The contact interface (CI) forms as the discontinuous surface between the outward moving detonation products and the shock-heated air and is RT unstable. Due to the over-expansion of the flow near the origin, the tail of the expansion wave typically turns into a secondary shock (SS) wave. The initially weak SS wave begins expanding outward, entrained within the detonation products. As it strengthens, the SS wave turns and implodes (Brode, 1959). The deceleration of the contact interface during this phase of the flow promotes the growth of RT structures. Eventually, the SS wave reflects off the origin and propagates out, interacting with the contact interface and initiating RM instability. This

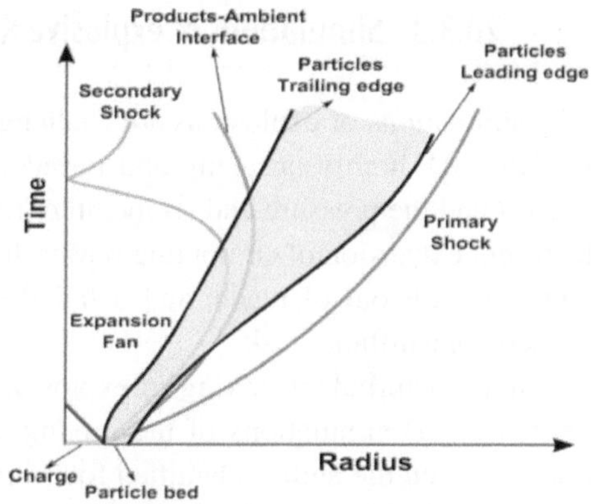

Fig. 20.20 Schematic wave diagram for the detonation of a heterogeneous explosive. Fig. 35 of Zhou et al. (2021).

interaction reverses the direction of travel of the CI, which begins to move outward. The SS continues as a transmitted shock and the interaction also generates a weaker, reflected, tertiary shock that travels inward and reflects from the origin. Meanwhile the CI slows and eventually heads back toward the origin, where it encounters the outward moving tertiary shock. This process continues until there is not enough energy left to repeat itself (Brode, 1959). When particles are initially emplaced around the high-pressure gas, the particle cloud presents an inner front and an outer front, across which the gas-particle density increases and decreases, respectively. As the BW, and later SS, propagates over the particle cloud, both the inner and outer fronts can undergo RM instability. Also, due to the deceleration of the particle bed, after the rapid initial acceleration, the outer front also becomes RT unstable. In contrast, when particles are initially distributed within the explosive (and within the high-pressure products of detonation) the process is like that of the single-phase limit, especially when the particles are very fine.

The instability of a CI in a detonation problem resulting from a spherically or cylindrically expanding material front is in several ways more complex than those considered in corresponding planar configurations (Crittenden and Balachandar, 2018; Mankbadi and Balachandar, 2012, 2013). For example, classical studies have considered the simplified problem of an isolated interface with a density jump, with uniform fields on either side of the interface, and with a controlled evolution of the interface. In contrast, in a spherical or cylindrical shock tube, the PS and SS remain very close to the CI, so that the CI cannot be considered to be spatially isolated. At later times, as the SS approaches and interacts with the CI, the evolution of the interface switches from RT-unstable to RM-unstable. Since the SS passes from heavy-to-light gas, the CI instability is expected to first decay, change phase, and then grow linearly due to the RM instability. After the passage of the SS, the instability again switches from RM-unstable to RT-unstable. Thus, the time evolution of perturbation at the CI is quite complex. Furthermore, in the detonation problem the compression rate on either side of the CI is not the same, nor is it spatially uniform.

20.3.3 Simulations of explosive RT and RM instabilities

High-fidelity simulations of explosions are challenging by nature – these flow are broadly multiscale, fully 3D, highly unsteady and turbulent; their characteristic time scales are extremely short and the pressure and temperature ranges reach extreme values. Therefore, despite the recent expansion of computing power, fully resolved, multiphysics simulations of explosions are still out of reach, and detailed studies of the embedded RT and RM instabilities remain a difficult task.

An early study of turbulent mixing in explosions was proposed by Kuhl (1993). The work consisted of 2D simulations of the mixing taking place in the fireball of an HE-driven blast wave, and the author identified four mixing phases (Anisimov and Zeldovich, 1977; Anisimov et al., 1983; Kuhl et al., 1993): (i) a strong blast wave phase, during which the fireball interface is subject to the RM instability when the detonation wave reaches the edge of the charge, causing the mixing width to grow linearly; (ii) an implosion phase in which the fireball is subject to the RT instability, and the mixing width grows according to a power-law; (iii) a reshock phase in which a secondary shock interacts with the density structures in the fireball, inducing another RM instability; and finally (iv) an asymptotic mixing phase at late time, during which the mixing layer achieves an asymptotically-constant width. It was found that a small residual of fluctuating kinetic energy, driven by the vorticity field, remained at late times.

Recently, Courtiaud et al. (2018) reported a numerical and experimental study of mixing inside fireballs using a non-reacting pressurized glass sphere experiment, illustrated in Fig. 20.15. Removing the numerical complication associated with modeling a high explosive (HE), they were able to concentrate on mixing. Their 2D and 3D LES confirmed that the glass sphere analog is a good approximation for the flow of a HE fireball, as all four characteristic mixing phases mentioned above (Kuhl, 1993; Balakrishnan et al., 2010) were captured. Also, their results suggested that the overall mixing layer development qualitatively parallels the classic RT theory.

Annamalai et al. (2014) isolated the RT instability within the explosion context. They performed 2D and 3D studies of an outward propagating, but decelerating, single-mode perturbation around the contact interface, following a finite source cylindrical blast wave. The RT instability in this context contrasts with classical studies. It involves a cylindrical geometry, time-dependent acceleration, radially inward and outward displacement of the contact interface-causing the wavelength under consideration to change in time and a time-dependent compressible background flow. Their simulations show that in the linear regime, the instability's perturbation growth rate is proportional to the exponential of the square-root of the wave number. In the nonlinear regime, they reported that the bubble height grows as the inverse of the square root of the wavenumber.

An observation widely reported in the explosive dispersal of particles surrounding a HE charge is the formation of late-time aerodynamically-stable particle jets. The mechanism for formation and evolution of these jetting instabilities is still unclear. However, the important role played by hydrodynamic instabilities in HE flows points to the RT and RM instabilities as natural candidates for contributing to late-time jetting. Numerical simulations showed that the HE detonation causes a spall layer, then an accretion layer that can break up, thus seeding the conditions for particle jet formation (Milne et al., 2010). These fragments follow a ballistic trajectory while leaving a trail of debris. The authors' analysis of 2D simulations, coupled with experimental data, lead them to conclude that the timescale of the RT instability was too long to successfully initiate the formation of particle jets. In a comparable study, Ripley et al. (2012) conjectured the jet formation follows the shock-particle front interaction near the charge's surface and that the number of jets is a function of shock pressure. In other words, the RM instability is responsible for seeding particle jets. The subsequent growth of these jets was linked to transverse displacement of particles induced by complex shock interactions, inelastic collisions, and particle-wake interactions that were qualitatively modeled with a particle attraction force.

20.3.4 Experiments of explosive RT and RM instabilities

Carefully controlled experiments of explosively expanding flows are difficult to design and hence remain scarce. To date, few have presented a detailed view of the hydrodynamic mixing that follows the primary blast wave.

Courtiaud et al. (2018) exercised the compressed balloon method to study mixing inside a HE explosion by analogy. This experimental setup consists of a pressurized glass sphere sealed to a metallic base with gas inlet enclosed in a steel vessel with optical access. Experiments using this apparatus have been successful at reproducing the characteristic features of blast-driven flows and have provided results consistent with numerical simulation results available in the literature.

Recently, a new facility to study blast-driven RT and RM at non-diffuse interfaces has been tested by Musci et al. (2018, 2020, 2023). The explosive charge is located in a reloadable cylindrical sleeve inserted at the bottom of the chamber. This sleeve is sealed prior to a test to ensure the upward propagation of the blast wave in the test chamber. The heavy gas is filled from the bottom and the light gas from the top of the test chamber. Finally, a loudspeaker is used to induce small disturbances at the interface. Initial testing of this facility has suggested good repeatability and controllability of the blast wave. In addition, the characteristic mixing phases of a HE-driven blast wave appear to have been observed.

An elegant experimental apparatus was designed by Rodriguez et al. (2017) to investigate the mechanisms of late-time particle jet formation following the explosive dispersal of a particle cloud by a blast wave. In this experiment, a weak blast wave generated in an open shock tube at the center of a packed bed of particles is confined to a Hele-Shaw cell. This allows for clear visualization of the solid particle jet formation in a quasi-2D geometry, as illustrated in Fig. 20.21(a). It appears that initial disturbances, referred to as filaments, are seeded at the inner surface of the particle cloud. These filaments continue to migrate to the outer particle cloud front and contribute to jet forming at late time. Figure 20.21(b) shows the result of a four-way coupled simulation inspired by Rodriguez et al.'s experiment, using a 42% mean initial particle volume fraction (Koneru et al., 2020). This study indicates that the deposition of vorticity through a multiphase analog of RM instability plays a key role in channeling the particles into well-defined jets at the outer edge of the particle bed.

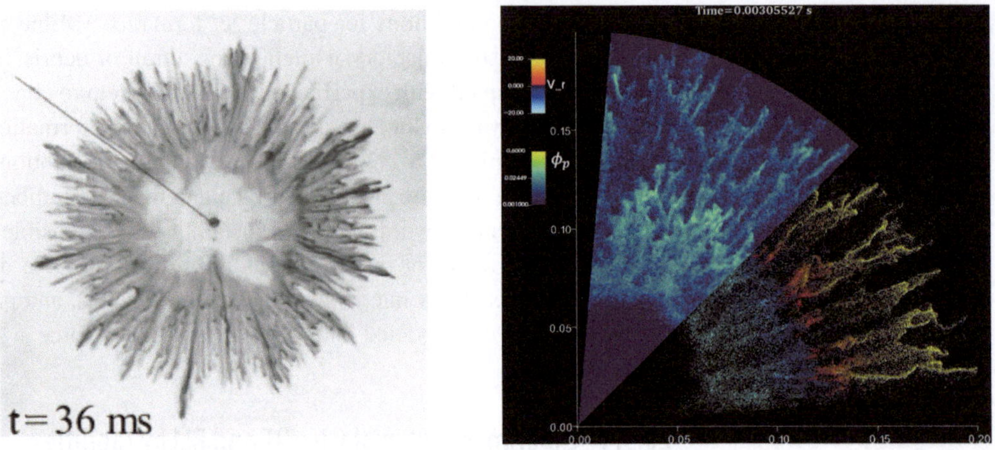

(a) (b)

Fig. 20.21 (a). Initially packed bed of flour particles dispersed by a blast wave in a hele-shaw cell at late time (experiment) Fig. 2 of (Rodriguez et al., 2017). Shock Waves. © Springer. (b). Four-way coupled point-particle particle simulation with discrete element method collision model of a blast-driven dense particle bed in a hele-shaw cell (42% mean initial volume fraction). The upper part of the figure displays particle volume fraction, while the lower part shows the particles colored by their radial velocity. Koneru et al. (2020), with permission from the AIP publishing.

20.4 Ejecta

Ejecta is a term used in a variety of fields for impulsively generated flows, often laden with particles, driven from an initial high density source into its environment. The impulsive driver can result from a variety of mechanisms: as a result of the rapid release of energy by phase change or violent chemical reaction, or by the impact of a body such as a bullet or a meteorite either directly onto a surface (Collins et al., 2005), or at the rear of a thin sample (Asay et al., 1976). At the time of writing, another topical example of relevance is the role of violent respiratory events, such as coughs and sneezes, in transferring respiratory diseases between infectious and susceptible individuals (Bourouiba et al., 2014; Scharfman et al., 2016), in the context of the coronavirus disease 2019 (COVID-19) pandemic (Bahl et al., 2022; Bourouiba, 2020; Chu et al., 2020) (Fig. 20.22).

In each case, the cloud of dense material generated will initially move ballistically, until any diffuse medium through which it passes starts to exert an aerodynamic effect. Once particles have swept through a column-density of diffuse material comparable to their own column-density, they will start to be subject to significant drag forces, which may even be strong enough to lead to the particles breaking up. The resulting particle-laden flows are often turbulent and subject to internal multiphase instabilities driven by the initial inter-component velocity, or by buoyancy differences between particles and the diffuse medium.

The initial creation of particles can result from a variety of processes. When volcanic ejecta form, the primary breakup mechanism is the result of a violent dissolution of gases as

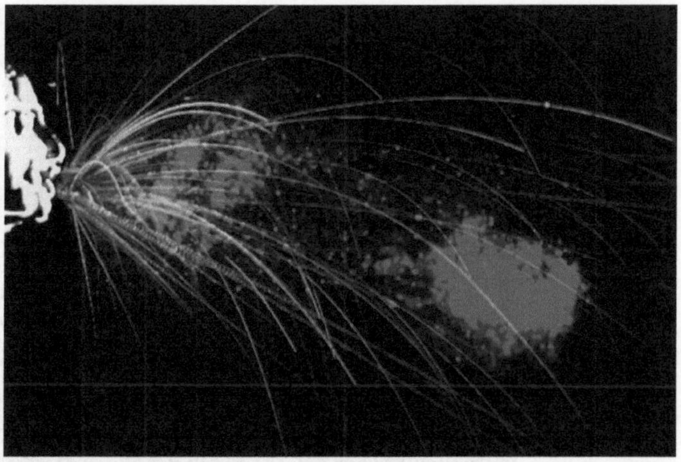

Fig. 20.22 Multiphase turbulent gas cloud from a human sneeze. The fluid dynamics data is relevant for evaluating social distancing strategies during the COVID-19 outbreak. Image reproduction showing the semi-ballistic largest drops, visible to the naked eye, and on the order of mm, which can overshoot the puff at its early stage of emission. The puff continues to propagate and entrain ambient air as it moves forward, carrying its payload of a continuum of drops (Bourouiba et al., 2014), over distances up to 8 meters for violent exhalations such as sneezes (Bourouiba, 2020). Figure 1 of Balachandar et al. (2020).

the geological pressure confining the lava is released. For impact and shock-driven ejecta, the mechanism is the formation of jets by processes akin to RM instability, where breakup ceases once constitutive forces such as strength and surface tension limit kinematic dispersion. For most astrophysical ejecta, the fluid flow becomes clumpy as a result of rapid cooling of material due to expansion and radiative losses, leading to its collapse into a thin sheet, and this thin sheet then fragmenting as it expands further (Vishniac, 1983; Ryu and Vishniac, 1988; Garcia-Segura and Mac Low, 1995).

The initial ballistic nature of the expansion once the dense ejecta have formed results in a characteristic kinematic signature, in which the velocity of the material varies linearly in position with the time of the initial impulsive driving event, the "kinematic age," being (to a good approximation) $\tau_{\text{kin}} \simeq 1/(dv/dx)$. This kinematic signature can be seen clearly in astrophysical spectroscopic observations, for example in the spectacular explosive ejecta around Eta Carinae, Fig. 20.23 (Meaburn et al., 1996).

Eventually, the ejecta particles will become entrained in the surrounding gaseous flow, a process which results in a variety of complex multi-phase dynamical phenomena. For volcanic ejecta (see Fig. 20.24), buoyancy is created by the dissolved gases and the heating of entrained air (Fisher, 1979; Woods, 1995). This often generates a plume of material which rises high into the atmosphere. This form of volcanic eruption is named after the Roman author Pliny the Younger, who observed the eruption of Vesuvius in A.D. 79 and accurately described the formation of the plume and associated flows in his letters. As the particle-laden plume rises, expands and cools, it eventually loses buoyancy and collapses,

Fig. 20.23 The explosively-driven ejection nebula around Eta Carinae. Image Credit: NASA, ESA, HST; Processing: Judy Schmidt. Reproduced under the Creative Commons Attribution 2.0 Generic License.

often generating pyroclastic surges that move rapidly across the landscape and can deposit the volcanic ejecta in deep layers.

20.4.1 Surface ejecta physics

Particulate ejection from the surface of a dense material is often observed when a shock wave is driven through it. Experiments to study this process (Buttler et al., 2017) have been performed using a variety of shock drivers, including gas guns (Asay et al., 1976; Rigg et al., 2006; Bell et al., 2017), high explosives (Buttler et al., 2012; Mikhailov et al., 2014; Hartsfield et al., 2022) and high-power lasers (Foster et al., 2005; Hartigan et al., 2009; Signor et al., 2009; Roland et al., 2018; Sollier and Lescoute, 2020; Saunders et al., 2020) (see Fig. 20.25). These different techniques offer distinct capabilities, both in the extent to which the drive pressure can be tuned to explore different scenarios, and the diagnostics which can be applied to the resulting ejecta.

It is now well recognized that this ejecta-forming process is an extreme limit of the RM instability, where the Atwood number $\mathcal{A}$ tends to -1 (Dimonte et al., 2013). The low density of the material ahead of the shocked interface has a substantial effect on the asymptotic behavior of the growing mixing layer. Rather than evolving to a fully self-similar state at late time (in which the growth law is controlled by the turbulence evolution), jets of dense material propagate away from the surface at nearly constant velocity. The bubbles slow as these quasi-ballistic jets remove the kinetic energy of the fluctuations from the surface. Indeed, the initial kinetic energy per unit area, $\mathcal{K}_0 \simeq \rho\lambda\dot{a}_k^2$ for a perturbation

Fig. 20.24 The particulate ejecta plume from the eruption of Mount St. Helens, May 18, 1980. Image Credit: USGS.

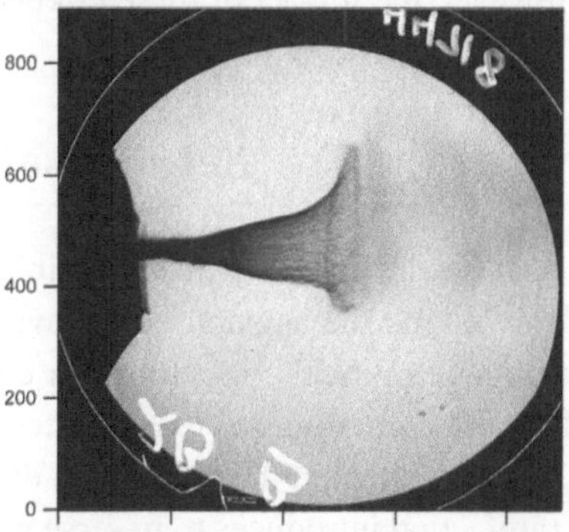

Fig. 20.25 Ejecta jet generated by a plane shock incident on the surface of a sample with a cylindrical "divot," expanding into vacuum. This high-resolution backlit radiograph shows the tip of the ejecta plume, which collapses to the jet expected for this perturbation geometry. Thin sheets break into thread-like structures oriented along the jet. The sheets likely break into particles, although the particles cannot be individually resolved. From (Williams, 2018), based on previously unpublished data from the experiments of (Hartigan et al., 2009). Reproduced with permission. UK Ministry of Defence © Crown Owned Copyright 2022/AWE.

of characteristic size scale $\lambda = 2\pi/k$ is one of the dominant characteristics of ejecta flows. Here the RMI growth rate in the fluid regime can be approximated by

$$\dot{a}_k = k\Delta u \mathcal{A}^+ \left(a_0^+ + a_0^-\right)/2, \tag{20.4}$$

where the superscripts $-$ and $+$ denote pre- and post-shock quantities, respectively (Richtmyer, 1960; Meyer and Blewett, 1972).

The process described above can be strongly affected by material properties such as strength and surface tension. As the movement of material into the spikes leads to loss of kinetic energy in the bulk perturbations, these material property effects can become increasingly important at later times. If the stress on the interfacial material due to RMI is sufficient to exceed the material's yield strength in the nonlinear phase, then the material will fail and perturbation growth will continue. However, the strength of the material will still play a role in reducing the bubble growth and mass transfer into the jets. Hence, the formation of a fully developed turbulent mixing cascade is prevented.

Buttler et al. (2012) performed experiments using a variety of initial surfaces to study the effect of strength on the growth rate of the bubble tips over time, as illustrated in Fig. 20.27, and related these results to a simple and effective theory.

In addition, when the ejecta is driven by an unsupported shock such as a blast wave, the growth of the RMI can be affected by spall, that is, internal brittle failure or rupture (cracking). This occurs when the material is placed under tensile loading due to the interaction of the rarefaction wave trailing the unsupported shock with the wave reflected from the material surface. Cracks initiated due to spall may be sufficiently large to cause a separation of thin surface layers from the bulk of the material, limiting the reservoir of material which can feed the growing ejecta jets.

The dominant mechanism for ejecta formation from shocked metal surfaces is believed to be the formation of jets from microscopic surface imperfections. The planar symmetry can also be broken up by other effects, such as the presence of grain structure or sub-surface inclusions. However, the impact of these effects scales inversely with their depth, limiting their overall effect.

20.4.2 Spike formation

As we have discussed previously, it is possible to estimate RM linear instability growth rates in fluids by taking the asymptotic behavior of impulsively-accelerated, incompressible RT, with minor empirical corrections for compressibility effects (Eq. 20.4). However, the details of the jet formation process can become more important in the context of ejecta formation. The direct effect of the interaction of the shock with the perturbed surface is that the reflected rarefaction introduces transverse motions in the material. Due to the relatively low Mach number of transverse motions, these soon result in the formation of jets of material (known as spikes) where they are compressive, or draw-back of the surface (bubbles) where they are expanding. The form of the spikes varies with the topology of the initial perturbation: an initial pit or divot in the surface leads to a localized jet, while a groove will lead to the formation of a sheet of ejecta. A surface with a raised feature will lead to the formation of a divot in the post-shock surface surrounded by an expanding ring of ejected

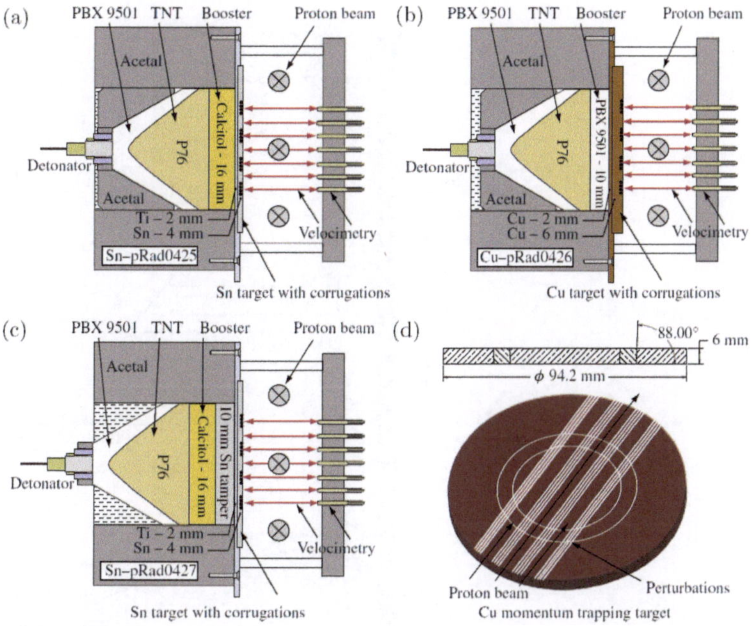

Fig. 20.26 Radiative RM experimental package geometries encase in acetal plastic a 76 mm diameter plane-wave HE lens used to uniformly detonate a HE booster cylinder in contact with a buffer plate; the sample/target is mounted onto the buffer plate on top of the acetal plastic as well. The target design incorporates momentum trapping concepts, as illustrated (d) together with the target perturbation schemes. The targets were machined to a diamond-turn finish that included four bands of sinusoidal corrugations, $\lambda = 500\ \mu m$, but with varying amplitudes. Each corrugation band was eight wavelengths wide and separated by flat regions 5 mm wide. Over each corrugation band and intervening flat region, a velocimetry probe measured the jump times and velocity histories throughout the experiment. The proton beam is aligned into the page, along the perturbations. Fig. 2 of Buttler et al. (2012).

material. This process is somewhat analogous to the impact of a droplet on a fluid surface, or on a larger scale, to a meteorite impact. A surface with regular machining grooves will generate an array of parallel sheets.

While the underlying equation, Eq. 20.4, is based on incompressible flow, Figs. 20.26 and 20.28 illustrate clearly that the sub-surface structures in real experiments can in fact have very high-density contrasts. The velocity field below the surface is imprinted by the reflected rarefaction, as usual, but in this case the rarefaction is strong enough that the material breaks up into sharply defined layers, as also observed by Liu et al. (2018b).

The form of the grooves on the initial surface can be quite variable, depending on the details of the surface preparation. However, the shape of the bubbles that form, and hence the mass source to the spikes after the initial transient, is rather weakly dependent on the initial surface profile (e.g., Cherne et al., 2015). In addition to the change in perturbation amplitude, or – if the material fails – peak expansion rate of the ejecta front, another important factor is the total mass of material ejected from the surface. As with other forms of RMI, the source of dense material into the ejecta sheets is controlled by the dynamics of the bubbles: The process of bubble competition can lead to sheets or jets pinching off, but

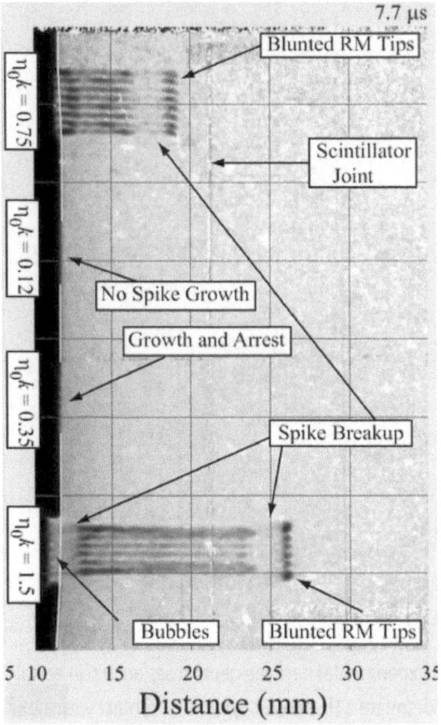

Fig. 20.27 Experimental data (from Buttler et al. (2012), with permission) illustrating the effect of varying the initial perturbation amplitude, and hence the RMI growth rate, on the growth of ejecta jets. Note that the authors used η_0, instead of a_0, for the initial amplitude. For low initial amplitudes, the post-shock material strength can suppress or arrest the spike growth. For higher initial amplitudes, ejecta jets can escape, but discrete clumps of material form at the jet tips, and the effects of strength and surface tension eventually act to cause the supply of material entering the jets to cease. Buttler et al. (2012).

for a sufficiently regular surface, this process of bubble competition will be suppressed. For real materials, subject to strength or surface tension, the spikes will tend to break up into discrete particles at late times.

20.4.3 Particle formation

It is important to understand the size distribution of the particles which the spikes break up into, as this affects their transport and reactivity. There is extensive literature on the break-up of fluid jets, which has been surveyed, for example, by Eggers (1997); however this literature tends to concentrate primarily on continuous rather than impulsive jets.

For an impulsive source, the process of particulate formation typically occurs in two phases. An initial process of rapid primary breakup occurs before the ejecta jets have interacted significantly with their environment, as a result of the balance between local kinematic perturbations within the dense fluid and restoring forces such as surface tension

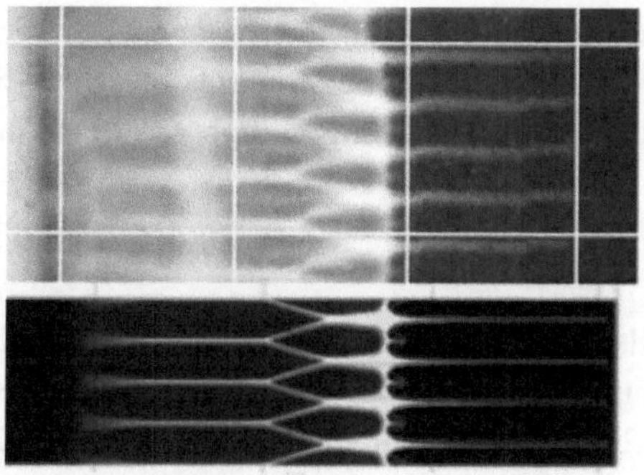

Fig. 20.28 Comparison of experimental pRad data from Buttler et al. (Buttler et al., 2012) (above) with 2D numerical calculations (Grieves, 2007) [Reproduced with permission. UK Ministry of Defence © Crown Owned Copyright 2022/AWE.]. The experimental data show both the growing spikes and the sub-surface structures driving them, which in this case have a far sharper structure than expected for an incompressible flow. Buttler et al. (2012).

and elastic strength. Except in the case where the ejecta is driven into a strict vacuum, an additional reservoir of kinetic energy becomes available once the interaction of the ejecta with the gas in its environment becomes significant. This leads to an additional secondary breakup process driven by drag on the ejecta particles. Of course, which process dominates the breakup at any given moment depends on the balance between the effect of the perturbations due to initial nonuniformities, and those generated by the unstable drag.

The primary breakup process can be enhanced by the effects of non-monotonic expansion velocities, non-uniform mass flux and capillary instability. The effect of the non-monotonic expansion velocity is analogous to the formation of internal working surfaces in high Mach number fluid jets (e.g., Raga et al., 1990), with additional breakup processes resulting from, for example, caustic formation in ejecta sheets with small lateral motions.

The process of secondary breakup is generally considered to be the result of a balance between aerodynamic forces on the particles, their inertia, and restoring forces such as material strength and surface tension. A variety of breakup models have been developed (O'Rourke and Amsden, 1987; Turner et al., 2012), which capture an essentially similar set of physical processes with varying fidelity.

A variety of techniques have been applied to the characterization of the particle size distribution in experiments. These techniques have complementary benefits and limitations, in terms of their ability to measure the properties of single particles against the population as a whole and the completeness of the sampling that is achieved.

Holography (Sorenson et al., 2017), shadowgraphy (Roland et al., 2017), and phase Doppler anemometry (Bell and Chapman, 2017) allow the properties of single particles to be characterized, but results are limited to rather small total numbers of particles, and particular experimental geometries. The Mie scattering diagnostic gives a measure of the ejecta particle population as a whole by measuring the angle by which light is diverted by diffraction around individual particles (Schauer et al., 2017), while multi-wavelength extinction measures the loss of light along the direct path (Steele et al., 2017). In both approaches, valuable additional information may be obtained by measuring at more than one probe wavelength. Such measurements provide statistics for the particle population as a whole and can be applied to cases in which the particle number would overwhelm a tracking diagnostic, but it adds complexity to the interpretation of the results.

20.4.4 Modeling

The details of the ejecta production process have been studied using codes which model either continuum flow (Grieves, 2007; Liu and Grieves, 2014; Roland et al., 2018; Karkhanis et al., 2018) or atomistic molecular dynamics (Zhakhovskii et al., 2006; Germann et al., 2009; Cherne et al., 2012; Durand and Soulard, 2012; Shao et al., 2014; Cherne et al., 2015; Durand and Soulard, 2015; Wu et al., 2018; Durand et al., 2020) (see Fig. 20.29). Comparisons between these techniques (Cherne et al., 2015; Durand et al., 2017; Dyachkov et al., 2017) tend to emphasize their good agreement on the overall properties of the jet formation and their complementary strengths. Molecular dynamical studies include the effects of constitutive properties modelled in an *ab initio* consistent manner (subject to the accuracy of the numerical scheme and the interatomic potential), but to date have been constrained to domain sizes far smaller than the most relevant experimental data. Continuum studies are not limited to small scales, but require supporting models for constitutive properties such as strength, surface tension and fragmentation.

Once ejecta particles have formed, the trajectories they follow in a vacuum will be determined purely by kinematics. However, in a medium of finite density, they will be subject to the usual hydrodynamical forces of drag and lift and may also be subject to interparticle collisions and breakup. Given the number of particles produced from a surface, it is generally impractical to model in detail the propagation of each individual particle, and so

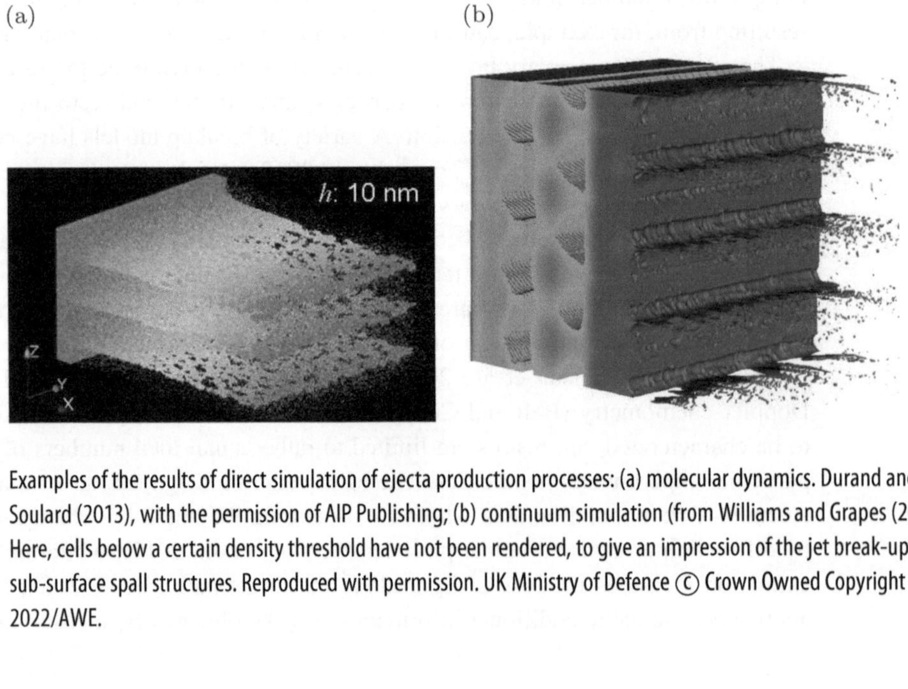

Fig. 20.29 Examples of the results of direct simulation of ejecta production processes: (a) molecular dynamics. Durand and Soulard (2013), with the permission of AIP Publishing; (b) continuum simulation (from Williams and Grapes (2017)) – Here, cells below a certain density threshold have not been rendered, to give an impression of the jet break-up and sub-surface spall structures. Reproduced with permission. UK Ministry of Defence © Crown Owned Copyright 2022/AWE.

some statistical model must be applied to the population. Many computational codes have been developed for the modeling of particle- and droplet-laden flows in a wide variety of applications, such as diesel sprays (O'Rourke and Amsden, 1987; Gosman and Clerides, 1997), fire suppression (Grant et al., 2000), fluidized beds (Chiesa et al., 2005), pyroclastic flows (Valentine and Wohletz, 1989; Dobran et al., 1993), spray irrigation, and inkjet printing. These approaches can be categorized broadly as (i) Eulerian–Lagrangian, in which the particle populations are modeled using representative Monte-Carlo particles (Cloutman, 1991; Fung et al., 2013; Mikhailov et al., 2014; McFarland et al., 2016) and (ii) Eulerian–Eulerian, in which the particle populations are modeled using a smoothed continuum phase field (Valentine and Wohletz, 1989; Dobran et al., 1993). In each case, many choices are available for the completeness with which processes such as drag are modeled, as well as the level of detail with which properties such as the particle size distribution are treated, whether by random sampling, binning at some resolution, or moment-based techniques.

It is often the case that differing approaches suit different applications, with Eulerian–Lagrangian techniques being particularly appropriate for modeling the initial ballistic expansion phase, while Eulerian–Eulerian approaches can capture cases where drag is dominant, or where particles are compacted into an ejecta bed. The above modeling methods may also be applied to flows in which a particle population is already present due to the seeding of the initial flow.

21 Astrophysical and space applications

In many space and astrophysical flows, including those in the Sun and its atmosphere, supernova remnants, P Cygni's nebula, binary neutron star mergers, equatorial tongue accretion, and star-forming galaxies, hydrodynamic instabilities are often observed (e.g., Bao et al., 2021; Blondin et al., 2022; Burdonov et al., 2022; Girichidis et al., 2021; González et al., 2018; Habib and Pierrehumbert, 2024; Magyar et al., 2019; Mishra and Srivastava, 2019; Müller, 2019; Peng et al., 2020; Popescu Braileanu et al., 2021a,b; Rast et al., 2021; Shen et al., 2022; Skoutnev et al., 2021; Wang et al., 2018).

The ejecta studied in the laboratory setting also found important applications in astrophysical planetary nebulae. In particular, astrophysical ejecta result from stellar wind interactions late in the life of intermediate mass stars, when the stars are subject to substantial mass loss rates and evolve rapidly through alternate phases in which the stellar winds are slow, cool, and dense or fast, hot, and diffuse. The internal collisions between these successive outflows give rise to shocks, and RT and RM instability within the ejecta (Balick and Frank, 2002). Other examples in which both RTI and RMI must be considered together are at the interface between the relativistic jets and their surrounding medium (Matsumoto and Masada, 2013) and in the uniform mixing of radioactive nuclides in the photo-solar nebula (Gritschneder et al., 2011).

21.1 Supernovae

Supernovae (SN or SNe) were recorded as "guest stars" that suddenly appeared in the sky as early as the second century A.D. by Chinese astronomers and were visible for a year or longer (Clark and Stephenson, 1977; Bethe, 1990; Murdin and Murdin, 1985). The Crab Nebula is the remnant of a supernova explosion (Woltjer, 1972) witnessed on July 4, 1054 A.D. Even at a distance of $\sim$2 kpc (Trimble, 1968), the Chinese astronomer Yei-Te Wang reported that the "guest star" was visible during the daytime for three weeks and was visible at night for 22 months (Clark and Stephenson, 1977; Hester, 2008). Supernova 1987A was discovered in Chile, February 23.316UTC, 1987 (Arnett et al., 1989b; Hillebrandt and Hoflich, 1989), when light and neutrinos from one of the brightest exploding stars in more than 400 years arrived on Earth. The circumstances of SN 1987A, in particular its proximity to Earth, are nearly ideal for studying supernovae (Arnett et al., 1989b). As a result, SN 1987A has become one of the most thoroughly studied objects outside the solar system and triggered a burst of publications in the scientific literature and popular press (McCray, 1993).

Broadly speaking, there are two types of supernovae. Type Ia supernovae are thermonuclear explosions of white dwarfs, whereas core-collapse supernovae occur when the iron core of a massive star collapses to form a neutron star (Type II). RT and RM instabilities play a major role in both types of explosions. In Type Ia supernovae, through a sequence of processes initiated by RMI or RTI, the transition of fusing plasma to turbulence creates a detonation front, a process termed deflagration-to-detonation transition (DDT) discussed in Chapter 20. At the astrophysical scale, RTI has been hypothesized to trigger the DDT process in type Ia supernovae explosions of white dwarfs in binary systems. The supernova is caused by runaway thermonuclear burn in a dense, electron-degenerate carbon–oxygen star. The star is driven over the Chandrasekhar stability limit by accretion from a binary companion. Recent simulations by Gamezo et al. (2003), summarized in Fig. 21.1, used high-resolution numerical simulations to investigate the deflagration process. The mass of a stable white dwarf can exceed the Chandrasekhar limit through accretion from a neighboring star, resulting in a sequence of processes that culminate in runaway ignition (Gamezo et al., 2003). The white dwarf undergoes core contraction, an compression of core material, an increase in temperature, and the acceleration of thermonuclear reactions. The ensuing thermonuclear flame propagates radially outward, initially with a laminar flame speed, but soon succumbs to the growth of RT unstable modes. The onset of RT on the flamefront signals a significant spike of the flame speed to a turbulent velocity (e.g., Khokhlov, 1993). The rising plumes of hot "burned" gas are interspersed with cold

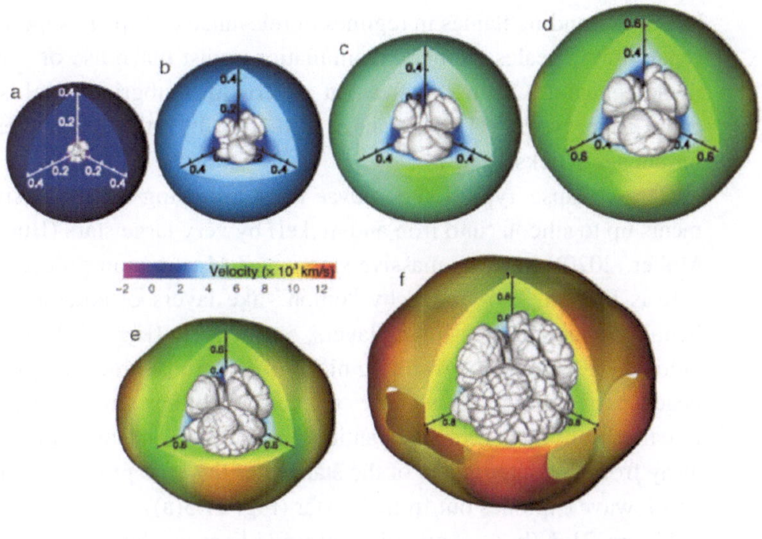

Fig. 21.1 Evolution of a carbon–oxygen white dwarf showing the RT-dominated thermonuclear flame (gray iso-surfaces) as well as the outer extent of the greatly expanded star. Spatial coordinates are scaled by the initial radius of the white dwarf. Panels (a)–(f) correspond to 1.26, 1.49, 1.57, 1.65, 1.76, and 1.9 s after ignition. Images are reproduced from Gamezo et al. (2003). Reprinted with permission from AAAS.

jets of "unburned" material that descend toward the core, as reported in Gamezo et al. (2003). For a detailed overview of RT flame simulations in the context of SN Ia objects, we point the reader to the references discussed in Hicks (2015). The detailed simulations of Gamezo et al. (2003) show a complex array of processes in which the RT instability corrugates the flame surface across an ever-widening range of scales. Secondary structures are observed on the flamefront. The counterflow of burned and unburned material triggers the KH instability, which further increases the mixing and the flame speed. Comparison of simulated carbon and oxygen spectra (Gamezo et al., 2003) with Ia observed spectra (Jeffery et al., 1992; Fisher et al., 1997) show a discrepancy in the velocities associated with these elements. The authors of Gamezo et al. (2003) hypothesize that the higher observed velocities imply these elements occur in the outer layers of the expanding star, and that the flame has undergone a transition to a detonation, likely instigated by RT. The higher detonation flame velocity and the more complete burning would explain the observation of C/O in the outer layers. The total energy output in detonation models is in broad agreement with the observations in Wheeler et al. (1995).

SN Ia are used as standard candles, since their light curves appear to satisfy a self-similar solution parametrized by the peak luminosity and the rate of decay (Oran, 2005). Awarded the 2011 Nobel Prize in Physics, Perlmutter et al. (1997, 1999) and Riess et al. (1998) recognized the use of Type Ia supernovae as standard candles to measure vast cosmic distances. They simultaneously discovered that not only was the Universe's expansion not slowing down, it was speeding up, contrary to previous cosmological theories. However, a significant source of uncertainty in the empirical relations used to collapse the different light curves is the precise explosion mechanism, including whether a DDT is observed and the mechanisms that drive it. A second source of uncertainty arises in the modeling of turbulent, expanding flames in regimes of relevance to Type Ia supernovae. Owing to the wide separation in scales, numerical simulations must make use of subgrid models to describe flame behavior. One parameter that governs the subgrid model is the flame speed, which could either be the RT flame speed or the result of the flame interaction with the ambient turbulence (Hicks, 2015).

Core-collapse Type II supernovae form following the successive fusion of heavier elements up to silicon (into iron and nickel) by very large stars (Burrows, 2000; Janka, 2012; Müller, 2020). In more massive stars ($\gtrsim 9\,M_\odot$), fusion proceeds until a degenerate iron core is formed, surrounded by "onion"-like layers of silicon-, oxygen-, neon-, carbon-, helium-, and hydrogen-fusing layers, respectively (Fig. 21.2). Eventually, the huge gravitational pressure on the resulting nickel-iron core is large enough to overcome the electron degeneracy pressure. This causes the core to collapse into itself, resulting in electron capture that emits an enormous quantity of neutrinos. The core continues to collapse into itself away from the outer layers of the star. After this (the precise mechanism is still debated) a shock wave explodes out from the star (Fig. 21.3(a)).

Figure 21.3 (b–d) shows the entropy slices in the z-y-plane at 80, 200, and 400 ms after bounce for the perturbed 3D progenitor model (Müller et al., 2018). The evolution through several hundred ms after core bounce demonstrates one possible mechanism for shock revival – namely, vigorous non-spherical motions that push the shock out to larger radii, thereby enhancing the neutrino heating efficiency of the post-shock material, leading

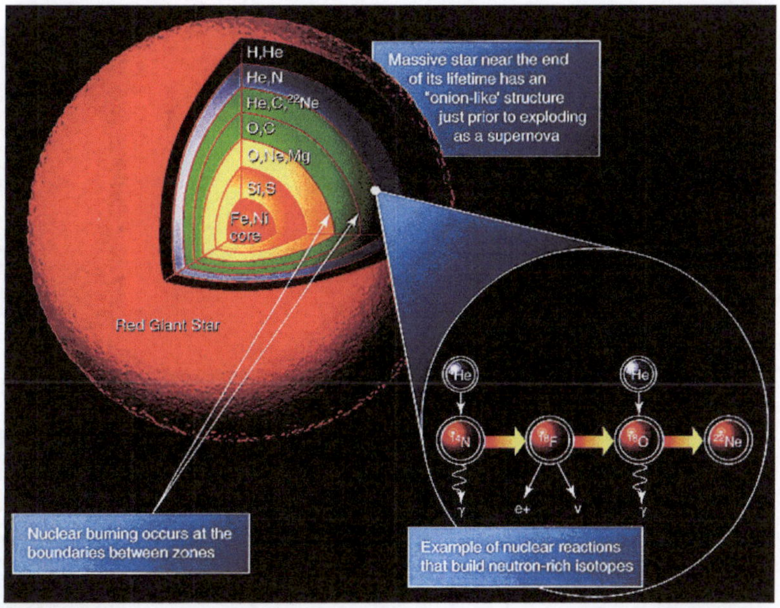

Fig. 21.2 Just before a supernova, a star resembles a many-layer, enormous onion as successively heavier layers of fuel ignite and burn. Layers' relative radii are enlarged for clarity and are not to scale. Image Credit: NASA.

eventually to explosion (Müller et al. 2018). The brightness of this event has a particular luminosity evolution over time, called a light curve.

From the particular smoothness of some light curves, especially from SN 1987 A,[1] as well as from supernova spectra (which reveal the distribution of elements in both velocity- and real-space), it has become evident that some kind of hydrodynamic mixing occurs inside the star during the core collapse and subsequent explosion (Arnett et al., 1989a; Hammer et al., 2010). During the shock transit phase of the core collapse, the RM instability is triggered at each discontinuity in the density profile of the star. As the explosion ensues, conditions arise in which dense material is accelerated by lighter material, the classic environment for the RTI. This is prominently the case at the boundary between the denser helium core ([C+O]/He, He/H) that is decelerated by the lower-density hydrogen envelope. Similar effects can occur with smaller amplitudes at inner-core composition boundaries (Si/O, for instance). Shear around RT fingers triggers KHI, and shocks propagating across steep density gradients near composition discontinuities induce RMI (Kifonidis et al., 2003; Branch and Wheeler, 2017; Wongwathanarat et al., 2015) (Fig. 21.4).

The outward mixing of higher-density, radioactive core material (e.g., ^{56}Ni, ^{56}Co, ^{44}Ti) brings a radioactive heat source toward the surface of the star. These explosion products decay via the emission of γ-rays, which Compton scatter off electrons in their vicinity. This reheats the photosphere and causes the secondary maximum in the light curve at 40–120

[1] Note that SN1987A was peculiar in showing a large effect of hydrodynamic mixing on the light curve. This is because it had a dim and short plateau, owing to the blue supergiant nature of its progenitor. In normal Type II supernovae, the effect of mixing on the light curve is not as pronounced.

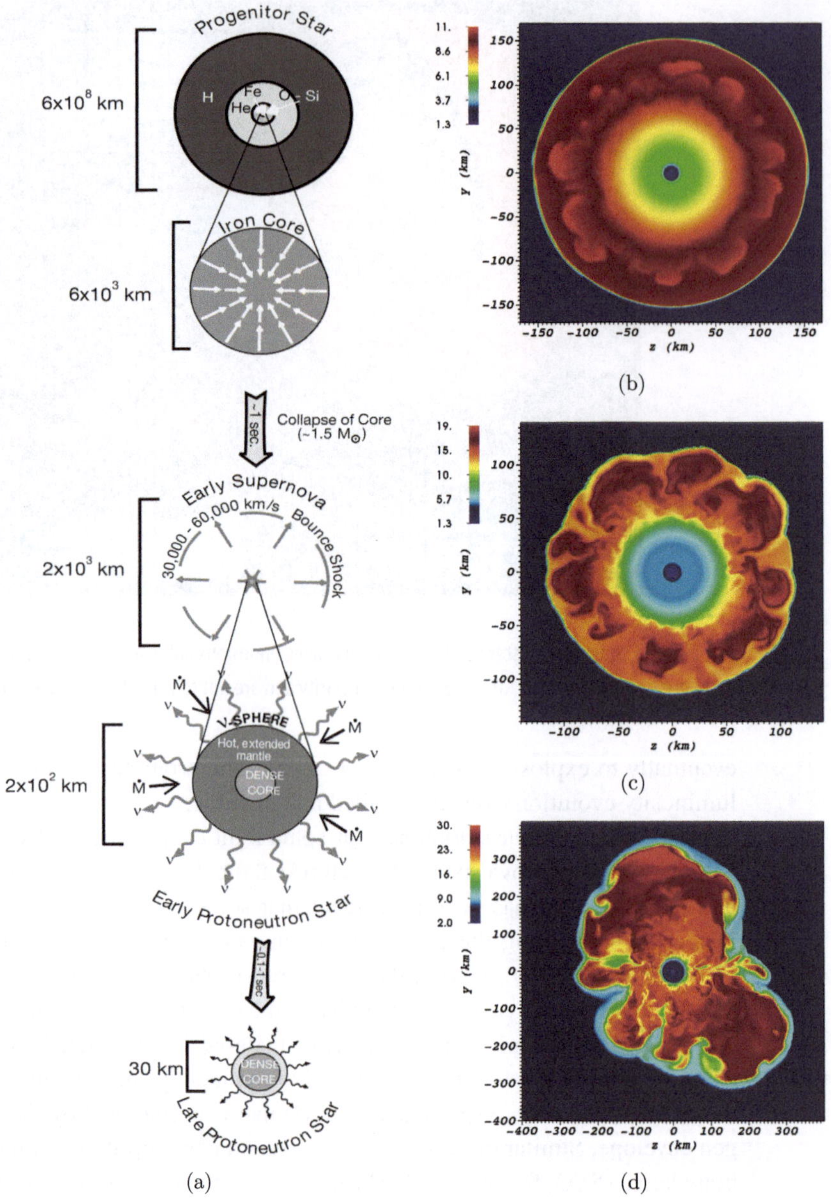

Fig. 21.3 (a) The sequence of events in the collapse of a stellar core to a nascent neutron star (This figure is not to scale). It begins with a massive star with an "onion-skin" structure, goes through a white-dwarf core implosion, a core bounce and shock-wave formation, and a protoneutron-star stage before exploding, and finally cooling to an isolated-neutron-star stage (From Burrows (2000), with permission from Nature). Entropy slices in the *z*-*y*-plane at 80 ms (b), 200 ms (c), and 400 ms (c) for the perturbed 3D progenitor model s18-3D using the CoCoNuT-GMT code (Müller and Janka, 2015). At 80 ms, the model begins to exhibit signs of neutrino-driven convection. At 200 ms, infalling perturbations begin to interact with large-scale oscillations, At 400 ms, the shock is deformed even more strongly and is likely on its way to shock revival and eventual explosion. Note the changing scale of the figures. Fig. 5 of Müller et al. (2018), with permission.

Fig. 21.4 Logarithm of the simulated density of SN 1987A. Top panel: From left to right (a) $t = 100$ s (left), (b) $t = 300$ s (middle), (c) $t = 1{,}500$ s (right). Bottom panel: From left to right (a) $t = 5{,}000$ s (left), (b) $t = 10{,}000$ s (middle), (c) $t = 20{,}000$ s (right). Note the change of the radial scale. The supernova shock is visible as the outermost circular discontinuity in the density plots. Kifonidis et al. (2003), with permission.

days. RT mixing hastens and broadens this secondary maximum. Light curve observations of SN1987A unambiguously showed this broadening of the secondary peak, suggesting enhanced transport from the core out to the photosphere (Arnett et al., 1989b; Hillebrandt and Hoflich, 1989; Remington et al., 2000).

Supernova explosions exist in complex electromagnetic environments, and fully understanding them requires, among other things, a proper accounting of the electromagnetic effects (Janka, 2012). The magnetorotational mechanism (Meier et al., 1976), is one candidate to explain the core-collapse supernova explosion. While the magnetic field inside the star is amplified to a great extent during gravitational collapse, this amplification alone is not enough to appreciably affect the hydrodynamics. However, as the star collapses and loses gravitational energy, its rotational energy increases due to the conservation of angular momentum. The increased rotation rate can boost the magnetic energy density further by "wrapping" the lines around the core in an increased number of windings (LeBlanc and Wilson, 1970; Wheeler et al., 2002). By incorporating such effects, MHD models have been used to develop propagation models of nuclear combustion fronts in Type Ia supernovae (Remming and Khokhlov, 2014, 2016).

In reality, RT, RM, and MHD effects all contribute to the complex dynamics of supernova remnants. The uneven shape of SNRs is due to RTI and RMI resulting from perturbations in the stellar structure during the initial explosion (e.g., Aschenbach et al., 1995). The SNRs move within a magnetic field, which significantly affects their dynamics (Bucciantini et al., 2004). Moreover, SNRs are known to greatly accelerate cosmic rays that pass through them. This can be explained using the MHD waves produced by these rays (Uchiyama et al., 2007), but the phenomenon may be best understood via the magnetic field amplification occurring in the MHD RM and RT instabilities (Jun et al., 1995; Sano et al., 2012).

21.2 Solar prominences

Having detailed the effect of magnetic fields on the RTI and RMI development in Chapter 17, we now turn our focus to some "real-world" examples. Our featured example is the development of RTI plumes in solar prominences.

The high density material that constitutes solar prominences is suspended against gravity by the magnetic field of the solar atmosphere. When buoyant bubbles of magnetic field rise beneath prominences, the high-density material of the prominence finds itself situated above material of much lower density. Recent observations have shown that, unsurprisingly, RTI develops in this situation. This section will present the information required to understand these observations and the peculiarities of RTI in this context.

21.2.1 Prominences and their dynamics

Prominences, an example of which is shown in Fig. 21.5, are one of the most striking features of the solar atmosphere. Seen clearly when the light from the Sun's disk is obscured during solar eclipses, prominences appear as colorful clouds in the solar corona. These "clouds" are composed of, what is relatively speaking, cool and dense plasma (in fact prominences are composed of neutral atoms, ions and electrons dominated by the neutral hydrogen component) surrounded by the hot and tenuous solar corona. To give a more precise statement of these quantities, the characteristic temperature and density of the solar corona are $\sim 10^6$ K and $\sim 10^{-12}\,\mathrm{kg\,m^{-3}}$, while for prominences we have $\sim 10^4$ K

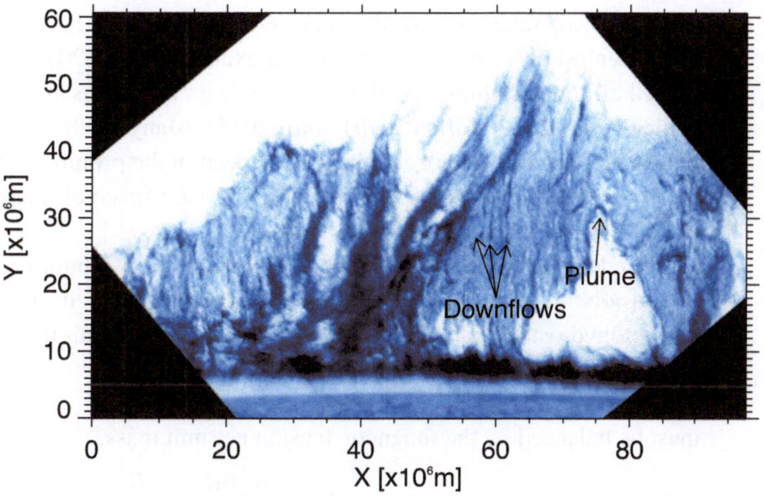

Fig. 21.5 A quiescent prominence observed by Hinode SOT on September 29, 2008 at 10:02UT showing the intensity in the Ca II H broadband filter, a plume and three downflowing knots are marked on the figure. Fig. 2 of Hillier (2018). This article is distributed under the terms of the Creative Commons Attribution 4.0 International License.

and $\sim 10^{-10}$ kg m^{-3} (Tandberg-Hanssen, 2013; Mackay et al., 2010) (note that these values can vary depending on where exactly in the solar atmosphere they are being measured).

Though prominences are observed at the solar limb, as the Sun rotates, they will rotate around from the solar limb onto the solar disk. On disk, the material that forms the prominence absorbs some of the light radiated from the solar surface below. As a result prominences can be observed on disk as dark features known as filaments. This multi-angle view of prominences means that characteristic dimensions can be measured, which are (note that there can be large variances depending on the prominence measured): length of 6×10^4 to 6×10^5 km, height of 1.5×10^4 to 10^5 km and thickness 5×10^3 to 1.5×10^4 km (Tandberg-Hanssen, 2013).

The low plasma β (defined in Eq.17.8, and is $\lesssim 0.2$ in the solar corona) means that the Lorentz force, in particular the magnetic tension component, see Eq. 17.7 and related discussion, becomes crucial for supporting the prominence material. Our current basic understanding is that prominence material collects in dipped regions of the coronal magnetic field, so that the tension force can work upward to support the material (for example, Aulanier and Démoulin, (1998)). This implies that the magnetic field in prominences is nearly horizontal (strengths of this horizontal field are regularly found to be between 3 and 30 G or 0.0003 to 0.003 T), as confirmed by measurements, for example, Leroy (1988) and Levens et al. (2016).

Due to the importance of magnetic fields in prominences, the type of dynamics seen can be correlated with the strength of the prominence magnetic field, and this field strength is determined by the magnetic field in the photosphere (the base of the solar atmosphere). For regions in which the photospheric magnetic field is weak, the visible characteristics of the prominence are more influenced by gravity. On scales $O(10^3$ km), these prominences display a wide range of flow dynamics (Al-Janabi et al., 2019), with strong motions both in and against the direction of gravity, for example, Chae (2010), and Hillier et al. (2011b), as well as flow instabilities and turbulence, (e.g., Leonardis et al., 2012; Freed et al., 2016; Hillier et al., 2017; Hillier and Polito, 2018). Many of the flows in prominences reach speeds of O(10 km/s), close to the sound speed in the prominence material (8 to 10 km/s), but lower than the Alfvén speed (the characteristic speed of magnetic waves), which for a $\beta = 0.1$ prominence becomes ∼30 km/s.

To understand where gravity becomes important in prominence dynamics, it is important to look at the relative importance of the key forces in the system. For a system in magnetohydrostatic balance, the gravitational force per unit mass,

$$F_{\rm G} = \rho g, \tag{21.1}$$

must be balanced by the magnetic tension per unit mass,

$$F_{\rm T} = \frac{B_x}{\mu_0}\frac{\partial B_z}{\partial x} \sim \frac{B^2}{\mu_0 L}, \tag{21.2}$$

where L is the magnetic field's radius of curvature. Taking the ratio of these two, and assuming an ideal gas ($p = \rho k_{\rm B} T/\mu_{\rm m} M_P$), gives:

$$\frac{F_{\rm G}}{F_{\rm T}} \sim \rho g \frac{\mu_0 L}{B^2} = \frac{p}{B^2/2\mu_0}\frac{g\mu_{\rm m} M_P}{k_{\rm B} T}\frac{L}{2} = \frac{\beta}{2}\frac{L}{H_{\rm p}}. \tag{21.3}$$

This expression highlights the fact that even for $\beta < 1$, gravity can be an important force if the system size is much larger than the pressure scale height (as is possible in prominences). The L value required to set this ratio to unity, given the pressure scale height of the prominence material ($H_{\rm p}$∼300 km) is ∼3×10^3 km for a prominence associated with weak photospheric fields (which is much smaller than the prominence global scale). This implies that gravity can be dynamically important in quiescent prominences.

21.2.2 Prominence plumes as observational evidence of the magnetic RTI

We next proceed to discuss observations of the magnetic Rayleigh–Taylor instability in quiescent prominences. These features are observed as dark plumes that rise upward through the prominence, an example of which is shown in Fig. 21.5 (note this is a negative image so the dark plumes appear light).

The first observations of plumes were presented by Stellmacher and Wiehr (1973). In these observations a large bubble (22,000 km in size) formed beneath the prominence, from which a plume broke off and then rose at approximately 12 km/s through the prominence before fragmenting. By analyzing the spectra, the authors observed that the lack of cool emission in the plume was a result of a dearth of cool, prominence material, and argued that the observed dynamics were created by an instability. These observations did not garner much attention at the time of publication, and interest in them grew only when the plumes were rediscovered by two research groups in 2008 (de Toma et al., 2008; Berger et al., 2008)

There are now a number of studies looking at prominences plumes, and these allow us to determine some general characteristics of the plumes. Plumes are found to have a wide range of widths from 500 to 5,000 km. Some of the observations show not only the rising plumes mentioned above, but also falling spikes of the dense prominence material between two rising plumes (Berger et al., 2010). Berger et al. (2008) noted that brightenings form at the plume head, which connect to downflows of prominence material. Further investigations into this bright material showed that it represented a Doppler-shift variation along the head of the plume (Berger et al., 2011; Suárez et al., 2014). This could be a signature of the plume pushing prominence material out the way. Berger et al. (2010) found that for most of its lifetime, the plume rose through the prominence material at a constant velocity with mean speeds of approximately 16 km/s (note that these speeds are supersonic, but sub-Alfvénic).

The majority of prominence plumes are observed to develop from large bubbles that form beneath the prominence. These are often called voids and cavities in the literature. It is from the boundary between the prominence and the bubble that the plumes develop and then rise up through the prominence. This has made the nature of the bubbles one of the key subjects for understanding how plumes are formed. Measurements show there is significantly less cool material in the bubbles than in the prominences (e.g., Heinzel et al. (2008)), though the presence of any excess hot material relative to the corona is still a matter of dispute (Berger et al., 2011; Gunár et al., 2014).

These observations can be summarized as follows: relatively dense prominence material is supported above a low density bubble. The boundary between these two systems can become unstable to the formation of rising plumes and falling spikes. As the prominence is threaded with a magnetic field, and the magnetic field is understood to be generally horizontal where the prominence material is found (Levens et al., 2016), the classic scenario for magnetic RTI is realized (e.g., Kruskal and Schwarzschild, 1954; Chandrasekhar, 1961). It was Ryutova et al. (2010) who was the first to realize that this instability could be key to plume formation, with Berger et al. (2010) developing the idea further.

21.2.3 Linear and nonlinear modeling of prominence plumes

For plumes formed by the magnetic Rayleigh–Taylor instability, both linear and nonlinear theory can be used to understand the dynamics of prominences. Observations of prominences can in turn be used to advance our understanding of the magnetic RT instability. In this section, we introduce some of the key theoretical results that have been developed to understand the plumes, and the physics that underpins these results.

There have been numerous attempts to use the linear dispersion relation of the magnetic RTI to investigate prominence plumes. In general, simple models for the linear stability assuming an incompressible plane-parallel system have been applied (Ryutova et al., 2010; Berger et al., 2010, 2017). These models use a classical dispersion relation similar to those found in, for example, Chandrasekhar (1961) (see also, Eq. 17.38).

By assuming a fixed angle between the magnetic field and the wave vector, Ryutova et al. (2010) used Eq. 17.38 to estimate the strength of the magnetic field of the prominence. By using observations to measure the growth rate of the instability as $\sim 8 \times 10^{-3}\,\mathrm{s}^{-1}$ and a wavelength in the plane-of-sky of $\sim 1.2 \times 10^{6}$ m observed in a prominence on November 30, 2006, Ryutova et al. (2010) determined the magnetic field strength to be 6 G. This value

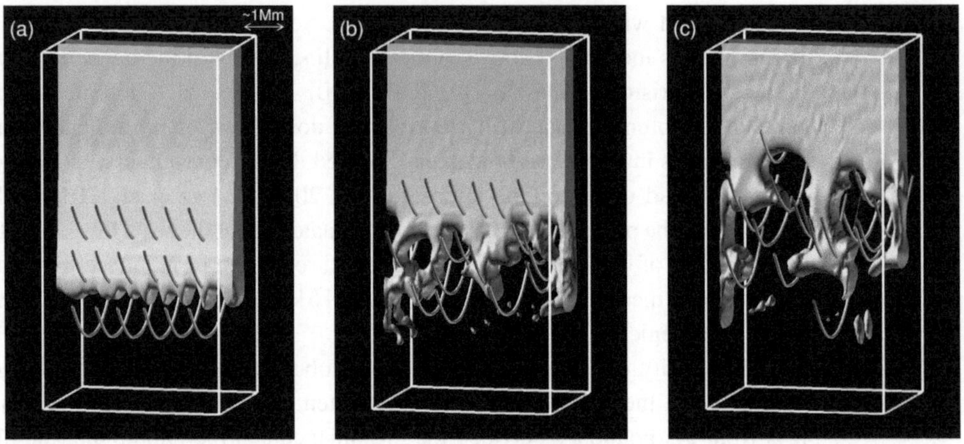

Fig. 21.6 Evolution of plumes driven by the magnetic Rayleigh–Taylor instability in the KS prominence model presented in Hillier et al. (2012). Reproduced with permission.

is consistent with those measured for quiescent prominences (Leroy, 1988). These ideas were further developed by Berger et al. (2017), and applied to a case where there was also a shear flow at the boundary (used as a potential additional source of instability). Again, a similar magnetic field strength was inferred.

There are a number of key components a nonlinear model of the instability occurring in prominences would have to include. The most basic of these is that there needs to be dense material supported against gravity by the Lorentz force, preferably the magnetic tension component of that force. These models must also include a low density bubble beneath the prominence material. Since this phenomena occurs in the solar corona, it should be expected that the plasma β of the model is less than unity, with reasonable values ranging between 0.01 to 0.2.

The first attempt at plume modeling was local – that is, it did not include the entire global prominence structure – model of Hillier et al. (2011a, 2012). They used the Kippenhahn–Schlüter (KS) model as the starting point, which provides an analytic solution for mass collecting in the dips of the magnetic field, so that it can be supported against gravity. The plasma β used for this model is <0.5. On perturbing the system, RT plumes rose from the bubble through the simulated prominence. Due to the strong, predominantly horizontal magnetic field, the plumes are forced to align with the direction of the magnetic field (see Fig. 21.6), so that when we observe the plumes in the prominence we are looking approximately along the direction of the magnetic field. This happens because the instability grows while working to minimize the bending of the magnetic field. The plumes rose at speeds of $\sim$8 km/s through the prominence body. Figure 21.6 shows a 3D rendering of the density, with magnetic field lines included, of the initial conditions (showing the buoyant bubble beneath the prominence) and at late times when rising plumes and falling spikes of dense prominence material had developed.

So-called global models, which represent the next level of complexity in modeling prominence plumes, can be seen as reproducing the plume dynamics. These models include

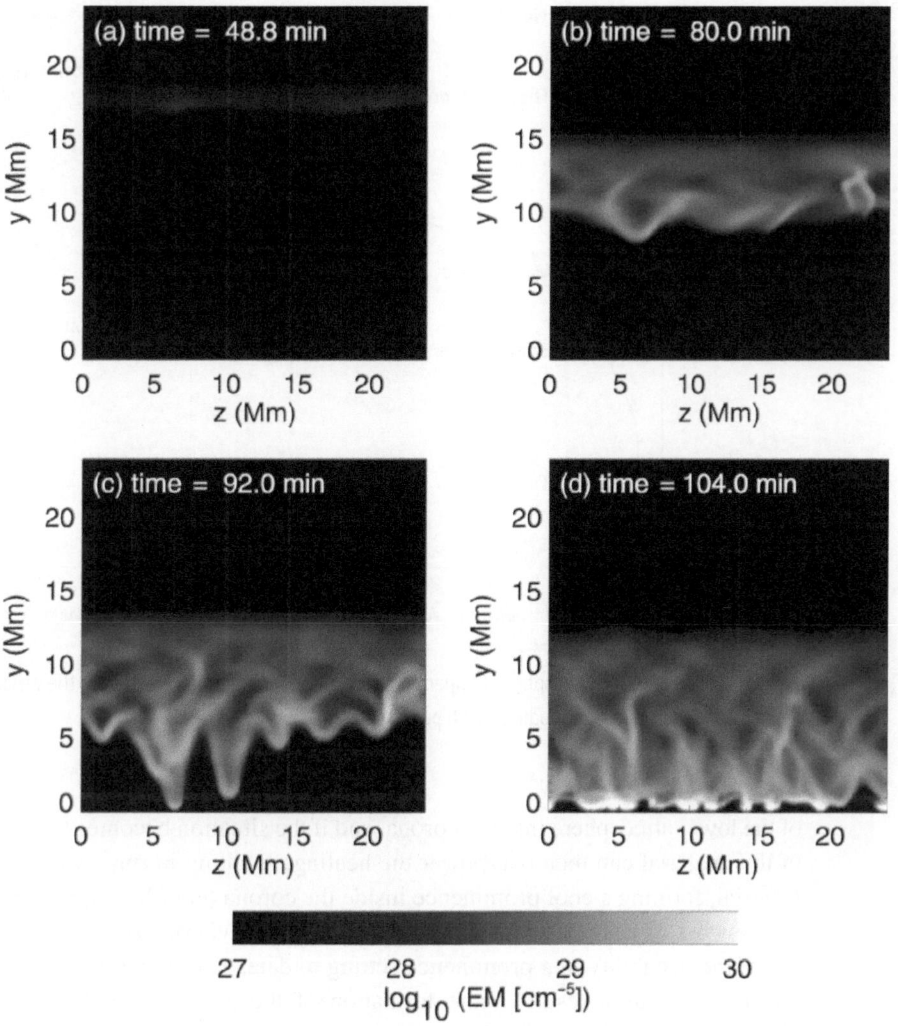

Fig. 21.7 Development of the RTI in a global prominence model (Kaneko and Yokoyama, 2018). Reproduced with permission.

the prominence plume dynamics by embedding the prominence in a hot, tenuous corona with the height and width (but not necessarily the length) of the prominence contained within the calculation domain. As seen in the simulations presented in Keppens et al. (2015) and Xia and Keppens (2016), the plume dynamics produced are similar to the local models.

To go beyond these "simple" global models, additional physics should be included to model the physical conditions of the solar corona in which a prominence might form. These include a heating source, energy losses from the atmosphere via radiation, and thermal conduction that is dominated by the conduction of heat along the magnetic field (as this is the direction in which the electrons have the greatest freedom to transport heat). The inclusion of these processes will allow for material to be evaporated from the dense layers

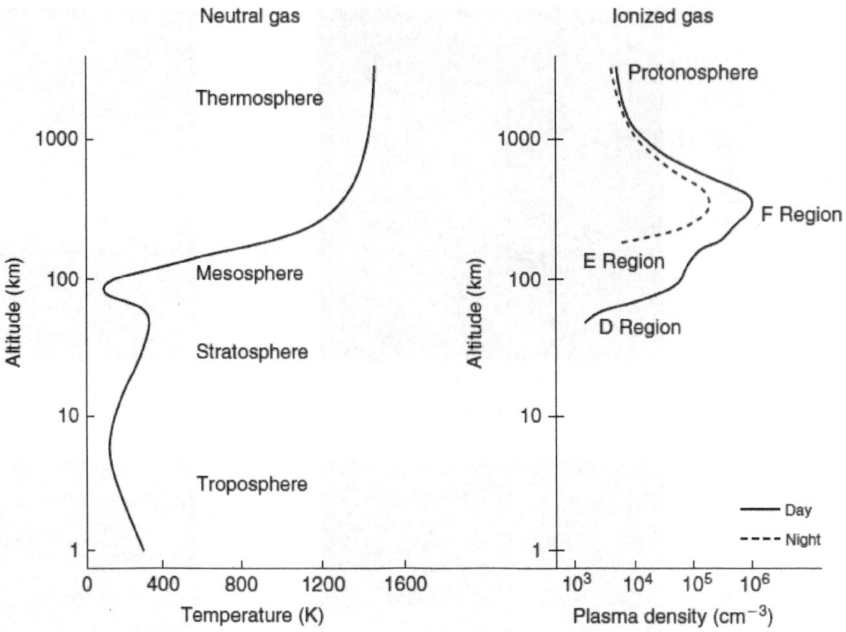

Fig. 21.8 Typical profiles of neutral atmospheric temperature and ionospheric plasma density with the various layers designated (Kelley, 2009). Reproduced with permission. © Elsevier.

of the lower atmosphere into the corona, and if the situation becomes favorable, the cooling of this material can then overpower the heating, resulting in run-away cooling of coronal material, forming a cool prominence inside the corona (e.g., Xia and Keppens, 2016).

The work of Kaneko and Yokoyama (2018) can be viewed as the most complete modeling of the instability in a prominence setting to date, as it was able to reproduce magnetic RT dynamics as a result of the formation of the prominence. In this model material condensed in the corona to form a prominence. However, due to mass of the material that condensed, the magnetic RTI developed, driving highly turbulent motions in the prominence (as shown in Fig. 21.7).

To summarize what has been learned from these studies, the conditions for the magnetic RT instability are satisfied in solar prominences, creating plumes that rise through the prominence body. Given the high resolution and temporal cadence of plume observations, these observations are quite possibly the highest quality observations of the magnetic RT instability in any astrophysical system. Note that observations of the RT instability in prominences are not only restricted to prominence plumes; there have been some clear observations of the instability in erupting prominences (e.g., Carlyle et al., 2014), which show the breakup of the dense prominence material as it falls back to the solar surface. These observations have also driven new areas of study, both for linear instability analysis and for nonlinear modeling. If the reader is interested in additional details of the phenomenon of prominence plumes and our current understanding of them, they are directed to Hillier (2018).

21.3 Ionospheric flows

Space physics encompasses the study and monitoring of Earth's space environment and solar-terrestrial relationships. The field is critical to advancing and protecting the nation's economic, defense, and scientific interests. Its strategic importance has led to the creation of the National Space Weather Program and the publication of The National Space Weather Strategy and Action Plan. Multi-scale physics-based predictive modeling, data assimilation and computation of ionospheric dynamics, and turbulence are major frontiers of space sciences and space weather research.

The ionosphere is a dynamic mixture of ions, electrons and neutral gases surrounding the Earth in the altitude range from approximately 90 km to beyond 1000 km. The ionosphere can also be viewed as a transition region from the earth's lower atmospheric regions (i.e., troposphere, stratosphere, mesosphere) to the outer space environment (i.e., the magnetosphere). As such, the ionosphere acts to mediate and transmit external forces and drivers from below and from above (Fig. 21.8(a)).

The ionosphere involves interactions between phenomena of varying scale sizes. Large-scale variations, including the solar cycle, seasonal and tidal effects, drive large-scale changes in global structure. Such changes define the mean state (climate) of the ionospheric system. These processes have characteristic spatial scales greater than a thousand kilometers and time scales from several hours to a few years. General circulation models have been developed to simulate these large-scale structures (e.g., Dickinson et al., 1981; Fuller-Rowell and Rees, 1980; Fuller-Rowell et al., 1987; Roble and Ridley, 1994). These global models, together with the large observational datasets that have been accumulated over the years, have led to a much greater understanding of large-scale structures in the ionosphere and the response of these structures to variations in geophysical inputs.

Space weather is the perturbation of the ionosphere and thermosphere from its long-term global mean state. These perturbations involve not only large-scale variations, but also mesoscale and small-scale processes that occur locally and may have short periods. Mesoscale and small-scale processes, such as RT instabilities and turbulence, affect not only local plasma and neutral distributions but also large-scale structures through dynamical and energetic coupling. Such coupling between processes of different scales occurs in the equatorial region, mid- and high latitudes. The dynamics of large and mesoscale winds are often characterized by the presence of zonal jets, layers and anisotropic turbulence (i.e., DeBoer et al., 2010; Gilman et al., 2017; Heelis et al., 1982; Hysell et al., 2004; Park et al., 2008; Raghavarao et al., 1992; Retterer, 2010a,b; Sekar et al., 1994; Sekar and Kherani, 2002; Steenburgh et al., 2008; Woodman and La Hoz, 1976; Sun et al., 1995; Xu et al., 2004). Improving our understanding of neutral/plasma interactions and RT processes at ionospheric meso/microscales is a challenge for ionospheric science studies.

The F region is the terrestrial plasma environment between an altitude of 120 and 800 km (Fig. 21.8(b)). The F region is bounded by the E region below and the exosphere above, and together these three regions form the terrestrial ionosphere (Sojka, 1989). The RTI was first proposed in Dungey (1956) as the process driving convective equatorial ionospheric storms. By analogy with the hydrodynamic RT instability when a light fluid supports

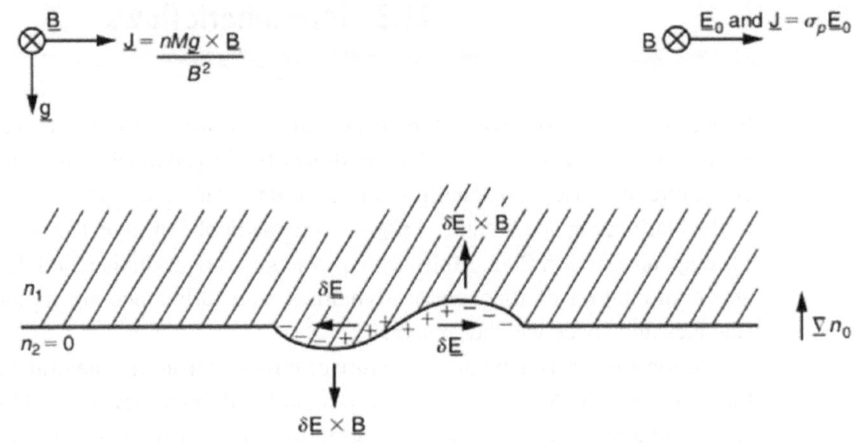

Fig. 21.9 Schematic diagram of the plasma analog of the Rayleigh–Taylor instability in the equatorial geometry (Kelley, 2009). Reproduced with permission. © Elsevier.

a heavier fluid against gravity, it was suggested that lower (higher) density ionospheric plasma is advected upward (downward), creating a larger perturbation, and rendering the system unstable. In case of the ionospheric, the "light fluid" is the low-density plasma, which carries a gravity-driven current providing the $\mathbf{J}\times\mathbf{B}$ force, preventing the plasma from freely falling. The system is unstable when the vectors $\mathbf{g}$ and ∇n are oppositely directed. Here, n is the electron number density of plasma (Kelley, 2009). See the illustration in Fig. 21.9.

The scale sizes of the density irregularities in the ionosphere range from a few meters to a few hundred kilometers, and can occur at all latitudes. At night, fully developed "spread F" is characterized by plasma bubbles, which are vertically elongated wedges of depleted plasma that drift upward from beneath the bottom F layer (Fig. 21.8(b)). A density perturbation can trigger the RT instability on the bottomside of the F layer under certain conditions. Once triggered, density irregularities develop, and the field-aligned depletions then bubble up through the F layer. The east-west extent of a disturbed region can be several thousand kilometers, with the horizontal distance between separate depleted regions being tens to hundreds of kilometers. Equatorial plasma bubbles (EPB) are irregular plasma density depletions in the post-sunset ionosphere that degrade communication and navigation signals.

The electron density in the ionosphere varies diurnally, geographically and seasonally with sunspot number and other solar phenomena. The total electron content (TEC) can vary by two orders of magnitude depending on the time and location of observations. Apart from the variation with altitude, the electron density varies with the Sun's activity level, time of the year, time of the day, and geographical position.

22 Mix models

In this chapter we focus on some of the methodologies currently utilized to tackle the practical problem of simulating hydrodynamic instabilities in engineering designs.

By necessity, experimental studies have been the key to advancement for centuries in fluid dynamics. However, with the rapid increase of computational capabilities, numerical approaches have become an acceptable surrogate for experiments. Calculations must resolve the Navier–Stokes equations or approximate models constructed from them. Despite intensive efforts to develop increased computational capabilities, Reynolds averaged Navier–Stokes (RANS) models remain the most viable approach for modeling many applications. The reason for this state of affairs becomes abundantly clear when one considers the difficulties of achieving the desired turnaround time for applied fluid dynamics calculations.

Let us begin by considering research activities relevant to the fluid dynamics of the natural sensing processes, which have attracted much attention recently (Settles, 2005; Reddy et al., 2021). A seemingly mundane, though ubiquitous case is when vertebrates aim their noses at regions of interest and sniff to acquire olfactory trace signals that carry information regarding food, kinship, danger, etc. Perhaps the best example is human's best friend, the dog, which recent demographic analysis (Wang et al., 2013a) indicates split from wolves 32,000 years ago. Dogs have an amazing ability to search for people who are lost in the wilderness (Jinn et al., 2020) as well as to detect explosive traces and landmines (Lazarowski and Dorman, 2014), and at the time of writing, another rather timely example, COVID-19 (Mendel et al., 2021; Otto et al., 2021). As Settles (2005) pointed out, the dog's nose is recognized as the gold standard of olfactory acuity. Our closest animal companions are the ultimate mobile, instinctive, intelligent sniffing platforms.

In fact, there is significant interest in the fluid dynamics of sniffing, in the hope that we might one day replicate nature's accomplishments and develop artificial ("electronic") noses. This task is extremely challenging, in part because the canine nasal airway is an impressively complex anatomical structure – inhospitable for methods built to model idealized flows. As an example, my Shih Tzu, Teddy, is a brachycephalic dog with a very short nose; his best friend, Ollie, a Maltipoo, has a much longer nose (both Teddy and Ollie (Fig. 22.1(a) and (b)) were adopted from an animal shelter). Thus, the fluid dynamics of dogs' olfactory trace detection must be examined using combined measurements (Greatbatch et al., 2015), computational fluid dynamics (Craven et al., 2010), and the Reynolds average Navier–Stokes (RANS) models (Zhao et al., 2006) amenable to tackling complex flows (Fig. 22.1(c)).

In this chapter, fundamental RANS models and numerical schemes for studying hydrodynamic instabilities and turbulence are explained. For mixing processes of practical

(a) Teddy, my Shih Tzu, is a brachycephalic dog with a very short nose. Photograph is provided by the author.

(b) Ollie, Teddy's best friend, is a Maltipoo who has a much longer nose. Photograph is provided by the author.

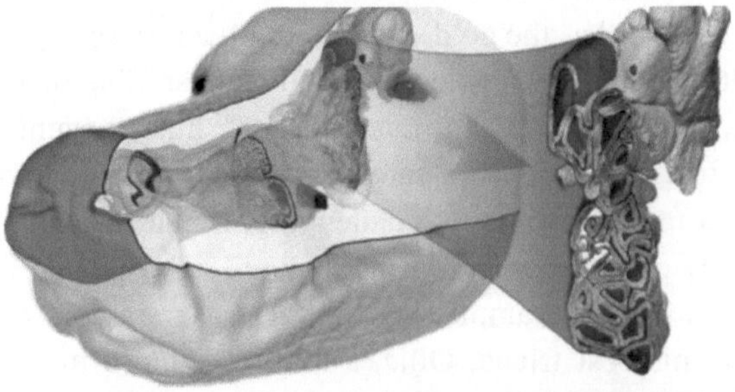

(c) Three-dimensional model of the left canine nasal airway, reconstructed from high-resolution MRI scans. The olfactory recess is located in the rear of the nasal cavity and contains scroll-like ethmoturbinates, which are lined with olfactory epithelium. Craven et al. (2010), with the permission of The Royal Society.

Fig. 22.1 The dog's nose is recognized as the gold standard of olfactory acuity, but the canine nasal airway is a complex anatomical structure. Reynolds averaged Navier–Stokes (RANS) models, along with computational fluid dynamics (CFD), are used for studying their sensing processes.

interest driven by density instabilities (e.g., RT and RM), excitations of spatial scales exceed the resolution possible in current and foreseeable direct numerical simulations. These practical problems constitute the major reason for pursuing the RANS approach. Transport models have been incorporated and integrated into most ICF and astrophysics computer codes. Note that these models implicitly assume one begins with fully developed turbulence, as opposed to transitioning to turbulence from a quiescent state.

It is sensible to begin our study of the fundamental RANS approximation by elaborating the closure problem caused by the strong nonlinearity of the Navier–Stokes equations. In particular, we will review two RANS approaches that are widely used in aerospace and mechanical engineering applications. These will pave the way for us to turn our attention to the Favre averaged mix models routinely employed for computations of instabilities, turbulence, and mixing.

22.1 Strong nonlinearity and closure problem

In RANS modeling, the flow is split into an appropriately defined mean and a fluctuating part (Reynolds, 1895). The Reynolds decomposition of an arbitrary function, $\mathfrak{f}$, is then denoted by

$$\mathfrak{f} = \bar{\mathfrak{f}} + \mathfrak{f}', \tag{22.1}$$

where "overbar" denotes an ensemble average. In this section, we will restrict ourselves to the incompressible Navier–Stokes system to illustrate the complications arising from its strong nonlinearity and the associated closure problem. However, the resulting models are not directly applicable to our problem of interest. When density variations are negligible, the averaged Navier–Stokes equation becomes (e.g., Speziale, 1991)

$$\frac{\partial \bar{u}_i}{\partial t} + \bar{u}_j \frac{\partial \bar{u}_i}{\partial x_j} = -\frac{\partial \bar{p}}{\partial x_i} + \nu \nabla^2 \bar{u}_i - \frac{\partial R_{ij}}{\partial x_j}, \tag{22.2}$$

where

$$R_{ij} = \overline{u'_i u'_j}. \tag{22.3}$$

Note that p is now redefined as the modified pressure for the purpose of simplicity (Speziale, 1991). The mean continuity equation is

$$\frac{\partial \bar{u}_j}{\partial x_j} = 0. \tag{22.4}$$

Meanwhile, the fluctuating velocity equation can be obtained by subtracting Eq. 22.2 from Eq. 2.7:

$$\frac{\partial u'_i}{\partial t} + \bar{u}_j \frac{\partial u'_i}{\partial x_j} = -u'_j \frac{\partial u'_i}{\partial x_j} - u'_j \frac{\partial \bar{u}_i}{\partial x_j} - \frac{\partial p'}{\partial x_i} + \nu \nabla^2 u'_i + \frac{\partial R_{ij}}{\partial x_j}, \tag{22.5}$$

while the transport equation for the Reynolds stress takes the form (Hinze, 1975)

$$\frac{\partial R_{ij}}{\partial t} + \bar{u}_k \frac{\partial R_{ij}}{\partial x_k} = -R_{ik} \frac{\partial \bar{u}_j}{\partial x_k} - R_{jk} \frac{\partial \bar{u}_i}{\partial x_k} + \wp_{ij} - \mathcal{E}_{ij} - \frac{\partial C_{ijk}}{\partial x_k} + \nu \nabla^2 R_{ij}. \tag{22.6}$$

In Eq. 22.6, the dissipation-rate correlation, third order diffusion correlation, and pressure-strain correlation are given by

$$\mathcal{E}_{ij} = 2\nu \overline{\frac{\partial u_i'}{\partial x_k}\frac{\partial u_j'}{\partial x_k}}, \tag{22.7}$$

$$C_{ijk} = \overline{u_i'u_j'u_k'} + \overline{p'u_i'}\delta_{jk} + \overline{p'u_j'}\delta_{ik}, \tag{22.8}$$

and

$$\wp_{ij} = \overline{p'\left(\frac{\partial u_i'}{\partial x_j} + \frac{\partial u_j'}{\partial x_i}\right)}. \tag{22.9}$$

The presence of the first and second terms on the right side of Eq. 22.6, which couples the mean velocity gradient to the Reynolds stress, reveals a mechanism for transferring energy from the mean flow field to the turbulent fluctuations (Reynolds, 1895). This insight was so far ahead of its time, as Jackson and Launder (2007) noted, that the corresponding equation for the mean square temperature fluctuations of Corrsin (1951) was not published until the 1950s.

Reynolds' work also illustrated the *closure problem* of turbulence. In Eq. 22.6, the second-order velocity correlation is not closed without the third-order one. Clearly, the equation for the latter requires the expression of a fourth-order correlation, and so on. While evolution equations for the correlation functions of the fluctuating velocity field can be formed, each i-th order equation so obtained demands the correlation function of i+1 order. This closure problem is the main challenge in turbulence and is a direct result of the nonlinearity of the Navier–Stokes equation.

22.2 Reynolds averaged Navier–Stokes closure approximations

The RANS method was largely developed for engineering calculations that require rapid turn-around times (Chou, 1940, 1945; Rotta, 1951; Launder et al., 1975). Over the years, many RANS models of increasing complexity have been proposed, with varying degrees of success (Wilcox, 1998). In principle, this approach attempts to model all of the fluctuating scales. Nevertheless, even under the best circumstances, RANS models can only provide information about first and second one-point moments (e.g., the mean velocity, turbulence intensity, and mean pressure), which is usually all that is needed for design purposes. The discussions in this section follow closely to those in Speziale (1991).

It is reasonable to begin with a conversation about two-equation models, which are among the most popular RANS models for scientific and engineering applications. By a contraction of equation 22.6, the transport equation for turbulent kinetic energy, $K = R_{ii}/2$, is obtained:

$$\frac{\partial K}{\partial t} - \nu\nabla^2 K + \bar{u}_j\frac{\partial K}{\partial x_j} = -R_{ij}\frac{\partial \bar{u}_i}{\partial x_j} - \mathcal{E} - \frac{\partial}{\partial x_i}\left(\frac{1}{2}\overline{u_i'u_k'u_k'} + \overline{p'u_i'}\right). \tag{22.10}$$

Similarly, the dissipation rate transport equation, $\mathcal{E} = \mathcal{E}_{ii}/2$, is obtained (Speziale, 1991):

$$\begin{aligned}
&\frac{\partial \mathcal{E}}{\partial t} - \nu \nabla^2 \mathcal{E} + \overline{u}_j \frac{\partial \mathcal{E}}{\partial x_j} = \\
&-2\nu \overline{\frac{\partial u_i'}{\partial x_j}\frac{\partial u_k'}{\partial x_j}} \frac{\partial \overline{u}_i}{\partial x_k} - 2\nu \overline{\frac{\partial u_j'}{\partial x_i}\frac{\partial u_j'}{\partial x_k}} \frac{\partial \overline{u}_i}{\partial x_k} - 2\nu \overline{u_k' \frac{\partial u_i'}{\partial x_j}} \frac{\partial^2 \overline{u}_i}{\partial x_k \partial x_j} - 2\nu \overline{\frac{\partial u_i'}{\partial x_k}\frac{\partial u_i'}{\partial x_m}\frac{\partial u_k'}{\partial x_m}} \\
&-\nu \frac{\partial}{\partial x_k}\left(\overline{u_k' \frac{\partial u_i'}{\partial x_m}\frac{\partial u_i'}{\partial x_m}} \right) - 2\nu \frac{\partial}{\partial x_k}\left(\overline{\frac{\partial p'}{\partial x_m}\frac{\partial u_k'}{\partial x_m}} \right) \\
&-2\nu^2 \overline{\frac{\partial^2 u_i'}{\partial x_k \partial x_m}\frac{\partial^2 u_i'}{\partial x_k \partial x_m}}. \qquad (22.11)
\end{aligned}$$

On the right-hand side of Eq. 22.11, the first four terms give rise to the production of dissipation, the next two terms represent the turbulent diffusion of dissipation, and the last term represents turbulent destruction of dissipation.

22.2.1 Two-equation models

In two equation models, two separate modeled transport equations are solved for the turbulent length and time scales (or for any two linearly independent combinations thereof) (Speziale, 1991). The length and time scales are built up from the turbulent kinetic energy and dissipation rate or integral length scale.

In the standard K–$\mathcal{E}$ model, Jones and Launder (1973) noted that much of the original ideas come from Harlow and Nakayama (1967). The dissipation rate equation is obtained by dividing each term of the kinetic energy equation by the turbulent time scale, $K/\mathcal{E}$. The final form of the equation is obtained by introducing appropriate dimensionless coefficients (see also, Daly and Harlow, 1970; Hanjalić and Launder, 1972; Launder and Spalding, 1974). The K–$\mathcal{E}$ model is probably the most popular two-equation model (Speziale, 1991).

From Eq. 22.10, the transport equation for the turbulent kinetic energy is given by

$$\frac{\partial K}{\partial t} - \nu \nabla^2 K + \overline{u}_j \frac{\partial K}{\partial x_j} = -R_{ij} \frac{\partial \overline{u}_i}{\partial x_j} - \mathcal{E} + \frac{\partial}{\partial x_i}\left(\frac{\nu_T}{\sigma_K}\frac{\partial K}{\partial x_i} \right), \qquad (22.12)$$

and the dissipation rate is provided,

$$\frac{\partial \mathcal{E}}{\partial t} - \nu \nabla^2 \mathcal{E} + \overline{u}_j \frac{\partial \mathcal{E}}{\partial x_j} = -C_{\mathcal{E}1} \frac{\mathcal{E}}{K} R_{ij} \frac{\partial \overline{u}_i}{\partial x_j} - C_{\mathcal{E}2} \frac{\mathcal{E}^2}{K} + \frac{\partial}{\partial x_i}\left(\frac{\nu_T}{\sigma_{\mathcal{E}}}\frac{\partial \mathcal{E}}{\partial x_i} \right), \qquad (22.13)$$

where $C_\mu, C_{\mathcal{E}1}, C_{\mathcal{E}2}, \sigma_K$ and $\sigma_{\mathcal{E}}$ are empirical constants obtained by comparison of the model predictions with the results of physical experiments or numerical simulations.

One particular challenge to obtaining closure of the governing hydrodynamic equations arises from the dependence of the Reynolds-stress tensor, Eq. 22.3, on the boundary and initial conditions of the fluctuating velocity field (Lumley, 1970). If one assumes that the boundary and initial conditions merely set the length and time scales of the flow, eddy viscosity models, which are the simplest of the RANS modeling, take the form

$$R_{ij} = \frac{2}{3}K\delta_{ij} - \nu_T\left(\frac{\partial \bar{u}_i}{\partial x_j} + \frac{\partial \bar{u}_j}{\partial x_i}\right). \tag{22.14}$$

Equation 22.14 can be further extended to second- or third-order nonlinearity, which are often termed anisotropic eddy-viscosity models, to incorporate additional physics (e.g., Speziale, 1991; Rubinstein and Barton, 1990; Zhou et al., 1994; Zhou, 2010).

In Eqs. 22.12 and 22.13, the eddy or turbulent viscosity is

$$\nu_T \propto L_T^2/\tau_T, \tag{22.15}$$

where L_T and τ_T are turbulent length and time scales, respectively. For the two-dimensional $K-\mathcal{E}$ model,

$$\nu_T = C_\mu \frac{K^2}{\mathcal{E}}, \tag{22.16}$$

where the constant C_μ assumes the approximate value of 0.09 (Speziale, 1991). The eddy viscosity is clearly inspired by molecular viscosity.

Another popular two-equation model is based on the solution of a transport equation for an integral length scale,

$$\begin{aligned}\frac{\partial(KL)}{\partial t} + \bar{u}_j\frac{\partial(KL)}{\partial x_j} &= \frac{\partial}{\partial x_i}\left[(\nu + C_{L1}K^{1/2}L)\frac{\partial(KL)}{\partial x_i} + C_{L2}K^{3/2}L\frac{\partial L}{\partial x_i}\right] \\ &\quad - C_{L3}LR_{ij}\frac{\partial \bar{u}_i}{\partial x_j} - C_{L4}K^{3/2},\end{aligned} \tag{22.17}$$

where $C_{L1}, \ldots, C_{L4}$ are empirical constants and, for the $K-L$ model,

$$\nu_T \propto K^{1/2}L. \tag{22.18}$$

The length-scale equation is first attributed to the classical work by Rotta (1951) and Wolfshtein (1970), and more recently, by Speziale (1991). Mellor and Herring (1973) and Mellor and Yamada (1982) have discussed the application of the $K-L$ model to either atmospheric or oceanic boundary layers. It has also been applied to the prediction of astrophysical phenomena (Brüggen et al., 2009; Scannapieco and Brüggen, 2008) and compared against measured mixing layer widths from ICF targets (Hurricane et al., 2012; Smalyuk, 2012).

22.2.2 Second order closure models

While attractive due to their relative simplicity, two-equation models have significant deficiencies that make their application to complex turbulent flows precarious. Most of these deficiencies are intimately tied to the assumption that there is a clear-cut separation of scales at the second-moment level (i.e., the level of the Reynold-stress tensor).

As articulated by Speziale, the *raison d'etre* of second-order closure modeling is as follows: since crude approximations for the second moments in eddy viscosity often yield adequate approximations in eddy viscosity for first-order moments, it may follow that crude approximation for the third-order moments can yield adequate approximations for the second-order moments in Reynolds-stress transport models.

Furthermore, second-order closure models are based on the Reynolds-stress transport equation (22.6). Since this equation automatically accounts for the convection and diffusion of Reynolds stresses, second-order closure models (unlike eddy-viscosity models) can account for strong nonlocal and history effects (Speziale, 1991).

Second-order closure models must be closed not only by the Reynolds stress model but also the transport equation for the dissipation rate (22.13) or integral length scale (22.17).

To develop a second-order closure, models must be provided for the higher-order correlations $\mathcal{C}_{ijk}$, $\wp_{ij}$, and $\mathcal{E}$. Such modeling efforts invariably introduce several additional empirical constants, all of which must be calibrated by physical experiments. For details, we refer the reader to Spezial (1991) and Wilcox (1998).

22.3 Favre averaged Navier–Stokes closure approximations

The governing equations for RANS simulations are developed by ensemble averaging the Navier–Stokes equations. Neglecting mean molecular diffusion, the Favre averaged continuity, momentum, total energy, and species mass fraction equations are the starting point of the model development (e.g., Schwarzkopf et al., 2011):

$$\frac{\partial \bar{\rho}}{\partial t} + (\bar{\rho}\tilde{u}_j)_{,j} = 0, \tag{22.19}$$

$$\frac{\partial \bar{\rho}\tilde{u}_i}{\partial t} + (\bar{\rho}\tilde{u}_i\tilde{u}_j + \bar{p}\delta_{ij} + \bar{\rho}\tilde{R}_{ij})_{,j} = \bar{\rho}g_i, \tag{22.20}$$

$$\begin{aligned}\frac{\partial \bar{\rho}\tilde{E}}{\partial t} + (\bar{\rho}\tilde{E}\tilde{u}_j + \bar{p}\tilde{u}_j + \bar{\rho}\tilde{u}_i\tilde{R}_{ij})_{,j} &= -(\overline{p'u_j'})_{,j} - (\overline{\rho e''u_j''})_{,j} \\ &\quad - \frac{1}{2}(\overline{\rho u_i''u_i''u_j''})_{,j} - \overline{q}_{dj,j},\end{aligned} \tag{22.21}$$

where $,j$ indicates a spatial differential in direction j and the last term on the right-hand side is the interdiffusional enthalpy flux. The mean of the mass-weighted velocity fluctuation is

$$a_i = -\overline{u_i''} = \overline{\rho' u_i'}/\bar{\rho}. \tag{22.22}$$

The averaged transport equation for the mass fraction of species m, Y_m, is given by

$$\frac{\partial \bar{\rho}\tilde{Y}_m}{\partial t} + (\bar{\rho}\tilde{u}_j\tilde{Y}_m + \overline{\rho Y_m''u_j''})_{,j} = 0, \tag{22.23}$$

where the turbulent flux of the species mass fraction is provided by

$$a_i^m = -\overline{\rho Y_m'' u_i''}. \tag{22.24}$$

The Favre average technique will be applied without warning in the remaining sections of this chapter. The resulting basic equations introduce additional unclosed terms, as expected; closure models are needed to complete the system of equations (Eqs. 22.19–22.24).

The rest of the chapter will be devoted to modeling efforts of different levels of sophistication to solve this system. As will became clear to the reader, researchers must decide what approach to take based on what physics is most pertinent for the application under consideration.

Before proceeding, a note on our upcoming presentation and notation is in order. Many models are closed slightly differently and written with different symbols. Since our objective is to illustrate the methodology, for the sake of brevity these variations will not be documented in detail here, and the reader is referred to the original publications.

22.4 Two-equation mix models

Andronov et al (1976) presented the first use of a two-equation mix model for RT/RM – not $K-\mathcal{E}$ or $K-L$ but $K-\nu_T$. Gauthier and Bonnet (1990) calibrated the $K-\mathcal{E}$ model with the shock tube experiments of Andronov et al. (1976) and Brouillette and Sturtevant (1989). For the model to be valid under strong gradients, the authors found that the classical linear closure relations must be supplemented with some algebraic relations. While most constants were standard, two sets of constants required careful tuning in order to match the experimental data.

The Favre averaged $K-L$ model has an attractive feature in the context of ICF applications. Namely, the initialization of the L profile can be related, at least conceptually, to the interface perturbation, which corresponds directly to the surface roughness of the capsule. Dimonte and Tipton (2006) developed a $K-L$ model for computing hydrodynamic mix aided by the LEM experimental data of Dimonte and Schneider (2000) and the shear layer mixing data obtained by Brown and Roshko (1974). Here, $K = \widetilde{u_k''u_k''}/2$ is the density weighted turbulence kinetic energy. The $K-L$ model is designed as a simple model capable of capturing the self-similar growth of RT and RM flows with a planar (extended) interface (Dimonte and Tipton, 2006; Kokkinakis et al., 2015, 2020a; Zhang et al., 2020b; Xiao et al., 2021; Xie et al., 2023).

The transport equation for K can be written as,

$$\frac{\partial \bar{\rho} K}{\partial t} + \frac{\partial \bar{\rho} K \tilde{u}_j}{\partial x_j} = S_k + \tilde{R}_{ij}\frac{\partial \tilde{u}_i}{\partial x_j} + \frac{\partial}{\partial x_j}\left(\frac{\mu_t}{\sigma_k}\frac{\partial K}{\partial x_j}\right) - D_k, \tag{22.25}$$

where the turbulent eddy viscosity is described by

$$\mu_t = C_\mu \bar{\rho} K^2/\mathcal{E}, \tag{22.26}$$

and $\nu_t = \mu_t/\bar{\rho}$.

The "dissipation" or "destruction" term can be modeled as

$$D_k = C_D \bar{\rho}(2K)^{3/2}/L, \tag{22.27}$$

and S_k is the production term,

$$S_k = \bar{\rho}(2K)^{1/2}\begin{cases} C_B \max(\mathcal{A}_{Li}g_i, 0), & \text{for RT} \\ C_B|\mathcal{A}_{Li}g_i|, & \text{for RM.}\end{cases} \tag{22.28}$$

Here, C_B, C_D, C_μ, and σ_k are model constants and $g_i = -(1/\bar{\rho})\partial\bar{p}/\partial x_i$ is the acceleration (Kokkinakis et al., 2015).

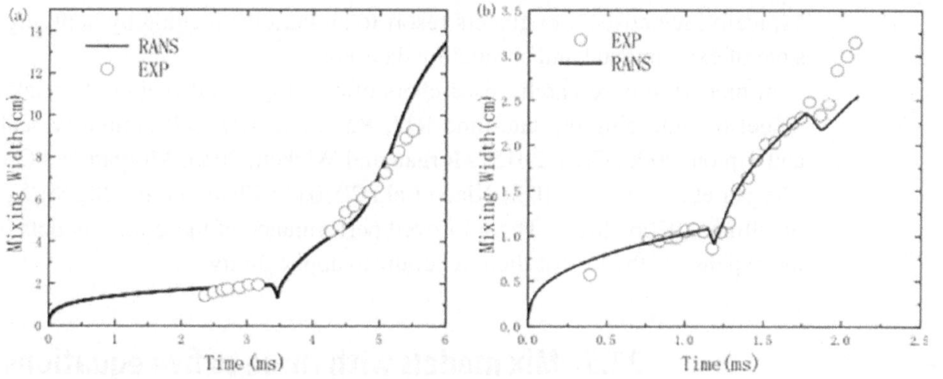

Fig. 22.2 Comparison of temporal evolutions of the total mixing width between experimental results and the two-equation ($K - \mathcal{E}$) RANS results at Atwood numbers (a) 0.67 and (b) −0.67. The line denotes the RANS results, where the circle denotes the experimental results of Vetter and Sturtevant (1995). Fig. 9 of Xie et al. (2021a). This figure is licensed under a Creative Commons Attribution (CC BY) license.

An important point regarding the Dimonte and Tipton model is that the source term distinguishes between impulsive (RM) and gradual acceleration (RT). Another key ingredient in S_k is an Atwood number, $\mathcal{A}_L$, that is calculated on a zone by zone basis (Dimonte and Tipton, 2006). Other forms of S_k can be found in Xiao et al. (2020a,b). Additionally, a realizable K–L model (Xiao et al., 2021) can be constructed utilizing a proposal by Shih et al. (1995).

The set of equations is completed by an equation for the turbulence length scale,

$$\frac{\partial \bar{\rho} L}{\partial t} + \frac{\partial \bar{\rho} L \tilde{u}_j}{\partial x_j} = C_L \rho \sqrt{2K} + C_C \bar{\rho} L \frac{\partial \tilde{u}_j}{\partial x_j} + \frac{\partial}{\partial x_j}\left(\frac{\mu_t}{\sigma_L}\frac{\partial L}{\partial x_j}\right), \tag{22.29}$$

where the right-hand side model production, compressibility effects, and turbulent diffusion, respectively (Kokkinakis et al., 2015). Note also that C_L, C_C, and σ_L are model constants. Equation 22.29 can be generalized further by the addition of a shear production term. While this term is neglected in the Dimonte-Tipton K–L model, it is very important to the correctly modeling KH growth rate.

The K–$\mathcal{E}$ model has also been applied to RTI, tilted RTI, reshocked RMI (Fig. 22.2), and other problems (e.g., Yanilkin et al., 1995; Gauthier and Bonnet, 1990; Gréa, 2015; Morán-López and Schilling, 2013, 2014; Xie et al., 2021a). Using Eqs. 22.25 and 22.29, the Favre averaged $\mathcal{E}$ model can be constructed using $\mathcal{E} \equiv K^{3/2}/L$. Fig. 22.3(a) compared the performance of a realizable two-equation K–L model with the reshocked RMI experimental results of Vetter and Sturtevant (1995) and the standard K–L model.

It is interesting to note that Johnson and Schilling (2011a,b) showed that two-equation RANS models can also describe the linear instability phase for buoyancy- and shear-driven mixing.

The hallmark of RANS is its associated empirical constants; the number of these parameters increases as the complexity of the model grows to incorporate more physics. Even at the simplest two-equation level, there are roughly ten coefficients to be determined.

Typically, scientists or engineers resort to an iterative method by applying the model to a suite of experimental and simulation datasets.

In modern mix research, researchers often adopt another line of attack to improve the model by enforcing the late-time RTI, RMI, and KHI self-similarity scalings (Dimonte and Tipton, 2006; Gréa, 2015; Morgan and Wickett, 2015; Morgan and Greenough, 2016; Morgan et al., 2018b, 2023; Xiao et al., 2020a,b; Zhang et al., 2020b; Xiao et al., 2021; Schilling, 2020b, 2021). The improved performance of these mix models is most likely at the expense of the loss of their generalized applicability.

22.5 Mix models with three to five equations

22.5.1 *K*–*L*–𝔞 family

To close the buoyancy production term, the K–L model relies on a gradient diffusion approximation to model the mass-flux velocity,[1] $\mathfrak{a}_j$. Such an approximation should break down for flows with significant countergradient transport (Denissen et al., 2014).

Inspired by Banerjee et al. (2010b), Morgan and Wickett (2015) provides the third equation to form a three-equation model (Kokkinakis et al., 2019),

$$\frac{\partial \bar{\rho}\mathfrak{a}_i}{\partial t} + \frac{\partial \bar{\rho}\tilde{u}_j\mathfrak{a}_i}{\partial x_j} = \mathcal{B}_{\mathfrak{a}j} + \frac{\tilde{R}_{ij}}{\bar{\rho}}\frac{\partial \bar{\rho}}{\partial x_j} + \frac{\partial}{\partial x_j}\left(\frac{\mu_t}{\sigma_\mathfrak{a}}\frac{\partial \mathfrak{a}_i}{\partial x_j}\right) - C_{D\mathfrak{a}}\bar{\rho}\mathfrak{a}_i\frac{\sqrt{2K}}{L}, \tag{22.30}$$

where $\sigma_\mathfrak{a}$ and $C_{D\mathfrak{a}}$ are constants. Here, the buoyancy production term is typically written in the form:

$$\mathcal{B}_{\mathfrak{a}i} = C_{\mathfrak{b}1}\mathfrak{b}\frac{\partial \bar{p}}{\partial x_i}, \tag{22.31}$$

where $C_{\mathfrak{b}1}$ is a constant and $\mathfrak{b} = -\overline{\rho'\left(\frac{1}{\rho}\right)'}$ is the dimensionless density-specific-volume covariance requiring closure (e.g., Stalsberg-Zarling and Gore, 2011; Kurien et al., 2020; Pal et al., 2021).

Given the possibility of unphysical behavior for models relying on Eq. 22.31, the following alternative approach is proposed, which follows a similar form to buoyancy production terms appearing in other model turbulence equations (Gauthier and Bonnet, 1990; Morgan, 2022b):

$$\mathcal{B}_{\mathfrak{a}i} = C_{\mathfrak{b}2}\mathfrak{a}_i\frac{\mathfrak{a}_j}{2K}\frac{\partial \bar{p}}{\partial x_j}, \tag{22.32}$$

where $C_{\mathfrak{b}2}$ is a model constant. Notice that use of Eq. (22.32) effectively decouples $\mathfrak{b}$ from the model.

As mentioned previously, the density-specific-volume, $\mathfrak{b}$, is not closed. A proposed closure of Eq. 22.31 is given by (Kokkinakis et al., 2019; Morgan et al., 2018b)

$$\mathfrak{b} = \bar{\rho}\frac{\sum_i f_i/\rho_i}{\sum_i f_i} - 1, \tag{22.33}$$

where f_i is the volume fraction of fluid i.

[1] The mass-flux velocity, $\mathfrak{a}_i$, should not be confused with the perturbation amplitude, a.

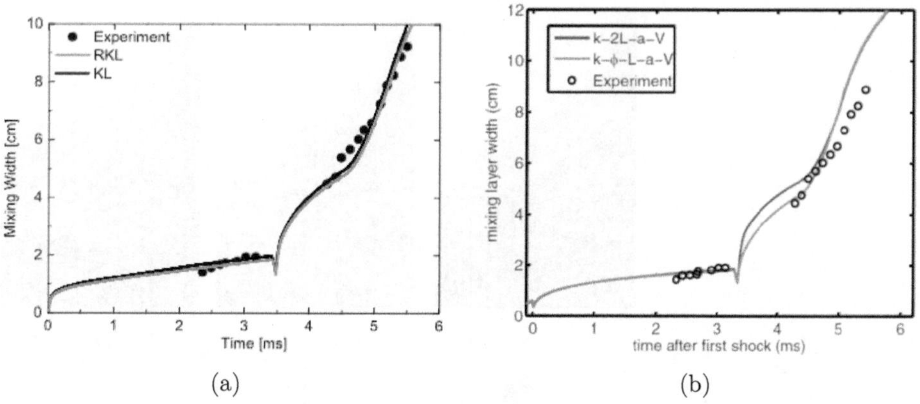

Fig. 22.3 Mixing width as a function of time for the Vetter and Sturtevant (1995) shock tube experiment compared against RANS models: (a) the $K{-}L$ models. Fig. 6 of Xiao et al. (2021), with the permission of AIP Publishing. (b) the $K{-}2L{-}\mathfrak{a}{-}V$ and $K{-}\phi{-}L{-}\mathfrak{a}{-}V$ models. Fig. 6 of Morgan (2021b), with the permission of APS.

Morgan (2021b) developed a $K{-}\phi{-}L{-}\mathfrak{a}{-}V$ model to achieve high-order spatial profiles that are self-consistent with a similarity analysis. This allows one to determine model coefficients that will exactly reproduce desired growth rates while maintaining high-order spatial profiles (Fig. 22.3(b)).

To simultaneously capture both the growth parameter and the intensity of a KH shear layer when modeling coefficients are constrained by similarity analysis, Morgan et al. (2018b) added a second length-scale equation, separating the turbulence destruction length scale from the turbulence transport scale ($K{-}2L{-}\mathfrak{a}$ model).

To model reacting flows, Morgan et al. (2018a) and Morgan (2021a) suggested the one- and two-length scale models, known as $K{-}L{-}\mathfrak{a}{-}V$ and $K{-}2L{-}\mathfrak{a}{-}V$ (Fig. 22.3(b)), and compared the results with LES data. It is also worth mentioning the existence of the $K{-}L{-}\mathfrak{a}{-}\mathfrak{C}$ and $K{-}\phi{-}L{-}\mathfrak{a}{-}\mathfrak{C}$ models introduced in Morgan (2022b) for multicomponent reacting flows.

22.5.2 $K{-}L{-}\mathfrak{a}{-}\mathfrak{b}$ family

Equation 22.30 shows that $\mathfrak{b}$ governs the primary production mechanism of the turbulent mass-flux (Stalsberg-Zarling and Gore, 2011). Thus, the density-specific volume covariance needs to be modeled accurately in order to reflect the effects of changes in the density fluctuations. The BHR-2 model (Banerjee et al., 2010b; Stalsberg-Zarling and Gore, 2011; Denissen et al., 2014; Xie et al., 2021b) employed a governing equation for $\mathfrak{b}$, replacing its model in the algebraic form of Eq. 22.33.

The governing equation for $\mathfrak{b}$ is written as

$$\frac{\partial \bar{\rho}\mathfrak{b}}{\partial t} + \frac{\partial \bar{\rho}\mathfrak{b}\tilde{u}_j}{\partial x_j} = -2(\mathfrak{b}+1)\mathfrak{a}_j\frac{\partial \overline{\rho}}{\partial x_j} + \bar{\rho}^2\frac{\partial}{\partial x_j}\left(\frac{\mu_t}{\bar{\rho}^2\sigma_{\mathfrak{b}}}\frac{\partial \mathfrak{b}}{\partial x_j}\right) - C_{D\mathfrak{b}}\frac{\sqrt{2K}}{L}\mathfrak{b}, \tag{22.34}$$

where $C_{D\mathfrak{b}}$ and $\sigma_{\mathfrak{b}}$ are model constants (Kokkinakis et al., 2019).

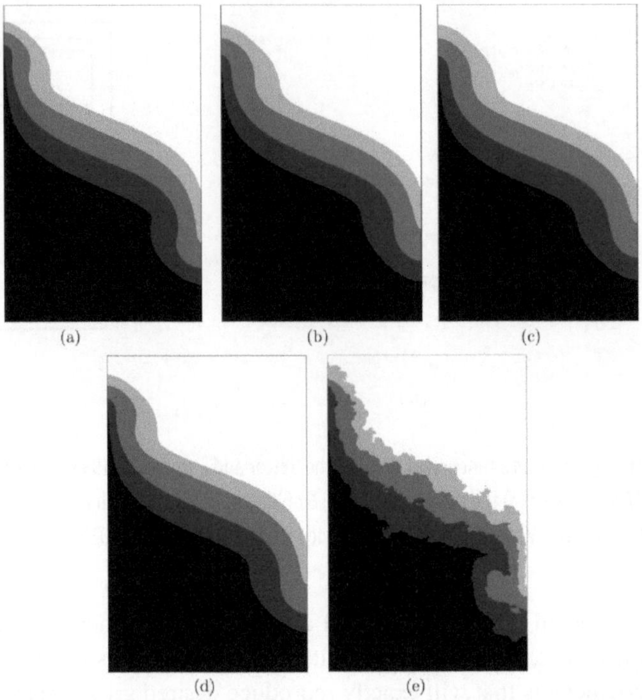

Fig. 22.4 Mass fraction contour plot for the tilted-rig test cases (at nondimensional time 1.741): (a) $K-L$, (b) $K-L-\mathfrak{a}$, (c) $K-L-\mathfrak{a}-\mathfrak{b}$, (d) BHR-2, and (e) ILES. Fig. 9 of Kokkinakis et al. (2019), with the permission of APS.

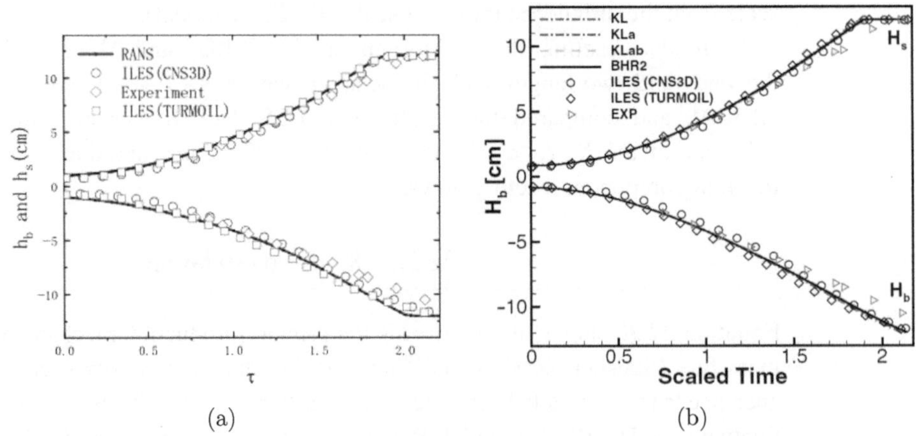

Fig. 22.5 Comparison of the bubble and spike height evolution from "tilted-rig" RTI experiments, implicit large eddy simulations (with the TURMOIL and CNS3D codes), and mix models. (a) $K-\mathcal{E}$ models. Fig. 10 of Xie et al. (2021a). This figure is licensed under a Creative Commons Attribution (CC BY) license. (b) $K-L$, $K-L-\mathfrak{a}$, $K-L-\mathfrak{a}-\mathfrak{b}$, and BHR-2 models. Fig. 7 of Kokkinakis et al. (2019), with the permission of APS.

Equation 22.34 is similar to the BHR-2 model, but with a few terms slightly modified, as pointed out in Kokkinakis et al. (2019). Concerning the tilted-rig case, the BHR-2 model performs better than $K{-}L{-}a{-}b$ in the spike, but worse in the bubble plume region. The authors also show that using a model transport equation for $\mathfrak{b}$, instead of the algebraic expression used by the $K{-}L{-}\mathfrak{a}$ model, provides some accuracy improvement – particularly in relation to the prediction of the turbulent mass-flux velocities and the turbulence kinetic energy (Figs. 22.4 and 22.5(b)).

More recently, Xie et al. (2021b) applied the methodology for determining the coefficients of turbulent mixing models (Zhang et al., 2020b), based on the self-similar analysis mentioned above. Good agreement between experiments, large-eddy simulations, and RANS results was obtained for tilted-RTI (Fig. 22.5(a)), re-shocked RMI, and spherical implosion mixing.

22.6 Second-order closure mix models

The second-order closure used to model mix problems involves evolving transport equations for the Reynolds stress, $\tilde{R}_{ij}$, which must be modelled in some fashion (BHR-3, Besnard et al. (1992), or GSG, Grégoire et al. (2005)).[2]

The Reynolds stress transport is modeled as (Schwarzkopf et al., 2011)

$$
\begin{aligned}
\frac{\partial \bar{\rho}\tilde{R}_{ij}}{\partial t} + (\bar{\rho}\tilde{u}_k\tilde{R}_{ij})_{,k} \;=\;& \overbrace{[\mathfrak{a}_i\bar{p}_{,j} + \mathfrak{a}_j\bar{p}_{,i}] - \bar{\rho}[\tilde{R}_{ik}\tilde{u}_{j,k} + \tilde{R}_{jk}\tilde{u}_{i,k}]}^{\text{Production}} \\
& \underbrace{+\, C_r\bar{\rho}\left(\frac{L}{\sqrt{K}}\tilde{R}_{kn}\tilde{R}_{ij,n}\right)_{,k}}_{\text{Diffusion}} \underbrace{-\, C_{r3}\bar{\rho}\frac{\sqrt{K}}{L}\left(\tilde{R}_{ij} - \frac{1}{3}\tilde{R}_{kk}\delta_{ij}\right)}_{\text{Return to Isotropy}} \\
& \overbrace{-C_{r1}[\mathfrak{a}_i\bar{p}_{,j} + \mathfrak{a}_j\bar{p}_{,i}] + C_{r2}\bar{\rho}[\tilde{R}_{ik}\tilde{u}_{j,k} + \tilde{R}_{jk}\tilde{u}_{i,k}]}^{\text{Rapid Distortion}} \\
& \underbrace{-\, C_{r2}\frac{2}{3}\bar{\rho}\tilde{R}_{mk}\tilde{u}_{m,k}\delta_{ij} + C_{r1}\frac{2}{3}\mathfrak{a}_k\bar{p}_{,k}\delta_{ij}}_{\text{Pressure Strain}} \underbrace{-\, \bar{\rho}\frac{2}{3}\frac{K^{3/2}}{L}\delta_{ij}}_{\text{Dissipation}}.
\end{aligned}
\tag{22.35}
$$

This equation is supplemented by several other governing equations already invoked for describing the evolution of:

(i) a length scale (BHR-3) or dissipation rate (GSG),
(ii) the turbulent mass flux, $\mathfrak{a}_j$, which arises in the unclosed equation for Reynolds stress,
(iii) a density-specific volume correlation, $\mathfrak{b}$ (BHR), or the density variance $\overline{\rho'^2}$ (GSG).

[2] Again, $\tilde{R}_{ij} = \widetilde{u''_i u''_j} = \overline{\rho u''_i u''_j}/\bar{\rho}$.

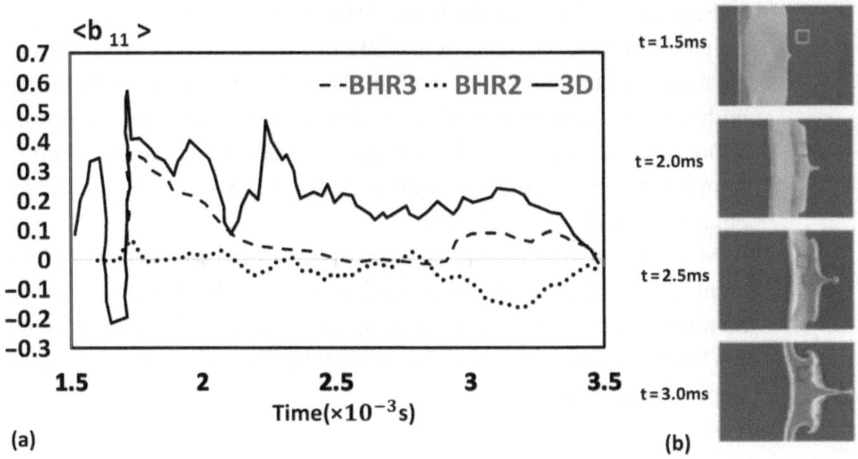

Fig. 22.6 (a) Anisotropy component in the direction of the shock as a function of time for the inverse chevron experiment of Section 15.1. At each time, b_{11} is spatially averaged over the box indicated in (b). b_{11} is based on z-averaged ILES. Fig. 10 of Haines et al. (2013). © Taylor & Francis.

The main motivation for second-order closures is to capture the anisotropy of the flow fields. The analysis of Wong et al. (2022) addressed the importance of two second-order moments: turbulent mass flux and density-specific-volume covariance, together with their transport equations. These quantities play an essential role in the development of Favre-averaged Reynolds stress and turbulent kinetic energy. Using simulations of the inverse chevron shock tube of Holder and Barton (2004), Haines et al. (2013) examine the evolution of the anisotropy tensor in the direction of the shock (b_{11}), spatially averaged over the sampling box indicated in Fig. 22.6(b). The ILES results highlight the fact that after first reshock the Reynolds stress in the direction of the shock remains anisotropic for some time, a fact not captured by the model based on algebraic stress closure, $K\text{–}L\text{–}\mathfrak{a}\text{–}\mathfrak{b}$ (e.g., BHR-2). Figure 22.6(a) shows that BHR-3 improves on BHR-2 by providing a better overall agreement with the ILES data; quantitatively, BHR-3 predicts the anisotropy quite reasonably compared to ILES after the first reshock (1.7 ms), but less so after the second reshock (2.25 ms).

The shock tube RMI data of Poggi et al. (1998) provides an opportunity to compare and contrast the performance of GSG (Figs. 13.24–13.25). Overall, the agreement between the GSG prediction and the data is reasonable. The two components of the Reynolds stress tensor also compare reasonably well with the data, although the model predictions in the transverse direction are somewhat broader than the data (Grégoire et al., 2005).

At positions 161 and 178.5 mm beyond the initial interface, which is in the decay region after the first shock, but before reshock, the BHR-3 model also compares reasonably well with the Poggi experimental data, Fig. 22.7(a). The peak in R_{11}, between 1.85 and 1.9 ms (at 178.5 mm), is due to the second reshock. Data for the Reynolds stress anisotropy at 169 mm are found in Grégoire et al. (2005). Here, the BHR-3 model is compared with the data

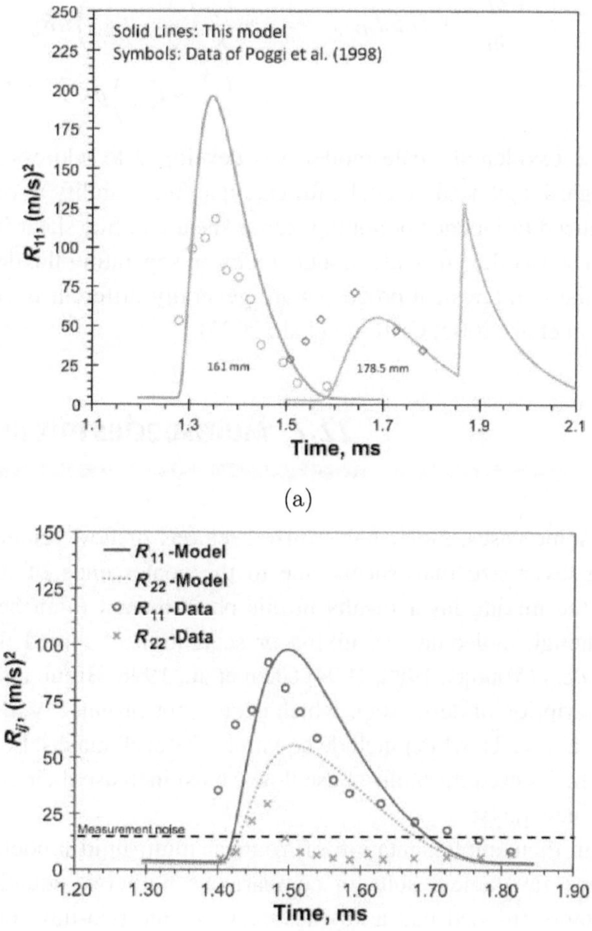

Fig. 22.7 BHR-3 second-moment model compared to measurements from the CEA shock tube (Poggi, 1997; Poggi et al., 1998); (a) the axial Reynolds stress, (b) the Reynolds stress components. Figs. 16 and 17 of Schwarzkopf et al. (2011). © Taylor & Francis.

and shown in Fig. 22.7(b). Once again, the two components compare reasonably well with the data (Schwarzkopf et al., 2011).

A related model, BHR-3.1, requires a pair of length scales – a transport length scale (L_T) satisfying

$$\frac{\partial \bar{\rho} L_T}{\partial t} + (\bar{\rho}\tilde{u}_j L_T)_{,j} = -\frac{L_T}{K}\left(\frac{3}{2} - C_1\right)\bar{\rho}\tilde{R}_{ij}\tilde{u}_{i,j} + \frac{L_T}{K}\left(\frac{3}{2} - C_4\right)a_j\bar{P}_{,j} - \left(\frac{3}{2} - C_2\right)\bar{\rho}\sqrt{K} + \frac{C_\mu}{\sigma_s}\left(\bar{\rho} L_T \sqrt{K} L_{T,j}\right)_{,j}, \qquad (22.36)$$

and a dissipation length scale (L_D) governed by

$$\frac{\partial \bar{\rho} L_D}{\partial t} + (\bar{\rho}\tilde{u}_j L_D)_{,j} = -\frac{L_D}{K}\left(\frac{3}{2} - C_{1\nu}\right)\bar{\rho}\tilde{R}_{ij}\tilde{u}_{i,j} + \frac{L_D}{K}\left(\frac{3}{2} - C_{4\nu}\right)a_j\bar{P}_{,j}$$
$$-\left(\frac{3}{2} - C_{2\nu}\right)\bar{\rho}\sqrt{K} + \frac{C_\mu}{\sigma_{2s}}\left(\bar{\rho}L_T\sqrt{\tilde{K}}L_{D,j}\right)_{,j}. \tag{22.37}$$

The two-length scale model was developed to address an inconsistency present in the single-length scale models, for example, the inability to match either variable density RTI induced turbulence or homogeneous shear and free shear flows (Schwarzkopf et al., 2016).

The two-length scale models focus on separating the decay and transport length scales, as the two physical processes are generally different in inhomogeneous turbulence (e.g., Zhou et al., 2002; Griffond et al., 2023).

22.7 Multispecies mix models

In some cases, molecularly mixed regions of flow remain mixed, but the apparent mixing layer size may shrink due to the coalescence of unmixed regions. The shrinking of the mixing layer results in this phenomenon often being referred to as "de-mixing," although molecular de-mixing or separation of mixed fluid does not occur. Multi-fluid models (Youngs, 1989, 1994; Chen et al., 1996; Braun and Gore, 2021) are needed for the description of de-mixing, which occurs, for instance, when the acceleration reverses sign. Such models, which include equations for each material component in a manner similar to methods used for multi-phase flows, must increase their complexity in order to incorporate the new physics.

In their implementation of Youngs' multi-fluid model, Kokkinakis et al. (2015) show some favorable results in comparisons with two-equation RANS models. Indeed, the authors stressed that a key advantage of the two-fluid model is that it can represent the degree of molecular mixing directly, by transferring mass between the two phases. In a more recent follow-up, Kokkinakis et al. (2020a) reported that to capture the highly localized, non-self-similar tip of a chevron, the more complex, and computationally expensive, multi-fluid model performs better than does a two-equation $K{-}L$ model.

In addition to the mean flow equations, multi-fluid models also solve a separate set of equations for each fluid, supplemented by other closure equations. Two more equations, for the turbulence kinetic energy, K, and the length scale, L, are used to close various terms of Youngs' models (Youngs, 1989, 1994; Kokkinakis et al., 2015, 2020a).

A recently developed multispecies turbulence model of Braun and Gore (2021) is based on the BHR family of models, but is extended to track the evolution of the turbulent fluxes and fluctuations in the material mass fractions for each species present (BHR-4). In addition to the equations of the two-length scale model BHR-3.1, the model includes equations for turbulent species mass-fraction flux, a_i^k, and the correlation between density and species mass fraction fluctuations, b_i^k. Hence, the BHR-4 model is perhaps the most complex and comprehensive one proposed to date (Figs. 22.8–22.10).

Figure 22.11 shows the mixing layer as a function of time for a classical two-component RT system, the width of which, as calculated with several LES and RANS models, is shown

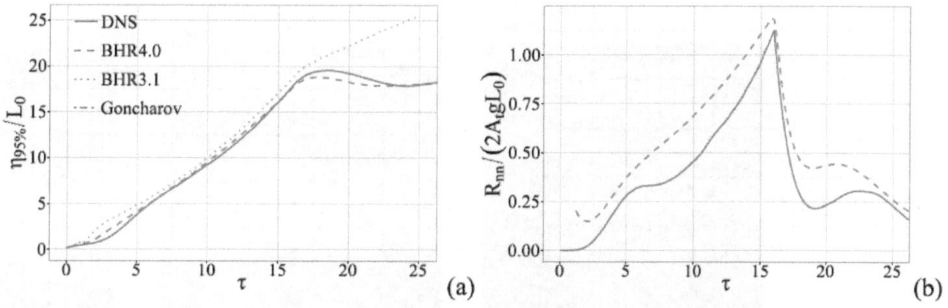

Fig. 22.8 Evolution of mixing layer statistics for an RTI with gravity reversal, with the mixing layer width based on 95% volume fraction, for an Atwood number of 0.5. Fig. 2 of Braun and Gore (2021). © Taylor & Francis.

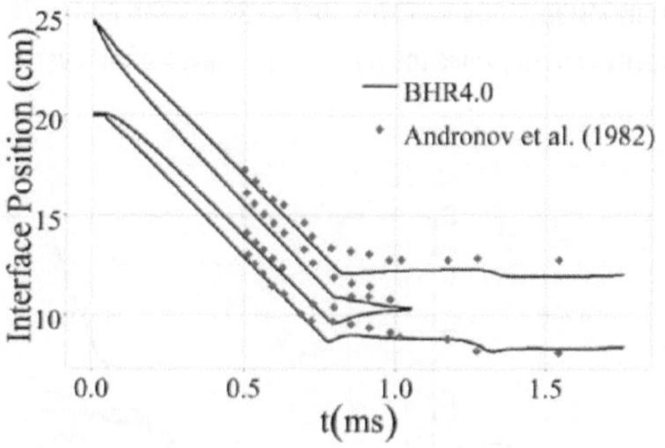

Fig. 22.9 Location of air-helium interfaces in the two-interface shock tube experiments of Andronov et al. The lines show the location of 1% and 99% volume fraction of air in BHR-4 simulations of the experiment, while the symbols show the interface locations reported by Andronov et al. (1982). Fig. 15 of Braun and Gore (2021). © Taylor & Francis.

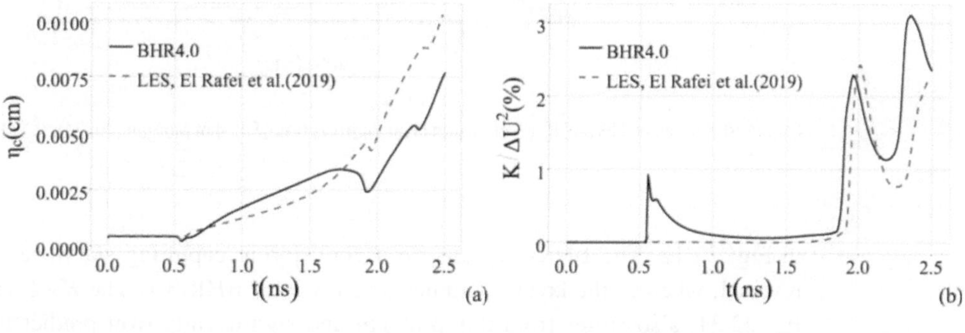

Fig. 22.10 Mixing layer width (a) and turbulent kinetic energy averaged over the mixing layer (b) on the interior surface of a dense shell of fluid during a spherical implosion. The solid line is BHR-4, and the dashed lines are LES (El Rafei et al., 2019). Fig. 17 of Braun and Gore (2021). © Taylor & Francis.

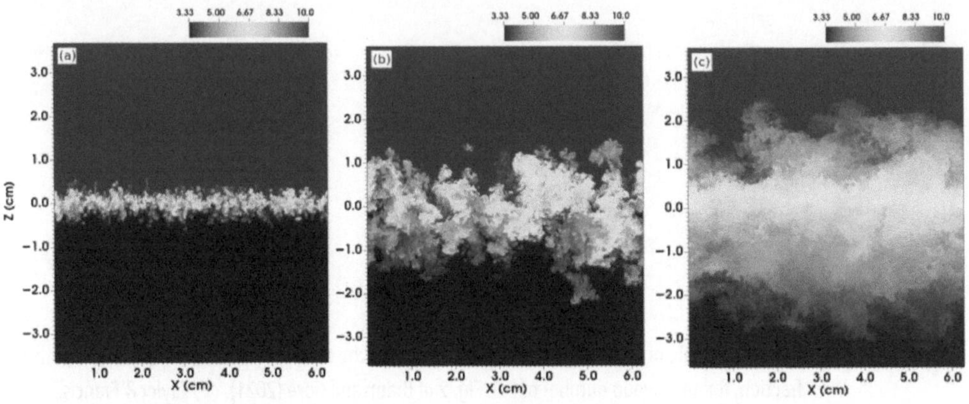

Fig. 22.11 Contours of density (mg/cm^3 in the y $= \pi$ plane for LES of a classical two-component RT mixing layer at three different time instants: (a) t $=$ 10 ms, (b) t $=$ 30 ms, and (c) t $=$ 70 ms. The left and middle images are from times prior to gravity reversal, while the right image is taken after reversal. Fig. 3 of Morgan (2022a). © APS.

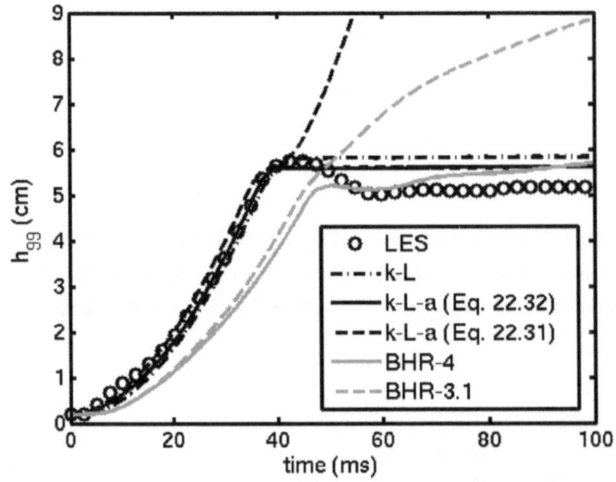

Fig. 22.12 Comparison between LES and RANS of mixing layer width vs time. Fig. 4 of Morgan (2022a). © APS.

in Fig. 22.12. The BHR-4 results do a good job at capturing stabilization after gravity reversal, whereas the layer continues to grow with BHR-3.1. The $K{-}L{-}a$ results, using Eq. 22.31, also suffer from this problem and significantly over-predict the layer width. Rather interestingly, the $K{-}L$ and $K{-}L{-}a$ with Eq. 22.32 models do much better (Morgan, 2022a). This example with gravity reversal emphasizes the improvement that Eq. 22.32 provides over Eq. 22.31.

Finally, Llor and Bailly (2003) introduced a two-structure two-field concept (so-called 2SFK model) for modeling RT, RM, and KH mixing layers (Gréa et al., 2016).

22.8 Other mix models of note

The buoyancy-drag model, which computes the evolution of the mixing region amplitudes, was the workhorse of many practical, first generation mix model calculations. Briefly, the buoyancy-drag model is essentially an equation of motion that balances the inertia, buoyancy, and Newtonian drag forces (Lighthill, 1986; Clift et al., 1978; Landau and Lifshitz, 1959). The model was extended to represent turbulent mixing where the growth of the mixing region depends on the density ratio (namely, the Atwood number) and the acceleration history (Eqs. 3.4–3.7).

Youngs (summarized in the Appendix of Hansom et al. (1990)), Shvarts et al. (2000), Dimonte (2000), Dimonte and Schneider (1996, 2000), Baker and Freeman (1981), Ramshaw (1998), Srebro et al. (2003), Zhou et al. (2002), and Schilling (2020a,b) are examples of papers describing buoyancy-drag models.

More recently, results of 3D turbulent mixing simulations have been used to construct the modifications required to represent the mixing layer width evolution due to single-shock RMI. The simulations used narrowband initial random perturbations (Youngs and Thornber, 2020b). Youngs and Thornber (2020a) extend this analysis to give separate equations for the bubble and spike distances (the depths to which the mixing zone penetrates the dense and light fluids). Moreover, Schilling and Mueschke (2017) and Thornber et al. (2019) should be mentioned for their studies of transport equation budgets of RT and RM, respectively.

Brown et al. (2014b) summarized the "initialization" practice put in place sometime ago at the UK's Atomic Weapons Establishment. At early times, a buoyancy-drag mix model is employed at material interfaces to calculate the expected interpenetration between light and heavy fluids. This model is "passive," in that it does not lead to any modification of the host fluid properties. Once the passively modeled mixing region has grown sufficiently large as to be accurately represented by the host mesh, calculated fluid penetration depths are used to instantiate a smooth interpenetrating volume-fraction profile at the material interface, directly modifying the properties of the host fluid. For subsequent hydrodynamic time steps, a mix model (in the form of a multi-fluid model described in Youngs, 1989, 1994, see the section above) is employed to model the turbulent mixing processes within the host fluids. This model is "dynamic" since it directly interacts with, and imprints upon, the fluid properties of the host code.

Rollin and Andrews (2013), Braun and Gore (2020), and Canfield et al. (2020) developed a two-step initialization strategy for RANS-based simulations. These authors were especially concerned with the BHR model. The first step uses a nonlinear modal model. It assumes small amplitude initial perturbations and incompressible, inviscid fluids. The model tracks the growth of modes that exist in each initial perturbation spectrum, as well as modes generated by modal interactions. As reviewed previously, Haan (1991) has derived such a second-order mode-coupling equation. However, Rollin and Andrews (2013) and Canfield et al. (2020) used Goncharov's (2002) nonlinear solution, rather than Haan's linear one, for the second-order coupling term in the RTI problem. Hence, the iterative integration of the model equations would develop fully nonlinear solutions, an improvement to iteratively solving only the linear solution. It is interesting to mention in passing that the multi-mode model presented in Rollin and Andrews (2013) is somewhat related to a model proposed by Ofer et al. (1996). Ofer extended Haan's model in order to make it applicable to the cases where nonlinear mode generation is important. The bubble and spike front positions and velocities, along with the simplified mixture fraction profiles, were used with a two-fluid approximation to derive profiles for the turbulence model variables.

Both RANS and LES require turbulence models for closure that are capable of capturing the complex features of the flow field. This includes features associated with the interfacial mixing and planar/convergent geometry. For practical applications, it can be argued that continuous models spanning the entire range from RANS to DNS are needed (Speziale, 1998b,a). This approach, theoretically speaking, has the capability of bridging the gap between DNS, LES, and RANS. It allows a DNS to go continuously to a RANS computation in the coarse mesh/infinite Reynolds number limit.

Recently, these methodologies have been refined to a suitable continuous RANS-LES model, applicable to RTI and RMI induced flows (e.g., Grinstein et al., 2020; Bin et al., 2021). In coarse-grained methodology, effective equations are derived for the observable fields, spatially-averaged at an arbitrary scale of resolution (Aluie, 2013). At large coarse-graining length scales, the direct dissipation of the invariants by microscopic mechanisms is shown to be negligible (Aluie, 2017).

An intermediate class of turbulence models, such as the local wavenumber model (LWN) (Besnard et al., 1996; Clark and Zemach, 1998; Pal et al., 2021; Kurien and Pal, 2022), may become more useful as computer power advances. This class of models can be shown to have some "heritage" from statistical closure theories (Zhou, 2021), but are much less costly. They maintain the multiplicity of dynamical scales, thus allowing a richer dynamical behavior. But they are more costly than the usual engineering RANS models. They inhabit a middle world, still too costly for most engineering computations, and lacking the rigor of the formal theories. But as computer power increases, they may warrant a new examination.[3]

[3] T.T. Clark, private communication, 2020).

23 Numerical simulations of mixing

23.1 Trade-off between tractability and complexity

In solving the governing equations of fluid dynamics, the trade-off between tractability and complexity is an important one, as it generally guides the choice of numerical methods for a particular study. First, assumptions are made in deriving the governing equations to improve tractability and compute time, but also place restrictions on the range of validity of the results. An additional level of approximation is introduced when the equations are discretized and solved numerically, which the researcher seeks to minimize by choosing an appropriate numerical method.

Generally, the first aspect to consider is whether the problem of interest contains multiple phases of matter. Historically, numerical studies of turbulence arising from RT/RM have focused almost exclusively on single-phase fluids, as this greatly reduces the range of physical phenomena that must be considered.

Perhaps the biggest distinction, however, from an algorithmic point of view, is whether advection of the fluids may be treated as compressible or incompressible. Assuming incompressibility greatly improves the numerical tractability of the problem, as now the pressure and velocity fields are guaranteed to remain smooth and continuous, which allows for the use of numerical methods that can take advantage of this assumption. Needless to say this also reduces the range of validity of the results, which will be limited to applications where acoustic effects are negligible.

If acoustic effects are important, as quantified by the Mach number, then a fully compressible formulation must be used, which allows for the possibility of discontinuous changes in the flow properties.

Another assumption commonly made within the context of incompressible formulations relates to the Atwood number. In the limit of $\mathcal{A} \to 0$, variations in density can be assumed to be small (and influential only through the buoyancy term), and the governing equations may be simplified, known as the Boussinesq approximation. This may be considered the simplest formulation that still permits the study of buoyancy-driven turbulence.

The specification of a fully defined system of equations is completed by the constitutive relations for viscous dissipation, thermal conductivity and molecular diffusion. Newton's law of viscosity gives the form of the viscous stress tensor $\boldsymbol{\tau}$ (note that Stokes' hypothesis of zero bulk viscosity is invoked), while the heat flux vector is given by Fourier's law of conductivity:

$$\mathbf{q}_c = -\kappa_{\mathrm{T}} \nabla T. \qquad (23.1)$$

Numerical methods must be able to robustly handle shocks and resolve acoustic waves, in addition to capturing fine-scale vortical motions and material interfaces. This represents the key challenge that compressible solvers must overcome, and in general, some form of trade-off must be made between stably capturing discontinuities and resolving fine-scale structures. For this reason, a more diverse range of numerical methods is available in the literature for compressible formulations, as each approach has its strengths and weaknesses.

Artificial fluid properties are used to regularize sharp gradients and discontinuities in the flow. This can include artificial viscosity or numerical smoothing methods. While causing loss of accuracy, these methods prevent the arisal of numerical artifacts in the solution.

An important consideration in going from incompressible to compressible formulations is preserving stability and high-order accuracy at non-periodic boundaries, while also avoiding unwanted wave reflections, since the objective often is to represent a very large (or even infinite) physical domain with a finite computational one. In incompressible simulations of RTI, the boundary conditions applied in the inhomogeneous direction are typically no-slip or free-slip conditions, placed sufficiently far away that the pressure and velocity fields are always uniform at the boundaries. For compressible simulations, the presence of acoustic phenomena means that more advanced treatments are required.

In incompressible solvers, the velocity divergence constraint is satisfied by deriving an elliptic equation for pressure. This equation is obtained by taking the divergence of the momentum equation, combined with the divergence constraint. The constant-density and Boussinesq formulations result in a constant-coefficient Poisson equation for pressure, which can be readily solved through a variety of techniques (Ku et al., 1987). The variable-density formulation leads to a Poisson equation with a factor of $1/\rho$ in the coefficients, making its solution more complicated and computationally demanding.

Livescu and Ristorcelli (2007) overcame the reduction in accuracy of the temporal discretization by deriving an exact nonlinear equation for pressure (i.e., no finite difference approximation required for $\mathbf{u}^{n+1}$). The trade-off for eliminating temporal discretization errors is that the pressure equation now requires an iterative solution, increasing the overall computational cost of the scheme. This approach also only works for triply periodic flows.

A third approach is given by Chung and Pullin (2010), who use a Helmholtz–Hodge decomposition on the pressure gradient terms in the momentum equation. This leads to a constant-coefficient Poisson equation for the scalar potential of this decomposition and a nonlinear equation for the divergence-free component that is solved by iteration. This approach ensures that temporal discretization errors are isolated to the divergence-free component. The iteration introduces additional computational cost, however, with the convergence rate depending on the Atwood number. In addition, $\ln(\rho)$ is used instead of ρ in the determining the time step, as this ensures ρ is always positive after dealiasing but at the cost of not discretely conserving mass. To date, computations have been run using this method up to a density ratio of 10 (Gat et al., 2017). A similar procedure was used by Livescu et al. (2011) to simulate planar RTI between fluids with density ratios as high as 19. Recent simulations have also been performed of shear-driven mixing layers between hydrogen and air (density ratio 16) using this method (Baltzer and Livescu, 2019). A key difference with the method of Chung and Pullin is that ρ is advanced in the timestepping, which ensures that mass is conserved, but at the cost of a smaller timestep (D. Livescu, private communication). Other approaches have also been presented, typically in the context of incompressible two-phase flows, such as the method of Dodd and Ferrante (2014) who use a pressure-correction technique coupled with a volume-of-fluid method. This approach is quite similar to that of Livescu and Ristorcelli (2007), but using only the first step of the iteration.

While the variable-density formulation allows for accurate results to be obtained for mixtures of miscible fluids at arbitrary Atwood numbers, its range of applicability is still limited to low-speed flows. Typically, departures from incompressibility are considered to occur starting at a Mach number (mean and/or fluctuating) of ≈ 0.3 (Anderson, 2007). In general, this precludes the study of the RMI using this formulation, as although the development of the instability at late time is virtually incompressible for small to moderate shock Mach numbers, the shock wave itself is an inherently compressible phenomenon. One approach, valid for initial perturbations that are entirely linear, is to use an equivalent velocity perturbation (Thornber et al., 2010), thus circumventing the need to initialize a shock wave.

23.2 Interface tracking

The above formulations and numerical approaches are strictly valid for simulating turbulence between miscible fluids. Thus, physical diffusion across a material interface is either explicitly modeled or treated using interface capturing schemes in which the interface is "smeared" across some finite width region within the computational domain. However, there are often compelling reasons for modeling an interface as exactly discontinuous, without numerical or physical diffusion. One such scenario is when the fluids being studied are immiscible or of different phases (multi-phase). Another scenario is when the interface is of negligible thickness compared to the size of the computational grid, which is the case for flame fronts or the early stages of RM/RT instabilities. Indeed, early research into RT and RM instabilities focused on understanding single-mode growth and typically, some form of interface tracking was used for this purpose, particularly due to the limited computational resources available at the time (Youngs, 1984). Such simulations were also usually inviscid. When examining turbulence induced by RTI/RMI, interface tracking was deemed to no longer be appropriate in situations where significant fine-scale breakup of the interface occurs. However, for situations in which the interface radius of curvature is greater than the grid size, such as those discussed above, some form of interface tracking remains a viable, and sometimes necessary, approach to modeling the interface evolution in time.

The use of interface tracking methods in simulations of turbulent mixing due to RT and RM instabilities has primarily been to study the effects of immiscibility. The use of some form of interface tracking to resolve gradients that are sharp with respect to the computational grid is also potentially beneficial for modeling early time development of RM and RT flows between miscible fluids.

23.3 Direct numerical simulation

The mathematical formulations of direct numerical simulations (DNS) describe completely, at the continuum level, the evolution of the flow field that ensues from a given initial condition (assuming well-posedness). Under suitable conditions, the flow will become turbulent. The wide range of scales accompanying turbulence is described in full by the governing equations. However, in order to obtain a numerical solution, these equations must be discretized without losing a significant amount of information about the fine-scale structure of the flow in the process. In particular, turbulent motions at scales smaller than the computational grid employed will be lost, while scales that are close to the grid scale will be severely impacted by the numerics. This will, in turn, affect the evolution of the grid-resolved flow field in time, since there exists a cascade of energy from large to small scales, and this energy is only dissipated at the very smallest scales by viscosity (Davidson, 2015). The handling of the impacts of finite grid resolution when simulating unsteady, turbulent flow is addressed in this section.

All the scales of motion must be resolved in the discretization (both spatially and temporally). In fully developed RT/RM turbulence, multiple scales must be adequately resolved: the initial interface thickness, the dominant wavenumber in the initial condition, the Kolmogorov scale, λ_K, at which viscous dissipation occurs and when mass diffusion is present, the Batchelor scale, λ_B, at which scalar dissipation occurs.

The following expression for λ_K can be obtained (e.g., Ristorcelli and Clark (2004)):

$$\lambda_K = \frac{(\frac{\nu}{2})^{0.75} t^{-0.25}}{\sqrt{\alpha \mathcal{A} g}}. \tag{23.2}$$

Thus, scale separation in self-similar RT implies a Kolmogorov scale nearly independent of time (while the large scales grow as $\sim t^2$), a result that is fortuitous to the design of numerical simulations, since the smallest flow scales dictate the grid resolution.

However, there are a few issues with the above definitions and analyses. First, the assumption that the Kolmogorov scale is representative of the smallest eddies in the flow is predicated upon the presence of an inertial range in the turbulent kinetic energy spectrum, that is, that the turbulence is fully developed. In a temporally developing flow, such as a mixing layer, this is not always the case. Second, mixing layers can be anisotropic as well as inhomogeneous, making the definition of a single Kolmogorov scale problematic.

For Schmidt numbers greater than one, it is also necessary to ensure that scalar dissipation is accurately resolved. In addition to requiring sufficient grid resolution at the small scales, the domain size must ensure the largest scales in the flow are accurately represented.

There are multiple ways in which numerical errors arise and affect the flow field during simulations. One major source of errors are truncation errors resulting from approximating gradients or interpolation, which reduce the level of fidelity with which fine-scale motions in the flow can be resolved. These errors correspond to numerical sources of dissipation and dispersion, due to similarities with their physical counterparts in their effect on the flow field. Errors also stem from the finite representation of continuous functions, known as aliasing errors. These errors arise through nonlinear interactions leading to the spurious generation of modes that are not in the set of modes being represented and whose contributions are added incorrectly to the flow field. Finally, errors due to finite grid resolution are also introduced at scales near the grid spacing, where triadic interactions that occur between scales (because of nonlinearities in the governing equations) are not represented properly.

Flows with shock waves present an additional complication, since the Navier–Stokes equations do not, in general, give the correct shock width. Other techniques, such as solution of the Boltzmann equation or molecular dynamics are needed for this (Sherman, 1955; Gombosi, 1994).

Some calculations have been performed with fully resolved shock waves, for example the simulations of early-time RMI by Margolin and Reisner (2017) or shock-turbulence interaction by Ryu and Livescu (2014). However, the typical approach, when simulating such flows, is to use a shock-capturing approach (Groom and Thornber, 2019), where the internal structure of the shock is lost but the jump conditions across the shock are still enforced. With this approach, it is important to minimize the impacts of the extra numerical dissipation that is introduced, in order to stably capture the shock.

23.4 Large-eddy simulation

For Reynolds numbers of practical interest, the modeling approach known as large-eddy simulation (LES) is most practical, where the unsteady dynamics of the largest scales are explicitly computed, while the influence of the smaller scales, which are statistically universal, is modeled. A fundamental assumption in LES subgrid modeling is that of scale locality, which justifies the existence of universal statistics and motivates the belief that models of subscale terms in the equations for large scales can be of general utility, independent of the particulars of the turbulent flows under study. Conceptually, LES involves a low-pass filtering operation, in which the flow field is decomposed into a filtered and residual component. The equations of motion for the filtered component are derived from the original governing equations and will contain contributions arising from residual motions, which are supplied by some form of closure model. It is important to note that the filtering operation must be independent of the grid resolution. In other words, the filtering operation is a convolution of f with some kernel function G, where G depends on a characteristic filter width (Δ) and potentially also the grid position ($\mathbf{x}$). If $\Delta = h$ then the residual component of the flow field may also be called the subgrid component, while the filtered component may also be called the resolved component (Pope, 2004).

As I have just discussed, an assumption that underlies LES modeling efforts is the notion of *scale locality* (Eyink, 2005; Zhou, 1993a,b). If the coupling is mostly between scales of a comparable size, then the flow interactions are said to be scale-local (Kraichnan, 1959). Scale-locality is central to devising LES subgrid models that are a function of the resolved flow. However, the conditions required for scale-locality to hold are may not be valid in transitional turbulent flows (Aluie, 2011).

During transition, the small scale properties may become under-resolved on a given grid, but may not yet be turbulent. In that period of mixing layer development, both explicit and implicit LES may not give accurate results for properties of the subgrid scales, such as mixing. This can happen when most of the subgrid flow exists at scales much smaller than that of the grid, resulting in a wide scale-gap between the resolved flow and the subgrid flow (Eyink and Aluie, 2009; Aluie and Eyink, 2009). Such a scenario leads to the dominant coupling between scales being long-range, which is challenging for existing LES models. Until models of unresolved mixing during transition are developed, one approach is to apply a combination of DNS of the initial, laminar and transitional growth with LES (implicit or explicit) of the late time self-similar growth. In RM in particular, the Reynolds number varies as $t^{2\theta-1}$, and thus may reduce in time for some initial conditions. In that case, DNS may also need to be employed at late-time if the initial Reynolds number is sufficiently low.

23.4.1 Explicit subgrid modeling

If a closure model for the residual stresses is provided, either in physical or wavenumber space, then the modeling approach is referred to as explicit subgrid modeling or explicit LES. In order to obtain grid-independent solutions, the filtering operation must be performed explicitly. This approach is sometimes referred to as pure LES (Pope, 2004). However, in order for the LES to generate useful results, a significant portion of the turbulent kinetic energy (TKE) in the flow should be resolved. Considering homogeneous turbulence as an example, assuming a Kolmogorov $k^{-5/3}$ model spectrum and requiring that 80% of the TKE be resolved gives $k_c L_{11} = 15$ (Pope, 2000), where k_c is the cutoff wavenumber and L_{11} the integral length scale. Calculating accurate statistics at large scales requires that $L \geq 20L_{11}$ (Thornber, 2016), implying that at least 48 modes need to be resolved, requiring a minimum grid resolution of 96^3. In addition, in order to obtain negligible numerical error, most schemes require a filter width to grid spacing ratio of at least $\Delta/h = 4$ (Chow and Moin, 2003), bringing the minimum required grid resolution in this example to 384^3. If instead, 90% of TKE must be resolved, then the grid resolution requirement becomes at least 960^3. Therefore, as noted by Pope (2004), the optimal value of Δ/h for a given grid resolution most likely corresponds to some non-negligible amount of numerical error being retained, so that the modeling error is reduced.

For flow at high Reynolds numbers, the requirement that the majority of TKE in the flow be resolved implies that the filter width be located in the inertial subrange of the energy spectrum. This represents the ideal case for the application of LES, as (i) it is in this regime that the various assumptions inherent in simple subgrid models (i.e., eddy viscosity models) are well justified and (ii) accurate estimates of the residual quantities are available.

In incompressible flows, there are only a handful of terms in the filtered governing equations requiring closure; namely the residual stress tensor as well as subgrid mass/scalar fluxes in the case of multicomponent flow. A wide range of closure models exist for these terms, and the most commonly used models are summarized in the review articles of Lesieur and Metais (1996), Meneveau and Katz (2000), Moser et al. (2021), and the book by Sagaut (2006). In contrast, for compressible flows there are more terms requiring more complicated closures due to the presence of the energy equation. Additional complications also occur in flows with discontinuities, such as shock waves, where the combination of a subgrid model and numerical dissipation can be excessively dissipative. The subgrid motion in a computational cell containing a shock is also significantly different from the canonical turbulence that subgrid models are typically constructed to represent.

For RTI flows, Amala et al. (1994), Amala and Rodrigue (1994), and Amala (1995) performed large-eddy simulations including a combined subgrid model: a modified Smagorinsky (1963) eddy viscosity and a stochastic backscatter (e.g., Leith, 1990). While it was an a priori study, Cabot et al. (2004) showed the potential utility of a spectral eddy viscosity for LES of RT mixing. Wang and Li (2005) and Wang et al. (2002) also performed 3D LES of RTI induced flows in which representation of the subgrid-scale flux was given by a simple eddy viscosity coefficient and a dynamical equation for the subgrid-scale turbulence intensity (Deardorff, 1980).

Burton (2011) computed RTI flows using his nonlinear large-eddy simulation (nLES) method (Burton, 2008). He demonstrated the utility of a convective subgrid model with a deterministic backscatter limiter, which applies neither an eddy-viscosity nor other artificial dissipation, in calculating ultra-high Atwood numbers RTI flows. His results indicated that the method may remain accurate for Atwood numbers beyond $\mathcal{A} = 0.96$, the highest test in that study. Yilmaz (2020) also performed LES of RTI for Atwood numbers up to 0.9, but used an algebraic eddy-viscosity model utilizing a differential operator based on the traceless symmetric part of the square of the velocity gradient tensor. This so-called WALE (wall adapting local eddy-viscosity) model is known for its applicability in laminar, transitional, and turbulent regimes (Nicoud and Ducros, 1999).

The hyperviscosity has been shown as a very effective method for a LES to avoid wasting numerical resolution by reducing the range of scales over which dissipation is effective. Essentially, the normal Laplacian dissipation is replaced by a higher power one (Frisch et al., 2008) so that the fluid transport coefficient is altered to provide the sharpest possible cutoff near the Nyquist wavenumber.

In addition, LES simulations have provided a detailed analysis of the 3D planar RMI flows with reshock (Hill et al., 2006; Lombardini et al., 2011) and turbulent mixing driven by spherical implosions (Lombardini et al., 2014a,b). These LES utilized the stretched-vortex subgrid-scale model developed by Misra and Pullin (1997). On the other hand, Bai et al. (2010b, 2012b,a) and Wang et al. (2016) have carried LES of RMI induced flows with reshock, but with the subgrid model proposed by Vreman (2014). Moreover, Wang et al. (2015c) have compared and contrasted the performance of the dynamic Smagorinsky model (Germano et al., 1991), the stretched-vortex model (Misra and Pullin, 1997), as well as the approach advanced by Vreman (2004).

23.4.2 Implicit subgrid modeling

Calculations that exclude the use of an explicit filter and subgrid model are referred to as Implicit LES (ILES) or Monotone Integrated LES (MILES), and can be used to obtain accurate statistics of the large scales in very high Reynolds number flows (Boris et al., 1992; Drikakis, 2003; Grinstein et al., 2007; Youngs, 1991). ILES aims to tackle several of the identified challenges for explicit LES with non-dissipative algorithms, namely the need to (i) ensure monotonic behavior at shock waves and contact surfaces, (ii) provide minimal dissipation in the early linear and nonlinear stages of instability growth, (iii) formally separate the filter and mesh scale, or ensure numerical dissipation and dispersion do not conflict with the explicit model, and (iv) consider non-equilibrium effects.

ILES uses the numerical dissipation of the underlying algorithm to provide the dissipation necessary to remove energy from the smallest resolved scales — acting as an implicit subgrid model. In principle, any sufficiently dissipative numerical method can be used to perform ILES, provided there is sufficient scale separation in the flow being simulated so that the largest scales of motion are insensitive to the specific mechanism by which energy is removed from the flow. This is guaranteed if the flow satisfies scale-locality (Kolmogorov, 1941; Zhou, 1993a,b; Eyink, 2005; Aluie, 2011), which is also required in explicit subgrid modeling. Although some numerical methods do aim to mimic specific subgrid models, or have explicit stabilization terms, in most algorithms, the dissipation rate is a complex function of the pressure, density and velocity fields and does not directly map to an explicit subgrid model. However, this does not preclude the use of ILES, since Kolmogorov's assumption of the independence of large scales from the dissipative scales, which was demonstrated for variable density flows in Aluie (2013) and Zhao and Aluie (2018), may be applied directly to any subgrid dissipation mechanism. Thus, as long as the specific numerical dissipation mechanism does not impact the resolved large scales, the large scales will evolve physically (Zhou, 2007). This fact was exhibited clearly in the very good agreement in large scale quantities between substantially different ILES codes when applied to an identical problem in Thornber et al. (2017).

For the majority of numerical methods that do not attempt to mimic a specific subgrid model, numerical dissipation is introduced to control ringing of the solution at shocks. Such numerical dissipation is required due to the fact that higher-order modeling approaches appropriate to incompressible flows with low density contrasts often lead to unphysical results around shocks and strong contact discontinuities. For the purposes of numerical stability, energy must be removed from the flow at high wavenumbers by the numerical scheme, using physically motivated limiting approaches, thus playing a similar function to an explicit subgrid model. In practice, this is typically achieved by using a non-oscillatory numerical method, such as the flux-corrected transport, piecewise parabolic or Godunov finite-volume methods. Most approaches employ at least second-order accurate schemes. The advantage of higher-order methods is that the implicit subgrid model is nonlinear and heavily weighted toward the smallest scales resolved by the grid, which improves the computational efficiency of the calculation since it increases scale separation. Use of these schemes also allows for discontinuities such as shock waves and material interfaces to be robustly handled in an ILES calculation. The implicit filtering operation is also naturally anisotropic wherever there is anisotropy in the grid (this is also true of explicit LES schemes with implicit filtering).

Given that each numerical method has a unique dissipation mechanism and magnitude, an ILES practitioner must pay careful attention to the properties of their numerical method in order to choose the appropriate grid resolution for a given problem. This is typically established by undertaking computations of single-mode instability development at very coarse grid resolution (e.g., 8–16 grid points per wavelength), equivalent to the resolution of very high wavenumber modes on a much larger grid for multi-mode perturbations. The total required grid size for a specific problem can then be computed once the minimum number of points to accurately evolve a single-mode is established, along with the expected integral scale and the requirement for the resolution of a given proportion of total kinetic energy (as already discussed).

There are three key caveats to consider in the application of ILES. The first is that the dissipation from the subgrid model may not enforce any expected universality of the unresolved scales. The second is that there is no formal point-wise grid independence, which may be gained in explicitly filtered LES. This means that for each grid, a different flow is being computed. While methods exist for defining the effective Reynolds number (and hence viscosity) in an ILES calculation (Aspden et al., 2008; Zhou et al., 2014; Zhou and Thornber, 2016; Domaradzki, 2021), it remains problematic to evaluate during the simulation itself, although this is also true of many explicit LES models. Third, when estimates of subgrid motions are required for models of other subgrid processes (such as reactions), it is only possible to use methods based on the resolved scales (see e.g., Thornber et al., 2011).

To expand on the third point above, it is assumed that the flows being solved with ILES have sufficiently high Reynolds numbers to have passed the mixing transition (Dimotakis, 2000) and minimum state (Zhou, 2007; Zhou et al., 2019) (as described in Chapter 8) and that the dissipative processes are controlled by an eddy cascade to high-wave numbers and are insensitive to specific values of viscosity and diffusivity. As for LES, the mesh resolution needs to be high enough to resolve a significant portion of the inertial range. The numerical Schmidt and Prandtl numbers are nominally equal to one (Livescu, 2013). With this in mind, most ILES calculations of compressible turbulent mixing are performed in the nominally inviscid limit, as the contributions from viscous terms to the resolved scales of motion will be negligible in comparison with numerical dissipation.

Given the considerations above, one particularly appropriate application of ILES is in computing the high Reynolds number limit of statistical quantities that depend on the energy containing scales, in flows in which the rate-controlling processes are also determined by these scales. Free shear flows, as well as RT- and RM-induced mixing layers, satisfy this requirement in the self-similar regime. Hence, ILES is expected to be effective at accurately computing integral quantities such as mix width for these flows in the high Reynolds number limit.

23.5 DNS and LES compared

For an RTI induced flow, Cook et al. (2004) confirmed that the LES converges to DNS at low Schmidt and Reynolds numbers. In this case, a DNS was first carried out for a case very similar to that described in Cook and Zhou (2002): $\mathcal{A} = 0.5$, with isotropic perturbations peaked at mode 16. An accompanying LES was subsequently performed with a set-up identical to the DNS, but with the filter and hyperviscosity subgrid-scale (SGS) models

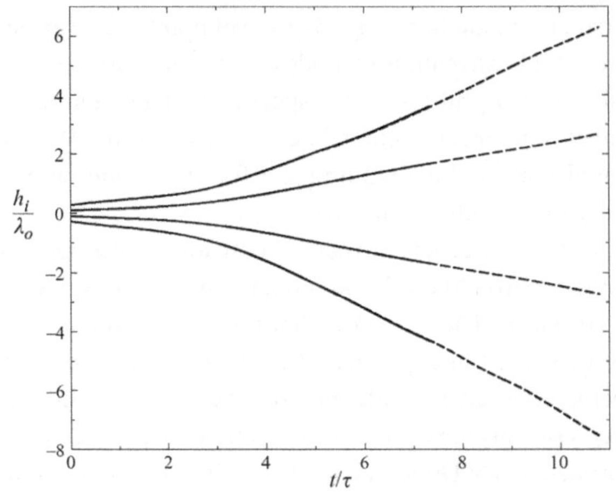

Fig. 23.1 Mixing heights from validation DNS (solid) and LES (dashed). Top and bottom curves are (1%) threshold or "visual" widths for the bubble and spike. Middle curves are the corresponding integrated mixing widths. Fig. 3 of Cook et al. (2004).

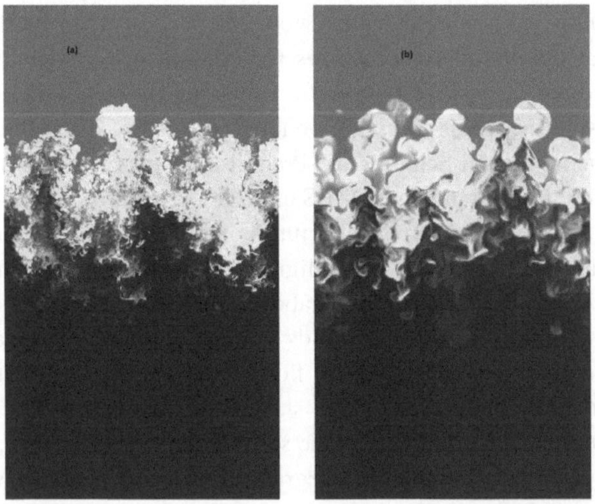

Fig. 23.2 Volume fraction distributions for plane sections for a dimensionless time approximately 3, denser fluid at top; (a) ILES, (b) DNS. Fig. 2 of Youngs (2017). UK Ministry of Defence © Crown Owned Copyright 2022/AWE.

active. Based on visual inspection, the flows appear identical, illustrating that convergence is achieved (Fig. 23.1).

One of the ILES calculations described in Youngs (2013) is repeated with viscosity and diffusivity included in Fig. 23.2. As described in Youngs (2017), the Atwood number is 0.9 and the resolution used for the DNS is the same as that used for the ILES, $1{,}550 \times 1{,}000^2$ meshes. The ILES and DNS give very similar values of late-time RTI growth $\alpha \sim 0.03$.

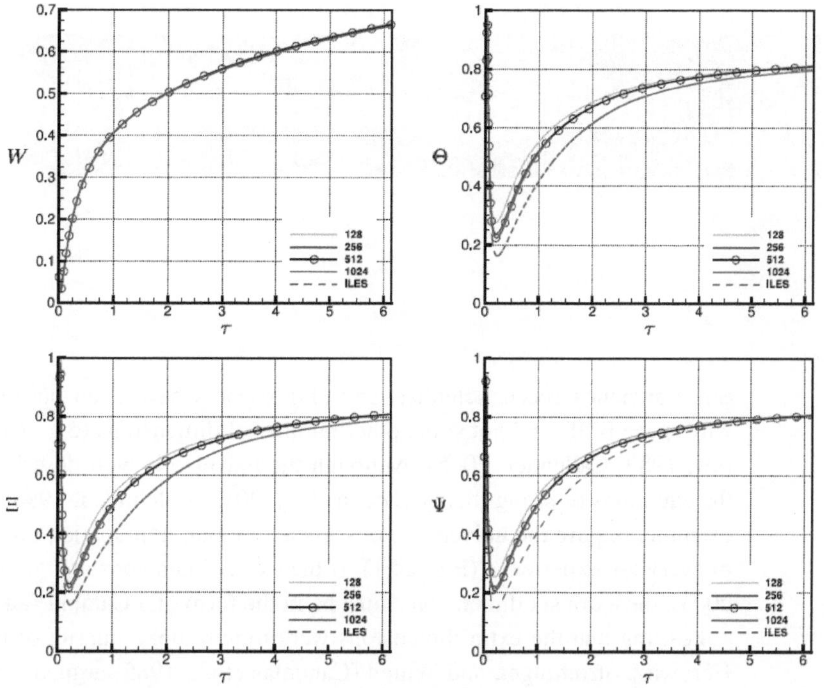

Fig. 23.3 Time histories of integral width and mixing parameters (mixedness, Θ and Ξ, and normalized mixed mass, Ψ) comparing the DNS and ILES results. Fig. 7 of Groom and Thornber (2019).

A much finer-scale structure is present in the ILES. Viscosity and diffusivity reduce the growth rate of the short-wavelength initial perturbations.

Latini et al. (2007a) showed the utility of both ILES and higher order methods for RM with reshock. Groom and Thornber (2019) compared their DNS of RMI initiated by a Mach 1.84 shock with ILES results from the θ-group collaboration (Thornber et al., 2017), generated using the same numerical algorithm. Overall, the agreement in the large scales between DNS and ILES of an identical initial condition is very good and hence the mixing width and growth rate exponent are very similar (Fig.23.3(a)). There are substantial differences in small-scale behavior, however (Fig. 23.3(b–d)). At early time, the noticeably larger values of mixedness and normalized mixed mass for the DNS, compared to the ILES, are brought about by the additional spreading of the layer due to molecular diffusion and the suppression of turbulence. Despite the inhibition of fine-scale motions in the DNS, the late time values of the DNS and ILES quantities (in Fig. 23.3) are very similar, indicating that the mixing is dominated by large-scale motions at late time (Groom and Thornber, 2019).

24 Does 2D turbulence resemble 3D turbulence?

Since ancient Greece, scientists and philosophers have been puzzled by the question of why space is 3D and not some other number of dimensions (e.g., Neugebauer, 1975; Barrow, 1983; Callender, 2005). Although the universe is well described by 4D space-time, the equations of string theory (Green et al., 2012; Polchinski, 1998; Zwiebach, 2004), for example, require the universe to possess additional, non-spatial, dimensions beyond those of everyday experience (Fig. 24.1). If the extra dimensions are "compactified" (Susskind, 2003), the extra six dimensions must be in the form of a Calabi–Yau manifold (Fig. 24.2).

Positing that the extra dimensions were minuscule, a quartet of physicists – Candelas, Horowitz, Strominger, and Witten (Candelas et al., 1985) argued, would not only explain why we haven't seen them, but could also provide the missing bridge to experimental verification (Green, 2015).

Even in fluid mechanics, several authors have studied homogeneous, isotropic turbulence in d spatial dimensions, looking for possible theoretical simplifications. This may seem counterintuitive, but a theory may become more mathematically tractable by modeling spacetime in arbitrary dimensions (e.g., Forster et al., 1977; Fournier and Frisch, 1978). Using the statistical closure theories, and DNS as an example, Gotoh et al. (2007) found that the energy transfer is more efficient in 4D than in 3D.

There are a number of situations in which theories in 2D are useful for describing phenomena in fluid dynamics. The relevance of idealized 2D turbulence to certain aspects of atmospheric motion has been emphasized since the late last century (Gage, 1979; Charney, 1971; Herring, 1980; Lilly, 1983), because large-scale motions (many hundreds of kilometers) in the atmosphere are approximately 2D. For instance, 2D turbulence has found success (Zhang et al., 2008) in describing the category 3 Hurricane Isidore, September 2002 (Fig. 24.3). Similarly, Figure 24.4 shows Jupiter's atmosphere, where eddies span a wide range of length scales from the 40,000-km-diameter Great Red Spot down to the images' resolution scale (Young and Read, 2017).

As a practical matter, it is desirable to run computations in 2D whenever possible, due to the high computational cost and time-consuming nature of fully 3D simulations of hydrodynamic instabilities and turbulence. The utility of 2D simulations has been questioned sometimes by well-known experts, however. For example, Pullin and Saffman (1998) memorably stated that "two-dimensional turbulence ... is a consequence of the construction of large computers."

All of this begs the question, does 2D turbulence resemble 3D turbulence?

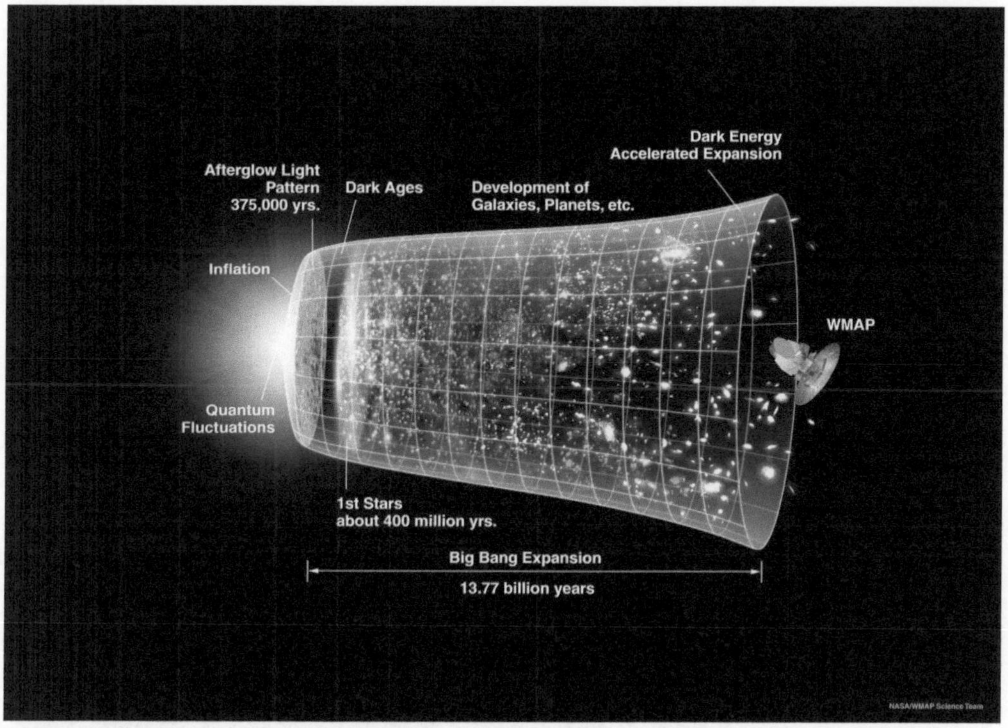

Fig. 24.1 The timeline of the universe, beginning with the Big Bang. According to String Theory, this is just one of many possible worlds. Credit: NASA. https://map.gsfc.nasa.gov/media/060915/060915_CMB_Timeline600.jpg.

Fig. 24.2 String space – superstring theory lives in 10 dimensions, which means that six of the dimensions have to be "compactified" in order to explain why we can only perceive four. The best way to do this is to use a complicated 6D geometry called a Calabi–Yau manifold, in which all the intrinsic properties of elementary particles are hidden. This image is a somewhat fair reconstruction of an image of the Calabi–Yau manifold that appears as a figure in the paper: L. Susskind (2003). To recreate the image's likeness without copying the image, the artist used the description on the home page of Prof. Andrew Hanson of the computer science department at the Indiana University, and the procedure he provides in the paper: A.J. Hanson (1994). "A construction for computer visualization of certain complex curves." This file is licensed under the Creative Commons Attribution-Share Alike 2.5 Generic license. https://commons.wikimedia.org/wiki/File:Calabi-Yau.png.

Fig. 24.3 Hurricane Isidore with winds of 120 mph on the Yuctan coast, on September 22, 2002. Satellite image via National Oceanic and Atmospheric Administration (NOAA) from www1.ncdc.noaa.gov/pub/data/images/hurr-isidore-20020922-1945-g81kmvis.gif.

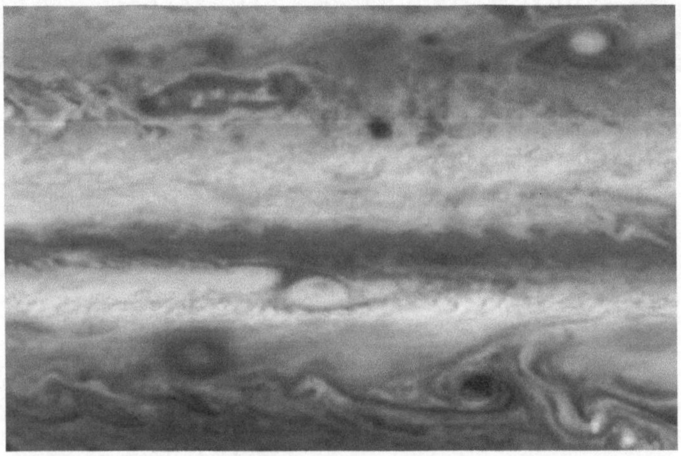

Fig. 24.4 Image of Jupiter's clouds with jet scale and deformation radius. This is part of a global mosaic of NASA Cassini Imaging Science Subsystem narrow-angle camera images combining the CB2 (750 nm) and BL1 (451 nm) filters. It shows a 45° × 30° region of Jupiter's clouds (Young and Read, 2017).

24.1 Enstrophy vortex stretching and direct or inverse energy cascade

The Navier–Stokes equation can be recast into an equation for enstrophy (Kraichnan and Montgomery, 1980; Tabeling, 2002; Boffetta and Ecke, 2012), $\mho = \boldsymbol{\zeta}^2/2\rho$, where the vorticity is obtained by taking the curl of the velocity field, $\boldsymbol{\zeta} = \nabla \times \mathbf{u}$:

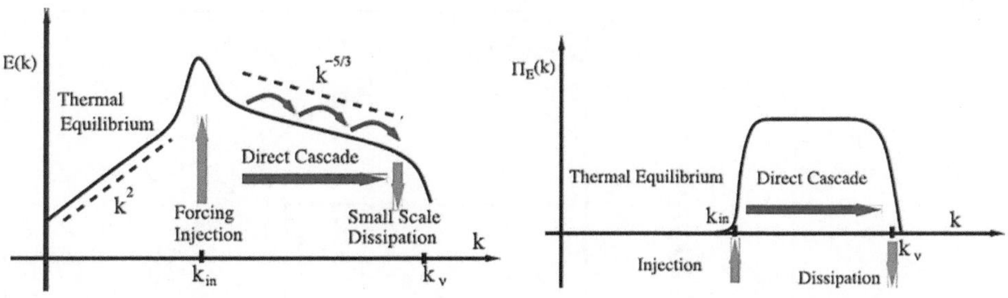

Fig. 24.5 Left: Schematic of the spectrum for the direct cascade regime, with the Kolmogorov slope in the direct cascade range. Right: Same as in the left panel but for the energy flux. Fig. 5 of Alexakis and Biferale (2018).

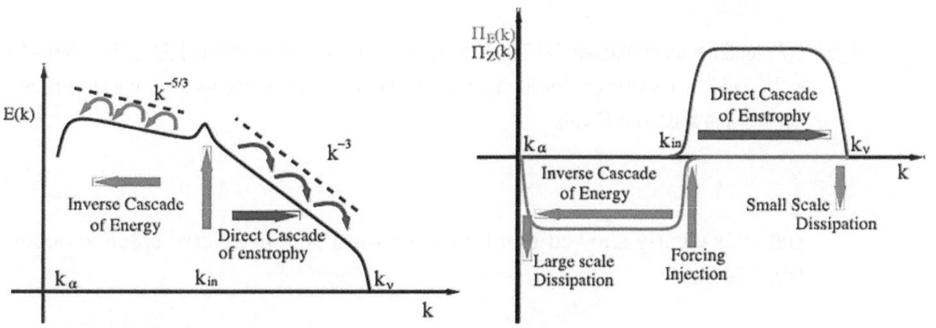

Fig. 24.6 Log–log schematic of the energy spectrum (left) and of energy and enstrophy fluxes (right) for the 2D Batchelor–Kraichnan–Leith theory. Fig. 6 of Alexakis and Biferale (2018).

$$\frac{\partial \mho}{\partial t} + \nabla \cdot (\mho \mathbf{u}) = \rho^{-1}\Big[\boldsymbol{\zeta}\boldsymbol{\zeta}{:}\mathbf{S} + \boldsymbol{\zeta} \cdot \mathbf{B} + \boldsymbol{\zeta} \cdot (\nabla \times \rho^{-1}\nabla \cdot \tilde{\boldsymbol{\sigma}})\Big]; \qquad (24.1)$$

where $\mathbf{B} = \rho^{-2}\nabla\rho\times\nabla p$ is the baroclinic vector, $\mathbf{S} = (\nabla\mathbf{u} + \mathbf{u}\nabla)/2$ is the velocity strain tensor, $\tilde{\boldsymbol{\sigma}}$ is the viscous stress tensor, and the first term on the right hand side is the contraction of two second-rank tensors (e.g., Tennekes and Lumley, 1972; Cabot, 2006).

The first term on the right-hand side of Eq. 24.1 is referred to as the enstrophy production from vortex stretching (which vanishes in 2D flow), while the second and third terms are the production from baroclinicity, and viscous dissipation and redistribution, respectively. In 3D, vortex stretching triggers the transition to a turbulent flow field.

The more developed a flow, the longer its inertial range, and the sharper the difference between 3D and 2D can be. This is best illustrated by comparing the energy spectra and flux functions in the two cases. It is an empirical fact that in 3D there is only a direct energy cascade (Fig. 24.5). On the other hand, the vortex-stretching term vanishes completely in a pure 2D flow (Lee, 1951). Indeed, the energy transfer in a 2D flow is inverse-cascading (Fig. 24.6), rather than forward cascading (Kraichnan, 1967; Leith, 1967; Batchelor, 1969).

Boffetta and Musacchio (2010) reported numerical results of 2D HIT turbulence on the basis of a very high resolution of 32768^2 with forcing at moderate wavenumbers. The

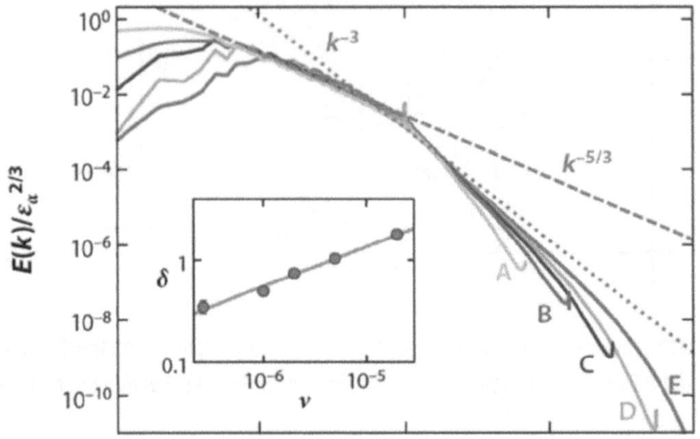

Fig. 24.7 $E(k)$ for different resolutions: (A) 2,048, (B) 4,096, (C) 8,192, (D) 16,384, and (E) 32,768. Dashed and dotted lines are $k^{-5/3}$ and k^{-3}, respectively. The inset shows the correction to the exponent -3 as a function of viscosity. Fig. 2 of Boffetta and Musacchio (2010).

statistics clearly showed that both a forward and an inverse cascade occur simultaneously (Fig. 24.7).

24.2 2D versus 3D in RTI

The generalized Layzer-type model (Layzer, 1955) found that 3D RTI perturbations grow at a rate $\sqrt{3}$ times faster than 2D perturbations in both the linear and nonlinear regimes (Goncharov, 2002; Mikaelian, 1998). In these analytical treatments, the 2D flow is between two "parallel walls" separated by a distance. In the 3D geometry, also called "tubular flow," a bubble rises in a cylindrical tube of diameter $\mathfrak{D}$.

For single-mode perturbations, bubble-merger models can be used to provide comparable rates of mixing layer growth for 2D and 3D perturbations. As described in Oron et al. (2001), while the 2D simulated growth rate value saturates at a dimensionless time of 0.5, the 3D rate saturates at 1.5 (Fig. 24.8), indicating growth of much more elongated bubbles in the 3D case. This factor of 3 enhancement is in very good agreement with the experimental value of Dimonte and Schneider (2000).

Numerically, Fig. 24.9 shows a direct comparison between the lengths of the bubble and spike of the 2D and 3D single-mode RTI with $\mathcal{A} = 0.5$ and $Re = 3000$. With the same initial perturbation magnitude, a 3D RT instability exhibits larger absolute penetration lengths of the bubble and spike compared to a 2D RT instability (Hamzehloo et al., 2021). Such 2D and 3D differences in the absolute penetration lengths have also been reported previously, for example, in simulation works of Anuchina et al. (2004) and Bian et al. (2020), who showed that compared to 2D, a single-mode RTI bubble in 3D accelerates more easily to speeds larger than those predicted by generalized Layzer-type models.

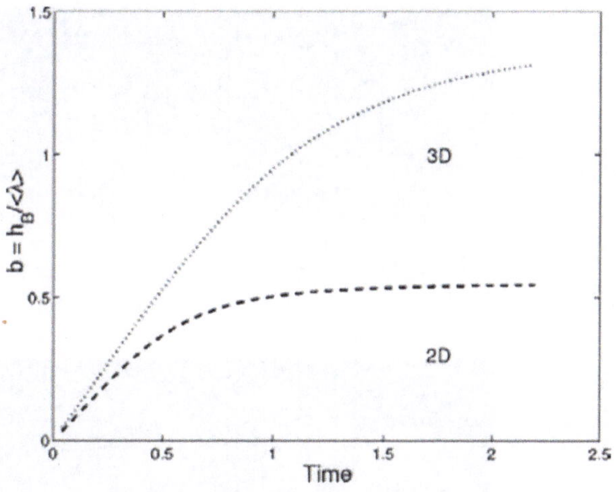

Fig. 24.8 Temporal evolution of the self-similarity parameter $h_B/\langle\lambda\rangle$ in the $\mathcal{A} = 0.5$, 2D and 3D simulations of RTI induced flows. (Oron et al., 2001, with the permission of AIP Publishing).

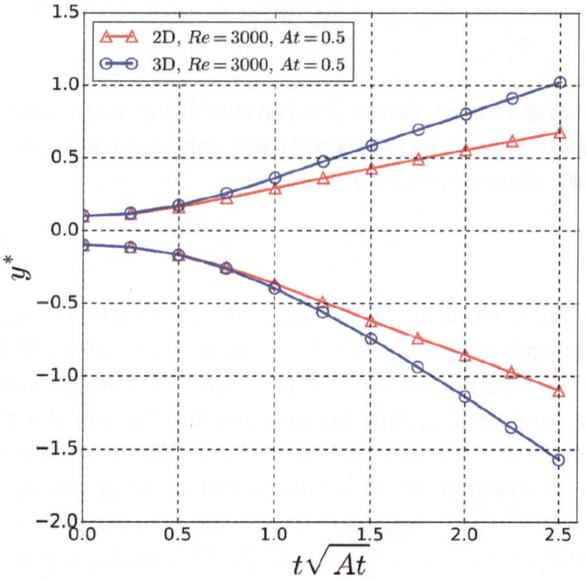

Fig. 24.9 Temporal evolution of the vertical coordinates of a spike and bubble for RTI induced flows. Direct comparison between 2D and 3D single-mode instabilities with the same maximum initial perturbation amplitude, Reynolds and Atwood numbers. Hamzehloo et al. (2021), with the permission of AIP Publishing.

Turning to small-scale multi-mode initial conditions, Cabot (2006) illustrated that the 2D flow develops larger structures (Fig. 24.10), is much less well mixed, and grows more quickly than its 3D counterpart at late times (Fig. 24.11). After the initial entrainment at $\tau = 3$, the level of 3D mixedness continues to increase, reaching a plateau of $\Xi = 0.79$.

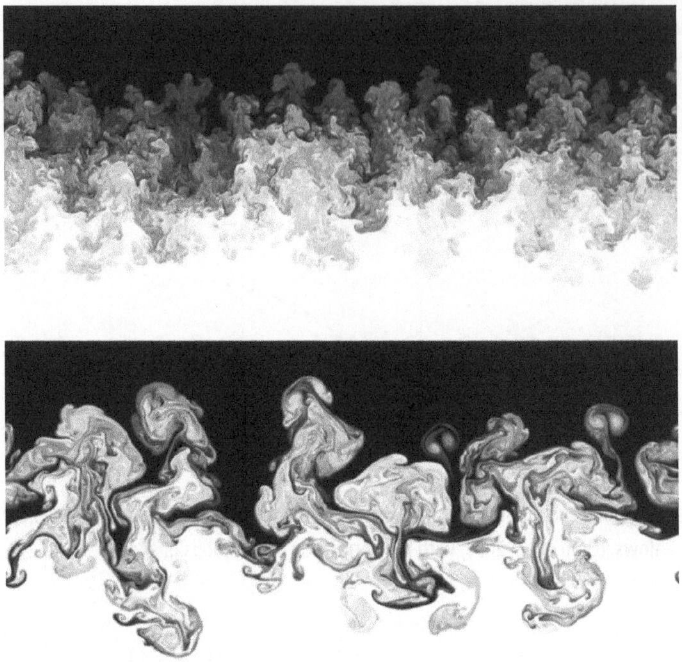

Fig. 24.10 Vertical slice of density from the 3D simulation of RTI (top) compared with the density field from a 2D RTI simulation (bottom), both at dimensionless time 25. Black represents the heavy fluid, and white the light fluid. Fig. 2 of Cabot (2006), with the permission of AIP Publishing.

On the other hand, the 2D mixedness, after a brief rise, declines to a level of $\Xi = 0.53$, independent of initial conditions (Fig. 5 of Cabot (2006), not shown). Kuchugov et al. (2014) also present the results of their 2D and 3D numerical simulations in support of the last point regarding the mixed width. Zhao et al. (2022) analyzed the energy transfer using coarse-graining and showed that 2D multi-mode RTI has a positive feedback loop that is absent in 3D. In 2D, the potential energy released cascades to larger scales, which in turn drives a large-scale quasi-isotropic overturning circulation (absent in 3D), which releases even more potential energy. This feedback loop in 2D results in a faster growth of the mixed width compared to 3D. However, the atomic mixing in 2D is lower than in 3D (Cabot, 2006).

The findings discussed above are consistent with the concept of an inverse cascade in 2D flows. To demonstrate this point, it is instructive to evaluate each term in Eq. 24.1 for 2D and 3D RTI flows. During the initial stages of the flow's development, baroclinicity is the dominant mechanism for generating enstrophy, while at late times in the 3D flow the vortex-stretching term dominants. Only the baroclinic production operates in the 2D flow and a substantially lower amount of enstrophy is generated (Fig. 24.12).

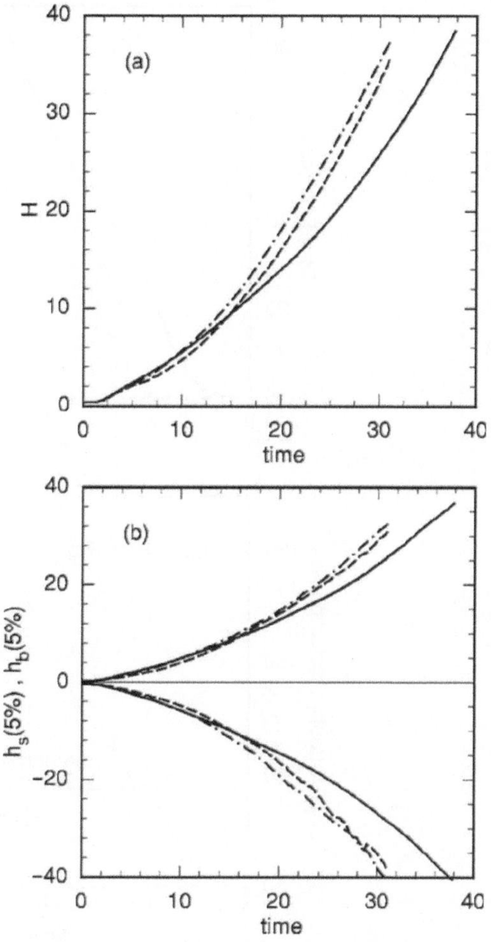

Fig. 24.11 (a) Integral product height and (b) 5% threshold values of mean density for the 3D (solid lines), and 2D (dashed and dot-dashed lines) simulations. The standard deviation in the 2D simulations is about 7% in (a) and 10% in (b) at late times. Fig. 3 of Cabot (2006), with the permission of AIP Publishing.

Figures 24.13–24.14 show that any given class of initial conditions probably leads to an "envelope" of values for the RTI growth parameter, α. Indeed, Clark (2003) indicated that as many as 104 independent realizations at 1000^2 grid resolution were required to achieve reasonable statistical convergence for 2D RTI simulations. Therefore, the statistics of the 2D case are probably more sensitive to outliers due to an exaggerated emphasis on bubble-merger and eddy-merger events.

Finally, Yu and Livescu (2008) found that 2D circular calculations of RTI growth do not seem to be a good model for the fully 3D cylindrical case.

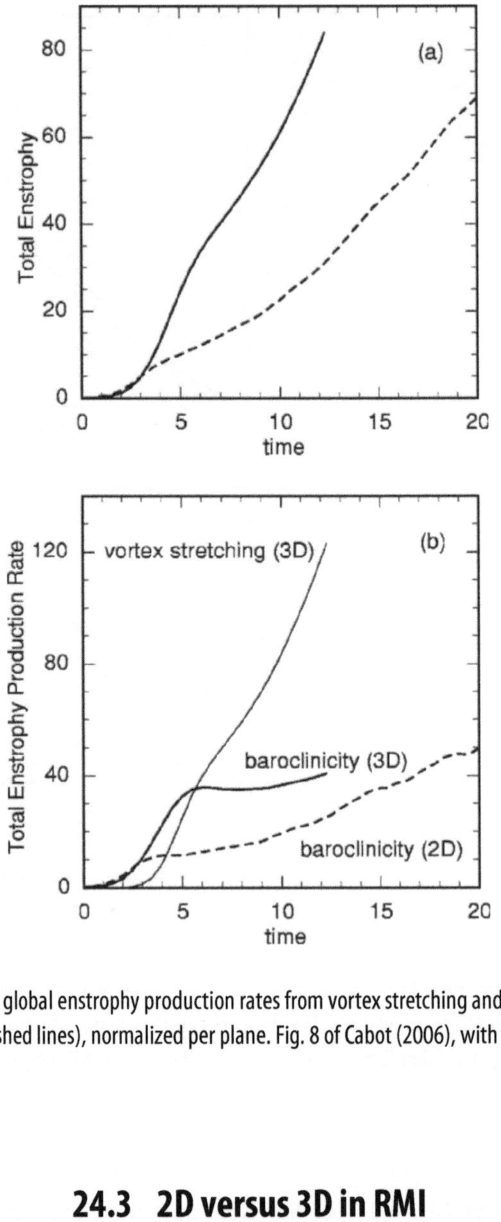

Fig. 24.12 (a) Global enstrophy and (b) global enstrophy production rates from vortex stretching and baroclinicity from 3D DNS (solid lines) and 2D DNS (dashed lines), normalized per plane. Fig. 8 of Cabot (2006), with the permission of AIP Publishing.

24.3 2D versus 3D in RMI

According to the linear analysis of Meyer and Blewett (1972) for a heavy/light interface, the ratio of 3D and 2D RMI growth rates is $\sqrt{2}$. By comparing to computation of a single-mode, single shock RMI, Li and Zhang (1997) studied the validity of nonlinear analysis for the perturbation amplitude in both two and three dimensions. Mikaelian (1998) also studied RMI in two and three dimensions using analytic formulas for the nonlinear evolution of the bubble amplitude. Here, the generalized Layzer-type model predicts that 3D perturbations grow at a rate $3/2$ times that of 2D perturbations.

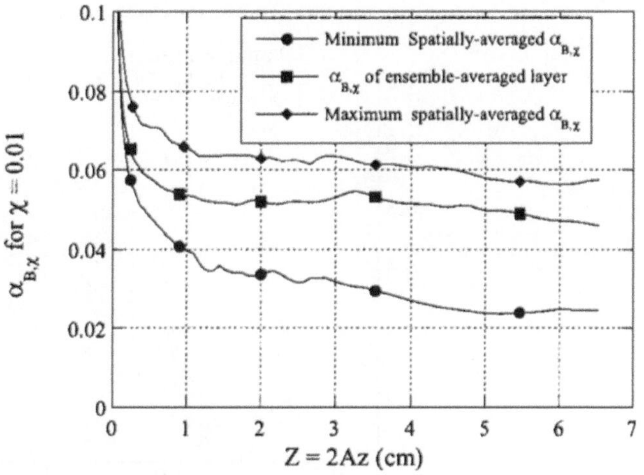

Fig. 24.13 Evolution of the bubble-growth parameter, showing extrema within the ensemble of simulations as well as the value for the ensemble-averaged layer. Fig. 3 of Clark (2003).

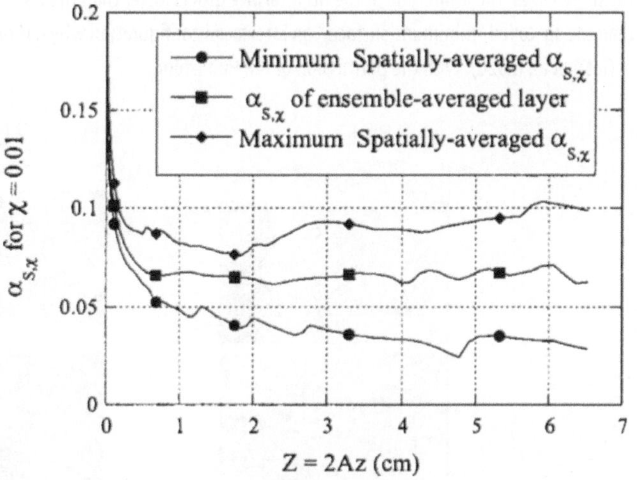

Fig. 24.14 Evolution of the spike-growth parameter, showing extrema within the ensemble of simulations as well as the value for the ensemble-averaged layer. Fig. 5 of Clark (2003), with the permission of AIP Publishing.

The numerical results shown in Fig. 24.15 confirm that the amplitude of the 3D interface grows faster than the 2D interface (Guan et al., 2020). The last case considered in Fig. 24.15, the so-called "minimum-surface" configuration defined by the soap film property, is inspired by Krechetnikov (2009), who emphasized the importance of the interfacial curvature on RMI (see also, Luo et al. 2016a).

Yosef-Hai et al. (2003) studied the late-time growth rate of the RMI experimentally in a shock tube with 2D and 3D single-mode initial perturbations (Fig. 24.16). To make the

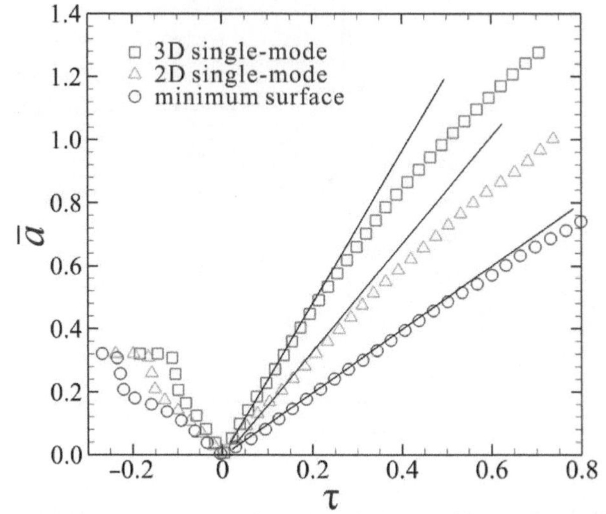

Fig. 24.15 Amplitude growth histories of a 3D single-mode interface (3D, squares), a 2D single-mode interface (2D, triangles) and the present minimum surface (3D, circles). Black solid lines denote the theoretical impulsive growth predictions of the three cases. The amplitude of the 3D interface grows faster than the 2D interface, which is in turn grows faster than the so-called "minimum-surface," an interfacial configuration which is defined by the soap film property. Fig. 9 of Guan et al. (2020), with the permission of AIP Publishing.

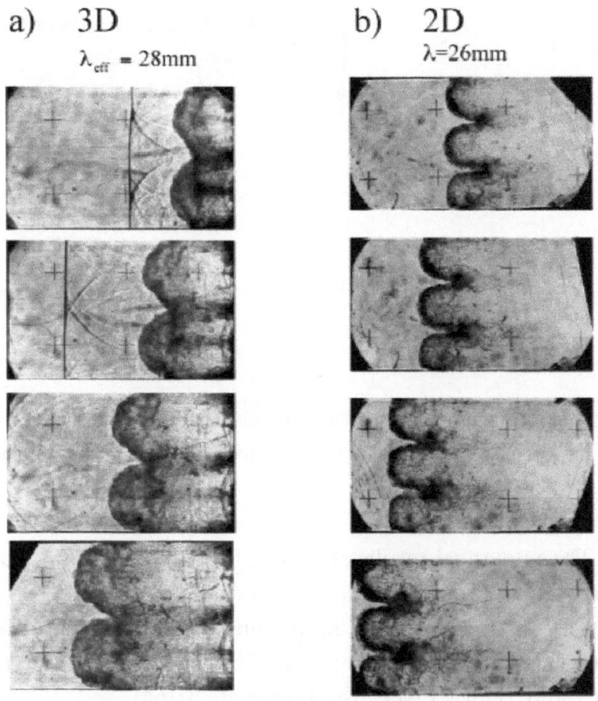

Fig. 24.16 Schlieren photographs for the dimensionality investigation for RMI induced flows. Fig. 5 of Yosef-Hai et al. (2003).

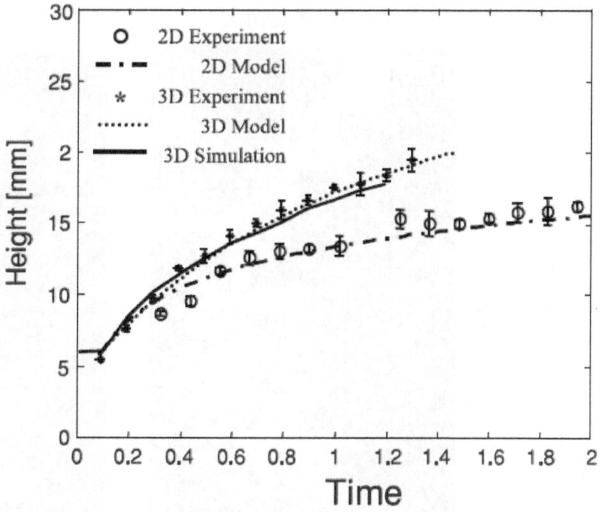

Fig. 24.17 Bubble height versus time for 2D and 3D experiments for RMI induced flows, along with the bubble-competition model and a 3D simulation. Fig. 6 of Yosef-Hai et al. (2003).

comparison clearer, the effective wavelengths were chosen to be similar in both experiments, with an Atwood number of 0.7 (air to SF_6) and an incident Mach number of 1.2. The experiments focused on the interaction between adjacent 3D bubbles in the nonlinear stage of a dual-mode initial perturbation, an interaction that is the basis for the bubble-merger picture of a multi-mode initial perturbation's evolution (Oron et al., 2001). The bubble height was measured from the Schlieren photographs and plotted versus time in Fig. 24.17. There is very good agreement between the experimental results, the predictions of the model, and the numerical results, confirming the different behavior of 2D and 3D bubbles as manifested in their asymptotic velocities.

A comparison of 2D and 3D RMI results for 1024^2 and 1024^3 simulations (Cohen et al., 2002) illustrates the difference in dynamics for one of the shock-tube experimental settings of Vetter and Sturtevant (1995). Here the shock passes from the low- to the high-density fluids and the membrane is initially on the same side of the mesh as the shock. Figure 24.18 shows a rendering of the entropy field for a 2D simulation and for a vertical slice of a 3D simulation. The contrast is evident: a predominantly fine-scale structure in the slice of the 3D simulation (left panel) versus predominance of sizable structures with sharp edges in the 2D results (right panel).

Indeed, the 2D simulations are characterized by coalescence into structures comparable to, or larger than, the initial short-wavelength scale, while the 3D simulations evolve toward fine scales. We see that the high-k value of the energy flux is negative in the former case and positive in the latter, consistent with the dominance of inverse cascade dynamics in two dimensions and forward cascade in three dimensions (Fig. 24.19) (Batchelor, 1969, Kraichnan, 1973, Leith, 1971, Kraichnan and Montgomery, 1980).

The RMI integral mix layer width for the initially narrowband and broadband perturbations is shown in Fig. 24.20. The 2D integral width is initially lower than that of the 3D

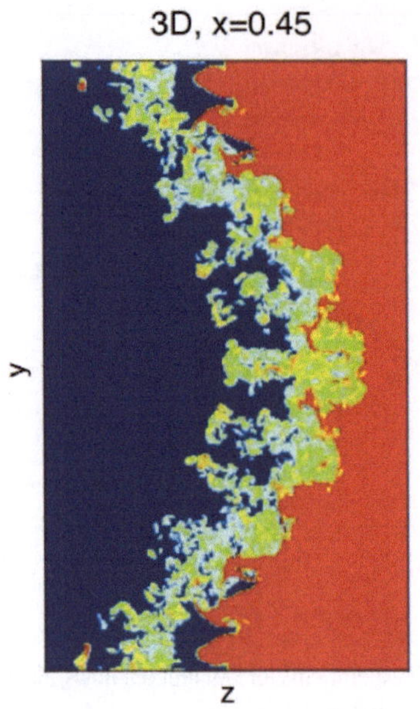

Fig. 24.18 Entropy from a 2D RMI simulation at 1024^2 resolution (right) versus a slice of a 3D simulation at 1024^3 resolution (left), both at dimensionless time 9. Fig. 11 of Cohen et al. (2002).

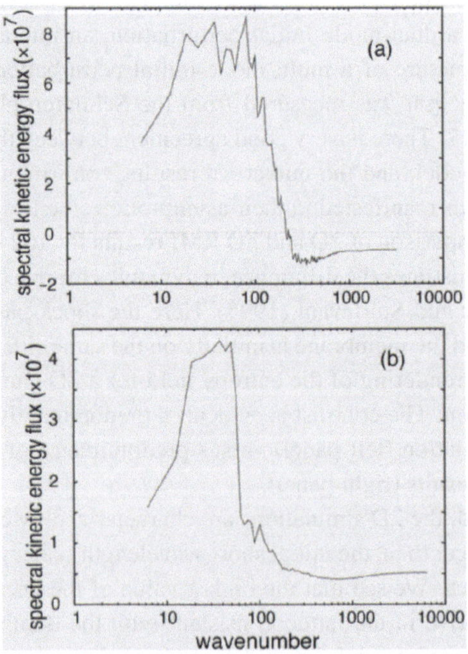

Fig. 24.19 Spectral flux in a 2D RMI simulation at 1024^2 resolution (a) versus 3D simulation at 1024^3 resolution (b), both at dimensionless time 9. Fig. 13 of Cohen et al. (2002).

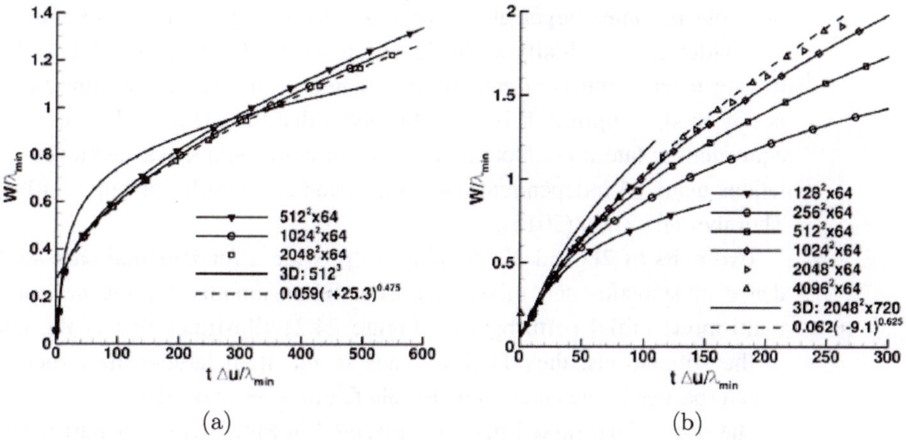

Fig. 24.20 The integral mix width plotted for narrow- (left) and broad- (right) band initial perturbations. Fig. 5 of Thornber and Zhou (2015).

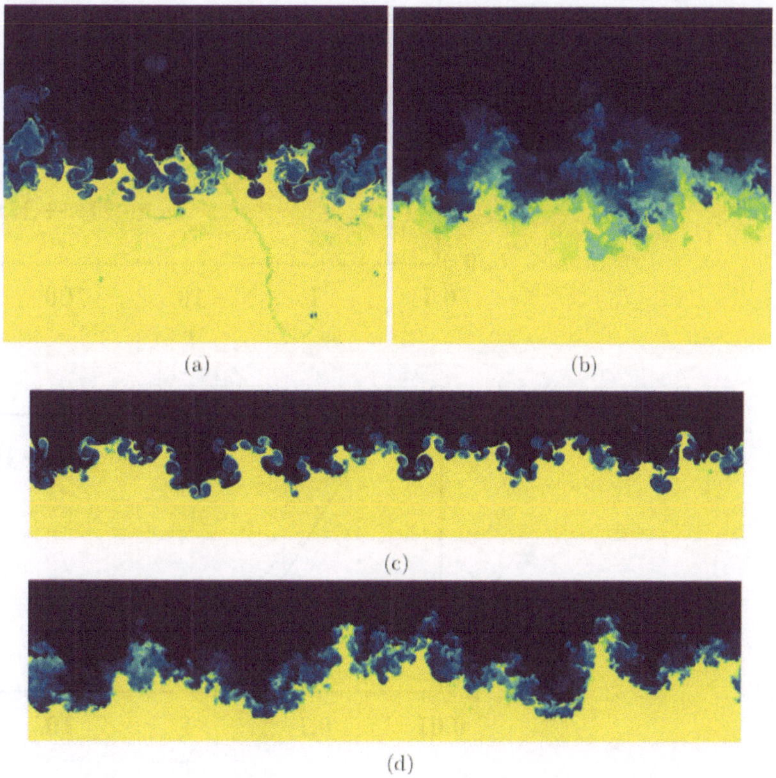

Fig. 24.21 Visualization of the latest time for the 2D and 3D cases, (a) and (b) are narrowband, and (c) and (d) are broadband initial perturbations. Fig. 1 of Zhou et al. (2020a).

flow, but becomes larger at late time due to its higher growth exponent. The broadband mix width grows initially at the same rate as the 3D case, which is to be expected as the mixing layer width is effectively dominated by the currently saturating nonlinear mode, not by mode coupling. It is clear, however, that the 2D flow does not converge to the 3D equivalent at late time. Note that statistical convergence was achieved for 2D RMI simulations using 64 independent realizations and grid resolutions up to 4096^2 in the plane by Thornber and Zhou (2015).

Two pairs of 2D and 3D simulations provide a fundamental case study for the dependence of statistics and mixed mass on dimensionality for the narrow- and broad-band multi-mode initial perturbations. Figure 24.21 illustrates that at the latest time reached in the simulations, the 2D flow tends to maintain large-scale structures, while its 3D counterpart generates more small-scale features, as expected.

The normalized mixed mass is presented in Fig. 24.22. For narrowband perturbations, the normalized mixed mass is substantially lower in two than three dimensions, consistent

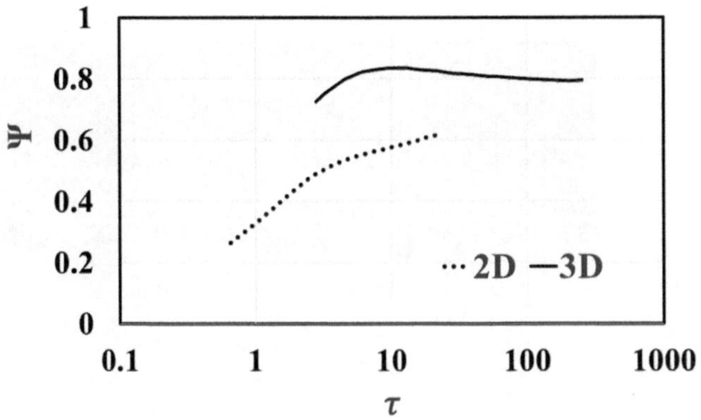

(a)

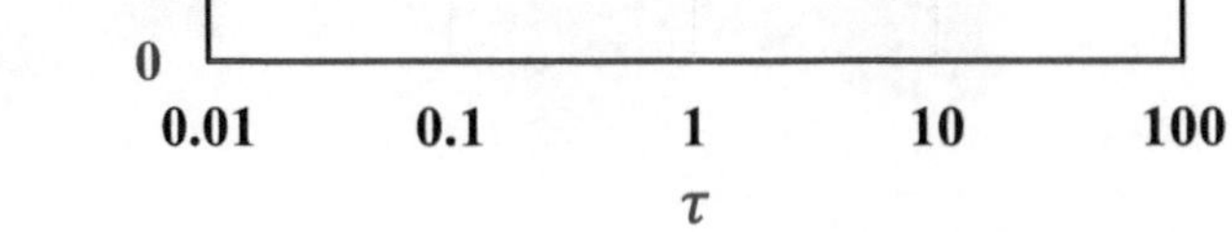

(b)

Fig. 24.22 2D and 3D normalized mixed mass (Ψ) for (a) narrowband and (b) broadband initial perturbations. 2D (dotted lines) and 3D (solid lines). Fig. 12 of Zhou, Groom, and Thornber, (2020a).

with previous observations on the mixing parameters. The broadband cases are closer (Zhou et al., 2020a).

24.4 2D versus 3D in reshocked RM

The numerical study of Probyn et al. (2014) details the development of the mixing layer width both before and after reshock for 2D and 3D single-mode RMI using an initial setup comparable to that of existing experimental work (Collins and Jacobs, 2002). In this experiment, a sinusoidal interface between the two gases is formed by oscillating the entire shock tube in the horizontal direction. The instability is generated by two incident shock strengths of Mach 1.11 and 1.21. The simulation end wall is an inviscid wall to ensure that a reshock occurs at the appropriate time, as in the experiment. The validated code is subsequently used to evaluate the growth rate for a range of different initial conditions.

In a related work, Tritschler et al. (2013) compared 2D simulations to experimental results of Jacobs and Krivets (2005). Note that Jacobs and Krivets used the same experimental setup to repeat the Collins' experiment, but with a stronger incident shock (Mach 1.3). As larger Mach numbers promote faster growth, they were able to obtain information on the late-time development of the single-mode instability.

More recently, the growth dynamics of the 2D and 3D single-mode reshocked RMI of Jacobs and Krivets (2005) was further investigated systematically by Latini and Schilling (2020), using data from high-resolution ILES. Figures 24.23 and 24.24 show the arrival of the reflected shock at 5.6 ms, the reflected shock refracting at the interface at 5.7 ms and 5.8 ms, and the inception of the inversion process at 5.9 ms, when bubbles transform into spikes and vice versa. The differences between the 2D and 3D cases is again attributable to the absence of vortex stretching in two dimensions, in which coherence of the flow structures persists (Latini and Schilling, 2020).

Figure 24.25 shows the 2D and 3D simulated mixing layer amplitudes and the predictions of the Mikaelian, Brouillette–Sturtevant, Charakhch'yan, and Lombardini et al. reshock models. The 3D preshock amplitude is adjusted so that the corresponding initial velocity is the same as in the 2D simulation, All of these models predict linear growth in time. The models are intended to apply immediately after reshock ($\approx$ 5.77 ms) until the arrival of the reflected rarefaction wave at $\approx$ 7.0 ms. The model predictions are in reasonable agreement with the simulation data.

Based on the experimental setups of Leinov et al. (2009), Ukai et al. (2011) numerically analyzed single- and multi-mode RMI in 2D and 3D domains. The comparisons of growth rates for each of the simulations are shown in Fig. 24.26. Each study shows distinctly different growth rates after reshock. In particular, the single-mode cases tend to have larger growth rates in both 2D and 3D.

Figure 24.26 shows that 3D multi-mode RMI predicts a growth rate in very good agreement with the experimental value (Leinov et al., 2009). Yet, 2D multi-mode RMI realizes very small growth rates compared to the 3D multi-mode RMI, as also reported by Grinstein et al. (2011).

Olson and Greenough (2014) provided another comparison between 2D and 3D reshocked RMI. The authors found that the growth rate of the 2D flow outpaces that of the 3D after reshock for the duration of the calculation (Fig. 24.27). The larger mixing width is easily observed in Figure 24.28, which shows large 2D structures nearly encompassing the entire width of the image. Note that for the 2D case, only 1/5th of the actual width (y-direction) is shown.

The development of the multi-mode mixed width in 2D and 3D is compared in Fig. 24.29. The mixed width initially grows at a slightly faster rate in 2D after the first shock, but after reshock, the 2D mixed width grows at a much higher rate than the 3D mixed width (Wong et al., 2019). This result is consistent with the finding of Olson and Greenough (2014) but differs from that of Ukai et al. (2011) and Grinstein et al. (2011). This difference

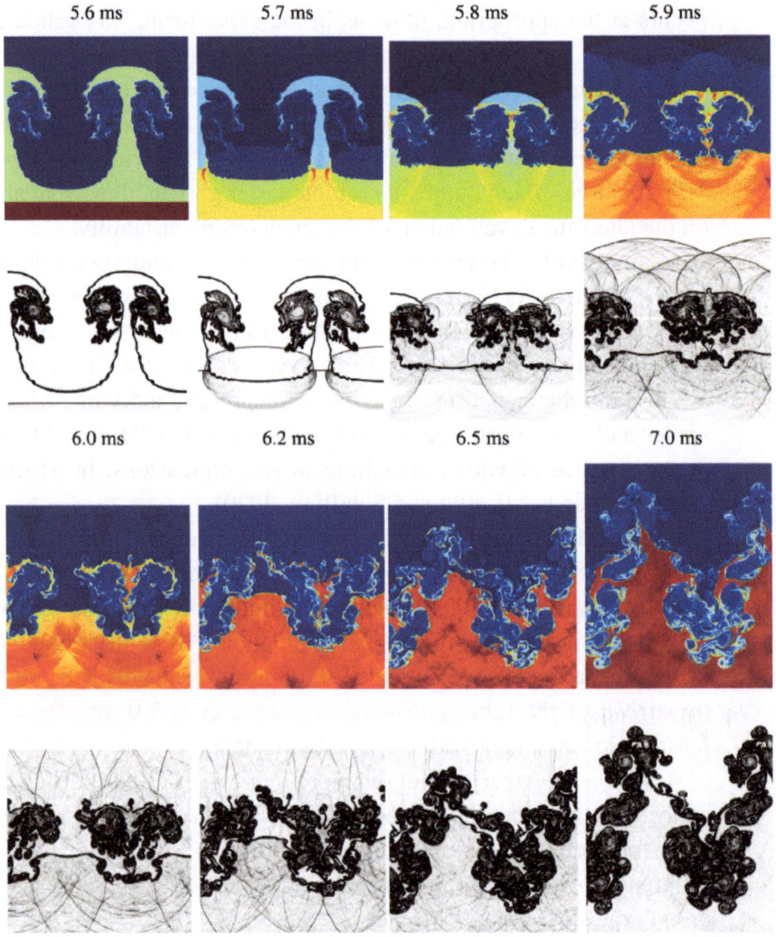

Fig. 24.23 Time-evolution of the density and simulated density Schlieren fields from the WENO simulation illustrating the reshock process at 5.6, 5.7, 5.8, 5.9, 6.0, 6.2, 6.5, and 7.0 ms. Reshock occurs at $\approx$ 5.65 ms. Fig. 9 of Latini and Schilling (2020).

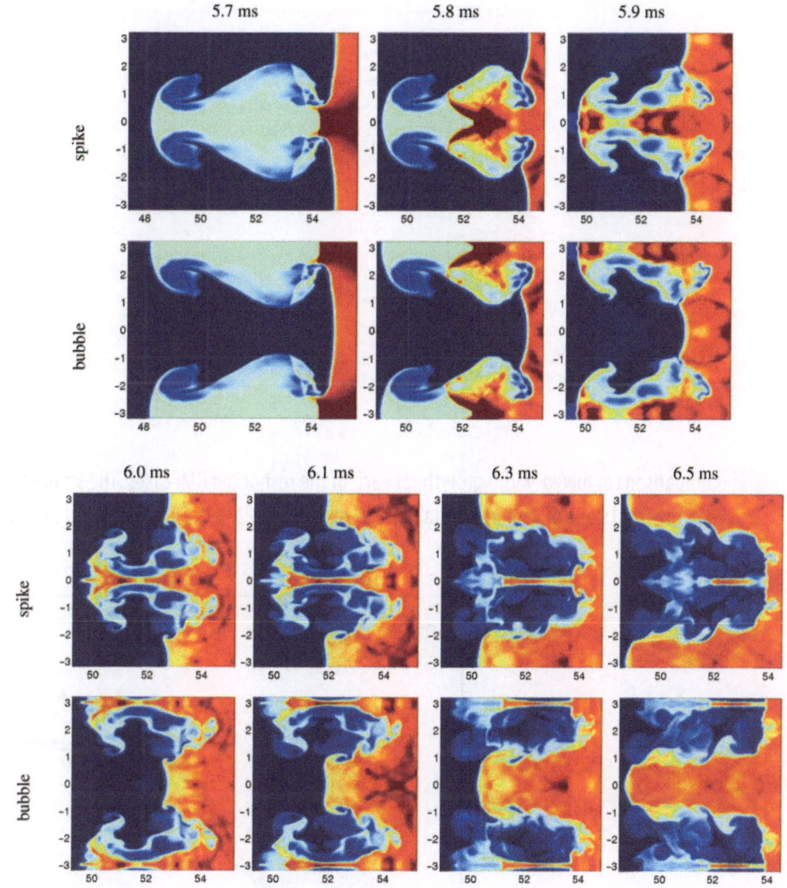

Fig. 24.24 Time-evolution of the density in the (x, y)-plane from a 3D simulation of Jacob reshock experiments. Both the evolution of the bubble and spike are shown. Fig. 21 Latini and Schilling (2020).

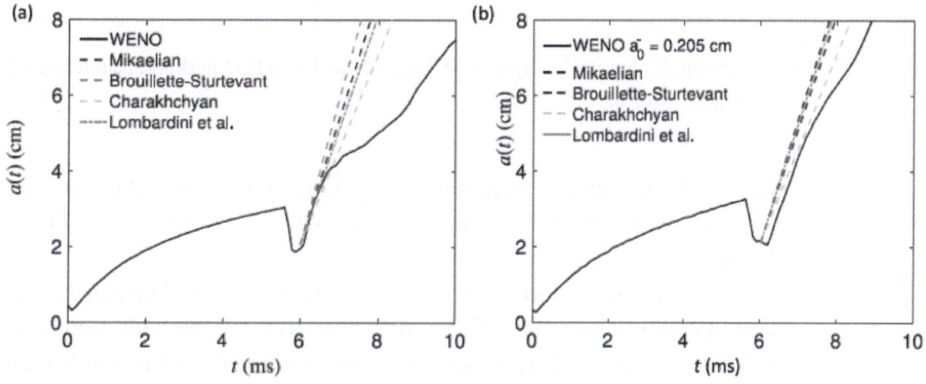

Fig. 24.25 The mixing layer amplitude versus time from the WENO simulations in (a) 2D and (b) 3D together with the predictions of the linear reshock models. Figs. 12 and 23 of Latini and Schilling (2020).

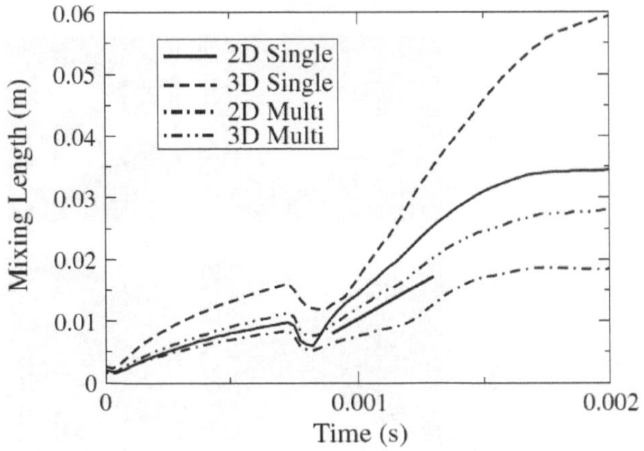

Fig. 24.26 Comparisons of mixed-width growth for each of the reshocked RMI cases. The straight-line represents the experimental slope of 23.1 m/s, obtained by Leinov et al. (2009). Fig. 8 of Ukai et al. (2011).

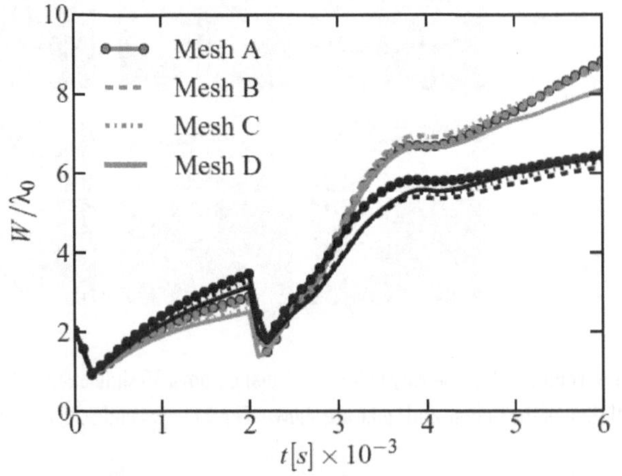

Fig. 24.27 Mixed widths vs. time in 2D (gray) and 3D (black) for the RMI with reshock at various resolutions. Fig. 2 of Olson and Greenough (2014).

may be due to differences in initial conditions or perhaps a dependence on the simulation box size, which tends to rapidly saturate with large 2D modes (Olson and Greenough, 2014).

Comparing the 3D cases with different transport coefficients on the same figure, it can be seen that although the growths of the normalized mixing widths are very similar just after first shock, they become smaller with decreasing Reynolds number until reshock. After reshock, the growth rates that are initially different become very similar near the end of the simulations among all 3D cases, as shown in Fig. 24.29(b).

The mixedness of the 2D case is smaller at all times than that of the 3D case with physical transport coefficients (Fig. 24.30). At late times after reshock, mixedness for all

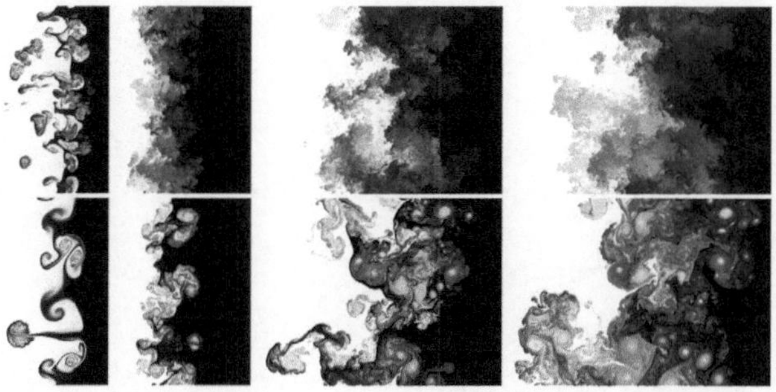

Fig. 24.28 Contours of the mass fraction of SF_6 from white (0.0) to black (0.8) in 3D (top) and 2D (bottom) showing the qualitative difference. Temporal evolution is from left to right for times $t = 2.0, 2.5, 4.0, 6.0$ ms. Reshock occurs at $t = 2.1$ ms and first shock at $t \approx 0.05$ ms. Fig. 1 of Olson and Greenough (2014).

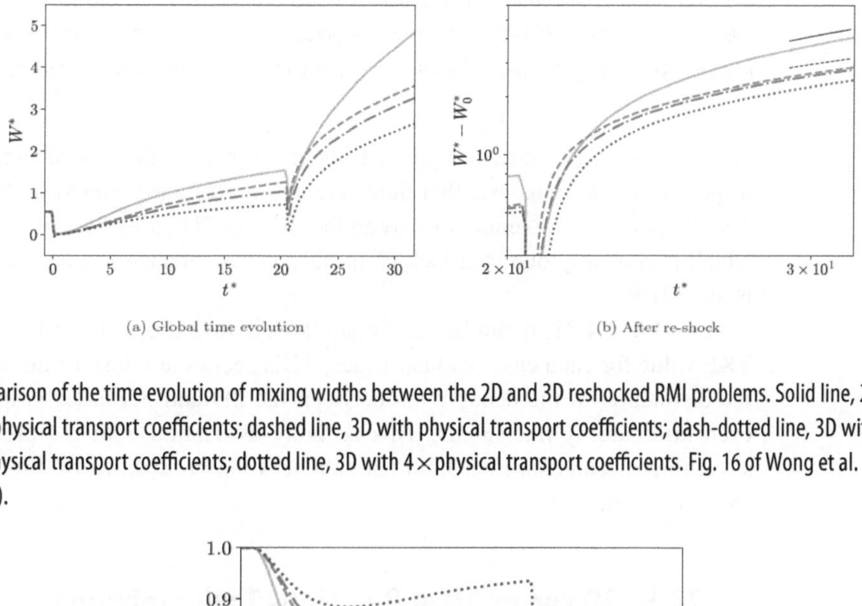

Fig. 24.29 Comparison of the time evolution of mixing widths between the 2D and 3D reshocked RMI problems. Solid line, 2D with physical transport coefficients; dashed line, 3D with physical transport coefficients; dash-dotted line, 3D with 2 × physical transport coefficients; dotted line, 3D with 4 × physical transport coefficients. Fig. 16 of Wong et al. (2019).

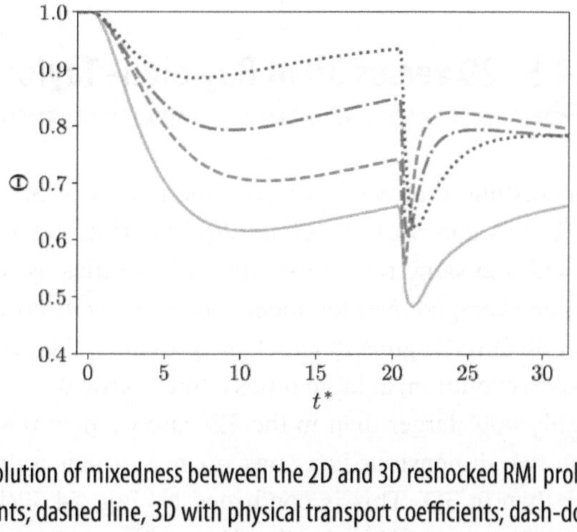

Fig. 24.30 Comparison of the time evolution of mixedness between the 2D and 3D reshocked RMI problems. Solid line, 2D with physical transport coefficients; dashed line, 3D with physical transport coefficients; dash-dotted line, 3D with 2 × physical transport coefficients; dotted line, 3D with 4 × physical transport coefficients. Fig. 17 of Wong et al. (2019).

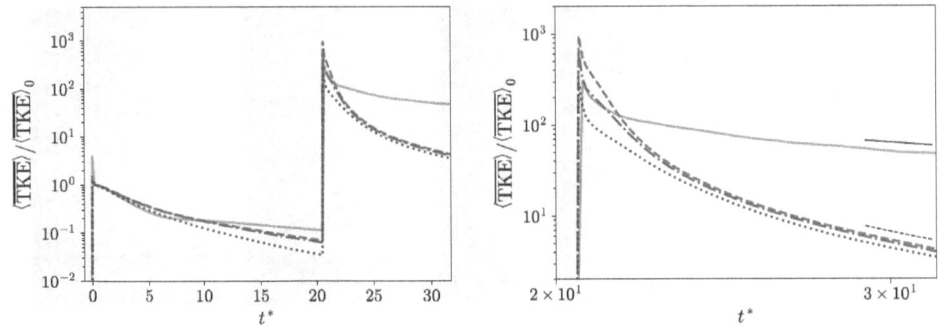

(a) Global time evolution (b) After re-shock

Fig. 24.31 Time evolution of the normalized means of turbulent kinetic energy, TKE, in the central part of mixing layer for 2D and 3D reshocked RMI problems. The mean TKE is normalized by its initial value. Solid line, 2D with physical transport coefficients; dashed line, 3D with physical transport coefficients; dash-dotted line, 3D with 2 × physical transport coefficients; dotted line, 3D with 4 × physical transport coefficients. The thin solid and dashed lines indicate the post-reshock decaying scalings, −0.5 and −1.4, respectively. Fig. 21 of Wong et al. (2019).

3D and 2D cases seems to approach the asymptotic values of around 0.8 and 0.65–0.7, respectively. This suggests that fluids are less molecularly mixed in the 2D configuration. The difference in mixedness between the 2D and 3D cases becomes larger after reshock, which is probably due to a lack of turbulence to enhance mixing in the 2D case (Wong et al., 2019).

From Fig. 24.31, it can be seen that after the first shock, there is a sudden jump in the TKE value for each case. At later times, TKE decays at a faster rate in 3D. After reshock, the distinction in the decay rates between the 2D and 3D cases is much larger, with an even faster decay rate in 3D. This is likely associated with the presence of the vortex stretching mechanism in 3D, which enhances the breakdown of large-scale features into smaller scales.

24.5 2D versus 3D of Rayleigh–Taylor mixing in porous media

It is interesting to directly compare the results of 3D simulations with corresponding 2D cases. To this aim, Boffetta et al. (2020) performed a set of numerical simulations for 2D flows with the same resolutions and diffusivities as those of 3D. The results in this case have been averaged over ten independent realizations (Fig. 24.32(a)).

After an initial regime in which the growth of the vertical scale $h(t)$ is faster than ballistic (diffusive) evolution, at large times a linear growth is observed in 2D, but with a coefficient α roughly 40% larger than in the 3D case ($\alpha_{2D} \approx 0.84$). The faster growth of the mixing layer in two dimensions is a consequence of density plumes that are more elongated and thinner than in 3D. This is confirmed by Fig. 24.32(b), which shows that the horizontal correlation scale in 2D is smaller than in 3D.

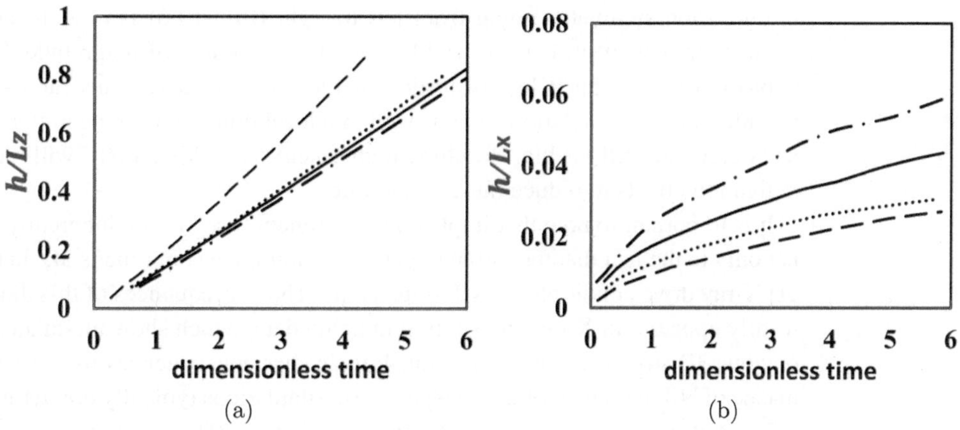

Fig. 24.32 Time evolution of the components of the mixing layer for 3D simulations of RT mixing in porous media at three Péclet numbers, $Pe = \mathcal{LU}/D$: 0.8×10^4 (dot-dashed lines), 1.6×10^4 (solid lines), and 3.1×10^4 (dotted lines). The dashed line represents the results for the 2D case with $Pe = 3.1 \times 10^4$. (a) the mixing layer width $h(t)$. (b) the horizontal correlation scale $h_x(t)$. Here, L_z and L_x are the vertical and horizontal sizes of the simulation box, respectively. Figs. 2 and 3 of Boffetta et al. (2020).

24.6 2D versus 3D in ICF

As noted previously in Section 19.2.4, satisfactory quantitative comparisons between simulation and experimental data from high-convergence NIF implosions have only proven feasible for fully 3D, multi-effect simulations (Clark et al., 2013, 2015, 2016, 2017). Examples of such 3D simulations of recent NIF implosion experiments are illustrated in Fig. 19.6. In each instance, the rendering shows a cutaway from the simulation at the time of peak neutron production ("bang time") where the left cutaway shows ion temperature, the right cutaway mass density, and the outer bounding surface is the ablation front of the imploding shell. Progressing from left to right are simulations of a low-foot implosion (Edwards et al., 2013), a high-foot implosion (Dittrich et al., 2014), and a recent implosion using high-density carbon (HDC) in place of the previous plastic ablators (Le Pape et al., 2018; Baker et al., 2018).

Evident in each of the simulations referenced above are deviations from sphericity due to the growth of hydrodynamic instabilities and also due to long-wavelength asymmetries in the X-ray drive. The defects seeded by the capsule support tents, the ring-like distortions surrounding the north and south poles, are particularly evident in the low- and high-foot implosions, as are the broadband ablation front modulations in the low-foot case. These are both a direct consequence of the coupled RM and RT instability growth in these implosions. Indeed, the under-performance of the low-foot implosions compared to expectations is now principally understood to be due to strong ablation front instabilities, in some cases leading to the mixing of ablator material deep into the hot spot, thereby poisoning the potential fusion burn (Regan et al., 2013; Ma et al., 2013). The substantial improvement in implosion symmetry going from left to right (Fig. 19.6) is obvious and corresponds to an improvement in neutron yield of nearly two orders of magnitude. While the HDC implosion shows much less distortion of the hot spot, it remains far from perfect, and modulations of the ablation front as well as a local thinning of the confining shell from the tent defect are still visible. Reaching higher neutron yields on NIF will require continued mitigation efforts to reduce these instabilities.

It is important to note that implosion experiments on NIF are inherently 3D. This is due not only to the fact that the initial target configuration at shot time is 3D, but also due to the 3D X-ray drive conditions in NIF implosions. The consequences of this dimensionality are readily apparent in X-ray and neutron imaging data, which show modulated hot spots with obvious 3D structure. The dimensionality also has consequences for the simulated performance of NIF targets, with 2D axisymmetric simulations typically predicting yields that are a factor of two or more higher than either analogous 3D simulations or experiments. This difference can be ascribed to the fundamental difference in stagnation dynamics between 2D axisymmetric flows and realistic 3D flows. In 2D flows, material reaching the symmetry axis is, by construction, met by an exactly opposing flow. This perfect momentum balance maximizes the conversion of shell kinetic energy to hot spot internal energy, and hence maximizes the compression and heating of the hot spot material, thereby enhancing fusion reactivity. In realistic 3D flows, this perfect momentum balancing is unrealized, rendering the conversion of kinetic to potential energy less efficient, with more energy left in useless, residual hot spot flows. The well-known higher growth rate of nonlinear RT instabilities (Layzer, 1955) in 3D versus 2D also contributes to the differences. Specifically, the development of weak spots in the confining shell at ignition time that can rupture and cause a loss of confinement happens more rapidly in realistic 3D geometries. This leads to shorter burn durations and lower hot spot pressures.

Figure 24.33 illustrates many sources of ICF instabilities (Smalyuk et al., 2017, 2018). One of these sources is the jet of ablator material injected into the hot spot via the fill tube. Indeed, it was found that interaction of target defects with the radiation drive can seed 3D flows. Contaminants prevent efficient compression and heating of the fuel. Again, fuel contamination in ICF involves complex 3D flows, as the added degree of freedom reduces the simulated neutron yield. The simulations that account for these effects more closely represent the experimental data, while 2D simulations under-predict the impact (see Fig. 24.34) (Haines et al., 2017, 2019).

24.7 2D versus 3D in supernovae

Janka et al. (2016) stated that "supernova theory and 2D simulations are not in a marriage of love but rather a partnership of convenience that was enforced by the limitations of computing resources." Importantly, for numerical simulations to capture the effects of turbulence properly, 3D models are required. Historically, 1D models routinely either fail to explode without ad hoc prescriptions such as a mixing-length algorithm (e.g., Wilson, 1985) or else explode weakly as in the lower-mass O-Ne-Mg progenitor case (e.g., Kitaura

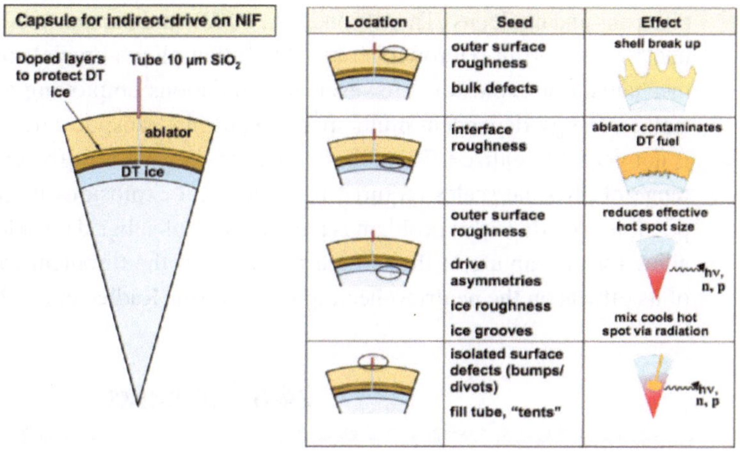

Fig. 24.33 Schematic of an ICF capsule showing the locations of unstable interfaces, seeds for instability growth, and effects on implosion performance. Reproduced with permission from Smalyuk et al. (2017). Copyright 2017 Institute of Physics.

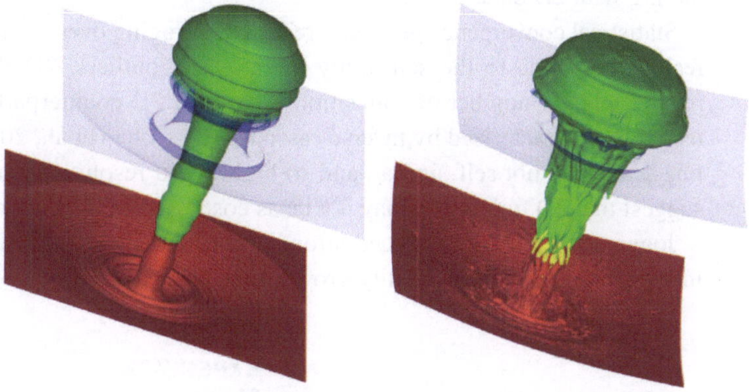

Fig. 24.34 Comparison of material distribution in the fill tube jet at 4.8 ns for both 2D (synthetically rotated; left) and 3D (right) simulations. Red: ablator interfaces; blue: DT wetted foam/DT gas interface; green: fill tube glass; and yellow: glue. Reproduced with permission from Haines et al. (2019). Copyright 2019 AIP Publishing.

et al., 2006). With 2D axisymmetry, turbulence exhibits a well-known inverse cascade from small scales to large (Kraichnan, 1967), which may artificially aid the model to explode. Moreover, a popular simplification of neutrino transport could lead to the misattribution of the standing accretion shock instability (SASI) as the aiding factor (Skinner et al., 2016). This so-called "ray-by-ray" approach (see, e.g., Marek and Janka, 2009) tends in 2D to exaggerate the axial sloshing observed in those simulations, aiding explosion.

Fully 3D numerical simulations of core-collapse supernovae (ccSNe) are now in their adolescence. Simulations like those of Vartanyan et al. (2019) are able to capture the multi-dimensional effects of turbulence and various other important physics such as general-relativistic gravitational effects (Rampp and Janka, 2002), inelastic neutrino scattering off electrons and nucleons (Thompson et al., 2003), and the many-body correction to neutrino–nucleon scattering (Horowitz et al., 2017) that play a crucial role in the neutrino-heating mechanism of explosion. However, in simulations employing all the relevant physics, as well as energy-dependent multi-species neutrino transport, the character of the turbulence is not yet fully realized. This usually owes to the lack of sufficient resolution to capture the range of physical scales required for successful explosions in global ccSNe models. One possible remedy to this problem is the inclusion of subgrid turbulence models, which would allow for the capture of the full inertial range of the turbulent cascade and the assessment of its effects on the neutrino-heating mechanism (Radice et al., 2018).

24.8 Remarks

It is worth noting that molecular mixing occurs at fluid interfaces, and the area-to-volume ratios in three-dimensions are very different from the length-to-area ratios in two-dimensions. Therefore, care must be taken when attempting to assess realistic 3D mixing with 2D data.

Statistical convergence generally requires averaging over a suitable number of flow field realizations. Due to the sensitivity to statistical outliers, 2D RTI and RMI simulations require a larger number of realizations than their 3D counterparts. Simulations in 2D also tend to be characterized by inverse cascades (e.g., Zhao et al., 2022), which means that the fine-scales are not self-similar, and so higher grid resolutions are required. These factors suggest that 2D turbulence may not be as cost-effective as first envisioned.

Joggerst et al. (2014) argued strongly that one should not take the results of any 2D investigation of fluid instability growth too seriously. 2D geometries experience artificial

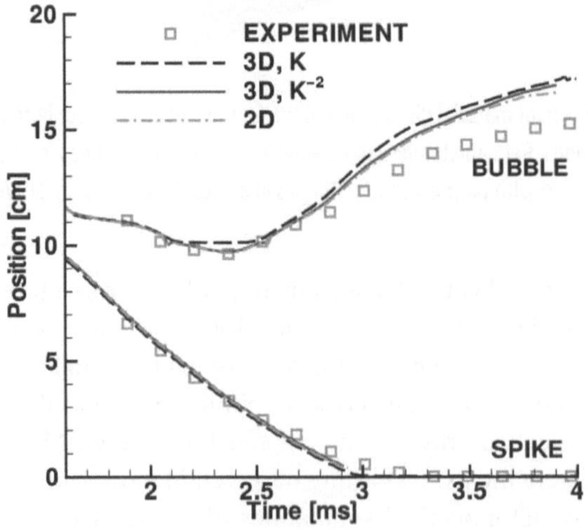

Fig. 24.35 Bubble and spike positions from experiment and simulations for inverse chevron experiments. Fig. 7 of Hahn et al. (2011).

drag forces that can slow the growth of instabilities (Hammer et al., 2010; Joggerst et al., 2010). The artificial drag forces will be different in 2D cylindrical and 2D axisymmetric calculations, as well. To that end, the authors stressed, these simulations should not be taken as realistic representations of the growth rate of instabilities in a 3D implosion geometry. As a result, 3D calculations are necessary to accurately determine the growth rate.

And yet, in some cases, 2D simulations can produce a reasonable representation of a 3D system for some integrated measurements. As shown in Fig. 15.2, large-scale features of the inverse chevron experiments (Section 15.1) are not very sensitive to the characteristics of the small-scale perturbations. If one is satisfied with the bubble and spike positions, Fig. 24.35 shows that by restricting the simulation to two dimensions and removing any additional perturbation produces almost identical curves to those from a fully 3D simulation. Hence, 2D simulations are useful in giving some indication of the amount of instability growth at an interface in such a case. The small-scale structures are, of course, very different. Thus, if one is interested in the characteristics generated by the small-scale perturbations, it is essential to consider the fully 3D flow (Hann et al., 2011).

It is our hope that this chapter further enhances the message in Zhou et al. (2019) that one must exercise caution and evaluate on a case-by-case basis the applicability of 2D RTI and RMI simulations for engineering applications such as inertial confinement fusion.

References

Abarzhi, S.I., Nishihara, K., and Glimm, J. 2003. Rayleigh–Taylor and Richtmyer–Meshkov instabilities for fluids with a finite density ratio. *Phys. Lett. A*, **317**, 470.

Abd-El-Fattah, A.M., and Henderson, L.F. 1978a. Shock waves at a fast-slow gas interface. *J. Fluid Mech.*, **86**, 15.

Abd-El-Fattah, A.M., and Henderson, L.F. 1978b. Shock waves at a slow-fast gas interface. *J. Fluid Mech.*, **89**, 79.

Abd-El-Fattah, A.M., Henderson, L.F., and Lozzi, A. 1976. Precursor shock waves at a slow – fast gas interface. *J. Fluid Mech.*, **76**, 157.

Abdikamalov, E., Ott, C.D., Radice, D., Roberts, L.F., Haas, R., Reisswig, C., Mösta, P., Klion, H., and Schnetter, E. 2015. Neutrino-driven turbulent convection and standing accretion shock instability in three-dimensional core-collapse supernovae. *Astrophys. J.*, **808**, 70.

Abu-Shawareb, H., Acree, R., Adams, P., et al. 2022. Lawson criterion for ignition exceeded in an inertial fusion experiment. *Phys. Rev. Lett.*, **129**, 075001.

Abu-Shawareb, H., et al. 2024. Achievement of target gain larger than unity in an inertial fusion experiment. *Phys. Rev. Lett.*, **132**, 065102.

Adkins, R., Shelton, E.M., Renoult, M.-C., Carles, P., and Rosenblatt, C. 2017. Interface coupling and growth rate measurements in multilayer Rayleigh–Taylor instabilities. *Phys. Rev. Fluids*, **2**, 062001.

Aglitskiy, Y., Velikovich, A.L., Karasik, M., Serlin, V., Pawley, C.J., Schmitt, A.J., Obenschain, S.P., Mostovych, A.N., Gardner, J.H., and Metzler, N. 2002. Direct observation of mass oscillations due to ablative Richtmyer–Meshkov instability and feedout in planar plastic targets. *Phys. Plasmas*, **9**, 2264.

Aglitskiy, Y., Velikovich, A., Karasik, M., et al. 2010. Basic hydrodynamics of Richtmyer-Meshkov type growth and oscillations in the inertial confinement fusion-relevant conditions. *Phil. Trans. R. Soc. A*, **368**, 1739.

Akkerman, V.Y., and Law, C.K. 2013. Flame dynamics and consideration of deflagration-to-detonation transition in central gravitational field. *Proc. Combust. Inst.*, **34**, 1921.

Akkerman, V.Y., Law, C.K., and Bychkov, V. 2011. Self-similar accelerative propagation of expanding wrinkled flames and explosion triggering. *Phys. Rev. E*, **83**, 026305.

Akula, B., and Ranjan, D. 2016. Dynamics of buoyancy-driven flows at moderately high Atwood numbers. *J. Fluid Mech.*, **795**, 313.

Akula, B., Andrews, M.J., and Ranjan, D. 2013. Effect of shear on Rayleigh–Taylor mixing at small Atwood number. *Phys. Rev. E*, **87**, 033013.

Akula, B., Suchandra, P., Mikhaeil, M., and Ranjan, D. 2017. Dynamics of unstably stratified free shear flows: an experimental investigation of coupled Kelvin–Helmholtz and Rayleigh–Taylor instability. *J. Fluid Mech.*, **816**, 619.

Al-Janabi, K., Antolin, P., Baker, D., et al. 2019. Achievements of Hinode in the first eleven years. *Publ. Astron. Soc. Jpn.*, **71**, R1.

Al'bikov, Z.A., Velikhov, E.P., Veretennikov, A.I., et al. 1990. Angara-5-1 experimental complex. *Soviet Atomic Energy*, **68**, 34.

Aleshin, A.N., Lazareva, E.V., Zaitsev, S.G., Rozanov, V.B., Gamalii, E.G., and Lebo, I.G. 1990. Linear, nonlinear, and transient stages in the development of the Richtmyer–Meshkov instability. *Sov. Phys. Doklady*, **35**, 159.

Aleshin, A.N., Lazareva, E.V., Chebotareva, E.I., Sergeev, S.V., and Zaytsev, S.G. 1997. Investigation of Richtmyer–Meshkov instability induced by the incident and reflected shock waves. Page 1 of: Jourdan, G., and Houas, L. (eds), *Proceedings of the Sixth International Workshop on the Physics of Compressible Turbulent Mixing*. IUSTI Universite de Provence, Marseille, France.

Alexakis, A., and Biferale, L. 2018. Cascades and transitions in turbulent flows. *Phys. Rep.*, **767**, 1.

Allred, J.C., Blount, G.H., and Miller III, J.H. 1953. *Experimental studies of Taylor instability*. Tech. rept. Los Alamos National Lab, Los Alamos, NM.

Alon, U., Shvarts, D., and Mukamel, D. 1993. Scale-invariant regime in Rayleigh–Taylor bubble-front dynamics. *Phys. Rev. E*, **48**, 1008.

Alon, U., Hecht, J., Mukamel, D., and Shvarts, D. 1994. Scale invariant mixing rates of hydrodynamically unstable interfaces. *Phys. Rev. Lett.*, **72**, 2867.

Alon, U., Hecht, J., Ofer, D., and Shvarts, D. 1995. Power laws and similarity of Rayleigh–Taylor and Richtmyer–Meshkov mixing fronts at all density ratios. *Phys. Rev. Lett.*, **74**, 534.

Aluie, H. 2011. Compressible turbulence: the cascade and its locality. *Phys. Rev. Lett.*, **106**, 174502.

Aluie, H. 2013. Scale decomposition in compressible turbulence. *Physica D*, **247**, 54.

Aluie, H. 2017. Coarse-grained incompressible magnetohydrodynamics: analyzing the turbulent cascades. *New J. Phys.*, **19**, 025008.

Aluie, H., and Eyink, G.L. 2009. Localness of energy cascade in hydrodynamic turbulence. II. Sharp spectral filter. *Phys. Fluids*, **21**, 115108.

Alvi, F.S., Krothapalli, A., and Washington, D. 1996. Experimental study of a compressible countercurrent turbulent shear layer. *AIAA J.*, **34**, 728.

Amala, P.A.K. 1995. *Large-eddy simulation of the Rayleigh–Taylor instability on a massively parallel computer*. Tech. rept. UCRL-LR-119748, Lawrence Livermore National Laboratory, Livermore, CA.

Amala, P.A.K., and Rodrigue, G.H. 1994. Programming models for three-dimensional hydrodynamics on the CM-5 (Part II). *Comput. Phys.*, **8**, 300.

Amala, P.A.K., Rodrigue, G.H., and Dubois, P.F. 1994. Programming models for three-dimensional hydrodynamics on the CM-5 (Part I). *Comput. Phys.*, **8**, 181.

Amendt, P., Colvin, J.D., Tipton, R.E., et al. 2002. Indirect-drive noncryogenic double-shell ignition targets for the National Ignition Facility: design and analysis. *Phys. Plasmas*, **9**, 2221.

Amendt, P., Cerjan, C., Hamza, A., Hinkel, D.E., Milovich, J.L., and Robey, H.F. 2007. Assessing the prospects for achieving double-shell ignition on the National Ignition Facility using vacuum hohlraums. *Phys. Plasmas*, **14**, 056312.

Amendt, P., Cerjan, C., Hinkel, D.E., Milovich, J.L., Park, H.S., and Robey, H.F. 2008. Rugby-like hohlraum experimental designs for demonstrating X-ray drive enhancement. *Phys. Plasmas*, **15**, 012702.

Amendt, P., Milovich, J., Perkins, L.J., and Robey, H. 2010. An indirect-drive non-cryogenic double-shell path to 1ω Nd-laser hybrid inertial fusion–fission energy. *Nucl. Fusion*, **50**, 105006.

Amendt, P., Ho, D., Ping, Y., et al. 2019. Ultra-high (>30%) coupling efficiency designs for demonstrating central hot-spot ignition on the National Ignition Facility using a Frustraum. *Phys. Plasmas*, **26**, 082707.

Amendt, P., Ho, D., Nora, R., Ping, Y., and Smalyuk, V. 2020. High-volume and-adiabat capsule ("HVAC") ignition: lowered fuel compression requirements using advanced Hohlraums. *Phys. Plasmas*, **27**, 122708.

Anderson, J. 2007. *Fundamentals of Aerodynamics*. 4th edn. McGraw-Hill, New York.

Anderson, M.H., Puranik, B.P., Oakley, J.G., Brooks, P.W., and Bonazza, R. 2000. Shock tube investigation of hydrodynamic issues related to inertial confinement fusion. *Shock Waves*, **10**, 377.

André, M.L. 1999. The French megajoule laser project (LMJ). *Fusion Eng. Design*, **44**, 43.

Andreopoulos, Y., Agui, J.H., and Briassulis, G. 2000. Shock wave-turbulence interactions. *Annu. Rev. Fluid Mech.*, **32**, 309.

Andrews, M.J., and Spalding, D.B. 1990. A simple experiment to investigate two-dimensional mixing by Rayleigh–Taylor instability. *Phys. Fluids A*, **2**, 922.

Andrews, M.J., Youngs, D.L., Livescu, D., and Wei, T. 2014. Computational studies of two-dimensional Rayleigh–Taylor driven mixing for a tilted-rig. *J. Fluids Eng.*, **136**, 091212.

Andronov, V.A., Bakhrakh, S.M., Meshkov, E.E., Mokhov, V.N., Nikiforov, V.V., Pevnitskii, A.V., and Tolshmyakov, A.I. 1976. Turbulent mixing at contact surface accelerated by shock waves. *Sov. Phys. JETP*, **44**, 424.

Andronov, V.A., Bakhrakh, S.M., Meshkov, E.E., Nikiforov, V.V., Pevnitskii, A.V., and Tolshmyakov, A.I. 1982. An experimental investigation and numerical modeling of turbulent mixing in one-dimensional flows. *Sov. Phys. Doklady*, **27**, 393.

Anisimov, S.I., and Zeldovich, Y.B. 1977. Rayleigh–Taylor instability of boundary between detonation products and gas in spherical explosion. *Pis'ma Zh. Eksp. Teor. Fiz.*, **3**, 5.

Anisimov, S.I., Zeldovich, Y. B., Inogamov, N.A., and Ivanov, M.F. 1983. The Taylor instability of contact boundary between expanding detonation products and a surrounding gas. *Prog. Astronaut. Aeronaut.*, **87**, 218.

Annamalai, S., Parmar, M.K., Ling, Y., and Balachandar, S. 2014. Nonlinear Rayleigh–Taylor instability of a cylindrical interface in explosion flows. *J. Fluids Eng.*, **136**, 060910.

Annamalai, S., Rollin, B., Ouellet, F., Neal, C., Jackson, T.L., and Balachandar, S. 2016. Effects of initial perturbations in the early moments of an explosive dispersal of particles. *J. Fluids Eng.*, **138**, 070903.

Anuchina, N.N., Kucherenko, Y.A., Neuvazhaev, V.E., Ogibina, V.N., Shibarshov, L.I., and Yakovlev, V.G. 1978. Turbulent mixing at an accelerating interface between liquids of different density. *Fluid Dyn.*, **13**, 916.

Anuchina, N.N., Volkov, V.I., Gordeychuk, V.A., Es'kov, N.S., Ilyutina, O.S., and Kozyrev, O.M. 2004. Numerical simulations of Rayleigh–Taylor and Richtmyer–Meshkov instability using MAH-3 code. *J. Comput. Appl. Math.*, **168**, 11.

Apffel, B., Novkoski, F., Eddi, A., and Fort, E. 2020. Floating under a levitating liquid. *Nature*, **585**, 48.

Aragón, J., Naumis, G.G., Bai, M., Torres, M., and Maini, P.K. 2008. Turbulent luminance in impassioned van Gogh paintings. *J. Math. Imaging Vis.*, **30**, 275.

Arnett, D., Fryxell, B., and Mueller, E. 1989a. Instabilities and nonradial motion in SN 1987A. *Astrophys. J. Lett.*, **341**, L63.

Arnett, W.D., Bahcall, J.N., Kirshner, R.P., and Woosley, S.E. 1989b. Supernova 1987A. *Annu. Rev. Astron. Astrophys.*, **27**, 629.

Asay, J.R., Mix, L.P., and Perry, F.C. 1976. Ejection of material from shocked surfaces. *Appl. Phys. Lett.*, **29**, 284.

Aschenbach, B., Egger, R., and Trümper, J. 1995. Discovery of explosion fragments outside the Vela supernova remnant shock-wave boundary. *Nature*, **373**, 587.

Aslangil, D., Banerjee, A., and Lawrie, A.G.W. 2016. Numerical investigation of initial condition effects on Rayleigh–Taylor instability with acceleration reversals. *Phys. Rev. E*, **94**, 053114.

Aslangil, D., Farley, Z., Lawrie, A.G.W., and Banerjee, A. 2020. Rayleigh–Taylor instability with varying periods of zero acceleration. *J. Fluids Eng.*, **142**, 121103.

Aslangil, D., Lawrie, A.G.W., and Banerjee, A. 2022. Effects of variable deceleration periods on Rayleigh–Taylor Instability with acceleration reversals. *Phys. Rev. E*, **105**, 065103.

Aslani, M., and Regele, J.D. 2018. Numerical simulation of finite disturbances interacting with laminar premixed flames. *Combust. Theory Model.*, **22**, 812.

Aspden, A., Nikiforakis, N., Dalziel, S., and Bell, J.B. 2008. Analysis of implicit LES methods. *Commun. Appl. Math. Comput. Sci.*, **3**, 103.

Atoyan, L., Hammer, D.A., Kusse, B.R., Byvank, T., Cahill, A.D., Greenly, J.B., Pikuz, S.A., and Shelkovenko, T.A. 2016. Helical plasma striations in liners in the presence of an external axial magnetic field. *Phys. Plasmas*, **23**, 022708.

Attal, N, and Ramaprabhu, P. 2015. Numerical investigation of a single-mode chemically reacting Richtmyer–Meshkov instability. *Shock Waves*, **25**, 307.

Attal, N., and Ramaprabhu, P. 2020. The stability of reacting single-mode Rayleigh–Taylor flames. *Physica D*, **404**, 132353.

Attal, N., Ramaprabhu, P., Hossain, J., Karkhanis, V., Uddin, M., Gord, J.R., and Roy, S. 2015. Development and validation of a chemical reaction solver coupled to the FLASH code for combustion applications. *Comput. Fluids*, **107**, 59.

Atzeni, S., and Meyer-ter-Vehn, J. 2004. *The Physics of Inertial Fusion: BeamPlasma Interaction, Hydrodynamics, Hot Dense Matter*. Clarendon Press, Oxford, UK.

Aulanier, G., and Démoulin, P. 1998. 3-D magnetic configurations supporting prominences. I. The natural presence of lateral feet. *Astron. Astrophys.*, **329**, 1125.

Awe, T.J., McBride, R.D., Jennings, C.A., Lamppa, D.C., Martin, M.R., Rovang, D.C., Slutz, S.A., Cuneo, M.E., Owen, A.C., Sinars, D.B., and Tomlinson, K. 2013. Observations of modified three-dimensional instability structure for imploding Z-pinch liners that are premagnetized with an axial field. *Phys. Rev. Lett.*, **111**, 235005.

Awe, T.J., Jennings, C.A., McBride, R.D., et al. 2014. Modified helix-like instability structure on imploding Z-pinch liners that are pre-imposed with a uniform axial magnetic field. *Phys. Plasmas*, **21**, 056303.

Awe, T.J., Peterson, K.J., Yu, E.P., McBride, R.D., Sinars, D.B., Gomez, M.R., Jennings, C.A., Martin, M.R., Rosenthal, S.E., Schroen, D.G., and Sefkow, A.B. 2016. Observations of modified three-dimensional instability structure for imploding Z-pinch liners that are premagnetized with an axial field. *Phys. Rev. Lett.*, **116**, 065001.

Azarova, O.A., and Gvozdeva, L.G. 2018. Control of triple-shock configurations in high-speed flows over a cylindrically blunted plate in gases for different Mach numbers. *Proc. Inst. Mech. Engineers, Part G: J. Aerospace Eng.*, **24**, 1.

Azarova, O.A., Krasnobaev, K.V., Kravchenko, O.V., Lapushkina, T.A., and Erofeev, A.V. 2020. Redistribution of energy in a viscous heat-conductive medium during the interaction of a shock wave with a temperature layered plasma region. *J. Phys. Conf. Ser.*, **1698**, 012004.

Baer, M.R. 2022. *Private communication*.

Bahl, P., Doolan, C., de Silva, C., Chughtai, A.A., Bourouiba, L., and MacIntyre, C.R. 2022. Airborne or droplet precautions for health workers treating COVID-19? *J. Infect. Dis.*, **225**, 1561.

Bai, J., Wang, T., Liu, K., Li, L., Zhong, M., Jiang, Y., Tang, M., Yu, J., Pei, X., and Li, P. 2012a. Large-eddy simulation of the three-dimensional experiment on Richtmyer–Meshkov instability induced turbulence. *Int. J. Astron. Astrophys.*, **2**, 28.

Bai, J., Wang, B., Wang, T., and Liu, K. 2012b. Numerical simulation of the Richtmyer–Meshkov instability in initially nonuniform flows and mixing with reshock. *Phys. Rev. E*, **86**, 066319.

Bai, J., Zou, L., Wang, T., Liu, K., Huang, W., Liu, J., Li, P., Tan, D., and Liu, C. 2010a. Experimental and numerical study of shock-accelerated elliptic heavy gas cylinders. *Phys. Rev. E*, **82**, 056318.

Bai, J., Liu, J.-H., Wang, T., Zou, L., Li, P., and Tan, D. 2010b. Investigation of the Richtmyer–Meshkov instability with double perturbation interface in nonuniform flows. *Phys. Rev. E*, **81**, 056302.

Bailey, M. 2018. *Starry Night: Van Gogh at the Asylum*. White Lion Publishing, London, UK.

Baker, G.R., Meiron, D.I., and Orszag, S.A. 1980. Vortex simulations of the Rayleigh–Taylor instability. *Phys. Fluids*, **23**, 1485.

Baker, K.L., Thomas, C.A., Casey, D.T., et al. 2018. High-performance indirect-drive cryogenic implosions at high adiabat on the National Ignition Facility. *Phys. Rev. Lett.*, **121**, 135001.

Baker, L., and Freeman, J.R. 1981. Heuristic model of the nonlinear Rayleigh–Taylor instability. *J. Appl. Phys.*, **52**, 655.

Balachandar, S., Zaleski, S., Soldati, A., Ahmadi, G., and Bourouiba, L. 2020. Host-to-host airborne transmission as a multiphase flow problem for science-based social distance guidelines. *Int. J. Multiph. Flow*, **132**, 103439.

Balakrishnan, K., Genin, F., Nance, D.V., and Menon, S. 2010. Numerical study of blast characteristics from detonation of homogeneous explosives. *Shock Waves*, **20**, 147.

Balakumar, B.J., Orlicz, G.C., Tomkins, C.D., and Prestridge, K.P. 2008. Simultaneous particle-image velocimetry-planar laser-induced fluorescence measurements of Richtmyer–Meshkov instability growth in a gas curtain with and without reshock. *Phys. Fluids*, **20**, 124103.

Balakumar, B.J., Orlicz, G.C., Ristorcelli, J.R., Balasubramanian, S., Prestridge, K.P., and Tomkins, C.D. 2012. Turbulent mixing in a Richtmyer–Meshkov fluid layer after reshock: velocity and density statistics. *J. Fluid Mech.*, **696**, 67.

Balasubramanian, S., Orlicz, G.C., Prestridge, K.P., and Balakumar, B.J. 2012. Experimental study of initial condition dependence on Richtmyer–Meshkov instability in the presence of reshock. *Phys. Fluids*, **24**, 034103.

Balasubramanian, S., Orlicz, G.C., and Prestridge, K.P. 2013. Experimental study of initial condition dependence on turbulent mixing in shock-accelerated Richtmyer–Meshkov fluid layers. *J. Turbul.*, **14**, 170.

Baldwin, K.A., Scase, M.M., and Hill, R.J.A. 2015. The inhibition of the Rayleigh–Taylor instability by rotation. *Sci. Rep.*, **5**, 11706.

Balescu, R. 1975. *Equilibrium and Nonequilibrium Statistical Mechanics*. John Wiley & Sons, New York.

Balick, B., and Frank, A. 2002. Shapes and shaping of planetary nebulae. *Annu. Rev. Astron. Astrophys.*, **40**, 439.

Balick, B., Huarte-Espinosa, M., Frank, A., Gomez, T., Alcolea, J., Corradi, R.L.M., and Vinković, D. 2013. Outflows from evolved stars: the rapidly changing fingers of CRL 618. *Astrophys. J.*, **772**, 20.

Baltzer, J.R., and Livescu, D. 2019. Low-speed turbulent shear-driven mixing layers with large thermal and compositional density variations. In: Livescu, D., Battaglia, F., and Givi, P. (eds), *Modeling and Simulation of Turbulent Mixing and Reaction: For Power, Energy, and Flight*. Springer, Singapore.

Bambauer, M., Hasslberger, J., and Klein, M. 2020. Direct numerical simulation of the Richtmyer–Meshkov instability in reactive and nonreactive flows. *Combustion Sci. Tech.*, **192**, 2010.

Bambauer, M., Chakraborty, N., Klein, M., and Hasslberger, J. 2021. Vortex dynamics and fractal structures in reactive and nonreactive Richtmyer–Meshkov instability. *Phys. Fluids*, **33**, 044114.

Bambauer, M., Hasslberger, J., Ozel-Erol, G., Chakraborty, N., and Klein, M. 2023. Surface topologies and self interactions in reactive and nonreactive Richtmyer–Meshkov instability. *Sci. Rep.*, **13**, 837.

Banerjee, A. 2020. Rayleigh–Taylor instability: a status review of experimental designs and measurement diagnostics. *J. Fluids Eng.*, **142**, 120801.

Banerjee, A., and Andrews, M.J. 2006. Statistically steady measurements of Rayleigh–Taylor mixing in a gas channel. *Phys. Fluids*, **18**, 035107.

Banerjee, A., and Andrews, M.J. 2009. 3D simulations to investigate initial condition effects on the growth of Rayleigh–Taylor mixing. *Int. J. Heat Mass Trans.*, **52**, 3906.

Banerjee, A., and Mutnuri, L.A.R. 2012. Passive and reactive scalar measurements in a transient high-Schmidt-number Rayleigh–Taylor mixing layer. *Exp. Fluids*, **53**, 717.

Banerjee, A., Kraft, W.N., and Andrews, M.J. 2010a. Detailed measurements of a statistically steady Rayleigh–Taylor mixing layer from small to high Atwood numbers. *J. Fluid Mech.*, **659**, 127.

Banerjee, A., Gore, R.A., and Andrews, M.J. 2010b. Development and validation of a turbulent-mix model for variable-density and compressible flows. *Phys. Rev. E*, **82**, 046309.

Banerjee, R., Mandal, L., Khan, M., and Gupta, M.R. 2012. Effect of viscosity and shear flow on the nonlinear two fluid interfacial structures. *Phys. Plasmas*, **19**, 122105.

Bao, B., Peng, Q., Yang, C., and Zhang, L. 2021. Evolutions of young Type Ia supernova remnants with two initial density profiles in a turbulent medium. *Astrophys. J.*, **909**, 173.

Barenblatt, G.-I. 1983. Self-similar turbulence propagation from an instantaneous point source. In: Barenblatt, G.I., Looss, G., and Joseph, D.D. (eds), *Non-Linear Dynamics and Turbulence*. Pitman, Boston, MA.

Barenblatt, G.I. 2001. George Keith Batchelor (1920–2000) and David George Crighton (1942–2000): Applied Mathematicians. *Not. Am. Math. Soc.*, **48**, 800.

Barnes, C.W., Batha, S.H., Dunne, A.M., et al. 2002. Observation of mix in a compressible plasma in a convergent cylindrical geometry. *Phys. Plasmas*, **9**, 4431.

Barnes, J.F., Blewett, P.J., McQueen, R.G., Meyer, K.A., and Venable, D. 1974. Taylor instability in solids. *J. Appl. Phys.*, **45**, 727.

Barrow, J.D. 1983. Dimensionality. *Philos. Trans. R. Soc. A*, **310**, 337.

Barton, N.R., and Rhee, M. 2013. A multiscale strength model for tantalum over an extended range of strain rates. *J. Appl. Phys.*, **114**, 123507.

Barton, N.R., Bernier, J.V., Becker, R., Arsenlis, A., Cavallo, R., Marian, J., Rhee, M., Park, H.S., Remington, B.A., and Olson, R.T. 2011. A multiscale strength model for extreme loading conditions. *J. Appl. Phys.*, **109**, 073501.

Basko, M.M. 1990. Spark and volume ignition of DT and D2 microspheres. *Nucl. Fusion*, **30**, 2443.

Basko, M.M., Kemp, A.J., and Meyer-ter Vehn, J. 2000. Ignition conditions for magnetized target fusion in cylindrical geometry. *Nucl. Fusion*, **40**, 59.

Batchelor, G.K. 1953. *The Theory of Homogeneous Turbulence*. Cambridge University Press. Cambridge, UK.

Batchelor, G.K. 1969. Computation of the energy spectrum in homogeneous two-dimensional turbulence. *Phys. Fluids*, **12**, II–233.

Batchelor, G.K. 1994. *The Life and Legacy of G.I. Taylor*. Cambridge University Press. Cambridge, UK

Batchelor, G.K., and Proudman, I. 1956. The large-scale structure of homogenous turbulence. *Phil. Trans. R. Soc. Lond. A*, **248**, 369.

Batchelor, G.K., Canuto, V.M., and Chasnov, J.R. 1992. Homogeneous buoyancy-generated turbulence. *J. Fluid Mech.*, **235**, 349.

Battimelli, G. 1986. On the history of the statistical theories of turbulence. *Rev. Mex. Fis.*, **32**, (Suppl. 1) S3.

Baus, M., and Hansen, J.-P. 1980. Statistical mechanics of simple Coulomb systems. *Phys. Rep.*, **59**, 1.

Begelman, M.C., Blandford, R.D., and Rees, M.J. 1984. Theory of extragalactic radio sources. *Rev. Mod. Phys.*, **56**, 255.

Belen'kii, S.Z., and Fradkin, E.S. 1965. Theory of turbulent mixing. *Tr. Fiz. Inst. Akad. Nauk SSSR*, **29**, 207.

Bell, D.J., and Chapman, D.J. 2017. Phase Doppler anemometry as an ejecta diagnostic. *AIP Conf. Proc.*, **1793**, 060011.

Bell, D.J., Routley, N.R., Millett, J.C.F., Whiteman, G., Collinson, M.A., and Keightley, P.T. 2017. Investigation of ejecta production from Tin at an elevated temperature and the eutectic alloy Lead–Bismuth. *J. Dyn. Behav. Mater.*, **3**, 208.

Bell, G.I. 1951. *Taylor instability on cylinders and spheres in the small amplitude approximation*. Los Alamos Scientific Lab. Report No. LA-1321. Los Alamos, NM.

Bell, J.H., and Mehta, R.D. 1990. Development of a two-stream mixing layer from tripped and untripped boundary layers. *AIAA J.*, **28**, 2034.

Bellan, P.M., You, S., and Hsu, S.C. 2005. Simulating astrophysical jets in laboratory experiments. In: Kyrala, G. (eds), *High Energy Density Laboratory Astrophysics*. Springer, Dordrecht, Netherlands.

Bellman, R., and Pennington, R.H. 1954. Effects of surface tension and viscosity on Taylor instability. *Quart. Appl. Math.*, **12**, 151.

Ben-Dor, G., Igra, O., and Elperin, T. 2000. *Handbook of Shock Waves, I–III*. Academic Press, New York.

Bender, C.M., and Orszag, S. 1978. *Advanced Mathematical Methods for Scientists and Engineers*. McGraw-Hill, New York.

Bender, J.D., Schilling, O., Raman, K.S., Managan, R.A., Olson, B.J., Copeland, S.R., Ellison, C.L., Erskine, D.J., Huntington, C.M., Morgan, B.E., Nagel, S.R., Prisbrey, S.T., Pudliner, B.S., Sterne, P.A., Wehrenberg, C.E., and Zhou, Y. 2021. Simulation and flow physics of a shocked and reshocked high-energy-density mixing layer. *J. Fluid Mech.*, **915**, A84.

Bennett, G.R., Wallace, J.M., Murphy, T.J., et al. 2000. Moderate-convergence inertial confinement fusion implosions in tetrahedral hohlraums at Omega. *Phys. Plasmas*, **7**, 2594.

Bennett, W.H. 1934. Magnetically self-focusing streams. *Phys. Rep.*, **45**, 890.

Bera, R.K., Song, Y., and Srinivasan, B. 2022. The effect of viscosity and resistivity on Rayleigh–Taylor instability induced mixing in magnetized high-energy-density plasmas. *J. Plasma Phys.*, **88**, 905880209.

Berger, T., Testa, P., Hillier, A., Boerner, P., Low, B.C., Shibata, K., Schrijver, C., Tarbell, T., and Title, A. 2011. Magneto-thermal convection in solar prominences. *Nature*, **472**, 197.

Berger, T., Hillier, A., and Liu, W. 2017. Quiescent prominence dynamics observed with the Hinode Solar Optical Telescope. II. Prominence bubble boundary layer characteristics and the onset of a coupled Kelvin–Helmholtz Rayleigh–Taylor instability. *Astrophys. J.*, **850**, 60.

Berger, T.E., Shine, R.A., Slater, G.L., Tarbell, T.D., Okamoto, T.J., Ichimoto, K., Katsukawa, Y., Suematsu, Y., Tsuneta, S., Lites, B.W., and Shimizu, T. 2008. Hinode SOT observations of solar quiescent prominence dynamics. *Astrophys. J. Lett.*, **676**, L89.

Berger, T.E., Slater, G., Hurlburt, N., Shine, R., Tarbell, T., Lites, B.W., Okamoto, T.J., Ichimoto, K., Katsukawa, Y., Magara, T., and Suematsu, Y. 2010. Quiescent prominence dynamics observed with the Hinode Solar Optical Telescope. I. Turbulent upflow plumes. *Astrophys. J.*, **716**, 1288.

Berry, M.V., and Geim, A.K. 1997. Of flying frogs and levitrons. *Eur. J. Phys.*, **18**, 307.

Besnard, D., Harlow, F., and R., Rauenzahn. 1992. *Turbulence transport equations for variable density turbulence and their relationship to two-field models*. Tech. rept. LA-12303-MS. Los Alamos National Laboratory, Los Alamos, NM.

Besnard, D.C., Haas, J.F., and Rauenzahn, R.M. 1989. Statistical modeling of shock-interface interaction. *Physica D*, **37**, 227.

Besnard, D.C., Harlow, F.H., Rauenzahn, R.M., and Zemach, C. 1996. Spectral transport model for turbulence. *Theor. Comput. Fluid Dyn.*, **8**, 1.

Bethe, H.A. 1990. Supernova mechanisms. *Rev. Mod. Phys.*, **62**, 801.

Betti, R., and Hurricane, O.A. 2016. Inertial-confinement fusion with lasers. *Nat. Phys.*, **12**, 435.

Betti, R., and Sanz, J. 2006. Bubble acceleration in the ablative Rayleigh–Taylor instability. *Phys. Rev. Lett.*, **97**, 205002.

Betti, R., McCrory, R.L., and Verdon, C.P. 1993. Stability analysis of unsteady ablation fronts. *Phys. Rev. Lett.*, **71**, 3131.

Betti, R., Goncharov, V.N., McCrory, R.L., and Verdon, C.P. 1995. Self-consistent cutoff wave number of the ablative Rayleigh–Taylor instability. *Phys. Plasmas*, **2**, 3844.

Betti, R., Zhou, C.D., Anderson, K.S., Perkins, L.J., Theobald, W., and Solodov, A.A. 2007. Shock ignition of thermonuclear fuel with high areal density. *Phys. Rev. Lett.*, **98**, 155001.

Biamino, L., Mariani, C., Jourdan, G., Houas, L., Vandenboomgaerde, M., and Souffland, D. 2014. Planar shock focusing through perfect gas lens: first experimental demonstration. *J. Fluids Eng.*, **136**, 091204.

Bian, X., Aluie, H., Zhao, D., Zhang, H., and Livescu, D. 2020. Revisiting the late-time growth of single-mode Rayleigh–Taylor instability and the role of vorticity. *Physica D*, **403**, 132250.

Biferale, L., Mantovani, F., Sbragaglia, M., Scagliarini, A., Toschi, F., and Tripiccione, R. 2010. High resolution numerical study of Rayleigh–Taylor turbulence using a thermal lattice Boltzmann scheme. *Phys. Fluids*, **22**, 115112.

Biferale, L., Boffetta, G., Mailybaev, A.A., and Scagliarini, A. 2018. Rayleigh–Taylor turbulence with singular nonuniform initial conditions. *Phys. Rev. Fluids*, **3**, 092601.

Bilger, R.W. 1975. A note on Favre averaging in variable density flows. *Combust. Sci. Technol.*, **11**, 215.

Billet, G., Giovangigli, V., and De Gassowski, G. 2008. Impact of volume viscosity on a shock–hydrogen-bubble interaction. *Combust. Theory Model.*, **12**, 221.

Billington, I.J. 1956. An experimental study of the one-dimensional refraction of a rarefaction wave at a contact surface. *J. Aero. Sci.*, **23**, 997.

Bin, Y., Xiao, M., Shi, Y., Zhang, Y., and Chen, S. 2021. A new idea to predict reshocked Richtmyer–Meshkov mixing: constrained large-eddy simulation. *J. Fluid Mech.*, **918**, R1.

Binnie, A.M. 1953. The stability of the surface of a cavitation bubble. *Math. Proc. Cambridge Philos. Soc.*, **49**, 151.

Bird, R.B., Stewart, W.E., and Lightfoot, E.N. 2006. *Transport Phenomena.* John Wiley & Sons, New York.

Birkhoff, G. 1954. *Taylor instability and laminar mixing*. Tech. rept. LA-1862. Los Alamos National Laboratory, Los Alamos, NM.

Birkhoff, G. 1962. Helmholtz and Taylor instability. Pages 55–76 of: *Proceedings of Symposia in Applied Mathematics*. Vol. 13. American Mathematical Society, Providence, RI.

Bishop, A.S. 1958. *Project Sherwood: The US Program in Control. Fusion.* Addison-Wesley, Reading, MA.

Blinova, A.A., Romanova, M.M., and Lovelace, R.V.E. 2016. Boundary between stable and unstable regimes of accretion. Ordered and chaotic unstable regimes. *Mon. Not. R. Astron. Soc.*, **459**, 2354.

Bliznetsov, M.V., Vlasov, Y.V., Dudin, V.I., Meshkov, E.E., Nikulin, A.A., Til'kunov, V.A., Tolshmyakov, A.I., and Kholkin, S.A. 1997. How the film may control the gas-gas turbulent mixing development in shock tube experiments. Page 90 of: *Proceedings the Sixth International Workshop on Compressible Turbulent Mixing*. Marseille, France.

Blondin, S., Bravo, E., Timmes, F.X., Dessart, L., and Hillier, D.J. 2022. Stable nickel production in Type Ia supernovae: a smoking gun for the progenitor mass? *Astron. Astrophys.*, **660**, A96.

Bodenschatz, E., and Eckert, M. 2011. Prandtl and the Göttingen school. In: P.A. Davidson, Y. Kaneda, K. Moffet, and K.R. Sreenivasan (eds), *A Voyage through Turbulence*. Cambridge University Press, Cambridge, UK.

Bodner, S.E. 1974. Rayleigh–Taylor instability and laser-pellet fusion. *Phys. Rev. Lett.*, **33**, 761.

Boehly, T.R., Brown, D.L., Craxton, R.S., et al. 1997. Initial performance results of the OMEGA laser system. *Opt. Commun.* **133**, 495.

Boffetta, G., and Ecke, R.E. 2012. Two-dimensional turbulence. *Annu. Rev. Fluid Mech.*, **44**, 427.

Boffetta, G., and Mazzino, A. 2017. Incompressible Rayleigh–Taylor turbulence. *Annu. Rev. Fluid Mech.*, **49**, 119.

Boffetta, G., and Musacchio, S. 2010. Evidence for the double cascade scenario in two-dimensional turbulence. *Phys. Rev. E*, **82**, 016307.

Boffetta, G, and Musacchio, S. 2022. Incompressible Rayleigh–Taylor mixing in circular and spherical geometries. *Phys. Rev. E*, **105**, 025104.

Boffetta, G., Mazzino, A., Musacchio, S., and Vozella, L. 2009. Kolmogorov scaling and intermittency in Rayleigh–Taylor turbulence. *Phys. Rev. E*, **79**, 065301.

Boffetta, G., Mazzino, A., Musacchio, S., and Vozella, L. 2010. Statistics of mixing in three-dimensional Rayleigh–Taylor turbulence at low Atwood number and Prandtl number one. *Phys. Fluids*, **22**, 035109.

Boffetta, G., Mazzino, A., and Musacchio, S. 2011. Effects of polymer additives on Rayleigh–Taylor turbulence. *Phys. Rev. E*, **83**, 056318.

Boffetta, G., De Lillo, F., Mazzino, A., and Musacchio, S. 2012. Bolgiano scale in confined Rayleigh–Taylor turbulence. *J. Fluid Mech.*, **690**, 426.

Boffetta, G., Mazzino, A., and Musacchio, S. 2016. Rotating Rayleigh–Taylor turbulence. *Phys. Rev. Fluids*, **1**, 054405.

Boffetta, G., Magnani, M., and Musacchio, S. 2019. Suppression of Rayleigh–Taylor turbulence by time-periodic acceleration. *Phys. Rev. E*, **99**, 033110.

Boffetta, G., Borgnino, M., and Musacchio, S. 2020. Scaling of Rayleigh–Taylor mixing in porous media. *Phys. Rev. Fluids*, **5**, 062501.

Bolgiano Jr, R. 1959. Turbulent spectra in a stably stratified atmosphere. *J. Geophys. Res.*, **64**, 2226.

Bonazza, R. 2022. Research activities at the Wisconsin shock tube laboratory. Pages 57–82 of: Takayama, K., and Igra, O. (eds), *Frontiers of Shock Wave Research*. Springer International Publishing, Cham, Switzerland.

Bonazza, R., and Sturtevant, B. 1996. X-ray measurements of growth rates at a gas interface accelerated by shock waves. *Phys. Fluids*, **8**, 2496.

Bond, D., Wheatley, V., Samtaney, R., and Pullin, D.I. 2017. Richtmyer–Meshkov instability of a thermal interface in a two-fluid plasma. *J. Fluid Mech.*, **833**, 332.

Borgnino, M., Boffetta, G., and Musacchio, S. 2021. Dimensional transition in Darcy-Rayleigh–Taylor mixing. *Phys. Rev. Fluids*, **6**, 074501.

Boris, J.P. 1989. Comment: the potential and limitations of direct and large-eddy simulations. Page 344 of: Lumley, J.L. (ed), *Whither Turbulence? Turbulence at the Crossroads*. Springer, Berlin, Germany.

Boris, J.P., Grinstein, F.F., Oran, E.S., and Kolbe, R.L. 1992. New insights into large eddy simulation. *Fluid Dyn. Res.*, **10**, 199.

Bourgade, J.L., Marmoret, R., Darbon, S., et al. 2008. Diagnostics hardening for harsh environment in Laser Mégajoule. *Rev. Sci. Instrum.*, **79**, 10F301.

Bourouiba, L. 2020. Turbulent gas clouds and respiratory pathogen emissions: potential implications for reducing transmission of COVID-19. *JAMA: J. Am. Med. Assoc.*, **323**, 1837.

Bourouiba, L., Dehandschoewercker, E., and Bush, J.W. 2014. Violent expiratory events: on coughing and sneezing. *J. Fluid Mech.*, **745**, 537.

Boussinesq, J. 1903. *Théorie Analytique de la Chaleur*. Vol. 2. Gauthier-Villars, Paris, France.

Bouzgarrou, G., Bury, Y., Jamme, S., Joly, L., and Haas, J.-F. 2014. Laser doppler velocimetry measurements in turbulent gaseous mixing induced by the Richtmyer–Meshkov instability: statistical convergence issues and turbulence quantification. *J. Fluids Eng.*, **136**, 091209.

Bradley, P.A. 2014. The effect of mix on capsule yields as a function of shell thickness and gas fill. *Phys. Plasmas*, **21**, 062703.

Bradshaw, P. 1994. Turbulence: the chief outstanding difficulty of our subject. *Exp. Fluids*, **16**, 203.

Braginskii, S.I. 1965. Transport processes in a plasma. *Rev. Plasma Phys.*, **1**, 205.

Branch, D., and Wheeler, J.C. 2017. *Supernova Explosions*. Springer, Berlin.

Branover, H., Eidelman, A., Golbraikh, E., and Moiseyev, S. 1999. *Turbulence and Structures: Chaos, Fluctuations, and Helical Self-Organization in Nature and the Laboratory*. Academic Press, San Diego, CA.

Braun, N.O., and Gore, R.A. 2020. A passive model for the evolution of subgrid-scale instabilities in turbulent flow regimes. *Physica D*, **404**, 132373.

Braun, N. O., and Gore, R. A. 2021. A multispecies turbulence model for the mixing and de-mixing of miscible fluids. *J. Turbul.*, **22**, 784.

Brenner, M.P., and Stone, H.A. 2000. Modern classical physics through the work of G. I. Taylor. *Phys. Today*, **53**, 30.

Briard, A., Gréa, B.-J., and Nguyen, F. 2022. Growth rate of the turbulent magnetic Rayleigh–Taylor instability. *Phys. Rev. E*, **106**, 065201.

Briard A, Gréa B.-J., and Nguyen, F. 2024. Turbulent mixing in the vertical magnetic Rayleigh–Taylor instability. *J. Fluid Mech.*, **979**, A8.

Brode, H.L. 1959. Blast wave from a spherical charge. *Phys. Fluids*, **2**, 217.

Brouillette, M. 2002. The Richtmyer–Meshkov instability. *Annu. Rev. Fluid Mech.*, **34**, 445.

Brouillette, M., and Sturtevant, B. 1989. Growth induced by multiple shock waves normally incident on plane gaseous interfaces. *Physica D*, **37**, 248.

Brouillette, M., and Sturtevant, B. 1994. Experiments on the Richtmyer–Meshkov instability: single-scale perturbations on a continuous interface. *J. Fluid Mech.*, **263**, 271.

Brown, G.L., and Roshko, A. 1974. On density effects and large structure in turbulent mixing layers. *J. Fluid Mech.*, **64**, 775.

Brown, J.L., Alexander, C.S., Asay, J.R., Vogler, T.J., and Ding, J.L. 2013. Extracting strength from high pressure ramp-release experiments. *J. Appl. Phys.*, **114**, 223518.

Brown, J.L., Alexander, C.S., Asay, J.R., Vogler, T.J., Dolan, D.H., and Belof, J.L. 2014a. Flow strength of tantalum under ramp compression to 250 GPa. *J. Appl. Phys.*, **115**, 043530.

Brown, M.A., Batha, C.A., and Williams, R.J.R. 2014b. Statistics for assessing mixing in a finite element hydrocode. *J. Fluids Eng.*, **136**, 091103.

Brüggen, M., Scannapieco, E., and Heinz, S. 2009. Evolution of X-ray cavities. *Mon. Not. R. Astron. Soc.*, **395**, 2210.

Bryson, A.E., and Gross, R.W.F. 1961. Diffraction of strong shocks by cones, cylinders, and spheres. *J. Fluid Mech.*, **10**, 1.

Bucciantini, N., Amato, E., Bandiera, R., Blondin, J.M., and Del Zanna, L. 2004. Magnetic Rayleigh–Taylor instability for pulsar wind nebulae in expanding supernova remnants. *Astron. Astrophys.*, **423**, 253.

Budil, K.S., Remington, B.A., Peyser, T.A., Mikaelian, K.O., Miller, P.L., Woolsey, N.C., Wood-Vasey, W.M., and Rubenchik, A.M. 1996. Experimental comparison of classical versus ablative Rayleigh–Taylor instability. *Phys. Rev. Lett.*, **76**, 4536.

Budil, K.S., Lasinski, B., Edwards, M.J., Wan, A.S., Remington, B.A., Weber, S.V., Glendinning, S.G., Suter, L., and Stry, P.E. 2001. The ablation-front Rayleigh–Taylor dispersion curve in indirect drive. *Phys. Plasmas*, **8**, 2344.

Burdonov, K., Yao, W., Sladkov, A., et al. 2022. Laboratory modelling of equatorial 'tongue' accretion channels in young stellar objects caused by the Rayleigh–Taylor instability. *Astron. Astrophys.*, **657**, A112.

Burlot, A., Gréa, B.-J., Godeferd, F.S., Cambon, C., and Soulard, O. 2015a. Large Reynolds number self-similar states of unstably stratified homogeneous turbulence. *Phys. Fluids*, **27**, 065114.

Burlot, A., Gréa, B.-J., Godeferd, F.S., Cambon, C., and Griffond, J. 2015b. Spectral modelling of high Reynolds number unstably stratified homogeneous turbulence. *J. Fluid Mech.*, **765**, 17.

Burrows, A. 2000. Supernova explosions in the universe. *Nature*, **403**, 727.

Burrows, K.D., Smeeton, V.S., and Youngs, D.L. 1984. Experimental investigation of turbulent mixing by Rayleigh–Taylor instability, II. *AWRE Report*, 22/84. Atomic Weapons Establishment, Aldermaston, UK.

Burton, G.C. 2008. The nonlinear large-eddy simulation method applied to $Sc \approx 1$ and $Sc \gg 1$ passive-scalar mixing. *Phys. Fluids*, **20**, 035103.

Burton, G.C. 2011. Study of ultrahigh Atwood-number Rayleigh–Taylor mixing dynamics using the nonlinear large-eddy simulation method. *Phys. Fluids*, **23**, 045106.

Buttler, W.T., Oró, D.M., Preston, D.L., et al. 2012. Unstable Richtmyer–Meshkov growth of solid and liquid metals in vacuum. *J. Fluid Mech.*, **703**, 60.

Buttler, W.T., Williams, R.J., and Najjar, F.M. (eds). 2017. The special issue on ejecta. *J. Dyn. Behav. Mater.*, **3**.

Cabot, W. 2006. Comparison of two-and three-dimensional simulations of miscible Rayleigh–Taylor instability. *Phys. Fluids*, **18**, 045101.

Cabot, W.H., and Cook, A.W. 2006. Reynolds number effects on Rayleigh–Taylor instability with possible implications for type Ia supernovae. *Nat. Phys.*, **2**, 562.

Cabot, W.H., and Zhou, Y. 2013. Statistical measurements of scaling and anisotropy of turbulent flows induced by Rayleigh–Taylor instability. *Phys. Fluids*, **25**, 015107.

Cabot, W.H., Schilling, O., and Zhou, Y. 2004. Influence of subgrid scales on resolvable turbulence and mixing in Rayleigh–Taylor flow. *Phys. Fluids*, **16**, 495.

Callender, C. 2005. Answers in search of a question: 'proofs' of the tri-dimensionality of space. *Stud. Hist. Philos. Mod. Phys.*, **36**, 113.

Campbell, E.M., Sangster, T.C., Goncharov, V.N., et al. 2020a. Direct-drive laser fusion: status, plans and future. *Phil. Trans. Roy. Soc. A*, **379**, 20200011.

Campbell, P.C., Jones, T.M., Woolstrum, J.M., et al. 2020b. Stabilization of liner implosions via a dynamic screw pinch. *Phys. Rev. Lett.*, **125**, 035001.

Canaud, B., Laffite, S., and Temporal, M. 2011. Shock ignition of direct-drive double-shell targets. *Nucl. Fusion*, **51**, 062001.

Candelas, P., Horowitz, G.T., Strominger, A., and Witten, E. 1985. Vacuum configurations for superstrings. *Nucl. Phys. B*, **258**, 46.

Canfield, J., Denissen, N., Francois, M., Gore, R., Rauenzahn, R., Reisner, J., and Shkoller, S. 2020. A comparison of interface growth models applied to Rayleigh–Taylor and Richtmyer–Meshkov instabilities. *J. Fluids Eng.*, **142**, 121108.

Cannone, M., and Friedlander, S. 2003. Navier: blow-up and collapse. *Not. Am. Math. Soc.*, **50**, 7.

Cao, C.Y., Sun, Y.B., Wang, C., Jia, X.Y., Zeng, R.H., and Yang, T.H. 2024. Coupled models for propagation of explosive shock waves in cylindrical and spherical geometries. *Phys. Plasmas*, **31**, 022706.

Cao, D., and Michaels, D. 2022. Vorticity transport in different regions of a strut-based scramjet. *AIAA J.*, **60**, 4532.

Cao, Y.G., Guo, H.Z., Zhang, Z.F., Sun, Z.H., and Chow, W.K. 2011. Effects of viscosity on the growth of Rayleigh–Taylor instability. *J. Physics A*, **44**, 275501.

Carlès, P., and Popinet, S. 2001. Viscous nonlinear theory of Richtmyer–Meshkov instability. *Phys. Fluids*, **13**, 1833.

Carlès, P., and Popinet, S. 2002. The effect of viscosity, surface tension and non-linearity on Richtmyer–Meshkov instability. *Eur. J. Mech. B/Fluids*, **21**, 511.

Carlès, P., Huang, Z., Carbone, G., and Rosenblatt, C. 2006. Rayleigh–Taylor instability for immiscible fluids of arbitrary viscosities: a magnetic levitation investigation and theoretical model. *Phys. Rev. Lett.*, **96**, 104501.

Carlyle, J., Williams, D.R., van Driel-Gesztelyi, L., Innes, D., Hillier, A., and Matthews, S. 2014. Investigating the dynamics and density evolution of returning plasma blobs from the 2011 June 7 eruption. *Astrophys. J.*, **782**, 87.

Carnevale, G.F., Purini, R., Orlandi, P., and Cavazza, P. 1995. Barotropic quasi-geostrophic f-plane flow over anisotropic topography. *J. Fluid Mech.*, **285**, 329.

Carnevale, G.F., Orlandi, P., Zhou, Y., and Kloosterziel, R.C. 2002. Rotational suppression of Rayleigh–Taylor instability. *J. Fluid Mech.*, **457**, 181.

Carter, J., Pathikonda, G., Jiang, N., Felver, J.J., Roy, S., and Ranjan, D. 2019. Time-resolved measurements of turbulent mixing in shock-driven variable-density flows. *Sci. Rep.*, **9**, 20315.

Case, K.M. 1960. Taylor instability of an inverted atmosphere. *Phys. Fluids*, **3**, 366.

Casey, D.T., Smalyuk, V.A., Tipton, R.E., Pino, J.E., Grim, G.P., Remington, B.A., Rowley, D.P., Weber, S.V., Barrios, M., Benedetti, L.R., and Bleuel, D.L. 2014. Development of the CD Symcap platform to study gas-shell mix in implosions at the National Ignition Facility. *Phys. Plasmas*, **21**, 092705.

Cashdollar, K.L. 1996. Coal dust explosibility. *J. Loss Prev. Process Ind.*, **9**, 65.

Casner, A., Caillaud, T., Darbon, S., et al. 2015. LMJ/PETAL laser facility: overview and opportunities for laboratory astrophysics. *High Energy Density Phys.*, **17**, 2.

Casner, A., Rigon, G., Albertazzi, B., et al. 2018. Turbulent hydrodynamics experiments in high energy density plasmas: scientific case and preliminary results of the TurboHEDP project. *High Power Laser Sci. Eng.*, **6**, e44.

Casner, A., Mailliet, C., Rigon, G., et al. 2019. From ICF to laboratory astrophysics: ablative and classical Rayleigh–Taylor instability experiments in turbulent-like regimes. *Nucl. Fusion*, **59**, 032002.

Castelvecchi, D. 2017. On the trail of turbulence. *Nature*, **548**, 382.

Catherall, A.T., Eaves, L., King, P.J., and Booth, S.R. 2003. Floating gold in cryogenic oxygen. *Nature*, **422**, 579.

Catherasoo, C.J., and Sturtevant, B. 1983. Shock dynamics in non-uniform media. *J. Fluid Mech.*, **127**, 539.

Cavailler, C. 2005. Inertial fusion with the LMJ. *Plasma Phys. Control. Fusion*, **47**, B389.

Celani, A., Cencini, M., Mazzino, A., and Vergassola, M. 2002. Active versus passive scalar turbulence. *Phys. Rev. Lett.*, **89**, 234502.

Celani, A., Cencini, M ., Mazzino, A., and Vergassola, M. 2004. Active and passive fields face to face. *New J. Phys.*, **6**, 72.

Chae, J.C. 2010. Dynamics of vertical threads and descending knots in a hedgerow prominence. *Astrophys. J.*, **714**, 618.

Champagne, F.H., Harris, V.G., and Corrsin, S. 1970. Experiments on nearly homogeneous turbulent shear flow. *J. Fluid Mech.*, **41**, 81.

Chandrasekhar, S. 1955. The character of the equilibrium of an incompressible fluid sphere of variable density and viscosity subject to radial acceleration. *Q. J. Mech. Appl. Math.*, **8**, 1.

Chandrasekhar, S. 1961. *Hydrodynamic and Hydromagnetic Stability*. Oxford University Press, London.

Chang, P.Y., Fiksel, G., Hohenberger, M., Knauer, J.P., Betti, R., Marshall, F.J., Meyerhofer, D.D., Séguin, F.H., and Petrasso, R.D. 2011. Fusion yield enhancement in magnetized laser-driven implosions. *Phys. Rev. Lett.*, **107**, 035006.

Chapman, D.R. 1979. Computational aerodynamics development and outlook. *AIAA J.*, **17**, 1293.

Chapman, P.R., and Jacobs, J.W. 2006. Experiments on the three-dimensional incompressible Richtmyer–Meshkov instability. *Phys. Fluids*, **18**, 074101.

Chapman, S., and Cowling, T.G. 1990. *The Mathematical Theory of Non-uniform Gases: An Account of the Kinetic Theory of Viscosity, Thermal Conduction and Diffusion in Gases*. 3rd edn. Cambridge University Press, Cambridge, UK.

Charakhch'an, A.A. 2000. Richtmyer–Meshkov instability of an interface between two media due to passage of two successive shocks. *J. Appl. Mech. Tech. Phys.*, **41**, 23.

Charakhch'yan, A.A. 2001. Reshocking at the non-linear stage of Richtmyer–Meshkov instability. *Plasma Phys. Control. Fusion*, **43**, 1169.

Charney, J.G. 1971. Geostrophic turbulence. *J. Atmos. Sci.*, **28**, 1087.

Chasnov, J.R. 1997. On the decay of inhomogeneous turbulence. *J. Fluid Mech.*, **342**, 335.

Chen, C., Xing, Y., Wang, H., Zhai, Z., and Luo, X. 2023. Experimental study on Richtmyer–Meshkov instability at a light–heavy interface over a wide range of Atwood numbers. *J. Fluid Mech.*, **975**, A29.

Chen, F., Xu, A., and Zhang, G. 2016. Viscosity, heat conductivity, and Prandtl number effects in the Rayleigh–Taylor instability. *Front. Phys.*, **11**, 1.

Chen, F., Xu, A., and Zhang, G. 2018. Collaboration and competition between Richtmyer–Meshkov instability and Rayleigh–Taylor instability. *Phys. Fluids*, **30**, 102105.

Chen, Q., Li, L., Zhang, Y., and Tian, B. 2019. Effects of the Atwood number on the Richtmyer–Meshkov instability in elastic-plastic media. *Phys. Rev. E*, **99**, 053102.

Chen, S., and Doolen, G.D. 1998. Lattice Boltzmann method for fluid flows. *Annu. Rev. Fluid Mech.*, **30**, 329.

Chen, Y., Glimm, J., Sharp, D.H., and Zhang, Q. 1996. A two-phase flow model of the Rayleigh–Taylor mixing zone. *Phys. Fluids*, **8**, 816.

Cheng, B., Glimm, J., and Sharp, D.H. 2000. Density dependence of Rayleigh–Taylor and Richtmyer–Meshkov mixing fronts. *Phys. Lett. A*, **268**, 366.

Cheng, B., Glimm, J., and Sharp, D.H. 2002a. Dynamical evolution of Rayleigh–Taylor and Richtmyer–Meshkov mixing fronts. *Phys. Rev. E*, **66**, 036312.

Cheng, B., Glimm, J., and Sharp, D.H. 2002b. A three-dimensional renormalization group bubble merger model for Rayleigh–Taylor mixing. *Chaos*, **12**, 267.

Cherfils, C., and Lafitte, O. 2000. Analytic solutions of the Rayleigh equation for linear density profiles. *Phys. Rev. E*, **62**, 2967.

Cherfils, C., and Mikaelian, K.O. 1996. Simple model for the turbulent mixing width at an ablating surface. *Phys. Fluids*, **8**, 522.

Cherne, F.J., Dimonte, G., and Germann, T.C. 2012. Richtmyer–Meshkov instability examinedwith large-scale molecular dynamics simulations. *AIP Conf. Proc.*, **1426**, 1307.

Cherne, F.J., Hammerberg, J.E., Andrews, M.J., Karkhanis, V., and Ramaprabhu, P. 2015. On shock driven jetting of liquid from non-sinusoidal surfaces into a vacuum. *J. Appl. Phys.*, **118**, 185901.

Chertkov, M. 2003. Phenomenology of Rayleigh–Taylor turbulence. *Phys. Rev. Lett.*, **91**, 115001.

Chertkov, M., Kolokolov, I., and Lebedev, V. 2005. Effects of surface tension on immiscible Rayleigh–Taylor turbulence. *Phys. Rev. E*, **71**, 055301.

Chiesa, M., Mathiesen, V., Melheim, J.A., and Halvorsen, B. 2005. Numerical simulation of particulate flow by the Eulerian–Lagrangian and the Eulerian–Eulerian approach with application to a fluidized bed. *Comput. Chem. Eng.*, **29**, 291.

Chizhkov, M.N., Karlykhanov, N.G., Lykov, V.A., Shushlebin, A.N., Sokolov, L.V., and Timakova, M.S. 2005. Computational optimization of indirect-driven targets for ignition on the Iskra-6 laser facility. *Laser Part. Beams*, **23**, 261.

Chou, P.Y. 1945. On velocity correlations and the solutions of the equations of turbulent fluctuation. *Quart. Appl. Math.*, **3**, 38.

Chou, P.Y. 1940. On an extension of Reynolds' method of finding apparent stress and the nature of turbulence. *Chin. J. Phys*, **4**, 1.

Chow, F.K., and Moin, P. 2003. A further study of numerical errors in large-eddy simulations. *J. Comput. Phys.*, **184**, 366.

Chu, B.-T., and Kovásznay, L.S.G. 1958. Non-linear interactions in a viscous heat-conducting compressible gas. *J. Fluid Mech.*, **3**, 494.

Chu, D.K., Akl, E.A., Duda, S., et al. 2020. Physical distancing, face masks, and eye protection to prevent person-to-person transmission of SARS-CoV-2 and COVID-19: a systematic review and meta-analysis. *Lancet*, **395**, 1973.

Chung, D., and Pullin, D.I. 2010. Direct numerical simulation and large-eddy simulation of stationary buoyancy-driven turbulence. *J. Fluid Mech.*, **643**, 279.

Ciardi, A., Ampleford, D.J., Lebedev, S.V., and Stehle, C. 2008. Curved Herbig-Haro jets: simulations and experiments. *Astrophys. J.*, **678**, 968.

Clark, D.H., and Stephenson, F.R. 1977. *The Historical Supernovae*. Pergamon, Oxford, UK.

Clark, D.S., and Tabak, M. 2005a. Acceleration-and deceleration-phase nonlinear Rayleigh–Taylor growth at spherical interfaces. *Phys. Rev. E*, **72**, 056308.

Clark, D.S., and Tabak, M. 2005b. Nonlinear Rayleigh–Taylor growth in converging geometry. *Phys. Rev. E*, **71**, 055302.

Clark, D.S., and Tabak, M. 2006. Linear and nonlinear Rayleigh–Taylor growth at strongly convergent spherical interfaces. *Phys. Fluids*, **18**, 064106.

Clark, D.S., Haan, S.W., Cook, A.W., Edwards, M.J., Hammel, B.A., Koning, J.M., and Marinak, M.M. 2011. Short-wavelength and three-dimensional instability evolution in National Ignition Facility ignition capsule designs. *Phys. Plasmas*, **18**, 082701.

Clark, D.S., Hinkel, D.E., Eder, D.C., et al. 2013. Detailed implosion modeling of deuterium-tritium layered experiments on the National Ignition Facility. *Phys. Plasmas*, **20**, 056318.

Clark, D.S., Milovich, J.L., Hinkel, D.E., et al. 2014. A survey of pulse shape options for a revised plastic ablator ignition design. *Phys. Plasmas*, **21**, 112705.

Clark, D.S., Marinak, M.M., Weber, C.R., et al. 2015. Radiation hydrodynamics modeling of the highest compression inertial confinement fusion ignition experiment from the National Ignition Campaign. *Phys. Plasmas*, **22**, 022703.

Clark, D.S., Weber, C.R., Milovich, J.L., et al. 2016. Three-dimensional simulations of low foot and high foot implosion experiments on the National Ignition Facility. *Phys. Plasmas*, **23**, 056302.

Clark, D.S., Kritcher, A.L., Milovich, J.L., et al. 2017. Capsule modeling of high foot implosion experiments on the National Ignition Facility. *Plasma Phys. Control. Fusion*, **59**, 055006.

Clark, T.T. 2003. A numerical study of the statistics of a two-dimensional Rayleigh–Taylor mixing layer. *Phys. Fluids*, **15**, 2413.

Clark, T.T., and Zemach, C. 1998. Symmetries and the approach to statistical equilibrium in isotropic turbulence. *Phys. Fluids*, **10**, 2846.

Clark, T.T., and Zhou, Y. 2003. Self-similarity of two flows induced by instabilities. *Phys. Rev. E*, **68**, 066305.

Clark, T.T., and Zhou, Y. 2006. Growth rate exponents of Richtmyer–Meshkov mixing layers. *J. Appl. Mech.*, **73**, 461.

Clérouin, J.G., Cherfi, M.H., and Zérah, G. 1998. The viscosity of dense plasmas mixtures. *Europhys. Lett.*, **42**, 37.

Clift, R., Grace, J.R., and Weber, M.E. 1978. *Bubbles, Drops, and Particles*. Academic Press, New York.

Cloutman, L.D. 1991. *A numerical model of particulate transport*. Lawrence Livermore National Lab. Report, Livermore, CA.

Coble, R.L. 1963. A model for boundary diffusion controlled creep in polycrystalline materials. *J. Appl. Phys.*, **34**, 1679.

Cobos-Campos, F., and Wouchuk, J.G. 2014. Analytical asymptotic velocities in linear Richtmyer–Meshkov-like flows. *Phys. Rev. E*, **90**, 053007.

Cobos-Campos, F., and Wouchuk, J.G. 2016. Analytical scalings of the linear Richtmyer–Meshkov instability when a shock is reflected. *Phys. Rev. E*, **93**, 053111.

Cobos-Campos, F, and Wouchuk, JG. 2017. Analytical scalings of the linear Richtmyer–Meshkov instability when a rarefaction is reflected. *Phys. Rev. E*, **96**, 013102.

Cohen, R.H., Dannevik, W.P., Dimits, A.M., Eliason, D.E., Mirin, A.A., Zhou, Y., Porter, D.H., and Woodward, P.R. 2002. Three-dimensional simulation of a Richtmyer–Meshkov instability with a two-scale initial perturbation. *Phys. Fluids*, **14**, 3692.

Colagrossi, A., Marrone, S., Colagrossi, P., and Le Touzé, D. 2021. Da Vinci's observation of turbulence: a French-Italian study aiming at numerically reproducing the physics behind one of his drawings, 500 years later. *Phys. Fluids*, **33**, 115122.

Cole, RH. 1948. *Underwater Explosions*. Princeton University Press, Princeton, NJ.

Cole, R.L., and Tankin, R.S. 1973. Experimental study of Taylor instability. *Phys. Fluids*, **16**, 1810.

Colgate, S.A., and White, R.H. 1966. The hydrodynamic behavior of supernovae explosions. *Astrophys. J.*, **143**, 626.

Collins, B.D., and Jacobs, J.W. 2002. PLIF flow visualization and measurements of the Richtmyer–Meshkov instability of an air/SF6 interface. *J. Fluid Mech.*, **464**, 113.

Collins, G.S., Melosh, H.J., and Marcus, R.A. 2005. Earth impact effects program: a web-based computer program for calculating the regional environmental consequences of a meteoroid impact on Earth. *Meteorit. Planet. Sci.*, **40**, 817.

Colvin, J.D., Legrand, M., Remington, B.A., Schurtz, G., and Weber, S.V. 2003. A model for instability growth in accelerated solid metals. *J. Appl. Phys.*, **93**, 5287.

Comte-Bellot, G., and Corrsin, S. 1966. The use of a contraction to improve the isotropy of grid-generated turbulence. *J. Fluid Mech.*, **25**, 657.

Comte-Bellot, G., and Corrsin, S. 1971. Simple Eulerian time correlation of full-and narrow-band velocity signals in grid-generated, 'isotropic' turbulence. *J. Fluid Mech.*, **48**, 273.

Cong, Z., Guo, X., Si, T., and Luo, X. 2022. Experimental and theoretical studies on heavy fluid layers with reshock. *Phys. Fluids*, **34**, 104108.

Cook, A.W. 2009. Enthalpy diffusion in multicomponent flows. *Phys. Fluids*, **21**, 055109.

Cook, A.W., and Cabot, W.H. 2004. A high-wavenumber viscosity for high-resolution numerical methods. *J. Comput. Phys.*, **195**, 594.

Cook, A.W., and Dimotakis, P.E. 2001. Transition stages of Rayleigh–Taylor instability between miscible fluids. *J. Fluid Mech.*, **443**, 69.

Cook, A.W., and Zhou, Y. 2002. Energy transfer in Rayleigh–Taylor instability. *Phys. Rev. E*, **66**, 026312.

Cook, A.W., Cabot, W.H., and Miller, P.L. 2004. The mixing transition in Rayleigh–Taylor instability. *J. Fluid Mech.*, **511**, 333.

Cook, R.C., Koziocziemski, B.J., Nikroo, A., et al. 2008. National Ignition Facility target design and fabrication. *Laser Part. Beams*, **26**, 479.

Corey, A.T. 1994. *Mechanics of Immiscible Fluids in Porous Media.* Water Resources Publication, Highlands Ranch, CO.

Coronel, S.A., Veilleux, J.-C., and Shepherd, J.E. 2021. Ignition of stoichiometric hydrogen-oxygen by water hammer. *Proc. Combust. Inst.*, **38**, 3537.

Corrsin, S. 1951. On the spectrum of isotropic temperature fluctuations in an isotropic turbulence. *J. Appl. Phys.*, **22**, 469.

Cottle, A.E., and Polanka, M.D. 2016. Numerical and experimental results from a common-source high-g ultra-compact combustor. In: *Proceedings of the ASME Turbo Expo 2016: Turbomachinery Technical Conference and Exposition. Volume 4A: Combustion, Fuels and Emissions.* Seoul, South Korea.

Courant, R., and Friedrichs, K.O. 1948. *Supersonic Flow and Shock Waves.* Interscience Publishers (Academic Press), New York.

Courtiaud, S., Lecysyn, N., Damamme, G., Poinsot, T., and Selle, L. 2018. Analysis of mixing in high-explosive fireballs using small-scale pressurised spheres. *Shock Waves*, **29**, 339.

Craik, A.D.D. 2005. George Gabriel Stokes on water wave theory. *Annu. Rev. Fluid Mech.*, **37**, 23.

Craik, A.D.D. 2012. Lord Kelvin on fluid mechanics. *Eur. Phys. J. H*, **37**, 75.

Craven, B.A., Paterson, E.G., and Settles, G.S. 2010. The fluid dynamics of canine olfaction: unique nasal airflow patterns as an explanation of macrosmia. *J. R. Soc. Interface*, **7**, 933.

Craxton, R.S., Anderson, K.S., Boehly, T.R., et al. 2015. Direct-drive inertial confinement fusion: a review. *Phys. Plasmas*, **22**, 110501.

Crittenden, P.E., and Balachandar, S. 2018. The stability of the contact interface of cylindrical and spherical shock tubes. *Phys. Fluids*, **30**, 064101.

Cuneo, M.E., Herrmann, M.C., Sinars, D.B., et al. 2012. Magnetically driven implosions for inertial confinement fusion at Sandia National Laboratories. *IEEE Trans. Plasma Sci.*, **40**, 3222.

Curzon, F.L., Folkierski, A., Latham, R., and Nation, J.A. 1960. Experiments on the growth rate of surface instabilities in a linear pinched discharge. *Proc. R. Soc. A*, **257**, 386.

Dai, J.L., Sun, Y.B., Wang, C., Zeng, R.H., and Zou, L.Y. 2023. Linear analytical model for magneto-Rayleigh–Taylor and sausage instabilities in a cylindrical liner. *Phys. Plasmas*, **30**, 022704.

Daly, B.J. 1969. Numerical study of the effect of surface tension on interface instability. *Phys. Fluids*, **12**, 1340.

Daly, B.J., and Harlow, F.H. 1970. Transport equations in turbulence. *Phys. Fluids*, **13**, 2634.

Dalziel, S.B., and Mouet, V. 2021. Rayleigh–Taylor instability between unequally stratified layers. *Physica D*, **423**, 132907.

Dalziel, S.B., Linden, P.F., and Youngs, D.L. 1999. Self-similarity and internal structure of turbulence induced by Rayleigh–Taylor instability. *J. Fluid Mech.*, **399**, 1.

Dalziel, S.B., Patterson, M.D., Caulfield, C.P., and Coomaraswamy, I.A. 2008. Mixing efficiency in high-aspect-ratio Rayleigh–Taylor experiments. *Phys. Fluids*, **20**, 065106.

Danckwerts, P.V. 1952. The definition and measurement of some characteristics of mixtures. *Appl. Sci. Res. A*, **3**, 279.

Danson, C.N., and Gizzi, L.A. 2023. Inertial confinement fusion ignition achieved at the National Ignition Facility – an editorial. *High Power Laser Sci. Eng.*, **11**, e40.

Darrieus, G. 1938. *Propagation d'un front de flame La Technique Moderne (Paris) and in 1945 at Congres de Mecanique Appliquee.* Unpublished work.

Darrigol, O. 2002. Between hydrodynamics and elasticity theory: the first five births of the Navier–Stokes equation. *Arch. Hist. Exact Sci*, **56**, 95.

Dastidar, A.G., Amyotte, P.R., and Pegg, M.J. 1997. Factors influencing the suppression of coal dust explosions. *Fuel*, **76**, 663.

Dávalos-Orozco, L.A., and Aguilar-Rosas, J.E. 1989a. Rayleigh–Taylor instability of a continuously stratified fluid under a general rotation field. *Phys. Fluids A*, **1**, 1192.

Dávalos-Orozco, L.A., and Aguilar-Rosas, J.E. 1989b. Rayleigh–Taylor instability of a continuously stratified magnetofluid under a general rotation field. *Phys. Fluids A*, **1**, 1600.

Davey, M.K., and Whitehead Jr, J.A. 1981. Rotating Rayleigh–Taylor instability as a model of sinking events in the ocean. *Geophys. Astrophys. Fluid Dyn.*, **17**, 237.

Davidson, P.A. 2015. *Turbulence: An Introduction for Scientists and Engineers*. Oxford University Press, Oxford, UK.

Davies, R.M., and Taylor, G.I. 1950. The mechanics of large bubbles rising through extended liquids and through liquids in tubes. *Proc. R. Soc. London. Ser. A.*, **200**, 375.

Davies Wykes, M.S., and Dalziel, S.B. 2014. Efficient mixing in stratified flows: experimental study of a Rayleigh–Taylor unstable interface within an otherwise stable stratification. *J. Fluid Mech.*, **756**, 1027.

De Gennes, P.-G., Brochard-Wyart, F., and Quéré, D. 2013. *Capillarity and Wetting Phenomena: Drops, Bubbles, Pearls, Waves*. Springer, New York.

de la Calleja, E.M., Zetina, S., and Zenit, R. 2014. Rayleigh–Taylor instability creates provocative images in painting. *Phys. Fluids*, **26**, 091102.

de Toma, G., Casini, R., Burkepile, J.T., and Low, B.C. 2008. Rise of a dark bubble through a quiescent prominence. *Astrophys. J. Lett.*, **687**, L123.

Deardorff, J.W. 1980. Stratocumulus-capped mixed layers derived from a three-dimensional model. *Bound.-Layer Meteorol.*, **18**, 495.

Debacq, M., Fanguet, V., Hulin, J.-P., Salin, D., and Perrin, B. 2001. Self-similar concentration profiles in buoyant mixing of miscible fluids in a vertical tube. *Phys. Fluids*, **13**, 3097.

Debacq, M., Hulin, J.-P., Salin, D., Perrin, B., and Hinch, E.J. 2003. Buoyant mixing of miscible fluids of varying viscosities in vertical tubes. *Phys. Fluids*, **15**, 3846.

DeBoer, J.D., Noël, J., and St-Maurice, J.P. 2010. The effects of mesoscale regions of precipitation on the ionospheric dynamics, electrodynamics and electron density in the presence of strong ambient electric field. *Ann. Geophys.*, **28**, 1345.

DeNeef, P. 1975. Two waves on a beam-plasma system. *Phys. Fluids*, **18**, 1209.

Deng, J., Xie, W., Feng, S., Wang, M., Li, H., Song, S., Xia, M., Ce, J., He, A., Tian, Q., and Gu, Y. 2016. From concept to reality – a review to the primary test stand and its preliminary application in high energy density physics. *Matter Radiat. Extremes*, **1**, 48.

Denissen, N.A., Rollin, B., Reisner, J.M., and Andrews, M.J. 2014. The tilted rocket rig: a Rayleigh–Taylor test case for RANS models. *J. Fluids Eng.*, **136**, 091301.

Desjardins, T.R., Di Stefano, C.A., Day, T., et al. 2019. A platform for thin-layer Richtmyer–Meshkov at OMEGA and the NIF. *High Energy Density Phys.*, **33**, 100705.

Dewald, E.L., Pino, J.E., Tipton, R.E., Salmonson, J.D., Ralph, J., Hartouni, E., Khan, S.F., Hatarik, R., Young, C.V., Thorn, D., and Smalyuk, V.A. 2019. Pushered single shell implosions for mix and radiation trapping studies using high-Z layers on National Ignition Facility. *Phys. Plasmas*, **26**, 072705.

Dewald, E.L., MacLaren, S.A., Martinez, D.A., et al. 2022. First graded metal pushered single shell capsule implosions on the National Ignition Facility. *Phys. Plasmas*, **29**, 052707.

Di Stefano, C.A., Malamud, G., Henry de Frahan, M.T., et al. 2014. Observation and modeling of mixing-layer development in high-energy-density, blast-wave-driven shear flow. *Phys. Plasmas*, **21**, 056306.

Di Stefano, C.A., Malamud, G., Kuranz, C.C., Klein, S.R., and Drake, R.P. 2015a. Measurement of Richtmyer–Meshkov mode coupling under steady shock conditions and at high energy density. *High Energy Density Phys.*, **17**, 263.

Di Stefano, C.A., Malamud, G., Kuranz, C.C., Klein, S.R., Stoeckl, C., and Drake, R.P. 2015b. Richtmyer–Meshkov evolution under steady shock conditions in the high-energy-density regime. *Appl. Phys. Lett.*, **106**, 114103.

Di Stefano, C.A., Rasmus, A.M., Doss, F.W., Flippo, K.A., Hager, J.D., Kline, J.L., and Bradley, P.A. 2017. Multimode instability evolution driven by strong, high-energy-density shocks in a rarefaction-reflected geometry. *Phys. Plasmas*, **24**, 052101.

Di Stefano, C.A., Doss, F.W., Merritt, E.C., et al. 2020. Experimental measurement of two copropagating shocks interacting with an unstable interface. *Phys. Rev. E*, **102**, 043212.

Dickinson, R.E., Ridley, E.C., and Roble, R.G. 1981. A three-dimensional general circulation model of the thermosphere. *J. Geophys. Res. Space Phys.*, **86**, 1499.

Diegelmann, F., Tritschler, V., Hickel, S., and Adams, N. 2016a. On the pressure dependence of ignition and mixing in two-dimensional reactive shock-bubble interaction. *Combust. Flame*, **163**, 414.

Diegelmann, Felix, Hickel, Stefan, and Adams, Nikolaus A. 2016b. Shock Mach number influence on reaction wave types and mixing in reactive shock–bubble interaction. *Combust. Flame*, **174**, 85.

Diegelmann, F., Hickel, S., and Adams, N.A. 2017. Three-dimensional reacting shock–bubble interaction. *Combust. Flame*, **181**, 300.

Dimonte, G. 1982. Experimental test of modulation theory and stochasticity of nonlinear oscillations. *Phys. Fluids*, **25**, 604.

Dimonte, G. 2000. Spanwise homogeneous buoyancy-drag model for Rayleigh–Taylor mixing and experimental evaluation. *Phys. Plasmas*, **7**, 2255.

Dimonte, G. 2004. Dependence of turbulent Rayleigh–Taylor instability on initial perturbations. *Phys. Rev. E*, **69**, 056305.

Dimonte, G., and Ramaprabhu, P. 2010. Simulations and model of the nonlinear Richtmyer–Meshkov instability. *Phys. Fluids*, **22**, 014104.

Dimonte, G., and Remington, B. 1993. Richtmyer–Meshkov experiments on the Nova laser at high compression. *Phys. Rev. Lett.*, **70**, 1806.

Dimonte, G., and Schneider, M. 1996. Turbulent Rayleigh–Taylor instability experiments with variable acceleration. *Phys. Rev. E*, **54**, 3740.

Dimonte, G., and Schneider, M. 1997. Turbulent Richtmyer–Meshkov instability experiments with strong radiatively driven shocks. *Phys. Plasmas*, **4**, 4347.

Dimonte, G., and Schneider, M. 2000. Density ratio dependence of Rayleigh–Taylor mixing for sustained and impulsive acceleration histories. *Phys. Fluids*, **12**, 304.

Dimonte, G., and Tipton, R. 2006. K-L turbulence model for the self-similar growth of the Rayleigh–Taylor and Richtmyer–Meshkov instabilities. *Phys. Fluids*, **18**, 085101.

Dimonte, G., Morrison, J., Hulsey, S., et al. 1996a. A linear electric motor to study turbulent hydrodynamics. *Rev. Sci. Instrum.*, **67**, 302.

Dimonte, G., Frerking, C.E., Schneider, M., and Remington, B. 1996b. Richtmyer–Meshkov instability with strong radiatively driven shocks. *Phys. Plasmas*, **3**, 614.

Dimonte, G., Youngs, D.L., Dimits, A., et al. 2004. A comparative study of the turbulent Rayleigh–Taylor instability using high-resolution three-dimensional numerical simulations: the Alpha-Group collaboration. *Phys. Fluids*, **16**, 1668.

Dimonte, G., Ramaprabhu, P., Youngs, D.L., Andrews, M.J., and Rosner, R. 2005. Recent advances in the turbulent Rayleigh–Taylor instability. *Phys. Plasmas*, **12**, 056301.

Dimonte, G., Ramaprabhu, P., and Andrews, M. 2007. Rayleigh–Taylor instability with complex acceleration history. *Phys. Rev. E*, **76**, 046313.

Dimonte, G., Terrones, G., Cherne, F.J., Germann, T.C., Dupont, V., Kadau, K., Buttler, W.T., Oro, D.M., Morris, C., and Preston, D.L. 2011. Use of the Richtmyer–Meshkov instability to infer yield stress at high-energy densities. *Phys. Rev. Lett.*, **107**, 264502.

Dimonte, G., Terrones, G., Cherne, F.J., and Ramaprabhu, P. 2013. Ejecta source model based on the nonlinear Richtmyer–Meshkov instability. *J. Appl. Phys.*, **113**, 024905.

Dimotakis, P.E. 2000. The mixing transition in turbulent flows. *J. Fluid Mech.*, **409**, 69.

Dimotakis, P.E. 2005. Turbulent mixing. *Annu. Rev. Fluid Mech.*, **37**, 329.

Dimotakis, P.E., and Samtaney, R. 2006. Planar shock cylindrical focusing by a perfect-gas lens. *Phys. Fluids*, **18**, 031705.

Ding, J., Si, T., Yang, J., Lu, X., Zhai, Z., and Luo, X. 2017. Measurement of a Richtmyer–Meshkov instability at an air-SF 6 interface in a semiannular shock tube. *Phys. Rev. Lett.*, **119**, 014501.

Ding, J., Deng, X., and Luo, X. 2021. Convergent Richtmyer–Meshkov instability on a light gas layer with perturbed inner and outer surfaces. *Phys. Fluids*, **33**, 102112.

Dittrich, T.R., Hurricane, O.A., Callahan, D.A., et al. 2014. Design of a high-foot high-adiabat ICF capsule for the National Ignition Facility. *Phys. Rev. Lett.*, **112**, 055002.

Do, A., Angulo, A.M., Nagel, S.R., Hall, G.N., Bradley, D.K., Hsing, W., Pickworth, L., Izumi, N., Robey, H., and Zhou, Y. 2022. High spatial resolution and contrast radiography of hydrodynamic instabilities at the National Ignition Facility. *Phys. Plasmas*, **29**, 080703.

Dobran, F., Neri, A., and Macedonio, G. 1993. Numerical simulation of collapsing volcanic columns. *J. Geophys. Res. Solid Earth*, **98**, 4231.

Dodd, M.S., and Ferrante, A. 2014. A fast pressure-correction method for incompressible two-fluid flows. *J. Comput. Phys.*, **273**, 416.

Dolai, B., and Prajapati, R.P. 2018. The rotating Rayleigh–Taylor instability in a strongly coupled dusty plasma. *Phys. Plasmas*, **25**, 083708.

Doludenko, A.N., and Fortova, S.V. 2015. Numerical simulation of Rayleigh–Taylor instability in inviscid and viscous media. *Comput. Math. Math. Phys.*, **55**, 874.

Domaradzki, J.A. 2021. Toward autonomous large eddy simulations of turbulence based on interscale energy transfer among resolved scales. *Phys. Rev. Fluids*, **6**, 104606.

Dong, M., Fan, Z.F., and Yu, C.X. 2019. Multiple eigenmodes of the Rayleigh–Taylor instability observed for a fluid interface with smoothly varying density. II. Asymptotic solution and its interpretation. *Phys. Rev. E*, **99**, 013109.

Donzis, D.A., and Sreenivasan, K.R. 2010. The bottleneck effect and the Kolmogorov constant in isotropic turbulence. *J. Fluid Mech.*, **657**, 171.

Dorofeev, S.B. 2011. Flame acceleration and explosion safety applications. *Proc. Combust. Inst.*, **33**, 2161.

Doss, F.W., Fincke, J.R., Loomis, E.N., Welser-Sherrill, L., and Flippo, K.A. 2013. The high-energy-density counterpropagating shear experiment and turbulent self-heating. *Phys. Plasmas*, **20**, 122704.

Doss, F.W., Kline, J.L., Flippo, K.A., et al. 2015. The Shock/Shear platform for planar radiation-hydrodynamics experiments on the National Ignition Facility. *Phys. Plasmas*, **22**, 056303.

Douglas, M.R., Deeney, C., and Roderick, N.F. 1997. Density and velocity statistics in variable density turbulent mixing. *Phys. Rev. Lett.*, **78**, 4577.

Douglas, M.R., De Groot, J.S., and Spielman, R.B. 2001. The magneto-Rayleigh–Taylor instability in dynamic Z pinches. *Laser Part. Beams*, **19**, 527.

Drake, R.P. 1999. Laboratory experiments to simulate the hydrodynamics of supernova remnants and supernovae. *J. Geophys. Res. Space Phys.*, **104**, 14505.

Drake, R.P. 2018. *High Energy Density Physics*. 2nd edn. Springer, Berlin.

Drake, R.P., Harding, E.C., and Kuranz, C.C. 2008. Approaches to turbulence in high-energy-density experiments. *Phys. Scr.*, **T132**, 014022.

Drazin, P.G., and Reid, W.H. 1981. *Hydrodynamic Stability*. Cambridge University Press, Cambridge, UK.

Drikakis, D. 2003. Advances in turbulent flow computations using high-resolution methods. *Prog. Aerosp. Sci.*, **39**, 405.

Drikakis, D., Grinstein, F., and Youngs, D. 2005. On the computation of instabilities and symmetry-breaking in fluid mechanics. *Prog. Aerosp. Sci.*, **41**, 609.

Duan, R., Jiang, F., and Yin, J. 2015. Rayleigh–Taylor instability for compressible rotating flows. *Acta Math. Scientia*, **35**, 1359.

Duff, R.E., Harlow, F.H., and Hirt, C.W. 1962. Effects of diffusion on interface instability between gases. *Phys. Fluids*, **5**, 417.

Duffell, P.C. 2016. A One-dimensional model for Rayleigh–Taylor instability in supernova remnants. *Astrophys. J.*, **821**, 76.

Dufreche, J.F., and Clerouin, J. 2000. Viscosity coefficient of dense fluid hydrogen. *J. de Physique-Colloques*, **10**, 303.

Dukler, Y., Ji, H., Falcon, C., and Bertozzi, A.L. 2020. Theory for undercompressive shocks in tears of wine. *Phys. Rev. Fluids*, **5**, 034002.

Dumitrescu, D. T. 1943. Strömung an einer Luftblase im senkrechten Rohr. *ZAMM – J. Appl. Math. and Mechanics/Zeitschrift für Angewandte Mathematik und Mechanik*, **23**, 139.

Dungey, J.W. 1956. Convective diffusion in the equatorial F region. *J. Atmos. Terr. Phys.*, **9**, 304.

Durand, O., and Soulard, L. 2012. Large-scale molecular dynamics study of jet breakup and ejecta production from shock-loaded copper with a hybrid method. *J. Appl. Phys.*, **111**, 044901.

Durand, O., and Soulard, L. 2013. Power law and exponential ejecta size distributions from the dynamic fragmentation of shock-loaded Cu and Sn metals under melt conditions. *J. Appl. Phys.*, **114**, 194902.

Durand, O., and Soulard, L. 2015. Mass-velocity and size-velocity distributions of ejecta cloud from shock-loaded tin surface using atomistic simulations. *J. Appl. Phys.*, **117**, 165903.

Durand, O., Jaouen, S., Soulard, L., Heuze, O., and Colombet, L. 2017. Comparative simulations of microjetting using atomistic and continuous approaches in the presence of viscosity and surface tension. *J. Appl. Phys.*, **122**, 135107.

Durand, O., Soulard, L., Colombet, L., and Prat, R. 2020. Influence of the phase transitions of shock-loaded tin on microjetting and ejecta production using molecular dynamics simulations. *J. Appl. Phys.*, **127**, 175901.

Düring, G., Josserand, C., Krstulovic, G., and Rica, S. 2019. Strong turbulence for vibrating plates: emergence of a Kolmogorov spectrum. *Phys. Rev. Fluids*, **4**, 064804.

Dávalos-Orozco, L.A. 1996. Rayleigh–Taylor stability of a two-fluid system under a general rotation field. *Dyn. Atmos. Oceans*, **23**, 247.

Dyachkov, S., Parshikov, A., and Zhakhovsky, V. 2017. Ejecta from shocked metals: comparative simulations using molecular dynamics and smoothed particle hydrodynamics. *AIP Conf. Proc.*, **1793**, 100024.

Eckart, C. 1948. An analysis of the stirring and mixing processes in compressible fluids. *J. Mar. Res.*, **7**, 265.

Eckhoff, R.K. 2005. Current status and expected future trends in dust explosion research. *J. Loss Prevent. Proc. Ind.*, **18**, 225.

Edwards, J., Lorenz, K.T., Remington, B.A., Pollaine, S., Colvin, J., Braun, D., Lasinski, B.F., Reisman, D., McNaney, J.M., Greenough, J.A., and Wallace, R. 2004. Laser-driven plasma loader for shockless compression and acceleration of samples in the solid state. *Phys. Rev. Lett.*, **92**, 075002.

Edwards, J., Marinak, M., Dittrich, T., Haan, S., Sanchez, J., Klingmann, J., and Moody, J. 2005. The effects of fill tubes on the hydrodynamics of ignition targets and prospects for ignition. *Phys. Plasmas*, **12**, 056318.

Edwards, M.J., Patel, P.K., Lindl, J.D., et al. 2013. Progress towards ignition on the National Ignition Facility. *Phys. Plasmas*, **20**, 070501.

Edwards, R. 2011. *The Heart of Ma Yuan: The Search for a Southern Song Aesthetic*. Hong Kong University Press, Hong Kong.

Eggers, J. 1997. Nonlinear dynamics and breakup of free-surface flows. *Rev. Mod. Phys.*, **69**, 865.

El Rafei, M., and Thornber, B. 2020. Numerical study and buoyancy–drag modeling of bubble and spike distances in three-dimensional spherical implosions. *Phys. Fluids*, **32**, 124107.

El Rafei, M., Flaig, M., Youngs, D.L., and Thornber, B. 2019. Three-dimensional simulations of turbulent mixing in spherical implosions. *Phys. Fluids*, **31**, 114101.

El Rafei, M., and Thornber, B. 2024. Turbulence statistics and transport in compressible mixing driven by spherical implosions with narrowband and broadband initial perturbations. *Physi. Rev. Fluids*, **9**, 034501.

Elbaz, Y., and Shvarts, D. 2018. Modal model mean field self-similar solutions to the asymptotic evolution of Rayleigh–Taylor and Richtmyer–Meshkov instabilities and its dependence on the initial conditions. *Phys. Plasmas*, **25**, 062126.

Elliott, L.A. 1965a. Calculation of the growth of interface instabilities by a Lagrangian Mesh Method. Pages 316–320 of: *Proceedings of the Fourth International Symposium on Detonation.* U.S. Naval Ordnance Lab., White Oak, MD.

Elliott, L.A. 1965b. Perturbations on fluid surfaces. *Proc. R. Soc. London. Ser. A*, **284**, 397.

Ellis, S., and Stöckhert, B. 2004. Elevated stresses and creep rates beneath the brittle–ductile transition caused by seismic faulting in the upper crust. *J. Geophys. Res. Solid Earth*, **109**, B05407.

Elyanov, A., Golub, V., and Volodin, V. 2018. Conditions for the development of Rayleigh–Taylor instability on the spherical flame front. *J. Phys. Conf. Series*, **1129**, 012011.

Emmons, H.W., Chang, C.T., and Watson, B.C. 1960. Taylor instability of finite surface waves. *J. Fluid Mech.*, **7**, 177.

Epstein, R. 2004. On the Bell–Plesset effects: the effects of uniform compression and geometrical convergence on the classical Rayleigh–Taylor instability. *Phys. Plasmas*, **11**, 5114.

Erez, L., Sadot, O., Oron, Da., Erez, G., Levin, L.A., Shvarts, D., and Ben-Dor, G. 2000. Study of the membrane effect on turbulent mixing measurements in shock tubes. *Shock Waves*, **10**, 241.

Evans, R.G. 1986. The influence of self-generated magnetic fields on the Rayleigh–Taylor instability. *Plasma Phys. Control. Fusion*, **28**, 1021.

Eyink, G.L. 2005. Locality of turbulent cascades. *Physica D*, **207**, 91.

Eyink, G.L., and Aluie, H. 2009. Localness of energy cascade in hydrodynamic turbulence. I. Smooth coarse graining. *Phys. Fluids*, **21**, 115107.

Falkovich, G. 2011. The Russian school. Pages 209–237 of: Davidson, P.A., Kaneda, Y., Moffatt, K., and Sreenivasan, K.R. (eds), *A Voyage through Turbulence*. Cambridge University Press, Cambridge, UK.

Falle, S.A., and Komissarov, S.S. 2001. On the inadmissibility of non-evolutionary shocks. *J. Plasma Phys.*, **65**, 29.

Fan, M., Zhai, Z., Si, T., Luo, X., Zou, L., and Tan, D. 2012. Numerical study on the evolution of the shock-accelerated SF 6 interface: influence of the interface shape. *Sci. China-Phys. Mech. Astron.*, **55**, 284.

Fan, Z., and Dong, M. 2020. Multiple eigenmodes of the Rayleigh–Taylor instability observed for a fluid interface with smoothly varying density. III. Excitation and nonlinear evolution. *Phys. Rev. E*, **101**, 063103.

Farley, D.R., Peyser, T.A., Logory, L.M., Murray, S.D., and Burke, E.W. 1999. High Mach number mix instability experiments of an unstable density interface using a single-mode, nonlinear initial perturbation. *Phys. Plasmas*, **6**, 4304.

Favre, A. 1965. Equations des gaz turbulents compressibles. I Formes générales. *J. Méc.*, **4**, 361.

Ferguson, K. 2022. *The Richtmyer–Meshkov instability in reshock in a dual driver vertical shock tube*. Ph.D. thesis, University of Arizona Tucson, AZ.

Ferguson, K., Wang, K.M., and Morgan, B.E. 2023. Mass and momentum transport in the Tilted Rocket Rig experiment. *Phys. Rev. Fluids*, **8**, 094502.

Fermi, E., and Von Neumann, J. 1955. *Taylor instability of incompressible liquids*. Tech. rept. 2979. United States Atomic Energy Commission, Technical Information Service, Washington DC.

Ferrari, A. 1998. Modeling extragalactic jets. *Annu. Rev. Astron. Astrophys.*, **36**, 539.

Feynman, R.P. 1963. *Feynman Lecture on Physics*. Addison-Wesley, Reading, MA.

Fincke, J.R., Lanier, N.E., Batha, S.H., et al. 2004. Postponement of saturation of the Richtmyer–Meshkov instability in a convergent geometry. *Phys. Rev. Lett.*, **93**, 115003.

Fincke, J.R., Lanier, N.E., Batha, S.H., et al. 2005. Effect of convergence on growth of the Richtmyer–Meshkov instability. *Laser Part. Beams*, **23**, 21.

Fisher, A., Branch, D., Nugent, P., and Baron, E. 1997. Evidence for a high-velocity carbon-rich layer in the Type Ia SN 1990N. *Astrophys. J. Lett.*, **481**, L89.

Fisher, R.V. 1979. Models for pyroclastic surges and pyroclastic flows. *J. Volcanol. Geotherm.*, **6**, 305.

Flaig, M., Clark, D., Weber, C., Youngs, D.L., and Thornber, B. 2018. Single-mode perturbation growth in an idealized spherical implosion. *J. Comput. Phys.*, **371**, 801.

Fleurot, N., Cavailler, C., and Bourgade, J.L. 2005. The Laser Megajoule (LMJ) Project dedicated to inertial confinement fusion: development and construction status. *Fusion Eng. Design*, **74**, 147.

Flippo, K.A., Doss, F.W., Kline, J.L., et al. 2016. Late-time mixing sensitivity to initial broadband surface roughness in high-energy-density shear layers. *Phys. Rev. Lett.*, **117**, 225001.

Flippo, K.A., Doss, F.W., Merritt, E.C., et al. 2018. Late-time mixing and turbulent behavior in high-energy-density shear experiments at high Atwood numbers. *Phys. Plasmas*, **25**, 056315.

Folgarait, L. 1987. *So Far from Heaven: David Alfaro Siqueiros' The March of Humanity and Mexican Revolutionary Politics*. Cambridge University Press, Cambridge, UK.

Forster, D., Nelson, D.R., and Stephen, M.J. 1977. Large-distance and long-time properties of a randomly stirred fluid. *Phys. Rev. A*, **16**, 732.

Foster, J.M., Wilde, B.H., Rosen, P.A., Williams, R.J.R., Blue, B.E., Coker, R.F., Drake, R.P., Frank, A., Keiter, P.A., Khokhlov, A.M., and Knauer, J.P. 2005. High-energy-density laboratory astrophysics studies of jets and bow shocks. *Astrophys. J. Lett.*, **634**, L77.

Fournier, J.B., and Cazabat, A.M. 1992. Tears of wine. *Europhys. Lett.*, **20**, 517.

Fournier, J.-D., and Frisch, U. 1978. d-Dimensional turbulence. *Phys. Rev. A*, **17**, 747.

Fraley, G. 1986. Rayleigh–Taylor stability for a normal shock wave–density discontinuity interaction. *Phys. Fluids*, **29**, 376.

Freed, M.S., McKenzie, D.E., Longcope, D.W., and Wilburn, M. 2016. Analysis of flows inside quiescent prominences as captured by Hinode/Solar Optical Telescope. *Astrophys. J.*, **818**, 57.

Frisch, U. 1995. *Turbulence: The Legacy of AN Kolmogorov*. Cambridge University Press, Cambridge, UK.

Frisch, U., Sulem, P.-L., and Nelkin, M. 1978. A simple dynamical model of intermittent fully developed turbulence. *J. Fluid Mech.*, **87**, 719.

Frisch, U., Kurien, S., Pandit, R., Pauls, W., Ray, S.S., Wirth, A., and Zhu, J.-Z. 2008. Hyperviscosity, Galerkin truncation, and bottlenecks in turbulence. *Phys. Rev. Lett.*, **101**, 144501.

Frost, D.L. 2018. Heterogeneous/particle-laden blast waves. *Shock Waves*, **28**, 439.

Frost, D.L., Zarei, Z., and Zhang, F. 2005. Instability of combustion products interface from detonation of heterogeneous explosives. In: *20th International Colloquium on the Dynamics of Explosions and Reactive Systems*. Montreal, Canada.

Frost, D.L., Gregoire, Y., Petel, O., Goroshin, S., and Zhang, F. 2012. Particle jet formation during explosive dispersal of solid particles. *Phys. Fluids*, **24**, 1109.

Fu, C., Zhao, Z., Xu, X., Wang, P., Liu, N., Wan, Z., and Lu, X. 2022. Nonlinear saturation of bubble evolution in a two-dimensional single-mode stratified compressible Rayleigh–Taylor instability. *Phys. Rev. Fluids*, **7**, 023902.

Fu, C., Zhao, Z., Wang, P., Liu, N., Wan, Z., and Lu, X. 2023. Bubble re-acceleration behaviours in compressible Rayleigh–Taylor instability with isothermal stratification. *J. Fluid Mech.*, **954**, A16.

Fu, Y., Yu, C., and Li, X. 2020. Energy transport characteristics of converging Richtmyer–Meshkov instability. *AIP Adv.*, **10**, 105302.

Fuller-Rowell, T.J., and Rees, D.A. 1980. A three dimensional time dependent global model of the thermosphere. *J. Atmos. Sci.*, **37**, 2545.

Fuller-Rowell, T.J., Rees, D., Quegan, S., Moffett, R.J., and Bailey, G.J. 1987. Interactions between neutral thermospheric composition and the polar ionosphere using a coupled ionosphere–thermosphere model. *J. Geophys. Res.*, **92**, 7744.

Fung, J., Harrison, A.K., Chitanvis, S., and Margulies, J. 2013. Ejecta source and transport modeling in the FLAG hydrocode. *Comput. Fluids*, **83**, 177.

Gad-el Hak, M. 1998. Fluid mechanics from the beginning to the third millennium. *Int. J. Eng. Edu.*, **14**, 177.

Gage, K.S. 1979. Evidence for a $k^{-5/3}$ law inertial range in mesoscale two-dimensional turbulence. *J. Atmos. Sci.*, **36**, 1950.

Galachov, I.V., Garanin, S.G., Eroshenko, V.A., Kirillov, G.A., Kochemasov, G.G., Murugov, V.M., Rukavishnikov, N.N., and Sukharev, S.A. 1999. Conception of the Iskra-6 Nd-laser facility. *Fusion Eng. Design*, **44**, 51.

Galmiche, D., and Gauthier, S. 1996. On the Reynolds number in laser experiments. *Jap. J. Appl. Phys.*, **35**, 4516.

Gamezo, V.N., Khokhlov, A.M., Oran, E.S., Chtchelkanova, A.Y., and Rosenberg, R.O. 2003. Thermonuclear supernovae: simulations of the deflagration stage and their implications. *Science*, **299**, 77.

Gancedo, F., Granero-Belinchón, R., and Scrobogna, S. 2020. Surface tension stabilization of the Rayleigh–Taylor instability for a fluid layer in a porous medium. *Annales de l'Institut Henri Poincaré C, Analyse non linéaire*, **37**, 1299.

García-Rubio, F., Betti, R., Sanz, J., and Aluie, H. 2021. Magnetic-field generation and its effect on ablative Rayleigh–Taylor instability in diffusive ablation fronts. *Phys. Plasmas*, **28**, 012103.

Garcia-Segura, Guillermo, and Mac Low, Mordecai-Mark. 1995. Wolf-Rayet bubbles. II. Gasdynamical simulations. *Astrophys. J.*, **455**, 160.

García-Senz, D., Cabezón, R., and Domínguez, I. 2018. Surface and core detonations in rotating white dwarfs. *Astrophys. J.*, **862**, 27.

Gardner, C.L., Glimm, J., McBryan, O., Menikoff, R., Sharp, D.H., and Zhang, Q. 1988. The dynamics of bubble growth for Rayleigh–Taylor unstable interfaces. *Phys. Fluids*, **31**, 447.

Gat, I., Matheou, G., Chung, D., and Dimotakis, P.E. 2017. Incompressible variable-density turbulence in an external acceleration field. *J. Fluid Mech.*, **827**, 506.

Gauthier, S. 2013. Compressibility effects in Rayleigh–Taylor flows: influence of the stratification. *Phys. Scr.*, **T155**, 014012.

Gauthier, S. 2017. Compressible Rayleigh–Taylor turbulent mixing layer between Newtonian miscible fluids. *J. Fluid Mech.*, **830**, 211.

Gauthier, S., and Bonnet, M. 1990. A K-ε model for turbulent mixing in shock-tube flows induced by Rayleigh–Taylor instability. *Phys. Fluids A*, **2**, 1685.

Ge, J., Zhang, X., Li, H., and Tian, B. 2020. Late-time turbulent mixing induced by multimode Richtmyer–Meshkov instability in cylindrical geometry. *Phys. Fluids*, **32**, 124116.

Ge, J., Li, H., Zhang, X., and Tian, B. 2022. Evaluating the stretching/compression effect of Richtmyer–Meshkov instability in convergent geometries. *J. Fluid Mech.*, **946**, A18.

Geim, A.K. 2011. Nobel Lecture: random walk to graphene. *Rev. Mod. Phys.*, **83**, 851.

George, E., Glimm, J., Li, X., Li, Y., and Liu, Xi. 2006. Influence of scale-breaking phenomena on turbulent mixing rates. *Phys. Rev. E*, **73**, 016304.

Gerashchenko, S., and Livescu, D. 2016. Viscous effects on the Rayleigh–Taylor instability with background temperature gradient. *Phys. Plasmas*, **23**, 072121.

Germann, T.C., Dimonte, G., Hammerberg, J.E., Kadau, K., Quenneville, J., and Zellner, M.B. 2009. Large-scale molecular dynamics simulations of particulate ejection and Richtmyer–Meshkov instability development in shocked copper. Page 1499 of: *DYMAT 2009–9th International Conference on the Mechanical and Physical Behavior of Materials under Dynamic Loading.* Brussels, Belgium.

Germano, M., Piomelli, U., Moin, P., and Cabot, W.H. 1991. A dynamic subgrid-scale eddy viscosity model. *Phys. Fluids A*, **3**, 1760.

Gharib, M., Kremers, D., Koochesfahani, M., and Kemp, M. 2002. Leonardo's vision of flow visualization. *Exp. Fluids*, **33**, 219.

Giacalone, J., and Jokipii, J.R. 2007. Magnetic field amplification by shocks in turbulent fluids. *Astrophys. J. Lett.*, **663**, L41.

Gillis, J. 1962. Book Review: hydrodynamic and hydromagnetic stability by S. Chandrasekhar. *Phys. Today*, **15**, 58.

Gilman, M., Smith, E., and Tsynkov, S. 2017. *Transionospheric Synthetic Aperture Imaging.* Springer, Cham, Switzerland.

Girichidis, P., Naab, T., Walch, S., and Berlok, T. 2021. The in situ formation of molecular and warm ionized gas triggered by hot galactic outflows. *Mon. Not. R. Astron. Soc.*, **505**, 1083.

Gittings, M., Weaver, R., Clover, M., et al. 2008. The RAGE radiation-hydrodynamic code. *Comput. Sci. Discov.* **1**, 015005.

Glendinning, S.G., Dixit, S.N., Hammel, B.A., et al. 1997. Measurement of a dispersion curve for linear-regime Rayleigh–Taylor growth rates in laser-driven planar targets. *Phys. Rev. Lett.*, **78**, 3318.

Glendinning, S.G., Colvin, J., Haan, S., et al. 2000. Ablation front Rayleigh–Taylor growth experiments in spherically convergent geometry. *Phys. Plasmas*, **7**, 2033.

Glendinning, S.G., Bolstad, J., Braun, D.G., Edwards, M.J., Hsing, W.W., Lasinski, B.F., Louis, H., Miles, A., Moreno, J., Peyser, T.A., Remington, B.A., Robey, H.F., Turano, E.J., Verdon, C.P., and Zhou, Y. 2003. Effect of shock proximity on Richtmyer–Meshkov growth. *Phys. Plasmas*, **10**, 1931.

Glimm, J., and Li, X.L. 1988. Validation of the Sharp–Wheeler bubble merger model from experimental and computational data. *Phys. Fluids*, **31**, 2077.

Glimm, J., Grove, J.W., Li, X.L., Oh, W., and Sharp, D.H. 2001. A critical analysis of Rayleigh–Taylor growth rates. *J. Comput. Phys.*, **169**, 652.

Glimm, J., Li, X.-L., and Lin, A.-D. 2002. Nonuniform approach to terminal velocity for single mode Rayleigh–Taylor instability. *Acta Math. Appl. Sin.*, **18**, 1.

Glimm, J., Sharp, D.H., Kaman, T., and Lim, H. 2013. New directions for Rayleigh–Taylor mixing. *Phil. Trans. R. Soc. A*, **371**, 20120183.

Glimm, J., Cheng, B., Sharp, D.H., and Kaman, T. 2020. A crisis for the verification and validation of turbulence simulations. *Physica D*, **404**, 132346.

Goedbloed, J.P., and Keppens, R. 2004. *Principles of Magnetohydrodynamics*. Cambridge University Press, Cambridge, UK.

Goedbloed, J.P., Keppens, R., and Poedts, S. 2010. *Advanced Magnetohydrodynamics: With Applications to Laboratory and Astrophysical Plasmas*. Cambridge University Press, Cambridge, UK.

Gombosi, T.I. 1994. *Gaskinetic Theory*. Cambridge University Press, Cambridge, UK.

Gomez, M.R., Slutz, S.A., Sefkow, A.B., Sinars, D.B., Hahn, K.D., Hansen, S.B., Harding, E.C., Knapp, P.F., Schmit, P.F., Jennings, C.A., and Awe, T.J. 2014. Experimental demonstration of fusion-relevant conditions in magnetized liner inertial fusion. *Phys. Rev. Lett.*, **113**, 155003.

Gomez, M.R., Slutz, S.A., Jennings, C.A., et al. 2020. Performance scaling in magnetized liner inertial fusion experiments. *Phys. Rev. Lett.*, **125**, 155002.

Goncharov, V.N. 2002. Analytical model of nonlinear, single-mode, classical Rayleigh–Taylor instability at arbitrary Atwood numbers. *Phys. Rev. Lett.*, **88**, 134502.

Goncharov, V.N., and Li, D. 2005. Effects of temporal density variation and convergent geometry on nonlinear bubble evolution in classical Rayleigh–Taylor instability. *Phys. Rev. E*, **71**, 046306.

Goncharov, V.N., Betti, R., McCrory, R.L., Sorotokin, P., and Verdon, C.P. 1996a. Self-consistent stability analysis of ablation fronts with large Froude numbers. *Phys. Plasmas*, **3**, 1402.

Goncharov, V.N., Betti, R., McCrory, R.L., and Verdon, C.P. 1996b. Self-consistent stability analysis of ablation fronts with small Froude numbers. *Phys. Plasmas*, **3**, 4665.

González, R.F., Cantó, J., and Raga, A.C. 2018. Analytical and numerical models of P Cygni's nebula. *Mon. Not. R. Astron. Soc.*, **480**, 5092.

Gorodnichev, K.E., Zakharov, P.P., Glazyrin, S.I., and Kuratov, S.E. 2024. The interface instability development induced by the bulk density perturbations in accelerated media. *Phys. Fluids*, **36**, 014115.

Gosman, A.D., and Clerides, D. 1997. Diesel spray modelling: a review. In: *Proceedings of ILASS-Europe*. The Institute for Liquid Atomization and Spray Systems, Florence, Italy.

Gotoh, T., Watanabe, Y., Shiga, Y., Nakano, T., and Suzuki, E. 2007. Statistical properties of four-dimensional turbulence. *Phys. Rev. E*, **75**, 016310.

Gou, J.N., Zeng, R.H., Wang, C., and Sun, Y.B. 2022. Analytical model for viscous and elastic Rayleigh–Taylor instabilities in convergent geometries at static interfaces. *AIP Adv.*, **12**, 075217.

Gowardhan, A.A., Ristorcelli, J.R., and Grinstein, F.F. 2011. The bipolar behavior of the Richtmyer–Meshkov instability. *Phys. Fluids*, **23**, 071701.

Grabovskii, E.V., Mitrofanov, K.N., Oleinik, G.M., and Porofeev, I.Yu. 2004. X-ray backlighting of the periphery of an imploding multiwire array in the Angara-5-1 facility. *Plasma Phys. Rep.*, **30**, 121.

Graham, M.J., and Zhang, Q. 2000. Numerical simulations of deep nonlinear Richtymer-Meshkov instability. *Astrophys. J. Suppl. Ser.*, **127**, 339.

Grant, G., Brenton, J., and Drysdale, D. 2000. Fire suppression by water sprays. *Progr. Energy Combust. Sci.*, **26**, 79.

Grant, H.L., Stewart, R.W., and Moilliet, A. 1962. Turbulence spectra from a tidal channel. *J. Fluid Mech.*, **12**, 241.

Gréa, B.-J. 2015. The dynamics of the $K - \mathcal{E}$ mix model toward its self-similar Rayleigh–Taylor solution. *J. Turbul.*, **16**, 184.

Gréa, B.J., and Briard, A. 2023. Inferring the magnetic field from the Rayleigh–Taylor instability. *Astrophys. J.*, **958**, 164.

Gréa, B.-J., and Soulard, O. 2019. Incompressible homogeneous buoyancy-driven turbulence. Pages 113–124 of: *Turbulent Cascades II*. Springer. Cham, Switzerland.

Gréa, B.-J., Burlot, A., Griffond, J., and Llor, A. 2016. Challenging mix models on transients to self-similarity of unstably stratified homogeneous turbulence. *J. Fluids Eng.*, **138**, 070904.

Greatbatch, I., Gosling, R.J., and Allen, S. 2015. Quantifying search dog effectiveness in a terrestrial search and rescue environment. *Wilderness Environ. Med.*, **26**, 327.

Green, B. 2015. Why string theory still offers hope we can unify physics. *Smithsonian Mag.*, www.smithsonianmag.com/science–nature/string–theory–about–unravel–180953637/.

Green, M.B., Schwarz, J.H., and Witten, E. 2012. *Superstring Theory: Vols. 1 and 2*. 25th anniversary edition. Cambridge University Press, Cambridge, UK.

Grégoire, O., Souffland, D., and Gauthier, S. 2005. A second-order turbulence model for gaseous mixtures induced by Richtmyer–Meshkov instability. *J. Turbul.*, **6**, N29.

Gregori, G., Reville, B., and Miniati, F. 2015. The generation and amplification of intergalactic magnetic fields in analogue laboratory experiments with high power lasers. *Phys. Rep.*, **601**, 1.

Grieves, B. 2007. 2D direct numerical simulation of ejecta production. In: *Proceedings of the 10th International Workshop on the Physics of Compressible Turbulent Mixing*. Paris, France.

Griffond, J., Gréa, B.-J., and Soulard, O. 2014. Unstably stratified homogeneous turbulence as a tool for turbulent mixing modeling. *J. Fluids Eng.*, **136**, 091201.

Griffond, J., Gréa, B.-J., and Soulard, O. 2015. Numerical investigation of self-similar unstably stratified homogeneous turbulence. *J. Turbul.*, **16**, 167.

Griffond, J., Haas, J.-F., Souffland, D., Bouzgarrou, G., Bury, Y., and Jamme, S. 2017. Experimental and numerical investigation of the growth of an air/SF6 turbulent mixing zone in a shock tube. *J. Fluids Eng.*, **139**, 091205.

Griffond, J., Soulard, O., and Gréa, B.-J. 2023. A modified dissipation equation for Reynolds stress turbulent mixing models. *J. Fluids Eng.*, **145**, 021502.

Grigoryev, S.Yu, Dyachkov, S.A., Parshikov, A.N., and Zhakhovsky, V.V. 2022. Limited and unlimited spike growth from grooved free surface of shocked solid. *J. Appl. Phys.*, **131**, 065104.

Grinstein, F.F., Margolin, L.G., and Rider, W.J. (eds). 2007. *Implicit Large Eddy Simulation*. Cambridge University Press, Cambridge, UK.

Grinstein, F.F., Gowardhan, A.A., and Wachtor, A.J. 2011. Simulations of Richtmyer–Meshkov instabilities in planar shock-tube experiments. *Phys. Fluids*, **23**, 034106.

Grinstein, F.F., Saenz, J.A., Rauenzahn, R.M., Germano, M., and Israel, D.M. 2020. Dynamic bridging modeling for coarse grained simulations of shock driven turbulent mixing. *Comput. Fluids*, **199**, 104430.

Gritschneder, M., Lin, D.N.C., Murray, S.D., Yin, Q.-Z., and Gong, M.-N. 2011. The supernova triggered formation and enrichment of our solar system. *Astrophys. J.*, **745**, 22.

Groom, M. 2020. *Direct numerical simulation of shock-induced turbulent mixing with high-resolution methods*. Ph.D. thesis, University of Sydney, Sydney, Australia.

Groom, M., and Thornber, B. 2019. Direct numerical simulation of the multimode narrowband Richtmyer–Meshkov instability. *Comput. Fluids*, **194**, 104309.

Groom, M., and Thornber, B. 2020. The influence of initial perturbation power spectra on the growth of a turbulent mixing layer induced by Richtmyer–Meshkov instability. *Physica D*, **407**, 132463.

Groom, M., and Thornber, B. 2021. Reynolds number dependence of turbulence induced by the Richtmyer–Meshkov instability using direct numerical simulations. *J. Fluid Mech.*, **908**, A31.

Gruber, H., and Gruber, V. 1956. Hermann von Helmholtz: nineteenth-century polymorph. *The Scientific Monthly*, **83**, 92.

Guan, B., Zhai, Z., Si, T., Lu, X., and Luo, X. 2017. Manipulation of three-dimensional Richtmyer–Meshkov instability by initial interfacial principal curvatures. *Phys. Fluids*, **29**, 032106.

Guan, B., Wang, D., Wang, G., Fan, E., and Wen, C.-Y. 2020. Numerical study of the Richtmyer–Meshkov instability of a three-dimensional minimum-surface featured SF6/air interface. *Phys. Fluids*, **32**, 024108.

Gull, S.F., and Longair, M.S. 1973. A numerical model of the structure and evolution of young supernova remnants. *Mon. Not. R. Astron. Soc.*, **161**, 47.

Gunár, S., Schwartz, P., Dudík, J., Schmieder, B., Heinzel, P., and Jurčák, J. 2014. Magnetic field and radiative transfer modelling of a quiescent prominence. *Astron. Astrophys.*, **567**, A123.

Guo, F., Li, S., Li, H., Giacalone, J., Jokipii, J.R., and Li, D. 2012. On the amplification of magnetic field by a supernova blast shock wave in a turbulent medium. *Astrophys. J.*, **747**, 98.

Guo, H.-Y., Yu, X.-J., Wang, L.-F., Ye, W.-H., Wu, J.-F., and Li, Y.-J. 2014. On the second harmonic generation through Bell–Plesset effects in cylindrical geometry. *Chin. Phys. Lett.*, **31**, 44702.

Guo, H.-Y., Wang, L.-F., Ye, W.-H., Wu, J.-F., and Zhang, W.-Y. 2017. Weakly nonlinear Rayleigh–Taylor instability in incompressible fluids with surface tension. *Chin. Phys. Lett.*, **34**, 045201.

Guo, H.-Y., Wang, L.-F., Ye, W.-H., Wu, J.-F., and Zhang, W.-Y. 2018. Weakly nonlinear Rayleigh–Taylor instability in cylindrically convergent geometry. *Chin. Phys. Lett.*, **35**, 055201.

Guo, W., and Zhang, Q. 2020. Universality and scaling laws among fingers at Rayleigh–Taylor and Richtmyer–Meshkov unstable interfaces in different dimensions. *Physica D*, **403**, 132304.

Guo, W., and Zhang, Q. 2022. Quantitative theory for spikes and bubbles in the Richtmyer–Meshkov instability at arbitrary density ratios in three dimensions. *Phys. Fluids*, **34**, 072115.

Guo, X., Zhai, Z., Si, T., and Luo, X. 2019. Bubble merger in initial Richtmyer–Meshkov instability on inverse-chevron interface. *Phys. Rev. Fluids*, **4**, 092001.

Guo, X., Zhai, Z., Ding, J., Si, T., and Luo, X. 2020. Effects of transverse shock waves on early evolution of multi-mode chevron interface. *Phys. Fluids*, **32**, 106101.

Guo, X., Si, T., Zhai, Z., and Luo, X. 2022a. Large-amplitude effects on interface perturbation growth in Richtmyer–Meshkov flows with reshock. *Phys. Fluids*, **34**, 082118.

Guo, X., Cong, Z., Si, T., and Luo, X. 2022b. Shock-tube studies of single-and quasi-single-mode perturbation growth in Richtmyer–Meshkov flows with reshock. *J. Fluid Mech.*, **941**, A65.

Guo, Y., and Tice, I. 2010. Linear Rayleigh–Taylor instability for viscous, compressible fluids. *SIAM J. Math. Anal.*, **42**, 1688.

Gupta, M.R., Roy, S., Khan, M., Pant, H.C., Sarkar, S., and Srivastava, M.K. 2009. Effect of compressibility on the Rayleigh–Taylor and Richtmyer–Meshkov instability induced nonlinear structure at two fluid interface. *Phys. Plasmas*, **16**, 032303.

Gupta, M.R., Banerjee, R., Mandal, L.K., Bhar, R., Pant, H.C., Khan, M., and Srivastava, M.K. 2012. Effect of viscosity and surface tension on the growth of Rayleigh–Taylor instability and Richtmyer–Meshkov instability induced two fluid interfacial nonlinear structure. *Indian J. Phys.*, **86**, 471.

Gupta, N.K., and Lawande, S.V. 1986. Rayleigh–Taylor instability in spherical geometry. *Phys. Rev. A*, **33**, 2813.

Haan, S.W. 1989. Onset of nonlinear saturation for Rayleigh–Taylor growth in the presence of a full spectrum of modes. *Phys. Rev. A*, **39**, 5812.

Haan, S.W. 1991. Weakly nonlinear hydrodynamic instabilities in inertial fusion. *Phys. Fluids B*, **3**, 2349.

Haan, S.W., Pollaine, S.M., Lindl, J.D., et al. 1995. Design and modeling of ignition targets for the National Ignition Facility. *Phys. Plasmas*, **2**, 2480.

Haas, J.-F., and Sturtevant, B. 1987. Interaction of weak shock waves with cylindrical and spherical gas inhomogeneities. *J. Fluid Mech.*, **181**, 41.

Habib, N., and Pierrehumbert, R.T. 2024. Modeling noncondensing compositional convection for applications to super-Earth and sub-Neptune atmospheres, *Astrophys. J.*, **961**, 35.

Hachisu, I., Matsuda, T., Nomoto, K., and Shigeyama, T. 1991. Rayleigh–Taylor instabilities and mixing in the helium star models for Type Ib/Ic supernovae. *Astrophys. J.*, **368**, L27.

Haehn, N., Ranjan, D., Weber, C., Oakley, J.G., Anderson, M.H., and Bonazza, R. 2010. Experimental investigation of a twice-shocked spherical density inhomogeneity. *Phys. Scr.*, **T142**, 014067.

Haehn, N, Weber, C, Oakley, J, Anderson, M, Ranjan, D, and Bonazza, R. 2011. Experimental investigation of a twice-shocked spherical gas inhomogeneity with particle image velocimetry. *Shock Waves*, **21**, 225.

Haehn, N., Ranjan, D., Weber, C., Oakley, J., Rothamer, D., and Bonazza, R. 2012. Reacting shock bubble interaction. *Combust. Flame*, **159**, 1339.

Haehn, N.S. 2012. *Experimental Investigation of the Reactive Shock-bubble Interaction*. Ph.D. thesis, The University of Wisconsin-Madison, Madison, WI.

Haines, B.M., Grinstein, F.F., and Schwarzkopf, J.D. 2013. Reynolds-averaged Navier–Stokes initialization and benchmarking in shock-driven turbulent mixing. *J. Turbul.*, **14**, 46.

Haines, B.M., Aldrich, C.H., Campbell, J.M., Rauenzahn, R.M., and Wingate, C.A. 2017. High-resolution modeling of indirectly driven high-convergence layered inertial confinement fusion capsule implosions. *Phys. Plasmas*, **24**, 052701.

Haines, B.M., Olson, R.E., Sweet, W., et al. 2019. Robustness to hydrodynamic instabilities in indirectly driven layered capsule implosions. *Phys. Plasmas*, **26**, 012707.

Haines, B.M., Sauppe, J.P., Keiter, P.A., et al. 2021. Constraining computational modeling of indirect drive double shell capsule implosions using experiments. *Phys. Plasmas*, **28**, 032709.

Haines, M.G. 2011. The past, present, and future of Z pinches. *Plasma Phys. Control. Fusion*, **53**, 093001.

Haines, M.G., Lebedev, S.V., Chittenden, J.P., Beg, F.N., Bland, S.N., and Dangor, A.E. 2000. The past, present, and future of Z pinches. *Phys. Plasmas*, **7**, 1672.

Hammel, B.A., Haan, S.W., Clark, D.S., Edwards, M.J., Langer, S.H., Marinak, M.M., Patel, M.V., Salmonson, J.D., and Scott, H.A. 2010. High-mode Rayleigh–Taylor growth in NIF ignition capsules. *High Energy Density Phys.*, **6**, 171.

Hammel, B.A., Tommasini, R., Clark, D.S., Field, J., Stadermann, M., and Weber, C. 2016. Simulations and experiments of the growth of the "tent" perturbation in NIF ignition implosions. *J. Physics: Conf. Ser.*, **717**, 012021.

Hammer, N.J., Janka, H.-Th., and Müller, E. 2010. Three-dimensional simulations of mixing instabilities in supernova explosions. *Astrophys. J.*, **714**, 1371.

Hamzehloo, A., Bartholomew, P., and Laizet, S. 2021. Direct numerical simulations of incompressible Rayleigh–Taylor instabilities at low and medium Atwood numbers. *Phys. Fluids*, **33**, 054114.

Han, L., Yuan, J., Dong, M., and Fan, Z. 2021. Secondary instability of the spike-bubble structures induced by nonlinear Rayleigh–Taylor instability with a diffuse interface. *Phys. Rev. E*, **104**, 035108.

Hanjalić, K., and Launder, B.E. 1972. A Reynolds stress model of turbulence and its application to thin shear flows. *J. Fluid Mech.*, **52**, 609.

Hansom, J.C.V., Rosen, P.A., Goldack, T.J., Oades, K., Fieldhouse, P., Cowperthwaite, N., Youngs, D.L., Mawhinney, N., and Baxter, A.J. 1990. Radiation driven planar foil instability and mix experiments at the AWE HELEN laser. *Laser Part. Beams*, **8**, 51.

Hanson, A.J., 1994. A construction for computer visualization of certain complex curves. *Not. Am. Math. Soc*, **41**, 1156.

Hao, Y., and Prosperetti, A. 1999. The dynamics of vapor bubbles in acoustic pressure fields. *Phys. Fluids*, **11**, 2008.

Harding E.C., Hansen J.F., Hurricane O.A., et al. 2009. Observation of a Kelvin–Helmholtz instability in a high-energy-density plasma on the Omega laser. *Phys. Rev. Lett.*, **103**, 045005.

Harig, C., Molnar, P., and Houseman, G.A. 2008. Rayleigh–Taylor instability under a shear stress free top boundary condition and its relevance to removal of mantle lithosphere from beneath the Sierra Nevada. *Tectonics*, **27**, TC6019.

Harlow, F.H., and Nakayama, P.I. 1967. Turbulence transport equations. *Phys. Fluids*, **10**, 2323.

Harris, E.G. 1962. Rayleigh–Taylor instabilities of a collapsing cylindrical shell in a magnetic field. *Phys. Fluids*, **5**, 1057.

Harrison, W.J. 1908. The influence of viscosity on the oscillations of superposed fluids. *Proc. London Math. Soc.*, **2**, 396.

Hartigan, P., Foster, J.M., Wilde, B.H., Coker, R.F., Rosen, P.A., Hansen, J.F., Blue, B.E., Williams, R.J.R., Carver, R., and Frank, A. 2009. Laboratory experiments, numerical simulations, and astronomical observations of deflected supersonic jets: application to HH 110. *Astrophys. J.*, **705**, 1073.

Hartsfield, T.M., Schulze, R.K., La Lone, B.M., et al. 2022. The temperatures of ejecta transporting in vacuum and gases. *J. Appl. Phys.*, **131**, 195104.

Hawley, J.F., and Zabusky, N.J. 1989. Vortex paradigm for shock-accelerated density-stratified interfaces. *Phys. Rev. Lett.*, **63**, 1241.

Haxhimali, T., Rudd, R.E., Cabot, W.H., and Graziani, F.R. 2015. Shear viscosity for dense plasmas by equilibrium molecular dynamics in asymmetric Yukawa ionic mixtures. *Phys. Rev. E*, **92**, 053110.

He, A., Liu, J., Liu, C., and Wang, P. 2018. Numerical and theoretical investigation of jet formation in elastic-plastic solids. *J. Appl. Phys.*, **124**, 185902.

He, Y., Hu, X., and Jiang, Z. 2008. Compressibility effects on the Rayleigh–Taylor instability growth rates. *Chin. Phys. Lett.*, **25**, 1015.

Hecht, J., Alon, U., and Shvarts, D. 1994. Potential flow models of Rayleigh–Taylor and Richtmyer–Meshkov bubble fronts. *Phys. Fluids*, **6**, 4019.

Heelis, R.A., Lowell, J.K., and Spiro, R.W. 1982. A model of the high-latitude ionospheric convection pattern. *J. Geophys. Res.*, **87**, 6339.

Heidt, L., Flaig, M., and Thornber, B. 2021. The effect of initial amplitude and convergence ratio on instability development and deposited fluctuating kinetic energy in the single-mode Richtmyer–Meshkov instability in spherical implosions. *Comput. Fluids*, **218**, 104842.

Heilig, W.H. 1969. Diffraction of a shock wave by a cylinder. *Phys. Fluids*, **12**, Suppl. 1, 154.

Heinzel, P., Schmieder, B., Fárník, F., Schwartz, P., Labrosse, N., Kotrč, P., Anzer, U., Molodij, G., Berlicki, A., DeLuca, E.E., and Golub, L. 2008. Hinode, TRACE, SOHO, and ground-based observations of a quiescent prominence. *Astrophys. J.*, **686**, 1383.

Heisenberg, W. 1948a. On the theory of statistical and isotropic turbulence. *Proc. R. Soc. A*, **195**, 402.

Heisenberg, W. 1948b. Zur statistischen Theorie der Turbulenz. *Zeitschrift für Physik*, **124**, 628.

Henry de Frahan, M.T., Belof, J.L., Cavallo, R.M., et al. 2015a. Experimental and numerical investigations of beryllium strength models using the Rayleigh–Taylor instability. *J. Appl. Phys.*, **117**, 225901.

Henry de Frahan, M.T., Varadan, S., and Johnsen, E. 2015b. A new limiting procedure for discontinuous Galerkin methods applied to compressible multiphase flows with shocks and interfaces. *J. Comput. Phys.*, **280**, 489.

Henry de Frahan, M.T., Movahed, P., and Johnsen, E. 2015c. Numerical simulations of a shock interacting with successive interfaces using the Discontinuous Galerkin method: the multilayered Richtmyer–Meshkov and Rayleigh–Taylor instabilities. *Shock Waves*, **25**, 329.

Herring, J.R. 1980. Statistical theory of quasi-geostrophic turbulence. *J. Atmos. Sci.*, **37**, 969.

Herrmann, M.C., Tabak, M., and Lindl, J.D. 2001. Ignition scaling laws and their application to capsule design. *Phys. Plasmas*, **8**, 2296.

Hester, J.J. 2008. The Crab Nebula: an astrophysical chimera. *Annu. Rev. Astron. Astrophys.*, **46**, 127.

Hewett, J.S., and Madnia, C.K. 1998. Flame–vortex interaction in a reacting vortex ring. *Phys. Fluids*, **10**, 189.

Hicks, E.P. 2015. Rayleigh–Taylor unstable flames – fast or faster? *Astrophys. J.*, **803**, 72.

Hide, R. 1955a. The character of the equilibrium of an incompressible heavy viscous fluid of variable density: an approximate theory. *Math. Proc. Cambridge Philos. Soc.*, **51**, 179.

Hide, R. 1955b. Waves a heavy, viscous, incompressible, electrically conducting fluid of variable density, in the presence of a magnetic field. *Proc. R. Soc. Lond. A*, **233**, 376.

Hide, R. 1956a. The character of the equilibrium of a heavy, viscous, incompressible, rotating fluid of variable density: I. General theory. *Q. J. Mech. Appl. Math.*, **9**, 22.

Hide, R. 1956b. The character of the equilibrium of a heavy, viscous, incompressible, rotating fluid of variable density: II. Two special cases. *Q. J. Mech. Appl. Math.*, **9**, 35.

Hill, D.J., Pantano, C., and Pullin, D.I. 2006. Large-eddy simulation and multiscale modelling of a Richtmyer–Meshkov instability with reshock. *J. Fluid Mech.*, **557**, 29.

Hillebrandt, W., and Hoflich, P. 1989. The supernova 1987a in the large magellanic cloud. *Rep. Prog. Phys.*, **52**, 1421.

Hillier, A. 2018. The magnetic Rayleigh–Taylor instability in solar prominences. *Rev. Mod. Plasma Phys.*, **2**, 1.

Hillier, A. 2019. Ideal MHD instabilities, With a focus on the Rayleigh–Taylor and Kelvin–Helmholtz instabilities. In: MacTaggart, D., and Hillier, A. (eds), *Topics in Magnetohydrodynamic Topology, Reconnection and Stability Theory*. Springer, Cham, Switzerland.

Hillier, A., and Polito, V. 2018. Observations of the Kelvin–Helmholtz instability driven by dynamic motions in a solar prominence. *Astrophys. J. Lett.*, **864**, L10.

Hillier, A., Isobe, H., Shibata, K., and Berger, T. 2011a. Numerical simulations of the magnetic Rayleigh–Taylor instability in the Kippenhahn-Schlüter prominence model. *Astrophys. J. Lett.*, **736**, L1.

Hillier, A., Isobe, H., and Watanabe, H. 2011b. Observations of plasma blob ejection from a quiescent prominence by Hinode solar optical telescope. *Publ. Astron. Soc. Jpn.*, **63**, L19.

Hillier, A., Berger, T., Isobe, H., and Shibata, K. 2012. Numerical simulations of the magnetic Rayleigh–Taylor instability in the Kippenhahn-Schlüter prominence model. I. Formation of upflows. *Astrophys. J.*, **746**, 120.

Hillier, A., Matsumoto, T., and Ichimoto, K. 2017. Investigating prominence turbulence with Hinode SOT Dopplergrams. *Astron. Astrophys.*, **597**, A111.

Hillier, A.S. 2016. On the nature of the magnetic Rayleigh–Taylor instability in astrophysical plasma: the case of uniform magnetic field strength. *Mon. Not. R. Astron. Soc.*, **462**, 2256.

Hinze, J.O. 1975. *Turbulence*. McGraw-Hill, New York.

Hirschfelder, J.O., Curtiss, C.F., and Bird, R.B. 1964. *Molecular Theory of Gases and Liquids*. Wiley, New York.

Ho, C.-M., and Huerre, P. 1984. Perturbed free shear layers. *Annu. Rev. Fluid Mech.*, **16**, 365.

Hohenberger, M., Chang, P.Y., Fiksel, G., Knauer, J.P., Betti, R., Marshall, F.J., Meyerhofer, D.D., Séguin, F.H., and Petrasso, R.D. 2012. Inertial confinement fusion implosions with imposed magnetic field compression using the OMEGA Laser. *Phys. Plasmas*, **19**, 056306.

Holder, D.A., and Barton, C.J. 2004. Shock tube Richtmyer–Meshkov experiments: inverse chevron and half height. In: *Proceedings of the Ninth International Workshop on the Physics of Compressible Turbulent Mixing*. University of Cambridge, Cambridge, UK.

Holder, D.A., Smith, A.V., Barton, C.J., and Youngs, D.L. 2003a. Mix experiments using a two-dimensional convergent shock-tube. *Laser Part. Beams*, **21**, 403.

Holder, D.A., Smith, A.V., Barton, C.J., and Youngs, D.L. 2003b. Shock-tube experiments on Richtmyer–Meshkov instability growth using an enlarged double-bump perturbation. *Laser Part. Beams*, **21**, 411.

Holford, J.M., Dalziel, S.B., and Youngs, D.L. 2003. Rayleigh–Taylor instability at a tilted interface in laboratory experiments and numerical simulations. *Laser Part. Beams*, **21**, 419.

Holmes, R.L., Dimonte, G., Fryxell, B., Gittings, M.L., Grove, J.W., Schneider, M., Sharp, D.H., Velikovich, A.L., Weaver, R.P., and Zhang, Q. 1999. Richtmyer–Meshkov instability growth: experiment, simulation and theory. *J. Fluid Mech.*, **389**, 55.

Hoogenboom, T., and Houseman, G.A. 2006. Rayleigh–Taylor instability as a mechanism for corona formation on Venus. *Icarus*, **180**, 292.

Hopps, N., Danson, C., Duffield, S., et al. 2013. Overview of laser systems for the Orion facility at the AWE. *Appl. Optics*, **52**, 3597.

Hopps, N., Oades, K., Andrew, J., et al. 2015. Comprehensive description of the Orion laser facility. *Plasma Phys. Control. Fusion*, **57**, 064002.

Horne, J.T., and Lawrie, A.G.W. 2020. Aspect-ratio-constrained Rayleigh–Taylor instability. *Physica D*, **406**, 132442.

Horowitz, C.J., Caballero, O.L., Lin, Z., O'Connor, E., and Schwenk, A. 2017. Neutrino-nucleon scattering in supernova matter from the virial expansion. *Phys. Rev. C*, **95**, 025801.

Hou, T.Y., Wu, X., Chen, S., and Zhou, Y. 1998. Effect of finite computational domain on turbulence scaling law in both physical and spectral spaces. *Phys. Rev. E*, **58**, 5841.

Houas, L., and Chemouni, I. 1996. Experimental investigation of Richtmyer–Meshkov instability in shock tube. *Phys. Fluids*, **8**, 614.

Houim, R.W., and Oran, E.S. 2016. A multiphase model for compressible granular–gaseous flows: formulation and initial tests. *J. Fluid Mech.*, **789**, 166.

Hsing, W.W., and Hoffman, N.M. 1997. Measurement of feedthrough and instability growth in radiation-driven cylindrical implosions. *Phys. Rev. Lett.*, **78**, 3876.

Hsing, W.W., Barnes, C.W., Beck, J.B., Hoffman, N.M., Galmiche, D., Richard, A., Edwards, J., Graham, P., Rothman, S., and Thomas, B. 1997. Rayleigh–Taylor instability evolution in ablatively driven cylindrical implosions. *Phys. Plasmas*, **4**, 1832.

Hu, X., Liang, H., and Wang, H. 2020. Lattice Boltzmann method simulations of the immiscible Rayleigh–Taylor instability with high Reynolds numbers. *Acta Physica Sinica*, **69**, 044701.

Hu, Z., Zhang, Y., Tian, B., He, Z., and Li, L. 2019. Effect of viscosity on two-dimensional single-mode Rayleigh–Taylor instability during and after the reacceleration stage. *Phys. Fluids*, **31**, 104108.

Huang, H., Liang, H., and Xu, J. 2021. Effect of surface tension on late-time growth of high-Reynolds-number Rayleigh–Taylor instability. *Acta Physica Sinica*, **70**, 114701.

Huang, Z., De Luca, A., Atherton, T.J., Bird, M., Rosenblatt, C., and Carles, P. 2007. Rayleigh–Taylor instability experiments with precise and arbitrary control of the initial interface shape. *Phys. Rev. Lett.*, **99**, 204502.

Hughes, K.T., Diggs, A., Park, C., Littrell, D., Haftka, R.T., Kim, N.H., and Balachandar, S. 2018. Simulation driven experiments of macroscale explosive dispersal of particles. Page 1545 of: *2018 AIAA Aerospace Sciences Meeting,* American Institute of Aeronautics and Astronautics, Kissimmee, FL.

Hunt, J., and Kevlahan, N. 1993. Rapid distortion theory and the structure of turbulence. Pages 285–316 of: *New Approaches and Concepts in Turbulence*. Birkhäuser, Basel, Switzerland.

Huntington, C.M., Shimony, A., Trantham, M., et al. 2018. Ablative stabilization of Rayleigh–Taylor instabilities resulting from a laser-driven radiative shock. *Phys. Plasmas*, **25**, 052118.

Huntington, C.M., Raman, K.S., Nagel, S.R., MacLaren, S.A., Baumann, T., Bender, J.D., Prisbrey, S.T., Simmons, L., Wang, P., and Zhou, Y. 2020. Split radiographic tracer technique to measure the full width of a high energy density mixing layer. *High Energy Density Phys.*, **35**, 100733.

Huppert, H.E., and Neufeld, J.A. 2014. The fluid mechanics of carbon dioxide sequestration. *Annu. Rev. Fluid Mech.*, **46**, 255.

Hurricane, O.A. 2008. Design for a high energy density Kelvin–Helmholtz experiment. *High Energy Density Phys.*, **4**, 97.

Hurricane, O.A., and Herrmann, M.C. 2017. High-energy-density physics at the National Ignition Facility. *Annu. Rev. Nucl. Part. Sci.*, **67**, 213.

Hurricane, O.A., Burke, E., Maples, S., and Viswanathan, M. 2000. Saturation of Richtmyer's impulsive model. *Phys. Fluids*, **12**, 2148.

Hurricane, O.A., Hansen, J.F., Robey, H.F., Remington, B.A., Bono, M.J., Harding, E.C., Drake, R.P., and Kuranz, C.C. 2009. A high energy density shock driven Kelvin–Helmholtz shear layer experiment. *Phys. Plasmas*, **16**, 056305.

Hurricane, O.A., Smalyuk, V.A., Raman, K., et al. 2012. Validation of a turbulent Kelvin–Helmholtz shear layer model using a high-energy-density omega laser experiment. *Phys. Rev. Lett.*, **109**, 155004.

Hurricane, O.A., Callahan, D.A., Casey, D.T., Celliers, P.M., Cerjan, C., Dewald, E.L., Dittrich, T.R., Döppner, T., Hinkel, D.E., Hopkins, L.B., and Kline, J.L. 2014. Fuel gain exceeding unity in an inertially confined fusion implosion. *Nature*, **506**, 343.

Hurricane, O.A., Patel, P.K., Betti, R., Froula, D.H., Regan, S.P., Slutz, S.A., Gomez, M.R. and Sweeney, M.A. 2023. Physics principles of inertial confinement fusion and US program overview. *Rev. Mod. Phys.*, **95**, 025005.

Hurricane, O.A., Callahan, D., Casey, D., et al. 2024. Energy principles of scientific breakeven in an inertial fusion experiment. *Phys. Rev. Lett.*, **132**, 06510.

Hysell, D.L., Larsen, M.F., and Zhou, Q.H. 2004. Common volume coherent and incoherent scatter radar observations of midlatitude sporadic E layers and QP echoes. *Ann. Geophys.*, **22**, 3277.

Ibragimov, N. 1999. *Elementary Lie Group Analysis and Ordinary Differential Equations*. Wiley, New York.

Ichimaru, S., Iyetomi, H., and Tanaka, S. 1987. Statistical physics of dense plasmas: thermodynamics, transport coefficients and dynamic correlations. *Phys. Rep.*, **149**, 91.

Ida, S., Nakagawa, Y., and Nakazawa, K. 1989. The Rayleigh–Taylor instability in a self-gravitating two-layer fluid sphere. *Earth, Moon, and Planets*, **44**, 149.

Igumenshchev, I.V., Goncharov, V.N., Marshall, F.J., et al. 2016. Three-dimensional modeling of direct-drive cryogenic implosions on OMEGA. *Phys. Plasmas*, **23**, 052702.

Ingraham, R.L. 1954. Taylor instability of the interface between superposed fluids-solution by successive approximations. *Proc. Phys. Soc. London B*, **67**, 748.

Inogamov, N.A. 1978. Turbulent stage of the Rayleigh–Taylor instability. *Sov. Tech. Phys. Lett.*, **4**, 743.

Inogamov, N.A. 1999. The role of Rayleigh–Taylor and Richtmyer–Meshkiv instabilities in astrophysics: an introduction. *Astrophys. Space Phys.*, **10**, 1.

Inogamov, N.A., and Oparin, A.M. 2003. Bubble motion in inclined pipes. *J. Exp. Theor. Phys.*, **97**, 1168.

Irving, T. 2013. *Deadly Dust: Changes in Both Technology and Culture are Needed to Reduce the Risks of Dust Explosions*. The Chemical Institute of Canada. Ottawa, Ontario, Canada. www.cheminst.ca/magazine/article/deadly-dust/.

Isenberg, C. 1992. *The Science of Soap Films and Soap Bubbles*. Dover, New York.

Ishihara, T., Morishita, K., Yokokawa, M., Uno, A., and Kaneda, Y. 2016. Energy spectrum in high-resolution direct numerical simulations of turbulence. *Phys. Rev. Fluids*, **1**, 082403.

Ishizaki, R., Nishihara, K., Sakagami, H., and Ueshima, Y. 1996. Instability of a contact surface driven by a nonuniform shock wave. *Phys. Rev. E*, **53**, R5592.

Jackson, D., and Launder, B. 2007. Osborne Reynolds and the publication of his papers on turbulent flow. *Annu. Rev. Fluid Mech.*, **39**, 19.

Jacobs, J.W. 1992. Shock-induced mixing of a light-gas cylinder. *J. Fluid Mech.*, **234**, 629.

Jacobs, J.W., and Catton, I. 1988a. Three-dimensional Rayleigh–Taylor instability Part 1. Weakly nonlinear theory. *J. Fluid Mech.*, **187**, 329.

Jacobs, J.W., and Catton, I. 1988b. Three-dimensional Rayleigh–Taylor instability Part 2. Experiment. *J. Fluid Mech.*, **187**, 353.

Jacobs, J.W., and Dalziel, S.B. 2005. Rayleigh–Taylor instability in complex stratifications. *J. Fluid Mech.*, **542**, 251.

Jacobs, J.W., and Krivets, V.V. 2005. Experiments on the late-time development of single-mode Richtmyer–Meshkov instability. *Phys. Fluids*, **17**, 034105.

Jacobs, J.W., and Sheeley, J.M. 1996. Experimental study of incompressible Richtmyer–Meshkov instability. *Phys. Fluids*, **8**, 405.

Jacobs, J.W., Krivets, V.V., Tsiklashvili, V., and Likhachev, O.A. 2013. Experiments on the Richtmyer–Meshkov instability with an imposed, random initial perturbation. *Shock Waves*, **23**, 407.

Jacobs, M., Kim, I., and Mészáros, A. 2021. Weak solutions to the Muskat problem with surface tension via optimal transport. *Arch. Ration. Mech. Anal.*, **239**, 389.

Jain, S.S., Mani, A., and Moin, P. 2020. A conservative diffuse-interface method for compressible two-phase flows. *J. Comput. Phys.*, **418**, 109606.

Janka, H.-T., Melson, T., and Summa, A. 2016. Physics of core-collapse supernovae in three dimensions: a sneak preview. *Annu. Rev. Nucl. Part. Sci.*, **66**, 341.

Janka, H.T. 2012. Explosion mechanisms of core-collapse supernovae. *Annu. Rev. Nucl. Part. Sci.*, **62**, 407.

Jeffery, D.J., Leibundgut, B., Kirshner, R.P., Benetti, S., Branch, D., and Sonneborn, G. 1992. Analysis of the photospheric epoch spectra of type 1a supernovae SN 1990N and SN 1991T. *Astrophys. J.*, **397**, 304.

Jeffreys, H. 1933. Evolution of hydrodynamics. *Nature*, **131**, 313.

Jensen, B.J., Cherne, F.J., Prime, M.B., et al. 2015. Jet formation in cerium metal to examine material strength. *J. Appl. Phys.*, **118**, 195903.

Jevons, W.S. 1857. II. On the cirrous form of cloud. *The London, Edinburgh, Dublin Philos. Mag. J. Sci.*, **14**, 22.

Jiang, R., Huang, X., Zou, L., Shi, H., and Wu, J. 2018a. Experimental investigation on the characteristics of unstability at Liquid- liquid tilted interface induced by Rayleigh–Taylor instability. *Chin. J. High Press. Phys.*, **32**, 054201.

Jiang, S., Wang, F., Ding, Y., et al. 2018b. Experimental progress of inertial confinement fusion based at the ShenGuang-III laser facility in China. *Nucl. Fusion*, **59**, 032006.

Jinn, J., Connor, E.G., and Jacobs, L.F. 2020. How ambient environment influences olfactory orientation in search and rescue dogs. *Chem. Senses*, **45**, 625.

Joggerst, C.C., Almgren, A., and Woosley, S.E. 2010. Three-dimensional simulations of Rayleigh–Taylor mixing in core-collapse supernovae. *Astrophys. J.*, **723**, 353.

Joggerst, C.C., Nelson, A., Woodward, Pa., Lovekin, C., Masser, T., Fryer, C.L., Ramaprabhu, P., Francois, M., and Rockefeller, G. 2014. Cross-code comparisons of mixing during the implosion of dense cylindrical and spherical shells. *J. Comput. Phys.*, **275**, 154.

Johnson, B.M., and Schilling, O. 2011a. Reynolds-averaged Navier–Stokes model predictions of linear instability. I: buoyancy-and shear-driven flows. *J. Turbul.*, **12**, N36.

Johnson, B.M., and Schilling, O. 2011b. Reynolds-averaged Navier–Stokes model predictions of linear instability. II: shock-driven flows. *J. Turbul.*, **12**, N37.

Johnson, J.R., Wing, S., and Delamere, P.A. 2014. Kelvin Helmholtz instability in planetary magnetospheres. *Space Sci. Rev.*, **184**, 1.

Jones, W.P., and Launder, B.E. 1973. The calculation of low-Reynolds-number phenomena with a two-equation model of turbulence. *Int. J. Heat Mass Transfer*, **16**, 1119.

Jourdan, G., and Houas, L. 1996. Experimental investigation of Richtmyer–Meshkov instability before and after the reflected shock compression. *Phys. Fluids*, **8**, 1353.

Jourdan, G., and Houas, L. 2005. High-amplitude single-mode perturbation evolution at the Richtmyer–Meshkov instability. *Phys. Rev. Lett.*, **95**, 204502.

Jun, B.I., Norman, M.L., and Stone, J.M. 1995. A numerical study of Rayleigh–Taylor instability in magnetic fluids. *Astrophys. J.*, **453**, 332.

Kadau, K., Germann, T.C., Hadjiconstantinou, N.G., Lomdahl, P.S., Dimonte, G., Holian, B.L., and Alder, B.J. 2004. Nanohydrodynamics simulations: an atomistic view of the Rayleigh–Taylor instability. *Proc. Natl. Acad. Sci.*, **101**, 5851.

Kadau, K., Rosenblatt, C., Barber, J.L., Germann, T.C., Huang, Z., Carlès, P., and Alder, B.J. 2007. The importance of fluctuations in fluid mixing. *Proc. Natl. Acad. Sci.*, **104**, 7741.

Kailasanath, K. 2000. Review of propulsion applications of detonation waves. *AIAA J.*, **38**, 1698.

Kaman, T., and Holley, R. 2022. Validation and verification of turbulence mixing due to Richtmyer–Meshkov instability of an air/SF_6 interface. *Int. J. Numer. Anal. Model.*, **19**, 822.

Kaman, T., Edwards, A., and McGarigal, J. 2021. Performance analysis of the parallel CFD code for turbulent mixing simulations. *J. Comput. Sci. Ed.*, **12**, 49.

Kaneko, T., and Yokoyama, T. 2018. Impact of dynamic state on the mass condensation rate of solar prominences. *Astrophys. J.*, **869**, 136.

Kapila, A.K., Menikoff, R., Bdzil, J.B., Son, S.F., and Stewart, D.S. 2001. Two-phase modeling of deflagration-to-detonation transition in granular materials: reduced equations. *Phys. Fluids*, **13**, 3002.

Karkhanis, V., Ramaprabhu, P., Cherne, F.J., Hammerberg, J.E., and Andrews, M.J. 2018. A numerical study of bubble and spike velocities in shock-driven liquid metals. *J. Appl. Phys.*, **123**, 025902.

Karlykhanov, N.G., Lykov, V.A., Timakova, M.S., and Chizhkov, M. 2004. 1D-simulation results of indirect-driven target optimization for ignition at Iskra-6 facility. *J. Exp. Theor. Phys. Lett.*, **79**, 25.

Kartoon, D., Oron, D., Arazi, L., and Shvarts, D. 2003. Three-dimensional multimode Rayleigh–Taylor and Richtmyer–Meshkov instabilities at all density ratios. *Laser Part. Beams*, **21**, 327.

Keele, K.D. 1952. *Leonardo Da Vinci on Movement of the Heart and Blood.* J. B. Lippincott, Philadelphia, PA.

Keele, K.D. 2014. *Leonardo da Vinci's Elements of the Science of Man.* Academic Press, New York.

Kelley, M.C. 2009. *Earth's Ionosphere.* 2nd edn. Elsevier, Amsterdam.

Keppens, R., Xia, C., and Porth, O. 2015. Solar prominences: double, double ... boil and bubble. *Astrophys. J. Lett.*, **806**, L13.

Khokhlov, A. 1993. Flame modeling in supernovae. *Astrophys. J.*, **419**, L77.

Khokhlov, A.M., Oran, E.S., Chtchelkanova, A.Y., and Wheeler, J.C. 1999. Interaction of a shock with a sinusoidally perturbed flame. *Combust. Flame*, **117**, 99.

Kida, S., and Orszag, S.A. 1990a. Energy and spectral dynamics in forced compressible turbulence. *J. Sci. Comput.*, **5**, 85.

Kida, S., and Orszag, S.A. 1990b. Enstrophy budget in decaying compressible turbulence. *J. Sci. Comput.*, **5**, 1.

Kidder, R.E. 1976. Laser-driven compression of hollow shells: power requirements and stability limitations. *Nucl. Fusion*, **16**, 3.

Kifonidis, K., Plewa, T., Janka, H.-Th., and Müller, E. 2003. Non-spherical core collapse supernovae-I. Neutrino-driven convection, Rayleigh–Taylor instabilities, and the formation and propagation of metal clumps. *Astron. Astrophys.*, **408**, 621.

Kilkenny, J.D., Glendinning, S.G., Haan, S.W., Hammel, B.A., Lindl, J.D., Munro, D., Remington, B.A., Weber, S.V., Knauer, J.P., and Verdon, C.P. 1994. A review of the ablative stabilization of the Rayleigh–Taylor instability in regimes relevant to inertial confinement fusion. *Phys. Plasmas*, **1**, 1379.

Kitaura, F.-S., Janka, H.-Th., and Hillebrandt, W. 2006. Explosions of O-Ne-Mg cores, the Crab supernova, and subluminous type II-P supernovae. *Astron. Astrophys.*, **450**, 345.

Kobayashi, K.U., and Kurita, R. 2022. Key connection between gravitational instability in physical gels and granular media. *Sci. Rep.*, **12**, 6290.

Kokkinakis, I.W., Drikakis, D., Youngs, D.L., and Williams, R.J.R. 2015. Two-equation and multi-fluid turbulence models for Rayleigh–Taylor mixing. *Int. J. Heat Fluid Flow*, **56**, 233.

Kokkinakis, I.W., Drikakis, D., and Youngs, D.L. 2019. Modeling of Rayleigh–Taylor mixing using single-fluid models. *Phys. Rev. E*, **99**, 013104.

Kokkinakis, I.W., Drikakis, D., and Youngs, D.L. 2020a. Two-equation and multi-fluid turbulence models for Richtmyer–Meshkov mixing. *Phys. Fluids*, **32**, 074102.

Kokkinakis, I.W., Drikakis, D., and Youngs, D.L. 2020b. Vortex morphology in Richtmyer–Meshkov-induced turbulent mixing. *Physica D*, **407**, 132459.

Kolmogorov, A.N. 1941. The local structure of turbulence in incompressible viscous fluid for very large Reynolds number. *Dokl. Akad. Nauk SSSR*, **30**, 301.

Koneru, R.B., Rollin, B., Durant, B., Ouellet, F., and Balachandar, S. 2020. A numerical study of particle jetting in a dense particle bed driven by an air-blast. *Phys. Fluids*, **32**, 093301.

Korycansky, D.G., Zahnle, K. J., and Law, M.-M. M. 2000. High-resolution calculations of asteroid impacts into the Venusian atmosphere. *Icarus*, **146**, 387.

Kraichnan, R.H. 1959. The structure of isotropic turbulence at very high Reynolds numbers. *J. Fluid Mech.*, **5**, 497.

Kraichnan, R.H. 1965. Inertial-range spectrum of hydromagnetic turbulence. *Phys. Fluids*, **8**, 1385.

Kraichnan, R.H. 1967. Inertial ranges in two-dimensional turbulence. *Phys. Fluids*, **10**, 1417.

Kraichnan, R.H., and Montgomery, D. 1980. Two-dimensional turbulence. *Rep. Progr. Phys.*, **43**, 547.

Krasheninnikova, N.S., Schmitt, M.J., Molvig, K., Hsu, S.C., Scheiner, B.S., Schmidt, D.W., Geppert-Kleinrath, V., McKenty, P.W., Michel, D.T., Edgell, D.H., and Marshall, F.J. 2020. Development of a directly driven multi-shell platform: laser drive energetics. *Phys. Plasmas*, **27**, 022706.

Krasny, R. 1986. Desingularization of periodic vortex sheet roll-up. *J. Comput. Phys.*, **65**, 292.

Krechetnikov, R. 2009. Rayleigh–Taylor and Richtmyer–Meshkov instabilities of flat and curved interfaces. *J. Fluid Mech.*, **625**, 387.

Krehl, P.O.K. 2008. *History of Shock Waves, Explosions and Impact: a Chronological and Biographical Reference*. Springer Science & Business Media, Berlin.

Kritcher, A. L., Zylstra, A. B., Callahan, D. A., et al. 2022. Design of an inertial fusion experiment exceeding the Lawson criterion for ignition. *Phys. Rev. E*, **106**, 025201.

Kritcher, A.L., Zylstra, A.B., Weber, C.R., Hurricane, O.A., et al. 2024. Design of the first fusion experiment to achieve target energy gain $G > 1$. *Phys. Rev. E*, **109**, 025204.

Krivets, V.V., Ferguson, K.J., and Jacobs, J.W. 2017. Turbulent mixing induced by Richtmyer–Meshkov instability. *AIP Conf. Proc.*, **1793**, 150003.

Kruskal, M.D., and Schwarzschild, M. 1954. Some instabilities of a completely ionized plasma. *Proc. R. Soc. Lond. Ser. A*, **223**, 348.

Krygier, A., Powell, P.D., McNaney, J.M., Huntington, C.M., Prisbrey, S.T., Remington, B.A., Rudd, R.E., Swift, D.C., Wehrenberg, C.E., Arsenlis, A., and Park, H.S. 2019. Extreme hardening of Pb at high pressure and strain rate. *Phys. Rev. Lett.*, **123**, 205701.

Ku, H.C., Taylor, T.D., and Hirsh, R.S. 1987. Pseudospectral methods for solution of the incompressible Navier–Stokes equations. *Comput. Fluids*, **15**, 195.

Kuang, L., Li, H., Jing, L., Lin, Z., Zhang, L., Li, L., Ding, Y., Jiang, S., Liu, J., and Zheng, J. 2016. A novel three-axis cylindrical hohlraum designed for inertial confinement fusion ignition. *Sci. Rep.*, **6**, 34636.

Kucherenko, Yu.A., Shibarshov, L.I., Chitaikin, V.I., Balabin, S.I., and Pylaev, A.P. 1991. Experimental study of the gravitational turbulent mixing self-similar mode. Pages 427–454 of: Dautray, R. (ed), *Third International Workshop on the Physics of Compressible Turbulent Mixing*. Cambridge University, Cambridge, UK.

Kucherenko, Yu.A., Balabzn, S.I., Ardashova, R.I., Kozelkov, O.E., I.A., Rornclnoc., Che'ret, R., and Haas, J.F. 1997. Experimental study of the gravitational turbulent mixing self-similar mode. Pages 258–265 of: Jourdan, G., and Houas, L. (eds), *Sixth International Workshop on the Physics of Compressible Turbulent Mixing*, Imprimerie Caractere, Marseille, France.

Kucherenko, Yu A., Shestachenko, O.E., Balabin, S.I., and Pylaev, A.P. 2003. RFNC-VNIITF multifunctional shock tube for investigating the evolution of instabilities in nonstationary gas dynamic flows. *Laser Part. Beams*, **21**, 381.

Kucherenko, Yu A., Pavlenko, A.V., Shestachenko, O.E., Balabin, S.I., Pylaev, A.P., and Tyaktev, A.A. 2010. Measurement of spectral characteristics of the turbulent mixing zone. *J. Appl. Mech. Tech. Phys.*, **51**, 299.

Kuchibhatla, S., and Ranjan, D. 2013. Effect of initial conditions on Rayleigh–Taylor mixing: modal interaction. *Phys. Scr.*, **2013**, 014057.

Kuchugov, P.A., Rozanov, V.B., and Zmitrenko, N.V. 2014. The differences in the development of Rayleigh–Taylor instability in 2D and 3D geometries. *Plasma Phys. Rep.*, **40**, 451.

Kuhl, A. 1993. *Mixing in Explosions*. Tech. Rep. Lawrence Livermore National Laboratory, Livermore, CA.

Kuhl, A. 1996. Spherical mixing layers in explosions. Page 291 of: Bowen, J.R. (ed), *Dynamics of Exothermicity*. Gordon and Breach Amsterdam, Netherlands.

Kuhl, A. 2015. On the structure of self-similar detonation waves in TNT charges. *Combust. Explos. Shock Waves*, **51**, 72.

Kuhl, A., Ferguson, R., Priolo, F., Chien, K., and Collins, J. 1993. *Baroclinic Mixing in He Fireballs*. Tech. rept. UCRL-JC-114982 Lawrence Livermore National Lab., El Segundo, CA.

Kuhl, A.L., Bell, J.B., Beckner, V.E., and Reichenbach, H. 2011. Gasdynamic model of turbulent combustion in TNT explosions. *Proc. Combust. Inst.*, **33**, 2177.

Kuhl, A.L., Bell, J.B., Beckner, V.E., Balakrishnan, K., and Aspden, A.J. 2013. Spherical combustion clouds in explosions. *Shock Waves*, **23**, 233.

Kuhlen, M, Woosley, SE, and Glatzmaier, GA. 2006. Carbon ignition in type Ia supernovae. II. A three-dimensional numerical model. *Astrophys. J.*, **640**, 407.

Kull, H.J., and Anisimov, S.I. 1986. Ablative stabilization in the incompressible Rayleigh–Taylor instability. *Phys. Fluids*, **29**, 2067.

Kulsrud, R.M. 2005. *Plasma Physics for Astrophysics*. Princeton University Press, Princeton, NJ.

Kundu, P.K., and Cohen, I.M. 2004. *Fluid Mechanics*. Elsevier, Boston, MA.

Kuranz, C.C., Park, H.-S., Huntington, C.M., et al. 2018. How high energy fluxes may affect Rayleigh–Taylor instability growth in young supernova remnants. *Nat. Commun.*, **9**, 1564.

Kurien, S., and Pal, N. 2022. The local wavenumber model for computation of turbulent mixing. *Philos. Trans. R. Soc. A*, **380**, 20210076.

Kurien, S., Doss, F.W., Livescu, D., and Flippo, K. 2020. Extracting a mixing parameter from 2D radiographic imaging of variable-density turbulent flow. *Physica D*, **405**, 132354.

La Fleche, M., Xiao, Q., Wang, Y., and Radulescu, M. 2017. Experimental study of the head-on interaction of a shock wave with a cellular flame. In: 26th *International Colloquium on the Dynamics of Explosions and Reactive Systems*. Boston, MA.

Lafay, M.-A., Le Creurer, B., and Gauthier, S. 2007. Compressibility effects on the Rayleigh–Taylor instability between miscible fluids. *Europhys. Lett.*, **79**, 64002.

Lagerstrom, P.A. 2013. *Matched Asymptotic Expansions: Ideas and Techniques*. Vol. 76. Springer Science & Business Media, New York.

Lamb, H. 1932. *Hydrodynamics*. 6th edn. Cambridge University Press, Cambridge, UK.

Lan, K., Liu, J., Li, Z., et al. 2016. Progress in octahedral spherical hohlraum study. *Mat. Rad. Extremes*, **1**, 8.

Landau, L.D. 1944. On the theory of slow combustion. *Acta Phys. USSR*, **19**, 77.

Landau, L.D., and Lifshitz, E.M. 1959. *Fluid Mechanics*. 2nd edn. Pergamon Press, Oxford, UK.

Landen, O.L., Baker, K.L., Clark, D.S., et al. 2016. Indirect-drive ablative Richtmyer–Meshkov node scaling. *J. Phys. Conf. Ser.*, **717**, 012034.

Laney, C.B. 1998. *Computational Gasdynamics*. Cambridge University Press, Cambridge, UK.

Lanier, N.E., Barnes, C.W., Batha, S.H., Day, R.D., Magelssen, G.R., Scott, J.M., Dunne, A.M., Parker, K.W., and Rothman, S.D. 2003. Multimode seeded Richtmyer–Meshkov mixing in a convergent, compressible, miscible plasma system. *Phys. Plasmas*, **10**, 1816.

Laplace, P.S. 1805. *Traite de Mechanique Celeste vol 4 supplements au Livre X*. Gauthier-Villars, Paris.

Lapushkina, T.A., Erofeev, A.V., Azarova, O.A., and Kravchenko, O.V. 2019. Interaction of a plane shock wave with an area of ionization instability of discharge plasma in air. *Aerospace Sci. Tech.*, **85**, 347.

Larmor, J. 1907. *Memoir and Scientific Correspondence of George Gabriel Stokes*. Cambridge University Press, Cambridge, UK.

Larroche, O., Rinderknecht, H.G., and Rosenberg, M.J. 2018. Nuclear yield reduction in inertial confinement fusion exploding-pusher targets explained by fuel-pusher mixing through hybrid kinetic-fluid modeling. *Phys. Rev. E*, **98**, 031201.

Latini, M., and Schilling, O. 2020. A comparison of two-and three-dimensional single-mode reshocked Richtmyer–Meshkov instability growth. *Physica D*, **401**, 132201.

Latini, M., Schilling, O., and Don, W.-S. 2007a. Effects of WENO flux reconstruction order and spatial resolution on reshocked two-dimensional Richtmyer–Meshkov instability. *J. Comput. Phys.*, **221**, 805.

Latini, M., Schilling, O., and Don, W.-S. 2007b. High-resolution simulations and modeling of reshocked single-mode Richtmyer–Meshkov instability: comparison to experimental data and to amplitude growth model predictions. *Phys. Fluids*, **19**, 024104.

Lau, Y.Y., Zier, J.C., Rittersdorf, I.M., Weis, M.R., and Gilgenbach, R.M. 2011. Anisotropy and feedthrough in magneto-Rayleigh–Taylor instability. *Phys. Rev. E*, **83**, 066405.

Launder, B. 2012. Horace Lamb and the circumstances of his appointment at Owens College. *Notes Rec. R. Soc.*, **67**, 139.

Launder, B.E. 2015. First steps in modelling turbulence and its origins: a commentary on Reynolds (1895) 'On the dynamical theory of incompressible viscous fluids and the determination of the criterion'. *Philos. Trans. R. Soc. A*, **373**, 20140231.

Launder, B.E., and Spalding, D.B. 1974. The numerical computation of turbulent flows. *Comput. Methods Appl. Mech. Eng.*, **3**, 269.

Launder, B.E., Reece, G. Jr, and Rodi, W. 1975. Progress in the development of a Reynolds-stress turbulence closure. *J. Fluid Mech.*, **68**, 537.

Law, C.K. 2006. *Combustion Physics*. Cambridge University Press, Cambridge, UK.

Lawrie, A.G.W., and Dalziel, S.B. 2011a. Rayleigh–Taylor mixing in an otherwise stable stratification. *J. Fluid Mech.*, **688**, 507.

Lawrie, A.G.W., and Dalziel, S.B. 2011b. Turbulent diffusion in tall tubes. I. Models for Rayleigh–Taylor instability. *Phys. Fluids*, **23**, 085109.

Lawrie, A.G.W., and Dalziel, S.B. 2011c. Turbulent diffusion in tall tubes. II. Confinement by stratification. *Phys. Fluids*, **23**, 085110.

Lax, P.D., and Richtmyer, R.D. 1956. Survey of the stability of linear finite difference equations. *Comm. Pure Appl. Math.*, **9**, 267.

Layes, G., and Le Métayer, O. 2007. Quantitative numerical and experimental studies of the shock accelerated heterogeneous bubbles motion. *Phys. Fluids*, **19**, 042105.

Layes, G., Jourdan, G., and Houas, L. 2003. Distortion of a spherical gaseous interface accelerated by a plane shock wave. *Phys. Rev. Lett.*, **91**, 174502.

Layes, G., Jourdan, G., and Houas, L. 2005. Experimental investigation of the shock wave interaction with a spherical gas inhomogeneity. *Phys. Fluids*, **17**, 028103.

Layes, G., Jourdan, G., and Houas, L. 2009. Experimental study on a plane shock wave accelerating a gas bubble. *Phys. Fluids*, **21**, 074102.

Layzer, D. 1955. On the instability of superposed fluids in a gravitational field. *Astrophys. J.*, **122**, 1.

Lazarowski, L., and Dorman, D.C. 2014. Explosives detection by military working dogs: olfactory generalization from components to mixtures. *Appl. Anim. Behav. Sci.*, **151**, 84.

Le Creurer, B., and Gauthier, S. 2008. A return toward equilibrium in a 2D Rayleigh–Taylor instability for compressible fluids with a multidomain adaptive Chebyshev method. *Theor. Comput. Fluid Dyn.*, **22**, 125.

Le Pape, S., Berzak Hopkins, L.F., Divol, L., et al. 2018. Fusion energy output greater than the kinetic energy of an imploding shell at the National Ignition Facility. *Phys. Rev. Lett.*, **120**, 245003.

Lebedev, A.I., Nizovtsev, P.N., Rayevsky, V.A., and Solovyov, V.P. 1996. Rayleigh–Taylor instability in strong media, experimental study. In: Young, R., Glimm, J., and Boston, B. (eds), *Proceedings of the Fifth International Workshop on Compressible Turbulent Mixing*. World Scientific, Singapore.

Lebedev, S.V., Beg, F.N., Bland, S.N., Chittenden, J.P., Dangor, A.E., Haines, M.G., Kwek, K.H., Pikuz, S.A., and Shelkovenko, T.A. 2001. Effect of discrete wires on the implosion dynamics of wire array Z pinches. *Phys. Plasmas*, **8**, 3734.

Lebedev, S.V., Ampleford, D.J., Bland, S.N., et al. 2005. Physics of wire array Z-pinch implosions: experiments at Imperial College. *Plasma Phys. Control. Fusion*, **47**, A91.

Lebedev, S.V., Frank, A., and Ryutov, D.D. 2019. Exploring astrophysics-relevant magnetohydrodynamics with pulsed-power laboratory facilities. *Rev. Mod. Phys.*, **91**, 025002.

LeBlanc, J.M., and Wilson, J.R. 1970. A numerical example of the collapse of a rotating magnetized star. *Astrophys. J.*, **161**, 541.

Lee, T.D. 1951. Difference between turbulence in a two-dimensional fluid and in a three-dimensional fluid. *J. Appl. Phys.*, **22**, 524.

Lee, Y.T., and More, R.M. 1984. An electron conductivity model for dense plasmas. *Phys. Fluids*, **27**, 1273.

Lei, F. 2017. *Experimental and theoretical study on converging Richtmyer–Meshkov instability*. Ph.D. thesis, University of Science and Technology of China, Hefei, China.

Lei, F., Ding, J., Si, T., Zhai, Z., and Luo, X. 2017. Experimental study on a sinusoidal air/SF interface accelerated by a cylindrically converging shock. *J. Fluid Mech.*, **826**, 819.

Leinov, E., Malamud, G., Elbaz, Y., Levin, L.A., Ben-Dor, G., Shvarts, D., and Sadot, O. 2009. Experimental and numerical investigation of the Richtmyer–Meshkov instability under re-shock conditions. *J. Fluid Mech.*, **626**, 449.

Leith, C.E. 1967. Diffusion approximation to inertial energy transfer in isotropic turbulence. *Phys. Fluids*, **10**, 1409.

Leith, C.E. 1986. *Development of a two-equation turbulent mix model*. Tech. rept. Lawrence Livermore National Laboratory, UCRL-96036, Livermore, CA.

Leith, CE. 1990. Stochastic backscatter in a subgrid-scale model: plane shear mixing layer. *Phys. Fluids A*, **2**, 297.

Lekner, J. 2017. Nurturing genius: the childhood and youth of Kelvin and Maxwell. *Substantia*, **1**, 133.

LeLevier, R., Lasher, G.J., and Bjorklund, F. 1955. *Effect of a density gradient on Taylor instability*. Tech. rept. Radiation Lab., University of California, Livermore, CA.

Leonardis, E., Chapman, S.C., and Foullon, C. 2012. Turbulent characteristics in the intensity fluctuations of a solar quiescent prominence observed by the Hinode Solar Optical Telescope. *Astrophys. J.*, **745**, 185.

Leroy, J.L. 1988. Observation of prominence magnetic fields. Pages 77–113 of: Priest, E.R. (ed), *Dynamics and Structure of Quiescent Solar Prominences*. Springer, Dordrecht, Netherlands.

Lesieur, M. 1987. *Turbulence in Fluids: Stochastic and Numerical Modelling*. Nijhoff, Boston, MA.

Lesieur, M., and Metais, O. 1996. New trends in large-eddy simulations of turbulence. *Annu. Rev. Fluid Mech.*, **28**, 45.

Lesieur, M., and Schertzer, D. 1978. Dynamique des gros tourbillons et décroissance de lénergie cinétique en turbulence tridimensionelle isotrope á grand nombre de Reynolds. *J. Méch.*, **17**, 607.

Levens, P.J., Schmieder, B., LópezAriste, A., Labrosse, N., Dalmasse, K., and Gelly, B. 2016. Magnetic field in atypical prominence structures: bubble, tornado, and eruption. *Astrophys. J.*, **826**, 164.

Lewis, D.J. 1950. The instability of liquid surfaces when accelerated in a direction perpendicular to their planes. II. *Proc. R. Soc. Lond. A*, **202**, 81.

Lherm, V., Deguen, R., Alboussière, T., and Landeau, M. 2022. Rayleigh–Taylor instability in impact cratering experiments. *J. Fluid Mech.*, **937**, A20.

Li, H., He, Z., Zhang, Y., and Tian, B. 2019a. On the role of rarefaction/compression waves in Richtmyer–Meshkov instability with reshock. *Phys. Fluids*, **31**, 054102.

Li, H., Tian, B., He, Z., and Zhang, Y. 2021a. Growth mechanism of interfacial fluid-mixing width induced by successive nonlinear wave interactions. *Phys. Rev. E*, **103**, 053109.

Li, J., Sun, Y., Pan, J., and Ren, Y. 2016. Instability and turbulent mixing of shocked "V" shaped interface. *Acta Phys. Sin.*, **65**, 245202.

Li, J., Ding, J., Luo, X., and Zou, L. 2022a. Instability of a heavy gas layer induced by a cylindrical convergent shock. *Phys. Fluids*, **34**, 042123.

Li, J., Yan, R., Zhao, B., Zheng, J., Zhang, H., and Lu, X. 2022b. Mitigation of the ablative Rayleigh–Taylor instability by nonlocal electron heat transport. *Matter Radiat. Extrem.*, **7**, 055902.

Li, M., Ding, J., Zhai, Z., Si, T., Liu, N., Huang, S., and Luo, X. 2020a. On divergent Richtmyer–Meshkov instability of a light/heavy interface. *J. Fluid Mech.*, **901**, A38.

Li, X., Zhang, J., Yang, S., Hou, Y., and Erdélyi, R. 2018a. Observing Kelvin–Helmholtz instability in solar blowout jet. *Sci. Rep.*, **8**, 8136.

Li, X., Fu, Y., Yu, C., and Li, L. 2021b. Statistical characteristics of turbulent mixing in spherical and cylindrical converging Richtmyer–Meshkov instabilities. *J. Fluid Mech.*, **928**, A10.

Li, X.L., and Zhang, Q. 1997. A comparative numerical study of the Richtmyer–Meshkov instability with nonlinear analysis in two and three dimensions. *Phys. Fluids*, **9**, 3069.

Li, Y., and Luo, X. 2014. Theoretical analysis of effects of viscosity, surface tension, and magnetic field on the bubble evolution of Rayleigh–Taylor instability. *Acta Phys. Sin.*, **63**, 85203.

Li, Y., Samtaney, R., and Wheatley, V. 2018b. The Richtmyer–Meshkov instability of a double-layer interface in convergent geometry with magnetohydrodynamics. *Matter Radiat. Extremes*, **3**, 207.

Li, Z., Wang, Z., Xu, R., Yang, J., Ye, F., Chu, Y., Xu, Z., Chen, F., Meng, S., Qi, J., and Hu, Q. 2019b. Experimental investigation of Z-pinch radiation source for indirect drive inertial confinement fusion. *Matter Radiat. Extremes*, **4**, 046201.

Li, Z., Wang, L., Wu, J., and Ye, W. 2020b. Numerical study on the laser ablative Rayleigh–Taylor instability. *Acta Mech. Sin.*, **36**, 789.

Liang, H., Li, Q.X., Shi, B.C., and Chai, Z.H. 2016. Lattice Boltzmann simulation of three-dimensional Rayleigh–Taylor instability. *Phys. Rev. E*, **93**, 033113.

Liang, H., Hu, X., Huang, X., and Xu, J. 2019. Direct numerical simulations of multi-mode immiscible Rayleigh–Taylor instability with high Reynolds numbers. *Phys. Fluids*, **31**, 112104.

Liang, H., Xia, Z., and Huang, H. 2021a. Late-time description of immiscible Rayleigh–Taylor instability: a lattice Boltzmann study. *Phys. Fluids*, **33**, 082103.

Liang, Y. 2022. The phase effect on the Richtmyer–Meshkov instability of a fluid layer. *Phys. Fluids*, **34**, 034106.

Liang, Y., and Luo, X. 2021a. On shock-induced heavy-fluid-layer evolution. *J. Fluid Mech.*, **920**, A13.

Liang, Y., and Luo, X. 2021b. Shock-induced dual-layer evolution. *J. Fluid Mech.*, **929**, R3.

Liang, Y., and Luo, X. 2022. On shock-induced light-fluid-layer evolution. *J. Fluid Mech.*, **933**, A10.

Liang, Y., and Luo, X. 2023. Hydrodynamic instabilities of two successive slow/fast interfaces induced by a weak shock. *J. Fluid Mech.*, **955**, A40.

Liang, Y., Liu, L., Zhai, Z., Si, T., and Wen, C. 2020. Evolution of shock-accelerated heavy gas layer. *J. Fluid Mech.*, **886**, A7.

Liang, Y., Liu, L., Zhai, Z., Ding, J., Si, T., and Luo, X. 2021b. Richtmyer–Meshkov instability on two-dimensional multi-mode interfaces. *J. Fluid Mech.*, **928**, A37.

Liang, Y., Liu, L., Zhai, Z., Si, T., and Luo, X. 2021c. Universal perturbation growth of Richtmyer–Meshkov instability for minimum-surface featured interface induced by weak shock waves. *Phys. Fluids*, **33**, 032110.

Liang, Y., Liu, L., Luo, X., and Wen, C.-Y. 2023. Hydrodynamic instabilities of a dual-mode air–SF6 interface induced by a cylindrically convergent shock. *J. Fluid Mech.*, **963**, A25.

Liao, S., Zhang, W., Chen, H., Zou, L., Liu, J., and Zheng, X. 2019. Atwood number effects on the instability of a uniform interface driven by a perturbed shock wave. *Phys. Rev. E*, **99**, 013103.

Liberatore, S., Gauthier, P., Willien, J.L., et al., 2023. First indirect drive inertial confinement fusion campaign on Laser Megajoule. *Phys. Plasmas*, **30**, 122707.

Lighthill, J. 1986. *An Informal Introduction to Theoretical Fluid Mechanics*. Clarendon Press, Oxford, UK.

Lilly, D.K. 1983. Stratified turbulence and the mesoscale variability of the atmosphere. *J. Atmos. Sci.*, **40**, 749.

Lin, C.C. 1956. Aspects of the problem of turbulent motion. *J. Aeronaut. Sci.*, **23**, 453.

Lindemuth, I.R., and Kirkpatrick, R.C. 1983. Parameter space for magnetized fuel targets in inertial confinement fusion. *Nucl. Fusion*, **23**, 263.

Linden, P.F., Redondo, J.M., and Youngs, D.L. 1994. Molecular mixing in Rayleigh–Taylor instability. *J. Fluid Mech.*, **265**, 97.

Lindl, J.D. 1995. Development of the indirect-drive approach to inertial confinement fusion and the target physics basis for ignition and gain. *Phys. Plasmas*, **2**, 3933.

Lindl, J.D. 1998. *Inertial Confinement Fusion: The Quest for Ignition and Energy Gain using Indirect Drive*. American Institute of Physics Press, New York.

Lindl, J.D., Amendt, P., Berger, R.L., Glendinning, S.G., Glenzer, S.H., Haan, S.W., Kauffman, R.L., Landen, O.L., and Suter, L.J. 2004. The physics basis for ignition using indirect-drive targets on the National Ignition Facility. *Phys. Plasmas*, **11**, 339.

Ling, Y., Wagner, J.L., Beresh, S.J., Kearney, S.P., and Balachandar, S. 2012. Interaction of a planar shock wave with a dense particle curtain: modeling and experiments. *Phys. Fluids*, **24**, 113301.

Liu, B., Zhang, C., Lou, Q., and Liang, H. 2022. Lattice Boltzmann study of three-dimensional immiscible Rayleigh–Taylor instability in turbulent mixing stage. *Front. Phys.*, **17**, 53506.

Liu, C., Zhang, Y., and Xiao, Z. 2023a. A unified theoretical model for spatiotemporal development of Rayleigh–Taylor and Richtmyer–Meshkov fingers. *J. Fluid Mech.*, **954**, A13.

Liu, H., and Xiao, Z. 2016. Scale-to-scale energy transfer in mixing flow induced by the Richtmyer–Meshkov instability. *Phys. Rev. E*, **93**, 053112.

Liu, H.-C., Yu, B., Chen, H., Zhang, B., Xu, H., and Liu, H. 2020. Contribution of viscosity to the circulation deposition in the Richtmyer–Meshkov instability. *J. Fluid Mech.*, **895**, A10.

Liu, J., Tan, D., Zhang, X., Huang, W., and Zou, L. 2012a. Experimental investigation of mixing at tilted interface induced by Rayleigh–Taylor instability. *Chin. J. High Press. Phys.*, **26**, 688.

Liu, J., Zou, L., Liao, S., Cao, R., Wang, Y., and Tan, D. 2014a. The Richtmyer–Meshkov instability by diffracted incident shock waves and reshock. *Sci. Sin.-Phys. Mech. Astron.*, **44**, 1203.

Liu, J.H., Tan, D.W., Bai, J.S., Wang, W.B., Zou, L.Y., and Zhang, X. 2012b. Experimental study of Richtmyer–Meshkov instability in nonuniform flow by shock tube. *J. Exp. Mech*, **27**, 160.

Liu, L., Liang, Y., Ding, J., Liu, N., and Luo, X. 2018a. An elaborate experiment on the single-mode Richtmyer–Meshkov instability. *J. Fluid Mech.*, **853** R2.

Liu, W.-H., Ma, W.F., and Wang, X.L. 2015a. Cylindrical effects in weakly nonlinear Rayleigh–Taylor instability. *Chin. Phys. B*, **24**, 015202.

Liu, W.-H., Chen, Y., Yu, C., and Li, X. 2015b. Harmonic growth of spherical Rayleigh–Taylor instability in weakly nonlinear regime. *Phys. Plasmas*, **22**, 112112.

Liu, W.-H., Yu, C., Jiang, H., and Li, X. 2017. Bell-Plessett effect on harmonic evolution of spherical Rayleigh–Taylor instability in weakly nonlinear scheme for arbitrary Atwood numbers. *Phys. Plasmas*, **24**, 022102.

Liu, W.B., Ma, D.J., He, A.M., and Wang, P. 2018b. Ejecta from periodic grooved Sn surface under unsupported shocks. *Chin. Phys. B*, **27**, 016202.

Liu, W.H., He, X.T., and Yu, C.P. 2012c. Cylindrical effects on Richtmyer–Meshkov instability for arbitrary Atwood numbers in weakly nonlinear regime. *Phys. Plasmas*, **19**, 072108.

Liu, W.H., Yu, C.P, and Li, X.L. 2014b. Effects of initial radius of the interface and Atwood number on nonlinear saturation amplitudes in cylindrical Rayleigh–Taylor instability. *Phys. Plasmas*, **21**, 112103.

Liu, W.H., Yu, C.-P., Ye, W.-H., and Wang, L.-F. 2014c. Nonlinear saturation amplitude of cylindrical Rayleigh–Taylor instability. *Chin. Phys. B*, **23**, 94502.

Liu, W.H., Yu, C.P., Chang, Ye, W.H., Wang, L.F., and He, X.T. 2014d. Nonlinear theory of classical cylindrical Richtmyer–Meshkov instability for arbitrary Atwood numbers. *Phys. Plasmas*, **21**, 062119.

Liu, X., George, E., Bo, W., and Glimm, J. 2006. Turbulent mixing with physical mass diffusion. *Phys. Rev. E*, **73**, 056301.

Liu, Y., and Grieves, B. 2014. Ejecta production and transport from a shocked Sn coupon. *J. Fluids Eng.*, **136**, 091202.

Liu, Y.X., Chen, Z., Wang, L.F., Li, Z.Y., Wu, J.F., Ye, W.H., and Li, Y.J. 2023b. Dynamic of shock–bubble interactions and nonlinear evolution of ablative hydrodynamic instabilities initialed by capsule interior isolated defects. *Phys. Plasmas*, **30**, 042302.

Livescu, D. 2004. Compressibility effects on the Rayleigh–Taylor instability growth between immiscible fluids. *Phys. Fluids*, **16**, 118.

Livescu, D. 2013. Numerical simulations of two-fluid turbulent mixing at large density ratios and applications to the Rayleigh–Taylor instability. *Phil. Trans. R. Soc. A*, **371**, 20120185.

Livescu, D. 2020. Turbulence with large thermal and compositional density variations. *Annu. Rev. Fluid Mech.*, **52**, 309.

Livescu, D., and Ristorcelli, J.R. 2007. Buoyancy-driven variable-density turbulence. *J. Fluid Mech.*, **591**, 43.

Livescu, D., Wei, T., and Petersen, M.R. 2011. Direct numerical simulations of Rayleigh–Taylor instability. *J. Phys. Conf. Ser*, **318**, 082007.

Livescu, D., Wei, T., and Brady, P.T. 2021. Rayleigh–Taylor instability with gravity reversal. *Physica D*, **417**, 132832.

Llor, A. 2003. Bulk turbulent transport and structure in Rayleigh–Taylor, Richtmyer–Meshkov, and variable acceleration instabilities. *Laser Part. beams*, **21**, 305.

Llor, A. 2005. *Statistical Hydrodynamic Models for Developed Mixing Instability Flows: Analytical "0D" Evaluation Criteria, and Comparison of Single-and Two-Phase Flow Approaches. Lecture Notes in Physics.* Vol. 681. Springer Science & Business Media, Berlin.

Llor, A. 2006. Invariants of free turbulent decay. *arXiv preprint physics/0612220.*

Llor, A., and Bailly, P. 2003. A new turbulent two-field concept for modeling Rayleigh–Taylor, Richtmyer–Meshkov, and Kelvin–Helmholtz mixing layers. *Laser Part. Beams*, **21**, 311.

Lobatchev, V., and Betti, R. 2000. Ablative stabilization of the deceleration phase Rayleigh–Taylor instability. *Phys. Rev. Lett.*, **85**, 4522.

Lombardini, M., Hill, D.J., Pullin, D.I., and Meiron, D.I. 2011. Atwood ratio dependence of Richtmyer–Meshkov flows under reshock conditions using large-eddy simulations. *J. Fluid Mech.*, **670**, 439.

Lombardini, M., Pullin, D.I., and Meiron, D.I. 2012. Transition to turbulence in shock-driven mixing: a Mach number study. *J. Fluid Mech.*, **690**, 203.

Lombardini, M., Pullin, D.I., and Meiron, D.I. 2014a. Turbulent mixing driven by spherical implosions. Part 1. Flow description and mixing-layer growth. *J. Fluid Mech.*, **748**, 85.

Lombardini, M., Pullin, D.I., and Meiron, D.I. 2014b. Turbulent mixing driven by spherical implosions. Part 2. Turbulence statistics. *J. Fluid Mech.*, **748**, 113.

Long, C.C., Krivets, V.V., Greenough, J.A., and Jacobs, J.W. 2009. Shock tube experiments and numerical simulation of the single-mode, three-dimensional Richtmyer–Meshkov instability. *Phys. Fluids*, **21**, 114104.

López Ortega, A., Lombardini, M., Pullin, D.I., and Meiron, D.I. 2014. Numerical simulations of the Richtmyer–Meshkov instability in solid-vacuum interfaces using calibrated plasticity laws. *Phys. Rev. E*, **89**, 033018.

López Ortega, A., Lombardini, M., Barton, P.T., Pullin, D.I., and Meiron, D.I. 2015. Richtmyer–Meshkov instability for elastic–plastic solids in converging geometries. *J. Mech. Phys. Solids*, **76**, 291.

Lugomer, S. 2019. Laser-generated Richtmyer–Meshkov and Rayleigh–Taylor instabilities in a semi-confined configuration: bubble dynamics in the central region of the Gaussian spot. *Phys. Scr.*, **94**, 015001.

Lugomer, S. 2020. Nano-wrinkles, compactons, and wrinklons associated with laser-induced Rayleigh–Taylor instability: I. Bubble environment. *Laser Part. Beams*, **38**, 101.

Lumley, J.L. 1970. Toward a turbulent constitutive relation. *J. Fluid Mech.*, **41**, 413.

Lumley, J.L. 1992. Some comments on turbulence. *Phys. Fluids A*, **4**, 203.

Lund, H.M., and Dalziel, S.B. 2014. Bursting water balloons. *J. Fluid Mech.*, **756**, 771.

Luo, K., Luo, Y., Jin, T., and Fan, J. 2017. Numerical analysis on shock-cylinder interaction using immersed boundary method. *Sci. China Technol. Sci.*, **60**, 1423.

Luo, T., and Wang, J. 2022. Mixing and energy transfer in compressible Rayleigh–Taylor turbulence for initial isothermal stratification. *Phys. Rev. Fluids*, **7**, 104608.

Luo, T., Wang, J., Xie, C., Wan, M., and Chen, S. 2020a. Effects of compressibility and Atwood number on the single-mode Rayleigh–Taylor instability. *Phys. Fluids*, **32**, 012110.

Luo, X., Wang, X., and Si, T. 2013. The Richtmyer–Meshkov instability of a three-dimensional air/SF6 interface with a minimum-surface feature. *J. Fluid Mech.*, **722**, R2.

Luo, X., Wang, M., Si, T., and Zhai, Z. 2015a. On the interaction of a planar shock with an polygon. *J. Fluid Mech.*, **773**, 366.

Luo, X., Ding, J., Wang, M., Zhai, Z., and Si, T. 2015b. A semi-annular shock tube for studying cylindrically converging Richtmyer–Meshkov instability. *Phys. Fluids*, **27**, 091702.

Luo, X., Guan, B., Zhai, Z., and Si, T. 2016a. Principal curvature effects on the early evolution of three-dimensional single-mode Richtmyer–Meshkov instabilities. *Phys. Rev. E*, **93**, 023110.

Luo, X., Dong, P., Si, T., and Zhai, Z. 2016b. The Richtmyer–Meshkov instability of a 'V' shaped air/interface. *J. Fluid Mech.*, **802**, 186.

Luo, X., Guan, B., Si, T., Zhai, Z., and Wang, X. 2016c. Richtmyer–Meshkov instability of a three-dimensional SF_6-air interface with a minimum-surface feature. *Phys. Rev. E*, **93**, 013101.

Luo, X., Li, M., Ding, J., Zhai, Z., and Si, T. 2019. Nonlinear behaviour of convergent Richtmyer–Meshkov instability. *J. Fluid Mech.*, **877**, 130.

Luo, X., Liu, L., Liang, Y., Ding, J., and Wen, C. 2020b. Richtmyer–Meshkov instability on a dual-mode interface. *J. Fluid Mech.*, **905**, A5.

Lutoschkin, E. 2013. *Pressure-gain combustion for gas turbines based on shock-flame interaction.* Ph.D. thesis, University Stuttgart, Stuttgart, Germany.

Ma, C., Liu, B., and Liang, H. 2022. Lattice Boltzmann simulation of three-dimensional fluid interfacial instability coupled with surface tension. *Acta Phys. Sin.*, **71**, 044701.

Ma, J., Chen, Y., Gan, B., and Yu, M.Y. 2006. Behavior of the Rayleigh–Taylor mode in a dusty plasma with rotational and shear flows. *Planet. Space Sci.*, **54**, 719.

Ma, T., Patel, P.K., Izumi, N., et al. 2013. Onset of hydrodynamic mix in high-velocity, highly compressed inertial confinement fusion implosions. *Phys. Rev. Lett.*, **111**, 085004.

Ma, T., Patel, P.K., Izumi, N., et al. 2017. The role of hot spot mix in the low-foot and high-foot implosions on the NIF. *Phys. Plasmas*, **24**, 056311.

Mackay, D.H., Karpen, J.T., Ballester, J.L., Schmieder, B., and Aulanier, G. 2010. Physics of solar prominences: II – magnetic structure and dynamics. *Space Sci. Rev.*, **151**, 333.

MacLaren, S.A., Ho, D.D.-M., Hurricane, O.A., et al. 2021. A pushered capsule implosion as an alternate approach to the ignition regime for inertial confinement fusion. *Phys. Plasmas*, **28**, 122710.

MacPhee, A., Casey, D., Clark, D., et al. 2017. X-ray shadow imprint of hydrodynamic instabilities on the surface of inertial confinement fusion capsules by the fuel fill tube. *Phys. Rev. E*, **95**, 031204.

MacPhee, A., Smalyuk, V., Landen, O., et al. 2018. Mitigation of x-ray shadow seeding of hydrodynamic instabilities on inertial confinement fusion capsules using a reduced diameter fuel fill-tube. *Phys. Plasmas*, **25**, 054505.

Maeder, A., Meynet, G., Lagarde, N., and Charbonnel, C. 2013. The thermohaline, Richardson, Rayleigh–Taylor, Solberg–Høiland, and GSF criteria in rotating stars. *Astron. Astrophys.*, **553**, A1.

Magyar, N., Van Doorsselaere, T., and Goossens, M. 2019. Understanding uniturbulence: self-cascade of MHD waves in the presence of inhomogeneities. *Astrophys. J.*, **882**, 50.

Malamud, G., Di Stefano, C.A., Elbaz, Y., Huntington, C.M., Kuranz, C.C., Keiter, P.A., and Drake, R.P. 2013a. A design of a two-dimensional, multimode RM experiment on OMEGA-EP. *High Energy Density Phys.*, **9**, 122.

Malamud, G., Shimony, A., Wan, W.C., Di Stefano, C.A., Elbaz, Y., Kuranz, C.C., Keiter, P.A., Drake, R.P., and Shvarts, D. 2013b. A design of a two-dimensional, supersonic KH experiment on OMEGA-EP. *High Energy Density Phys.*, **9**, 672.

Malamud, G., Leinov, E., Sadot, O., Elbaz, Y., Ben-Dor, G., and Shvarts, D. 2014. Reshocked Richtmyer–Meshkov instability: numerical study and modeling of random multi-mode experiments. *Phys. Fluids*, **26**, 084107.

Mandal, L., Roy, S., Banerjee, R., Khan, M., and Gupta, M.R. 2011. Evolution of nonlinear interfacial structure induced by combined effect of Rayleigh–Taylor and Kelvin–Helmholtz instability. *Nucl. Instrum. Methods Phys. Res. A*, **653**, 103.

Mankbadi, M., and Balachandar, S. 2013. Viscous effects on the non-classical Rayleigh–Taylor instability of spherical material interfaces. *Shock Waves*, **23**, 603.

Mankbadi, M.R., and Balachandar, S. 2012. Compressible inviscid instability of rapidly expanding spherical material interfaces. *Phys. Fluids*, **24**, 034106.

Mansoor, M. M., Dalton, S. M., Martinez, A. A., Desjardins, T., Charonko, J. J., and Prestridge, K. P. 2020. The effect of initial conditions on mixing transition of the Richtmyer–Meshkov instability. *J. Fluid Mech.*, **904**, A3.

Manuel, M.J.E., Li, C.K., Séguin, F.H., Frenje, J., Casey, D.T., Petrasso, R.D., Hu, S.X., Betti, R., Hager, J.D., Meyerhofer, D.D., and Smalyuk, V.A. 2012. First measurements of Rayleigh–Taylor-induced magnetic fields in laser-produced plasmas. *Phys. Rev. Lett.*, **108**, 255006.

Manuel, M.J.E., Khiar, B., Rigon, G., et al. 2021. On the study of hydrodynamic instabilities in the presence of background magnetic fields in high-energy-density plasmas. *Matter Radiat. Extrem.*, **6**, 026904.

Marble, F.E., Hendricks, G.J., and Zukoski, E.E. 1989. Progress toward shock enhancement of supersonic combustion processes. Pages 932–950 of: Borghi, R., Murthy, S.N.B. (eds), *Turbulent Reactive Flows*. Springer, New York.

Marble, F.E., Zukoski, E.E., Jacobs, J., Hendricks, G., and Waitz, I. 1990. Shock enhancement and control of hypersonic mixing and combustion. *AIAA paper*, 26th Joint Propulsion Conference, Orlando, FL. https://arc.aiaa.org/doi/abs/10.2514/6.1990-1981.

Marek, A., and Janka, H.-Th. 2009. Delayed neutrino-driven supernova explosions aided by the standing accretion-shock instability. *Astrophys. J.*, **694**, 664.

Margolin, L.G., and Reisner, J.M. 2017. Fully compressible solutions for early stage Richtmyer–Meshkov instability. *Comput. Fluids*, **151**, 46.

Mariani, C., Vandenboomgaerde, M., Jourdan, G., Souffland, D., and Houas, L. 2008. Investigation of the Richtmyer–Meshkov instability with stereolithographed interfaces. *Phys. Rev. Lett.*, **100**, 254503.

Marinak, M.M., Tipton, R.E., Landen, O.L., Murphy, T.J., Amendt, P., Haan, S.W., Hatchett, S.P., Keane, C.J., McEachern, R., and Wallace, R. 1996. Three-dimensional simulations of Nova high growth factor capsule implosion experiments. *Phys. Plasmas*, **3**, 2070.

Marinak, M.M., Glendinning, S.G., Wallace, R.J., Remington, B.A., Budil, K.S., Haan, S.W., Tipton, R.E., and Kilkenny, J.D. 1998. Nonlinear Rayleigh–Taylor evolution of a three-dimensional multimode perturbation. *Phys. Rev. Lett.*, **80**, 4426.

Marinak, M.M., Kerbel, G.D., Gentile, N.A., Jones, O., Munro, D., Pollaine, S., Dittrich, T.R., and Haan, S.W. 2001. Three-dimensional HYDRA simulations of National Ignition Facility targets. *Phys. Plasmas*, **8**, 2275.

Markstein, G.H. 1951. Experimental and theoretical studies of flame-front stability. *J. Aeronaut. Sci.*, **18**, 199.

Markstein, G.H. 1957a. Flow disturbances induced near a slightly wavy contact surface, or flame front, traversed by a shock wave. *J Aerosp. Sci.*, **24**, 238.

Markstein, G.H. 1957b. A shock-tube study of flame front-pressure wave interaction. *Symposium (International) on Combustion*, **6**, 387.

Markstein, G.H., and Schwartz, D. 1955. Interaction between pressure waves and flame fronts. *Jet Propulsion*, **24**, 173.

Marrow, T.J., Buffiere, J.Y., Withers, P.J., Johnson, G., and Engelberg, D. 2004. High resolutionX-ray tomography of short fatigue crack nucleation in austempered ductile cast iron. *Int. J. Fatigue*, **26**, 717.

Martinez, D.A., Smalyuk, V.A., Kane, J.O., Casner, A., Liberatore, S., and Masse, L.P. 2015. Evidence for a bubble-competition regime in indirectly driven ablative Rayleigh–Taylor instability experiments on the NIF. *Phys. Rev. Lett.*, **114**, 215004.

Marusic, I., and Broomhall, S. 2021. Leonardo da Vinci and fluid mechanics. *Annu. Rev. Fluid Mech.*, **53**, 1.

Massa, L., and Jha, P. 2012. Linear analysis of the Richtmyer–Meshkov instability in shock-flame interactions. *Phys. Fluids*, **24**, 056101.

Matalon, M. 2018. The Darrieus-Landau instability of premixed flames. *Fluid Dyn. Res.*, **50**, 051412.

Matsumoto, J., and Masada, Y. 2013. Two-dimensional numerical study for Rayleigh–Taylor and Richtmyer–Meshkov instabilities in relativistic jets. *Astrophys. J. Lett.*, **772**, L1.

Matsuo, K., Sano, T., Nagatomo, H., Somekawa, T., Law, K.F.F., Morita, H., Arikawa, Y., and Fujioka, S. 2021. Enhancement of ablative Rayleigh–Taylor instability growth by thermal conduction suppression in a magnetic field. *Phys. Rev. Lett.*, **127**, 165001.

Matsuoka, C. 2009. Vortex sheet motion in incompressible Richtmyer–Meshkov and Rayleigh–Taylor instabilities with surface tension. *Phys. Fluids*, **21**, 092107.

Matsuoka, C. 2020. Nonlinear dynamics of double-layer unstable interfaces with non-uniform velocity shear. *Phys. Fluids*, **32**, 102109.

Matsuoka, C., and Nishihara, K. 2006a. Analytical and numerical study on a vortex sheet in incompressible Richtmyer–Meshkov instability in cylindrical geometry. *Phys. Rev. E*, **74**, 066303.

Matsuoka, C., and Nishihara, K. 2006b. Fully nonlinear evolution of a cylindrical vortex sheet in incompressible Richtmyer–Meshkov instability. *Phys. Rev. E*, **73**, 055304.

Matsuoka, C., and Nishihara, K. 2006c. Vortex core dynamics and singularity formations in incompressible Richtmyer–Meshkov instability. *Phys. Rev. E*, **73**, 026304.

Matsuoka, C., and Nishihara, K. 2020. Nonlinear interaction between bulk point vortices and an unstable interface with nonuniform velocity shear such as Richtmyer–Meshkov instability. *Phys. Plasmas*, **27**, 052305.

Matsuoka, C., Nishihara, K., and Fukuda, Y. 2003. Nonlinear evolution of an interface in the Richtmyer–Meshkov instability. *Phys. Rev. E*, **67**, 036301.

Matthaeus, W.H., and Zhou, Y. 1989. Extended inertial range phenomenology of magnetohydrodynamic turbulence. *Phys. Fluids B*, **1**, 1929.

Matzen, M.K. 1997. Z pinches as intense X-ray sources for high-energy density physics applications. *Phys. Plasmas*, **4**, 1519.

Maxwell, J.C. 1891. *Theory of Heat*. Longmans, Green and Co., London, UK.

McBride, R.D., Slutz, S.A., Jennings, C.A., et al. 2012. Penetrating radiography of imploding and stagnating beryllium liners on the Z accelerator. *Phys. Rev. Lett.*, **109**, 135004.

McBride, R.D., Martin, M.R., Lemke, R.W., et al. 2013. Beryllium liner implosion experiments on the Z accelerator in preparation for magnetized liner inertial fusion. *Phys. Plasmas*, **20**, 056309.

McComb, D. 2008. Scale-invariance in three-dimensional isotropic turbulence: a paradox and its resolution. *J. Phys. A*, **41**, 075501.

McCray, R. 1993. Supernova 1987A revisited. *Annu. Rev. Astron. Astrophys.*, **31**, 175.

McFarland, J.A., Greenough, J.A., and Ranjan, D. 2013. Investigation of the initial perturbation amplitude for the inclined interface Richtmyer–Meshkov instability. *Phys. Scr.*, **T155**, 014014.

McFarland, J.A., Reilly, D., Creel, S., McDonald, C., Finn, T., and Ranjan, D. 2014. Experimental investigation of the inclined interface Richtmyer–Meshkov instability before and after reshock. *Exp. Fluids*, **55**, 1640.

McFarland, J.A., Reilly, D., Black, W., Greenough, J.A., and Ranjan, D. 2015. Modal interactions between a large-wavelength inclined interface and small-wavelength multimode perturbations in a Richtmyer–Meshkov instability. *Phys. Rev. E*, **92**, 013023.

McFarland, J.A., Black, W.J., Dahal, J., and Morgan, B.E. 2016. Computational study of the shock driven instability of a multiphase particle-gas system. *Phys. Fluids*, **28**, 024105.

McLaren, C.P., Kovar, T.M., Penn, A., Müller, C.R., and Boyce, C.M. 2019. Gravitational instabilities in binary granular materials. *Proc. Nat. Acad. Sci.*, **116**, 9263.

Meaburn, J., Boumis, P., Walsh, J.R., Steffen, W., Holloway, A.J., Williams, R.J.R., and Bryce, M. 1996. Highly supersonic motions within the outer features of the η Carinae nebulosity. *Mon. Not. R. Astron. Soc.*, **282**, 1313.

Meier, D.L., Epstein, R.I., Arnett, W.D., and Schramm, D.N. 1976. Magnetohydrodynamic phenomena in collapsing stellar cores. *Astrophys. J.*, **204**, 869.

Meinecke, J., Tzeferacos, P., Bell, A., et al. 2015. Developed turbulence and nonlinear amplification of magnetic fields in laboratory and astrophysical plasmas. *Proc. Natl. Acad. Sci.*, **112**, 8211.

Mejia-Alvarez, R., Wilson, B., Leftwich, M.C., Martinez, A.A., and Prestridge, K.P. 2015. Design of a fast diaphragmless shock tube driver. *Shock Waves*, **25**, 635.

Mellado, J.P., Sarkar, S., and Zhou, Y. 2005. Large-eddy simulation of Rayleigh–Taylor turbulence with compressible miscible fluids. *Phys. Fluids*, **17**, 076101.

Mellor, G.L., and Herring, H.J. 1973. A survey of the mean turbulent field closure models. *AIAA J.*, **11**, 590.

Mellor, G.L., and Yamada, T. 1982. Development of a turbulence closure model for geophysical fluid problems. *Rev. Geophys.*, **20**, 851.

Melvin, J., Rao, P., Kaufman, R., Lim, H., Yu, Y., Glimm, J., and Sharp, D.H. 2013. Atomic scale mixing for inertial confinement fusion associated hydro instabilities. *High Energy Density Phys.*, **9**, 288.

Melvin, J., Rao, P., Kaufman, R., Lim, H., Yu, Y., Glimm, J., and Sharp, D.H. 2014. Turbulent transport at high Reynolds numbers in an inertial confinement fusion context. *J. Fluids Eng.*, **136**, 091206.

Mendel, J., Frank, K., Edlin, L., Hall, K., Webb, D., Mills, J., Holness, H.K., Furton, K.G., and Mills, D. 2021. Preliminary accuracy of COVID-19 odor detection by canines and HS-SPME-GC-MS using exhaled breath samples. *Forensic Sci. Int.*, **3**, 100155.

Meneveau, C., and Katz, J. 2000. Scale-invariance and turbulence models for large-eddy simulation. *Annu. Rev. Fluid Mech.*, **32**, 1.

Menikoff, R., and Zemach, C. 1983. Rayleigh–Taylor instability and the use of conformal maps for ideal fluid flow. *J. Comput. Phys.*, **51**, 28.

Menikoff, R., Mjolsness, R.C., Sharp, D.H., and Zemach, C. 1977. Unstable normal mode for Rayleigh–Taylor instability in viscous fluids. *Phys. Fluids*, **20**, 2000.

Merritt, E.C., Doss, F.W., Loomis, E.N., Flippo, K.A., and Kline, J.L. 2015. Modifying mixing and instability growth through the adjustment of initial conditions in a high-energy-density counter-propagating shear experiment on OMEGA. *Phys. Plasmas*, **22**, 062306.

Merritt, E.C., Sauppe, J.P., Loomis, E.N., et al. 2019. Experimental study of energy transfer in double shell implosions. *Phys. Plasmas*, **26**, 052702.

Meshkov, E.E. 1969. Instability of the interface of two gases accelerated by a shock wave. *Fluid Dyn.*, **4**, 101.

Meshkov, E.E. 2013. Some peculiar features of hydrodynamic instability development. *Phil. Trans. R. Soc. A*, **371**, 20120288.

Meyer, K.A., and Blewett, P.J. 1972. Numerical investigation of the stability of a shock-accelerated interface between two fluids. *Phys. Fluids*, **15**, 753.

Mikaelian, K.O. 1982. Normal modes and symmetries of the Rayleigh–Taylor instability in stratified fluids. *Phys. Rev. Lett.*, **48**, 1365.

Mikaelian, K.O. 1983. Time evolution of density perturbations in accelerating stratified fluids. *Phys. Rev. A*, **28**, 1637.

Mikaelian, K.O. 1985. Richtmyer–Meshkov instabilities in stratified fluids. *Phys. Rev. A*, **31**, 410.

Mikaelian, K.O. 1986. Approximate treatment of density gradients in Rayleigh–Taylor instabilities. *Phys. Rev. A*, **33**, 1216.

Mikaelian, K.O. 1987. Comment on "Rayleigh–Taylor instability in spherical geometry." *Phys. Rev. A*, **36**, 411.

Mikaelian, K.O. 1989. Turbulent mixing generated by Rayleigh–Taylor and Richtmyer–Meshkov instabilities. *Physica D*, **36**, 343.

Mikaelian, K.O. 1990a. Rayleigh–Taylor and Richtmyer–Meshkov instabilities and mixing in stratified spherical shells. *Phys. Rev. A*, **42**, 3400.

Mikaelian, K.O. 1990b. Rayleigh–Taylor and Richtmyer–Meshkov instabilities in multilayer fluids with surface tension. *Phys. Rev. A*, **42**, 7211.

Mikaelian, K.O. 1990c. Stability and mix in spherical geometry. *Phys. Rev. Lett.*, **65**, 992.

Mikaelian, K.O. 1990d. Turbulent energy at accelerating and shocked interfaces. *Phys. Fluids A*, **2**, 592.

Mikaelian, K.O. 1991. Kinetic energy of Rayleigh–Taylor and Richtmyer–Meshkov instabilities. *Phys. Fluids A*, **3**, 2625.

Mikaelian, K.O. 1993a. Effect of viscosity on Rayleigh–Taylor and Richtmyer–Meshkov instabilities. *Phys. Rev. E*, **47**, 375.

Mikaelian, K.O. 1993b. Growth rate of the Richtmyer–Meshkov instability at shocked interfaces. *Phys. Rev. Lett.*, **71**, 2903.

Mikaelian, K.O. 1994a. Freeze-out and the effect of compressibility in the Richtmyer–Meshkov instability. *Phys. Fluids*, **6**, 356.

Mikaelian, K.O. 1994b. Oblique shocks and the combined Rayleigh–Taylor, Kelvin–Helmholtz, and Richtmyer–Meshkov instabilities. *Phys. Fluids*, **6**, 1943.

Mikaelian, K.O. 1995. Rayleigh–Taylor and Richtmyer–Meshkov instabilities in finite-thickness fluid layers. *Phys. Fluids*, **7**, 888.

Mikaelian, K.O. 1996. Connection between the Rayleigh and the Schrödinger equations. *Phys. Rev. E*, **53**, 3551.

Mikaelian, K.O. 1998. Analytic approach to nonlinear Rayleigh–Taylor and Richtmyer–Meshkov instabilities. *Phys. Rev. Lett.*, **80**, 508.

Mikaelian, K.O. 2003. Explicit expressions for the evolution of single-mode Rayleigh–Taylor and Richtmyer–Meshkov instabilities at arbitrary Atwood numbers. *Phys. Rev. E*, **67**, 026319.

Mikaelian, K.O. 2005a. Rayleigh–Taylor and Richtmyer–Meshkov instabilities and mixing in stratified cylindrical shells. *Phys. Fluids*, **17**, 094105.

Mikaelian, K.O. 2005b. Richtmyer–Meshkov instability of arbitrary shapes. *Phys. Fluids*, **17**, 034101.

Mikaelian, K.O. 2008. Limitations and failures of the Layzer model for hydrodynamic instabilities. *Phys. Rev. E*, **78**, 015303.

Mikaelian, K.O. 2009a. Nonlinear hydrodynamic interface instabilities driven by time-dependent accelerations. *Phys. Rev. E*, **79**, 065303(R).

Mikaelian, K.O. 2009b. Reshocks, rarefactions, and the generalized Layzer model for hydrodynamic instabilities. *Phys. Fluids*, **21**, 024103.

Mikaelian, K.O. 2010. Analytic approach to nonlinear hydrodynamic instabilities driven by time-dependent accelerations. *Phys. Rev. E*, **81**, 016325.

Mikaelian, K.O. 2011. Extended model for Richtmyer–Meshkov mix. *Physica D*, **240**, 935.

Mikaelian, K.O. 2013. Shock-induced interface instability in viscous fluids and metals. *Phys. Rev. E*, **87**, 031003.

Mikaelian, K.O. 2014a. Boussinesq approximation for Rayleigh–Taylor and Richtmyer–Meshkov instabilities. *Phys. Fluids*, **26**, 054103.

Mikaelian, K.O. 2014b. Comment on "The effect of viscosity, surface tension and non-linearity on Richtmyer–Meshkov instability"[Eur. J. Mech. B Fluids 21 (2002) 511–526]. *Eur. J. Mech. B Fluids*, **43**, 183.

Mikaelian, K.O. 2014c. Solution to Rayleigh–Taylor instabilities: bubbles, spikes, and their scalings. *Phys. Rev. E*, **89**, 053009.

Mikaelian, K.O. 2015. Testing an analytic model for Richtmyer–Meshkov turbulent mixing widths. *Shock Waves*, **25**, 35.

Mikaelian, K.O. 2019. Exact, approximate, and hybrid treatments of viscous Rayleigh–Taylor and Richtmyer–Meshkov instabilities. *Phys. Rev. E*, **99**, 023112.

Mikaelian, K.O., and Lindl, J.D. 1984. Density gradients to reduce fluid instabilities in multishell inertial-confinement-fusion targets. *Phys. Rev. A*, **29**, 290.

Mikaelian, K.O., and Olson, B.J. 2020. On modeling Richtmyer–Meshkov turbulent mixing widths. *Physica D*, **402**, 132243.

Mikhailov, A.L., Ogorodnikov, V.A., Sasik, V.S., Raevskii, V.A., Lebedev, A.I., Zotov, D.E., Erunov, S.V., Syrunin, M.A., Sadunov, V.D., Nevmerzhitskii, N.V., and Lobastov, S.A. 2014. Experimental-calculation simulation of the ejection of particles from a shock-loaded surface. *Sov. Phys. JETP*, **118**, 785.

Miles, A.R., Edwards, M.J., and Greenough, J.A. 2004. Effect of initial conditions on two-dimensional Rayleigh–Taylor instability and transition to turbulence in planar blast-wave-driven systems. *Phys. Plasmas*, **11**, 5278.

Miller, G.H., Moses, E.I., and Wuest, C.R. 2004. The national ignition facility. *Optical Eng.*, **43**, 2841.

Milne, A., Parrish, C., and Worland, I. 2010. Dynamic fragmentation of blast mitigants. *Shock Waves*, **20**, 41.

Milovich, J.L., Amendt, P., Marinak, M., and Robey, H. 2004. Multimode short-wavelength perturbation growth studies for the National Ignition Facility double-shell ignition target designs. *Phys. Plasmas*, **11**, 1552.

Mima, K., Kato, Y., Azechi, H., et al. 1996. Recent progress of implosion experiments with uniformity-improved GEKKO XII laser facility at the Institute of Laser Engineering, Osaka University. *Phys. Plasmas*, **3**, 2077.

Mirabel, I.F., and Rodriguez, L.F. 1999. Sources of relativistic jets in the galaxy. *Annu. Rev. Astron. Astrophys.*, **37**, 409.

Mishra, S.K., and Srivastava, A.K. 2019. The evolution of magnetic Rayleigh–Taylor unstable plumes and hybrid KH-RT instability into a loop-like eruptive prominence. *Astrophys. J.*, **874**, 57.

Misra, A., and Pullin, D.I. 1997. A vortex-based subgrid stress model for large-eddy simulation. *Phys. Fluids*, **9**, 2443.

Mizuno, Y., Pohl, M., Niemiec, J., Zhang, B., Nishikawa, K.I., and Hardee, P.E. 2010. Magnetic-field amplification by turbulence in a relativistic shock propagating through an inhomogeneous medium. *Astrophys. J.*, **726**, 62.

Mizuno, Y., Pohl, M., Niemiec, J., Zhang, B., Nishikawa, K.I., and Hardee, P.E. 2014. Magnetic field amplification and saturation in turbulence behind a relativistic shock. *Mon. Not. R. Astron. Soc.*, **439**, 3490.

Moffatt, H.K. 2002. G.K. Batchelor and the homogenization of turbulence. *Annu. Rev. Fluid Mech.*, **34**, 19.

Moffatt, H.K. 2008. Vortex dynamics: the legacy of Helmholtz and Kelvin. Page 1 of: Borisov, A.V., Kozlov, V.V., Mamaev, I.S., and Sokolovskiy, M.A. (eds), *IUTAM Symposium on Hamiltonian Dynamics, Vortex Structures, Turbulence*. Springer, Dordrecht, Netherlands.

Moffatt, H.K. 2010. George Batchelor: A personal tribute, ten years on. *J. Fluid Mech.*, **663**, 2.

Mogilner, A., and Manhart, A. 2018. Intracellular fluid mechanics: coupling cytoplasmic flow with active cytoskeletal gel. *Annu. Rev. Fluid Mech.*, **50**, 347.

Mohaghar, M., Carter, J., Musci, B., Reilly, D., McFarland, J., and Ranjan, D. 2017. Evaluation of turbulent mixing transition in a shock-driven variable-density flow. *J. Fluid Mech.*, **831**, 779.

Mohaghar, M., Carter, J., Pathikonda, G., and Ranjan, D. 2019. The transition to turbulence in shock-driven mixing: effects of Mach number and initial conditions. *J. Fluid Mech.*, **871**, 595.

Mohaghar, M., McFarland, J., and Ranjan, D. 2022. Three-dimensional simulations of reshocked inclined Richtmyer–Meshkov instability: effects of initial perturbations. *Phys. Rev. Fluids*, **7**, 093902.

Moin, P., and Mahesh, K. 1998. Direct numerical simulation: a tool in turbulence research. *Annu. Rev. Fluid Mech.*, **30**, 539.

Molvig, K., Schmitt, M.J., Albright, B.J., Dodd, E.S., Hoffman, N.M., McCall, G.H., and Ramsey, S.D. 2016. Low fuel convergence path to direct-drive fusion ignition. *Phys. Rev. Lett.*, **116**, 255003.

Molvig, K., Schmitt, M.J., Betti, R., Campbell, E.M., and McKenty, P. 2018. Stable and confined burn in a Revolver ignition capsule. *Phys. Plasmas*, **25**, 082708.

Mondal, P., and Korenaga, J. 2017. The Rayleigh–Taylor instability in a self-gravitating two-layer viscous sphere. *Geophys. J. Int.*, **212**, 1859.

Monin, A.S., and Yaglom, A.M. 1971. *Statistical Fluid Mechanics, Volume I*. MIT Press, Cambridge, MA.

Monin, A.S., and Yaglom, A.M. 1975. *Statistical Fluid Mechanics, Volume II*. MIT Press, Cambridge, MA.

Montgomery, D.S., Daughton, W.S., Albright, B.J., Simakov, A.N., Wilson, D.C., Dodd, E.S., Kirkpatrick, R.C., Watt, R.G., Gunderson, M.A., Loomis, E.N., and Merritt, E.C. 2018. Design considerations for indirectly driven double shell capsules. *Phys. Plasmas*, **25**, 092706.

Moore, A.S., Meezan, N.B., Thomas, C.A., et al. 2020a. Experimental demonstration of the reduced expansion of a laser-heated surface using a low density foam layer, pertaining to advanced hohlraum designs with less wall-motion. *Phys. Plasmas*, **27**, 082706.

Moore, A.S., Meezan, N.B., Milovich, J., et al. 2020b. Foam-lined hohlraum, inertial confinement fusion experiments on the National Ignition Facility. *Phys. Rev. E*, **102**, 051201.

Morán-López, J. T., and Schilling, O. 2013. Multicomponent Reynolds-averaged Navier–Stokes simulations of reshocked Richtmyer–Meshkov instability-induced mixing. *High Energy Density Phys.*, **9**, 112.

Morán-López, J.T., and Schilling, O. 2014. Multi-component Reynolds-averaged Navier–Stokes simulations of Richtmyer–Meshkov instability and mixing induced by reshock at different times. *Shock Waves*, **24**, 325.

Morgan, B.E. 2021a. Scalar mixing in a Kelvin–Helmholtz shear layer and implications for Reynolds-averaged Navier–Stokes modeling of mixing layers. *Phys. Rev. E*, **103**, 053108.

Morgan, B.E. 2021b. Self-consistent, high-order spatial profiles in a model for two-fluid turbulent mixing. *Phys. Rev. E*, **104**, 015107.

Morgan, B.E. 2022a. Large-eddy simulation and Reynolds-averaged Navier–Stokes modeling of three Rayleigh–Taylor mixing configurations with gravity reversal. *Phys. Rev. E*, **106**, 025101.

Morgan, B.E. 2022b. Simulation and Reynolds-averaged Navier–Stokes modeling of a three-component Rayleigh–Taylor mixing problem with thermonuclear burn. *Phys. Rev. E*, **105**, 045104.

Morgan, B.E., and Black, W.J. 2020. Parametric investigation of the transition to turbulence in Rayleigh–Taylor mixing. *Physica D*, **402**, 132223.

Morgan, B.E., and Greenough, J.A. 2016. Large-eddy and unsteady RANS simulations of a shock-accelerated heavy gas cylinder. *Shock Waves*, **26**, 355.

Morgan, B.E., and Wickett, M.E. 2015. Three-equation model for the self-similar growth of Rayleigh–Taylor and Richtmyer–Meskov instabilities. *Phys. Rev. E*, **91**, 043002.

Morgan, B.E., Olson, B.J., Black, W.J., and McFarland, J.A. 2018a. Large-eddy simulation and Reynolds-averaged Navier–Stokes modeling of a reacting Rayleigh–Taylor mixing layer in a spherical geometry. *Phys. Rev. E*, **98**, 033111.

Morgan, B.E., Schilling, O., and Hartland, T.A. 2018b. Two-length-scale turbulence model for self-similar buoyancy-, shock-, and shear-driven mixing. *Phys. Rev. E*, **97**, 013104.

Morgan, B.E., Ferguson, K., and Olson, B.J. 2023. Two self-similar Reynolds-stress transport models with anisotropic eddy viscosity. *Phys. Rev. E*, **108**, 055104.

Morgan, R.V., and Jacobs, J.W. 2020. Experiments and simulations on the turbulent, rarefaction wave driven Rayleigh–Taylor instability. *J. Fluids Eng.*, **142**, 121101.

Morgan, R.V., Aure, R., Stockero, J.D., Greenough, J.A., Cabot, W., Likhachev, O.A., and Jacobs, J.W. 2012. On the late-time growth of the two-dimensional Richtmyer–Meshkov instability in shock tube experiments. *J. Fluid Mech.*, **712**, 354.

Morgan, R.V., Likhachev, O.A., and Jacobs, J.W. 2016. Rarefaction-driven Rayleigh–Taylor instability. Part 1. Diffuse-interface linear stability measurements and theory. *J. Fluid Mech.*, **791**, 34.

Morgan, R.V., Cabot, W.H., Greenough, J.A., and Jacobs, J.W. 2018c. Rarefaction-driven Rayleigh–Taylor instability. Part 2. Experiments and simulations in the nonlinear regime. *J. Fluid Mech.*, **838**, 320.

Moser, R.D., Haering, S.W., and Yalla, G.R. 2021. Statistical properties of subgrid-scale turbulence models. *Annu. Rev. Fluid Mech.*, **53**, 255.

Moses, E.I., and Wuest, C.R. 2003. The National Ignition Facility: status and plans for laser fusion and high-energy-density experimental studies. *Fusion Sci. Tech.*, **43**, 420.

Mostert, W., Pullin, D.I., Wheatley, V., and Samtaney, R. 2017. Magnetohydrodynamic implosion symmetry and suppression of Richtmyer–Meshkov instability in an octahedrally symmetric field. *Phys. Rev. Fluids*, **2**, 013701.

Motl, B., Oakley, J., Ranjan, D., Weber, C., Anderson, M., and Bonazza, R. 2009. Experimental validation of a Richtmyer–Meshkov scaling law over large density ratio and shock strength ranges. *Phys. Fluids*, **21**, 126102.

Movahed, P., and Johnsen, E. 2013. A solution-adaptive method for efficient compressible multifluid simulations, with application to the Richtmyer–Meshkov instability. *J. Comput. Phys.*, **239**, 166.

Mueschke, N.J. 2010. *Experimental and numerical study of molecular mixing dynamics in Rayleigh–Taylor unstable flows*. Ph.D. thesis, Texas A&M, College Station, TX.

Mueschke, N.J., and Schilling, O. 2009a. Investigation of Rayleigh–Taylor turbulence and mixing using direct numerical simulation with experimentally measured initial conditions. I. Comparison to experimental data. *Phys. Fluids*, **21**, 014106.

Mueschke, N.J., and Schilling, O. 2009b. Investigation of Rayleigh–Taylor turbulence and mixing using direct numerical simulation with experimentally measured initial conditions. II. Dynamics of transitional flow and mixing statistics. *Phys. Fluids*, **21**, 014107.

Mueschke, N.J., Andrews, M.J., and Schilling, O. 2006. Experimental characterization of initial conditions and spatio-temporal evolution of a small-Atwood-number Rayleigh–Taylor mixing layer. *J. Fluid Mech.*, **567**, 27.

Mueschke, N.J., Schilling, O., Youngs, D.L., and Andrews, M. 2007. Measurements of molecular mixing in a high Schmidt number Rayleigh–Taylor mixing layer. *J. Fluid Mech.*, **632**, 17.

Mügler, C., and Gauthier, S. 2000. Two-dimensional Navier–Stokes simulations of gaseous mixtures induced by Richtmyer–Meshkov instability. *Phys. Fluids*, **12**, 1783.

Müller, B. 2019. A critical assessment of turbulence models for 1D core-collapse supernova simulations. *Mon. Not. R. Astron. Soc.*, **487**, 5304.

Müller, B. 2020. Hydrodynamics of core-collapse supernovae and their progenitors. *Living Rev. Comput. Astrophys.*, **6**, 1.

Müller, B., and Janka, H.-Th. 2015. Non-radial instabilities and progenitor asphericities in core-collapse supernovae. *Mon. Not. R. Astron. Soc.*, **448**, 2141.

Müller, B., Gay, D.W., Heger, A., Tauris, T.M., and Sim, S.A. 2018. Multidimensional simulations of ultrastripped supernovae to shock breakout. *Mon. Not. R. Astron. Soc.*, **479**, 3675.

Munro, D.H. 1988. Analytic solutions for Rayleigh–Taylor growth rates in smooth density gradients. *Phys. Rev. A*, **38**, 1433.

Munson, B.R., Young, D.F., and Okiishi, T.H. 1998. *Fundamentals of Fluid Mechanics*. 3rd edn. Wiley & Sons, New York.

Murdin, P., and Murdin, L. 1985. *Supernovae*. Cambridge University Press, Cambridge, UK.

Murillo, M.S. 2008. Viscosity estimates of liquid metals and warm dense matter using the Yukawa reference system. *High Energy Density Phys.*, **4**, 49.

Musci, B., Petter, S., Pathikonda, G., Ranjan, D., and Denissen, N. 2018. An experimental study of the blast driven Rayleigh–Taylor and Richtmyer–Meshkov instabilities: preliminary results. Page 202 of: *Proceedings of the 16th International Workshop on the Physics of Compressible Turbulent Mixing*. Marseille, France.

Musci, B., Petter, S., Pathikonda, G., Ochs, B., and Ranjan, D. 2020. Supernova hydrodynamics: a lab-scale study of the blast-driven instability using high-speed diagnostics. *Astrophys. J.*, **896**, 92.

Musci, B., Olson, B., Petter, S., Pathikonda, G., and Ranjan, D. 2023. Multifidelity validation of digital surrogates using variable-density turbulent mixing models. *Phys. Rev. Fluids*, **8**, 014501.

Muskat, M. 1937. *The Flow of Homogeneous Fluids through Porous Media: Analogies with Other Physical Problems*. McGraw-Hill, New York.

Nagel, S.R., Haan, S.W., Rygg, J.R., et al. 2015. Effect of the mounting membrane on shape in inertial confinement fusion implosions. *Phys. Plasmas*, **22**, 022704.

Nagel, S.R., Raman, K.S., Huntington, C.M., MacLaren, S.A., Wang, P., Barrios, M.A., Baumann, T., Bender, J.D., Benedetti, L.R., Doane, D.M., Felker, S., Fitzsimmons, P., Flippo, K.A., Holder, J.P., Kaczala, D.N., Perry, T.S., Seugling, R. M., Savage, L., and Zhou, Y. 2017. A platform for studying the Rayleigh–Taylor and Richtmyer–Meshkov instabilities in a planar geometry at high energy density at the National Ignition Facility. *Phys. Plasmas*, **24**, 072704.

Nagel, S.R., Raman, K.S., Huntington, C.M., MacLaren, S.A., Wang, P., Bender, J.D., Prisbrey, S.T., and Zhou, Y. 2022. Experiments on the single-mode Richtmyer–Meshkov instability with reshock at high energy densities. *Phys. Plasmas*, **29**, 032308.

Nakaya, S., Hikichi, Y., Nakazawa, Y., Sakaki, K., Choi, M., Tsue, M., Kono, M., and Tomioka, S. 2015. Ignition and supersonic combustion behavior of liquid ethanol in a scramjet model combustor with cavity flame holder. *Proc. Combust. Inst.*, **35**, 2091.

Narkis, J., Conti, F., and Beg, F.N. 2022. Material effects on dynamics in triple-nozzle gas-puff Z pinches. *Phys. Rev. E*, **105**, 045205.

Navier, C.L.M.H. 1827. Sur les lois du mouvement des fluids. *Memoir Acad. Roy. Sci. Inst. Fr*, **6**, 389.

Nayfeh, A.H. 1969. On the non-linear Lamb-Taylor instability. *J. Fluid Mech.*, **38**, 619.

Nayfeh, A.H. 1973. *Perturbation Methods*. Wiley, New York.

Nelkin, M. 1994. Universality and scaling in fully developed turbulence. *Adv. Phys.*, **43**, 143.

Nelson, N.J., and Grinstein, F.F. 2015. Effects of initial condition spectral content on shock-driven turbulent mixing. *Phys. Rev. E*, **92**, 013014.

Neugebauer, O. 1975. *A History of Ancient Mathematical Astronomy*. Vol. 1. Springer Science & Business Media, Berlin.

Neuvazhaev, V.E., and Yakovlev, V.G. 1976a. Theory of turbulent mixing at the interface of fluids in a gravity field. *J. Appl. Mech. Tech. Phys.*, **17**, 513.

Neuvazhaev, V.E., and Yakovlev, V.G. 1976b. Turbulent mixing of an interface in a numerical gasdynamic calculation. *Zh. Vychisl. Mat. Mat. Fiz*, **16**, 154.

Nguyen, Q.M., Oza, A.U., Abouezzi, J., Sun, G., Childress, S., Frederick, C., and Ristroph, L. 2021. Flow rectification in loopy network models of bird lungs. *Phys. Rev. Lett.*, **126**, 114501.

Nicoud, F., and Ducros, F. 1999. Subgrid-scale stress modelling based on the square of the velocity gradient tensor. *Flow Turbul. Combust.*, **62**, 183.

Niederhaus, C.E., and Jacobs, J.W. 2003. Experimental study of the Richtmyer–Meshkov instability of incompressible fluids. *J. Fluid Mech.*, **485**, 243.

Nield, D.A., and Bejan, A. 2006. *Convection in Porous Media*. 5th edn. Springer, Cham, Switzerland.

Nikiforov, V. V., Andronov, V. A., Razin, A. N., and Trutnev, Yu. A. 1995. Development of a turbulent mixing zone driven by a shock wave. *Physics – Doklady*, **40**, 333.

Nikolov, A., Wasan, D., and Lee, J. 2018. Tears of wine: the dance of the droplets. *Adv. Colloid Interface Sci.*, **256**, 94.

Nishihara, K., Wouchuk, J.G., Matsuoka, C., Ishizaki, R., and Zhakhovsky, V.V. 2010. Richtmyer–Meshkov instability: theory of linear and nonlinear evolution. *Phil. Trans. R. Soc. A*, **368**, 1769.

Nobile, A., Nikroo, A., Cook, R.C., et al. 2006. Status of the development of ignition capsules in the US effort to achieve thermonuclear ignition on the National Ignition Facility. *Laser Part. Beams*, **24**, 567.

Noble, C.D., Herzog, J.M., Ames, A.M., Oakley, J., Rothamer, D.A., and Bonazza, R. 2020a. High speed PLIF study of the Richtmeyer-Meshkov instability upon re-shock. *Physica D*, **401**, 132519.

Noble, C.D., Herzog, J.M., Rothamer, D.A., Ames, A.M., Oakley, J., and Bonazza, R. 2020b. Scalar power spectra and scalar structure function evolution in the Richtmyer–Meshkov instability upon reshock. *J. Fluids Eng.*, **142**, 121102.

Noble, C., Ames, A., McConnell, R., Oakley, J., Rothamer, D., and Bonazza, R. 2023. Simultaneous measurements of kinetic and scalar energy spectrum time evolution in the Richtmyer–Meshkov instability upon reshock. *J. Fluid Mech.*, **975**, A39.

Nuckolls, J., Wood, L., Thiessen, A., and Zimmerman, G. 1972. Laser compression of matter to super-high densities: thermonuclear (CTR) applications. *Nature*, **239**, 139.

Obenschain, S.P., Bodner, S.E., Colombant, D., et al. 1996. The Nike KrF laser facility: performance and initial target experiments. *Phys. Plasmas*, **3**, 2098.

Obukhov, AM. 1949. Temperature field structure in a turbulent flow. *Izv. Acad. Nauk SSSR Ser. Geog. Geofiz*, **13**, 58.

Ofer, D., Alon, U., Shvarts, D., McCrory, R.L., and Verdon, C.P. 1996. Modal model for the nonlinear multimode Rayleigh–Taylor instability. *Phys. Plasmas*, **3**, 3073.

Oggian, T., Drikakis, D., Youngs, D.L., and Williams, R.J.R. 2015. Computing multi-mode shock-induced compressible turbulent mixing at late times. *J. Fluid Mech.*, **779**, 411.

Olson, B.J., and Cook, A.W. 2007. Rayleigh–Taylor shock waves. *Phys. Fluids*, **19**, 128108.

Olson, B.J., and Greenough, J.A. 2014. Comparison of two-and three-dimensional simulations of miscible Richtmyer–Meshkov instability with multimode initial conditions. *Phys. Fluids*, **26**, 101702.

Olson, B.J., Larsson, J., Lele, S.K., and Cook, A.W. 2011. Nonlinear effects in the combined Rayleigh–Taylor/Kelvin–Helmholtz instability. *Phys. Fluids*, **23**, 114107.

Olson, D.H., and Jacobs, J.W. 2009. Experimental study of Rayleigh–Taylor instability with a complex initial perturbation. *Phys. Fluids*, **21**, 034103.

Olsthoorn, J., Tedford, E.W., and Lawrence, G.A. 2019. Diffused-interface Rayleigh–Taylor instability with a nonlinear equation of state. *Phys. Rev. Fluids*, **4**, 094501.

Olver, P.J. 2000. *Applications of Lie Groups to Differential Equations*. Springer Science & Business Media New York.

Onsager, L. 1945. The distribution of energy in turbulence. *Phys. Rev.*, **68**, 286.

Onsager, L. 1949. Statistical hydrodynamics. *Il Nuovo Cimento (1943–1954)*, **6**, 279.

Oran, E.S. 2005. Astrophysical combustion. *Proc. Comb. Inst.*, **30**, 1823.

Oren, G.H., and Terrones, G. 2022. Finite boundary effects on the spherical Rayleigh–Taylor instability between viscous fluids. *AIP Adv.*, **12**, 045009.

Orlicz, G.C., Balasubramanian, S., and Prestridge, K.P. 2015a. Investigation of Mach number dependence on the Richtmyer–Meshkov mixing transition for a shocked heavy-gas curtain. Pages

1101–1106 of: Bonazza, R., and Ranjan, D. (eds), *29th International Symposium on Shock Waves 2*. Springer International Publishing, Cham, Switzerland.

Orlicz, G.C., Balasubramanian, S., and Prestridge, K.P. 2013. Incident shock Mach number effects on Richtmyer–Meshkov mixing in a heavy gas layer. *Phys. Fluids*, **25**, 114101.

Orlicz, G.C., Balasubramanian, S., Vorobieff, P., and Prestridge, K.P. 2015b. Mixing transition in a shocked variable-density flow. *Phys. Fluids*, **27**, 114102.

Oron, D., Arazi, L., Kartoon, D., Rikanati, A., Alon, U., and Shvarts, D. 2001. Dimensionality dependence of the Rayleigh–Taylor and Richtmyer–Meshkov instability late-time scaling laws. *Phys. Plasmas*, **8**, 2883.

O'Rourke, P.J., and Amsden, A.A. 1987. *The TAB method for numerical calculation of spray droplet breakup*. Tech. rept. SAE technical paper.

Orszag, S.A. 1970a. Analytical theories of turbulence. *J. Fluid Mech.*, **41**, 363.

Orszag, S.A. 1970b. Transform method for the calculation of vector-coupled sums: application to the spectral form of the vorticity equation. *J. Atmos. Sci.*, **27**, 890.

Orszag, S.A. 1977. Statistical theory of turbulence. In *Les Houcher Summer School in Physics*, Gordon and Breach, New York.

Orszag, S.A., and Patterson Jr, G.S. 1972. Numerical simulation of three-dimensional homogeneous isotropic turbulence. *Phys. Rev. Lett.*, **28**, 76.

Osin, D., Kroupp, E., Starobinets, A., Rosenzweig, G., Alumot, D., Maron, Y., Fisher, A., Yu, E., Giuliani, J.L., and Deeney, C. 2011. Evolution of MHD instabilities in plasma imploding under magnetic field. *IEEE Trans. Plasma Sci.*, **39**, 2392.

Otto, C.M., Sell, T.K., Veenema, T.G., Hosangadi, D., Vahey, R.A., Connell, N.D., and Privor-Dumm, L. 2021. The promise of disease detection dogs in pandemic response: lessons learned from COVID-19. *Disaster Med. Public Health Prep.* https://doi.org/10.1017/dmp.2021.183.

O'Neill, P.L., Nicolaides, D., Honnery, D., Soria, J., et al. 2004. Autocorrelation functions and the determination of integral length with reference to experimental and numerical data. Behnia, M., Lin, W., and McBain, G. D. (eds), *15th Australasian Uid Mechanics Conference*. Vol. 1. University of Sydney, Sydney, Australia.

Pacitto, G., Flament, C., Bacri, J.-C., and Widom, M. 2000. Rayleigh–Taylor instability with magnetic fluids: experiment and theory. *Phys. Rev. E*, **62**, 7941.

Pak, A., Zylstra, A.B., et al. 2024. Observations and properties of the first laboratory fusion experiment to exceed target gain of unity. *Phys. Rev. E*, **109**, 025203.

Pal, N., Boureima, I., Braun, N., Kurien, S., Ramaprabhu, P., and Lawrie, A. 2021. Local wave-number model for inhomogeneous two-fluid mixing. *Phys. Rev. E*, **104**, 025105.

Palmer, C.A.J., Schreiber, J., Nagel, S.R., et al. 2012. Rayleigh–Taylor instability of an ultra-thin foil accelerated by the radiation pressure of an intense laser. *Phys. Rev. Lett.*, **108**, 225002.

Papamoschou, D. 1995. Evidence of shocklets in a counterflow supersonic shear layer. *Phys. Fluids*, **7**, 233.

Paquette, C., Pelletier, C., Fontaine, G., and Michaud, G. 1986. Diffusion coefficients for stellar plasmas. *Astrophys. J. Suppl. Ser.*, **61**, 177.

Park, H.S., Lorenz, K.T., Cavallo, R.M., Pollaine, S.M., Prisbrey, S.T., Rudd, R.E., Becker, R.C., Bernier, J.V., and Remington, B.A. 2010. Viscous Rayleigh–Taylor instability experiments at high pressure and strain rate. *Phys. Rev. Lett.*, **104**, 135504.

Park, H.S., Rudd, R.E., Cavallo, R.M., Barton, N.R., Arsenlis, A., Belof, J.L., Blobaum, K.J.M., El-dasher, B.S., Florando, J.N., Huntington, C.M., and Maddox, B.R. 2015. Grain-size-independent plastic flow at ultrahigh pressures and strain rates. *Phys. Rev. Lett.*, **114**, 065502.

Park, J., Min, K.W., Kim, V.P., Kil, H., Su, S.Y., Chao, C.K., and Lee, J.J. 2008. Equatorial plasma bubbles with enhanced ion and electron temperatures. *J. Geophys. Res. Space Phys.*, **113**, A09318.

Parker, K., Horsfield, C.J., Rothman, S.D., Batha, S.H., Balkey, M.M., Delamater, N.D., Fincke, J.R., Hueckstaedt, R.M., Lanier, N.E., and Magelssen, G.R. 2004. Observation and simulation of plasma mix after reshock in a convergent geometry. *Phys. Plasmas*, **11**, 2696.

Pathikonda, G., Petter, S.J., Wall, I., and Ranjan, D. 2022. Temporal evolution of scalar modes in Richtmyer–Meshkov instability of inclined interface using high-speed PIV and PLIF measurements at 60 kHz. *Meas. Sci. Tech.*, **33**, 105206.

Pellone, Sa., Di Stefano, C.A., Rasmus, A.M., Kuranz, C.C., and Johnsen, E. 2021. Vortex-sheet modeling of hydrodynamic instabilities produced by an oblique shock interacting with a perturbed interface in the HED regime. *Phys. Plasmas*, **28**, 022303.

Peltier, W.R., and Caulfield, C.P. 2003. Mixing efficiency in stratified shear flows. *Ann. Rev. Fluid Mech.*, **35**, 135.

Peng, N., Yang, Y., and Xiao, Z. 2021a. Effects of the secondary baroclinic vorticity on the energy cascade in the Richtmyer–Meshkov instability. *J. Fluid Mech.*, **925**, A39.

Peng, N., Yang, Y., Wu, J., and Xiao, Z. 2021b. Mechanism and modelling of the secondary baroclinic vorticity in the Richtmyer–Meshkov instability. *J. Fluid Mech.*, **911**, A56.

Peng, Q., Bao, B., Yang, C., and Zhang, L. 2020. Simulations of young type Ia supernova remnants undergoing shock acceleration in a turbulent medium *Astrophys. J.*, **891**, 75.

Penney, W.G., and Price, A.T. 1942. On the changing form of a nearly spherical submarine bubble. Page 145 of: Hartmann, G.K. and Hill, E. G. (eds), *Underwater Explosion Research*. Vol. 2. Office of Naval Research, Washington, DC.

Pereira, F.S., Grinstein, F.F., Israel, D.M., and Rauenzahn, R. 2021. Molecular viscosity and diffusivity effects in transitional and shock-driven mixing flows. *Phys. Rev. E*, **103**, 013106.

Perez, F., Koenig, M., Batani, D., et al. 2009. Fast-electron transport in cylindrically laser-compressed matter. *Plasma Phys. Control. Fusion*, **51**, 124035.

Perkins, L.J., Logan, B.G., Zimmerman, G.B., and Werner, C.J. 2013. Two-dimensional simulations of thermonuclear burn in ignition-scale inertial confinement fusion targets under compressed axial magnetic fields. *Phys. Plasmas*, **20**, 072708.

Perkins, L.J., Ho, D.M., Logan, B.G., Zimmerman, G.B., Rhodes, M.A., Strozzi, D.J., Blackfield, D.T., and Hawkins, S.A. 2017. The potential of imposed magnetic fields for enhancing ignition probability and fusion energy yield in indirect-drive inertial confinement fusion. *Phys. Plasmas*, **24**, 062708.

Perlmutter, S., Gabi, S., Goldhaber, G., et al. 1997. Measurements of the cosmological parameters Ω and Λ from the first seven supernovae at $Z \geq 0.35$. *Astrophys. J.*, **483**, 565.

Perlmutter, S., Aldering, G., Goldhaber, G., et al. 1999. Measurements of Ω and Λ from 42 high-redshift supernovae. *Astrophys. J.*, **517**, 565.

Peterson, D.L., Bowers, R.L., Brownell, J.H., Greene, A.E., McLenithan, K.D., Oliphant, T.A., Roderick, N.F., and Scannapieco, A.J. 1996. Two-dimensional modeling of magnetically driven Rayleigh–Taylor instabilities in cylindrical Z pinches. *Phys. Plasmas*, **3**, 368.

Peterson, J.L., Clark, D.S., Masse, L.P., and Suter, L.J. 2014. The effects of early time laser drive on hydrodynamic instability growth in National Ignition Facility implosions. *Phys. Plasmas*, **21**, 092710.

Peyser, T.A., Miller, P.L., Stry, P.E., Budil, K.S., Burke, E.W., Wojtowicz, D.A., Griswold, D.L., Hammel, B.A., and Phillion, D.W. 1995. Measurement of radiation-driven shock-induced mixing from nonlinear initial perturbations. *Phys. Rev. Lett.*, **75**, 2332.

Philippe, F., Casner, A., Caillaud, T., et al. 2010. Experimental demonstration of X-ray drive enhancement with rugby-shaped hohlraums. *Phys. Rev. Lett.*, **104**, 035004.

Phillion, D.W., and Pollaine, S.M. 1994. Dynamical compensation of irradiation nonuniformities in a spherical hohlraum illuminated with tetrahedral symmetry by laser beams. *Phys. Plasmas*, **1**, 2963.

Picone, J.M., Oran, E.S., Boris, J.P., and Young, T.R. 1985. Theory of vorticity generation by shock wave and flame interactions. Pages 429–448 of: *Dynamics of Shock Waves, Explosions, and Detonations*. American Institute of Aeronautics and Astronautics, New York.

Ping, Y., Smalyuk, V.A., Amendt, P., et al. 2019. Enhanced energy coupling for indirectly driven inertial confinement fusion. *Nat. Phys.*, **15**, 138.

Piriz, A.R., López Cela, J.J., Tahir, N.A., and Hoffmann, D.H.H. 2008. Richtmyer–Meshkov instability in elastic-plastic media. *Phys. Rev. E*, **78**, 056401.

Piriz, A.R., López Cela, J.J., and Tahir, N.A. 2009. Richtmyer–Meshkov instability as a tool for evaluating material strength under extreme conditions. *Nucl. Instrum. Methods Phys. Res. A*, **606**, 139.

Piriz, A.R., Piriz, S.A., and Tahir, N.A. 2022. Cylindrical convergence effects on the Rayleigh–Taylor instability in elastic and viscous media. *Phys. Rev. E*, **106**, 015109.

Plesset, M.S. 1954. On the stability of fluid flows with spherical symmetry. *J. Appl. Phys.*, **25**, 96.

Poggi, F. 1997. *Analyse par vélocimétrie d'un mélange gazeux créé par instabilité de Richtmyer–Meshkov*. Ph.D. thesis, Université de Poitiers, Poitiers, France.

Poggi, F., Thorembey, M.-H., and Rodriguez, G. 1998. Velocity measurements in turbulent gaseous mixtures induced by Richtmyer–Meshkov instability. *Phys. Fluids*, **10**, 2698.

Poinsot, T., and Veynante, D. 2005. *Theoretical and Numerical Combustion*. R.T. Edwards, Inc. Philadelphia, PA.

Polavarapu, R., Roach, P., and Banerjee, A. 2019. Rayleigh–Taylor-instability experiments with elastic-plastic materials. *Phys. Rev. E*, **99**, 053104.

Polchinski, J.G. 1998. *String Theory*. Vols. I and II. Cambridge University Press, Cambridge, UK.

Pope, S.B. 2000. *Turbulent Flows*. Cambridge University Press, Cambridge, UK.

Pope, S.B. 2004. Ten questions concerning the large-eddy simulation of turbulent flows. *New J. Phys.*, **6**, 35.

Popescu Braileanu, B., Lukin, V.S., Khomenko, E., and de Vicente, A. 2021a. Two-fluid simulations of Rayleigh–Taylor instability in a magnetized solar prominence thread. I. Effects of prominence magnetization and mass loading. *Astron. Astrophys.*, **646**, A93.

Popescu Braileanu, B., Lukin, V.S., Khomenko, E., and de Vicente, A. 2021b. Two-fluid simulations of Rayleigh–Taylor instability in a magnetized solar prominence thread. II. Effects of collisionality. *Astron. Astrophys.*, **650**, A181.

Popil, R., and Curzon, F.L. 1979. Production of reproducible Rayleigh–Taylor instabilities. *Rev. Sci. Instrum.*, **50**, 1291.

Porter, D.H., Jones, T.W., and Ryu, D. 2015. Vorticity, shocks, and magnetic fields in subsonic, ICM-like turbulence. *Astrophys. J.*, **810**, 93.

Poujade, O., and Peybernes, M. 2010. Growth rate of Rayleigh–Taylor turbulent mixing layers with the foliation approach. *Phys. Rev. E*, **81**, 016316.

Pozzi, A. 1994. *Applications of Padé Approximation Theory in Fluid Dynamics*. Vol. 14. World Scientific, Singapore.

Praskovsky, A.A., Gledzer, E.B., Karyakin, M.Yu, and Zhou, Y. 1993. The sweeping decorrelation hypothesis and energy–inertial scale interaction in high Reynolds number flows. *J. Fluid Mech.*, **248**, 493.

Preston, D. L, Tonks, D. L, and Wallace, D. C. 2003. Model of plastic deformation for extreme loading conditions. *J. Appl. Phys.*, **93**, 211.

Prestridge, K., Balasubramanian, S., and Orlicz, G. 2012. Effects of initial conditions on mixing in Richtmyer–Meshkov turbulence experiments. Pages 321–327 of: Kontis, K., (ed), *28th International Symposium on Shock Waves*. Springer, Berlin Heidelberg.

Prime, M.B., Buttler, W.T., Buechler, M.A., et al. 2017. Estimation of metal strength at very high rates using free-surface Richtmyer–Meshkov instabilities. *J. Dyn. Behavior Mater.*, **3**, 189.

Prime, M.B., Buttler, W.T., Fensin, S.J., et al. 2019. Tantalum strength at extreme strain rates from impact-driven Richtmyer–Meshkov instabilities. *Phys. Rev. E*, **100**, 053002.

Pringle, J.E. 1981. Accretion discs in astrophysics. *Annu. Rev. Astron. Astrophys.*, **19**, 137.

Probyn, M., Thornber, B., Drikakis, D., Youngs, D., and Williams, R.J.R. 2014. An investigation into nonlinear growth rate of two-dimensional and three-dimensional single-mode Richtmyer–Meshkov instability using an arbitrary-Lagrangian-Eulerian algorithm. *J. Fluids Eng.*, **136**, 091208.

Probyn, M.G., Williams, R.J.R., Thornber, B., Drikakis, D., and Youngs, D.L. 2021. 2D single-mode Richtmyer–Meshkov instability. *Physica D*, **418**, 132827.

Pu, Y., Huang, T., Ge, F., Zou, S., Wang, F., Yang, J., Jiang, S., and Ding, Y. 2018. First integrated implosion experiments on the SG-III laser facility. *Plasma Phys. Control. Fusion*, **60**, 085017.

Pullin, D.I., and Saffman, P.G. 1998. Vortex dynamics in turbulence. *Annu. Rev. Fluid Mech.*, **30**, 31.

Puranik, P.B., Oakley, J.G., Anderson, M.H., and Bonazza, R. 2004. Experimental study of the Richtmyer–Meshkov instability induced by a Mach 3 shock wave. *Shock Waves*, **13**, 413.

Purcell, E.M. 1977. Life at low Reynolds number. *Amer. J. Phys.*, **45**, 3.

Radice, D., Abdikamalov, E., Ott, C.D., Mösta, P., Couch, S.M., and Roberts, L.F. 2018. Turbulence in core-collapse supernovae. *J. Phys. G*, **45**, 053003.

Raga, A.C., Cantó, J., Binette, L., and Calvet, N. 1990. Stellar jets with intrinsically variable sources. *Astrophys. J.*, **364**, 601.

Raghavarao, R., Sekar, R., and Suhasini, R. 1992. Nonlinear numerical simulation of equatorial spread *F* – Effects of winds and electric fields. *Adv. Space Res.*, **12**, 227.

Raman, K.S., Hurricane, O.A., Park, H.-S., Remington, B.A., et al. 2012. Three-dimensional modeling and analysis of a high energy density Kelvin–Helmholtz experiment. *Phys. Plasmas*, **19**, 092112.

Ramaprabhu, P., and Andrews, M.J. 2004. On the initialization of Rayleigh–Taylor simulations. *Phys. Fluids*, **16**, L59.

Ramaprabhu, P., Dimonte, G., and Andrews, M.J. 2005. A numerical study of the influence of initial perturbations on the turbulent Rayleigh–Taylor instability. *J. Fluid Mech.*, **536**, 285.

Ramaprabhu, P., Dimonte, G., Young, Y.-N., Calder, A.C., and Fryxell, B. 2006. Limits of the potential flow approach to the single-mode Rayleigh–Taylor problem. *Phys. Rev. E*, **74**, 066308.

Ramaprabhu, P., Dimonte, G., Woodward, P., Fryer, C., Rockefeller, G., Muthuraman, K., Lin, P.-H., and Jayaraj, J. 2012. The late-time dynamics of the single-mode Rayleigh–Taylor instability. *Phys. Fluids*, **24**, 074107.

Ramaprabhu, P., Karkhanis, V., and Lawrie, A.G.W. 2013. The Rayleigh–Taylor Instability driven by an accel-decel-accel profile. *Phys. Fluids*, **25**, 115104.

Ramaprabhu, P., Karkhanis, V., Banerjee, R., Varshochi, H., Khan, M., and Lawrie, A.G.W. 2016. Evolution of the single-mode Rayleigh–Taylor instability under the influence of time-dependent accelerations. *Phys. Rev. E*, **93**, 013118.

Ramis, R. 2013. Hydrodynamic analysis of laser-driven cylindrical implosions. *Phys. Plasmas*, **20**, 082705.

Rampp, M., and Janka, H.-T. 2002. Radiation hydrodynamics with neutrinos-Variable Eddington factor method for core-collapse supernova simulations. *Astron. Astrophys.*, **396**, 361.

Ramshaw, J.D. 1998. Simple model for linear and nonlinear mixing at unstable fluid interfaces with variable acceleration. *Phys. Rev. E*, **58**, 5834.

Ramshaw, J.D. 1999. Simple model for linear and nonlinear mixing at unstable fluid interfaces in spherical geometry. *Phys. Rev. E*, **60**, 1775.

Ramshaw, J.D. 2000. Effect of slow compression on the linear stability of an accelerated shear layer. *Phys. Rev. E*, **61**, 1486.

Ranjan, D., Anderson, M., Oakley, J., and Bonazza, R. 2005. Experimental investigation of a strongly shocked gas bubble. *Phys. Rev. Lett.*, **94**, 184507.

Ranjan, D., Niederhaus, J., Motl, B., Anderson, M., Oakley, J., and Bonazza, R. 2007. Experimental investigation of primary and secondary features in high-Mach-number shock-bubble interaction. *Phys. Rev. Lett.*, **98**, 024502.

Ranjan, D., Niederhaus, J.H.J., Oakley, J.G., Anderson, M.H., Bonazza, R., and Greenough, J.A. 2008. Shock-bubble interactions: features of divergent shock-refraction geometry observed in experiments and simulations. *Phys. Fluids*, **20**, 036101.

Ranjan, D., Oakley, J., and Bonazza, R. 2011. Shock-bubble interactions. *Annu. Rev. Fluid Mech.*, **43**, 117.

Rasmus, A.M., Di Stefano, C.A., Flippo, K.A., et al. 2018. Shock-driven discrete vortex evolution on a high-Atwood number oblique interface. *Phys. Plasmas*, **25**, 032119.

Rasmus, A.M., Di Stefano, C.A., Flippo, K.A., Doss, F.W., Kawaguchi, C.F., Kline, J.L., Merritt, E.C., Desjardins, T.R., Cardenas, T., Schmidt, D.W., and Donovan, P.M. 2019. Shock-driven hydrodynamic instability of a sinusoidally perturbed, high-Atwood number, oblique interface. *Phys. Plasmas*, **26**, 062103.

Rast, M.P., Bello González, N., Bellot Rubio, L., et al. 2021. Critical science plan for the Daniel K. Inouye solar telescope (DKIST). *Solar Phys.*, **296**, 70.

Rasteiro dos Santos, M., Bury, Y., Jamme, S., and Griffond, J. 2023. On the effect of characterised initial conditions on the evolution of the mixing induced by the Richtmyer–Meshkov instability. *Shock Waves*, **33**, 117.

Rasteiro dos Santos, M., Bury, Y., and Jamme, S. 2022. Designing multi-shape dual-gas initial conditions for the study of hydrodynamic instabilities. *J. Fluids Eng.*, **144**, 041503.

Ravid, A., Citron, R.I., and Jeanloz, R. 2021. Hydrodynamic instability at impact interfaces and planetary implications. *Nat. Comm.*, **12**, 2104.

Rayleigh, Lord. 1883. Investigation of the character of the equilibrium of an incompressible heavy fluid of variable density. *Proc. Lond. Math. Soc.*, **14**, 170.

Rayleigh, Lord. 1916. On convection currents in a horizontal layer of fluid, when the higher temperature is on the under side. *The London, Edinburgh, and Dublin Philos. Mag. and J. Sci.*, **32**, 529.

Read, K.I. 1984. Experimental investigation of turbulent mixing by Rayleigh–Taylor instability. *Physica D*, **12**, 45.

Read, K.I., and Youngs, D.L. 1983. Experimental investigation of turbulent mixing by Rayleigh–Taylor instability. *AWRE Report*, 11/83. Atomic Weapons Establishment, Aldermaston, UK.

Reckinger, S.J., Livescu, D., and Vasilyev, O.V. 2010. Adaptive wavelet collocation method simulations of Rayleigh–Taylor instability. *Phys. Scr.*, **2010**, 014064.

Reckinger, S.J., Livescu, D., and Vasilyev, O.V. 2016. Comprehensive numerical methodology for direct numerical simulations of compressible Rayleigh–Taylor instability. *J. Comput. Phys.*, **313**, 181.

Reddy, G., Murthy, V.N., and Vergassola, M., 2021. Olfactory sensing and navigation in turbulent environments. *Annu. Rev. Condensed Matter Phys.*, **13**, 191.

Reese, D.T., Ames, A.M., Noble, C.D., Oakley, J.G., Rothamer, D.A., and Bonazza, R. 2018. Simultaneous direct measurements of concentration and velocity in the Richtmyer–Meshkov instability. *J. Fluid Mech.*, **849**, 541.

Regan, S.P., Epstein, R., Hammel, B.A., et al. 2013. Hot-spot mix in ignition-scale inertial confinement fusion targets. *Phys. Rev. Lett.*, **111**, 045001.

Reid, W.H. 1961. The effects of surface tension and viscosity on the stability of two superposed fluids. *Proc. Cambridge Phil. Soc.*, **57**, 415.

Reipurth, B., and Bally, J. 2001. Herbig-Haro flows: probes of early stellar evolution. *Annu. Rev. Astron. Astrophys.*, **39**, 403.

Remington, B.A., Arnett, D., Drake, P.R., and Takabe, H. 1999. Modeling astrophysical phenomena in the laboratory with intense lasers. *Science*, **284**, 1488.

Remington, B.A., Drake, R.P., Takabe, H., and Arnett, D. 2000. A review of astrophysics experiments on intense lasers. *Phys. Plasmas*, **7**, 1641.

Remington, B.A., Bazan, G., Belak, J., et al. 2004. Materials science under extreme conditions of pressure and strain rate. *Metall. Mater. Trans. A*, **35**, 2587.

Remington, B.A., Drake, R.P., and Ryutov, D.D. 2006. Experimental astrophysics with high power lasers and Z pinches. *Rev. Mod. Phys.*, **78**, 755.

Remington, B.A., Park, H.-S., Casey, D.T., et al. 2019. Rayleigh–Taylor instabilities in high-energy density settings on the National Ignition Facility. *Proc. Natl. Acad. Sci.*, **116**, 18233.

Remming, I.S., and Khokhlov, A.M. 2014. The classification of magnetohydrodynamic regimes of thermonuclear combustion. *Astrophys. J.*, **794**, 87.

Remming, I.S., and Khokhlov, A.M. 2016. The internal structure and propagation of magnetohydrodynamical thermonuclear flames. *Astrophys. J.*, **831**, 162.

Renoult, M.-C., Carles, P., Ferjani, S., and Rosenblatt, C. 2013. 2D Rayleigh–Taylor instability: interfacial arc-length vs. deformation amplitude. *Europhys. Lett.,* **101**, 54001.

Renshaw, C.E., and Schulson, E.M. 2001. Universal behaviour in compressive failure of brittle materials. *Nature*, **412**, 897.

Retterer, J.M. 2010a. Forecasting low-latitude radio scintillation with 3-D ionospheric plume models: 1. Plume model. *J. Geophys. Res.*, **115**, A03306.

Retterer, J.M. 2010b. Forecasting low-latitude radio scintillation with 3-D ionospheric plume models: 2. Scintillation calculation. *J. Geophys. Res.*, **115**, A03307.

Reynolds, O. 1883. An experimental investigation of the circumstances which determine whether the motion of water shall be direct or sinuous, and of the law of resistance in parallel channels. *Philos. Trans. R. Soc.*, **174**, 935.

Reynolds, O. 1895. On the dynamical theory of incompressible viscous fluids and the determination of the criterion. *Philos. Trans. R. Soc.*, **186**, 123.

Richardson, L.F. 1922. *Weather Prediction by Numerical Process.* Cambridge University Press, Cambridge, UK.

Richtmyer, R.D. 1960. Taylor instability in shock acceleration of compressible fluids. *Commun. Pure Appl. Math.*, **13**, 297.

Riess, A.G., Filippenko, A.V., Challis, P., et al. 1998. Observational evidence from supernovae for an accelerating universe and a cosmological constant. *Astron. J.*, **116**, 1009.

Rigg, P.A., Anderson, W.W., Olson, R.T., Buttler, W.T., and Hixson, R.S. 2006. Investigation of ejecta production in tin using plate impact experiments. *AIP Conf. Proc.*, **845**, 1283.

Rigon, G., Casner, A., Albertazzi, B., et al. 2019. Rayleigh–Taylor instability experiments on the LULI2000 laser in scaled conditions for young supernova remnants. *Phys. Rev. E*, **100**, 021201.

Rigon, G., Albertazzi, B., Mabey, P., Michel, Th., Falize, E., Bouffetier, V., Ceurvorst, L., Masse, L., Koenig, M., and Casner, A. 2021a. Exploring the Atwood-number dependence of the highly nonlinear Rayleigh–Taylor instability regime in high-energy-density conditions. *Phys. Rev. E*, **104**, 045213.

Rigon, G., Albertazzi, B., Pikuz, T., et al. 2021b. Micron-scale phenomena observed in a turbulent laser-produced plasma. *Nat. Commun.*, **12**, 2679.

Rigon, G., Albertazzi, B., Mabey, P., Michel, Th., Barroso, P., Faenov, A., Kumar, R., Michaut, C., Pikuz, T., Sakawa, Y., Sano, T., Shimogawara, H., Tamatani, H., Casner, A., and Koenig, M. 2022. Hydrodynamic instabilities in a highly radiative environment. *Phys. Plasmas*, **29**, 072106.

Rikanati, A., Oron, D., Alon, U., and Shvarts, D. 2000. Statistical mechanics merger model for hydrodynamic instabilities. *Astrophys. J. Suppl. Ser.*, **127**, 451.

Rikanati, A., Oron, D., Sadot, O., and Shvarts, D. 2003. High initial amplitude and high Mach number effects on the evolution of the single-mode Richtmyer–Meshkov instability. *Phys. Rev. E*, **67**, 026307.

Ripley, R., Donahue, L., and Zhang, F. 2012. Jetting instabilities of particles from explosive dispersal. *AIP Conf. Proc.*, **1426**, 1615.

Ristorcelli, J.R., and Clark, T.T. 2004. Rayleigh–Taylor turbulence: self-similar analysis and direct numerical simulations. *J. Fluid Mech.*, **507**, 213.

Roberts, M.S., and Jacobs, J.W. 2016. The effects of forced small-wavelength, finite-bandwidth initial perturbations and miscibility on the turbulent Rayleigh–Taylor instability. *J. Fluid Mech.*, **787**, 50.

Robey, H.F. 2004. Effects of viscosity and mass diffusion in hydrodynamically unstable plasma flows. *Phys. Plasmas*, **11**, 4123.

Robey, H.F., Zhou, Y., Buckingham, A.C., Keiter, P., Remington, B.A., and Drake, R.P. 2003. The time scale for the transition to turbulence in a high Reynolds number, accelerated flow. *Phys. Plasmas*, **10**, 614.

Robey, H.F., Celliers, P.M., Moody, J.D., Sater, J., Parham, T., Kozioziemski, B., Dylla-Spears, R., Ross, J.S., LePape, S., Ralph, J.E., and Hohenberger, M. 2014. Shock timing measurements and analysis in deuterium-tritium-ice layered capsule implosions on NIF. *Phys. Plasmas*, **21**, 022703.

Robey, H.F., Berzak Hopkins, L., Milovich, J.L., and Meezan, N.B. 2018. The I-Raum: a new shaped hohlraum for improved inner beam propagation in indirectly-driven ICF implosions on the National Ignition Facility. *Phys. Plasmas*, **25**, 012711.

Roble, R.G., and Ridley, E.C. 1994. A thermosphere-ionosphere-mesosphere-electrodynamics general circulation model (TIME-GCM): equinox solar cycle minimum simulations (30–500 km). *Geophys. Res. Lett.*, **21**, 417.

Rodriguez, V., Saurel, R., Jourdan, G., and Houas, L. 2017. Impulsive dispersion of a granular layer by a weak blast wave. *Shock Waves*, **27**, 187.

Rodriguez Azara, J.L., and Emanuel, G. 1989. Compressible rotational flows generated by the substitution principle. II. *Phys. Fluids A*, **1**, 600.

Rogallo, R.S. 1981. *Numerical experiments in homogeneous turbulence*. Rep. 81315. National Aeronautics and Space Administration, Moffett field, CA.

Roland, C., de Rességuier, T., Sollier, A., Lescoute, E., Loison, D., and Soulard, L. 2017. Ejection of micron-scale fragments from triangular grooves in laser shock-loaded copper samples. *J. Dynam. Behav. Mater*, **3**, 156.

Roland, C., de Rességuier, T., Sollier, A., Lescoute, E., Tandiang, D., Toulminet, M., and Soulard, L. 2018. Ballistic properties of ejecta from a laser shock-loaded groove: SPH versus experiments. *AIP Conf. Proc.*, 080012.

Rollin, B., and Andrews, M.J. 2013. On generating initial conditions for turbulence models: the case of Rayleigh–Taylor instability turbulent mixing. *J. Turbul.*, **14**, 77.

Romero, B., Poroseva, S.V., Vorobieff, P., and Reisner, J.M. 2021. Simulations of the shock-driven Kelvin–Helmholtz instability in inclined gas curtains. *Phys. Fluids*, **33**, 064103.

Rose, H.A., and Sulem, P.L. 1978. Fully developed turbulence and statistical mechanics. *J. de Physique*, **39**, 441.

Rosen, M.D. 1999. The physics issues that determine inertial confinement fusion target gain and driver requirements: a tutorial. *Phys. Plasmas*, **6**, 1690.

Rosensweig, R.E., Hirota, Y., Tsuda, S., and Raj, K. 2008. Study of audio speakers containing ferrofluid. *J. Phys. Condens. Matter*, **20**, 204147.

Rott, N. 1956. Diffraction of a weak shock with vortex generation. *J. Fluid Mech.*, **1**, 111.

Rotta, J.C. 1951. Statistische theorie nichthomogener turbulenz. *Zeitschrift für Physik*, **129**, 547.

Rotty, R.M. 1962. *Introduction to Gas Dynamics*. Wiley, New York.

Roy, A., Ghosh, D., and Mandal, N. 2024. Dampening effect of global flows on Rayleigh–Taylor instabilities: Implications for deep-mantle plumes vis-à-vis hotspot distributions. *Geophys. J. Int.*, **236**, 119.

Roy, S., Mandal, L.K., Khan, M., and Gupta, M.R. 2014. Combined effect of viscosity, surface tension and compressibility on Rayleigh–Taylor bubble growth between two fluids. *J. Fluids Eng.*, **136**, 091101.

Rozanov, V.B., Kuchugov, P.A., Zmitrenko, N.V., and Yanilkin, Yu V. 2015. Effect of initial conditions on the development of Rayleigh–Taylor instability. *J. Russian Laser Res.*, **36**, 139.

Ruan, Y., Zhang, Y., Tian, B., and Zhang, X. 2020. Density-ratio-invariant mean-species profile of classical Rayleigh–Taylor mixing. *Phys. Rev. Fluids*, **5**, 054501.

Rubinstein, R., and Barton, J.M. 1990. Nonlinear Reynolds stress models and the renormalization group. *Phys. Fluids A*, **2**, 1472.

Ruderman, M.S., Terradas, J., and Ballester, J.L. 2014. Rayleigh–Taylor instabilities with sheared magnetic fields. *Astrophys. J.*, **785**, 110.

Rudinger, G. 1958. Shock wave and flame interactions, *Third AGARD Coll.* Pergamon, London.

Rudinger, G., and Somers, L.M. 1960. Behaviour of small regions of different gases carried in accelerated gas flows. *J. Fluid Mech.*, **7**, 161.

Rudyak, V., and Minakov, A. 2014. Modeling and optimization of Y-type micromixers. *Micromachines*, **5**, 886.

Ruev, G.A., Fedorov, A.V., and Fomin, V.M. 2006. Development of the Rayleigh–Taylor instability due to interaction of a diffusion mixing layer of two gases with compression waves. *Shock Waves*, **16**, 65.

Ruiz, D. E., Schmit, P. F., Yager-Elorriaga, D. A., Jennings, C. A., and Beckwith, K. 2023a. Exploring the parameter space of MagLIF implosions using similarity scaling. I. Theoretical framework. *Phys. Plasmas*, **30**, 032707.

Ruiz, D. E., Schmit, P. F., Yager-Elorriaga, D. A., Gomez, M. R., Weis, M. R., Jennings, C. A., Harvey-Thompson, A. J., Knapp, P. F., Slutz, S. A., Ampleford, D. J., Beckwith, K., and Matzen, M. K. 2023b. Exploring the parameter space of MagLIF implosions using similarity scaling. II. Current scaling. *Phys. Plasmas*, **30**, 032708.

Ruiz, D.E., Yager-Elorriaga, D.A., Peterson, K.J., Sinars, D.B., Weis, MR, Schroen, D.G., Tomlinson, K., Fein, J.R., and Beckwith, K. 2022. Harmonic generation and inverse cascade in the Z-pinch driven, preseeded multimode, magneto-Rayleigh–Taylor instability. *Phys. Rev. Lett.*, **128**, 255001.

Ryu, D., and Vishniac, E.T. 1988. A linear stability analysis for wind-driven bubbles. *Astrophys. J.*, **331**, 350.

Ryu, J., and Livescu, D. 2014. Turbulence structure behind the shock in canonical shock–vortical turbulence interaction. *J. Fluid Mech.*, **756**, R1.

Ryutov, D., Drake, R.P., Kane, J., Liang, E., Remington, B.A., and Wood-Vasey, W.M. 1999. Similarity criteria for the laboratory simulation of supernova hydrodynamics. *Astrophys. J.*, **518**, 821.

Ryutov, D.D. 2015. Characterizing the plasmas of dense *Z*-pinches. *IEEE Trans. Plasma Sci.*, **43**, 2363.

Ryutov, D.D., Drake, R.P., and Remington, B.A. 2000a. Criteria for scaled laboratory simulations of astrophysical MHD phenomena. *Astrophys. J. Suppl. Ser.*, **127**, 465.

Ryutov, D.D., Derzon, M.S., and Matzen, M.K. 2000b. The physics of fast Z pinches. *Rev. Mod. Phys.*, **72**, 167.

Ryutov, D.D., Remington, B.A., Robey, H.F., and Drake, R.P. 2001. Magnetohydrodynamic scaling: from astrophysics to the laboratory. *Phys. Plasmas*, **8**, 1804.

Ryutov, D.D., Awe, T.J., Hansen, S.B., McBride, R.D., Peterson, K.J., Sinars, D.B., and Slutz, S.A. 2014. Effect of axial magnetic flux compression on the magnetic Rayleigh–Taylor instability (theory). *AIP Conf. Proc.*, **1639**, 63.

Ryutova, M., Berger, T., Frank, Z., and Tarbell, T. 2010. Observation of plasma instabilities in quiescent prominences. *Solar Phys.*, **267**, 75.

Sabet, N., Hassanzadeh, H., De Wit, A., and Abedi, J. 2021. Scalings of Rayleigh–Taylor instability at large viscosity contrasts in porous media. *Phys. Rev. Lett.*, **126**, 094501.

Saddoughi, S.G., and Veeravalli, S.V. 1994. Local isotropy in turbulent boundary layers at high Reynolds number. *J. Fluid Mech.*, **268**, 333.

Sadek, M., and Aluie, H. 2018. Extracting the spectrum of a flow by spatial filtering. *Phys. Rev. Fluids*, **3**, 124610.

Sadler, J.D., Li, H., and Haines, B.M. 2020. Magnetization around mix jets entering inertial confinement fusion fuel. *Phys. Plasmas*, **27**, 072707.

Sadler, J.D., Green, S., Li, S., Zhou, Y., Flippo, K.A., and Li, H. 2022a. Faster ablative Kelvin–Helmholtz instability growth in a magnetic field. *Phys. Plasmas*, **29**, 052708.

Sadler, J.D., Walsh, C.A., Zhou, Y., and Li, H. 2022b. Role of self-generated magnetic fields in the inertial fusion ignition threshold. *Phys. Plasmas*, **29**, 072701.

Sadler, J.D., Louie, C., and Zhou, Y. 2023. Intricate structure of the plasma Rayleigh–Taylor instability in shock tubes. *Phys. Plasmas*, **30**, 022709.

Sadot, O., Erez, L., Alon, U., Oron, D., Levin, L.A., Erez, G., Ben-Dor, G., and Shvarts, D. 1998. Study of nonlinear evolution of single-mode and two-bubble interaction under Richtmyer–Meshkov instability. *Phys. Rev. Lett.*, **80**, 1654.

Sadot, O., Rikanati, A., Oron, D., Ben-Dor, G., and Shvarts, D. 2003. An experimental study of the high Mach number and high initial-amplitude effects on the evolution of the single-mode Richtmyer–Meshkov instability. *Laser Part. Beams*, **21**, 341.

Saenz, J.A., Aslangil, D., and Livescu, D. 2021. Filtering, averaging, and scale dependency in homogeneous variable density turbulence. *Phys. Fluids*, **33**, 025115.

Saffman, P.G. 1967. The large-scale structure of homogeneous turbulence. *J. Fluid Mech.*, **27**, 581.

Saffman, P.G., and Meiron, D.I. 1989. Kinetic energy generated by the incompressible Richtmyer–Meshkov instability in a continuously stratified fluid. *Phys. Fluids A*, **1**, 1767.

Sagaut, P. 2006. *Large Eddy Simulation for Incompressible Flows: An Introduction*. Springer Science & Business Media, Berlin.

Sagaut, P., and Cambon, C. 2019. *Homogeneous Turbulence Dynamics*. 2nd edn. Springer Cham, Switzerland.

Saigo, T, and Hamaguchi, S. 2002. Shear viscosity of strongly coupled Yukawa systems. *Phys. Plasmas*, **9**, 1210.

Sakagami, H., and Nishihara, K. 1990. Rayleigh–Taylor instability on the pusher–fuel contact surface of stagnating targets. *Phys. Fluids B*, **2**, 2715.

Samtaney, R. 2003. Suppression of the Richtmyer–Meshkov instability in the presence of a magnetic field. *Phys. Fluids*, **15**, L53.

Samtaney, R., and Zabusky, N.J. 1994. Circulation deposition on shock-accelerated planar and curved density-stratified interfaces: models and scaling laws. *J. Fluid Mech.*, **269**, 45.

Samtaney, R., Ray, J., and Zabusky, N.J. 1998. Baroclinic circulation generation on shock accelerated slow/fast gas interfaces. *Phys. Fluids*, **10**, 1217.

Sandoval, D. 1995. *The dynamics of variable-density turbulence*. Ph.D. thesis, University of Washington, Seattle, WA.

Sano, T. 2021. Alfvén Number for the Richtmyer–Meshkov Instability in Magnetized Plasmas. *Astrophys. J.*, **920**, 29.

Sano, T., Nishihara, K., Matsuoka, C., and Inoue, T. 2012. Magnetic field amplification associated with the Richtmyer–Meshkov instability. *Astrophys. J.*, **758**, 126.

Sano, T., Inoue, T., and Nishihara, K. 2013. Critical magnetic field strength for suppression of the Richtmyer–Meshkov instability in plasmas. *Phys. Rev. Lett.*, **111**, 205001.

Sano, T., Ishigure, K., and Cobos-Campos, F. 2020. Suppression of the Richtmyer–Meshkov instability due to a density transition layer at the interface. *Phys. Rev. E*, **102**, 013203.

Sanz, J. 1994. Self-consistent analytical model of the Rayleigh–Taylor instability in inertial confinement fusion. *Phys. Rev. Lett.*, **73**, 2700.

Sardina, G., Brandt, L., Boffetta, G., and Mazzino, A. 2018. Buoyancy-driven flow through a bed of solid particles produces a new form of Rayleigh–Taylor turbulence. *Phys. Rev. Lett.*, **121**, 224501.

Sasaki, T., and Abe, Y. 2007. Rayleigh–Taylor instability after giant impacts: imperfect equilibration of the Hf-W system and its effect on the core formation age. *Earth, Planets and Space*, **59**, 1035.

Saunders, A.M., Ali, S.J., Park, H.-S., Eggert, J., Najjar, F., Huntington, C., Haxhimali, T., Morgan, B., Ping, Y., and Rinderknecht, H.G. 2020. Development of high power laser platforms to study metal ejecta interactions. *AIP Conf. Proc.*, **2272**, 120025.

Sauppe, J.P., Palaniyappan, S., Tobias, B.J., et al. 2020. Demonstration of scale-invariant Rayleigh–Taylor instability growth in laser-driven cylindrical implosion Experiments. *Phys. Rev. Lett.*, **124**, 185003.

Scannapieco, E., and Brüggen, M. 2008. Subgrid modeling of AGN-driven turbulence in galaxy clusters. *Astrophys. J.*, **686**, 927.

Scase, M.M., and Sengupta, S. 2021. Cylindrical rotating Rayleigh–Taylor instability. *J. Fluid Mech.*, **907**.

Scase, M.M., Baldwin, K.A., and Hill, R.J.A. 2017. Rotating Rayleigh–Taylor instability. *Phys. Rev. Fluids*, **2**, 024801.

Scase, M.M., Baldwin, K.A., and Hill, R.J.A. 2020. Magnetically induced Rayleigh–Taylor instability under rotation: comparison of experimental and theoretical results. *Phys. Rev. E*, **102**, 043101.

Scharfman, B.E., Techet, A.H., Bush, J.W.M., and Bourouiba, L. 2016. Visualization of sneeze ejecta: steps of fluid fragmentation leading to respiratory droplets. *Exp. Fluids*, **57**, 24.

Schauer, M.M., Buttler, W.T., Frayer, D.K., Grover, M., Lalone, B.M., Monfared, S.K., Sorenson, D.S., Stevens, G.D., and Turley, W.D. 2017. Ejected particle size distributions from shocked metal surfaces. *J. Dynam. Behav. Mater.*, **3**, 217.

Scheiner, B., Schmitt, M.J., Hsu, S.C., Schmidt, D., Mance, J., Wilde, C., Polsin, D.N., Boehly, T.R., Marshall, F.J., Krasheninnikova, N., and Molvig, K. 2019. First experiments on Revolver shell collisions at the OMEGA laser. *Phys. Plasmas*, **26**, 072707.

Schill, W.J., Armstrong, M.R., Nguyen, J.H., Sterbentz, D.M., White, D.A., Benedict, L.X., Rieben, R.N., Hoff, A, Lorenzana, H.E., Belof, J.L., La Lone, B.M., and Staska, M.D. 2024. Suppression of Richtmyer–Meshkov instability via special pairs of shocks and phase transitions. *Phys. Rev. Lett.*, **132**, 024001.

Schilling, O. 2020a. A buoyancy–shear–drag-based turbulence model for Rayleigh–Taylor, reshocked Richtmyer–Meshkov, and Kelvin–Helmholtz mixing. *Physica D*, **402**, 132238.

Schilling, O. 2020b. Progress on understanding Rayleigh–Taylor flow and mixing using synergy between simulation, modeling, and experiment. *J. Fluids Eng.*, **142**.

Schilling, O. 2021. Self-similar Reynolds-averaged mechanical–scalar turbulence models for Rayleigh–Taylor, Richtmyer–Meshkov, and Kelvin–Helmholtz instability-induced mixing in the small Atwood number limit. *Phys. Fluids*, **33**, 085129.

Schilling, O., and Latini, M. 2010. High-order WENO simulations of three-dimensional reshocked Richtmyer–Meshkov instability to late times: dynamics, dependence on initial conditions, and comparisons to experimental data. *Acta Math. Scientia*, **30**, 595.

Schilling, O., and Mueschke, N.J. 2010. Analysis of turbulent transport and mixing in transitional Rayleigh–Taylor unstable flow using direct numerical simulation data. *Phys. Fluids*, **22**, 105102.

Schilling, O., and Mueschke, N.J. 2017. Turbulent transport and mixing in transitional Rayleigh–Taylor unstable flow: a priori assessment of gradient-diffusion and similarity modeling. *Phys. Rev. E*, **96**, 063111.

Schilling, O., Latini, M., and Don, W.S. 2007. Physics of reshock and mixing in single-mode Richtmyer–Meshkov instability. *Phys. Rev. E*, **76**, 026319.

Schils, R. 2012. *How James Watt Invented the Copier: Forgotten Inventions of Our Great Scientists*. Springer Science & Business Media, New York.

Schmit, P.F., Velikovich, A.L., McBride, R.D., and Robertson, G.K. 2016. Rayleigh–Taylor instabilities in magnetically driven solid metal shells by means of a dynamic screw pinch. *Phys. Rev. Lett.*, **117**, 205001.

Schmitt, F.G. 2017. Turbulence from 1870 to 1920: the birth of a noun and of a concept. *Comptes Rendus Mécanique*, **345**, 620.

Schneider, N. and Gauthier, S., 2015. Asymptotic analysis of Rayleigh–Taylor flow for Newtonian miscible fluids. *J. Eng. Math.*, **92**, 55.

Schnittman, J.D., and Craxton, R.S. 1996. Indirect-drive radiation uniformity in tetrahedral hohlraums. *Phys. Plasmas*, **3**, 3786.

Schowalter, D.G., Van Van Atta, C.W., and Lasheras, J.C. 1994. A study of streamwise vortex structure in a stratified shear layer. *J. Fluid Mech.*, **281**, 247.

Schwarzkopf, J.D., Livescu, D., Gore, R.A., Rauenzahn, R.M., and Ristorcelli, J.R. 2011. Application of a second-moment closure model to mixing processes involving multicomponent miscible fluids. *J. Turbul.*, **12**, N49.

Schwarzkopf, J.D., Livescu, D., Baltzer, J.R., Gore, R.A., and Ristorcelli, J.R. 2016. A two-length scale turbulence model for single-phase multi-fluid mixing. *Flow, Turbul. Combust.*, **96**, 1.

Schwarzschild, M. 2015. *Structure and Evolution of Stars*. Princeton University Press, Princeton, NJ.

Sefkow, A.B. 2016. *On the helical instability and efficient stagnation pressure production in thermonuclear magnetized inertial fusion*. Tech. rept. Sandia National Lab., Albuquerque, NM.

Sekar, R., and Kherani, E.A. 2002. Effects of molecular ions on the collisional Rayleigh–Taylor instability: nonlinear evolution. *J. Geophys. Res. Space Phys.*, **107**, 1139.

Sekar, R., Suhasini, R., and Raghavarao, R. 1994. Effects of vertical winds and electric fields in the nonlinear evolution of equatorial spread F. *J. Geophys. Res. Space Phys.*, **99**, 2205.

Sembian, S., Liverts, M., and Apazidis, N. 2018. Plane blast wave interaction with an elongated straight and inclined heat-generated inhomogeneity. *J. Fluid Mech.*, **851**, 245.

Sen, S., and Storer, R.G. 1997. Suppression of Rayleigh–Taylor instability by flow curvature. *Phys. Plasmas*, **4**, 3731.

Séon, T, Hulin, J-P, Salin, D, Perrin, B, and Hinch, EJ. 2004. Buoyant mixing of miscible fluids in tilted tubes. *Phys. Fluids*, **16**, L103.

Séon, T, Znaien, J, Salin, D, Hulin, JP, Hinch, EJ, and Perrin, B. 2007. Transient buoyancy-driven front dynamics in nearly horizontal tubes. *Phys. Fluids*, **19**, 123603.

Settles, G.S. 2005. Sniffers: fluid-dynamic sampling for olfactory trace detection in nature and homeland security. *J. Fluids Eng.*, **127**, 189.

Sewell, E.G., Ferguson, K.J., Krivets, V.V., and Jacobs, J.W. 2021. Time-resolved particle image velocimetry measurements of the turbulent Richtmyer–Meshkov instability. *J. Fluid Mech.*, **917**, A41.

Seyler, C.E., Martin, M.R., and Hamlin, N.D. 2018. Helical instability in MagLIF due to axial flux compression by low-density plasma. *Phys. Plasmas*, **25**, 062711.

Shakura, N.I., and Sunyaev, R.A. 1973. Black holes in binary systems. Observational appearance. *Astron. Astrophys.*, **24**, 337.

Shankar, S.K., and Lele, S.K. 2014. Numerical investigation of turbulence in reshocked Richtmyer–Meshkov unstable curtain of dense gas. *Shock Waves*, **24**, 79.

Shao, J.L., Wang, P., and He, A.M. 2014. Microjetting from a grooved Al surface under supported and unsupported shocks. *J. Appl. Phys.*, **116**, 073501.

Sharp, D.H. 1984. Overview of Rayleigh–Taylor instability. *Physica D*, **12**, 3.

Sharp, D.H., and Wheeler, J.A. 1961. *Late stage of Rayleigh–Taylor instability*. Tech. rept. Institute for Defense Analyses, Alexandria, VA.

Shen, C., Chen, B., Reeves, K.K., Yu, S., Polito, V., and Xie, X. 2022. The origin of underdense plasma downflows associated with magnetic reconnection in solar flares. *Nat. Astron.*, **6**, 317.

Sherman, F. 1955. *A low-density wind-tunnel study of shock-wave structure and relaxation phenomena in gases*. Tech. rept. Tech. Note 3298. National Advisory Committee for Aeronautics, Washington DC.

Shih, T.-H., Liou, W.W., Shabbir, A., Yang, Z., and Zhu, J. 1995. A new K-$\mathcal{E}$ eddy viscosity model for high reynolds number turbulent flows. *Comput. Fluids*, **24**, 227.

Shimony, A., Shvarts, D., Malamud, G., Di Stefano, C.A., Kuranz, C.C., and Drake, R.P. 2016. The effect of a dominant initial single mode on the Kelvin–Helmholtz instability evolution: new insights on previous experimental results. *J. Fluids Eng.*, **138**, 070902.

Shin, S., Sohn, S.-I., and Hwang, W. 2022. Numerical simulation of single-and multi-mode Rayleigh–Taylor instability with surface tension in two dimensions. *European J. Mech.-B/Fluids*, **91**, 141.

Shipley, G.A., Jennings, C.A., and Schmit, P.F. 2019. Design of dynamic screw pinch experiments for magnetized liner inertial fusion. *Phys. Plasmas*, **26**, 102702.

Shvarts, D., Alon, U., Ofer, D., McCrory, R.L., and Verdon, C.P. 1995. Nonlinear evolution of multimode Rayleigh–Taylor instability in two and three dimensions. *Phys. Plasmas*, **2**, 2465.

Shvarts, D., Sadot, O., Oron, D., Rikanati, A., and Alon, U. 2000. Shock-induced instability of interfaces. Page 489 of: Ben-Dor, G., Igra, O., and Elperin, T. (eds), *Handbook of Shock Waves*, Vol. 2. Academic, San Diego, CA.

Si, T., Zhai, Z., Yang, J., and Luo, X. 2012. Experimental investigation of reshocked spherical gas interfaces. *Phys. Fluids*, **24**, 054101.

Si, T., Long, T., Zhai, Z., and Luo, X. 2015. Experimental investigation of cylindrical converging shock waves interacting with a polygonal heavy gas cylinder. *J. Fluid Mech.*, **784**, 225.

Signor, L., Roy, G., Chanal, P.Y., Héreil, P.L., Buy, F., Voltz, C., Llorca, F., de Rességuier, T., and Dragon, A. 2009. Debris cloud ejection from shock-loaded tin melted on release or on compression. *AIP Conf. Proc.*, **1195**, 1065.

Simakov, A.N., and Molvig, K. 2014. Electron transport in a collisional plasma with multiple ion species. *Phys. Plasmas*, **21**, 024503.

Simakov, A.N., and Molvig, K. 2016a. Hydrodynamic description of an unmagnetized plasma with multiple ion species. I. General formulation. *Phys. Plasmas*, **23**, 032115.

Simakov, A.N., and Molvig, K. 2016b. Hydrodynamic description of an unmagnetized plasma with multiple ion species. II. Two and three ion species plasmas. *Phys. Plasmas*, **23**, 032116.

Sinars, D.B., Slutz, S.A., Herrmann, M.C., McBride, R.D., Cuneo, M.E., Peterson, K.J., Vesey, R.A., Nakhleh, C., Blue, B.E., Killebrew, K., and Schroen, D. 2010. Measurements of magneto-Rayleigh–Taylor instability growth during the implosion of initially solid Al tubes driven by the 20-MA, 100-ns Z facility. *Phys. Rev. Lett.*, **105**, 185001.

Sinars, D.B., Slutz, S.A., Herrmann, et al. 2011. Measurements of magneto-Rayleigh–Taylor instability growth during the implosion of initially solid metal liners. *Phys. Plasmas*, **18**, 056301.

Sinars, D.B., Sweeney, M.A., Alexander, C.S., et al. 2020. Review of pulsed power-driven high energy density physics research on Z at Sandia. *Phys. Plasmas*, **27**, 070501.

Singh, S. 2021. Contribution of Mach number to the evolution of the Richtmyer–Meshkov instability induced by a shock-accelerated square light bubble. *Phys. Rev. Fluids*, **6**, 104001.

Singh, S., Battiato, M. and Myong, R.S., 2021. Impact of bulk viscosity on flow morphology of shock-accelerated cylindrical light bubble in diatomic and polyatomic gases. *Phys. Fluids*, **33**, 066103.

Siqueiros, D.A. 1936a. *Collective suicide*. Museum of Modern Art, New York. Available at www.moma.org/collection/object.php?object_id=79146.

Siqueiros, D.A. 1936b. *Cosmos and disaster*. Tate Modern Museum, London, United Kingdom. Available at www.tate.org.uk/art/artworks/siqueiros-cosmos-and-disaster-102487.

Siqueiros, D.A. 1969. *Landscape in red*. Los Angeles County Museum of Art. Available at http://collections.lacma.org/node/193844.

Skinner, M. A., Burrows, A., and Dolence, J.C. 2016. Should one use the ray-by-ray approximation in core-collapse supernova simulations? *Astrophys. J.*, **831**, 81.

Skoutnev, V., Most, E.R., Bhattacharjee, A., and Philippov, A.A. 2021. Scaling of Small-scale Dynamo Properties in the Rayleigh–Taylor Instability. *Astrophys. J.*, **921**, 75.

Slutz, S.A., and Vesey, R.A. 2012. High-gain magnetized inertial fusion. *Phys. Rev. Lett.*, **108**, 025003.

Slutz, S.A., Herrmann, M.C., Vesey, R.A., Sefkow, A.B., Sinars, D.B., Rovang, D.C., Peterson, K.J., and Cuneo, M.E. 2010. Pulsed-power-driven cylindrical liner implosions of laser preheated fuel magnetized with an axial field. *Phys. Plasmas*, **17**, 056303.

Smagorinsky, J. 1963. General circulation experiments with the primitive equations: I. The basic experiment. *Monthly Weather Rev.*, **91**, 99.

Smalyuk, V.A. 2012. Experimental techniques for measuring Rayleigh–Taylor instability in inertial confinement fusion. *Phys. Scr.*, **86**, 058204.

Smalyuk, V.A., Boehly, T.R., Bradley, D.K., Goncharov, V.N., Delettrez, J.A., Knauer, J.P., Meyerhofer, D.D., Oron, D., and Shvarts, D. 1998. Saturation of the Rayleigh–Taylor growth of broad-bandwidth laser-imposed nonuniformities in planar targets. *Phys. Rev. Lett.*, **81**, 5342.

Smalyuk, V.A., Sadot, O., Delettrez, J.A., Meyerhofer, D.D., Regan, S.P., and Sangster, T.C. 2005. Fourier-space nonlinear Rayleigh–Taylor growth measurements of 3D laser-imprinted modulations in planar targets. *Phys. Rev. Lett.*, **95**, 215001.

Smalyuk, V.A., Sadot, O., Betti, R., Goncharov, V.N., Delettrez, J.A., Meyerhofer, D.D., Regan, S.P., Sangster, T.C., and Shvarts, D. 2006. Rayleigh–Taylor growth measurements of three-dimensional modulations in a nonlinear regime. *Phys. Plasmas*, **13**, 056312.

Smalyuk, V.A., Tipton, R.E., Pino, J.E., et al. 2014. Measurements of an ablator-gas atomic mix in indirectly driven implosions at the National Ignition Facility. *Phys. Rev. Lett.*, **112**, 025002.

Smalyuk, V.A., Robey, H.F., Casey, D.T., et al. 2017. Mix and hydrodynamic instabilities on NIF. *J. Instrum.*, **12**, C06001.

Smalyuk, V.A., Robey, H.F., Alday, C.L., et al. 2018. Review of hydro-instability experiments with alternate capsule supports in indirect-drive implosions on the National Ignition Facility. *Phys. Plasmas*, **25**, 072705.

Smeeton, V.S., and Youngs, D.L. 1987. Experimental investigation of turbulent mixing by Rayleigh–Taylor instability, III. *AWRE Report*, 35/87. Atomic Weapons Establishment, Aldermaston, UK.

Smith, A.V., Holder, D.A., Barton, C.J., Morris, A.P., and Youngs, D.L. 2001. Shock tube experiments on Richtmyer–Meshkov instability across a chevron profiled interface. In: *Proceedings of the Eighth International Workshop on the Physics of Compressible Turbulent Mixing*. Pasedena, CA.

Snider, D.M., and Andrews, M.J. 1994. Rayleigh–Taylor and shear driven mixing with an unstable thermal stratification. *Phys. Fluids*, **6**, 3324.

Sohn, S.-I. 2003. Simple potential-flow model of Rayleigh–Taylor and Richtmyer–Meshkov instabilities for all density ratios. *Phys. Rev. E*, **67**, 026301.

Sohn, S.-I. 2004a. Density dependence of a Zufiria-type model for Rayleigh–Taylor bubble fronts. *Phys. Rev. E*, **70**, 045301.

Sohn, S.-I. 2004b. Vortex model and simulations for Rayleigh–Taylor and Richtmyer–Meshkov instabilities. *Phys. Rev. E*, **69**, 036703.

Sohn, S.-I. 2009. Effects of surface tension and viscosity on the growth rates of Rayleigh–Taylor and Richtmyer–Meshkov instabilities. *Phys. Rev. E*, **80**, 055302(R).

Sohn, S.-I., and Baek, S. 2017. Bubble merger and scaling law of the Rayleigh–Taylor instability with surface tension. *Phys. Lett. A*, **381**, 3812.

Sojka, J.J. 1989. Global scale, physical models of the *F* region ionospere. *Rev. Geophys.*, **27**, 371.

Sollier, A., and Lescoute, E. 2020. Characterization of the ballistic properties of ejecta from laser shock-loaded samples using high resolution picosecond laser imaging. *Int. J. Impact Eng.*, **136**, 103429.

Song, Y., Wang, P., Wang, L., Ma, D., He, A., Chen, D., Fan, Z., Ma, Z., and Wang, J. 2021a. The early-time dynamics of Rayleigh–Taylor mixing with a premixed layer. *Comput. Fluids*, **229**, 105061.

Song, Y., Wang, P., and Wang, L. 2021b. Numerical investigations of Rayleigh–Taylor instability with a density gradient layer. *Comput. Fluids*, **220**, 104869.

Sorenson, D.S., Capelle, G.A., Grover, M., Johnson, R.P., Kaufman, M.I., LaLone, B.M., Malone, R.M., Marshall, B.F., Minich, R.W., Pazuchanics, P.D., and Smalley, D.D. 2017. Measurements of Sn ejecta particle-size distributions using ultraviolet in-line Fraunhofer holography. *J. Dynam. Behav. Mater.*, **3**, 233.

Soulaine, C., Quintard, M., Baudouy, B., and Van Weelderen, R. 2017. Numerical investigation of thermal counterflow of He II past cylinders. *Phys. Rev. Lett.*, **118**, 074506.

Soulard, O., and Griffond, J. 2022. Permanence of large eddies in Richtmyer–Meshkov turbulence for weak shocks and high Atwood numbers. *Phys. Rev. Fluids*, **7**, 014605.

Soulard, O., Griffond, J., and Gréa, B.-J. 2015. Large-scale analysis of unconfined self-similar Rayleigh–Taylor turbulence. *Phys. Fluids*, **27**, 095103.

Soulard, O., Guillois, F., Griffond, J., Sabelnikov, V., and Simoëns, S. 2018. Permanence of large eddies in Richtmyer–Meshkov turbulence with a small Atwood number. *Phys. Rev. Fluids*, **3**, 104603.

Soulard, O., Guillois, F., Griffond, J., Sabelnikov, V., and Simoëns, S. 2020. A two-scale Langevin PDF model for Richtmyer–Meshkov turbulence with a small Atwood number. *Physica D*, **403**, 132276.

Speziale, C.G. 1991. Analytical methods for the development of Reynolds-stress closures in turbulence. *Annu. Rev. Fluid Mech.*, **23**, 107.

Speziale, C.G. 1998a. A combined large-eddy simulation and time-dependent RANS capability for high-speed compressible flows. *J. Sci. Comput.*, **13**, 253.

Speziale, C.G. 1998b. Turbulence modeling for time-dependent RANS and VLES: a review. *AIAA J.*, **36**, 173.

Spielman, R.B., Deeney, C., Chandler, G.A., et al. 1998. Tungsten wire-array Z-pinch experiments at 200 TW and 2 MJ. *Phys. Plasmas*, **5**, 2105.

Spitzer, L. 2006. *Physics of Fully Ionized Gases*. 2nd edn. Dover Publications, New York.

Srebro, Y., Elbaz, Y., Sadot, O., Arazi, L., and Shvarts, D. 2003. A general buoyancy–drag model for the evolution of the Rayleigh–Taylor and Richtmyer–Meshkov instabilities. *Laser Part. Beams*, **21**, 347.

Sreenivasan, K.R. 1995. On the universality of the Kolmogorov constant. *Phys. Fluids*, **7**, 2778.

Sreenivasan, K.R. 2011. G.I. Taylor: the inspiration behind the Cambridge school. Pages 127–186 of: Davidson, P.A., Kaneda, Y., Moffatt, K., and Sreenivasan, K.R. (eds), *A Voyage through Turbulence*. Cambridge University Press, Cambridge, UK.

Srinivasan, B., and Tang, X.Z. 2012. Mechanism for magnetic field generation and growth in Rayleigh–Taylor unstable inertial confinement fusion plasmas. *Phys. Plasmas*, **19**, 082703.

Srinivasan, B., and Tang, X.Z. 2013. The mitigating effect of magnetic fields on Rayleigh–Taylor unstable inertial confinement fusion plasmas. *Phys. Plasmas*, **20**, 056307.

Srinivasan, B., Dimonte, G., and Tang, X.-Z. 2012. Magnetic field generation in Rayleigh–Taylor unstable inertial confinement fusion plasmas. *Phys. Rev. Lett.*, **108**, 165002.

Stalsberg-Zarling, K., and Gore, R.A. 2011. *The BHR-2 Turbulence Model: Incompressible Isotropic Decay, Rayleigh–Taylor, Kelvin–Helmholtz, and Homogeneous Variable Density Turbulence*. Tech. Rept. LA-UR-11-04773. Los Alamos National Laboratory, Los Alamos, NM.

Stanton, L.G., and Murillo, M.S. 2016. Ionic transport in high-energy-density matter. *Phys. Rev. E*, **93**, 043203.

Statsenko, V.P., Yanilkin, Yu V., and Zhmaylo, V.A. 2013. Direct numerical simulation of turbulent mixing. *Phil. Trans. R. Soc. Lond. A*, **371**, 20120216.

Steele, P.T., Jacoby, B.A., Compton, S.M., and Sinibaldi, J.O. 2017. Advances in ejecta diagnostics at LLNL. *J. Dynam. Behav. Mater.*, **3**, 253.

Steenburgh, R.A., Smithtro, C.G., and Groves, K.M. 2008. Ionospheric scintillation effects on single frequency GPS. *Space Weather*, **6**, S04D02.

Steinberg, D.J., and Lund, C.M. 1989. A constitutive model for strain rates from 10^{-4} to 10^{6} s^{-1}. *J. Appl. Phys.*, **65**, 1528.

Steinberg, D.J., Cochran, S.G., and Guinan, M.W. 1980. A constitutive model for metals applicable at high-strain rate. *J. Appl. Phys.*, **51**, 1498.

Stellmacher, G., and Wiehr, E. 1973. Observation of an instability in a quiescent prominence. *Astron. Astrophys.*, **24**, 321.

Sterbentz, D.M., Jekel, C.F., White, D.A., Aubry, S., Lorenzana, H.E., and Belof, J.L. 2022. Design optimization for Richtmyer–Meshkov instability suppression at shock-compressed material interfaces. *Phys. Fluids*, **34**, 082109.

Stokes, G.G. 1845. On the theories of the internal friction of fluids in motion, and of the equilibrium and motion of pendulums. *Trans. Camb. Phil. Soc.* **8**, 287.

Strykowski, P.J., Krothapalli, A., and Jendoubi, S. 1996. The effect of counterflow on the development of compressible shear layers. *J. Fluid Mech.*, **308**, 63.

Sturtevant, B., and Kulkarny, V.A. 1976. The focusing of weak shock waves. *J. Fluid Mech.*, **73**, 651.

Suárez, D.O., Díaz, A.J., Ramos, A.A., and Bueno, J.T. 2014. Time evolution of plasma parameters during the rise of a solar prominence instability. *Astrophys. J. Lett.*, **785**, L10.

Suchandra, P., and Ranjan, D. 2023. Dynamics of multilayer Rayleigh–Taylor instability at moderately high Atwood numbers. *J. Fluid Mech.*, **974**, A35.

Sun, M., and Takayama, K. 2003. Vorticity production in shock diffraction. *J. Fluid Mech.*, **478**, 237.

Sun, P., Ding, J., Huang, S., Luo, X., and Cheng, W. 2020a. Microscopic Richtmyer–Meshkov instability under strong shock. *Phys. Fluids*, **32**, 024109.

Sun, R., Ding, J., Zhai, Z., Si, T., and Luo, X. 2020b. Convergent Richtmyer–Meshkov instability of heavy gas layer with perturbed inner surface. *J. Fluid Mech.*, **902**, A3.

Sun, Y.B., Zeng, R.H., and Tao, J.J. 2021. Effects of viscosity and elasticity on Rayleigh–Taylor instability in a cylindrical geometry. *Phys. Plasmas*, **28**, 062701.

Sun, Z.P., Turco, R.P., Walterscheid, R.L., Venkateswaran, S.V., and Jones, P.W. 1995. Thermospheric response to morningside diffuse aurora: high-resolution three-dimensional simulations. *J. Geophys. Res. Space Phys.*, **100**, 23779.

Susskind, L. 2003. Superstrings. *Phys. World*, **16**, 29.

Svetsov, V.V., Nemtchinov, I.V., and Teterev, A.V. 1995. Disintegration of large meteoroids in Earth's atmosphere: theoretical models. *Icarus*, **116**, 131.

Sykes, J.P., Gallagher, T.P., and Rankin, B.A. 2021. Effects of Rayleigh–Taylor instabilities on turbulent premixed flames in a curved rectangular duct. *Proc. Combust. Inst.*, **38**, 6059.

Tabeling, P. 2002. Two-dimensional turbulence: a physicist approach. *Phys. Rep.*, **362**, 1.

Takabe, H. 2004. A historical perspective of developments in hydrodynamic instabilities, integrated codes and laboratory astrophysics. *Nucl. Fusion*, **44**, S149.

Takabe, H., Mima, K., Montierth, L., and Morse, R.L. 1985. Self-consistent growth rate of the Rayleigh–Taylor instability in an ablatively accelerating plasma. *Phys. Fluids*, **28**, 3676.

Tandberg-Hanssen, E. 2013. *The Nature of Solar Prominences*. Springer Science & Business Media. Dordrecht, Netherlands.

Tang, J., Zhang, F., Luo, X., and Zhai, Z. 2021a. Effect of Atwood number on convergent Richtmyer–Meshkov instability. *Acta Mech. Sin.*, **37**, 434.

Tang, K., Mostert, W., Fuster, D., and Deike, L. 2021b. Effects of surface tension on the Richtmyer–Meshkov instability in fully compressible and inviscid fluids. *Phys. Rev. Fluids*, **6**, 113901.

Tao, J.J., He, X.T., Ye, W.H., and Busse, F.H. 2013. Nonlinear Rayleigh–Taylor instability of rotating inviscid fluids. *Phys. Rev. E*, **87**, 013001.

Tapinou, K.C., Wheatley, V., Bond, D., and Jahn, I. 2022. The Richtmyer–Meshkov instability of thermal, isotope and species interfaces in a five-moment multi-fluid plasma. *J. Fluid Mech.*, **951**, A11.

Tapinou, K.C., Wheatley, V., Bond, D., and Jahn, I. 2023. The effect of collisions on the multi-fluid plasma Richtmyer–Meshkov instability. *Phys. Plasmas*, **30**, 022707.

Tassart, J. 2004. Overview of inertial fusion and high-intensity laser plasma research in Europe. *Nucl. Fusion*, **44**, S134.

Tavares, H.S., Biferale, L., Sbragaglia, M., and Mailybaev, A.A. 2021a. Immiscible Rayleigh–Taylor turbulence using mesoscopic lattice Boltzmann algorithms. *Phys. Rev. Fluids*, **6**, 054606.

Tavares, H.S., Biferale, L., Sbragaglia, M., and Mailybaev, A.A. 2021b. Validation and application of the lattice Boltzmann algorithm for a turbulent immiscible Rayleigh–Taylor system. *Philos. Trans. R. Soc. A*, **379**, 20200396.

Taylor, G.I. 1935. Sir Horace Lamb, FRS. *Nature*, **135**, 255.

Taylor, G.I. 1950. The instability of liquid surfaces when accelerated in a direction perpendicular to their planes. I. *Proc. R. Soc. London. A*, **201**, 192.

Taylor, G.I. 1974. The interaction between experiment and theory in fluid mechanics. *Annu. Rev. Fluid Mech.*, **6**, 1.

Temporal, M., Jaouen, S., Masse, L., and Canaud, B. 2006. Hydrodynamic instabilities in ablative tamped flows. *Phys. Plasmas*, **13**, 122701.

Tennekes, H., and Lumley, J.L. 1972. *A First Course in Turbulence*. MIT Press, Cambridge, MA.

Terrones, G., and Carrara, M.D. 2015. Rayleigh–Taylor instability at spherical interfaces between viscous fluids: fluid/vacuum interface. *Phys. Fluids*, **27**, 054105.

Theofanous, T.G. 2011. Aerobreakup of Newtonian and viscoelastic liquids. *Annu. Rev. Fluid Mech.*, **43**, 661.

Thomas, G., Bambrey, R., and Brown, C. 2001. Experimental observations of flame acceleration and transition to detonation following shock-flame interaction. *Combust. Theory Model.*, **5**, 573.

Thompson, S.P. 1907. Lord Kelvin. *Nature*, **77**, 175.

Thompson, T.A., Burrows, A., and Pinto, P.A. 2003. Shock breakout in core-collapse supernovae and its neutrino signature. *Astrophys. J.*, **592**, 434.

Thomson, J. 1855. XLII. On certain curious motions observable at the surfaces of wine and other alcoholic liquors. *The London, Edinburgh, and Dublin Philos. Mag. J. Sci.*, **10**, 330.

Thomson, W. 1887a. On the propagation of laminar motion through a turbulently moving inviscid liquid. *The London, Edinburgh, Dublin Philos. Mag. J. Sci.*, **24**, 342.

Thomson, W. 1887b. Stability of motion (continued from the May, June, and August Numbers).–Broad river flowing down an inclined plane bed. *The London, Edinburgh, Dublin Philos. Mag. J. Sci.*, **24**, 272.

Thornber, B. 2016. Impact of domain size and statistical errors in simulations of homogeneous decaying turbulence and the Richtmyer–Meshkov instability. *Phys. Fluids*, **28**, 045106.

Thornber, B., and Zhou, Y. 2012. Energy transfer in the Richtmyer–Meshkov instability. *Phys. Rev. E*, **86**, 056302.

Thornber, B., and Zhou, Y. 2015. Numerical simulations of the two-dimensional multimode Richtmyer–Meshkov instability. *Phys. Plasmas*, **22**, 032309.

Thornber, B., Drikakis, D., Youngs, D.L., and Williams, R.J.R. 2010. The influence of initial conditions on turbulent mixing due to Richtmyer–Meshkov instability. *J. Fluid Mech.*, **654**, 99.

Thornber, B., Drikakis, D., Youngs, D.L., and Williams, R.J.R. 2011. Growth of a Richtmyer–Meshkov turbulent layer after reshock. *Phys. Fluids*, **23**, 095107.

Thornber, B., Drikakis, D., Youngs, D.L., and Williams, R.J.R. 2012. Physics of the single-shocked and reshocked Richtmyer–Meshkov instability. *J. Turbul.*, **13**, N10.

Thornber, B., Griffond, J., Poujade, O., et al. 2017. Late-time growth rate, mixing, and anisotropy in the multimode narrowband Richtmyer–Meshkov instability: the θ-group collaboration. *Phys. Fluids*, **29**, 105107.

Thornber, B., Griffond, J., Bigdelou, P., Boureima, I., Ramaprabhu, P., Schilling, O., and Williams, R.J.R. 2019. Turbulent transport and mixing in the multimode narrowband Richtmyer–Meshkov instability. *Phys. Fluids*, **31**, 096105.

Tian, B., Fu, D., and Ma, Y. 2006. Numerical investigation of Richtmyer–Meshkov instability driven by cylindrical shocks. *Acta Mech. Sin.*, **22**, 9.

Tian, B.L., Zhang, X.T., Qi, J., and Wang, S.H. 2011. Effects of a Premixed Layer on the Richtmyer–Meshkov Instability. *Chin. Phys. Lett.*, **28**, 114701.

Ticknor, C., Kress, J.D., Collins, L.A., Clérouin, J., Arnault, P., and Decoster, A. 2016. Transport properties of an asymmetric mixture in the dense plasma regime. *Phys. Rev. E*, **93**, 063208.

Tomkins, C.D., Balakumar, B.J., Orlicz, G., Prestridge, K.P., and Ristorcelli, J.R. 2013. Evolution of the density self-correlation in developing Richtmyer–Meshkov turbulence. *J. Fluid Mech.*, **735**, 288.

Tommasini, R., Field, J.E., Hammel, B.A., et al. 2015. Tent-induced perturbations on areal density of implosions at the National Ignition Facility. *Phys. Plasmas*, **22**, 056315.

Tonks, L. 1937. Theory and phenomena of high current densities in low pressure arcs. *Trans. Electrochem. Soc.*, **72**, 167.

Townsend, A.A.R. 1980. *The Structure of Turbulent Shear Flow*. University Press, Cambridge, UK.

Trimble, V. 1968. Motions and structure of the filamentary envelope of the Crab Nebula. *Astron. J.*, **73**, 535.

Tritschler, V.K., Hu, X.Y., Hickel, S., and Adams, N.A. 2013. Numerical simulation of a Richtmyer–Meshkov instability with an adaptive central-upwind sixth-order WENO scheme. *Phys. Scr.*, **T155**, 014016.

Tritschler, V.K., Zubel, M., Hickel, S., and Adams, N.A. 2014a. Evolution of length scales and statistics of Richtmyer–Meshkov instability from direct numerical simulations. *Phys. Rev. E*, **90**, 063001.

Tritschler, V.K., Olson, B.J., Lele, S.K., Hickel, S., Hu, X.Y., and Adams, N.A. 2014b. On the Richtmyer–Meshkov instability evolving from a deterministic multimode planar interface. *J. Fluid Mech.*, **755**, 429.

Tritton, D.J. 1988. *Physical Fluid Dynamics*. Clarendon, Oxford, UK.

Tryggvason, G., and Unverdi, S.O. 1990. Computations of three-dimensional Rayleigh–Taylor instability. *Phys. Fluids A*, **2**, 656.

Tsiklashvili, V., Romero Colio, P.E., Likhachev, O.A., and Jacobs, J.W. 2012. An experimental study of small Atwood number Rayleigh–Taylor instability using the magnetic levitation of paramagnetic fluids. *Phys. Fluids*, **24**, 052106.

Tubbs, D.L., Barnes, C.W., Beck, J.B., Hoffman, N.M., Oertel, J.A., Watt, R.G., Boehly, T., Bradley, D., Jaanimagi, P., and Knauer, J. 1999. Cylindrical implosion experiments using laser direct drive. *Phys. Plasmas*, **6**, 2095.

Turner, J.S. 1997. G.I. Taylor in his later years. *Annu. Rev. Fluid Mech.*, **29**, 1.

Turner, M.R., Sazhin, S.S., Healey, J.J., Crua, C., and Martynov, S.B. 2012. A breakup model for transient diesel fuel sprays. *Fuel*, **97**, 288.

Uchiyama, Y., Aharonian, F.A., Tanaka, T., Takahashi, T., and Maeda, Y. 2007. Extremely fast acceleration of cosmic rays in a supernova remnant. *Nature*, **449**, 576.

Ukai, S., Balakrishnan, K., and Menon, S. 2011. Growth rate predictions of single-and multi-mode Richtmyer–Meshkov instability with reshock. *Shock Waves*, **21**, 533.

Vadivukkarasan, M. 2021. Temporal instability characteristics of Rayleigh–Taylor and Kelvin–Helmholtz mechanisms of an inviscid cylindrical interface. *Meccanica*, **56**, 117.

Vadivukkarasan, M., and Panchagnula, M.V. 2017. Combined Rayleigh–Taylor and Kelvin–Helmholtz instabilities on an annular liquid sheet. *J. Fluid Mech.*, **812**, 152.

Valentine, G.A., and Wohletz, K.H. 1989. Numerical models of Plinian eruption columns and pyroclastic flows. *J. Geophys. Res. Solid Earth*, **94**, 1867.

van Haren, H. 2015. Instability observations associated with wave breaking in the stable-stratified deep-ocean. *Physica D*, **292**, 62.

Van Haren, H., and Gostiaux, L. 2010. A deep-ocean Kelvin–Helmholtz billow train. *Geophys. Res. Lett.*, **37**, L03605.

Vandenboomgaerde, M., and Aymard, C. 2011. Analytical theory for planar shock focusing through perfect gas lens and shock tube experiment designs. *Phys. Fluids*, **23**, 016101.

Vandenboomgaerde, M., Mügler, C., and Gauthier, S. 1998. Impulsive model for the Richtmyer–Meshkov instability. *Phys. Rev. E*, **58**, 1874.

Vandenboomgaerde, M., Gauthier, S., and Mügler, C. 2002. Nonlinear regime of a multi-mode Richtmyer–Meshkov instability: a simplified perturbation theory. *Phys. Fluids*, **14**, 1111.

Vandenboomgaerde, M., Bastian, J .and Casner, A., Galmiche, D., Jadaud, J.-P., Laffite, S., Liberatore, S., Malinie, G., and Philippe, F. 2007. Prolate-spheroid ("rugby-shaped") hohlraum for inertial confinement fusion. *Phys. Rev. Lett.*, **99**, 065004.

Vandenboomgaerde, M., Souffland, D., Mariani, C., Biamino, L., Jourdan, G., and Houas, L. 2014. An experimental and numerical investigation of the dependency on the initial conditions of the Richtmyer–Meshkov instability. *Phys. Fluids*, **26**, 024109.

Vartanyan, D., Burrows, A., Radice, D., Skinner, M. A., and Dolence, J. 2019. A successful 3D core-collapse supernova explosion model. *Mon. Not. R. Astron. Soc.*, **482**, 351.

Velikovich, A.L., and Dimonte, G. 1996. Nonlinear perturbation theory of the incompressible Richtmyer–Meshkov instability. *Phys. Rev. Lett.*, **76**, 3112.

Velikovich, A.L., Herrmann, M., and Abarzhi, S.I. 2014. Perturbation theory and numerical modelling of weakly and moderately nonlinear dynamics of the incompressible Richtmyer–Meshkov instability. *J. Fluid Mech.*, **751**, 432.

Velikovich, A.L., Schmitt, A.J., Zulick, C., Aglitskiy, Y., Karasik, M., Obenschain, S.P., Wouchuk, J.G., and Cobos Campos, F. 2020. Multi-mode hydrodynamic evolution of perturbations seeded by isolated surface defects. *Phys. Plasmas*, **27**, 102706.

Venerus, D.C., and Simavilla, D.N. 2015. Tears of wine: new insights on an old phenomenon. *Sci. Rep.*, **5**, 16162.

Veraar, R., Mayer, A., Verreault, J., Stowe, R., Farinaccio, R., and Harris, P. 2009. Proof-of-principle experiment of a shock-induced combustion ramjet. Page 7432 of: *16th AIAA/DLR/DGLR*

International Space Planes and Hypersonic Systems and Technologies Conference. American Institute of Aeronautics and Astronautics, Rhode St. Genèse, Belgium.

Versluis, M., Schmitz, B., Von der Heydt, A., and Lohse, D. 2000. How snapping shrimp snap: through cavitating bubbles. *Science*, **289**, 2114.

Vetter, M., and Sturtevant, B. 1995. Experiments on the Richtmyer–Meshkov instability of an air/SF_6 interface. *Shock Waves*, **4**, 247.

Villermaux, E., Rehab, H., and Hopfinger, E.J. 1998. Shear instabilities in the near field of coaxial jets. *Phys. Fluids*, **10**, S2.

Vishniac, E.T. 1983. The dynamic and gravitational instabilities of spherical shocks. *Astrophys. J.*, **274**, 152.

Vladimirova, N., and Chertkov, M. 2009. Self-similarity and universality in Rayleigh–Taylor, Boussinesq turbulence. *Phys. Fluids*, **21**, 015102.

Vogler, T.J., and Hudspeth, M.C. 2021. Tamped Richtmyer–Meshkov instability experiments to probe high-pressure material strength. *J. Dyn. Behavior Mater.*, **7**, 262.

Vold, E., Yin, L., and Albright, B.J. 2021. Plasma transport simulations of Rayleigh–Taylor instability in near-ICF deceleration regimes. *Phys. Plasmas*, **28**, 092709.

von Helmholtz, H. 1868. On the discontinuous movements of fluids. *Monder Königlichen Preussische Akademie der Wissenschaften zu Berlin*, **23**, 215.

Vreman, A.W. 2004. An eddy-viscosity subgrid-scale model for turbulent shear flow: algebraic theory and applications. *Phys. Fluids*, **16**, 3670.

Vuilleumier, R, Ego, V, Neltner, L, and Cazabat, AM. 1995. Tears of wine: the stationary state. *Langmuir*, **11**, 4117.

Wadas, M.J., and Johnsen, E. 2020. Interactions of two bubbles along a gaseous interface undergoing the Richtmyer–Meshkov instability in two dimensions. *Physica D*, **409**, 132489.

Wadas, M.J., Khieu, L.H., Cearley, G.S., LeFevre, H.J., Kuranz, C.C., and Johnsen, E. 2023. Saturation of vortex rings ejected from shock-accelerated interfaces. *Phys. Rev. Lett.*, **130**, 194001.

Waddell, J.T., Niederhaus, C.E., and Jacobs, J.W. 2001. Experimental study of Rayleigh–Taylor instability: low Atwood number liquid systems with single-mode initial perturbations. *Phys. Fluids*, **13**, 1263.

Wade, N.J. 1994. Hermann von Helmholtz (1821–1894). *Perception*, **23**, 981.

Waitz, I., Marble, F., and Zukoski, E. 1993. Investigation of a contoured wall injector for hypervelocity mixing augmentation. *AIAA J.*, **31**, 1014.

Walchli, B., and Thornber, B. 2017. Reynolds number effects on the single-mode Richtmyer–Meshkov instability. *Phys. Rev. E*, **95**, 013104.

Wallace, J.M., Murphy, T.J., Delamater, N.D., et al. 1999. Inertial confinement fusion with tetrahedral hohlraums at OMEGA. *Phys. Rev. Lett.*, **82**, 3807.

Walsh, C.A. 2022. Magnetized ablative Rayleigh–Taylor instability in three dimensions. *Phys. Rev. E*, **105**, 025206.

Walsh, C.A., and Clark, D.S. 2023. Nonlinear ablative Rayleigh–Taylor instability: increased growth due to self-generated magnetic fields. *Phys. Rev. E*, **107**, L013201.

Walsh, C.A., Chittenden, J.P., McGlinchey, K., Niasse, N.P.L., and Appelbe, B.D. 2017. Self-generated magnetic fields in the stagnation phase of indirect-drive implosions on the National Ignition Facility. *Phys. Rev. Lett.*, **118**, 155001.

Wan, W.C., Malamud, G., Shimony, A., Di Stefano, C.A., Trantham, M.R., Klein, S.R., Shvarts, D., Kuranz, C.C., and Drake, R.P. 2015. Observation of single-mode, Kelvin–Helmholtz instability in a supersonic flow. *Phys. Rev. Lett.*, **115**, 145001.

Wan, W.C., Malamud, G., Shimony, A., Di Stefano, C.A., Trantham, M.R., Klein, S.R., Shvarts, D., Drake, R.P., and Kuranz, C.C. 2017. Observation of dual-mode, Kelvin–Helmholtz instability vortex merger in a compressible flow. *Phys. Plasmas*, **24**, 055705.

Wang, G., Zhai, W., Yang, H., et al. 2013a. The genomics of selection in dogs and the parallel evolution between dogs and humans. *Nat. Commun.*, **4**, 1860.

Wang, H., Cao, Q., Chen, C., Zhai, Z., and Luo, X. 2022a. Experimental study on a light–heavy interface evolution induced by two successive shock waves. *J. Fluid Mech.*, **953**, A15.

Wang, H., Wang, H., Zhai, Z., and Luo, X. 2023a. High-amplitude effect on single-mode Richtmyer–Meshkov instability of a light–heavy interface. *Phys. Fluids*, **35**, 016106.

Wang, H., Wang, H., Zhai, Z., and Luo, X. 2023b. High-amplitude effect on Richtmyer–Meshkov instability at a single-mode heavy-light interface. *Phys. Fluids*, **35**, 126107.

Wang, L., and Li, J. 2005. Fluid mixing due to Rayleigh–Taylor instability in a time-dependent acceleration field. *Commun. Nonlinear. Sci. Numer. Simul.*, **10**, 571.

Wang, L., Li, J., and Xie, Z. 2002. Large-eddy-simulation of 3-dimensional Rayleigh–Taylor instability in incompressible fluids. *Sci. China Math.*, **45**, 95.

Wang, L.F., Ye, W.H., and Li, Y.J. 2010. Combined effect of the density and velocity gradients in the combination of Kelvin–Helmholtz and Rayleigh–Taylor instabilities. *Phys. Plasmas*, **17**, 042103.

Wang, L.F., Ye, W.H., He, X.T., Zhang, W.Y., Sheng, Z.M., and Yu, M.Y. 2012. Formation of jet-like spikes from the ablative Rayleigh–Taylor instability. *Phys. Plasmas*, **19**, 100701.

Wang, L.F., Wu, J.F., Ye, W.H., Zhang, W.Y., and He, X.T. 2013b. Weakly nonlinear incompressible Rayleigh–Taylor instability growth at cylindrically convergent interfaces. *Phys. Plasmas*, **20**, 042708.

Wang, L.F., Wu, J.F., Guo, H.Y., Ye, W.H., Liu, J., Zhang, W.Y., and He, X.T. 2015a. Weakly nonlinear Bell-Plesset effects for a uniformly converging cylinder. *Phys. Plasmas*, **22**, 082702.

Wang, M., Si, T., and Luo, X. 2013c. Generation of polygonal gas interfaces by soap film for Richtmyer–Meshkov instability study. *Exp. Fluids*, **54**, 1.

Wang, P., Zhou, Y., MacLaren, S.A., Huntington, C.M., Raman, K.S., Doss, F.W., and Flippo, K.A. 2015b. Three-and two-dimensional simulations of counter-propagating shear experiments at high energy densities at the National Ignition Facility. *Phys. Plasmas*, **22**, 112701.

Wang, R., Song, Y., Ma, Z., Ma, D., Wang, L., and Wang, P. 2022b. The transition to turbulence in rarefaction-driven Rayleigh–Taylor mixing: effects of diffuse interface. *Phys. Fluids*, **34**, 015125.

Wang, R., Song, Y., Ma, Z., Zhang, Y., Wang, J., Xu, Y., Wang, L., and Wang, P. 2023c. Scale-to-scale energy transfer in rarefaction-driven Rayleigh–Taylor instability induced transitional mixing. *Phys. Fluids*, **35**, 025136.

Wang, T, Tao, G, Bai, JS, Li, P, and Wang, B. 2015c. Numerical comparative analysis of Richtmyer–Meshkov instability simulated by different SGS models. *Can. J. Phys.*, **93**, 519.

Wang, T, Bai, JS, Li, P, Wang, B, Du, L, and Tao, G. 2016. Large-eddy simulations of the multi-mode Richtmyer–Meshkov instability and turbulent mixing under reshock. *High Energy Density Phys.*, **19**, 65.

Wang, X., Yang, D., Wu, J., and Luo, X. 2015d. Interaction of a weak shock wave with a discontinuous heavy-gas cylinder. *Phys. Fluids*, **27**, 064104.

Wang, X.-G., Sun, S.K., Xiao, D.L., Wang, G.Q., Zhang, Y., Zhou, S.T., Ren, X.D., Xu, Q., Huang, X.B., Ding, N., and Shu, X.J. 2019a. Numerical study on magneto-Rayleigh–Taylor instabilities for thin liner implosions on the primary test stand facility. *Chin. Phys. B*, **28**, 035201.

Wang, X.-X. 2019b. Research at Tsinghua University on electrical explosions of wires. *Matter Radiat. Extremes*, **4**, 017201.

Wang, Y., and Dong, G. 2020. Interface evolutions and growth predictions of mixing zone on premixed flame interface during RM instability. *Chin. J. Theor. Appl. Mech.*, **52**, 1655.

Wang, Y., Bao, B., Yang, C., and Zhang, L. 2018. The impact of different interstellar medium structures on the dynamical evolution of supernova remnants. *Mon. Not. R. Astron. Soc.*, **478**, 2948.

Wang, Z., Wang, T., Bai, J., and Xiao, J. 2019. Numerical study of the Richtmyer–Meshkov instability induced by non-planar shock wave in non-uniform flows. *J. Turbul.*, **20**, 481.

Wang, Z., Xue, K., and Han, P. 2021. Bell–Plesset effects on Rayleigh–Taylor instability at cylindrically divergent interfaces between viscous fluids. *Phys. Fluids*, **33**, 034118.

Warhaft, Z. 1997. *An Introduction to Thermal-Fluid Engineering: The Engine and the Atmosphere.* Cambridge University Press, Cambridge, UK.

Weber, C., Haehn, N., Oakley, J., Anderson, M., and Bonazza, R. 2012. Richtmyer–Meshkov instability on a low Atwood number interface after reshock. *Shock Waves*, **22**, 317.

Weber, C.R., Cook, A.W., and Bonazza, R. 2013. Growth rate of a shocked mixing layer with known initial perturbations. *J. Fluid Mech.*, **725**, 372.

Weber, C.R., Haehn, N.S., Oakley, J.G., Rothamer, D.A., and Bonazza, R. 2014a. An experimental investigation of the turbulent mixing transition in the Richtmyer–Meshkov instability. *J. Fluid Mech.*, **748**, 457.

Weber, C.R., Clark, D.S., Cook, A.W., Busby, L.E., and Robey, H.F. 2014b. Inhibition of turbulence in inertial-confinement-fusion hot spots by viscous dissipation. *Phys. Rev. E*, **89**, 053106.

Weber, C.R., Clark, D.S., Cook, A.W., et al. 2015. Three-dimensional hydrodynamics of the deceleration stage in inertial confinement fusion. *Phys. Plasmas*, **22**, 032702.

Weber, C.R., Döppner, T., Casey, D.T., et al. 2016. First measurements of fuel-ablator interface instability growth in inertial confinement fusion implosions on the National Ignition Facility. *Phys. Rev. Lett.*, **117**, 075002.

Weber, C.R., Casey, D.T., Clark, D.S., et al. 2017. Improving ICF implosion performance with alternative capsule supports. *Phys. Plasmas*, **24**, 056302.

Weber, C.R., Clark, D.S., Pak, A., et al. 2020. Mixing in ICF implosions on the National Ignition Facility caused by the fill-tube. *Phys. Plasmas*, **27**, 032703.

Wei, T., and Livescu, D. 2012. Late-time quadratic growth in single-mode Rayleigh–Taylor instability. *Phys. Rev. E*, **86**, 046405.

Wei, Y., Li, Yu., Wang, Z., Yang, H., Zhu, Z., Qian, Y.H., and Luo, K.H. 2022. Small-scale fluctuation and scaling law of mixing in three-dimensional rotating turbulent Rayleigh–Taylor instability. *Phys. Rev. E*, **105**, 015103.

Weir, S.T., Chandler, E.A., and Goodwin, B.T. 1998. Rayleigh–Taylor instability experiments examining feedthrough growth in an incompressible, convergent geometry. *Phys. Rev. Lett.*, **80**, 3763.

Weis, M.R., Zhang, P., Lau, Y.Y., Schmit, P.F., Peterson, K.J., Hess, M., and Gilgenbach, R.M. 2015. Coupling of sausage, kink, and magneto-Rayleigh–Taylor instabilities in a cylindrical liner. *Phys. Plasmas*, **22**, 032706.

Weizsäcker, C.F. v. 1948. Das spektrum der turbulenz bei grossen Reynoldsschen zahlen. *Zeitschrift für Physik*, **124**, 614.

Wells, F. 2014. *The Heart of Leonardo*. Springer Science & Business Media, London.

Welser-Sherrill, L., Mancini, R.C., Haynes, D.A., et al. 2007. Development of two mix model postprocessors for the investigation of shell mix in indirect drive implosion cores. *Phys. Plasmas*, **14**, 072705.

Welser-Sherrill, L., Fincke, J., Doss, F., Loomis, E., Flippo, K., Offermann, D., Keiter, P., Haines, B., and Grinstein, F. 2013. Two laser-driven mix experiments to study reshock and shear. *High Energy Density Phys.*, **9**, 496.

Wheatley, V., Pullin, D.I., and Samtaney, R. 2005a. Regular shock refraction at an oblique planar density interface in magnetohydrodynamics. *J. Fluid Mech.*, **522**, 179.

Wheatley, V., Pullin, D.I., and Samtaney, R. 2005b. Stability of an impulsively accelerated density interface in magnetohydrodynamics. *Phys. Rev. Lett.*, **95**, 125002.

Wheatley, V., Gehre, R.M., Samtaney, R., and Pullin, D.I. 2013. The magnetohydrodynamic Richtmyer–Meshkov instability: the oblique field case. In: Bonazza, R., Ranjan, D. (eds), *29th International Symposium on Shock Waves 2. ISSW 2013*. Springer, Cham, Switzerland.

Wheeler, J.C., Harkness, R.P., Khokhlov, A.M., and Höflich, P. 1995. Stirling's supernovae: a survey of the field. *Phys. Rep.*, **256**, 211.

Wheeler, J.C., Meier, D.L., and Wilson, J.R. 2002. Asymmetric supernovae from magnetocentrifugal jets. *Astrophys. J.*, **568**, 807.

White, D.A. 2009. *Siqueiros: Biography of a Revolutionary Artist*. BookSurge, Charleston, SC.

White, F.M. 1994. *Fluid Mechanics*. 3rd edn. McGraw-Hill, New York.

White, J., Oakley, J., Anderson, M., and Bonazza, R. 2010. Experimental measurements of the nonlinear Rayleigh–Taylor instability using a magnetorheological fluid. *Phys. Re. E*, **81**, 026303.

Whitney, K.G. 1999. Momentum and heat conduction in highly ionizable plasmas. *Phys. Plasmas*, **6**, 816.

Wieland, S.A., Hamlington, P.E., Reckinger, S.J., and Livescu, D. 2019. Effects of isothermal stratification strength on vorticity dynamics for single-mode compressible Rayleigh–Taylor instability. *Phys. Rev. Fluids*, **4**, 093905.

Wilcox, D.C. 1998. *Turbulence Modeling for CFD*. DCW Industries, La Canada, CA.

Wilkinson, J.P., and Jacobs, J.W. 2007. Experimental study of the single-mode three-dimensional Rayleigh–Taylor instability. *Phys. Fluids*, **19**, 124102.

Willey, T.M., Champley, K., Hodgin, R., Lauderbach, L., Bagge-Hansen, M., May, C., Sanchez, N., Jensen, B.J., Iverson, A., and Van Buuren, T. 2016. X-ray imaging and 3D reconstruction of in-flight exploding foil initiator flyers. *J. Appl. Phys.*, **119**, 235901.

Williams, R.J.R. 2016. The late time structure of high density contrast, single mode Richtmyer–Meshkov flow. *Phys. Fluids*, **28**, 074108.

Williams, R.J.R. 2018. Ejecta sources and scalings. *AIP Conf. Proc.*, **1979**, 080015.

Williams, R.J.R., and Grapes, C.C. 2017. Simulation of double-shock ejecta production. *J. Dynam. Behav. Mater.*, **3**, 291.

Wilson, D.B. (ed). 1990. *The Correspondence between Sir George Gabriel Stokes and Sir William Thomson, Baron Kelvin of Largs*, 2 vols. Cambridge University Press, Cambridge, UK.

Wilson, D.C., Bradley, P.A., Goldman, S.R., Hoffman, N.M., Margevicius, R.W., Stephens, R.B., and Olson, R.E. 2000. Developments in NIF beryllium capsule design. *Fusion Techno.*, **38**, 16.

Wilson, J.R. 1985. Supernovae and post-collapse behavior. In: Centrella, J.M., LeBlanc, J.M., Bowers, R.L. and Wheeler, J.A. (eds), *Numerical astrophysics*, Jones and Bartlett Publ., Boston, MA.

Wolfshtein, M. 1970. Length-scale-of-turbulence equation. *Israel J. Tech.*, **8**, 87.

Woltjer, L. 1972. Supernova remnants. *Annu. Rev. Astron. Astrophys.*, **10**, 129.

Wong, M.L., Livescu, D., and Lele, S.K. 2019. High-resolution Navier–Stokes simulations of Richtmyer–Meshkov instability with reshock. *Phys. Rev. Fluids*, **4**, 104609.

Wong, M.L., Baltzer, J.R., Livescu, D., and Lele, S.K. 2022. Analysis of second moments and their budgets for Richtmyer–Meshkov instability and variable-density turbulence induced by reshock. *Phys. Rev. Fluids*, **7**, 044602.

Wongwathanarat, A., Mueller, E., and Janka, H.-Th. 2015. Three-dimensional simulations of core-collapse supernovae: from shock revival to shock breakout. *Astron. Astrophys.*, **577**, A48.

Wood, B.D., He, X., and Apte, S.V. 2020. Modeling turbulent flows in porous media. *Annu. Rev. Fluid Mech.*, **52**, 171.

Woodman, R.F., and La Hoz, C. 1976. Radar observations of F region equatorial irregularities. *J. Geophys. Res.*, **81**, 5447.

Woods, A.W. 1995. The dynamics of explosive volcanic eruptions. *Rev. Geophys.*, **33**, 495.

Woods, J.D. 1968. Wave-induced shear instability in the summer thermocline. *J. Fluid Mech.*, **32**, 791.

Woolstrum, J.M., Yager-Elorriaga, D.A., Campbell, P.C., Jordan, N.M., Seyler, C.E., and McBride, R.D. 2020. Extended magnetohydrodynamics simulations of thin-foil Z-pinch implosions with comparison to experiments. *Phys. Plasmas*, **27**, 092705.

Woosley, S.E., Wunsch, S., and Kuhlen, M. 2004. Carbon ignition in type Ia supernovae: an analytic model. *Astrophys. J.*, **607**, 921.

Wouchuk, J.G., and Nishihara, K. 1996. Linear perturbation growth at a shocked interface. *Phys. Plasmas*, **3**, 3761.

Wouchuk, J.G., and Nishihara, K. 1997. Asymptotic growth in the linear Richtmyer–Meshkov instability. *Phys. Plasmas*, **4**, 1028.

Wouchuk, JG, and Sano, T. 2015. Normal velocity freeze-out of the Richtmyer–Meshkov instability when a rarefaction is reflected. *Phys. Rev. E*, **91**, 023005.

Wu, C.C. 1987. On MHD intermediate shocks. *Geophys. Res. Lett.*, **14**, 668.

Wu, C.C. 1990. Formation, structure, and stability of MHD intermediate shocks. *J. Geophys. Res. Space Phys.*, **95**, 8149.

Wu, J., Lu, Y., Sun, F., Jiang, X., Wang, Z., Zhang, D., Li, X., and Qiu, A. 2019. Researches on preconditioned wire array Z pinches in Xi'an Jiaotong University. *Matter Radiat. Extremes*, **4**, 036201.

Wu, J., Liu, H., and Xiao, Z. 2021. Refined modelling of the single-mode cylindrical Richtmyer–Meshkov instability. *J. Fluid Mech.*, **908**.

Wu, Z., Huang, S., Ding, J., Wang, W., and Luo, X. 2018. Molecular dynamics simulation of cylindrical Richtmyer–Meshkov instability. *Sci. China Phys. Mech. Astron.*, **61**, 114712.

Wu-Wang, H., Liu, C., and Xiao, Z. 2024. Parametric effects on Richtmyer–Meshkov instability of a V-shaped gaseous interface within linear stage. *Phys. Fluids*, **36**, 024114.

Wykes, M.S.D., Hughes, G.O., and Dalziel, S. 2015. On the meaning of mixing efficiency for buoyancy driven mixing in stratified turbulent flows. *J. Fluid Mech.*, **781**, 261.

Xia, C., and Keppens, R. 2016. Internal dynamics of a twin-layer solar prominence. *Astrophys. J. Lett.*, **825**, L29.

Xiao, J.X., Bai, J.S., and Wang, T. 2016. Numerical study of initial perturbation effects on Richtmyer–Meshkov instability in nonuniform flows. *Phys. Rev. E*, **94**, 013112.

Xiao, M., Zhang, Y., and Tian, B. 2020a. Modeling of turbulent mixing with an improved K–L model. *Phys. Fluids*, **32**, 092104.

Xiao, M., Zhang, Y., and Tian, B. 2020b. Unified prediction of reshocked Richtmyer–Meshkov mixing with KL model. *Phys. Fluids*, **32**, 032107.

Xiao, M., Zhang, Y., and Tian, B. 2021. A *K–L* model with improved realizability for turbulent mixing. *Phys. Fluids*, **33**, 022104.

Xie, C., Tao, J., and Li, J. 2017a. Viscous Rayleigh–Taylor instability with and without diffusion effect. *Appl. Math. Mech.*, **38**, 263.

Xie, C.Y., Tao, J.J., Sun, Z.L., and Li, J. 2017b. Retarding viscous Rayleigh–Taylor mixing by an optimized additional mode. *Phys. Rev. E*, **95**, 023109.

Xie, H., Xiao, M., and Zhang, Y. 2021a. Predicting different turbulent mixing problems with the same $K - \mathcal{E}$ model and model coefficients. *AIP Adv.*, **11**, 075213.

Xie, H., Xiao, M., and Zhang, Y. 2021b. Unified prediction of turbulent mixing induced by interfacial instabilities via Besnard- Harlow- Rauenzahn-2 model. *Phys. Fluids*, **33**, 105123.

Xie, H., Zhao, Y., and Zhang, Y. 2023. Data-driven nonlinear K-L turbulent mixing model via gene expression programming method. *Acta Mech. Sin.*, **39**, 322315.

Xin, J., Yan, R., Wan, Z.-H., Sun, D.-J., Zheng, J., Zhang, H., Aluie, H., and Betti, R. 2019. Two mode coupling of the ablative Rayleigh–Taylor instabilities. *Phys. Plasmas*, **26**, 032703.

Xu, Z.W., Wu, J., and Wu, Z.S. 2004. A survey of ionospheric effects on space-based radar. *Waves Random Complex Media*, **14**, S189.

Yabe, T., Hoshino, H., and Tsuchiya, T. 1991. Two-and three-dimensional behavior of Rayleigh–Taylor and Kelvin–Helmholtz instabilities. *Phys. Rev. A*, **44**, 2756.

Yager-Elorriaga, D.A., Zhang, P., Steiner, A.M., Jordan, N.M., Campbell, P.C., Lau, Y.Y., and Gilgenbach, R.M. 2016. Discrete helical modes in imploding and exploding cylindrical, magnetized liners. *Phys. Plasmas*, **23**, 124502.

Yager-Elorriaga, D.A., Gomez, M.R., Ruiz, D.E., et al. 2022. An overview of magneto-inertial fusion on the Z Machine at Sandia National Laboratories. *Nucl. Fusion*, **62**, 042015.

Yakovenko, S.N. 2014. The effects of density difference and surface tension on the development of Rayleigh–Taylor instability of an interface between fluid media. *Fluid Dyn.*, **49**, 748.

Yamanaka, C., Kato, Y., Izawa, Y., Yoshida, K., Yamanaka, T., Sasaki, T., Nakatsuka, M., Mochizuki, T., Kuroda, J., and Nakai, S. 1981. Nd-doped phosphate glass laser systems for laser-fusion research. *IEEE J. Quantum Electron.*, **17**, 1639.

Yan, R., Betti, R., Sanz, J., Aluie, H., Liu, B., and Frank, A. 2016. Three-dimensional single-mode nonlinear ablative Rayleigh–Taylor instability. *Phys. Plasmas*, **23**, 022701.

Yan, Z., Fu, Y., Wang, L., Yu, C., and Li, X. 2022. Effect of chemical reaction on mixing transition and turbulent statistics of cylindrical Richtmyer–Meshkov instability. *J. Fluid Mech.*, **941**, A55.

Yang, H., and Radulescu, M.I. 2021. Dynamics of cellular flame deformation after a head-on interaction with a shock wave: reactive Richtmyer–Meshkov instability. *J. Fluid Mech.*, **923**, A36.

Yang, J., Kubota, T., and Zukoski, E.E. 1993. Applications of shock-induced mixing to supersonic combustion. *AIAA J.*, **31**, 854.

Yang, J., Kubota, T., and Zukoski, E.E. 1994a. A model for characterization of a vortex pair formed by shock passage over a light-gas inhomogeneity. *J. Fluid Mech.*, **258**, 217.

Yang, Q., Chang, J., and Bao, W. 2014. Richtmyer–Meshkov instability induced mixing enhancement in the scramjet combustor with a central strut. *Adv. Mech. Eng.*, **6**, 614189.

Yang, X.C., and Cao, Y.G. 2021. Effects of head loss, surface tension, viscosity and density ratio on the Kelvin–Helmholtz instability in different types of pipelines. *Physica D*, **424**, 132950.

Yang, Y., Zhang, Q., and Sharp, D.H. 1994b. Small amplitude theory of Richtmyer–Meshkov instability. *Phys. Fluids*, **6**, 1856.

Yanilkin, Yu V, Nikiforov, VV, Bondarenko, Yu A, Gubkov, EV, Zharova, GV, Statsenko, VP, and Tarasov, VI. 1995. Two-parameter model and method for computations of turbulent mixing in 2D compressible flows. In: Young, R., Glimm, J. & Boston, B. (eds), *Proceedings of 5th International Workshop on the Physics of Compressible Turbulent Mixing*. World Scientific, Singapore.

Ye, W.H., Wang, L.F., and He, X.T. 2010. Spike deceleration and bubble acceleration in the ablative Rayleigh–Taylor instability. *Phys. Plasmas*, **17**, 122704.

Ye, W.H., Wang, L.F., Xue, C., Fan, Z.F., and He, X.T. 2011. Competitions between Rayleigh–Taylor instability and Kelvin–Helmholtz instability with continuous density and velocity profiles. *Phys. Plasmas*, **18**, 022704.

Yeung, P.K., and Zhou, Y. 1997. Universality of the Kolmogorov constant in numerical simulations of turbulence. *Phys. Rev. E*, **56**, 1746.

Yih, C.-S. 1960. A transformation for non-homentropic flows, with an application to large-amplitude motion in the atmosphere. *J. Fluid Mech.*, **9**, 68.

Yilmaz, I. 2020. Analysis of Rayleigh–Taylor instability at high Atwood numbers using fully implicit, non-dissipative, energy-conserving large eddy simulation algorithm. *Phys. Fluids*, **32**, 054101.

Yin, L., Albright, B.J., Vold, E.L., Nystrom, W.D., Bird, R.F., and Bowers, K.J. 2019. Plasma kinetic effects on interfacial mix and burn rates in multispatial dimensions. *Phys. Plasmas*, **26**, 062302.

Yosef-Hai, A., Sadot, O., Kartoon, D., Oron, D., Levin, L.A., Sarid, E., Elbaz, Y., Ben-Dor, G., and Shvarts, D. 2003. Late-time growth of the Richtmyer–Meshkov instability for different Atwood numbers and different dimensionalities. *Laser Part. Beams*, **21**, 363.

Young, R.M.B., and Read, P.L. 2017. Forward and inverse kinetic energy cascades in Jupiter's turbulent weather layer. *Nat. Phys.*, **13**, 1135.

Young, T. 1805. III. An essay on the cohesion of fluids. *Phil. Trans. R. Soc. London*, **95**, 65.

Young, Y.-N., and Ham, F.E. 2006. Surface tension in incompressible Rayleigh–Taylor mixing flow. *J. Turbul.*, **7**, N71.

Youngs, D.L. 1984. Numerical simulation of turbulent mixing by Rayleigh–Taylor instability. *Physica D*, **12**, 32.

Youngs, D.L. 1989. Modelling turbulent mixing by Rayleigh–Taylor instability. *Physica D*, **37**, 270.

Youngs, D.L. 1991. Three-dimensional numerical simulation of turbulent mixing by Rayleigh–Taylor instability. *Phys. Fluids A*, **3**, 1312.

Youngs, D.L. 1994. Numerical simulation of mixing by Rayleigh–Taylor and Richtmyer–Meshkov instabilities. *Laser Part. Beams*, **12**, 725.

Youngs, D.L. 2004. Effect of initial conditions on self-similar turbulent mixing. In: *Proceedings of the 9th International Workshop on the Physics of Compressible Turbulent Mixing*, Cambridge, UK.

Youngs, D.L. 2013. The density ratio dependence of self-similar Rayleigh–Taylor mixing. *Phil. Trans. R. Soc. Lond.*, **371**, 20120173.

Youngs, D.L. 2017. Rayleigh–Taylor mixing: direct numerical simulation and implicit large eddy simulation. *Phys. Scr.*, **92**, 074006.

Youngs, D.L., and Thornber, B. 2020a. Buoyancy-Drag modelling of bubble and spike distances for single-shock Richtmyer–Meshkov mixing. *Physica D*, **410**, 132517.

Youngs, D.L., and Thornber, B. 2020b. Early time modifications to the Buoyancy-Drag model for Richtmyer–Meshkov mixing. *J. Fluids Eng.*, **142**, 121107.

Youngs, D.L., and Williams, R.J.R. 2008. Turbulent mixing in spherical implosions. *Int. J. Numer. Methods Fluids*, **56**, 1597.

Yu, B., He, M., Zhang, B., and Liu, H. 2020. Two-stage growth mode for lift-off mechanism in oblique shock-wave/jet interaction. *Phys. Fluids*, **32**, 116105.

Yu, E.P., Awe, T.J., Cochrane, K.R., Peterson, K.J., Yates, K.C., Hutchinson, T.M., Hatch, M.W., Bauer, B.S., Tomlinson, K., and Sinars, D.B. 2023a. Seeding the Electrothermal Instability through a Three-Dimensional, Nonlinear Perturbation. *Phys. Rev. Lett.*, **130**, 255101.

Yu, E.P., Awe, T.J., Cochrane, K.R., Peterson, K.J., Yates, K.C., Hutchinson, T.M., Hatch, M.W., Bauer, B.S., Tomlinson, K., and Sinars, D.B. 2023b. Three-dimensional feedback processes in current-driven metal. *Phys. Rev. E*, **107**, 065209.

Yu, H., and Livescu, D. 2008. Rayleigh–Taylor instability in cylindrical geometry with compressible fluids. *Phys. Fluids*, **20**, 104103.

Yuan, D., Shen, Y., Liu, Y., Li, H., Feng, X., and Keppens, R. 2019. Multilayered Kelvin–Helmholtz instability in the solar corona. *Astrophy. J. Lett.*, **884**, L51.

Zabusky, N.J. 1999. Vortex paradigm for accelerated inhomogeneous flows: visiometrics for the Rayleigh–Taylor and Richtmyer–Meshkov environments. *Annu. Rev. Fluid Mech.*, **31**, 495.

Zanella, R., Tegze, G., Le Tellier, R., and Henry, H. 2020. Two-and three-dimensional simulations of Rayleigh–Taylor instabilities using a coupled Cahn–Hilliard/Navier–Stokes model. *Phys. Fluids*, **32**, 124115.

Zaytsev, S.G., Krivets, V.V., Mazilin, I.M., Titov, S.N., Chebotareva, E.I., Nikishin, V.V., Tishkin, V.F., Bouquet, S., and Haas, J.F. 2003. Evolution of the Rayleigh–Taylor instability in the mixing zone between gases of different densities in a field of variable acceleration. *Laser Part. Beams*, **21**, 393.

Zel'dovich, Ya B., and Raizer, Yu P. 1966. *Physics of Shock Waves and High-Temperature Hydrodynamic Phenomena, I & II*. Academic Press, New York.

Zelina, J., Shouse, D.T., and Hancock, R.D. 2004. Ultra-compact combustors for advanced gas turbine engines. Pages 53–62 of: *ASME Turbo Expo 2004: Power for Land, Sea, and Air*. American Society of Mechanical Engineers, Vienna, Austria.

Zeng, R.H., Tao, J.J., and Sun, Y.B. 2020. Three-dimensional viscous Rayleigh–Taylor instability at the cylindrical interface. *Phys. Rev. E*, **102**, 023112.

Zeng, W., Pan, J., Ren, Y., and Sun, Y. 2018a. Numerical study on the turbulent mixing of planar shock-accelerated triangular heavy gases interface. *Acta Mech. Sin.*, **34**, 855.

Zeng, W., Pan, J., Sun, Y., and Ren, Y. 2018b. Turbulent mixing and energy transfer of reshocked heavy gas curtain. *Phys. Fluids*, **30**, 064106.

Zenit, R. 2019. Some fluid mechanical aspects of artistic painting. *Phys. Rev. Fluids*, **4**, 110507.

Zetina, S., Godínez, F.A., and Zenit, R. 2015. A hydrodynamic instability is used to create aesthetically appealing patterns in painting. *PloS One*, **10**, e0126135.

Zeytounian, R.K. 2003. Joseph Boussinesq and his approximation: a contemporary view. *Comptes Rendus Mecanique*, **331**, 575.

Zhai, Z., Si, T., Luo, X., and Yang, J. 2011. On the evolution of spherical gas interfaces accelerated by a planar shock wave. *Phys. Fluids*, **23**, 084104.

Zhai, Z., Wang, M., Si, T., and Luo, X. 2014. On the interaction of a planar shock with a light polygonal interface. *J. Fluid Mech.*, **757**, 800.

Zhai, Z., Dong, P., Si, T., and Luo, X. 2016. The Richtmyer–Meshkov instability of a "V" shaped air/helium interface subjected to a weak shock. *Phys. Fluids*, **28**, 082104.

Zhai, Z., Liang, Y., Liu, L., Ding, J., Luo, X., and Zou, L. 2018. Interaction of rippled shock wave with flat fast-slow interface. *Phys. Fluids*, **30**, 046104.

Zhai, Z., Ou, J., and Ding, J. 2019. Coupling effect on shocked double-gas cylinder evolution. *Phys. Fluids*, **31**, 096104.

Zhai, Z., Guo, X., and Si, T. 2020. Experimental study on bubble competition of shock-accelerated saw-tooth interface. *Acta Aerodynamica Sin.*, **38**, 339.

Zhakhovskii, V.V., Zybin, S.V., Abarzhi, S.I., and Nishihara, K. 2006. Atomistic dynamics of the Richtmyer–Meshkov instability in cylindrical and planar geometries. *AIP Conf. Proc.*, **845**, 433.

Zhang, C., Gao, C., Yu, C., Xu, X., Fan, Z., and Wang, P. 2023a. Numerical study of the high-intensity heat conduction effect on turbulence induced by the ablative Rayleigh–Taylor instability. *Phys. Fluids*, **35**, 054106.

Zhang, D., Ding, J., Si, T., and Luo, X. 2023b. Divergent Richtmyer–Meshkov instability on a heavy gas layer. *J. Fluid Mech.*, **959**, A37.

Zhang, H., Kaman, T., She, D., Cheng, B., Glimm, J., and Sharp, D.H. 2018a. V&V for turbulent mixing in the intermediate asymptotic regime. *Pure Appl. Math. Q.*, **14**, 193.

Zhang, H., Betti, R., Yan, R., and Aluie, H. 2020a. Nonlinear bubble competition of the multimode ablative Rayleigh–Taylor instability and applications to inertial confinement fusion. *Phys. Plasmas*, **27**, 122701.

Zhang, J., Wang, L.F., Ye, W.H., Wu, J.F., Guo, H.Y., Zhang, W.Y., and He, X.T. 2017. Weakly nonlinear incompressible Rayleigh–Taylor instability in spherical geometry. *Phys. Plasmas*, **24**, 062703.

Zhang, J.A., Katsaros, K.B., Black, P.G., Lehner, S., French, J.R., and Drennan, W.M. 2008. Effects of roll vortices on turbulent fluxes in the hurricane boundary layer. *Bound. Layer Meteorol.*, **128**, 173.

Zhang, L., Ding, Y., Jiang, S., et al. 2015. Reducing wall plasma expansion with gold foam irradiated by laser. *Phys. Plasmas*, **22**, 110703.

Zhang, L., Ding, Y., Lin, Z., et al. 2016a. Demonstration of enhancement of x-ray flux with foam gold compared to solid gold. *Nucl. Fusion*, **56**, 036006.

Zhang, Q. 1998. Analytical solutions of Layzer-type approach to unstable interfacial fluid mixing. *Phys. Rev. Lett.*, **81**, 3391.

Zhang, Q., and Graham, M.J. 1997. Scaling laws for unstable interfaces driven by strong shocks in cylindrical geometry. *Phys. Rev. Lett.*, **79**, 2674.

Zhang, Q., and Graham, M.J. 1998. A numerical study of Richtmyer–Meshkov instability driven by cylindrical shocks. *Phys. Fluids*, **10**, 974.

Zhang, Q., and Guo, W. 2016. Universality of finger growth in two-dimensional Rayleigh–Taylor and Richtmyer–Meshkov instabilities with all density ratios. *J. Fluid Mech.*, **786**, 47.

Zhang, Q., and Guo, W. 2022. Quantitative theory for spikes and bubbles in the Richtmyer–Meshkov instability at arbitrary density ratios. *Phys. Rev. Fluids*, **7**, 093904.

Zhang, Q., and Sohn, S.-I. 1996. An analytical nonlinear theory of Richtmyer–Meshkov instability. *Phys. Lett. A*, **212**, 149.

Zhang, Q., and Sohn, S.-I. 1997. Nonlinear theory of unstable fluid mixing driven by shock wave. *Phys. Fluids*, **9**, 1106.

Zhang, Q., and Sohn, S.-I. 1999. Quantitative theory of Richtmyer–Meshkov instability in three dimensions. *Zeitschrift für angewandte Mathematik und Physik ZAMP*, **50**, 1.

Zhang, Q., Deng, S., and Guo, W. 2018b. Quantitative theory for the growth rate and amplitude of the compressible Richtmyer–Meshkov instability at all density ratios. *Phys. Rev. Lett.*, **121**, 174502.

Zhang, W., Wu, Q., Zou, L., Zheng, X., Li, X., Luo, X., and Ding, J. 2018c. Mach number effect on the instability of a planar interface subjected to a rippled shock. *Phys. Rev. E*, **98**, 043105.

Zhang, X. 2013. Single-mode bubble evolution simulations of Rayleigh Taylor instability with spectral element method and a viscous model. *Comput. Fluids*, **88**, 813–821.

Zhang, X., Zheng, W., Wei, X., et al. 2005. Preliminary experimental results of Shenguang III technical integration experiment line. Page 6 of: *High-Power Lasers and Applications III*. Vol. 5627. International Society for Optics and Photonics, Bellingham, WA.

Zhang, X., Zhang, S., Yan, Z., Duan, H., Ding, Y., and Kang, W. 2024. Mode-coupled perturbation growth on the interfaces of cylindrical implosion: A comparison between theory and experiment. *Phys. Rev. E*, **109**, 035203.

Zhang, Y., He, Z., Gao, F., Li, X., and Tian, B. 2016b. Evolution of mixing width induced by general Rayleigh–Taylor instability. *Phys. Rev. E*, **93**, 063102.

Zhang, Y., He, Z., Xie, H., Xiao, M., and Tian, B. 2020b. Methodology for determining coefficients of turbulent mixing model. *J. Fluid Mech.*, **905**, A26.

Zhang, Y., Ruan, Y., Xie, H., and Tian, B. 2020c. Mixed mass of classical Rayleigh–Taylor mixing at arbitrary density ratios. *Phys. Fluids*, **32**, 011702.

Zhang, Y., Ni, W., Ruan, Y., and Xie, H. 2020d. Quantifying mixing of Rayleigh–Taylor turbulence. *Phys. Rev. Fluids*, **5**, 104501.

Zhang, Y.-T., Shi, J., Shu, C.-W., and Zhou, Y. 2003. Numerical viscosity and resolution of high-order weighted essentially nonoscillatory schemes for compressible flows with high Reynolds numbers. *Phys. Rev. E*, **68**, 046709.

Zhang, Y.-T., Shu, C.-W., and Zhou, Y. 2006. Effects of shock waves on Rayleigh–Taylor instability. *Phys. Plasmas*, **13**, 062705.

Zhao, D., and Aluie, H. 2018. Inviscid criterion for decomposing scales. *Phys. Rev. Fluids*, **3**, 054603.

Zhao, D., Betti, R., and Aluie, H. 2022. Scale interactions and anisotropy in Rayleigh–Taylor turbulence. *J. Fluid Mech.*, **930**, A29.

Zhao, K., Dalton, P., Yang, G.C., and Scherer, P.W. 2006. Numerical modeling of turbulent and laminar airflow and odorant transport during sniffing in the human and rat nose. *Chem. Senses*, **31**, 107.

Zhao, Y., Xia, M., and Cao, Y. 2020a. A study of bubble growth in the compressible Rayleigh–Taylor and Richtmyer–Meshkov instabilities. *AIP Adv.*, **10**, 015056.

Zhao, Z., Wang, P., Liu, N.-S., and Lu, X. 2020b. Analytical model of nonlinear evolution of single-mode Rayleigh–Taylor instability in cylindrical geometry. *J. Fluid Mech.*, **904**, A24.

Zhao, Z., Liu, N.-S., and Lu, X.-Y. 2020c. Kinetic energy and enstrophy transfer in compressible Rayleigh–Taylor turbulence. *J. Fluid Mech.*, **904**, A37.

Zhao, Z., Wang, P., Liu, N.-S., and Lu, X. 2021. Scaling law of mixing layer in cylindrical Rayleigh–Taylor turbulence. *Phys. Rev. E*, **104**, 055104.

Zheng, J.G., Lee, T.S., and Winoto, S.H. 2008a. Numerical simulation of Richtmyer–Meshkov instability driven by imploding shocks. *Math. Comput. Simul.*, **79**, 749.

Zheng, W., Zhang, X., Wei, X., et al. 2008b. Status of the SG-III solid-state laser facility. *J. Phys. Conf. Ser.*, 112, 032009.

Zhou, Y. 1993a. Degrees of locality of energy transfer in the inertial range. *Phys. Fluids A*, **5**, 1092.

Zhou, Y. 1993b. Interacting scales and energy transfer in isotropic turbulence. *Phys. Fluids A*, **5**, 2511.

Zhou, Y. 1995. A phenomenological treatment of rotating turbulence. *Phys. Fluids*, **7**, 2092.

Zhou, Y. 2001. A scaling analysis of turbulent flows driven by Rayleigh–Taylor and Richtmyer–Meshkov instabilities. *Phys. Fluids*, **13**, 538.

Zhou, Y. 2007. Unification and extension of the similarity scaling criteria and mixing transition for studying astrophysics using high energy density laboratory experiments or numerical simulations. *Phys. Plasmas*, **14**, 082701.

Zhou, Y. 2010. Renormalization group theory for fluid and plasma turbulence. *Phys. Rep.*, **488**, 1.

Zhou, Y. 2017a. Rayleigh–Taylor and Richtmyer–Meshkov instability induced flow, turbulence, and mixing. I. *Phys. Rep.*, **720–722**, 1.

Zhou, Y. 2017b. Rayleigh–Taylor and Richtmyer–Meshkov instability induced flow, turbulence, and mixing. II. *Phys. Rep.*, **723–725**, 1.

Zhou, Y. 2021. Turbulence theories and statistical closure approaches. *Phys. Rep.*, **935**, 1.

Zhou, Y., and Cabot, W.H. 2019. Time-dependent study of anisotropy in Rayleigh–Taylor instability induced turbulent flows with a variety of density ratios. *Phys. Fluids*, **31**, 084106.

Zhou, Y., and Matthaeus, W.H. 1990. Models of inertial range spectra of interplanetary magnetohydrodynamic turbulence. *J. Geophys. Res. Space Phys.*, **95**, 14881.

Zhou, Y., and Oughton, S. 2011. Nonlocality and the critical Reynolds numbers of the minimum state magnetohydrodynamic turbulence. *Phys. Plasmas*, **18**, 072304.

Zhou, Y., and Speziale, C.G. 1998. Advances in the fundamental aspects of turbulence: energy transfer, interacting scales, and self-preservation in isotropic decay. *Appl. Mech. Rev.*, **51**, 267.

Zhou, Y., and Thornber, B. 2016. A comparison of three approaches to compute the effective Reynolds number of the implicit large-eddy simulations. *J. Fluids Eng.*, **138**, 70905.

Zhou, Y., Vahala, G., and Thangam, S. 1994. Development of a turbulence model based on recursion renormalization group theory. *Phys. Rev. E*, **49**, 5195.

Zhou, Y., Zimmerman, G.B., and Burke, E.W. 2002. Formulation of a two-scale transport scheme for the turbulent mix induced by Rayleigh–Taylor and Richtmyer–Meshkov instabilities. *Phys. Rev. E*, **65**, 056303.

Zhou, Y., Robey, H.F., and Buckingham, A.C. 2003a. Onset of turbulence in accelerated high-Reynolds-number flow. *Phys. Rev. E*, **67**, 056305.

Zhou, Y., Remington, B.A., Robey, H.F., Cook, A.W., Glendinning, S.G., Dimits, A., Buckingham, A.C., Zimmerman, G.B., Burke, E.W., Peyser, T.A., Cabot, W., and Eliason, D. 2003b. Progress in understanding turbulent mixing induced by Rayleigh–Taylor and Richtmyer–Meshkov instabilities. *Phys. Plasmas*, **10**, 1883.

Zhou, Y., Matthaeus, W.H., and Dmitruk, P. 2004. Colloquium: magnetohydrodynamic turbulence and time scales in astrophysical and space plasmas. *Rev. Mod. Phys.*, **76**, 1015.

Zhou, Y., Grinstein, F.F., Wachtor, A.J., and Haines, B.M. 2014. Estimating the effective Reynolds number in implicit large-eddy simulation. *Phys. Rev. E*, **89**, 013303.

Zhou, Y., Cabot, W.H., and Thornber, B. 2016. Asymptotic behavior of the mixed mass in Rayleigh–Taylor and Richtmyer–Meshkov instability induced flows. *Phys. Plasmas*, **23**, 052712.

Zhou, Y., Clark, T.T, Clark, D.S., Glendinning, S.G., Skinner, M.A., Huntington, C.M., Hurricane, O.A., Dimits, A.M., and Remington, B.A. 2019. Turbulent mixing and transition criteria of flows induced by hydrodynamic instabilities. *Phys. Plasmas*, **26**, 080901.

Zhou, Y., Groom, M., and Thornber, B. 2020a. Dependence of enstrophy transport and mixed mass on dimensionality and initial conditions in the Richtmyer–Meshkov instability induced flows. *J. Fluids Eng.*, **142**, 121104.

Zhou, Y., Williams, R.J.R., Ramaprabhu, P., Groom, M., Thornber, B., Hillier, A., Mostert, W., Rollin, B., Balachandar, S., Powell, P.D., Mahalov, A., and Attal, N. 2021. Rayleigh–Taylor and Richtmyer–Meshkov instabilities: a journey through scales. *Physica D*, **423**, 132838.

Zhou, Y., Sadler, J.D., and Hurricane, O.A. 2025. Instabilities and mixing in inertial confinement fusion, *Annu. Rev. Fluid Mech.*, **57**, https://doi.org/10.1146/annurev-fluid-022824-110008.

Zhou, Y., Zou, S., Pu, Y., Xue, Q., and Liu, H. 2022. Terminal velocities and vortex dynamics of weakly compressible Rayleigh–Taylor instability. *AIP Adv.*, **12**, 015325.

Zhou, Z., Zhang, Y., and Tian, B. 2018. Dynamic evolution of Rayleigh–Taylor bubbles from sinusoidal, W-shaped, and random perturbations. *Phys. Rev. E*, **97**, 033108.

Zhou, Z., Ding, J., Zhai, Z., Cheng, W., and Luo, X. 2020b. Mode coupling in converging Richtmyer–Meshkov instability of dual-mode interface. *Acta Mech. Sin.*, **36**, 356.

Zhou, Z., Ding, J., Cheng, W., and Luo, X. 2023. Scaling law of structure function of Richtmyer–Meshkov turbulence. *J. Fluid Mech.*, **972**, A18.

Zou, L., Liu, J., Liao, S., Zheng, X., Zhai, Z., and Luo, X. 2017. Richtmyer–Meshkov instability of a flat interface subjected to a rippled shock wave. *Phys. Rev. E*, **95**, 013107.

Zufiria, J.A. 1988. Bubble competition in Rayleigh–Taylor instability. *Phys. Fluids*, **31**, 440.

Zulick, C., Aglitskiy, Y., Karasik, M., Schmitt, A.J., Velikovich, A.L., and Obenschain, S.P. 2020a. Isolated defect evolution in laser accelerated targets. *Phys. Plasmas*, **27**, 072706.

Zulick, C., Aglitskiy, Y., Karasik, M., Schmitt, A.J., Velikovich, A.L., and Obenschain, S.P. 2020b. Multimode hydrodynamic instability growth of preimposed isolated defects in ablatively driven foils. *Phys. Rev. Lett.*, **125**, 055001.

Zwiebach, B. 2004. *A First Course in String Theory*. Cambridge University Press, Cambridge, UK.

Zylstra, A. B., Kritcher, A. L., Hurricane, O. A., et al. 2022. Experimental achievement and signatures of ignition at the National Ignition Facility. *Phys. Rev. E*, **106**, 025202.

Index